DTV Handbook

The Revolution in Digital Video
Third Edition

On-Line Updates

Additional updates relating to DTV in general, and this book in particular, can be found at the *Standard Handbook of Video and Television Engineering* web site:

www.tvhandbook.com

The tvhandbook.com web site supports the professional video community with news, updates, and product information relating to the broadcast, post production, and business/industrial applications of digital video.

Check the site regularly for news, updated chapters, and special events related to video engineering. The technologies encompassed by *DTV: The Revolution in Digital Video* are changing rapidly, with new standards proposed and adopted each month. Changing market conditions and regulatory issues are adding to the rapid flow of news and information in this area.

Specific services found at **www.tvhandbook.com** include:

- **Video Technology News**. News reports and technical articles on the latest developments in digital television, both in the U.S. and around the world. Check in at least once a month to see what's happening in the fast-moving area of digital television.

- **Television Handbook Resource Center**. Check for the latest information on professional and broadcast video systems. The Resource Center provides updates on implementation and standardization efforts, plus links to related web sites.

- **tvhandbook.com Update Port**. Updated material for *DTV: The Revolution in Digital Video* is posted on the site each month. Material available includes updated sections and chapters in areas of rapidly advancing technologies.

- **tvhandbook.com Book Store**. Check to find the latest books on digital video and audio technologies. Direct links to authors and publishers are provided. You can also place secure orders from our on-line bookstore.

In addition to the resources outlined above, detailed information is available on other books in the McGraw-Hill Video/Audio Series.

DTV Handbook

The Revolution in Digital Video
Third Edition

Jerry C. Whitaker

McGraw-Hill
New York   San Francisco   Washington D.C.   Auckland   Bogotá
Caracas   Lisbon   London   Madrid   Mexico City   Milan
Montreal   New Delhi   San Juan   Singapore
Sydney   Tokyo   Toronto

Cataloging-in-Publication Data is on file with the Library of Congress.

McGraw-Hill

*A Division of The **McGraw·Hill** Companies*

Copyright © 2001, by The McGraw-Hill Companies, Inc. All rights reserved. Printed in the United States of America. Except as permitted under the United States Copyright Act of 1976, no part of this publication may be reproduced or distributed in any form or by any means, or stored in a data base or retrieval system, without the prior written permission of the publisher.

1 2 3 4 5 6 7 8 9 0   DOC/DOC   0 6 5 4 3 2 1 0

P/N 0-07-137171-0

Part of

ISBN 0-07-137170-2

The sponsoring editor for this book was Steve Chapman and the production supervisor was Sherri Souffrance. The book was set in Times New Roman and Helvetica by Technical Press, Morgan Hill, CA.

Printed and bound by R. R. Donnelley & Sons Company.

McGraw-Hill books are available at special quantity discounts to use as premiums and sales promotions, or for use in corporate training programs. For more information, please write to the Director of Special Sales, McGraw-Hill, Two Penn Plaza, New York, NY 10121-2298. Or contact your local bookstore.

 This book is printed on recycled, acid-free paper containing a minimum of 50% recycled, de-inked fiber.

Introduction

There are few more beloved appliances in the home than the television set. Television is a window to the world through which we can be ceaselessly entertained and informed. It encompasses a tremendous variety of programming for consumption by an ever-enthusiastic consumer audience.

Consider for a moment the ubiquity of television broadcasting in the U.S. marketplace:

- There are more than 1,550 broadcast television stations in the U.S.

- 98 percent of U.S. households own television receivers.

- 67 percent of U.S. households own two or more receivers.

- Average daily television household viewing is over seven hours.

- Television is the main news source for 70 percent of the U.S. public.

Television is certainly a pervasive medium.

However, the technical system upon which this marvel of engineering is based is nearing the limits of its useful life. Like all other communications services, television and television broadcasting are moving toward a digital delivery platform. As this revolution gets under-way, the migration to digital technology will be harmonious with, and is perhaps even a natural consequence of, the increasing technical sophistication of the modern television viewing audience. Once exposed to the technology, one is easily convinced that television viewers will be rapidly swayed by the quality improvements and service flexibility offered by digital technology and HDTV.

It is particularly gratifying to us that the terrestrial broadcasting industry has done so much to bring this new service to reality. As you read the story of how the DTV phenomenon has unfolded, you will see the central and recurring role that broadcasters have played throughout the process. We are unabashedly proud of this role and equally enthusiastic that the onset of digital television broadcasting service in 1998 did indeed mark the beginning of a new "golden era" of television.

Lynn D. Claudy

Senior Vice President, Science and Technology
National Association of Broadcasters
Washington, D.C.

October, 2000

For 79 years, the National Association of Broadcasters has provided the resources and leadership necessary to advance the interests of radio and television broadcasters everywhere. NAB is a valuable resource for information on technical and regulatory matters. For more information visit the NAB World Wide Web site at **www.nab.org**.

Contents

Introduction by Lynn D. Claudy — v

Foreword by Dr. Joseph A. Flaherty — xix

Preface by Jerry C. Whitaker — xxvii

Dedication — xxxiii

Chapter 1: The Road to DTV — **1**

  Introduction — 1
    The Pioneers — 2
    Transmission Standards — 4
    Conventional TV Systems — 8
  Early Development of HDTV — 12
    1125/60 Equipment Development — 13
    The 1125/60 System — 16
    European HDTV Systems — 17
    A Perspective on HDTV Standardization — 18
    Digital Systems Emerge — 19
    Digital Video Broadcasting — 21
    Involvement of the Film Industry — 22
    Political Considerations — 23
    Terminology — 24
  Proposed Routes to HDTV — 25
    System Testing: Round 2 — 27
    Formation of the Grand Alliance — 29
    Testing the Grand Alliance System — 31
    The Home Stretch for the Grand Alliance — 32
    Digital Broadcasting Begins — 34
    Continuing Work on the ATSC Standard — 34
    A Surprising Challenge — 37
    FCC Reviews the Digital Television Conversion Process — 39
  Compatibility in HDTV Systems — 41
    Compromises in Compatible Systems — 42
    Adapters for Compatible Service — 42
    Transcoding Functions — 43
    Standardization Issues — 43
    Harmonized Standards — 45
  References — 46
  Bibliography — 47

Chapter 2: Applications for HDTV — **49**

  Introduction — 49

Resolution	49
Production Systems vs. Transmission Systems	50
Defining Terms	50
HDTV Applications	50
Business and Industrial Applications	51
Broadcast Applications	52
Computer Applications	54
ATSC Datacasting	57
Characteristics of the Video Signal	58
Critical Implications for the Viewer and Program Producer	58
Image Size	60
Format Development	61
Aural Component of Visual Realism	62
Hearing Perception	63
Matching Audio to Video	64
Making the Most of Audio	65
Ideal Sound System	66
Dolby AC-3	66
Motion Picture Film Origination	67
Film Types	68
Synergy Between Film and HDTV Program Origination	69
Standards Conversion	69
Film-to-Video Transfer Systems	71
Video-to-Film Laser-Beam Transfer	75
References	78
Bibliography	79

Chapter 3: Fundamental Imaging System Principles — **81**

Introduction	81
Visual Fields of Television	81
Foveal and Peripheral Vision	82
Vertical Detail and Viewing Distance	82
Horizontal Detail and Picture Width	84
Detail Content of the Image	86
Perception of Depth	86
Contrast and Tonal Range	86
Luminance and Chrominance	88
Chromatic Aspects of Vision	89
Acuity in Color Vision	91
Temporal Factors in Vision	92
Temporal Aspects of Illumination	92
Smear and Related Effects	93
Flicker	94
Visual Basis of Video Bandwidth	95
Gamma	96
Color Signal Transformations	100
Overview of the Grand Alliance System	101
Scanning Formats	101
Relative Horizontal and Vertical Spacing	102

Luminance and Chrominance Components   102
Pixel Transmission Rates and Digital Modulation   104
Video Compression   104
References   106

Chapter 4: Digital Coding of Video Signals   109

Introduction   109
Digital Signal Conversion   109
The Nyquist Limit and Aliasing   109
The A/D Conversion Process   110
The D/A Conversion Process   114
Converter Performance Criteria   115
Space and Time Components of Video Signals   116
Dimensions of the Nyquist Volume   116
Significant Frequencies in the Nyquist Volume   121
Signal Distinctions in the Nyquist Volume   122
Digital Modulation   125
QPSK   127
Signal Analysis   127
Digital Coding   128
Error-Correction Coding   130
8-VSB Modulation System   130
Digital Filters   131
FIR Filters   132
Infinite Impulse Response Filters   136
Digital Signal Processing and Distribution   138
Video Sampling   139
Serial Digital Interface   141
IEEE 1394   145
Fibre Channel   148
Gigabit Ethernet   149
References   151
Bibliography   153

Chapter 5: Video and Audio Compression   155

Introduction   155
Transform Coding   155
Planar Transform   156
Interframe Transform Coding   158
The JPEG Standard   159
Compression Techniques   160
DCT and JPEG   161
The MPEG Standard   163
Basic Provisions   163
Motion Compensation   164
Putting It All Together   166
Profiles and Levels   168
Studio Profile   170
MPEG-2 Features of Importance for DTV   170

MPEG-2 Layer Structure   171
Slices   172
Pictures, Groups of Pictures, and Sequences   172
Vector Search Algorithm   174
Motion-Vector Precision   175
Motion-Vector Coding   175
Encoder Prediction Loop   175
Dual Prime Prediction Mode   179
Adaptive Field/Frame Prediction Mode   179
Image Refresh   180
Discrete Cosine Transform   181
Entropy Coding of Video Data   183
Spatial and S/N Scalability   185
Concatenation   186
Compression Artifacts   187
Video Encoding Process   189
Encoding Tools   189
Signal Conditioning   190
SMPTE RP 202   191
Digital Audio Data Compression   192
PCM Versus Compression   192
Audio Bit-Rate Reduction   193
Prediction and Transform Algorithms   196
Processing and Propagation Delay   198
Bit Rate and Compression Ratio   199
Editing Compressed Data   200
Common Compression Techniques   200
Dolby E Coding System   208
Objective Quality Measurements   210
Perspective on Audio Compression   211
References   212
Bibliography   213

Chapter 6: High-Definition Production Systems   **215**

Introduction   215
The 1125/60 System   215
Fundamental Principles of 1125/60   216
The Standardization of 1125/60 HDTV   217
HDTV Production Standard   219
Technical Aspects of SMPTE 240M   219
Bandwidth and Resolution Considerations   223
Interlace and Progressive Scanning   227
Digital Representation of SMPTE 240M   228
SMPTE 260M   229
Sampling and Encoding   230
Principal Operating Parameters   233
Production Aperture   235
SMPTE 240M-1995   237
DTV-Related Raster-Scanning Standards   240

1920 x 1080 Scanning Standard   240
1280 x 720 Scanning Standard   242
720 x 483 Scanning Standard   244
MPEG-2 4:2:2 Profile at High Level   246
High-Definition Serial Digital Interface   247
A Practical Implementation   247
Audio Interface Provisions   254
Data Services   255
Time Division Multiplexing on SMPTE 292M   256
Packet Transport   257
540 Mbits/s Interface   259
Serial Data Transport Interface   259
SDTI Data Structure   260
SDTI in Computer-Based Systems   262
SDTI Content Package Format   263
Camera Systems   266
The Optical System   268
Digital Signal Processing   268
Camera Specifications   270
Camera-Related Standards   270
HDTV Camera/Film Production Issues   273
References   275
Bibliography   277

Chapter 7: DTV Audio Encoding and Decoding   279

Introduction   279
AES Audio   279
AES3 Data Format   281
SMPTE 324M   282
Audio Compression   282
Encoding   284
Decoding   285
Implementation of the AC-3 System   286
Audio-Encoder Interface   287
Output Signal Specification   288
Operational Details of the AC-3 Standard   288
Transform Filterbank   289
Coded Audio Representation   290
Bit Allocation   291
Rematrixing   292
Coupling   292
Bit Stream Elements and Syntax   293
Loudness and Dynamic Range   294
Encoding the AC-3 Bit Stream   295
AC-3/MPEG Bit Stream   297
Decoding the AC-3 Bit Stream   297
Algorithmic Details   300
Bit Allocation   301
Audio System Level Control   302

Dialogue Normalization 302
Dynamic Range Compression 303
Heavy Compression; COMPR, COMPR2 305
Audio System Features 306
Complete Main Audio Service (CM) 306
Main Audio Service, Music and Effects (ME) 306
Visually Impaired (VI) 307
Hearing Impaired (HI) 307
Dialogue (D) 307
Commentary (C) 308
Emergency (E) 308
Voice-Over (VO) 308
Multilingual Services 309
Channel Assignments and Levels 311
References 312
Bibliography 312

Chapter 8. The ATSC DTV System   313

Introduction 313
System Overview 313
Video Systems Characteristics 316
Transport System Characteristics 316
Overview of Video Compression and Decompression 318
MPEG-2 Levels and Profiles 319
Overview of the DTV Video System 319
Color Component Separation and Processing 320
Number of Lines Encoded 321
Film Mode 321
Pixels 321
Transport Encoder Interfaces and Bit Rates 324
Concatenated Sequences 324
Guidelines for Refreshing 325
Transmission Characteristics for Terrestrial Broadcast 326
Channel Error Protection and Synchronization 328
Modulation 333
Service Multiplex and Transport Systems 334
Overview of the Transport Subsystem 335
Higher-Level Multiplexing Functionality 340
The PES Packet Format 345
High Data-Rate Mode 346
Compatibility With Other Transport Systems 347
Program and System Information Protocol 352
Elements of PSIP 353
A/65 Technical Corrigendum and Amendment 355
Conditional Access System 357
Transport Stream Identification 360
Closed Captioning 362
SMPTE 333M 364
Data Broadcasting 364

PSIP and SDF for Data Broadcasting 365
Opportunistic Data 366
RP 203 368
ATSC A/90 Standard 368
References 369
Bibliography 371

Chapter 9: DTV Transmission Issues 373

Introduction 373
   Real World Conditions 374
   Bit-Rate Considerations 374
Performance Characteristics of the Terrestrial Broadcast Mode 375
   Transmitter Signal Processing 375
   Upconverter and RF Carrier Frequency Offsets 377
   Performance Characteristics of High-Data-Rate Mode 378
Spectrum Issues 378
   UHF Taboos 378
   Co-Channel Interference 380
   Adjacent-Channel Interference 381
   Power Ratings of NTSC and DTV 384
   Sixth Report and Order 386
   ATSC Standards for Satellite 387
Transmitter Considerations 392
   Operating Power 392
   Technology Options 393
   Digital Signal Pre-Correction 404
   FCC Emissions Mask 406
   Implementation Issues 407
   Channel-Combining Considerations 408
   Antenna Systems 409
References 409

Chapter 10. Receiver Systems and Display Devices 413

Introduction 413
   Noise Figure 413
Receiver System Overview 414
   Tuner 415
   Channel Filtering and VSB Carrier Recovery 417
   Segment Sync and Symbol Clock Recovery 418
   Noncoherent and Coherent AGC 418
   Data Field Synchronization 419
   Interference-Rejection Filter 420
   Channel Equalizer 424
   Phase-Tracking Loop 425
   Trellis Decoder 426
   Data De-Interleaver 429
   Other Receiver Functional Blocks 429
   Receiver Equalization Issues 431
   Receiver Performance in the Field 434

Compatibility Standards	439
Digital Receiver Advancements	440
HDTV Display Considerations	440
Color Space Issues in Digital Video	441
Display Technology Trends	447
Color CRT Display Devices	451
Projection Systems	473
Light-Valve Systems	485
LCD Projection Systems	493
Projection Requirements for Cinema Applications	508
Consumer Computer and Networking Issues	509
IEEE 1394	510
Digital Home Network	512
Advanced Television Enhancement Forum	513
Digital Application Software Environment	514
DTV Product Classification	515
Cable/DTV Receiver Labeling	517
References	518
Bibliography	522

Chapter 11: The DVB Standard 523

Introduction	523
European System	523
D-MAC/D2-MAC Systems	524
Enhanced Television Objectives and Constraints	524
Eureka Program	525
The End of Eureka	528
Digital Video Broadcasting (DVB)	529
Technical Background of the DVB System	529
DVB Services	530
The DVB Conditional-Access Package	534
Multimedia Home Platform	535
DVB Data Broadcast Standards	536
DVB and the ATSC DTV System	538
COFDM Technical Principles	538
DVB-T and ATSC Modulation Systems Comparison	544
Operating Parameters	544
Multipath Distortion	546
Mobile Reception	546
Spectrum Efficiency	547
HDTV Capability	547
Single Frequency Network	548
Impulse Noise	549
Tone Interference	549
Co-Channel Analog TV Interference	549
Co-Channel DTV Interference	550
Phase Noise Performance	550
Noise Figure	550
Indoor Reception	551

Scaling for Different Channel Bandwidth 552
References 552
Bibliography 554

Chapter 12: Video Measurement Techniques 557

Introduction 557
The Video Spectrum 559
    Minimum Video Frequency 560
    Maximum Video Frequency 560
    Horizontal Resolution 562
    Video Frequencies Arising From Scanning 562
Measurement of Color Displays 564
    Assessment of Color Reproduction 567
    Chromatic Adaptation and White Balance 568
    Overall Gamma Requirements 568
    Perception of Color Differences 568
    Display Resolution and Pixel Format 570
    Contrast Ratio 571
    Color Bar Test Patterns 571
    Conventional Video Measurements 574
    Automated Video Signal Measurement 583
    Applications of the Zone Plate Signal 583
    Display Measurement Techniques 593
    Subjective CRT Measurements 594
    Objective CRT Measurements 597
    Viewing Environment Considerations 599
Video Camera Performance Characterization and Verification 602
    Visual Inspection and Mechanical Check 602
    Confirmation of the Camera Encoder 602
    Confirmation of Auto Black 603
    Lens Back-Focus 603
    Black Shading 604
    Detail Circuit 604
    Optional Tests 604
    Color Reference Pattern 606
Picture-Quality Measurements for Digital Television 607
    Signal/Picture Quality 608
    Automated Picture-Quality Measurement 610
Serial Digital Bit Stream Analysis 614
    SMPTE RP 259M 615
    Jitter 615
    The Serial Digital Cliff 616
    Pathological Testing 617
    Eye Diagram 622
Transmission Issues 622
    Transmission System Measurements 623
    In-Band Signal Characterization 627
    Power Specification and Measurement 627
References 627

Bibliography   629

Chapter 13: DTV Implementation Issues   631

Introduction   631
MPEG Bit Stream Splicing   631
    Splice Flags   634
    SMPTE 312M   635
    Transition Clip Generator   637
    SMPTE 328M   637
MPEG-2 Recoding   638
    SMPTE 327M   639
    SMPTE 329M   639
    SMPTE 319M   640
    SMPTE 351M   640
    SMPTE 353M   641
Planning the DTV Infrastructure   642
    Considerations Regarding Interlaced and Progressive Scanning   642
    Network Contribution Options   643
    DTV Implementation Scenarios   645
    Top Down System Analysis   648
    Advanced System Control Architecture   650
24-Frame Mastering   653
    SMPTE 211   655
    Global HDTV Program Origination Standard   656
References   657
Bibliography   658

Chapter 14: Glossary   659

Terms Employed   659
Acronyms and Abbreviations   664
References   667

Chapter 15: Reference Documents   669

General   669
    Video   669
    Audio   669
ATSC DTV Standard   670
    Service Multiplex and Transport Systems   670
    System Information Standard   670
    Receiver Systems   671
    Program Guide   671
    Program/Episode/Version Identification   671
DVB   671
    General   672
    Multipoint Distribution Systems   672
    Interactive Television   672
    Conditional Access   672
    Interfaces   672

SMPTE Documents Relating to Digital Television 673
 General Topics 673
 Ancillary 673
 Digital Control Interfaces 674
 Edit Decision Lists 674
 Image Areas 674
 Interfaces and Signals 675
 Monitors 676
 MPEG-2 676
 Test Patterns 677
 Video Recording and Reproduction 677
SCTE Standards 679
References Cited in this Book 681

Index of Tables and Figures **703**
 Tables 703
 Figures 705

Subject Index **719**

Afterword by Dr. Robert Hopkins **735**

About the Author **741**

On the CD-ROM **743**

Foreword

Digital TV and HDTV: Reinventing Broadcasting

Dr. Joseph A. Flaherty (FIEE, FRTS), CBS, New York

At an earlier time, on June 15, 1936, when television itself was emerging from its radio foundations, David Sarnoff, then President of the Radio Corporation of America, presented a lecture to the Federal Communications Commission (FCC) entitled "The Future of Radio and the Public Interest, Convenience, and Necessity." In that lecture, the General said of the development of television:

"Of the future industries now visible on the horizon, television has gripped the public imagination most firmly. To bring television to the perfection needed for public service, our work proceeds under high pressure at great cost and with encouraging technical results."

"Such experiments call for ... imagination of the highest order and for the courage to follow where that imagination leads. It is in this spirit that our laboratories and our scientists are diligently and devotedly engaged in a task of the highest service to humanity."

From such work television was born, and today, the same genius and dedication in the service of mankind gave birth to the Advanced Television Systems Committee (ATSC) digital television (DTV) and high-definition television (HDTV) standard.

Television, this 20th century phenomenon, is so far advanced that today—the World over—more people watch television than are literate. Yet, along the way, every significant advance in television quality since John Logie Baird's 28 line mechanical scanning system has been heralded as "high-definition." High-definition has always been the best that could be done—the full state of the art.

Thus it was in 1970, 34 years after General Sarnoff's lecture, that modern HDTV began. NHK launched the modern development of high-definition television, and carried out extensive research and psychophysical testing as groundwork for choosing a scanning format, an aspect ratio, and an entirely new electronic imaging system. From 1970 through 1977, technical papers were published around the World covering the subjective evaluation of picture quality, the response of the human visual system to increased line scanning, the visual effects of interlace scanning, and the chromatic spatial frequency response of the human visual system. Then, in 1973, NHK described the original wide screen 1125 line HDTV system.

In 1974 the International Telecommunications Union, through the CCIR, adopted an HDTV Study Question stating:

"Considering: That high-definition television systems will require a resolution which is approximately equivalent to that of 35 mm film and corresponds to at least twice the horizontal and twice the vertical resolution of present television systems:"

"The CCIR UNANIMOUSLY DECIDES that this question should be studied: What standards should be recommended for high-definition television systems intended for broadcasting to the general public?"

By 1977, the Society of Motion Picture and Television Engineers (SMPTE) Study Group on High-Definition Television was formed, and in 1980 the *SMPTE Journal* published that group's report on HDTV. The report stated:

"The appropriate standard of comparison (for HDTV) is the current and prospective optimum performance of the 35 mm release print as projected on a wide screen."

The SMPTE HDTV Study Group concluded: "The appropriate line rate for HDTV is approximately 1100 lines-per-frame and the frame rate should be 30 frames-per-second, interlaced 2-to-1 ..."

Indeed, rather than being better than necessary, HDTV was to finally put television on a par with cinema quality! We're not too good, we're just catching up—catching up to a quality widely accepted by the creative community and by the American public. Through full HDTV, television would finally achieve its technical maturity.

The first glimpse of this maturity came in 1981 when CBS and NHK presented the first HDTV demonstration in America at the SMPTE Winter Television Conference in San Francisco. Two weeks later, an HDTV demonstration was presented in Washington, D.C. for the FCC and other government entities. These demonstrations were quickly followed by packed demonstrations in New York and Hollywood, and viewers were captivated.

There was no turning back. Television would never again be the same!

On November 21st, 1985, with apologies to Arthur C. Clarke for plagiarizing his title, I delivered a lecture entitled, "2001, A Broadcasting Odyssey." In that lecture I said:

"As we evaluate tomorrow's TV and HDTV and plan for its implementation, we must bear in mind that today's *standard of service* enjoyed by the viewer will not be his or her *level of expectation* tomorrow. *Good enough*, is no longer *perfect*, and may become wholly unsatisfactory."

"Quality is a moving target, both in programs and in technology. Our judgements as to the future must not be based on today's performance, nor on minor improvements thereto."

Television's quality target continued to move, and in 1987 the FCC sought private sector advice and formed the Advisory Committee on Advanced Television Service (ACATS), under the chairmanship of Richard E. Wiley, and charged it to study the problems of the terrestrial broadcasting of advanced television, to test proposed systems, and to make a recommendation to the FCC by the second quarter of 1993 for selecting a single terrestrial HDTV transmission standard for the U.S.

In the ACATS process, advanced television (ATV) system proposals peaked at twenty one, but by 1990 they had shrunk to nine. Only two of these were HDTV simulcast systems, and all were analog designs.

In April of 1990, then FCC Chairman Sikes had announced:

".. the Commission's intent is to select a simulcast high-definition television standard that is compatible with the current 6 MHz channelization plan but employing new design principles independent of NTSC technology. We do not envision ... that the Commission would adopt an enhanced definition standard, if at all, prior to reaching a final decision on an HDTV standard."

America had a goal—full quality terrestrial HDTV.

Two years and eight months into the U.S. FCC Advisory Committee Advanced TV process, on June 1, 1990, General Instrument proposed an all digital terrestrial HDTV system, and television was forever changed. The digital era had begun and analog television was doomed worldwide! Television was to make its most fundamental technological change since its invention and its subsequent colorization.

By 1991, only five HDTV systems remained, and, of these, only one was a hybrid analog/digital system—NHK's Narrow MUSE. The other four were all-digital systems, and the digital changeover had extended the Advisory Committee schedule by about six months.

The systems were:

- NHK's hybrid analog/digital *Narrow MUSE* system employing frequency split pulse amplitude modulation.

- General Instrument's *DigiCipher* system employing digital DCT compression algorithms and 16 or 32 state quadrature amplitude modulation or 16/32 QAM.

- AT&T and Zenith's *Digital Spectrum Compatible HDTV* or *DSC-HDTV* employing progressive scanning, digital DCT compression algorithms, and four level vestigial sideband modulation (4VSB/2-VSB).

- Thomson, Philips, Sarnoff Labs, and NBC's *Advanced Digital HDTV* or *AD-HDTV* employing digital DCT compression algorithms and spectrally shaped quadrature amplitude modulation.

- General Instrument and MIT's *Channel Compatible DigiCipher* or *CC-DigiCipher* employing progressive scan, digital DCT compression algorithms, and 16 or 32 state quadrature amplitude modulation (16/32 QAM).

In September of 1990, in its First Report and Order, the FCC decided:

"We do not find it useful to give further consideration to systems that use additional spectrum to 'augment' an existing 6 MHz television channel to provide NTSC compatible service."

"Consistent with our goal of ensuring excellence in ATV service, we intend to select a simulcast high-definition television system."

"A simulcast system also will be spectrum efficient and facilitate the implementation of advanced television service. Such a system will transmit the increased information of an HDTV signal in the same 6 MHz channel space used in the current television channel plan."

Thus, as of 1990, the FCC and the private sector Advisory Committee had abandoned "enhanced" and "augmentation" systems from consideration, focused further work on incompatible HDTV simulcast systems, and ensured that complete and objective tests would be made on all proponent systems before the approval of any HDTV system.

This was best expressed by FCC Chairman Sikes when he said:

"I understand the concerns of those who believe in an incremental, step-by-step progression toward full HDTV.. .but pursuing extended-definition options would tend to maximize transition costs for both industry and consumers. Stations presumably would need to make a series of sequential investments, as they inched toward full HDTV. At the same time, however, consumers almost certainly would be confused, and would probably resist buying equipment which, in relatively short order, might be rendered obsolete." ... "The FCC cares enough about broadcasting and the service it provides the public to want the very best—full HDTV—and not some incremental solution to this formidable challenge."

The private sector Advanced Television Test Center (ATTC) laboratory completed the objective laboratory tests and "expert viewer" psychophysical tests of the five systems by August 1992. The "nonexpert" psychophysical tests were completed at the Advanced Television Evaluation Lab (ATEL) in Canada in October 1992. The laboratory test reports were issued by December 1992 in preparation for a meeting of the Special Panel of the FCC Advisory Committee in February 1993.

In parallel with this work, the FCC issued its second Notice of Proposed Rule Making, or NPRM. In it, the FCC proposed how the HDTV service would be defined, and the time schedule for its implementation and for the replacement of the NTSC service.

This second FCC NPRM stated:

"We envision HDTV ... will eventually replace existing NTSC. In order to make a smooth transition to this technology, we earlier decided to permit delivery of advanced television on a separate 6 MHz (simulcast) channel. In order to continue to promote spectrum efficiency, we intend to require broadcasters to "convert" entirely to HDTV—i.e., to surrender one 6 MHz frequency and broadcast only in HDTV once HDTV becomes the prevalent medium."

In May of 1992, in its Second Report and Order for implementing the HDTV service, the Commission decided to make a block allotment of frequencies for HDTV, and broadcasters would have the first option on these frequencies.

In its third Notice of Proposed Rule Making, the FCC proposed to transition broadcasting to an all HDTV service, and to require broadcasters to surrender one of their two paired channels in 15 years from the date an HDTV standard is set and a final table of HDTV channel allotments is effective. At this time the NTSC service would be abandoned, but this schedule would be periodically reviewed.

Thus, the FCC was finalizing the regulatory procedures and rules that would govern HDTV terrestrial broadcasting.

The ACATS recommendation of an HDTV system was to have been made early in 1993, but in November 1992, a funny thing happened on the way to the recommendation. Toward the end of the testing process, and based on the test results, each of the system proponents identified a series of "improvements" for their systems, and requested permission of the FCC Advisory Committee to implement the improvements.

A Technical Sub-Group of the Advisory Committee chaired by Dr. Irwin Dorros and myself was appointed to review the improvement proposals, and approve those that were considered appropriate. This Technical Sub-group met on November 18, 1993 and approved many of the proposals.

The Special Panel of the Advisory Committee met the week of February 8, 1993 to consider the test results and the system improvements with a view toward selecting a final HDTV system to recommend to the Advisory Committee.

While all the systems produced good HDTV pictures in a 6 MHz channel, none of the systems were judged to have performed sufficiently well to be selected as the single standard at that time. But, since the all-digital systems performed significantly better than the hybrid analog/digital Narrow Muse system, this system was dropped from further consideration. The Special Panel approved making the improvements for the four all-digital systems, and recommended expeditious re-testing of the systems.

Meanwhile, the final four digital system proponents began to examine the possibility of combining their systems into a single best-of-the-best HDTV system through a consortium that came to be known as the "Grand Alliance". The Grand Alliance was formed on May 24, 1993 by the four digital HDTV system proponents—AT&T/Zenith, General Instrument, DSRC (Sarnoff)/Thomson/Philips, and MIT.

The FCC Advisory Committee assigned its Technical Subgroup, including Official Observers from Canada, Mexico, the EBU, and NHK—and still under the chairmanship of Dr. Dorros and myself—the task of reviewing the Grand alliance proposal, modifying it as necessary, selecting final specifications, and approving the system for prototype construction.

Following detailed system review and modification, the Grand Alliance and the Technical Subgroup recommended the basic system parameters:

- The system would support two, and only two, scanning rates of 1080 active lines with 1920 square pixels-per-line interlace scanned at 59.94 and 60 fields/second and 720 active lines with 1280 pixels-per-line progressively scanned at 59.94 and 60 frames/second. Both formats would also would operate in the progressive scanning mode at 30 and 24 frames/second.

- The system would employ MPEG-2 compatible video compression and transport systems.

- The system would use the Dolby AC-3, 384 Kbits/s audio system.

Following the subsystem transmission tests of the vestigial sideband (VSB) system and the quadrature amplitude modulation (QAM) system, the 8VSB system was approved on February 24, 1994.

The Grand Alliance system was built, tested, and field tested to verify that it performed better than any of the four separate proponent HDTV systems. It was the-best-of-the-best.

At this late date, in the Spring of 1995, FCC Chairman Reed Hundt required the Advisory Committee to include several standard-definition or SDTV formats in the DTV standard, and—without further SDTV tests—the SDTV formats were added to the ATSC scanning formats, and the planned transition to HDTV in America became a transition to SDTV and HDTV.

The ACATS digital TV and HDTV standard was recommended to the FCC by the Advisory Committee on November 28, 1995.

In another last minute change, the FCC promoted a series of meetings among the broadcasters, the consumer equipment manufacturers, and members of the computer industry to agree on a compromise in the scanning formats of the ATSC standard. Thus, the scanning formats were privatized and are now private sector ATSC standards. Following this compromise, the ATSC standard was finally approved by the FCC on December 24th, 1996 and was mandated for terrestrial DTV/HDTV broadcasting. 81 days later, on April 3, 1997, the FCC adopted a digital channel assignment plan and the DTV service rules.

The FCC's fifth Report and Order in the proceeding on digital television was summarized by the FCC as follows:

"The overarching goal in this proceeding is to provide for the success of free, local digital broadcast television. To bolster DTV's chance for success, the Commission's decisions allow broadcasters to use their channels according to their best business judgement, as long as they continue to offer free programming on which the public has come to rely. Specifically, broadcasters must provide a free digital video programming service that is at least comparable in resolution to today's service and aired during the same time periods as today's analog service."

"Broadcasters will be able to put together whatever package of digital product they believe will best attract customers and to develop partnerships with others to help make the most productive and efficient use of their channels. Giving broadcasters the flexibility in their use of their digital channel will allow them to put together the best mix of services and programming to stimulate consumer acceptance of digital technology and the purchase of digital receivers."

"The Commission requires the affiliates of the top four networks (ABC, CBS, Fox, and NBC) in the top 10 markets to be on-the-air with a digital signal by May 1, 1999. Affiliates of the top four networks in markets 11–30 must be on-the-air by November 1, 1999."

"An important goal in this proceeding is the return of the analog (NTSC) spectrum at the end of the DTV transition period. The Commission has set a target of 2006 as a reasonable end-date for NTSC service. The Commission will review that date in its periodic reviews, which will be conducted every two years to allow evaluation of the progress of DTV and changes in Commission rules, if necessary."

Under a separate voluntary agreement made at the urging of the FCC, some group owners—including CBS—agreed to have some top-ten market digital stations on-the-air by November 1, 1998—an on-air date that was successfully met.

Thus, after 9 years, 3 months and 22 days of study, debate, design, construction, testing, and rulemaking, America's terrestrial broadcasters have the ATSC digital TV and HDTV system that, if used promptly, will assure their competitive parity with other 21st century distribution media.

The FCC has posed an aggressive rollout, a short transition period, and outlined a difficult DTV broadcasting schedule. Moreover, this digital transition will not occur in a "free marketplace" environment. The Federal Government wants to recover the present NTSC spectrum in the shortest possible time to auction and re-use it, and this will result in constant government pressure to complete the digital transition in the shortest possible time. This, in turn, will foreshorten the transition period over that which would normally occur in a fully "free marketplace".

In the past, all competition in the TV marketplace was based on the same general technical quality. It was 525 line NTSC from the camera on stage to the receiver in the home, and senior managers never had to make decisions on program presentation quality. Hereafter, there will be a wide range of technical qualities delivered to viewers from SDTV, through "SDTV multiplex" programming, to full quality HDTV. Technical quality will become an increasing factor in the competition for viewers. Wide screen HDTV will be offered by various television media and, thus, HDTV will always be just a *channel click* away.

Since the previous edition of this book appeared, enormous progress has been made in the implementation of DTV and HDTV in the U.S. The transition from analog to digital techniques in all phases of the television process now advances at an explosive pace. As we move into the 21st century, a flash picture of the situation is in order.

The conversion of television stations to digital operation continues on schedule. Today, 1615 stations out of a total of 1619, have applied for a license to construct their DTV/HDTV transmission facilities. 525 stations have been granted licenses, and 150 stations are now on the air, reaching 64 percent of all American television households.

We now have the world's only HDTV production standard: the ITU-R Recommendation BT. 709-3, based upon a Common Image Format of 1920 pixels per line, an aspect ratio of 16:9, and with 1080 lines per picture progressively scanned at 24, 25, and 30 frames per second and both interlace and progressively scanned at 50 and 60 pictures per second.

In the last year there has been a massive launch of digital HDTV programming in the U.S. Virtually every major sporting event is now covered in HDTV, while episodic drama for television is moving rapidly to the new medium. At one television network, CBS, nearly all the primetime entertainment schedule, 18 hours per week, is broadcast in HDTV. Overall, more than 150 hours per week of HDTV programming is being delivered to American television audiences by broadcasters, cable operators, and direct-to-home satellite services.

In the world of feature movies for the cinema, the 24 frames per second progressive version of the Common Image Format standard is making possible the electronic production of film-style programs for both television and the cinema. Thus, the "Episode 2" of the Lucas Films' "Star Wars" is produced entirely electronically with the camera photography using the 16:9 wide screen Common Image Format of 1920/1080 at 24 frames per second. The same technique has been used by Crest Entertainment's production of 21 one-hour episodes of "The Secret Adventures of Jules Verne", and the current production of nine full-length features for cinema exhibition.

While the number of programs produced using electronic image capture is increasing steadily, it is to be noted that up to 90 percent of all commercial primetime programs in the U.S. have been produced routinely in high definition—namely, 35 mm film. In most of the world, 35 mm film means cinema. In the U.S., it also means television. Thus, in starting an HDTV service, programs can be shot in the wide screen format on the same film and using the same cameras as those used for current NTSC productions, with the film negative being transferred to HDTV videotape for post production and broadcast.

Since a large inventory of programming is required to reach the critical mass required to motivate consumer response, the sale of receivers is necessarily the final step in the transition to DTV and HDTV. It is now projected that the number of US TV households with wide screen digital TV and HDTV receivers will exceed 500,000 by the end of this year, and achieve a 30 to 50 percent market penetration by 2006.

It is important to remember that HDTV is not just pretty pictures for today's small screen television sets. Rather, it is a wholly new digital platform that will support the larger and vastly improved displays now coming to market. HDTV viewed on such large wide screen displays will create an entirely new viewing experience in the home, and the *home theater* will at last become a reality.

In accepting this reality, we do well to heed the advice of Alexander Pope, who, in his "Essay on Criticism," warned:

"Be not the first by whom the new are tried,
Nor yet the last to lay the old aside."

Joseph A. Flaherty

October, 2000

Preface

DTV—and more importantly, HDTV—are on the air. Stations coast-to-coast are broadcasting regular HDTV programming and beginning to fashion new business strategies made possible by this new technology. While broadcasters are the most visible early adopters of DTV, many other industries are gearing up and—indeed—are already making commercials, television programs, and motion pictures using HDTV equipment made possible by the ATSC DTV standard. Put side-by-side, the breadth of the HDTV applications now finding commercial success is staggering—from NASA to postproduction to medical imaging.

This is an exciting time for video professionals and for consumers. Our new digital television system, the product of more than a decade of work by scientists and engineers around the world, is on the air. It works. It makes beautiful pictures. It provides capabilities never before possible. Indeed, the ATSC DTV system offers features thought to be little more than fantasy just 10 years ago.

In the course of preparing this third edition, I discussing the status of DTV implementation with a number of television station engineers working to put their digital signals on the air. The word from the trenches was that implementation problems are common. Some stations are running into tower licensing roadblocks. Others have horror stories to tell about the difficulties they have found in interfacing new digital gear and getting it to work. Still others are baffled by the abundance of scanning formats.

These are, of course, all important issues. But, in the greater scheme of things, I've got to say they're *just details*. Important details to be sure. But just details.

The launch of every new technology has been fraught with growing pains. Consider what television engineers of the 1950s faced when their management decided that it was good business to convert to color. In 1954, getting a black-and-white signal on the air was an accomplishment, let alone holding the frequency and waveform tolerances sufficiently tight that the NTSC color system would look halfway decent. And then there were the receivers: big, expensive, and unstable. Round picture tubes with a long necks. The sets commandeered living rooms across America. And service these things? There were two dozen convergence adjustments alone in the RCA CTC-15 chassis, the first design that really worked—and it was released nearly a decade after the first color broadcasts began.

These problems were, in the final analysis, only details. Obstacles to be overcome. Challenges to be met. Such is the case with DTV. As this book goes to press, uncertainties over the performance of DTV receivers under certain conditions (mostly urban multipath) continues. A number of initiatives underway by the ATSC and other organizations were seeking to identify what issues—if any—could (or should) be addressed. If the history of color television tells us anything, however, it is that a well-conceived standard will enable enormous improvements in end-to-end performance as receiver manufacturers in partnership with transmission system engineers fine-tune the overall system.

The ATSC digital television standard is, of course, just technology—a technical scheme that permits certain tasks to be accomplished. As such, it is neutral insofar as intrinsic benefits. DTV will, instead, become whatever we choose make it.

At this point in the transition to DTV, many challenges exist. Many more will be uncovered as the early-adopters learn by doing. The bottom line, however, is that DTV is the future of the professional video industry.

The DTV train has left the station, and there's no turning back. It promises to be a splendid ride.

About the Book

DTV: The Revolution in Digital Video, Third Edition, examines the technology of digital high-definition video for a wide range of applications. The underlying technology of DTV is examined, and examples are provided of typical uses. New developmental efforts also are explained and the benefits of the work are outlined.

This publication is directed toward technical and engineering personnel involved in the design, specification, installation, and maintenance of broadcast television systems and non-broadcast professional imaging systems. The basic principles of digital television—in general—and HDTV—in particular—are discussed, with emphasis on how the underlying technologies influence the ultimate applications.

The author has made every effort to cover the subject of digital television comprehensively. Extensive references are included at the end of each chapter to direct readers to sources of additional information.

The entire DTV system is based—of course—on the work of the Advanced Television Systems Committee (ATSC). The ATSC has published a comprehensive set of documents that describe the DTV service. Several chapters in this book draw heavily upon the landmark work of the ATSC, and the author gratefully acknowledges this extensive contribution to *DTV: The Revolution in Digital Video*, Third Edition.

For those who are interested in acquiring a complete set of the ATSC documents, the applicable standards are available from the ATSC World Wide Web site (www.atsc.org). A printed version of the documents is also available for purchase from the SMPTE, White Plains, N.Y. (www.smpte.org), and the National Association of Broadcasters, Washington, D.C. (www.nab.org).

Because of the rapidly changing nature of DTV implementation, readers are encouraged to check in regularly with the Internet sites listed previously. Another resource for DTV developments is the author's web site, www.technicalpress.com. Visitors can find articles, background information, and links to DTV-related organizations.

Another valuable resource is the SMPTE Television Standards on CD-ROM. This product, available for purchase from the SMPTE, contains all existing and proposed Standards, Engineering Guidelines, and Recommended Practices for television work. In this era of digital video, this product is indispensable.

In the Third Edition

A new edition of *DTV: The Revolution in Digital Video* is being released because of the rapid changes in digital television implementation plans and industry standards that have occurred within the past two years. More than a dozen new standards have been adopted or proposed since the second edition was released that directly impact the DTV transition now underway. In the third edition, every chapter has been updated with new information.

Technology for Tomorrow

Advances in imaging resolution and clarity are the driving forces behind many new product offerings. In computer applications, increased imaging resolution is an important area of current and future work. As computer concepts merge with those of the video industry, HDTV will gain even more importance.

The field of science encompassed by DTV and HDTV is broad and exciting. It is an area of growing importance to broadcast, professional video, military, and industrial customers, and it is soon to be of keen interest to consumers as well. It is the consumer market, of course, that will make digital television a household word.

It is the goal of *DTV: The Revolution in Digital Video*, Third Edition, to bring these diverse concepts and technologies together in an understandable form.

Jerry C. Whitaker

October, 2000

DTV Handbook

The Revolution in Digital Video
Third Edition

Dedicated to the memory of

Howard Miller

one of the pioneers of television engineering

The Road to DTV

"Standardization at the present stage is dangerous. It is extremely difficult to change a standard, however undesirable it may prove, after the public has invested thousands of dollars in equipment. The development goes on, and will go on. There is no question that the technical difficulties will be overcome."

This warning does not address the problems faced by high-definition television or the delivery of video by fiber to the home. The writer is addressing the problems faced by *television* in a book published in April 1929. Technology changes, but the problems faced by the industry do not.

1.1   Introduction

After decades of research by companies and organizations around the world, and after years of standardization efforts and testing, a digital transmission system for high-definition television (HDTV) for the United States is now on the air. The historic action by private industry and the Federal Communications Commission (FCC) paved the way for consumers to enjoy the benefits of top-quality video—and accompanying audio—from a diverse group of industries, including:

- Terrestrial television broadcasting

- Cable television

- Direct-broadcast satellite television

- Personal computers

- Video-on-demand services

The formal process, which began in 1987, and continues (after a fashion) to this day, has led to the establishment of a new television broadcasting standard. It is, in fact, the third American TV broadcasting standard, with monochrome and color being the other two.

From a historical viewpoint, the battle over HDTV was simply a repeat of the two previous U.S. standardization fights, with a few new twists. Until the formation of the Grand

Alliance in 1993, which brought together the competing digital HDTV systems, the fight—and it was a fight—represented politics on a grand scale. The process evolved from a discussion of technical merits to one centered on national pride and national security.

The road to HDTV has been long and difficult. To fully appreciate the accomplishment, it is necessary to review the pioneering efforts of the past.

1.1.1 The Pioneers

The list of inventors whose contributions made television possible is lengthy and distinguished. Two names, however, stand out:

- Philo Farnsworth

- Vladimir Zworykin

The work of a nontechnical pioneer—David Sarnoff—also must be recognized.

Philo Farnsworth

Legend has it that Philo Farnsworth conceived of electronic television as a 15-year-old high school sophomore in Rigby, Idaho, a small town about 200 miles north of Salt Lake City. When he was 19 years old, he met a financial expert by the name of George Everson in Salt Lake City. Farnsworth persuaded Everson to try to secure venture capital for an all-electronic television system.

The potential financial backers of this unproven young man with unorthodox ideas had one main concern: whether anyone else was investigating an electronic method of television. Obviously, many people were interested in capturing control over patents of a vast new field for profit. If no one else was working on this method, then Farnsworth had a clear field. If, on the other hand, other companies were working in secret without publishing their results, then Farnsworth would have little chance of receiving the patent awards and the royalty income that surely would result. Farnsworth and Everson were able to convince their financial investors that they alone were on the trail of a total electronic television system.

Farnsworth established his laboratory first in Los Angeles and later in San Francisco, at the foot of Telegraph Hill. He was the proverbial lone basement experimenter. It was in 1927 at his Green Street laboratory in San Francisco that Farnsworth gave the first public demonstration of the television system he had dreamed of for 6 years. He was not yet 21 years of age.

Because he was quick to develop the basic concepts of an electronic television system, he had an edge on most other inventors in the race for patents. His patents included the principle of blacker-than-black synchronizing signals, linear sweep, and the ratio of forward sweep to retrace time.

Farnsworth's original "broadcast" included the transmission of graphic images, film clips of a Jack Dempsey-Gene Tunney fight, and scenes of Mary Pickford combing her hair (from her role in *Taming of the Shrew*). In his early systems, Farnsworth could transmit pictures with 100- to 150-line definition at a repetition rate of 30 lines/s. This pioneering demonstration set in motion the progression of technology that would lead to commercial broadcast television a decade later.

Vladimir Zworykin

Russian-born Vladimir Zworykin immigrated to the United States after World War I and went to work for the Westinghouse company in Pittsburgh. Zworykin had left Russia for America to develop his dream: television. During his stay at Westinghouse, from 1920 to1929, Zworykin performed some of his early experiments in television. His conception of the first practical TV camera tube, the *Iconoscope* (1923), and his development of the *kinescope* picture tube formed the basis for subsequent advances in the field. Zworykin is credited by most historians as "the father of television."

Zworykin's Iconoscope (from Greek for "image" and "to see") consisted of a thin aluminum-oxide film that was supported by a thin aluminum film, then coated with a photosensitive layer of potassium hydride. With this crude camera tube and a cathode-ray tube (CRT) as the picture reproducer, he had the essential elements for electronic television.

Continuing his research, Zworykin developed an improved Iconoscope 6 years later that employed a relatively thick, 1-sided target area. He had, in the meantime, continued work on improving the quality of the CRT and presented a paper on his efforts to the Eastern Great Lakes District Convention of the Institute of Radio Engineers (IRE) on November 18, 1929. The presentation attracted the attention of another Russian immigrant, David Sarnoff, then vice president and general manager of the Radio Corporation of America (RCA). Sarnoff persuaded Zworykin to join RCA Victor in Camden, New Jersey, where he was made director of RCA's electronics research laboratory. The company provided the management and financial backing that enabled Zworykin and the RCA scientists working with him to develop television into a practical system.

By 1931, with the development of the Iconoscope and CRT well under way, the launch of electronic television was imminent—and Sarnoff and RCA were ready for the new industry of television.

Farnsworth held many patents for television, and through the mid-1930s remained RCA's fiercest competitor in developing new technology. Indeed, Farnsworth's thoughts seemed to be directed toward cornering patents for the field of television and protecting his ideas. In the late 1930s, fierce patent conflicts between RCA and Farnsworth flourished. They were settled in September 1939 when RCA capitulated and agreed to pay continuing royalties to Farnsworth for the use of his patents.

David Sarnoff and Contemporaries

The radio networks of the 1930s (NBC and CBS) took early leads in paving the way for commercial television. NBC, through the visionary eyes of David Sarnoff and the resources of RCA, stood ready to undertake pioneering efforts to advance the new technology. Sarnoff accurately reasoned that television could establish an industrywide dominance only if all television set manufacturers and broadcasters were using the same standards. He knew this would occur only if the FCC adopted suitable standards and allocated the needed frequency spectrum. Toward this end, in April 1935, Sarnoff made a dramatic announcement that RCA would put millions of dollars into television development. One year later, RCA began field-testing TV transmission methods from a transmitter atop the Empire State Building.

In a parallel move, CBS (after years of deliberation) was ready to make its views public. In 1937, the company announced a $2 million experimental program that consisted of field-testing various television systems. This was a complete change of direction from CBS's previous stance. Several years earlier, in 1931, the network had put an experimental TV station on the air in New York City and transmitted programs for more than a year before becoming disillusioned with the commercial aspects of the new medium.

The Allen B. DuMont Laboratories also made significant contributions to early television. Although DuMont is best known for CRT development and synchronization techniques, the company's major historical contribution was its production of early electronic television sets for the public beginning in 1939.

It was during the 1939 World's Fair in New York, and the Golden Gate International Exposition in San Francisco the same year, that live and filmed television was demonstrated on a large scale for the first time. Franklin Roosevelt's World's Fair speech (April 30, 1939) marked the first use of television by a U.S. president. The public was fascinated by the new technology.

TV sets were available for sale at the World's Fair RCA pavilion. Prices ranged from $200 to $600, a huge sum at the time. Screen sizes ranged from 5 to 12 in, with the 12-in set being the "big screen" model. Because CRT technology at that time did not permit wide deflection angles, the pictures tubes were long. So long, in fact, that the devices were mounted vertically (in the larger-sized models). A hinge-mounted mirror at the top of the receiver cabinet permitted viewing.

At the San Francisco exposition, RCA had another large exhibit that featured live television. The actors and models used in the demonstrations could stand the hot lights for only a limited period. The studio areas were small and suitable only for interviews and commentary. People were allowed to walk through the television studio and stand in front of the camera for a few seconds. Friends and family members watched on monitors outside the booth. It was great fun; the lines were always long and the crowds enthusiastic. The interest caused by these first mass demonstrations of television sparked a keen interest in the commercial potential of TV broadcasting. Both expositions ran for a second season in 1940, but the war had started in Europe, and television development was about to grind to a halt.

1.1.2 Transmission Standards

In the late 1930s, as television was nearing the point of commercialization, the FCC insisted that the standards for television, as well as for other services, be set only when the industry was in substantial agreement on the form that the standards should take. Following this direction, the Radio Manufacturers Association (RMA)—the forerunner of the Electronics Industries Association (EIA)—set up a committee in 1936 to recommend standards for a commercial TV broadcasting service. In December 1937, the committee advised the FCC to adopt the RCA 343-line/30-frame system, which had been undergoing intensive development since 1931. The RCA system was the only one tested under both laboratory and field conditions. A majority of the RMA membership, however, objected to the RCA system because of the expectation that rapidly advancing technology would soon render this marginal system obsolete and, perhaps more importantly, it would place members at a competitive disadvantage. RCA had been prepared to immediately start manufacturing TV studio equipment and sets. Commercial development of television was, thus, put on hold.

At an FCC hearing in January 1940, a majority of the RMA membership was willing to embrace the RCA system, now improved to 441 lines. However, a strong dissenting minority (Zenith, Philco, and DuMont) still was able to block any action.

The FCC, chaired by James Lawrence Fly, sent the standards-setting process back to the drawing board. The result was the establishment of the *National Television System Committee* (NTSC). The concept of the NTSC arose in a meeting between Fly and Dr. W. R. G. Baker, a General Electric company executive and director of engineering of the RMA. The NTSC was formed as a private sector organization and placed under the sponsorship of the RMA. The deliberations were open to all members of the industry who were technically qualified to participate, whether or not they were members of the RMA.

The result was that the NTSC functioned essentially as a forum to investigate various options. DuMont proposed a 625-line/15-frame/4-field interlaced system. Philco advocated a 605-line/24-frame system. Zenith took the stance that it was still premature to adopt a national standard.

The original record of the NTSC was 11 volumes, totaling approximately 2000 pages. The first meeting was chaired by Dr. Baker on July 31, 1940. The work of the NTSC was organized into nine panels responsible for various areas of study.

A progress report presented to the FCC on January 27, 1941, stated that the members had reached substantial agreement on all parts of the new standard except for two points: the specification of 441 scanning lines per frame and amplitude modulation for the synchronization signals. At its final meeting on March 8, 1941, the NTSC agreed to specify 525 scanning lines per frame and rewrote the portion of the standard concerning synchronization to permit the use of frequency modulation.

The final report of the NTSC was delivered to the FCC on March 20, 1941, recommending adoption of the NTSC monochrome standard. The only opposition to the document at that time was put forward by the DuMont Laboratories, which urged that a variable number of lines and frames be used. Effective April 30, 1941, the FCC officially adopted the standard and ruled that commercial television broadcasting based on the format would be permitted as of July 1, 1941. Key elements of the standard included:

- The use of a 6 MHz radio frequency (RF) channel with the picture carrier 1.25 MHz above the bottom of the channel and the sound carrier 4.5 MHz above the picture carrier

- Vestigial sideband modulation of the picture carrier with negative modulation and preservation of the dc component

- Frequency modulation of the sound carrier

- 525 scanning lines per frame with 2:1 interlace at 30 frames (60 fields) per second (see Figure 1.1)

- 4:3 aspect ratio

In July 1941, the FCC authorized the first two commercial TV stations to be constructed in the United States, but the growth of early television was ended by the licensing freeze that accompanied World War II. By the end of 1945 there were just nine commercial TV stations authorized, with six of them on the air. The first postwar full-service commercial license was issued to WNBW, the NBC-owned station in Washington, D.C.

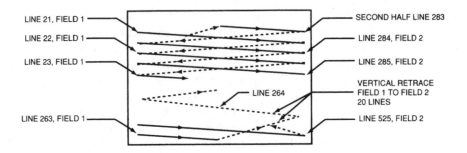

Figure 1.1 The interlaced-scanning pattern (raster) of the television image. (*Source: Electronic Industries Association.*)

As the number of television stations on the air began to grow, the newest status symbol became a TV antenna on the roof. Sets were expensive and not always reliable. Sometimes there was a waiting list to get one. Nobody cared; it was all very exciting—pictures through the air. People would stand in front of a department store window just to watch a test pattern.

Color Standard

During early development of commercial television systems—even as early as the 1920s—it was assumed that color would be demanded by the public. Primitive field-sequential systems were demonstrated in 1929. Peter Goldmark of CBS showed a field-sequential (*color filter wheel*) system in the early 1940s and promoted it vigorously during the postwar years. Despite the fact that it was incompatible with existing receivers, had limited picture-size possibilities, and was mechanically noisy, the CBS system was adopted by the FCC as the national color television standard in October 1950.

The engineering community (CBS excepted) felt betrayed. Monochrome television was little more than 9 years old, but with a base of 10 to 15 million receivers. Broadcasters and the public were faced with the possibility that much of their new, expensive equipment would become obsolete. The general wisdom was that color must be an adjunct to the 525/30 monochrome system so that existing terminal equipment and receivers could accept color transmissions.

Was the decision to accept the CBS color wheel approach simply a political one? Not entirely, because it was based on engineering tests presented to the FCC in early 1950. Contenders were the RCA dot-sequential, the CTI (Color Television Incorporated) line-sequential, and the CBS field-sequential systems. The all-electronic compatible approach was in its infancy, and there were no suitable display devices. Thus, for a decision made in 1950 based on the available test data, the commission's move to embrace the color wheel system was reasonable. CBS, however, had no support from other broadcasters or manufacturers; indeed, the company had to purchase the Hytron-Air King company to produce color TV sets (which would also receive black and white NTSC). Two hundred sets were manufactured for public sale.

Programming commenced on a 5-station east coast network on June 21, 1951, presumably to gain experience. Color receivers went on sale in September, but only 100 or so were sold. The last broadcast took place on October 21, 1951. The final curtain fell in November of that year when the National Production Authority (an agency created during the Korean war) imposed a prohibition on the manufacture of color sets for public use. Because the production of monochrome sets was not restricted, cynics interpreted this action as a design to get CBS off the hook and save face in the process.

The proponents of compatible, all-electronic color systems, meanwhile, were making significant advances. RCA had demonstrated a tricolor delta-delta kinescope. Hazeltine demonstrated the constant luminance principle, as well as the "shunted monochrome" idea. General Electric introduced the frequency interlaced color system. Philco showed a color signal composed of wideband luminance and two color-difference signals encoded by a quadrature-modulated subcarrier.

Because of the controversy surrounding the CBS color standard and the technical progress being made on various fronts, Dr. Baker reactivated the NTSC in January 1950. The panel structure was reorganized to address the particular problems of color television, and the membership was greatly expanded. The work of the second NTSC was contained in 18 volumes totaling more than 4000 pages. With Dr. Baker again serving as chairman, the second NTSC divided the work among eight panels. The first meeting was held in June 1951 to reorganize the committee for the purpose of pooling resources in the development of a compatible system. By November, a system employing the basic concepts of today's NTSC color system was demonstrated.

Field tests showed certain defects, such as sound interference caused by the choice of color subcarrier. This was corrected by the selection of a slightly different frequency, but at the expense of lowering the frame rate to 29.97 Hz. Finally, RCA demonstrated the efficacy of unequal I and Q color-difference bandwidths. Following further field tests, the proposal was forwarded to the FCC on July 22, 1953. All traces of the earlier CBS color system controversy had disappeared, and the industry was able to present a truly united front.

Still, a major problem remained: the color kinescope. It was expensive and yielded only a 9×12-in picture. Without the promise of an affordable large-screen display, the future of color television was uncertain. Then came the development of a method of applying the phosphor dots directly on the faceplate together with a curved shadow mask mounted directly on the faceplate, a breakthrough developed by the CBS-Hytron company.

The new color system was demonstrated for the FCC on October 15, 1953. On December 17 of the same year, the FCC approved the color standard. Color service was authorized to begin on January 23, 1954. The second NTSC was officially disbanded on February 4, 1954.

Only a few changes were made to the monochrome standard to include color. Of the key elements noted previously for the monochrome standard, only the frame/field rate changed, and that by the ratio of 1000/1001. The color subcarrier was specified to be 1/2 an odd multiple (455) of the horizontal frequency to minimize the visibility of the subcarrier in the picture. To minimize beats between the sound carrier and the color subcarrier, the sound carrier was specified to be 1/2 an even multiple of the horizontal frequency. To ensure compatibility with monochrome receivers, the sound carrier remained the same as for monochrome: 4.5 MHz. For monochrome, the ratio of the sound carrier to the horizontal frequency was close to, but not precisely, 286. For color, it was set precisely to 286. As a result, the horizontal frequency was changed slightly, thus the vertical frequency was changed slightly. This gave rise to the NTSC color vertical frequency of 59.94 Hz, or 1000/1001 × 60 Hz. A modulated subcarrier containing the color information was added, and the color burst was added to the synchronizing waveform. Some signal tolerances also were made tighter than the monochrome standard.

As a matter of fact, the *phase alternation line* (PAL) principle was tried, but practical hardware to implement the scheme would not be available until 10 years later.

Those who suspect conspiracy behind every happening suggested that the field-sequential system was the best and that RCA forced an inferior system on the public. However, the facts show that many of the prime features of the NTSC system were advanced by other research organizations. After more than 4 decades of use, NTSC still possesses a remarkable aptitude for improvement and manipulation. Even with the advantage of compatibility for monochrome viewers, however, it took 10 years of equipment development and programming support from RCA and NBC before the general public started buying color receivers in significant numbers.

In France, Germany, and other European countries, engineers began work in the mid-1950s aimed at improving upon the NTSC system. France developed the SECAM (*Sequentiel Couleur avec Mémoire*) system, and Germany produced the PAL system. Proponents of each of the three systems still debate their advantages today.

1.1.3 Conventional TV Systems

TV transmitters for NTSC service in the United States operate in three frequency bands:

- *Low-band VHF*—channels 2 through 6 (54 to 72 MHz and 76 to 88 MHz)

- *High-band VHF*—channels 7 through 13 (174 to 216 MHz)

- *UHF*—channels 14 through 69 (470 to 806 MHz).

Table 1.1 shows the frequencies used by TV broadcast stations. Maximum power output limits are specified by the FCC for each type of service. The maximum *effective radiated power* (ERP) for low-band VHF is 100 kW, for high-band VHF it is 316 kW, and for UHF it is 5 MW.

Aside from output power, the second major factor that affects the coverage area of a TV station is antenna height, known as *height above average terrain* (HAAT). HAAT takes into consideration the effects of the geography in the vicinity of the transmitting tower. The maximum HAAT permitted by the FCC for a low- or high-band VHF station is 1000 ft (305 m) east of the Mississippi River, and 2000 ft (610 m) west of the Mississippi. UHF stations are permitted to operate with a maximum HAAT of 2000 ft (610 m) anywhere in the United States, including Alaska and Hawaii.

The ratio of visual output power to aural power can vary from one installation to another, but the aural typically is operated at 10 to 20 percent of the visual power. This difference is the result of the reception characteristics of the two signals. It requires much greater signal strength at the consumer's receiver to recover the visual portion of the transmission than it does to recover the aural portion The aural power output is intended to be sufficient for

Table 1.1 Channel Designations for VHF- and UHF-TV Stations in the United States

Channel Designation	Frequency Band, MHz	Channel Designation	Frequency Band, MHz	Channel Designation	Frequency Band, MHz
2	54–60	30	566–572	57	728–734
3	60–66	31	572–578	58	734–740
4	66–72	32	578–584	59	740–746
5	76–82	33	584–590	60	746–752
6	82–88	34	590–596	61	752–758
7	174–180	35	596–602	62	758–764
8	180–186	36	602–608	63	764–770
9	186–192	37	608–614	64	776–782
10	192–198	38	614–620	65	782–788
11	198–204	39	620–626	66	788–794
12	204–210	40	626–632	67	794–800
13	210–216	41	632–638	68	800–806
14	470–476	42	638–644	69	806–812
15	476–482	43	644–650	70	812–818
16	482–488	44	650–656	71	818–824
17	488–494	45	656–662	72	824–830
18	494–500	46	662–668	73	830–836
19	500–506	47	668–674	74	836–842
20	506–512	48	674–680	75	842–848
21	512–518	49	680–686	76	848–864
22	518–524	50	686–692	77	854–860
23	524–530	51	692–698	78	860–866
24	530–536	52	698–704	79	866–872
25	536–542	53	704–710	80	872–878
26	542–548	54	710–716	81	878–884
27	548–554	55	716–722	82	878–884
28	554–560	56	722–728	83	884–890
29	560–566				

good reception at the fringe of the station's coverage area, but not beyond. It is pointless for a consumer to be able to receive a TV station's audio signal, but not the video.

In addition to the full-power stations discussed previously, two classifications of low-power TV stations have been established by the FCC to meet certain community needs. They are:

- *Translators*, low-power systems that rebroadcast the signal of another station on a different channel. Translators are designed to provide "fill-in" coverage for a station that cannot reach a particular community because of the local terrain. Translators operating in the VHF band are limited to 100 W (ERP), and UHF translators are limited to 1 kW.

- *Low-power television* (LPTV), a service established by the FCC to meet the special needs of particular communities. LPTV stations operating on VHF frequencies are limited to 100 W ERP, and UHF stations are limited to 1 kW. LPTV stations originate their own programming and can be assigned by the FCC to any channel, as long as full protection against interference to a full-power station is afforded.

TV Transmission Standards

Conventional television broadcast signals transmitted throughout the world have the following similarities:

- All systems use two fields interlaced to create a complete frame.

- All contain luminance, chrominance, sync, and sound components.

- All use amplitude modulation to put picture information onto the visual carrier.

- Modulation polarity, in most cases, is negative (greatest power output from the transmitter occurs during the sync interval; least power output occurs during peak white).

- The sound is transmitted on an aural carrier that is offset on a higher frequency than the visual carrier, using frequency modulation in most cases.

- All systems use a vestigial lower sideband approach.

- All systems derive a luminance and two color-difference signals from red, green, and blue components.

At this point the similarities stop, and the differences begin. Three primary color transmission standards are in use:

- *NTSC*—used in the United States, Canada, Central America, most of South America and Japan. In addition, NTSC has been accepted for use in various countries or possessions heavily influenced by the United States. The major components of the NTSC signal are shown in Figures 1.2 and 1.3.

- *PAL*—used in England, most countries and possessions influenced by the British Commonwealth, many western European countries, and China. Several variation exists in PAL systems.

- *SECAM*—used in France, countries and possessions influenced by France, Russia, most of the former Soviet Bloc nations, and other areas influenced by Russia.

The three standards are incompatible for the following reasons:

- Channel assignments are made in different frequency spectra in many parts of the world. Some countries have VHF only, some have UHF only, others have both. Assignments with VHF and UHF are not necessarily the same from one country to the next.

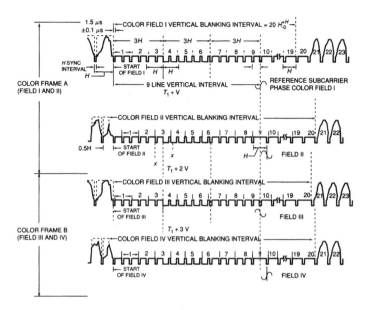

Figure 1.2 The principal components of the NTSC color system. (*Source: Electronic Industries Association.*)

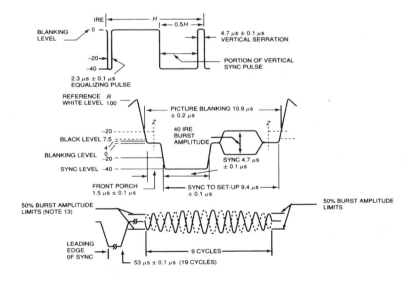

Figure 1.3 Sync pulse widths for the NTSC color system. (*Source: Electronic Industries Association.*)

- Channel bandwidths are different. NTSC uses a 6 MHz channel width. Versions of PAL exist with 6, 7, and 8 MHz bandwidths. SECAM channels are 8 MHz wide.

- Video signal bandwidths are different; NTSC uses 4.2 MHz; PAL uses 4.2, 5, and 5.5 MHz; and SECAM uses 6 MHz.

- The line structure of the signals varies. NTSC uses 525 lines per frame, 30 frames (60 fields) per second. PAL and SECAM use 625 lines per frame, 25 frames (50 fields) per second. As a result, the scanning frequencies also vary.

- The color subcarrier signals are incompatible. NTSC uses 3.579545 MHz. PAL uses 4.43361875 MHz. SECAM utilizes two subcarriers, 4.40625 and 4.250 MHz. The color subcarrier values are derived from the horizontal frequencies in order to interleave color information into the luminance signal without causing undue interference.

- The color encoding systems of all three standards differ.

- The offset between visual and aural carriers varies. In NTSC, it is 4.5 MHz. In PAL, the separation is 5.5 or 6 MHz, depending upon the PAL type. SECAM uses 6.5 MHz separation.

- One form of SECAM uses positive polarity visual modulation (peak white produces greatest power output of transmitter) with amplitude modulation for sound.

- Channels transmitted on UHF frequencies may differ from those on VHF in some forms of PAL and SECAM. Differences include channel bandwidth and video bandwidth.

It is possible to convert from one television standard to another electronically. The most difficult part of the conversion process results from the differing number of scan lines. In general, the signal must be disassembled in the input section of the standards converter, then placed in a large dynamic memory. Complex computer algorithms compare information on pairs of lines to determine how to create the new lines required (for conversion to PAL or SECAM) or how to remove lines (for conversion to NTSC). Motionless elements in the picture present no great difficulties, but motion in the picture may produce objectionable artifacts as the result of the sampling system.

1.2 Early Development of HDTV

The first developmental work on a high-definition television system began in 1968 at the technical research laboratories of Nippon Hoso Kyokai (NHK) in Tokyo [1]. In the initial effort, film transparencies were projected to determine the levels of image quality associated with different numbers of lines, aspect ratios, interlace ratios, and image dimensions. Image diagonals of 5 to 8 ft, aspect ratios of 5:3 to 2:1, and interlace ratios from 1:1 to 5:1 were used. The conclusions reached in the study were that a 1500-line image with 1.7:1 aspect ratio and a display size of 6.5 ft diagonal would have a subjective image quality improvement of more than 3.5 grades on a 7-grade scale, compared to the conventional 525-line, 4:3 image on a 25-in display. To translate these findings to television specifications, the following parameters were thus adopted:

- 2-to-1 interlaced scanning at 60 fields/s

- 0.7 Kell factor

- 0.5 to 0.7 interlace factor

- 5:3 aspect ratio

- Visual acuity of 9.3 cycles per degree of arc

Using these values, the preferred viewing distance for an 1125-line image was found to be 3.3 times the picture height. A display measuring 3×1.5 ft was constructed in 1972 using half mirrors to combine the images of three 26-in color tubes. Through the use of improved versions of wide-screen displays, the bandwidth required for luminance was established to be 20 MHz and, for wideband and narrowband chrominance, 7.0 and 5.5 MHz, respectively. The influence of viewing distance on sharpness also was investigated.

Additional tests were conducted to determine the effect of a wide screen on the realism of the display. In an elaborate viewing apparatus, motion picture film was projected on a spherical surface. As the aspect ratio of the image was shifted, the viewers' reactions were noted. The results showed that the realism of the presentation increased when the viewing angle was greater than 20°. The horizontal viewing angles for NTSC and PAL/SECAM were determined to be 11° and 13°, respectively. The horizontal viewing angle of the NHK system was set at 30°.

The so-called provisional standard for the NHK system was published in 1980. Because the NHK HDTV standard of 1125 lines and 60 fields/s was, obviously, incompatible with the conventional (NTSC) service used in Japan, adoption of these parameters raised a number of questions. No explanations appear in the literature, but justification for the values can be found in the situation faced by NHK prior to 1980. At that time, there was widespread support for a single worldwide standard for HDTV service. Indeed, the International Radio Consultative Committee (CCIR) had initiated work toward such a standard as early as 1972. If this were achieved, the NTSC and PAL/SECAM field rates of 59.94 and 50 Hz would prevent one (or all) of these systems from participation in such a standard. The 50 Hz rate was conceded to impose a severe limit on display brightness, and the 59.94 (vs. 60) Hz field rate posed more difficult problems in transcoding. Thus, a 60-field rate was proposed for the world standard. The choice of 1125 lines also was justified by the existing operating standards. The midpoint between 525 and 625 lines is 575 lines. Twice that number would correspond to 1150 lines for a *midpoint* HDTV system. This even number of lines could not produce alternate-line interlacing, then thought to be essential in any scanning standard. The nearest odd number having a common factor with 525 and 625 was the NHK choice: 1125 lines. The common factor—25—would make line-rate transcoding among the NHK, NTSC, and PAL/SECAM systems comparatively simple.

1.2.1   1125/60 Equipment Development

The 1970s saw intense development of HDTV equipment at the NHK Laboratories. By 1980, when the NHK system was publicly demonstrated, the necessary camera tubes and cameras, picture tubes and projection displays, telecines, and videotape recorders were available. Also, the choices of transmission systems, signal formats, and modulation/

demodulation parameters had been made. Work with digital transmission and fiber optics had begun, and a prototype tape recorder had been designed. Much of the HDTV equipment built to the 1125/60 NHK standard and brought to market by various vendors can be traced to these early developments.

In 1973, the NHK cameras used three 1-½-in Vidicons, then commercially available. The devices, however, lacked the necessary resolution and sensitivity. To solve this problem, NHK developed the *return-beam Saticon* (RBS), which had adequate resolution and sensitivity, but about 30 percent lag. Cameras using three RBS tubes came into production in 1975 and were used during much of the NHK HDTV system development. By 1980, another device—the *diode-gun impregnated-cathode Saticon* (DIS) tube—was ready for public demonstration. This was a 1-in tube, having a resolution of 1600 lines (1200 lines outside the 80 percent center circle), lag of less than 1 percent, and 39 dB signal-to-noise ratio (S/N) across a 30 MHz band. Misregistration among the primary color images, about 0.1 percent of the image dimensions in the earlier cameras, was reduced to 0.03 percent in the DIS camera. When used for sporting events, the camera was fitted with a 14-× zoom lens of *f*/2.8 aperture. The performance of this camera established the reputation of the NHK system among industry experts, including those from the motion picture business.

The task of adapting a conventional television display to high definition began in 1973, when NHK developed a 22-in picture tube with a shadow-mask hole pitch of 310 μm (compared with 660 μm for a standard tube) and an aspect ratio of 4:3. In 1978, a 30-in tube with hole pitch of 340 μm and 5:3 aspect ratio was produced. This tube had a peak brightness of 30 foot-lamberts (100 cd/m^2). This was followed in 1979 by a 26- in tube with 370 μm hole pitch, 5:3 aspect ratio, and peak brightness of 45 foot-lamberts (ft-L).

The need for displays larger than those available in picture tubes led NHK to develop projection systems. A system using three CRTs with Schmidt-type focusing optics produced a 55-in image (diagonal) on a curved screen with a peak brightness of about 30 ft-L. A larger image (67 in) was produced by a light-valve projector, employing Schlieren optics, with a peak brightness of 100 ft-L at a screen gain of 10.

Development by NHK of telecine equipment was based on 70 mm film to assure a high reserve of resolution in the source material. The first telecine employed three Vidicons, but they had low resolution and high noise. It was decided that these problems could be overcome through the use of highly monochromatic laser beams as light sources: helium-neon at 632.8 nm for red, argon at 514.5 nm for green, and helium-cadmium at 441.6 nm for blue. The basic elements of the laser-beam telecine are shown in Figure 1.4. To avoid variation in the laser output levels, each beam was passed though an acoustic-optical modulator with feedback control. The beams then were combined by dichroic mirrors and scanned mechanically. Horizontal scanning was provided by a 25-sided mirror, rotating at such a high speed (81,000 rpm) as to require aerostatic bearings. This speed was required to scan at 1125 lines, 30 frames/s, with 25 mirror facets. The deflected beam was passed through relay lenses to another mirror mounted on a galvanometer movement, which introduced the vertical scanning. The scanned beam then passed through relay lenses to a second mirror-polygon of 48 sides, rotating at 30 rpm, in accurate synchronism with the continuous movement of the 70 mm film.

The resolution provided by this telecine was limited by the mechanical scanning elements to 35 percent modulation transfer at 1000 lines, 2:1 aspect ratio. This level was

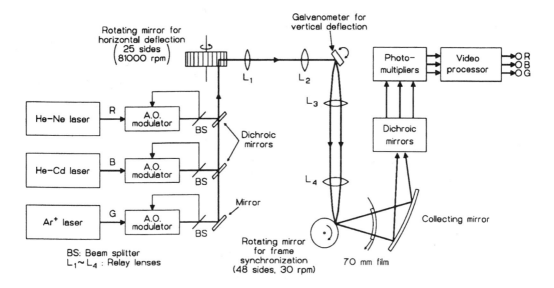

Figure 1.4 Major elements of a mechanically scanned laser-beam telecine. (*From* [1]. *Used with permission.*)

achieved only by maintaining high precision in the faces of the horizontal scanning mirror and in the alignment of successive faces. To keep the film motion and the frame-synchronization mirror rotation in precise synchronism, an elaborate electronic feedback control was used between the respective motor drives. In all other respects, the performance was more than adequate.

The processes involved in producing a film version of a video signal were, essentially, the reverse of those employed in the telecine, the end point being exposure of the film by laser beams controlled by the R, G, and B signals. A prototype system was shown in 1971 by CBS Laboratories, Stamford, Connecticut. The difference lay in the power required in the beams. In the telecine, with highly sensitive phototubes reacting to the beams, power levels of approximately 10 mW were sufficient. To expose 35 mm film, power approaching 100 mW is needed. With the powers available in the mid-1970s, the laser-beam recorder was limited to the smaller area of 16 mm film. A prototype 16 mm version was constructed using the basic mechanical scanning system of the telecine shown in Figure 1.5. In the laser recorder, the R, G, and B video signals were fed to three acoustic-optical modulators, and the resulting modulated beams were combined by dichroic mirrors into a single beam that was mechanically deflected. The scanned beam moved in synchronism with the moving film, which was exposed line by line. Color negative film typically was used in such equipment, but color duplicate film, having higher resolution and finer grain, was preferred for use with the 16 mm recorder. This technique is only of historical significance; the laser system has been discarded for the *electron-beam recording* system.

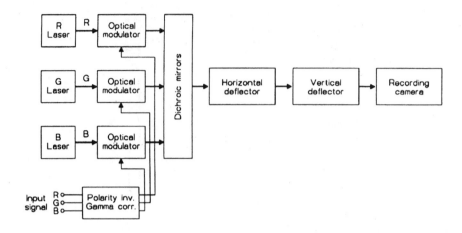

Figure 1.5 Block diagram of a mechanically scanned laser-beam video-to-film recorder. (*From* [1]. *Used with permission.*)

1.2.2 The 1125/60 System

In the early 1980s, NHK initiated a development program directed toward providing a high-definition television broadcasting service to the public. The video signal format was based upon a proposed 1125/60 standard published in 1980 by NHK. Experimental broadcasts were transmitted to prototype receivers in Japan over the MUSE (multiple sub-Nyquist encoding) satellite system. The 1984 Olympic Games, held in the United States and televised by NHK for viewers in Japan, was the first event of worldwide interest to be covered using high-definition television. The HDTV signals were *pan-scanned* to a 1.33 aspect ratio and transcoded to 525 lines for terrestrial transmission and reception over regular TV channels. On June 3, 1989, NHK inaugurated regular HDTV program transmissions for about an hour each day using its MS-2 satellite.

Various engineering committees within the Society of Motion Picture and Television Engineers (SMPTE), as well as other engineering committees in the United States, had closely studied the 1125/60 format since the 1980 NHK publication of the system details. Eventually, a production specification describing the 1125/60 format was proposed by SMPTE for adoption as an American National Standard. The document, SMPTE 240M, was published in April 1987 by the Society and forwarded to the American National Standards Institute (ANSI). Because of objections by several organizations, however, the document was not accepted as a national standard. One argument against the proposed standard was that it would be difficult to convert to the NTSC system, particularly if a future HDTV version of NTSC was to be compatible with the existing service.

NBC, for one, recommended that the HDTV studio standard be set at 1050 lines (525 × 2) and that the field rate of 59.94 Hz be retained, rather than the 60 Hz value of SMPTE 240M. Philips Laboratories and the David Sarnoff Laboratories concurred and based their proposed HDTV systems on the 1050-line, 59.94-field specifications.

Despite the rejection as a national standard for production equipment, SMPTE 240M remained a viable *recommended practice* for equipment built to the 1125/60 system.

The first full-scale attempt at international HDTV standardization was made by the CCIR in Dubrovnik, Yugoslavia, in May 1986. Japan and the United States pushed for a 1125-line/60 Hz production standard. The Japanese, of course, already had a working system. The 50 Hz countries, which did not have a working HDTV system of their own, demurred, asking for time to perfect and demonstrate a non-60 Hz (i.e., 50 Hz) system. Because of this objection, the CCIR took no action on the recommendation of its Study Group, voting to delay a decision until 1990, pending an examination of alternative HDTV production standards. There was strong support of the Study Group recommendations in the United States, but not among key European members of the CCIR.

The HDTV standardization fight was by no means simply a matter of North America and Japan vs. Europe. Of the 500 million receivers then in use worldwide, roughly half would feel the effect of any new frame rate. The Dubrovnik meeting focused mainly on a production standard. Television material can, of course, be produced in one standard and readily converted to another for transmission. Still, a single universal standard would avoid both the bother and degradation of the conversion process.

During this developmental period, the commercial implications of HDTV were inextricably intertwined with the technology. Even more in Europe than elsewhere, commercial considerations tended to dominate thinking. The 1125/60 system was, basically, a Japanese system. The United States came in late and jumped on Japan's coattails, aided greatly by the fact that the two countries have identical television standards (CCIR System *M*). But the Europeans did not want to have to buy Japanese or U.S. equipment; did not want to pay any Japanese or U. S. royalties; and did not want to swallow their NIH ("not invented here") pride. This feeling also emerged in the United States during the late 1980s and early 1990s for the same reasons—with American manufacturers not wanting to be locked into the Japanese and/or European HDTV systems.

1.2.3 European HDTV Systems

Early hopes of a single worldwide standard for high-definition television began a slow dissolve to black at the 1988 International Broadcasting Convention (IBC) in Brighton, England. Brighton was the public debut of the HDTV system developed by the European consortium *Eureka EU95*, and supporters made it clear that their system was intended to be a direct competitor of the 1125/60 system developed by NHK.

The Eureka project was launched in October 1986 (5 months after the Dubrovnik CCIR meeting) with the goal of defining a European HDTV standard of 1250-lines/50 Hz that would be compatible with existing 50 Hz receivers. EU95 brought together 30 television-related organizations, including major manufacturers, broadcasters, and universities. The Brighton showing included products and technology necessary for HDTV production, transmission, and reception. HD-MAC (high-definition multiplexed analog component) was the transmission standard developed under the EU95 program. HD-MAC was an extension of the MAC-packet family of transmission standards.

The primary movers in EU95 were the hardware manufacturers Bosch, Philips, and Thomson. The aim of the Eureka project was to define a 50 Hz HDTV standard for submission to the plenary assembly of the CCIR in 1990. The work carried out under this effort involved defining production, transmission, recording, and projection systems that would bring high-definition pictures into viewers' homes.

Supporters of the 1125/60 system also were planning to present their standard to the CCIR in 1990 for endorsement. The entry of EU95 into the HDTV arena significantly changed the complexion of the plenary assembly. For one thing, it guaranteed that no world-wide HDTV production standard—let alone a broadcast transmission standard—would be developed.

The 1250/50 Eureka HDTV format was designed to function as the future direct-broadcast satellite transmission system to the entire Western European Community. Although some interest had been expressed in the former Eastern Bloc countries, work there was slowed by more pressing economic and political concerns.

1.2.4 A Perspective on HDTV Standardization

HDTV production technology was seen from the very beginning as an opportunity to simplify program exchange, bringing together the production of programs for television and for the cinema [2]. Clearly, the concept of a single production standard that could serve all regions of the world *and* have application in the film community would provide benefits to both broadcasting organizations and program producers. All participants stated their preference for a single worldwide standard for HDTV studio production and international program exchange.

The work conducted in the field of studio standards showed that the task of creating a recommendation for HDTV studio production and international program exchange was made somewhat difficult by the diversity of objectives foreseen for HDTV in different parts of the world. There were differences in approach in terms of technology, support systems, and compatibility. It became clear that, for some participants, the use of HDTV for production of motion pictures and their subsequent distribution via satellites was the most immediate need. For others, there was a greater emphasis on satellite broadcasting, with a diversity of opinion on both the time scale for service introduction and the frequency bands to be used. For still others, the dominant consideration was terrestrial broadcasting services.

The proposal for a draft recommendation for an HDTV studio standard based on a 60 Hz field rate was submitted to the CCIR, initially in 1985, with the basic parameters outlined in Table 1.2. The proposal for a draft recommendation for an HDTV studio standard based on a 50 Hz field rate was submitted to the CCIR in 1987, with characteristics given in Table 1.3. Unfortunately, neither set of parameters in those drafts brought a consensus within the CCIR as a single worldwide standard. However, both had sufficient support for practical use in specific areas to encourage manufacturers to produce equipment.

Despite the lack of an agreement on HDTV, a great deal of progress was made during this effort in the area of an HDTV source standard. The specific parameters agreed upon included:

- Picture aspect ratio of 16:9

- Color rendition

- Equation for luminance

Table 1.2 Basic Characteristics of Video Signals Based on an 1125/60 System (After [3].)

Item	Characteristics	Value
1	Number of lines per frame	1125
2	Number of picture lines per frame	1035
3	Interlace ratio	2:1
4	Picture aspect ratio (H:V)	16:9
5	Field frequency (fields/s)	60.00
6	Line frequency (Hz)	33750

Table 1.3 Basic Characteristics of Video Signals Based on a 1250/50 System (After [4].)

Item	Characteristics	Value
1	Number of lines per frame	1250
2	Number of picture lines per frame	1152
3	Interlace ratio 1:1	(Proscan)
4	Picture aspect ratio (H:V)	16:9
5	Field frequency (fields/s)	50.00
6	Line frequency (Hz)	62500

Thus, for the first time in the history of television, all countries of the world agreed on the technical definition of a basic tristimulus color system for display systems. Also agreed upon, in principle, were the digital HDTV bit-rate values for the studio interface signal, which was important in determining both the interface for HDTV transmission and the use of digital recording. All of these agreements culminated in Recommendation 709, adopted by the XVII Plenary Assembly of the CCIR in 1990 in Dusseldorf [5].

1.2.5 Digital Systems Emerge

By mid-1991, publications reporting developments in the United States, the United Kingdom, France, the Nordic countries, and other parts of the world showed that bit-rate reduction schemes on the order of 60:1 could be applied successfully to HDTV source images. The results of this work implied that HDTV image sequences could be transmitted in a relatively narrowband channel in the range of 15 to 25 Mbits/s. Using standard proven modulation technologies, it would therefore be possible to transmit an HDTV program within the existing 6, 7, and/or 8 MHz channel bandwidths provided for in the existing VHF- and UHF-TV bands.

One outgrowth of this development was the initiation of studies into how—if at all—the existing divergent broadcasting systems could be included under a single unifying cover. Thus was born the *HDTV-6-7-8* program. HDTV-6-7-8 was based on the following set of assumptions [2]:

First, the differences between the bandwidths of the 6, 7, and 8 MHz channels might give rise to the development of three separate HDTV scenarios that would fully utilize the bandwidth of the assigned channels. It was assumed that the 6 MHz implementation would have the potential to provide pictures of sufficiently high quality to satisfy viewers' wishes for a "new viewing experience." The addition of a 1 or 2 MHz increment in the bandwidth, therefore, would not be critical for further improvement for domestic reception. On this basis, there was a possibility of adopting, as a core system, a single 6 MHz scheme to provide a minimum service consisting of video, two multichannel sound services, and appropriate support data channels for conditional access, and—where appropriate—closed captioning, program identification, and other user-oriented services.

Second, given the previous assumption, the 1 or 2 MHz channel bandwidth that could be saved in a number of countries might be used for transmission of a variety of additional information services, either within a 7 or 8 MHz composite digital signal or on new carriers in the band above the HDTV-6 signal. Such additional information might include narrowband TV signals that provide for an HDTV stereoscopic service, enhanced television services, multiprogram TV broadcasting, additional sound services, and/or additional data services.

Third, it would be practical to combine audio and video signals, additional information (data), and new control/test signals into a single HDTV-6-7-8 signal, in order to avoid using a secondary audio transmitter. In combination with an appropriate header/descriptor protocol and appropriate signal processing, the number of frequency channels could be increased, and protection ratio requirements would be reduced. In the course of time, this could lead to a review of frequency plans at the national and international levels for terrestrial TV transmission networks and cable television. This scheme, therefore, could go a considerable distance toward meeting the growing demand for frequency assignments.

This digital television system offered the prospect of considerably improved sound and image quality, while appreciably improving spectrum utilization (as compared to the current analog services). It was theorized that one way of exploiting these possibilities would be to use the bit stream available in digital terrestrial or satellite broadcasting to deliver to the public a certain number of digitally compressed conventional television programs instead of a single conventional, enhanced-, or high-definition program. These digitally compressed TV signals would be accompanied by digital high-quality sound, coded conditional access information, and ancillary data channels. Furthermore, the same approach could be implemented in the transmission of multiprogram signals over existing digital satellite or terrestrial links, or cable TV networks.

Despite the intrinsic merits of the HDTV-6-7-8 system, it was quickly superseded by the Digital Video Broadcasting project in Europe and the Grand Alliance project in the United States.

HD-DIVINE

HD-DIVINE began as a developmental project in late 1991. The aim was to prove that a digital HDTV system was possible within a short time frame without an intermediate step based on analog technology. The project was known as "Terrestrial Digital HDTV," but later was renamed "HD-DIVINE" (*digital video narrowband emission*). Less than 10 months after the project started, HD-DIVINE demonstrated a digital terrestrial HDTV system at the 1992 IBC show in Amsterdam. It was a considerable success, triggering discussion in the Nordic countries on cooperation for further development.

HD-DIVINE subsequently was developed to include four conventional digital television channels—as an alternative to HDTV—contained within an 8 MHz channel. Work also was done to adapt the system to distribution via satellite and cable, as demonstrated at the Montreux Television Symposium in 1993.

Meanwhile, in the spring of 1992, a second coordinated effort began with the goal of a common system for digital television broadcasting in Europe. Under the auspices of a European Launching Group composed of members of seven countries representing various organizations involved in the television business, the Working Group for Digital Television Broadcasting (WGDTB) defined approaches for a digital system. Two basic schemes were identified for further study, both multilayer systems that offered multiple service levels, either in a hierarchical or multicast mode. The starting point for this work was the experience with two analog TV standards, D2-MAC and HD-MAC, which had been operational for some time.

Eureka ADTT

On June 16, 1994, the leaders of the Eureka project approved the start of a new research effort targeted for the development of future television systems in Europe based on digital technologies. The Eureka 1187 *Advanced Digital Television Technologies* (ADTT) project was formed with the purpose of building upon the work of the Eureka 95 HDTV program. The effort, which was to run for 2½ years, included as partners Philips, Thomson Consumer Electronics, Nokia, and English and Italian consortia. The initial objectives of Eureka ADTT were to address design issues relating to production, transmission, reception, and display equipment and their key technologies.

A prototype high-definition broadcast system was developed and tested, based on specifications from—and in close consultation with—the European Project for Digital Video Broadcasting. The Eureka ADTT project also was charged with exploring basic technologies and the development of key components for products, such as advanced digital receivers, recording devices, optical systems, and multimedia hardware and software.

The underlying purpose of the Eureka ADTT effort was to translate, insofar as possible, the work done on the analog-based Eureka 95 effort to the all-digital systems that emerged from 1991 to 1993.

1.2.6 Digital Video Broadcasting

The European DVB project began in the closing months of 1990. Experimental European projects such as SPECTRE showed that the digital video-compression system known as *motion-compensated hybrid discrete cosine transform coding* was highly effective in reducing the transmission capacity required for digital television [6]. Until then, digital TV broadcasting was considered impractical to implement.

In the United States, the first proposals for digital terrestrial HDTV were made. In Europe, Swedish Television suggested that fellow broadcasters form a concerted pan-European platform to develop digital terrestrial HDTV. During 1991, broadcasters and consumer equipment manufacturers discussed how this could be achieved. Broadcasters, consumer electronics manufacturers, and regulatory bodies agreed to come together to discuss the formation of a pan-European group that would oversee the development of digital television in Europe—the European Launching Group (ELG). Over the course of about a year, the ELG expanded to include the major European media interest groups—both public and private—consumer electronics manufacturers, and common carriers.

The program officially began in September 1993, and the European Launching Group became the DVB (*Digital Video Broadcasting*) Project. Developmental work in digital television, already under way in Europe, then moved forward under this new umbrella. Meanwhile, a parallel activity, the *Working Group on Digital Television*, prepared a study of the prospects and possibilities for digital terrestrial television in Europe.

By 1999, a bit of a watershed year for digital television in general, the Digital Video Broadcasting Project had grown to a consortium of over 200 broadcasters, manufacturers, network operators, and regulatory bodies in more than 30 countries worldwide. Numerous broadcast services using DVB standards were operational in Europe, the Americas, Africa, Asia, and Australia.

At the '99 NAB Convention in Las Vegas, mobile and fixed demonstrations of the DVB system were made using a variety of equipment in various typical situations. Because mobile reception is the most challenging environment for television, the mobile system received a good deal of attention. DVB organizers used the demonstrations to point out the strengths of their chosen modulation method, the multicarrier *coded orthogonal frequency division multiplexing* (COFDM) technique.

In trials held in Germany beginning in 1997, DVB-T, the terrestrial transmission mode, has been tested in slow-moving city trams and at speeds in excess of 170 mph.

1.2.7 Involvement of the Film Industry

From the moment it was introduced, HDTV was the subject of various aims and claims. Among them was the value of video techniques for motion picture production, making it possible to bypass the use of film in part or completely. Production and editing were said to be enhanced, resulting in reduced costs to the producer. However, the motion picture industry was in no hurry to discard film, the medium that had served it well for the better part of a century. Film quality continues to improve, and film is unquestionably *the* universal production standard. Period. Although HDTV has made inroads in motion picture editing and special effects production, Hollywood has not rushed to hop on board the HDTV express.

Nevertheless, it is certainly true that the film industry has embraced elements of high-definition imaging. As early as 1989, Eastman Kodak unveiled the results of a long-range program to develop an *electronic-intermediate* (EI) digital video postproduction system. The company introduced the concept of an HDTV system intended primarily for use by large-budget feature film producers to provide new, creative dimensions for special effects without incurring the quality compromises of normal edited film masters.

In the system, original camera negative 35 mm film input to the electronic-intermediate system was transferred to a digital frame store at a rate substantially slower than real time, initially about 1 frame/s. Sequences could be displayed a frame at a time for unfettered image manipulation and compositing. This system was an electronic implementation of the time-standing-still format in which film directors and editors have been trained to exercise their creativity.

Kodak established a consortium of manufacturers and software developers to design and produce elements of the EI system. Although the announced application was limited strictly to the creation of artistic high-resolution special effects on film, it was hoped that the EI system eventually would lead to the development of a means for electronic real-time distribution of theatrical films.

1.2.8   Political Considerations

In an ideal world, the design and direction of a new production or transmission standard would be determined by the technical merits of the proposed system. However, as demonstrated in the past by the controversy surrounding the NTSC monochrome and color standards, technical considerations are not always the first priority. During the late 1980s, when concern over foreign competition was at near-frenzy levels in the United States and Europe, the ramifications of HDTV moved beyond just technology and marketing, and into the realm of politics. In fact, the political questions raised by the push for HDTV promised to be far more difficult to resolve than the technical ones.

The most curious development in the battle over HDTV was the interest politicians took in the technology. In early 1989, while chairing the House Telecommunications subcommittee, Representative Ed Markey (D-Massachusetts) invited comments from interested parties on the topic of high-definition television. Markey's subcommittee conducted two days of hearings in February 1989. There was no shortage of sources of input, including:

- *The American Electronics Association*, which suggested revised antitrust laws, patent policy changes, expanded exports of high-tech products, and government funding of research on high-definition television.

- *Citizens for a Sound Economy*, which argued for a relaxation of antitrust laws.

- *Committee to Preserve American Television*, which encouraged strict enforcement of trade laws and consideration of government research and development funds for joint projects involving semiconductors and advanced display devices.

- *Maximum Service Telecasters*, which suggested tax credits, antitrust exemptions, and low-interest loans as ways of encouraging U.S. development of terrestrial HDTV broadcasting.

- *Committee of Corporate Telecommunications Users*, which suggested the creation of a "Technology Corporation of America" to devise an open architecture for the production and transmission of HDTV and other services.

Representative Don Ritter (R-Pennsylvania) expressed serious concern over the role that U.S.-based companies would play—or more to the point, might not play—in the development of high-definition television. Ritter believed it was vital for America to have a piece of the HDTV manufacturing pie. Similar sentiments were echoed by numerous other lawmakers.

The Pentagon, meanwhile, expressed strong interest in high-definition technology for two reasons: direct military applications and the negative effects that the lack of domestic HDTV expertise could have on the American electronics industry. The Department of Defense, accordingly, allocated money for HDTV research.

Not everyone was upset, however, about the perceived technological edge the Japanese (and the Europeans to a lesser extent) had at that time over the United States. A widely read, widely circulated article in *Forbes* magazine [7] described HDTV as a technology that would be "obsolete" by the time of its introduction. The author argued, "The whole issue is phony," and maintained that HDTV products would hit the market "precisely at the time when the U.S. computer industry will be able to supply far more powerful video products at a lower price."

Although many of the concerns relating to HDTV that were voiced during the late 1980s now seem rather baseless and even hysterical, this was the atmosphere that drove the pioneering work on the technology. The long, twisting road to HDTV proved once again that the political implications of a new technology may be far more daunting than the engineering issues.

1.2.9 Terminology

During the development of HDTV, a procession of terms was used to describe performance levels between conventional NTSC and "real" HDTV (1125/60-format quality). Systems were classified in one or more of the following categories:

- *Conventional systems*: The NTSC, PAL, and SECAM systems as standardized prior to the development of advanced systems.

- *Improved-definition television* (IDTV) systems: Conventional systems modified to offer improved vertical and/or horizontal definition, also known as *advanced television* (ATV) or *enhanced-definition television* (EDTV) systems.

- *Advanced systems*: In the broad sense, all systems other than conventional ones. In the narrow sense, all systems other than conventional and "true" HDTV.

- *High-definition television* (HDTV) systems: Systems having vertical and horizontal resolutions approximately twice those of conventional systems.

- *Simulcast systems*: Systems transmitting conventional NTSC, PAL, or SECAM on existing channels and HDTV of the same program on one or more additional channels.

- *Production systems*: Systems intended for use in the production of programs, but not necessarily in their distribution.

- *Distribution systems*: Terrestrial broadcast, cable, satellite, videocassette, and videodisc methods of bringing programs to the viewing audience.

1.3   Proposed Routes to HDTV

Although HDTV production equipment had been available since 1984, standardization for broadcast service was slowed by lack of agreement on how the public could best be served. The primary consideration was whether to adopt a system compatible with NTSC or a simulcast system requiring additional transmission spectrum and equipment.

On November 17, 1987, at the request of 58 U.S. broadcasters, the FCC initiated a rule-making on ATV service and established a blue ribbon Advisory Committee on Advanced Television Service (ACATS) for the purpose of recommending a broadcast standard. Former FCC Chairman Richard E. Wiley was appointed to chair ACATS. At that time, it was generally believed that HDTV could not be broadcast using 6 MHz terrestrial broadcasting channels. Broadcasting organizations were concerned that alternative media would be used to deliver HDTV to the viewing public, placing terrestrial broadcasters at a severe disadvantage. The FCC agreed that this was a subject of importance and initiated a proceeding (MM Docket No. 87-268) to consider the technical and public policy issues of ATV.

The first interim report of the ACATS, filed on June 16, 1988, was based primarily on the work of the Planning Subcommittee. The report noted that proposals to implement improvements in the existing NTSC television standard ranged from simply enhancing the current standard all the way to providing full-quality HDTV. The spectrum requirements for the proposals fell into three categories: 6 MHz, 9 MHz, and 12 MHz. Advocates of a 12 MHz approach suggested using two channels in one of two ways:

- An existing NTSC-compatible channel supplemented by a 6 MHz *augmentation channel* (either contiguous or noncontiguous)

- An existing NTSC-compatible channel, unchanged, and a separate 6 MHz channel containing an independent non-NTSC-compatible HDTV signal

It was pointed out that both of these methods would be "compatible" in the sense that existing TV receivers could continue to be serviced by an NTSC signal.

The first interim report stated that, "based on current bandwidth-compression techniques, it appears that full HDTV will require greater spectrum than 6 MHz." The report went on to say that the Advisory Committee believed that efforts should focus on establishing, at least ultimately, an HDTV standard for terrestrial broadcasting. The report concluded that one advantage to simulcasting was that at some point in the future—after the NTSC standard and NTSC-equipped receivers were retired—part of the spectrum being utilized might be reemployed for other purposes. On the basis of preliminary engineering studies, the Advisory Committee stated that it believed sufficient spectrum capacity in the current television allocations table might be available to allow all existing stations to provide ATV through either an augmentation or simulcast approach.

With this launch, the economic, political, and technical implications of HDTV caused a frenzy of activity in technical circles around the world; proponents came forward to offer their ideas. The Advanced Television Test Center (ATTC) was set up to consider the proposals and evaluate their practicality. In the first round of tests, 21 proposed methods of transmitting some form of ATV signals (from 15 different organizations) were considered. The ATTC work was difficult for a number of reasons, but primarily because the 21 systems were in various stages of readiness. Most, if not all, were undergoing continual refinement.

Only a few systems existed as real, live black boxes, with "inputs" and "outputs". Computer simulation made up the bulk of what was demonstrated in the first rounds. The ATTC efforts promised, incidentally, to be as much a test of computer simulation as a test of hardware. Of the 21 initial proposals submitted to ACATS in September 1988, only six actually were completed in hardware and tested.

The race begun, engineering teams at various companies began assembling the elements of an ATV service. One of the first was the Advanced Compatible Television (ACTV) system, developed by the Sarnoff Research Center. On April 20, 1989, a short ACTV program segment was transmitted from the center, in New Jersey, to New York for broadcast over a WNBC-TV evening news program. The goal was to demonstrate the NTSC compatibility of ACTV. Consisting of two companion systems, the scheme was developed to comply with the FCC's *tentative decision* of September 1988, which required an HDTV broadcast standard to be compatible with NTSC receivers.

The basic signal, ACTV-I, was intended to provide a wide-screen picture with improved picture and sound quality on new HDTV receivers, while being compatible with NTSC receivers on a single 6 MHz channel. A second signal, ACTV-II, would provide full HDTV service on a second augmentation channel when such additional spectrum might be available.

In the second interim report of the ACATS (April 26, 1989), the committee suggested that its life be extended from November 1989 to November 1991. It also suggested that the FCC should be in a position to establish a single terrestrial ATV standard sometime in 1992. The Advisory Committee noted that work was ongoing in defining tests to be performed on proponent systems. An issue was raised relating to subjective tests and whether source material required for testing should be produced in only one format and transcoded into the formats used by different systems to be tested, or whether source material should be produced in all required formats. The Advisory Committee also sought guidance from the FCC on the minimum number of audio channels that an ATV system would be expected to provide.

The large number of system proponents, and delays in developing hardware, made it impossible to meet the aggressive timeline set for this process. It was assumed by experts at the time that consumers would be able to purchase HDTV, or at least advanced television, sets for home use by early 1992.

The FCC's tentative decision on compatibility, although not unexpected, laid a false set of ground rules for the early transmission system proponents. The requirement also raised the question of the availability of frequency spectrum to accommodate the added information of the ATV signal. Most if not all of the proposed ATV systems required total bandwidths of one, one and a half, or two standard TV channels (6 MHz, 9 MHz, or 12 MHz). In some cases, the added spectrum space that carried the ATV information beyond the basic 6 MHz did not have to be contiguous with the main channel.

Any additional use of the present VHF- and UHF-TV broadcast bands would have to take into account co-channel and adjacent-channel interference protection. At UHF, an additional important unanswered question was the effect of the UHF "taboo channels" on the availability of extra frequency space for ATV signals. These "taboos" were restrictions on the use of certain UHF channels because of the imperfect response of then-existing TV receivers to unwanted signals, such as those on image frequencies, or those caused by local oscillator radiation and front-end intermodulation.

The mobile radio industry had been a longtime combatant with broadcasters over the limited available spectrum. Land mobile had been asking for additional spectrum for years, saying it was needed for public safety and other worthwhile purposes. At that time, therefore, the chances of the FCC allocating additional spectrum to television broadcasters in the face of land mobile demands were not thought to be very good. Such was the case; the land mobile industry (and other groups) made a better case for the spectrum.

In any event, the FCC informally indicated that it intended to select a simulcast standard for HDTV broadcasting in the United States using existing television band spectrum and would not consider any augmentation-channel proposals.

1.3.1   System Testing: Round 2

With new groundwork clearly laid by the FCC, the second round of serious system testing was ready to begin. Concurrent with the study of the various system proposals, the ACATS began in late 1990 to evaluate the means for transmission of seven proposed formats for the purpose of determining their suitability as the U.S. standard for VHF and UHF terrestrial broadcasting. The initial round of tests were scheduled for completion by September 30, 1992.

The FCC announced on March 21, 1990 that it favored a technical approach in which high-definition programs would be broadcast on existing 6 MHz VHF and UHF channels separate from the 6 MHz channels used for conventional (NTSC) program transmission, but the commission did not specifically address the expected bandwidth requirements for HDTV. However, the implication was that only a single channel would be allocated for transmission of an HDTV signal. It followed that this limitation to a 6 MHz channel would require the use of video-compression techniques. In addition, it was stated that no authorization would be given for any enhanced TV system, so as not to detract from development of full high-definition television. The spring of 1993 was suggested by the FCC as the time for a final decision on the selection of an HDTV broadcasting system.

Under the simulcast policy, broadcasters would be required to transmit NTSC simultaneously on one channel of the VHF and UHF spectra and the chosen HDTV standard on another 6 MHz TV broadcast channel. This approach was similar to that followed by the British in their introduction of color television, which required monochrome programming to continue on VHF for about 20 years after 625/50 PAL color broadcasting on UHF was introduced. Standards converters working between 625-line PAL color and 405-line monochrome provided the program input for the simultaneous black-and-white network transmitters. The British policy obviously benefited the owners of old monochrome receivers who did not wish to invest in new color receivers; it also permitted program production and receiver sales for the new standard to develop at a rate compatible with industry capabilities. All television transmission in Great Britain now is on UHF, with the VHF channels reassigned to other radio services.

For the development of HDTV, the obvious advantage of simulcasting to viewers with NTSC receivers is that they may continue to receive all television broadcasts in either the current 525-line standard or the new HDTV standard—albeit the latter without the benefit of wide-screen and double resolution—without having to purchase a dual-standard receiver or a new HDTV receiver. Although it was not defined by the FCC, it was presumed that the HDTV channels also would carry the programs available only in the NTSC standard. Ideally for the viewer, these programs would be converted to the HDTV transmission standard from the narrower 1.33 aspect ratio and at the lower resolution of the 525-line format. A less desirable solution would be to carry programs available only in the NTSC standard without conversion to HDTV and require HDTV receivers to be capable of switching automatically between standards. A third choice would be not to carry NTSC-only programs on the HDTV channel and to require HDTV receivers to be both HDTV/NTSC channel and format switchable.

The development of HDTV involved exhaustive study of how to improve the existing NTSC system. It also meant the application of new techniques and the refinement of many others, including:

- Receiver enhancements, such as higher horizontal and vertical resolution, digital processing, and implementation of large displays.

- Transmission enhancements, including new camera technologies, image enhancement, adaptive prefilter encoding, digital recording, and advanced signal distribution.

- Signal compression for direct satellite broadcasting.

- Relay of auxiliary signals within conventional TV channels.

- Allocation and optimization of transmission frequency assignments.

Concurrently, an extensive study was undertaken concerning the different characteristics of the major systems of program distribution: terrestrial broadcasting, cable distribution by wire or fiber optics, satellite broadcasting, and magnetic and optical recorders. The major purposes of this study were to determine how the wide video baseband of HDTV could be accommodated in each system, and whether a single HDTV standard could embrace the needs of all systems. This work not only provided many of the prerequisites of HDTV, but by advancing the state of the conventional art, it established a higher standard against which the HDTV industry must compete.

In the third interim report (March 21, 1990), the Advisory Committee approved the proposed test plans and agreed that complete systems, including audio, would be required for testing. It also was agreed that proposed systems must be precertified by June 1, 1990. That date, naturally, became quite significant to all proponents. The pace of work was accelerated even further.

It is noteworthy that the first all-digital proposal was submitted shortly before the June deadline. The third interim report also stated that psychophysical tests of ATV systems would be conducted and that the Planning Subcommittee, through its Working Party 3 (PS/WP3), would undertake the development of preliminary ATV channel allotment plans and assignment options.

In the fourth interim report (April 1, 1991), the Advisory Committee noted changes in proponents and proposed systems. Most significant was that several all-digital systems had been proposed. Testing of proponent systems was scheduled to begin later that year. Changes had been required in the test procedures because of the introduction of the all-digital systems. It was reported also that the System Standards Working Party had defined a process for recommending an ATV system, and that PS/WP3 was working toward the goal of providing essentially all existing broadcasters with a simulcast channel whose coverage characteristics were equivalent to NTSC service.

By the time the fifth interim report was issued (March 24, 1992), there were five proponent systems, all simulcast—one analog and four all-digital. The Planning Subcommittee reported that it had reconstituted its Working Party 4 to study issues related to harmonizing an ATV broadcast transmission standard with other advanced imaging and transmission schemes that would be used in television and nonbroadcast applications. The Systems Subcommittee reported that its Working Party 2 had developed procedures for field-testing an ATV system. It was noted that the intent of the Advisory Committee was to field-test only the system recommended to the FCC by the Advisory Committee based on the laboratory tests. It also was reported that the Systems Subcommittee Working Party 4 had developed a process for recommending an ATV system and had agreed to a list of 10 selection criteria.

Hundreds of companies and organizations worked together within the numerous subcommittees, working parties, advisory groups, and special panels of ACATS during the 8-year existence of the organization. The ACATS process became a model for international industry-government cooperation. Among its accomplishments was the development of a competitive process by which proponents of systems were required to build prototype hardware that would then be thoroughly tested. This process sparked innovation and entrepreneurial initiative.

1.3.2 Formation of the Grand Alliance

Although the FCC had said in the spring of 1990 that it would determine whether all-digital technology was feasible for a terrestrial HDTV transmission standard, most observers viewed that technology as being many years in the future. Later the same year, however, General Instrument became the first proponent to announce an all-digital system. Later, all-digital systems were announced by MIT, the Philips-Thomson-Sarnoff consortium, and Zenith-AT&T.

The FCC anticipated the need for interoperability of the HDTV standard with other media. Initially, the focus was on interoperability with cable television and satellite delivery; both were crucial to any broadcast standard. But the value of interoperability with computer and telecommunications applications became increasingly apparent with the advent of all-digital systems.

Proponents later incorporated packetized transmission, headers and descriptors, and composite-coded surround sound in their subsystems. (The Philips-Thomson-Sarnoff consortium was the first to do so.) These features maximized the interoperability of HDTV with computer and telecommunications systems. The introduction of all-digital systems had made such interoperability a reality.

The all-digital systems set the stage for another important step, which was taken in February 1992, when the ACATS recommended that the new standard include a flexible, adaptive data-allocation capability (and that the audio also be upgraded from stereo to surround sound). Following testing, the Advisory Committee decided in February 1993 to limit further consideration only to those proponents that had built all-digital systems: two systems proposed by General Instrument and MIT; one proposed by Zenith and AT&T; and one proposed by Sarnoff, Philips, and Thomson. The Advisory Committee further decided that although all of the digital systems provided impressive results, no single system could be proposed to the FCC as the U.S. HDTV standard at that time. The committee ordered a round of supplementary tests to evaluate improvements to the individual systems.

At its February 1993 meeting, the Advisory Committee also adopted a resolution encouraging the digital HDTV groups to try to find a way to merge the four remaining all-digital systems. The committee recognized the merits of being able to combine the best features of those systems into a single "best of the best" system. With this encouragement, negotiations between the parties heated up, and on May 24, the seven companies involved announced formation of the Digital HDTV Grand Alliance.

By the spring of 1994, significant progress had been made toward the final HDTV system proposal. Teams of engineers and researchers had finished building the subsystems that would be integrated into the complete HDTV prototype system for testing later in the year. The subsystems—scanning formats, digital video compression, packetized data, audio, and modulation—all had been approved by the ACATS. Key features and specifications for the system included:

- **Support of two fundamental arrays of pixels** (picture elements): 1920 × 1080 and 1280 × 720. Each of these pixel formats supported a wide-screen 16:9 aspect ratio and square pixels, important for computer interoperability. Frame rates of 60, 30, and 24 Hz were supported, yielding a total of six different possible scanning formats—two different pixel arrays, each having three frame rates. The 60 and 30 Hz frame rates were important for video source material and 24 Hz for film. A key feature of the system was the Grand Alliance's commitment to using progressive scanning, also widely used in computer displays. Entertainment television traditionally had used interlaced scanning, which was efficient but subject to various unwanted artifacts. Of the six video formats, progressive scanning was used in all three 720-line formats and in the 30 and 24 Hz 1080-line formats. The sixth video format was a 60 Hz 1080-line scheme. It was neither technically or economically feasible to initially provide this as a progressive format, although it was a longer-term goal for the Grand Alliance. The 1080-line, 60-Hz format was handled in the initial standard by using interlaced rather than progressive scanning.

- **Video compression**: Utilizing the MPEG-2 (Moving Picture Experts Group)-proposed international standard allowed HDTV receivers to interoperate with MPEG-2 and MPEG-1 computer, multimedia, and other media applications.

- **Packetized data transport**: Also based on MPEG-2, this feature provided for the flexible transmission of virtually any combination of video, audio, and data.

- **Compact-disc-quality digital audio**: This feature was provided in the form of the 5.1-channel Dolby AC-3 surround sound system.

- **8-VSB** (8-level vestigial sideband): The modulation system selected for transmission provided maximum coverage area for terrestrial digital broadcasting.

The Grand Alliance format employed principles that made it a highly interoperable system. It was designed with a layered digital architecture that was compatible with the international Open Systems Interconnect (OSI) model of data communications that forms the basis of virtually all modern digital systems. This compatibility allowed the system to interface with other systems at any layer, and it permitted many different applications to make use of various layers of the HDTV architecture. Each individual layer of the system was designed to be interoperable with other systems at corresponding layers.

Because of the interoperability of the system between entertainment television and computer and telecommunications technologies, the Grand Alliance HDTV standard was expected to play a major role in the establishment of the *national information infrastructure* (NII). It was postulated that digital HDTV could be an engine that helped drive deployment of the NII by advancing the development of receivers with high-resolution displays and creating a high-data-rate path to the home for a multitude of entertainment, education, and information services.

1.3.3   Testing the Grand Alliance System

Field tests of the 8-VSB digital transmission subsystem began on April 11, 1994, under the auspices of the ACATS. The 8-VSB transmission scheme, developed by Zenith, had been selected for use in the Grand Alliance system two months earlier, following comparative laboratory tests at the ATTC. The field tests, held at a site near Charlotte, North Carolina, were conducted on channel 53 at a maximum effective radiated power of 500 kW (peak NTSC visual) and on channel 6 at an ERP of 10 kW (peak NTSC visual).

The tests, developed by the Working Party on System Testing, included measurements at approximately 200 receiving sites. Evaluations were based solely on a pseudorandom data signal as the input source; pictures and audio were not transmitted. The 8-VSB measurements included carrier-to-noise ratio (C/N), error rate, and *margin tests*, performed by adding noise to the received signal until an agreed-upon threshold of performance error rate occurred and noting the difference between the C/N and the C/N without added noise. Testing at the Charlotte facility lasted for about 3 months, under the direction of the Public Broadcasting System (PBS).

In 1995, extensive follow-up tests were conducted, including:

- Laboratory tests at the Advanced Television Test Center in Alexandria, Virginia

- Lab tests at Cable Television Laboratories, Inc. (CableLabs) of Boulder, Colorado

- Subjective viewer testing at the Advanced Television Evaluation Laboratory in Ottawa, Canada

- Continued field testing in Charlotte, North Carolina, by PBS, the Association for Maximum Service Television (MSTV), and CableLabs

The laboratory and field tests evaluated the Grand Alliance system's four principal subsystems: scanning formats, video and audio compression, transport, and transmission. Test results showed that:

- Each of the proposed HDTV scanning formats exceeded targets established for static and dynamic luminance and chrominance resolution.

- Video-compression testing, using 26 different HDTV sequences, showed that the Grand Alliance MPEG-2 compression algorithm was "clearly superior" to the four original ATV systems in both the 1080 interlaced- and 720 progressive-scanning modes. Significantly, the testing also showed little or no deterioration of the image quality while transmitting 3 Mbits/s of ancillary data.

- The 5.1-channel digital surround sound audio subsystem of the Grand Alliance system, known as Dolby AC-3, performed better than specifications in multichannel audio testing and met the expectations in long-form entertainment listening tests.

- The packetized data transport subsystem performed well when tested to evaluate the switching between compressed data streams, robustness of headers and descriptors, and interoperability between the compression and transport layers. Additional testing also demonstrated the feasibility of carrying the ATV transport stream on an *asynchronous transfer mode* (ATM) telecommunications network.

- Field and laboratory testing of the 8-VSB digital transmission subsystem reinforced test results achieved in the summer of 1994 in Charlotte. Testing for spectrum utilization and transmission robustness again proved that the Grand Alliance system would provide broadcasters significantly better transmission performance than the current analog transmission system, ensuring HDTV service "in many instances where NTSC service is unacceptable." Extensive testing on cable systems and fiber optic links of the 16-VSB subsystem also showed superior results.

The final technical report, approved on November 28, 1995, by the Advisory Committee, concluded that—based on intensive laboratory and field testing—the Grand Alliance digital television system was superior to any known alternative system in the world, better than any of the four original digital HDTV systems, and had surpassed the performance objectives of the ACATS.

Marking one of the last steps in an 8-year process to establish a U.S. ATV broadcasting standard, the 25-member blue-ribbon ACATS panel recommended the new standard to the FCC on November 28, 1995. Richard E. Wiley, ACATS chairman commented, "This is a landmark day for many communications industries and, especially, for American television viewers."

1.3.4 The Home Stretch for the Grand Alliance

With the technical aspects of the Grand Alliance HDTV system firmly in place, work proceeded to step through the necessary regulatory issues. Primary among these efforts was the establishment of a table of DTV assignments, a task that brought with it a number of significant concerns on the part of television broadcasters. Questions raised at the time involved whether a station's DTV signal should be equivalent to its present NTSC signal, and if so, how this should be accomplished.

Approval of the DTV standard by the FCC was a 3-step process:

- A *notice of proposed rulemaking* (NPRM) on policy matters, issued on August 9, 1995.

- Official acceptance of the Grand Alliance system. On May 9, 1996, the commission voted to propose that a single digital TV standard be mandated for over-the-air terrestrial broadcast of digital television. The standard chosen was that documented under the auspices of the Advanced Television Systems Committee (ATSC).

- Final acceptance of a table of assignments for DTV service.

During the comment period for the NPRM on Advanced Television Service (MM Docket 87-268), a number of points of view were expressed. Some of the more troublesome—from the standpoint of timely approval of the Grand Alliance standard, at least—came from the computer industry. Among the points raised were:

- **Interlaced scanning**. Some computer interests wanted to ban the 1920×1080 interlaced format.

- **Square pixels**. Computer interests recommended banning the use of formats that did not incorporate square pixels.

- **60 Hz frame rate**. Computer interests recommended a frame rate of 72 Hz and banning of 60 Hz.

Meanwhile, certain interests in the motion picture industry rejected the 16:9 (1.78:1) wide-screen aspect ratio in favor of a 2:1 aspect ratio.

One by one, these objections were dealt with. Negotiations between the two primary groups in this battle—broadcasters and the computer industry—resulted in a compromise that urged the FCC to adopt a standard that does not specify a single video format for digital television, but instead lets the various industries and companies choose formats they think will best suit consumers. The lack of a mandated video format set the stage for a lively competition between set makers and personal computer (PC) manufacturers, who were expected to woo consumers by combining sharp pictures with features peculiar to computers.

By early December, a more-or-less unified front had again been forged, clearing the way for final action by the FCC. With approval in hand, broadcasters then faced the demands of the commission's timetable for implementation, which included the following key points:

- By late 1998, 26 TV stations in the country's largest cities—representing about 30 percent of U.S. television households—would begin broadcasting the Grand Alliance DTV system.

- By mid-1999, the initial group would expand to 40; by 2000, it would expand to 120 stations.

- By 2006, every TV station in the country would be broadcasting a digital signal or risk losing its FCC license.

Fierce debates about the wisdom of this plan—and whether such a plan even could be accomplished—then ensued.

1.3.5   Digital Broadcasting Begins

If HDTV truly was going to be the "next big thing," then it was only fitting to launch it with a bang. The ATSC system received just such a sendoff, playing to excited audiences from coast to coast during the launch of Space Shuttle mission STS-95.

The first nationwide broadcast of a high-definition television program using the ATSC DTV system, complete with promos and commercials, aired on October 29, 1998. The live HDTV broadcast of Senator John Glenn's historic return to space was transmitted by ABC, CBS, Fox, NBC, and PBS affiliates from coast to coast.

The feed was available free for any broadcaster who could receive the signal. The affiliates and other providers transmitted the broadcast to viewing sites in Washington, D.C., New York, Atlanta, Chicago, Los Angeles, and 15 other cities. Audiences in those cities watched the launch on new digital receivers and projectors during special events at museums, retail stores, broadcast stations, and other locations. Many of the stations moved their on-air dates ahead of schedule in order to show the Glenn launch broadcast. The largest scheduled viewing site was the Smithsonian's National Air and Space Museum in Washington, D.C., where viewers watched the launch on an IMAX theatre screen and four new digital receivers.

Beyond the technical details was an even more important story insofar as HDTV production is concerned. All of the cameras used in the coverage provided an HD signal except for one NASA pool feed of the launch control center at the Kennedy Space Center, which was upconverted NTSC. On occasion, the director would switch to the launch center feed, providing a dramatic "A/B" comparison of high-definition versus standard-definition. The difference was startling. It easily convinced the industry observers present at the Air and Space Museum of the compelling power of the HDTV image.

The second production issue to come into focus during the broadcast was the editorial value of the wide aspect ratio of HDTV. At one point in the coverage, the program anchor described to the television audience how the Shuttle was fueled the night before. In describing the process, the camera pulled back from the launch pad shot to reveal a fuel storage tank off to one side of the pad. As the reporter continued to explain the procedure, the camera continued to pull back to reveal a second fuel storage tank on the other side of the launch pad. Thanks in no small part to the increased resolution of HDTV and—of course—the 16:9 aspect ratio, the television audience was able to see the entire launch area in a single image. Such a shot would have been wholly impossible with standard-definition imaging.

The STS-95 mission marked a milestone in space, and a milestone in television.

1.3.6   Continuing Work on the ATSC Standard

The creation by the Advanced Television Systems Committee of the DTV standard in 1995, and the FCC's subsequent adoption of the major elements of the standard into the FCC Rules in 1996, represented landmark achievements in the history of broadcast television. While these events represented the culmination of the ATSC's work in one sense, they also marked the beginning of a whole new effort to take the DTV standard as developed and turn it into a functioning, and profitable, system that would be embraced by both industry and consumers alike. To that end, the ATSC organized and continues to support the work of two primary technical standards-setting groups, each focusing on a different aspect of DTV deployment. The groups are:

- The *Technology Group on Distribution* (T3), which has as its mission the development and recommendation of voluntary, international technical standards for the distribution of television programs to the public using advanced imaging technology.

- *DTV Implementation Subcommittee* (IS), established to investigate and report on the requirements for implementation of advanced television. The subcommittee evaluates technical requirements, operational impacts, preferred operating methods, time frames, and cost impacts of implementation issues.

The ATSC is composed of more than 200 member corporations, industry associations, standards bodies, research laboratories, and educational institutions. It is an international group whose charter is to develop voluntary standards for the entire range of advanced television systems. Major efforts have included a certification program for television sets, computers, and other consumer video devices in cooperation with the Consumer Electronics Manufacturers Association. Another element of the ATSC work has been to explain and demonstrate the DTV system to international groups, the goal being adoption of the ATSC DTV standard in other countries.

ATSC Documents

The following is a partial list of ATSC Standards and technical activities.

- **ATSC Digital Television Standard**, *Document A/53*. The Digital Television Standard describes the system characteristics of the advanced television (ATV) system. The document and its normative annexes provide detailed specification of the parameters of the system including the video encoder input scanning formats and the pre-processing and compression parameters of the video encoder, the audio encoder input signal format and the pre-processing and compression parameters of the audio encoder, the service multiplex and transport layer characteristics and normative specifications, and the VSB RF/ transmission subsystem. The system is modular in concept and the specifications for each of the modules are provided in the appropriate annex. This document includes Amendment No. 1 to Doc. A/53.

- **Guide to the use of the ATSC Digital Television Standard**, *Document A/54*. This guide provides an overview and tutorial of the system characteristics of the advanced television (ATV) system defined by ATSC Standard A/53, *ATSC Digital Television*.

- **Digital Audio Compression (AC-3)**, *Document A/52*. This document specifies coded representation of audio information and the decoding process, as well as information on the encoding process. The coded representation specified is suitable for use in digital audio transmission and storage applications, and may convey from 1 to 5 full bandwidth audio channels, along with a low frequency enhancement channel. A wide range of encoded bit-rates is supported by this specification. Typical applications of digital audio compression are in satellite or terrestrial audio broadcasting, delivery of audio over metallic or optical cables, or storage of audio on magnetic, optical, semiconductor, or other storage media.

- **Standard for Coding 25/50 Hz Video**, *Document A/63*. This document describes the characteristics for the video subsystem of a digital television system operating at 25 Hz and 50 Hz frame rates.

- **Transmission Measurement and Compliance Standard for DTV**, *Document A/64 Rev. A*. This document describes methods for testing, monitoring, and measurement of the transmission subsystem intended for use in the digital television (DTV) system, including specifications for maximum out-of-band emissions, parameters affecting the quality of the inband signal, symbol error tolerance, phase noise and jitter, power, power measure, frequency offset, and stability. In addition, it describes the condition of the RF symbol stream upon loss of MPEG packets. (The ATSC approved a revision to this document on May 30, 2000, that includes the revised FCC DTV emission mask.

- **PSIP for Terrestrial Broadcast and Cable**, *Document A/65 Rev A with Amendment No. 1*. The Program and System Information Protocol Standard provides a methodology for transporting digital television system information and electronic program guide data. The standard includes an amendment that provides new functionality known as *Directed Channel Change* (DCC). This new feature will allow broadcasters to tailor programming or advertising based upon parameters defined by the viewer such as: postal, zip or location code, program identifier, demographic category, and content subject category. Potential applications include customized programming services, commercials based upon demographics, and localized weather and traffic reports.

- **Conditional Access For Terrestrial Broadcast**, *Document A/70*. This document defines a standard for the Conditional Access system for ATSC terrestrial broadcasting to enable broadcasters to fully utilize the capabilities of digital broadcasting. This standard is based, whenever possible, on existing open standards and defines the building blocks necessary to ensure interoperability. The ATSC CA module is replaceable; to ensure that ATSC hosts are protected against obsolescence as security is upgraded. This standard applies to all CA vendors that supply CA service on behalf of an ATSC service provider. An overview of the CA standard is given in Annex C. (This document includes an Amendment that the ATSC approved on May 30, 2000.)

- **Modulation and Coding Requirements for DTV Applications Over Satellite**, *Document A/80*. This document defines a standard for modulation and coding of data delivered over satellite for digital television contribution and distribution applications. The data can be a collection of program material including video, audio, data, multimedia, or other material. It includes the ability to handle multiplexed bit streams in accordance with the MPEG-2 systems layer, but it is not limited to this format and makes provision for arbitrary types of data as well. QPSK, 8PSK and 16 QAM modulation modes are included, as well as a range of forward error correction techniques.

- **Data Broadcast**, *Document A/90*. The ATSC Data Broadcast Standard defines protocols for data transmission compatible with digital multiplex bit streams constructed in accordance with ISO/IEC 13818-1 (MPEG-2 systems). The standard supports data services that are both TV program related and non-program related. Applications may include enhanced television, webcasting, and streaming video services. Data broadcasting receivers may include PCs, televisions, set-top boxes, or other devices. The standard provides mechanisms for download of data, delivery of datagrams, and streaming data.

1.3.7 A Surprising Challenge

As of the 1999 NAB Convention, over 50 stations were transmitting DTV signals. It was no surprise, then, that a proposal by one broadcaster—Sinclair Broadcast Group—that the chosen modulation method for the ATSC system be reevaluated, created shock waves throughout the convention, and indeed, throughout the industry. Sinclair's argument was that early production model DTV receivers performed poorly—in their estimation—relative to NTSC, and that the end result would be less reliance by the consumer on over-the-air signals and more reliance on cable television signals, thus putting broadcasters at a disadvantage. Sinclair suggested that COFDM, the method chosen by the European DVB consortium for terrestrial transmission, would perform better than 8-VSB.

Broadcast industry leaders were, generally speaking, in shock over the proposal. Many felt betrayed by one of their own. A number of theories were issued by leaders and observers to explain the Sinclair idea, and few had anything to do with the technical issues of signal coverage. Nevertheless, Sinclair went ahead with plans to conduct side-by-side tests of COFDM and 8-VSB later in the year.

It is important to note that in the early days of the Grand Alliance program, COFDM was tested against 8-VSB. It was, in fact, tested several times. On the basis of technical merit, 8-VSB was chosen by the Grand Alliance, and ultimately endorsed by the FCC.

The Sinclair-sponsored tests did, indeed, take place during the Summer of 1999. The reviews from stations, networks, and industry groups were mixed. In general, observers stated that the tests did show the benefits of COFDM for operation with inside receive antennas in areas that experience considerable multipath. Multipath tolerance, of course, is one of the strengths of the COFDM system. Whether any significant progress was made to change industry minds, however, was unclear.

The ITU also found itself being drawn into the modulation wars of Summer 1999. A report issued on May 11 by an ITU Radiocommunication Study Group presented an objective, scientifically valid comparison of the modulation schemes under a variety of conditions. The report, titled "Guide for the Use of Digital Television Terrestrial Broadcasting Systems Based on Performance Comparison of ATSC 8-VSB and DVB-T COFDM Transmission Systems," provided considerable detail of the relative merits of each system. The report, interestingly enough, was used by both the COFDM and 8-VSB camps to support their positions.

In essence, the report concluded that the answer to the question of which system is better is: "it depends." According to the document, "Generally speaking, each system has its unique advantages and disadvantages. The ATSC 8-VSB system is more robust in an *additive white Gaussian noise* (AWGN) channel, has a higher spectrum efficiency, a lower peak-to-average power ratio, and is more robust to impulse noise and phase noise. It also has comparable performance to DVB-T on low level ghost ensembles and analog TV interference into DTV. Therefore, the ATSC 8-VSB system could be more advantageous for *multi-frequency network* (MFN) implementation and for providing HDTV service within a 6 MHz channel."

"The DVB-T COFDM system has performance advantages with respect to high level (up to 0 dB), long delay static and dynamic multipath distortion. It could be advantageous for services requiring large scale *single frequency network* (SFN) (8k mode) or mobile reception (2k mode). However, it should be pointed out that large scale SFN, mobile reception and HDTV service cannot be achieved concurrently with any existing DTTB system over any channel spacing, whether 6, 7 or 8MHz."

The ITU report concluded, "DTV implementation is still in its early stage. The first few generations of receivers might not function as well as anticipated. However, with the technical advances, both DTTB systems will accomplish performance improvements and provide a much improved television service."

"The final choice of a DTV modulation system is based on how well the two systems can meet the particular requirements or priorities of each country, as well as other non-technical (but critical) factors, such as geographical, economical, and political connections with surrounding countries and regions. Each country needs to clearly establish their needs, then investigates the available information on the performances of different systems to make the best choice."

Petition Filed

Despite minimal, and sometimes conflicting, data on the relative performance of 8-VSB-versus-COFDM, Sinclair formally put the question before the FCC on October 8, 1999. The petition requested that the Commission allow COFDM transmissions *in addition to* 8-VSB for DTV terrestrial broadcasting. If granted, the petition would, in effect, require receiver manufacturers to develop and market dual mode receivers, something that receiver manufacturers were not predisposed to do. It was estimated that when the Sinclair petition was filed, over 5,000 DTV receivers had been sold to consumers.

Petition Rejected

On February 4, 2000, the FCC released a letter denying a Petition for Expedited Rulemaking, filed by Sinclair requesting that the Commission modify its rules to allow broadcasters to transmit DTV signals using COFDM modulation in addition to the current 8-VSB modulation standard. The Commission said that numerous studies supported the conclusion that NTSC replication is attainable under the 8-VSB standard. It said that the concerns raised in the Sinclair petition had "done no more than to demonstrate a shortcoming of early DTV receiver implementation." The Commission pointed out that receiver manufacturers were aware of problems cited by Sinclair and were aggressively taking steps to resolve multipath problems exhibited in some first-generation TV sets.

The Commission noted that the FCC Office of Engineering and Technology had analyzed the relative merits of the two standards, and concluded that the benefits of changing the DTV transmission standard to COFDM would not outweigh the costs of making such a revision. The Commission reiterated its view that allowing more than one standard could result in compatibility problems that could cause consumers and licensees to postpone purchasing DTV equipment and lead to significant delays in the implementation and provision of DTV services to the public. It said that development of a COFDM standard would result in a multi-year effort, rather than the "unrealistic" 120 days suggested in the Sinclair petition.

At the same time it dismissed the petition, the Commission recognized the importance of the issues raised by Sinclair. The Commission, stated, however, that the issue of the adequacy of the DTV standard is more appropriately addressed in the context of its biennial review of the entire DTV transition effort.

1.3.8   FCC Reviews the Digital Television Conversion Process

On March 8, 2000, the FCC began its first periodic review of the progress of conversion of the U.S. television system from analog technology to DTV. In a Notice of Proposed Rulemaking (NPRM 00-83), the Commission invited public comment on a number of issues that it said required resolution to insure continued progress in the DTV conversion and to eliminate potential sources of delay. The Commission said its goal was to insure that the DTV transition went smoothly for American consumers, broadcasters, and other interested parties. This periodic review followed through on the conclusion, adopted by the Commission as part of its DTV construction schedule and service rules in the May 1997 5th Report and Order, that it should undertake a periodic review every two years until the cessation of analog service to help the Commission insure that the DTV conversion fully served the public interest.

In the NPRM, the Commission noted that broadcast stations were facing relatively few technical problems in building digital facilities, and that problems encountered by some stations with tower availability and/or local zoning issues did not seem to be widespread. However, it asked for comment on whether broadcasters were able to secure necessary tower locations and construction resources, and to what extent any zoning disputes, private negotiations with tower owners, and the availability of tower construction resources affected the DTV transition.

In the NPRM, the Commission asked for comments on whether to adopt a requirement that DTV licensees replicate their NTSC service area, and whether a replication requirement should be based on the population or the area served by the station. The Commission noted that several licensees had sought authority to move their DTV station to a more central location in their market—or toward a larger market—or had asked to change their DTV allotment—including their assumed transmitter site and/or technical facilities—and it asked for comments on the effect that these situations had on the general replication requirements. In addition, the Commission asked for comments on a proposed requirement that DTV stations' principal community be served by a stronger signal level than that specified for the general DTV service contour.

The Commission asked for comments on what date stations with both their NTSC and DTV channels within the DTV *core* (channels 2–51) would have to choose the channel they intend to keep following the transition. It said that with the target date for the end of DTV transition set for December 31, 2006, it would be reasonable for stations to identify the DTV channels they will be using not later than 2004. It asked for comment on whether this date represented the proper balance between the goals of allowing DTV stations enough time to gain experience with DTV operation, and allowing stations that must move enough time to plan for their DTV channel conversion.

The Commission also invited comments on DTV application processing procedures, including whether to establish DTV application cut-off procedures, how to resolve conflicts between DTV applications to implement "initial" allotments, and the order of priority between DTV and NTSC applications. The Commission said it was seeking comment on whether to adopt a cut-off procedure for DTV area-expansion applications to minimize the number of mutually exclusive applications and to facilitate applicants' planning, and how to resolve any mutual exclusive applications that arise.

In the NPRM, the Commission noted that concerns had arisen regarding the 8-VSB DTV transmission standard. It invited comment on the status of this standard, including information on any additional studies conducted regarding NTSC replication using the 8-VSB standard. It specifically asked for comments on progress that was being made to improve indoor DTV reception, and manufacturers' efforts to implement DTV design or chip improvements.

The Commission noted certain industry agreements relating to cable compatibility, but said these agreements did not cover labeling of digital receivers, and asked whether a failure to reach agreement on the labeling issue would hinder the DTV transition. The Commission also asked for comments on the extent to which a lack of agreement on copy protection technology licensing and related issues would also hinder the DTV transition. Noting that some broadcasters had recommended that the Commission address over-the-air signal reception by setting receiver standards, the Commission asked for comments on whether the FCC had authority to set minimum performance levels for DTV receivers, whether it would be desirable to do so, and, if so, how such requirements should be structured.

In this "periodic review," the Commission said it was not asking for comment on issues that were the subject of separate proceedings—such as the issue of digital broadcast signal carriage on cable systems—or requests for reconsideration of already-decided issues—such as eligibility issues, certain issues relating to public television, and channel allotment or change requests. The Commission also said it is too early in the transition process to address other issues, including reconsidering the flexible approach to ancillary or supplementary services and the application of the simulcast requirement.

Perhaps most importantly—from the standpoint of broadcasters at least—the Commission said it would be inappropriate to review the 2006 target date for complete conversion to DTV because of Congressional action in the Balanced Budget Act of 1997, which confirmed December 31, 2006 as the transition completion date and established procedures and standards for stations seeking an extension of that date.

ATSC Comments Filed

The Advanced Television Systems Committee filed comments on the FCC's DTV review, stating, "The ATSC fully endorses and supports the Commission's DTV transmission standard, based on the ATSC DTV standard, and we urge the Commission to take all appropriate action to support and promote the rapid transition to digital television. Even so, the ATSC continues to seek ways to improve its standards and the implementation thereof in order to better meet the existing and evolving service requirements of broadcasters. To this end, and also in response to the specific concerns of some broadcasters regarding RF system performance (modulation, transmission and reception), the Executive Committee of the ATSC (has) formed a task force on RF system performance."

This task force was charged with examining a variety of technical issues that had been raised regarding the theoretical and realized performance of the DTV RF system—including receivers—and based on its findings, to make recommendations to the ATSC Executive Committee as to what, if any, technical initiatives should be undertaken by the ATSC.

At the time of the filing (May, 2000), the task force had identified three areas of focus:

- 8-VSB performance

- Broadcaster requirements

- Field test methodology

The task force was charged with evaluating the extent to which 8-VSB enables broadcasters to achieve reliable DTV reception within their current NTSC service areas. The group then was to assess the current performance of 8-VSB, including 8-VSB receiving equipment, as well as expected improvements. In addition, the task force was to look into potential backward-compatible enhancements to VSB, i.e., enhancements that could be implemented without affecting the operation of existing receivers that are not capable of delivering the enhanced features.

The task force also hoped to recommend potential actions that could be taken by the ATSC and its member organizations that might hasten improvements in the performance of the DTV RF system. The final report on the ATSC Task Force on RF Performance was nearing completion as this book went to press.

1.4   Compatibility in HDTV Systems

Whenever a new system is proposed, the question of its impact on the utility and value of existing equipment and services is raised [1]. A new system of television is defined as *compatible* with an old one if the receivers of the old system retain their utility when used with the new system. Few systems are totally compatible; some impairment in the performance of old receivers usually occurs because the new system is designed to meet different objectives.

A *compatible system* is fundamentally concerned with preserving the performance of receivers, that is, with the format of the transmitted signal. But there are many functions in the production and distribution of program material, prior to transmission, that contribute to, or are affected by, the compatibility requirement.

There is general agreement that the old system should be protected while the new service establishes its technical validity and shows its economic strength in the marketplace. The alternative—a sudden change from the old to the new—would have economic and political consequences that no one cares to face. Cutting off NTSC service in the United States would render useless nearly 200 million receivers whose purchase value exceeds $35 billion, a loss that would cause the removal from office of any governmental body that authorized it.

Preserving the existing service has been, without exception, the preferred procedure whenever a new television service has been introduced. However, shutting down of the older service has been known to be delayed so long that its diminished value no longer warranted the cost of continuing it. An instructive example lies in the history of the television service in the British Isles. Public broadcasting began there in 1936, using the 405-line monochrome system. When the 625-line color system was introduced in Britain in 1967, it was incompatible with the 405-line service. To preserve the existing service, new means of distribution had to be found for the color transmissions. This was found in the then largely unused UHF channels. The color and monochrome systems continued as separate services operating in parallel.

The 405-line service had a life span of 50 years. When it was discontinued in 1986, so few 405-line receivers remained in use and the cost of serving them was so high that it was proposed that Parliament provide the 405-line audience with 625-line receivers free of charge. But this economy was deemed politically unacceptable.

A similar, if less extreme, situation occurred in France, where the 819-line monochrome service continued for 37 years after it was introduced to the public in 1948. In both cases, the introduction of an incompatible color service depended entirely on the availability of the UHF channels.

1.4.1 Compromises in Compatible Systems

The impairment of the old service by a compatible system is well illustrated by the NTSC color system. When this scheme was introduced in 1954, many monochrome receivers (those responding to video frequencies at or above 3.58 MHz) could not discriminate against the color subcarrier, and this produced a pattern of fine dots, superimposed on the monochrome version of the color image. The NTSC designers attempted to minimize this effect by arranging for the subcarrier to have opposite phase on successive scans of each line, so that the dot pattern averaged out to zero during each frame scan. The eye was not completely fooled by this artifact, and some vestige of the dots remained in the form of an upward crawl of the scanning lines. In the enthusiasm for the new color service, this effect was tolerated. In time, the line crawl was eliminated by the decision of manufacturers to limit the bandwidth of color and monochrome receivers to less than 3.0 MHz, well below the 3.58 MHz (approximately) subcarrier. This action lowered the horizontal resolution to about 250 lines, 25 percent less than the 330 lines of the broadcast signal.

The low definition was not a matter of wide concern until improved cameras and signal enhancement provided additional detail, hidden by the existing receivers. Receiver designers reasoned that a new market could be established by restoring the lost resolution. This led to the introduction of the comb filter, which permitted, in the more expensive receivers, use of the full 4.2 MHz bandwidth in the broadcast signal. This is an example of how advances in technology can overcome problems once thought to be inherent in the system design.

1.4.2 Adapters for Compatible Service

Questions concerning the utility of the television service have risen with each new advance in video engineering, notably the wide spectrum of protected channels provided by the cable service, the storage of programs on videotape, and the reception of programs from satellites. In each case, when the service was first offered, it was necessary for the viewer to purchase an adapter or recorder at a price often greater than that of the receiver to which it was attached. This requirement was thought to limit the market to enthusiasts who had the necessary resources. But this judgment was wide of the mark.

Today in the United States, most homes enjoying television have video recorders and/or cable service, for which fees are charged. This violation of the "free television" doctrine of the broadcasters testifies that most viewers have found access to cable programs and the freedom to view them at a convenient time well worth the cost of the adapter, recorder, and fees.

1.4.3 Transcoding Functions

To convert otherwise incompatible signals to compatible ones, it is necessary to encode or decode (*transcode*) them. Digital video transcoding methods have reached a point at which, given unlimited resources, it should be possible to render a signal format into and out of any other, with little or no loss of signal quality. But resources are never unlimited, and the complications of transcoding are such that some deterioration of signal quality occurs in most practical applications.

The demands of compatibility extend to program origination and distribution, with specifications that exceed those of the broadcast signal. The need for excess bandwidth in studio equipment, for example, has been long established in conventional operations and is equally desired for HDTV systems. Most prime-time programs currently originate from film, in the 35 mm format. Because 35 mm film as projected has definition equivalent to that of the HDTV image, *excess definition* is available only in the 70 mm film format. Although several major film producers offer 70 mm prints, most producers use 35 mm film, and nearly all of the older films that are currently available are in that format. New film productions with the HDTV market in mind are well advised to consider recording on 70 mm prints.

Allied with the size of the film is its aspect ratio. The compatibility issue arises when the HDTV image is displayed on the conventional 4×3 screen with the left and right portions of the HDTV image cut off. To maintain the center of interest on the 4×3 screen, the portion of the 16×9 image contained within the 4×3 limits often must be shifted laterally during scanning.

1.4.4 Standardization Issues

Video standards have a complex structure [8]. They provide a detailed description of video signals compatible with an intended receiver. Strictly speaking, a broadcast television standard is a set of technical specifications defining the method of on-air radio-frequency transmission of a picture with accompanying sound. However, in the video program production environment, the RF parameters are usually irrelevant and sound can be handled in many different ways. For this reason, the term *video standard* is more appropriate.

It is possible, and many times useful, to divide a video standard into five component parts, as follows:

- The *scanning standard*, which determines how the picture is sampled in space and in time (i.e., the number of lines in the picture, the number of pictures per second, and whether interlace or progressive scanning is used). Only the conversion of video signals between different scanning standards—for example, NTSC to PAL—can truly be called *standards conversion.*

- *Color information representation*, which determines how color information is conveyed. Color formats are divided into two basic categories: *component*, primarily used for production, and *composite*, traditionally used for conventional television broadcasting.

- *Aspect ratio*, which describes how the picture fits into a screen of a particular proportion.

- *Signal levels*, which determine how a receiver will interpret a video waveform in the voltage domain. In the digital domain, it is necessary to standardize the relationship between analog voltages and the digital codes that they signify.

- *Format*, which is an agreed way of packaging the picture information for transmission or recording.

Historically, the most significant boundary in the television world was the one between different frame rates. At the dawn of black-and-white television, it was reasonable to link the field repetition rate with the frequency of the ac power line. This prevented slow scroll of the horizontal hum bar on the received picture. When the color era began, with crystal-referenced operating frequencies and improved filtering of the dc power supply, this link became more or less irrelevant. The only justifiable reason for preserving the relationship was to reduce the visibility of unpleasant low-frequency beating between studio lighting and the video camera field rate. Unfortunately, the economic (and political) necessity for backward compatibility did not permit significant changes to be made to the field and frame rates.

There are a number of organizations responsible for standardization at global, regional, and national levels. In the broadcast television field, one of the most prominent is the former CCIR—the International Radio Consultative Committee. CCIR was a branch of the International Telecommunication Union (ITU), which, in turn, is part of the United Nations Organization. In 1993, the CCIR was renamed as ITU-R (the Radiocommunication Sector of the ITU). Television is a main topic for ITU-R Study Group 11. The crucial documents issued by ITU-R are Recommendations and Reports. They contain explicit information, but in reasonably general form, often leaving room for different variants in practical implementation.

Digital Television System Model

In April 1997, the ITU adopted a series of Recommendations (standards) defining a *digital terrestrial television broadcasting* (DTTB) system [9]. The set of Recommendations and associated descriptive Reports represents a four-year effort by ITU Task Group 11/3. The Task Group began with the premise that it should establish an infrastructure that enabled a communication environment in which a continuum of television and other data services could be brought to the consumer via wire, recorded media, and through the air. Terrestrial broadcasting presents the most challenging set of constraints of all of the forms of media. Therefore, a system that works well in the terrestrial broadcasting environment should suffice for other media. The first step in harmonizing the delivery of these services was the development of a suitable service model.

The Task Group work was based in many ways on both the work of and the philosophy behind the MPEG-2 standard. The MPEG-2 document provides a set of tools that can be used to describe a system. The set of Recommendations developed by Task Group 11/3 defined a constrained set of tools that can be used to provide a DTTB service. This set of tools provides for a single, low-cost decoder that can deliver both ATSC- and DVB-coded images and sound.

Task Group 11/3 also established a harmonized subset of MPEG-2 that allowed for a single decoder that can translate the service multiplex and transport layer, and decode the audio and video compression and coding layers for any system that conforms to the DTTB set of Recommendations.

The set of harmonized Recommendations fully meets the request of the World Broadcasting Union for unique global broadcasting systems leading to single universal consumer appliances. A report on the Task Group's efforts can be found in [9].

1.4.5 Harmonized Standards

The convergence of the television, computer, and communications industries is well under way, having been anticipated for quite some time [10]. Video and audio compression methods, server technology, and digital networking are all making a significant impact on television production, post-production, and distribution. Accompanying these technological changes are potential benefits in reduced cost, improved operating efficiencies and creativity, and increased marketability of material. Countering the potential benefits are threats of confusion, complexity, variable technical performance, and increased costs if not properly managed. The technological changes now unfolding will dramatically alter the way in which video is produced and distributed in the future.

In this context, the Society of Motion Picture and Television Engineers (SMPTE) and the European Broadcasting Union (EBU) jointly formed the *Task Force for Harmonized Standards for the Exchange of Program Material as Bitstreams*. The Task Force, with the participation by approximately 200 experts from around the world, produced two reports. The first, published in April 1997, was called *User Requirements* for the systems and techniques that will implement new technologies. The second report, published in July 1998, provided *Analyses and Results* from the deliberations of the Task Force. Taken together, the two documents are meant to guide the converging industries in their decisions regarding specific implementations of the applicable technologies, and to steer the future development of standards that are intended to maximize the benefits and minimize the detriments of implementing such systems.

The goals of the Task Force were to look into the future a decade or more, to determine the requirements for systems in that time frame, and to identify the technologies that could be implemented within the coming years in order to meet the specified requirements over the time period. This approach recognized that it takes many years for new technologies to propagate throughout the industries implicated in the changes. Because of the large and complex infrastructures involved, choices must be made of the methods that can be applied in the relatively near future, but which will still be viable over the time period contemplated by the Task Force's efforts.

To meet its objectives, the Task Force partitioned its work among six separate Sub-Groups, each of which was responsible for a portion of the investigation. These Sub-Groups included the following general subject areas:

- Systems

- Compression

- Wrappers and File Formats

- Metadata

- File Transfer Protocols

- Physical Link and Transport Layers for Networks

Some of the Sub-Groups found that their areas of interest were inextricably linked with one another and, consequently, performed their work jointly (and produced a common report).

The Task Force made significant contributions in identifying the technologies and standards required to carry converging industries with an interest in television to the next plane of cooperation and interoperation. A detailed report of the Task Force can be found in [10].

1.5   References

1. Benson, K. B., and D. G. Fink: *HDTV: Advanced Television for the 1990s*, McGraw-Hill, New York, N.Y., 1990.

2. Krivocheev, Mark I., and S. N. Baron: "The First Twenty Years of HDTV: 1972–1992," *SMPTE Journal*, SMPTE, White Plains, N.Y., pg. 913, October 1993.

3. CCIR Report 801-3: "The Present State of High-Definition Television," pg. 37, June 1989.

4. CCIR Report 801-3: "The Present State of High-Definition Television," pg. 46, June 1989.

5. CCIR Document PLEN/69-E (Rev. 1): "Minutes of the Third Plenary Meeting," pp. 2–4, May 29, 1990.

6. Based on information supplied by the DVB Project on its Web site: www.dvb.com.

7. Gilder, George: "IBM-TV?," *Forbes*, February 20, 1989.

8. Pank, Bob (ed.): *The Digital Fact Book*, 9th ed., Quantel Ltd, Newbury, England, 1998.

9. Baron, Stanley: "International Standards for Digital Terrestrial Television Broadcast: How the ITU Achieved a Single-Decoder World," *Proceedings of the 1997 BEC*, National Association of Broadcasters, Washington, D.C., pp. 150–161, 1997.

10. SMPTE and EBU, "Task Force for Harmonized Standards for the Exchange of Program Material as Bitstreams," *SMPTE Journal*, SMPTE, White Plains, N.Y., pp. 605–815, July 1998.

1.6   Bibliography

ATSC: "Comments of The Advanced Television Systems Committee, MM Docket No. 00-39," ATSC, Washington, D.C., May, 2000.

"ATV System Recommendation," *1993 NAB HDTV World Conference Proceedings*, National Association of Broadcasters, Washington, D.C., pp. 253–258, 1993.

Appelquist, P.: "The HD-Divine Project: A Scandinavian Terrestrial HDTV System," *1993 NAB HDTV World Conference Proceedings*, National Association of Broadcasters, Washington, D.C., p. 118, 1993.

Battison, John: "Making History," *Broadcast Engineering*, Intertec Publishing, Overland Park, Kan., June 1986.

Benson, K. B., and Jerry C. Whitaker (eds.): *Television Engineering Handbook*, rev. ed., McGraw-Hill, New York, N.Y., 1992.

Benson, K. B., and J. C. Whitaker (eds.): *Television and Audio Handbook for Engineers and Technicians*, McGraw-Hill, New York, N.Y., 1989.

Federal Communications Commission: Notice of Proposed Rule Making 00-83, FCC, Washington, D.C., March 8, 2000.

Hopkins, R.: "Advanced Television Systems," *IEEE Transactions on Consumer Electronics*, vol. 34, pp. 1–15, February 1988.

Lincoln, Donald: "TV in the Bay Area as Viewed from KPIX," *Broadcast Engineering*, Intertec Publishing, Overland Park, Kan., May 1979.

McCroskey, Donald: "Setting Standards for the Future," *Broadcast Engineering*, Intertec Publishing, Overland Park, Kan., May 1989.

Reimers, U. H.: "The European Perspective for Digital Terrestrial Television, Part 1: Conclusions of the Working Group on Digital Terrestrial Television Broadcasting," *1993 NAB HDTV World Conference Proceedings*, National Association of Broadcasters, Washington, D.C., p. 117, 1993.

Schow, Edison: "A Review of Television Systems and the Systems for Recording Television," *Sound and Video Contractor*, Intertec Publishing, Overland Park, Kan., May 1989.

Schreiber, W. F., A. B. Lippman, A. N. Netravali, E. H. Adelson, and D. H. Steelin: "Channel-Compatible 6-MHz HDTV Distribution Systems," *SMPTE Journal*, SMPTE, White Plains, N.Y., vol. 98, pp. 5–13, January 1989.

Schreiber, W. F., and A. B. Lippman: "Single-Channel HDTV Systems—Compatible and Noncompatible," Report ATRP-T-82, Advanced Television Research Program, MIT Media Laboratory, Cambridge, Mass., March 1988.

"Sinclair Seeks a Second Method to Transmit Digital-TV Signals," *Wall Street Journal*, Dow Jones, New York, N.Y., October 7, 1999.

Whitaker, Jerry C.: *Electronic Displays: Technology, Design, and Applications*, McGraw-Hill, New York, N.Y., 1994.

2

Applications for HDTV

2.1 Introduction

The term *high-definition television* applies more to a class of technology than to a single system. HDTV can be logically divided into two basic applications:

- Closed-loop production systems

- Broadcast systems

Each system has its own applications, its own markets, and its own basic technology.

Japanese professional video manufacturers, under the direction of NHK (the Japanese national broadcasting company), launched a major HDTV development program long before either European or American organizations gave HDTV serious consideration. Early Japanese efforts to establish common international standards for HDTV were largely responsible for stimulating development projects both in North America and in Europe.

2.1.1 Resolution

The resolution of the displayed picture is the most basic attribute of any HDTV system. Generally speaking, the HDTV image has approximately twice as much luminance definition horizontally and vertically as the 525-line NTSC system or the 625-line PAL and SECAM systems. The total number of luminance picture elements (*pixels*) in the image, therefore, is 4 times as great. The wider aspect ratio of the HDTV system adds even more visual information. This increased detail in the image is achieved by employing a video bandwidth approximately 5 times that of conventional (NTSC) systems.

The HDTV image is 25 percent wider than the conventional video image, for a given image height. The ratio of image width to height in HDTV systems is 16:9, or 1.777. The conventional NTSC image has a 4:3 aspect ratio.

As discussed in Section 1.2, the HDTV image may be viewed more closely than is customary in conventional television systems. Full visual resolution of the detail of conventional television is available when the image is viewed at a distance equal to about 6 or 7 times the height of the display. The HDTV image may be viewed from a distance of about 3 times picture height for the full detail of the scene to be resolved.

2.1.2 Production Systems vs. Transmission Systems

Bandwidth is perhaps the most basic factor that separates production HDTV systems from transmission-oriented systems for broadcasting. A closed-circuit system does not suffer the same restraints imposed upon a video image that must be transported by radio frequency means from an origination center to consumers. It is this distinction that has led to the development of widely varied systems for production and transmission applications. Terrestrial broadcasting of NTSC video, for example, is restricted to a video baseband that is 4.2 MHz wide. The required bandwidth for full resolution HDTV, however, is on the order of 30 MHz. Fortunately, video-compression algorithms are available that can reduce the required bandwidth without noticeable degradation and still fit within the restraints of a standard NTSC, PAL, or SECAM channel. The development of efficient compression systems was, in fact, the breakthrough that made the all-digital HDTV system possible.

Video compression involves a number of compromises. In one case, a tradeoff is made between higher definition and precise rendition of moving objects. It is possible, for example, to defer the transmission of image detail, spreading the signal over a longer time period, thus reducing the required bandwidth. If motion is present in the scene over this longer interval, however, the deferred detail may not occupy its proper place. Smearing, ragged edges, and other types of distortion can occur.

2.1.3 Defining Terms

The following terms commonly are used to describe and evaluate high-definition imaging systems:

- *Aspect ratio*. The ratio of picture width to picture height.

- *Contrast*. The range of brightness in the displayed image.

- *Horizontal resolution*. The number of elements separately present in the picture width.

- *Vertical resolution*. The number of picture elements separately present in the picture height.

- *Color depth*. The range and accuracy of colors that can be reproduced by the imaging system.

2.2 HDTV Applications

Serious work on high-definition television began in Japan in the late 1960s. (See Section 1.2.) NHK undertook an in-depth, long-term examination of what might constitute a totally "new viewing experience" using the electronic medium of television. A research team began a search for the technical parameters that would define a television system of radically improved capability. This TV system was envisaged to attain a quality level sufficient to address the needs of a future sophisticated "information society" [1]. To that end, it would embrace:

- Industrial applications

- Motion picture production

- Broadcasting (future applications and services)

- High-quality information capture, storage, and retrieval

- Education, medical, and cultural applications

- Community viewing, including electronic theaters

The term "new viewing experience"—from the standpoint of the human viewer—is worthy of elaboration. As targeted by the NHK research group, this new experience would comprise: 1) a visual image of substantially higher quality, and 2) high-quality multichannel sound. The separate qualities to realize a higher-performance video image would, in turn, be an ensemble of:

- A larger (and wider) picture

- Increased horizontal and vertical resolution

- Considerably enhanced color rendition and resolution

- Elimination of current television system artifacts

2.2.1 Business and Industrial Applications

Although the primary goal of the early thrust in HDTV equipment development centered on those system elements essential to support program production, a considerably broader view of HDTV anticipated important advances in its wider use. The application of video imaging has branched out in many directions from the original exclusive over-the-air broadcast system intended to bring entertainment programming to the home. Throughout the past 3 decades, television has been applied increasingly to a vast array of teaching, training, scientific, corporate, and industrial applications. The same era has seen the emergence of an extensive worldwide infrastructure of independent production and postproduction facilities (more than 1000 in the United States alone) to support these needs.

As the Hollywood film industry became increasingly involved in supplying prime-time programming for television (on an international basis) via 35 mm film origination, video technology was harnessed to support off-line editing of these film originals, and to provide creative special effects. Meanwhile, as the world of computer-generated imaging grew at an explosive pace, it too began penetrating countless industrial, scientific, and corporate applications, including film production. The video industry has, in essence, splintered into disparate (although at certain levels, overlapping) specialized industries, as illustrated in Figure 2.1. Any of these video application sectors is gigantic in itself. It was into this environment that HDTV was born.

Apart from the issues of terrestrial (or cable and/or direct-broadcast satellite, DBS) distribution of HDTV entertainment programs, there exists today another crucial issue: elec-

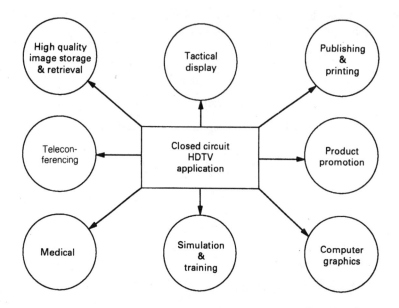

Figure 2.1 Applications for high-definition imaging in business and industrial facilities. (*After* [1].)

tronic imaging as a whole. Video technology is being applied to a vast diversity of applications. These include medical, teleconferencing, science, art, corporate communication, industrial process control, and surveillance. For some of these applications, 525 NTSC has been adequate, for some barely adequate, and for others woefully inadequate.

2.2.2 Broadcast Applications

As with many things technical, the plans of the designers and those of the end users do not always coincide. It was assumed from the beginning of the standardization process for HDTV that the result would be a system specifically for the delivery of pictures and sound of superb quality—a quantum leap beyond NTSC. The reality unfolding as the ATSC DTV format enters the implementation stage, however, is shaping up to be a bit different. The decision-makers at TV stations and networks are asking, "Do we really want to transmit HDTV or would we rather transmit more of the same signals that we send out now?" The flexible nature of the ATSC DTV system permits broadcasters to decide whether they would like to send to viewers one superquality HDTV program or several "NTSC-equivalent" programs.

The arguments for full-quality HDTV are obvious: great pictures and sound—a completely new viewing experience. Numerous programs could benefit greatly from the increased resolution and viewing angle thus provided; sporting events and feature films immediately come to mind. However, many programs, such as news broadcasts and situation comedies, would gain little or nothing from HDTV. As with most issues in commercial broadcasting, the programming drives the technology.

Arguments for the multiple-stream approach to digital broadcasting follow along similarly predictable lines. Broadcasters long have felt constrained by the characteristics of the NTSC signal: one channel, one program. Cable companies, obviously, have no such constraint. As with "true HDTV," programming will drive the multistream business model. For example, a station might allocate its channels as follows:

- Regular broadcast schedule channel

- 24-hour news channel (which would simply feed off—or feed to—the main channel operation)

- Special events channel (public affairs shows during the day and movies at night)

- Text-based informational channel

It is reasonable to assume that such a model could be successful for a station in a major market. But the question is: How many such services can a single market support? Furthermore, aside from news—which is expensive to do, if done right—programming will make or break these come-along channels.

A significant component of the FCC decision on DTV was, of course, the timetable for implementation. Few industry observers believe that the timetable, which calls for every television station in the United States to be broadcasting the ATSC DTV standard by 2006, can be met. Even fewer believe that the deadline really will stick. In any event, the most basic question for television stations is what to do with the information-carrying capacity of the DTV system. Obviously, the choice of HDTV programming or multiple-stream standard-definition programming has an immense impact on facility design and budget requirements. Once that decision has been made, the focus moves to implementation issues, such as:

- Signal coverage requirements vs. facility costs

- Tower space availability for a DTV antenna

- Transmitter tradeoffs and choices

- Studio-to-transmitter (STL), intercity relay (IRC), and satellite links

- Master control switching and routing

- Production equipment (cameras, switchers, special effects systems, recorders, and related hardware)

- Studios and sets for wide-screen presentations

The conversion from NTSC to DTV often is compared with the long-ago conversion from black and white to color. Many of the lessons learned in the late 1950s and early 1960s, however, are of little use today because the broadcast paradigm has shifted. But the most important lesson from the past still is valid: build the technology around the programming, not the other way around.

2.2.3 Computer Applications

One of the characteristics that set the ATSC effort apart from the NTSC efforts of the past was the inclusion of a broad range of industries—not just broadcasters and receiver manufacturers, but all industries that had an interest in imaging systems. It is, of course, fair to point out that during the work of the NTSC for the black-and-white and color standards, there were no other industries involved in imaging. Be that as it may, the broad-based effort encompassed by the ATSC system has ensured that the standard will have applications in far more industries than simply broadcast television. The most visible—and vocal—of these allied industries is the computer business.

As detailed in Chapter 1, computer hardware and software manufacturers lobbied hard for adjustments to the Grand Alliance system that would optimize the standard for use on personal computers. Heavy-hitter vendors such as Microsoft explained that the future of television would be computers. With computers integrated into television receivers, consumers would have a host of new options and services at their fingertips, hence facilitating the public interest, convenience, and necessity.

In an industry that has seen successive waves of hype and disappointment, it is not surprising that such visions of the video future were treated skeptically, at least by broadcasters who saw these predictions by computer companies as an attempt to claim a portion of their turf.

HDTV and Computer Graphics

The core of HDTV production is the creation of high-quality images. As HDTV was emerging in the early 1980s, a quite separate and initially unrelated explosion in electronic imaging was also under way in the form of high-resolution computer graphics. This development was propelled by broad requirements within a great variety of business and industrial applications, including:

- Computer-aided design (CAD)

- Computer-aided manufacturing (CAM)

- Printing and publishing

- Creative design (such as textiles and decorative arts)

- Scientific research

- Medical diagnosis

A natural convergence soon began between the real-time imagery of HDTV and the non-real-time high-resolution graphic systems. A wide range of high-resolution graphic display systems is commonly available today. This range addresses quite different needs for resolution within a broad spectrum of industries. Some of the more common systems are listed in Table 2.1.

The ATSC DTV system thus enjoys a good fit within an expanding hierarchy of computer graphics. This hierarchy has the range of resolutions that it does because of the varied needs of countless disparate applications. HDTV further offers an important wide-screen display organization that is eminently suited to certain critical demands. The 16:9 display,

Table 2.1 Common Computer System Display Formats

Horizontal Pixels	Vertical Pixels (Active Lines)
640	480
800	600
1024	768
1280	1024
1280	1536
2048	1536
2048	2048

for example, can encompass two side-by-side 8 × 11-in pages, important in many print applications. The horizontal form factor also lends itself to many industrial design displays that favor a horizontally oriented rectangle, such as automobile and aircraft portrayal.

This convergence will become increasingly important in the future. The use of computer graphics within the broadcast television industry has seen enormous growth during this decade. Apart from this trend, however, there is also the potential offered by computer graphics techniques and HDTV imagery for the creation of special effects within the motion picture production industry, already demonstrated convincingly in countless major releases.

Resolution Considerations

With few exceptions, computers were developed as stand-alone systems using proprietary display formats [2]. Until recently, computers remained isolated with little need to exchange video information with other systems. As a consequence, a variety of specific display formats were developed to meet computer industry needs that are quite different from the broadcast needs.

Among the specific computer industry needs are bright and flickerless displays of highly detailed pictures. To achieve this, computers use progressive vertical scanning with rates varying from 56 Hz to 75 Hz, referred to as *refresh rates*, and an increasingly high number of lines per picture. High vertical refresh rates and number of lines per picture result in short line durations and associated wide video bandwidths.

Because the video signals are digitally generated, certain analog resolution concepts do not directly apply to computer displays. To begin with, all *analog-to-digital* (A/D) conversion concerns related to sampling frequencies and associated anti-aliasing filters are nonexistent. Furthermore, there is no vertical resolution ambiguity, and the vertical resolution is equal to the number of active lines. The only limiting factor affecting the displayed picture resolution is the CRT dot structure and spacing, as well as the driving video amplifiers.

The computer industry uses the term *vertical resolution* when referring to the number of active lines per picture and *horizontal resolution* when referring to the number of active pixels per line. This resolution concept has no direct relationship to the television resolution concept and can be misleading (or at least confusing). Table 2.2 summarizes the basic characteristics of common IBM-type computer graphics displays.

Table 2.2 Basic Characteristics Of IBM-Type Computer Graphics Displays (*After* [2])

Horizontal Resolution	640	800	800	1024	1280
Vertical Resolution	480	600	600	768	1024
Active Lines/Frame	480	600	600	768	1024
Total Lines/Frame	525	628	666	806	1068
Active Line Duration, μs	20.317	20.00	16.00	13.653	10.119
f_h (Hz)	37.8	37.879	48.077	56.476	76.02
f_v (MHz)	72.2	60.316	72.188	70.069	71.18
Pixel Clock, MHz	31.5	40	50	75	126.5
Video Bandwidth, MHz	15.75	20	25	37.5	63.24

Video/Computer Optimization

In recognition of the interest on the part of the computer industry in television in general, and DTV in particular, detailed guidelines were developed by Intel and Microsoft to provide for interoperability of the ATSC DTV system with personal computers of the future. Under the industry guidelines known as *PC99*, design goals and interface issues for future devices and systems were addressed. To this end, the PC99 guidelines referred to existing industry standards or performance goals (benchmarks), rather than prescribing fixed hardware implementations. The video guidelines were selected for inclusion in the guide based on an evaluation of possible system and device features. Some guidelines are defined to provide clarification of available system support, or design issues specifically related to the Windows 98 and Windows NT operating system architectures (and their successor systems, Windows Millennium Edition and Windows 2000).

The requirements for digital broadcast television apply for any type of computer system that implements a digital broadcast subsystem, whether receiving satellite, cable, or terrestrial broadcasts. Such capabilities were recommended, but not required, for all system types. The capabilities were strongly recommended, however, for entertainment PC systems.

Among the specific recommendations was that systems be capable of simultaneously receiving two or more broadcast frequencies. The ability to tune to multiple frequencies results in better concurrent data and video operation. For example, with two tuners/decoders, the viewer could watch a video on one frequency and download web pages on the other. This also enables picture-in-picture or multiple data streams on different channels or transponders. Receiver were also recommended to support conditional access mechanisms for subscription services, pay-per-view events, and other network-specific access-control mechanisms available on the broadcast services for which they were designed.

As this book went to press, the second generation of products meeting these guidelines were showing up on retail store shelves.

2.2.4 ATSC Datacasting

Although the primary focus of the ATSC system is the conveyance of entertainment programming, datacasting is a practical and viable additional feature of the standard. The concept of datacasting is not new; it has been tried with varying degrees of success for years using the NTSC system in the U.S. and PAL and SECAM elsewhere. The tremendous data throughput capabilities of DTV, however, permit a new level of possibilities for broadcasters and cable operators.

In general, the industry has defined two major categories of datacasting [3]:

- *Enhanced television*—data content related to and synchronized with the video program content. For example, a viewer watching a home improvement program might be able to push a button on the remote control to find more information about the product being used or where to buy it.

- *Data broadcast*—data services not related to the program content. An example would be current traffic conditions, stock market activity, or even subscription services that utilize ATSC conditional access capabilities.

The ATSC Digital Television Application Software Environment (DASE) specialist group has worked to define the software architecture for digital receivers. DASE has identified the following critical issues:

- **Open architecture**. Receiver manufacturers want independence from any particular vendor of hardware or software subsystems, freeing them from the PC industry model where a small number of companies dictate product specifications.

- **JAVA**. The DASE standard will likely support the JAVA language.

- **Wide range of datacasting services**. Receiver manufacturers want to be able to offer products that support different levels of datacasting and browsing features.

- **Use of existing Web authoring tools**. The standard is being based on HTML (Hyper-Text Mark-up Language), the programming language used for the Internet. Consequently ATSC datacasting services will be able to make use of the pool of experienced authors already creating content for the Web, as well as support reuse of existing Web content.

- **Web links**. The datacasting content can automatically link to a Web site if the receiver is equipped with an Internet browser.

- **Synchronize data to program content**.The datacasting standard provides techniques to synchronize data content to specific segments of a video stream and provide precise layout control for the data content to coexist with the video image.

- **Extensible**. The standard supports new media types by using content decoders that are extensible with downloadable software code.

Effective use of datacasting could have far reaching effects on advertising and commercial broadcaster business models. A new generation of intelligent ATSC receivers with built in Internet browsers and reverse communications channels will be able to seamlessly integrate Internet services with broadcast television.

2.3 Characteristics of the Video Signal

High-definition television has improved on earlier techniques primarily by calling more fully upon the resources of human vision [1]. The primary objective of HDTV has been to enlarge the visual field occupied by the video image. This has called for larger, wider pictures that are intended to be viewed more closely than conventional video. To satisfy the viewer upon this closer inspection, the HDTV image must possess proportionately finer detail and sharper outlines.

2.3.1 Critical Implications for the Viewer and Program Producer

In its search for a "new viewing experience," NHK conducted an extensive psychophysical research program in which a large number of attributes were studied. Viewers with nontechnical backgrounds were exposed to a variety of electronic images, whose many parameters were then varied over a wide range. A definition of those imaging parameters was being sought, the aggregate of which would satisfy the average viewer that the TV image portrayal produced an emotional stimulation similar to that of the large-screen film cinema experience.

Central to this effort was the pivotal fact that the image portrayed would be large—considerably larger than current NTSC television receivers. Some of the key definitions being sought by NHK were precisely how large, how wide, how much resolution, and the optimum viewing distance of this new video image.

A substantial body of research gathered over the years has established that the average U.S. consumer views the TV receiver from a distance of approximately seven picture heights. This translates to perhaps a 27-in NTSC screen viewed from a distance of about 10 ft. At this viewing distance, most of the NTSC artifacts are invisible, with perhaps the exception of cross color. Certainly the scanning lines are invisible. The luminance resolution is satisfactory on camera close-ups. A facial close-up on a modern high-performance 525-line NTSC receiver, viewed from a distance of 10 ft, is quite a realistic and pleasing portrayal. But the system quickly fails on many counts when dealing with more complex scene content.

Wide-angle shots (such as jersey numbers on football players) are one simple and familiar example. TV camera operators, however, have adapted to this inherent restriction of 525-line NTSC, as witnessed by the continual zooming in for close-ups during most sporting events. The camera operator accommodates for the technical shortcomings of the conventional television system and delivers an image that meets the capabilities of NTSC, PAL, and SECAM quite reasonably. There is a penalty, however, as illustrated in Figure 2.2. The average home viewer is presented with a very narrow angle of view—on the order of 10°. The video image has been rendered "clean" of many inherent disturbances by the 10-ft viewing distance and made adequate in resolution by the action of the camera operator; but, in the process, the scene has become a small "window". The now "acceptable" television image pales in comparison with the sometimes awesome visual stimulation of the cinema. The primary limitation of conventional TV systems is, therefore, image size. A direct consequence is further limitation of image content; the angle of view constantly is constricted by the need to provide adequate resolution. There is significant, necessary, and unseen inter-

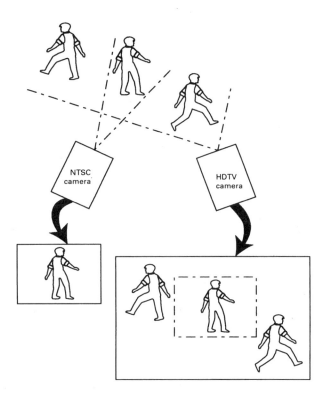

Figure 2.2 An illustration of the differences in the scene capture capabilities of conventional video and HDTV. (*After* [1].)

vention by the TV program director in the establishment of image content that can be passed on to the home viewer with acceptable resolution.

Compared with the 525-line NTSC signal (or the marginally better PAL and SECAM systems), the ATSC DTV system and the North American HDTV studio standard (SMPTE 240M) and its digital representation (SMPTE 274M) offer a vast increase in total information contained within the visual image. If all this information is portrayed on an appropriate HDTV studio monitor (commonly available in 19-, 28-, and 38-in diagonal sizes), the dramatic technical superiority of HDTV over conventional technology easily can be seen. The additional visual information, coupled with the elimination of composite video artifacts, portrays an image almost totally free (subjectively) of visible distortions, even when viewed at a close distance.

On a direct-view CRT monitor, HDTV displays a technically superb picture. The *information density* is high; the picture has a startling clarity. However, when viewed from a distance of approximately seven picture heights, it is virtually indistinguishable from a good NTSC portrayal. The wider aspect ratio is the most dramatic change in the viewing experience at normal viewing distances.

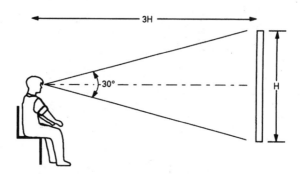

Figure 2.3 Viewing angle as a function of screen distance for HDTV. (*After* [1].)

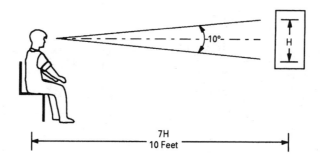

Figure 2.4 Viewing angle as a function of screen distance for conventional video systems. (*After* [1].)

2.3.2 Image Size

If HDTV is to find a home with the consumer, it will find it in the living room. If consumers are to retain the average viewing distance of 10 ft, then the minimum image size required for an HDTV screen for the average definition of a totally new viewing experience is about a 75-in diagonal. This represents an image area considerably in excess of present "large" 27-in NTSC (and PAL/SECAM) TV receivers. In fact, as indicated in Figure 2.3, the viewing geometry translates into a viewing angle close to 30° and a distance of only three picture heights between the viewer and the HDTV screen. Compare this with the viewing angle for conventional systems at 10°, as shown in Figure 2.4.

HDTV Image Content

There is more to the enhanced viewing experience than merely increasing picture size [1]. Unfortunately, this fundamental premise has been ignored in some audience surveys. The larger artifact-free imaging capability of HDTV allows a new image portrayal that capitalizes on the attributes of the larger screen. As mentioned previously, as long as the camera operator appropriately fills the 525 (or 625) scanning system, the resulting image (from a resolution viewpoint) is actually quite satisfactory on conventional systems. If, however, the same football game is shot with an HDTV camera the angle of view of the lens is adjusted to portray the same resolution (in the picture center) as the 525 camera when capturing a close-up of a player on its 525 screen, a vital difference between the two pictures emerges: the larger HDTV image contains considerably more information, as illustrated in Figure 2.2.

The HDTV picture shows more of the football field—more players, more of the total action. Thus, the HDTV image is radically different from the NTSC portrayal. The individual players are portrayed with the same resolution on the retina—at the same viewing distance—but a totally different viewing experience is provided for the consumer. The essence of HDTV imaging is this greater sensation of reality.

The real, dramatic impact of HDTV on the consumer will be realized only when two key ingredients are included:

- Presentation of an image size of approximately 75 in diagonal (minimum).

- Presentation of image content that capitalizes on new camera freedom in formatting larger, wider, and more true-to-life angles of view.

2.3.3 Format Development

Established procedures in the program production community provide for the 4:3 aspect ratio of video productions and motion picture films shot specifically for video distribution. This format convention has, by and large, been adopted by the computer industry for desktop computer systems.

In the staging of motion picture films intended for theatrical distribution, no provision generally is made for the limitations of conventional video displays. Instead, the full screen, in wide aspect ratios—such as CinemaScope—is used by directors for maximum dramatic and sensory impact. Consequently, cropping of essential information may be encountered more often than not on the video screen This problem is particularly acute in wide-screen features where cropping of the sides of the film frame is necessary to produce a print for video transmission. This objective is met in one of the following ways:

- *Letter-box* transmission with blank areas above and below the wide-screen frame. Audiences in North America and Japan have not generally accepted this presentation format, primarily because of the reduced size of the picture images and the aesthetic distraction of the blank screen areas.

- Printing the full frame height and cropping equal portions of the left and right sides to provide a 4:3 aspect ratio. This process frequently is less than ideal because, depending upon the scene, important visual elements may be eliminated.

- Programming the horizontal placement of a 4:3 aperture to follow the essential picture information. Called *pan and scan*, this process is used in producing a print or in making a film-to-tape transfer for video viewing. Editorial judgment is required for determining the scanning cues for horizontal positioning and, if panning is used, the rate of horizontal movement. This is an expensive and laborious procedure and, at best, it compromises the

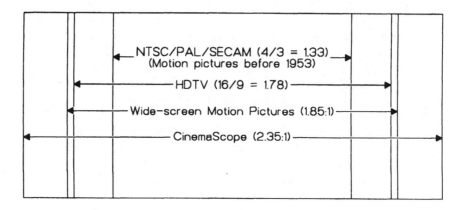

Figure 2.5 Comparison of the aspect ratios of television and motion pictures. (*After* [4].)

artistic judgments made by the director and the cinematographer in staging and shooting, and by the film editor in postproduction.

These considerations are also of importance to the computer industry, which is keenly interested in multimedia technology.

One of the reasons for moving to a 16:9 format is to take advantage of consumer acceptance of the 16:9 aspect ratio commonly found in motion picture films. Actually, however, motion pictures are produced in several formats, including:

- 4:3 (1.33)

- 2.35, used for 35 mm anamorphic CinemaScope film

- 2.2 in a 70 mm format

Still, the 16:9 aspect ratio generally is supported by the motion picture industry. Figure 2.5 illustrates some of the more common aspect ratios.

The SMPTE has addressed the mapping of pictures in various aspect ratios to 16:9 in Recommended Practice 199-1999. The Practice describes a method of mapping images originating in aspect ratios different from 16:9 into a 16:9 scanning structure in a manner that retains the original aspect ratio of the work. Ratios of 1.33 to 2.39 are described in RP199-1999 [5].

2.4 Aural Component of Visual Realism

The realism of a television image depends to a great degree on the realism of the accompanying sounds. Particularly in the close viewing of HDTV images, if the audio system is monophonic, the sounds seem to be confined to the center of the screen. The result is that the visual and aural senses convey conflicting information. From the beginning of HDTV system design, it has been clear that stereophonic sound must be used. The generally accepted quality standard for high-fidelity audio has been set by the digital compact disc (CD). This medium covers audio frequencies from below 30 Hz to above 20 kHz, with a dynamic range of 90 dB or greater.

Sound is an important element in the viewing environment. To provide the greatest realism for the viewer, the picture and the sound should be complementary, both technically and editorially. The sound system should match the picture in terms of positional information and offer the producer the opportunity to use the spatial field creatively. The sound field can be used effectively to enlarge the picture. A *surround sound* system can further enhance the viewing experience.

2.4.1 Hearing Perception

There is a large body of scientific knowledge on how humans localize sound. Most of the research has been conducted with subjects using earphones to listen to monophonic signals to study *lateralization. Localization* in stereophonic listening with loudspeakers is not as well understood, but the research shows the dominant influence of two factors: *interaural amplitude* differences and *interaural time delay.* Of these two properties, time delay is the more influential factor. Over intervals related to the time it takes for a sound wave to travel around the head from one ear to the other, interaural time clues determine where a listener will perceive the location of sounds. Interaural amplitude differences have a lesser influence. An amplitude effect is simulated in stereo music systems by the action of the stereo balance control, which adjusts the relative gain of the left and right channels. It is also possible to implement stereo balance controls based on time delays, but the required circuitry is more complex.

A listener positioned along the line of symmetry between two loudspeakers will hear the center audio as a phantom or *virtual image* at the center of the stereo stage. Under such conditions, sounds—dialogue, for example—will be spatially coincident with the on-screen image. Unfortunately, this coincidence is lost if the listener is not positioned properly with respect to the loudspeakers. Figure 2.6 illustrates the sensitivity of listener positioning to aural image shift. As illustrated, if the loudspeakers are placed 6 ft apart with the listener positioned 10 ft back from the speakers, an image shift will occur if the listener changes position (relative to the centerline of the speakers) by just 16 in. The data shown in the figure is approximate and will yield different results for different types and sizes of speakers. Also, the effects of room reverberation are not factored into the data. Still, the sensitivity of listener positioning can be seen clearly. Listener positioning is most critical when the loudspeakers are spaced widely, and less critical when they are spaced closely. To limit loudspeaker spacing, however, runs counter to the purpose of wide-screen displays. The best solution is to add a third audio channel dedicated exclusively to the transmission of center-channel signals for reproduction by a center loudspeaker positioned at the video display, and to place left and right speakers apart from the display to emphasize the wide-screen effect. The addition of *surround sound* speakers further improves the realism of the aural component of the production.

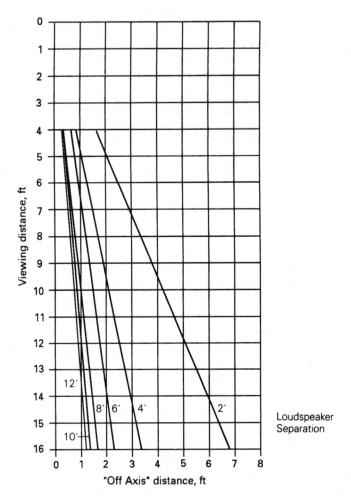

Figure 2.6 The effects of listener positioning on center image shift. (*After* [6].)

2.4.2 Matching Audio to Video

It has been demonstrated that even with no picture to provide visual cues, the ear/brain combination is sensitive to the direction of sound, particularly in an arc in front of and immediately in back of the listener. Even at the sides, listeners are able to locate direction cues with reasonable accuracy. With a large-screen display, visual cues make the accuracy of sound positioning even more important.

If the number of frontal loudspeakers and the associated channels is increased, the acceptable viewing/listening area can be enlarged. Three-channel frontal sound using three loudspeakers provides good stereo listening for three or four viewers, and a 4-channel presentation increases the area even more. The addition of one or more rear channels permits surround sound effects.

Surround sound presentations, when done correctly, significantly improve the viewing experience. For example, consider the presentation of a concert or similar performance in a public hall. Members of the audience, in addition to hearing the direct performance sound from the stage, also receive reflected sound, usually delayed slightly and perhaps diffused, from the building surfaces. These acoustic elements give a hall its tonal quality. If the spatial quality of the reflected sound can be made available to the home viewer, the experience will be enhanced greatly. The home viewer will see the stage performance in high definition and hear both the direct and indirect sound, all of which will add to the feeling of being present at the performance.

In sports coverage, much use can be made of positional information. In a tennis match, for example, the umpire's voice would be located in the center sound field—in line with his or her observed position—and crowd and ambient sounds would emanate from left and right.

Several methods have been used to successfully convey the surround sound channel(s) in conventional NTSC broadcasts. The Dolby AC-3 sound system is used in the ATSC DTV system, offering 5.1 channels of audio information to accompany the HDTV image.

2.4.3   Making the Most of Audio

In any video production, there is a great deal of sensitivity to the power of the visual image portrayed through elements such as special effects, acting, and directing that build the scene. All too often, however, audio tends to becomes separated from the visual element. Achieving a good audio product is difficult because of its subjective content. There are subtleties in the visual area, understood and manipulated by video specialists, that an audio specialist might not be aware of. By the same token, there are psychoacoustic subtleties relating to how humans hear and experience the world around them that audio specialists can manipulate to their advantage.

Reverb, for example, is poorly understood; it is more than just echo. This tool can be used creatively to trigger certain psychoacoustic responses in an audience. The brain will perceive a voice containing some reverb to be louder. Echo has been used for years to effectively change positions and dimensions in audio mixes.

To use such psychoacoustic tools is to work in a delicate and specialized area, and audio is a subjective discipline that is short on absolute answers. One of the reasons it is difficult to achieve good quality sound is because it is hard to define what that is. It is usually easier to quantify video than audio. Most people, given the same video image, come away with the same perception of it. With audio, however, accord is not so easy to come by. Musical instruments, for example, are harmonically rich and distinctive devices. A violin is not a pure tone; it is a complex balance of textures and harmonics. Audio offers an incredible palette, and it is acceptable to be different. Most video images have any number of absolute references by which images can be judged. These references, by and large, do not exist in audio.

When an audience is experiencing a program—be it a television show or an aircraft simulator training session—there is a balance of aural and visual cues. If the production is done right, the audience members will be drawn into the program, putting themselves into the

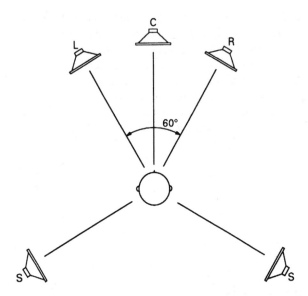

Figure 2.7 Optimum system speaker placement for HDTV viewing. (*After*[4].)

events occurring on the screen. This *suspension of disbelief* is the key to effectively reaching the audience.

2.4.4 Ideal Sound System

Based on the experience of the film industry, it is clear that for the greatest impact, HDTV sound should incorporate, at minimum, a 4-channel system with a center channel and surround sound. Figure 2.7 illustrates the optimum speaker placement for enhancement of the viewing experience. This viewpoint was taken into consideration by the ATSC in its study of the Grand Alliance system.

2.4.5 Dolby AC-3

Under the ATSC DTV sound system, complete audio programs are assembled at the user's receiver from various services sent by the broadcaster. The concept of assembling services at the user's end was intended to provide for greater flexibility, including various-language multichannel principal programs supplemented with optional services for the hearing impaired and visually impaired.

A variety of multichannel formats for the main audio services also is provided, adapting program by program to the best stereo presentation for a particular offering. An important idea that emerged during the process of writing the recommendation was that the principal sound for a program should take up only the digital bit space required by that program. The idea was born that programs fall into production categories and may be classified by the utilization of loudspeaker channels [7]. The categories include:

- **1/0—one front center channel, no surround**. 1/0 is most likely to be used in programs such as news, which have exceedingly strict production time requirements. The advantage in having a distinct monaural mode is that those end users having a center channel loudspeaker will hear the presentation over only that one loudspeaker, with an attendant improvement over hearing mono presented over two loudspeakers [8].

- **2/0—conventional 2-channel stereo**. 2/0 is intended principally for preexisting 2-channel program material. It is also useful for film production recorded in the Dolby Stereo or Ultra Stereo formats with a 4:2:4 amplitude-phase matrix (for which there is an accompanying indicator flag to toggle surround decoding on at the receiver).

- **3/0—left, center, and right front channels**. 3/0 was expected to be used for programs in which stereo is useful but surround sound effects are not, such as an interview program with a panel of experts.

- **3/2/.1—left, center, right front, left and right surround, and a low-frequency effects channel**. 3/2/.1 was expected to be used primarily for films and entertainment programming, matching the current motion picture production practice.

Monitoring

Aural monitoring of program production is a critical element in the production chain [7]. Although the monitor system—with its equalizers, power amplifiers, and loudspeakers—is not in the signal path, monitoring under highly standardized conditions has helped the film industry to make an extremely interchangeable product for many years. With strict monitor standards, there is less variation in program production, and the differences that remain are the result of the director's creative intent. Monitor standards must address the following criteria:

- Room acoustics for monitor spaces

- Physical loudspeaker arrangement

- Loudspeaker and electronics requirements

- Electroacoustics performance

2.5 Motion Picture Film Origination

To appreciate the role that HDTV production systems can play—and, indeed, are playing—in the film industry, it is necessary to first examine some of the important parameters of 35 mm film.

Color motion picture film consists of three photosensitive layers sensitized to red, green, and blue light [9]. Exposure and processing produce cyan, magenta, and yellow dye images in these layers that, when projected on a large screen or scanned in a telecine, produce the color pictures. Although the 16 mm width was the most common format for distribution of television programs for many years, 35 mm is currently the most used format ("super 35" and "super 16" also are utilized).

35 mm film is used extensively for the production of prime-time television programs and commercials. Manufacturers continue to improve film for speed, grain, and sharpness. It has been estimated that another 10× improvement still is possible with silver halide technology [9]. The quality of film images, versatility of production techniques, and worldwide standardization keep 35 mm film an important and valuable tool for the production of images for television broadcast. New versions of both flying-spot type and charge-coupled device film scanners continue to be developed for direct broadcast or transfer of film to video.

Motion pictures, most prime-time TV programs, and TV commercials are originated on color negative film. From these originals, prints can be made for broadcast and distribution in either 35 or 16 mm formats. Duplicate negatives can be made and large numbers of prints prepared for theatrical release. Telecines are capable of scanning either color positives or color negatives. A program originated on color negative can be transferred directly to tape, edited electronically, then broadcast. The same negative can be edited, then broadcast from either film or a transferred tape.

There are many reasons why film remains the medium of choice for origination. One of the main reasons is an undefined phenomenon called the "film look." This characteristic has defied quantification by performance parameters but continues to be a major consideration. Another advantage of film origination is that it is a standard format worldwide, and it has been for many decades. Programs originated on film can be readily syndicated for distribution in any of the conventional video standards.

2.5.1 Film Types

In addition to the higher level of technical performance required of an HDTV film-to-video transfer system, a wide variety of film formats must be accommodated. These include:

- Conventional television aspect ratio of 1.33

- Wide-screen nonanamorphic aspect ratio of 1.85

- Wide-screen anamorphic aspect ratio of 2.35

- Positive and negative films

- 16 mm films (optional)

The major source of 35 mm program material consists of films produced for television and prints of features made for theatrical release. In anticipation of standards approval for HDTV, some filmed television programs have been produced in wide-screen formats. On the other hand, most theatrical productions are in wide-screen formats either horizontally compressed—*anamorphic CinemaScope* (2.35:1)—or nonanamorphic (1.85:1). Because of this variation in screen size, some form of image truncation (pan and scan, discussed previously) or variable-size letterbox presentation is necessary.

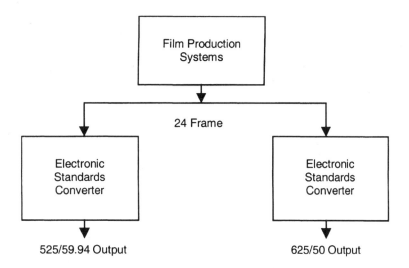

Figure 2.8 Conventional telecine conversion process. (*After* [10].)

2.5.2   Synergy Between Film and HDTV Program Origination

HDTV is not intended to replace film as a primary source of program production. This is a vital point in the marriage of film and video. HDTV, instead, provides a powerful new synergism that offers the choice of two mediums (of approximately comparable picture quality) by which a given program may be produced [10]. The result is greater flexibility, in the form of a new imaging tool, for program producers.

The producers will make their choices quite independently of the viewpoints of technologists. And they will make their choices based upon specific imaging requirements, which, in turn, usually will be based upon the particular script under consideration. The requirements of that script may clearly dictate 35 mm film as the medium of choice, or it may suggest an electronic HDTV medium offering unique picture-making advantages. The script may even suggest the use of both mediums, each applied to those scenes in which a particular form of imagery is sought, or a certain logistical convenience is required.

This freedom is possible because of the capability to transfer between mediums. This was part of the overall planning of the HDTV production standard (SMPTE 240M) from the beginning. Just as important was careful protection of the capability of program producers to release their programs via any distribution medium of choice, anywhere in the world. Figure 2.8 illustrates the methodology of program origination and television distribution in widespread use today.

2.5.3   Standards Conversion

Prime-time television programs typically are produced on 35 mm 24-frame film. By means of well-known standard converters (telecine machines), they are downconverted to the 525-

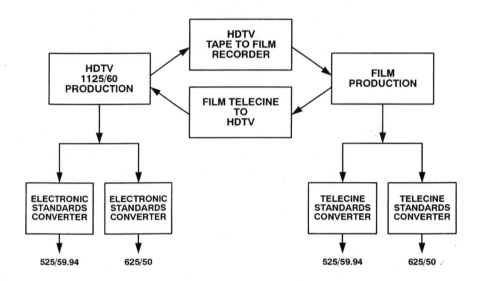

Figure 2.9 Standards conversion between 35 mm 24-frame film, HDTV, and conventional broadcast systems. (*After* [10]. *Courtesy of Sony.*)

or 625-line broadcast television formats. Essential to an understanding of the process is this fact: Film is a program production standard with *no direct compatible relationship* with any TV standard. Indeed, it has no technical relationship whatsoever with television [10]. The process of standards conversion is, thus, a fundamental philosophy inherent to the total scheme of TV programming today. A 24-frame celluloid medium is converted to a 25-frame or a 29.97-frame TV medium—depending on the region of the globe—for conventional television broadcast.

Figure 2.9 shows the same basic concept applied to the all-electronic system of HDTV studio origination followed by conversion to the existing 525/625 television media. As illustrated in the figure, the telecine has been replaced with electronic digital standards converters. The principle, however, remains identical. This concept is the very essence of the long search for a single worldwide HDTV format for studio origination and international program exchange. All HDTV studios the world over would produce programs to a single television format. This key point was the original driving force for standardization of the 1125/60 format by SMPTE and other organizations. The fact that a consensus failed to materialize does not diminish the importance of the concept.

It is fair to point out, however, that the increasing use of MPEG-2 compression is leading to the environment envisioned by the early proponents of 1125/60 standardization for program production. MPEG-2 is rapidly becoming the de facto standard for professional video, and related signal parameters tend to fall in line behind it. Indeed, program exchange capability was one of the reasons MPEG-2 was chosen for the Grand Alliance DTV system and for the European DVB project. Furthermore, the video standards harmonization efforts of

ITU-R Study Group 11/3 and SMPTE/EBU detailed in Chapter 1 have—for all intents and purposes—brought the worldwide HDTV production standard issue full-circle.

High-quality transfer between any two mediums is important because it allows easy intercutting of separately captured images. Regardless of the medium of production, the means exist to convert to any of the present television broadcast systems, conventional resolution or high-definition. Where such services are in place, 35 mm film programming can be released as HDTV. Likewise, HDTV programming can be released as 35 mm film. Furthermore, HDTV techniques are being integrated rapidly into the motion picture industry, adding a new flexibility there.

2.5.4 Film-to-Video Transfer Systems

Two types of film-to-video transfer systems presently are used for 525- and 625-line television service. Both are adaptable to operation for HDTV applications. The systems are:

- Continuous-motion film transport with CRT flying-spot light source and three photoelectric transducers.

- Continuous-motion film transport with three channels of CCD (charge-coupled device) line sensors.

The application of telecine systems is varied, but generally tends to divide between two basic functions:

- *Transfer telecine*—used to transfer feature films to videotape masters for subsequent duplication to videocassettes and videodisc formats, as well as for broadcast. Flying-spot and CCD telecines are used by film-to-tape transfer houses for these applications. A skilled *colorist* provides scene-to-scene color correction as necessary, using programmable color-correction controls. The transfer telecine is designed to accommodate negative film, print film, low-contrast print film, duplicate negative, or interpositive film. Some TV programs are edited on film and transferred to videotape in a similar manner with scene-to-scene color correction.

- *Production telecine*—used in the production of TV programs and commercials, typically flying-spot and CCD systems. Selected camera shots ("circle takes") are transferred from the camera negative film to videotape with a colorist providing scene-to-scene color correction.

Telecine Principles

The basic function of a telecine is to convert an optical image on motion picture film into a video signal [9]. This conversion involves an opto-electronic transducer and a scanning operation, as illustrated in Figure 2.10. Two basic telecine designs are available commercially: 1) cathode-ray tube flying-spot scanner (FSS), and 2) CCD line array. A laser telecine for HDTV use also has been produced.

Another design, which was actually the original system used for telecine operation, is the photoconductive telecine. This design involves the combination of a synchronized motion picture projector with a television camera. As the design evolved, specialized video-signal-

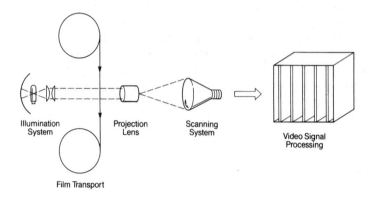

Figure 2.10 The primary components of a telecine. (*After* [9].)

processing circuitry was developed to improve the picture quality of the transferred or broadcast images. The photoconductive (camera tube) telecine is no longer manufactured, and probably only a few still exist. With the development of the flying-spot and the CCD line array systems, the photoconductive technique could not compete with the quality of the reproduced images or with the ease of operation and maintenance.

As shown in Figure 2.10, the major components of a generic telecine design include:

- Film transport
- Illumination system
- Projection lens
- Scanning system
- Video-signal-processing system

The implementation of these basic components is a function of the telecine technology and design.

CRT Flying-Spot Scanner

The CRT FSS produces a video signal by scanning the film images with a small spot of light and collecting the light transmitted through the film with a photomultiplier tube (PMT) [9]. A high-intensity CRT is scanned by an unblanked electron beam.

The CRT flying-spot scanner (developed by Rank Cintel) originally was designed for 24 frames/s film transfer to 25 frames/s European television standards (PAL and SECAM). Early attempts at 30 frames/s NTSC designs involved a complicated "jump-scan" approach, controlling the scan to implement both interlace and 3:2 frame-rate conversion. The development of the digital frame store—which permitted the film frame to be progressively scanned, then interlaced, and frame-rate-converted by controlling the output (read) rate—made the flying-spot scanner design practical for NTSC.

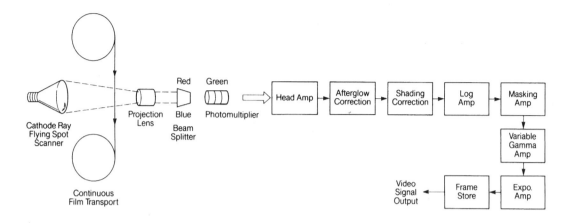

Figure 2.11 Block diagram of cathode-ray tube flying-spot scanner. (*After* [9].)

One of the fundamental innovations of the flying-spot scanner design was the development of a continuous-motion capstan-driven film transport. With this mechanism, the velocity of the film is monitored by a shaft encoder on a free-running timing sprocket that tracks the film perforations. The timing pulses then are used to control the capstan velocity via a servo loop. This transport has proved to be gentle enough to handle negative and intermediate film stocks in addition to print film.

A basic block diagram of a CRT FSS is shown in Figure 2.11. The scanning spot is divided into three color channels by a dichroic beam splitter, and each signal is picked up by a PMT. The signal from each PMT is buffered by the head amplifier and applied to the *afterglow correction* circuitry. Afterglow correction is a high-pass filtering operation that compensates for the persistence (afterglow) of the phosphor. The next step is *shading correction*, which compensates for uneven illumination, optical losses, and uneven tube sensitivity. Color masking and gamma correction are implemented on logarithmic signals, and the resulting signal is exponentiated for display.

A digital frame store is used to provide both interlace conversion and frame-rate conversion. This is accomplished by independent control of the input (write) clock and the output (read) clock. Aperture correction is also implemented digitally.

Of the recent advances in CRT flying-spot scanners, those of particular interest for high-definition systems include:

- All-digital video-signal-processing channel

- Development of a pin-registered gate for image compositing

- Development of an *electronic pin registration* (EPR) system for real-time steadiness correction

The primary advantage of the CRT flying-spot scanner is the scan flexibility; zoom, pan, and anamorphic expansion are easily implemented by changing the scanning raster. The

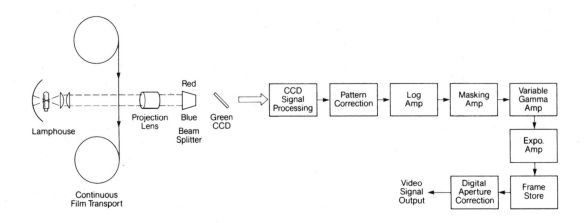

Figure 2.12 Block diagram of CCD line array telecine. (*After* [9].)

continuous-motion transport handles film gently, making the transfer of camera negative film viable.

The primary limitation of the CRT design is the short life of the tube (2000 to 5000 hours) before it must be replaced because of phosphor burn or spot size deterioration. Although no significant problem occurred with NTSC or PAL transfers, early CRT HDTV telecines exhibited limited sharpness and signal-to-noise performance (particularly when scanning negative film) as the CRT aged.

CCD Line Array

The CCD line array telecine was introduced in the early 1980s [9]. As its name implies, the heart of the system is a CCD line array imager that converts the optical image to a video signal by transferring the charge accumulated in each photosite of the line array through a charge-coupled output register. The CCD telecine design also utilizes the digital frame store and continuous-motion transport, first implemented in the CRT flying-spot scanner. The illumination system is a high-energy blue-rich tungsten halogen lamp, which is important in achieving high signal-to-noise (S/N) performance when transferring orange-masked color negative films.

Sensor clocks are generated to control the integration time of each line and the pixel readout rate. The commonly used CCD element photosensor has dual-channel readout with alternate samples interleaved. Pattern correction removes any stripe patterns resulting from photosite sensitivity variations and output shift register mismatches.

The video-signal-processing system in a CCD telecine is much the same as that of the CRT FSS telecine described previously. A basic block diagram is shown in Figure 2.12.

Programmable Color Correction

Telecines typically are controlled by programmable color-correction devices [9]. With these devices, the controls on the telecine can be programmed on a scene-by-scene or frame-by-frame basis. With a programmable color-correction unit, the telecine can be programmed to compensate for film differences contained within a roll or create special effects such as color enhancement, day for night, rotation, zoom, pan, noise reduction, and sharpness enhancement.

Programmable color correctors can perform *primary* and *secondary* color corrections. Primary control refers to lift, gamma, and gain controls, which adjust the blacks, midscale, and whites of an image respectively. Primaries also allow for red, green, and blue interactions in the black, midscale, and white portions of the image. Primaries are used to make overall color adjustments to the film transfer for the purpose of matching various elements or creating a color effect. Secondary color correction refers to additional, precise manipulations. For example, with secondary correction, a specific shade of red in a particular region of the image may be altered. Primary controls on a color corrector can be applied to both the signal processing within the telecine and the video processor of the color-correction unit. Secondary controls are applied to the video processor of the color-correction unit and usually cannot affect the internal workings of the telecine.

2.5.5   Video-to-Film Laser-Beam Transfer

Film exposure by laser beams can be accomplished through the use of converged high-resolution beams to expose color film directly. The vertical scan can be provided by a moving mirror, and the line scan by a mirrored rotating polygon. This technique was demonstrated as early as 1971 by CBS Laboratories in the United States [4]. (See Section 1.2.1.) Subsequently, NHK—together with other companies under its guidance in Japan—developed prototype 16 mm recorders using a similar system. This led to the design of an improved system for 35 mm film recording of 1125-line high-definition TV signals.

In the NHK system, the red (He-Ne), green (Ar^+), and blue (He-Cd) laser beams are varied in intensity, in accordance with the corresponding video signals, by *acoustic-optical modulators* (AOM) with a center frequency of approximately 200 MHz and a modulation bandwidth of about 30 MHz. The three color signals, converged by dichroic mirrors, are deflected horizontally by a 25-sided, 40-mm-diameter polygon mirror rotating at 81,000 rpm.

Direct recording on 35 mm color print film, such as EK 5383 (Kodak), meets the resolution and noise-level requirements of high-definition television and is superior to a conventional 35 mm print. A comparison of the modulation transfer functions (MTF) for various color films and a scanned laser beam is shown in Figure 2.13.

Higher resolution and less granular noise, in exchange for a 100:1 lower exposure speed, can be obtained by the use of color internegative film. This slower speed, on the order of ANSI 1 to 5, is easily within the light-output capability of a laser-beam recorder. Furthermore, sharpness can be improved significantly in laser recording by the use of electronic contour correction, a technique not possible in conventional negative-positive film photography.

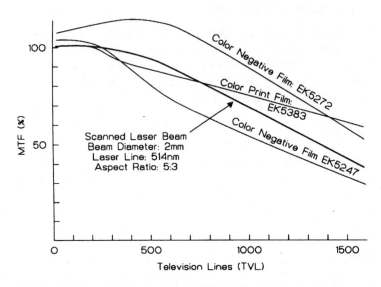

Figure 2.13 Modulation transfer function (MTF) of typical films and scanned laser beams. (*After* [4].)

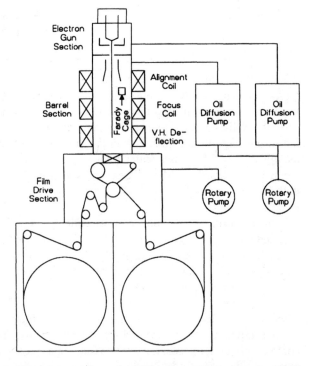

Figure 2.14 Block diagram of the basic tape-to-film electronic-beam recording system. (*After* [4].)

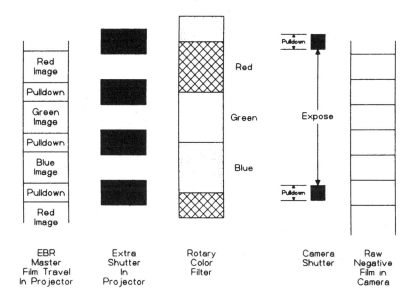

Figure 2.15 Relative timing of components in the EBR step printing process. (*After* [2].)

As with the FSS telecine, recording can be performed using a continuously moving—rather than an intermittent—film-transport mechanism. This requires a scanning standards converter to translate the video signal from interlaced to a noninterlaced sequential scan.

Electron-Beam Video-to-Film Transfer

The 1988 introduction of 1125/60 HDTV (SMPTE 240M) as a practical signal format for program production prompted intensive study of *electron-beam recording* (EBR) technology to meet the system specifications of HDTV program production [4]. The EBR system operates in non-real time to expose a monochrome (black and white) film master. A block diagram of the basic recording system is shown in Figure 2.14. In this scheme, the high-definition red, green, and blue video signals are processed for sequential application to the recorder. The progressive-scan mode was chosen for maximum resolution and a minimum of artifacts.

The sharply focused electron beam scans the film surface in a vacuum, producing a latent image in the emulsion. The conductivity of raw film stock is lowered substantially by evaporation of moisture from the film in the vacuum chamber necessary for the electron-beam scanning process. A reduction in conductivity can result in the buildup of a static charge and accompanying distortion of the scanning beam. Therefore, the scanning operation is limited to one scan per frame. The film is moved through the aperture gate in the EBR chamber by an intermittent claw pull-down mechanism.

The virtually still-frame film speed of the recording permits considerable latitude in the choice of black-and-white film for the EBR master recording. The sequential black-and-white exposures of red, green, and blue video signals produced by the EBR process are transferred to a color internegative in an optical step printer, as shown in Figure 2.15. Two shutters are used:

- The first to blank the pull-down of the master projector

- The second, at one-third the speed of the first, to blank the pull-down for the color camera

Each frame of the color negative film is exposed sequentially to the black-and-white master frames corresponding to the red, green, and blue video signals through appropriate red, green, and blue color filters mounted on a rotating disk, thus providing a color composite.

2.6   References

1. Thorpe, Laurence J.: "Applying High-Definition Television," *Television Engineering Handbook*, rev. ed., K. B. Benson and Jerry C. Whitaker (eds.), McGraw-Hill, New York, p. 23.4, 1991.

2. Robin, Michael: "Digital Resolution," *Broadcast Engineering*, Intertec Publishing, Overland Park, Kan., pp. 44–48, April 1998.

3. Venkat, Giri, "Understanding ATSC Datacasting—A Driver for Digital Television," *Proceedings of the NAB Broadcast Engineering Conference*, National Association of Broadcasters, Washington, D.C., pp. 113–116, 1999.

4. Benson, K. B., and D. G. Fink: *HDTV: Advanced Television for the 1990s*, McGraw-Hill, New York, 1990.

5. SMPTE Recommended Practice RP 199-1999, "Mapping of Pictures in Wide-Screen (16:9) Scanning Structure to Retain Original Aspect Ratio of the Work," SMPTE, White Plains, N.Y., 1999.

6. Torick, Emil L.: "HDTV: High Definition Video—Low Definition Audio?," *1991 HDTV World Conference Proceedings*, National Association of Broadcasters, Washington, D.C., April 1991.

7. Holman, Tomlinson: "The Impact of Multi-Channel Sound on Conversion to ATV," *Perspectives on Wide Screen and HDTV Production*, National Association of Broadcasters, Washington, D.C., 1995.

8. Holman, Tomlinson: "Psychoacoustics of Multi-Channel Sound Systems for Television," *Proceedings of HDTV World*, National Association of Broadcasters, Washington, D.C., 1992.

9. Bauer, Richard W.: "Film for Television," *NAB Engineering Handbook*, 9th ed., Jerry C. Whitaker (ed.), National Association of Broadcasters, Washington, D.C., 1998.

10. Thorpe, Laurence J.: "High Definition Production Systems," *Television Engineering Handbook*, rev. ed., K. B. Benson and Jerry C. Whitaker (eds.), McGraw-Hill, New York, p. 23.4, 1991.

2.7   Bibliography

Baldwin, M. Jr.: "The Subjective Sharpness of Simulated Television Images," *Proceedings of the IRE*, vol. 28, July 1940.

Belton, J.: "The Development of the CinemaScope by Twentieth Century Fox," *SMPTE Journal*, vol. 97, SMPTE, White Plains, N.Y., September 1988.

Fink, D. G.: "Perspectives on Television: The Role Played by the Two NTSCs in Preparing Television Service for the American Public," *Proceedings of the IEEE*, vol. 64, IEEE, New York, September 1976.

Fink, D. G: *Color Television Standards*, McGraw-Hill, New York, 1986.

Fink, D. G, et. al.: "The Future of High Definition Television," *SMPTE Journal*, vol. 9, SMPTE, White Plains, N.Y., February/March 1980.

Fujio, T., J. Ishida, T. Komoto, and T. Nishizawa: "High-Definition Television Systems— Signal Standards and Transmission," *SMPTE Journal*, vol. 89, SMPTE, White Plains, N.Y., August 1980.

Hamasaki, Kimio: "How to Handle Sound with Large Screen," *Proceedings of the ITS*, International Television Symposium, Montreux, Switzerland, 1991.

Hubel, David H.: *Eye, Brain and Vision*, Scientific American Library, New York, 1988.

Judd, D. B.: "The 1931 C.I.E. Standard Observer and Coordinate System for Colorimetry," *Journal of the Optical Society of America*, vol. 23, 1933.

Keller, Thomas B.: "Proposal for Advanced HDTV Audio," *1991 HDTV World Conference Proceedings*, National Association of Broadcasters, Washington, D.C., April 1991.

Kelly, R. D., A. V. Bedbord, and M. Trainer: "Scanning Sequence and Repetition of Television Images," *Proceedings of the IRE*, vol. 24, April 1936.

Kelly, K. L.: "Color Designation of Lights," *Journal of the Optical Society of America*, vol. 33, 1943.

Lagadec, Roger, Ph.D.: "Audio for Television: Digital Sound in Production and Transmission," *Proceedings of the ITS*, International Television Symposium, Montreux, Switzerland, 1991.

Miller, Howard: "Options in Advanced Television Broadcasting in North America," *Proceedings of the ITS*, International Television Symposium, Montreux, Switzerland, 1991.

Pitts, K. and N. Hurst: "How Much Do People Prefer Widescreen (16 × 9) to Standard NTSC (4 × 3)?," *IEEE Transactions on Consumer Electronics*, IEEE, New York, August 1989.

Pointer, R. M.: "The Gamut of Real Surface Colors, *Color Res. App.*, vol. 5, 1945.

Slamin, Brendan: "Sound for High Definition Television," *Proceedings of the ITS*, International Television Symposium, Montreux, Switzerland, 1991.

Suitable Sound Systems to Accompany High-Definition and Enhanced Television Systems: Report 1072. Recommendations and Reports to the CCIR, 1986. Broadcast Service—Sound. International Telecommunications Union, Geneva, 1986.

Fundamental Imaging System Principles

3.1   Introduction

High-definition television has improved on earlier techniques primarily by calling more fully on the resources of natural vision [1]. This chapter is concerned with those aspects of vision that are basic to the design of the digital TV service. Particular attention is devoted to the techniques required for HDTV, compared with the methods that are used in conventional NTSC, PAL, and SECAM services. The primary objective in HDTV has been to enlarge the visual field occupied by the television image. This has called for larger, wider pictures that are intended to be viewed closely. To satisfy the viewer upon this closer inspection, the HDTV image must possess proportionately finer detail and sharper outlines.

3.2   Visual Fields of Television

A principal objective of the television service is to offer the viewer a sense of presence in the scene and of participation in the events portrayed. To meet this objective, the televised image should convey as much of the spatial and temporal content of the scene as is economically and technically feasible. One important limitation of the conventional services, now clearly recognized, is that their images occupy too small a portion of the visual field of view. Experience in the motion picture industry has shown that a larger, wider picture, viewed closely, contributes greatly to the viewer's sense of presence and participation.

Ongoing development of the HDTV service is directed toward the same end. From the visual standpoint, the term "high-definition television" is, to some extent, a misnomer in that the primary objective of the system is not to produce images having fine, sharp detail, but to provide an image that occupies a larger part of the visual field. Higher definition is secondary; the definition need be no higher than is just adequate for closer scrutiny of the image.

3.2.1 Foveal and Peripheral Vision

There are two areas of the retina to be satisfied by television: the *fovea* and the areas peripheral to the fovea. A small central portion of the retina, the fovea perceives the fine detail and edges of an image [2]. Foveal vision extends over only about 1° of the visual field, whereas the total field to the periphery of vision extends about 160° horizontally and 80° vertically. Motions of the eye and head are necessary to ensure that the fovea is positioned on that part of the retinal image where the detailed structure of the scene is to be discerned.

The portion of the visual field outside the foveal region provides the remaining visual information. Thus, a large part of visual reality is conveyed by this *extrafoveal region*. As television engineering has developed, however, most of the attention has been paid to satisfying the needs of foveal vision. The vital perceptions of extrafoveal vision, notably motion and flicker, have received less attention. This is because a system capable of resolving fine detail presents the major technical challenge—transmission channels that offer essentially no discrimination in the amplitude or time delay among the signals they carry over a very wide band of frequencies. The difficulty encountered in trying to extend the channel bandwidths used by the NTSC, PAL, and SECAM services had, in fact, been the principal impediment to the introduction of HDTV service.

Attention to the properties of peripheral vision has led to a number of constraints. Peripheral vision is greatly sensitive to even modest changes in brightness and position. Thus, the bright portions of a wide image viewed closely are much more subject to flicker at the left and right edges than is the narrower image of the conventional systems. One result of this effect is that the 50 Hz image repetition rate of the PAL and SECAM services is inadequate for HDTV versions of these systems.

3.2.2 Vertical Detail and Viewing Distance

Figure 3.1 illustrates the geometry of the field occupied by the video image [1]. The ratio of the picture width W to its height H is the aspect ratio. The viewing distance D determines the angle h subtended by the picture height. This angle (2 tan – 1 $H/2D$) usually is measured by the ratio of the viewing distance to the picture height D/H. The smaller this ratio, the more fully the image fills the field of view. The useful limit to the viewing distance is that point at which the eye can just perceive the finest details in the image. Closer viewing serves no purpose in resolving detail, and more distant viewing prevents the eye from resolving all the detailed content of the image.

The preferred value of the viewing distance, expressed in picture heights, is the *optimal viewing ratio*, commonly referred to as the *optimal viewing distance*. It defines the distance at which the viewer with normal vision would prefer to see the image, when picture clarity is the criterion.

The optimal viewing ratio is not a definite value because it varies with the subject matter, the viewing conditions, and the acuity of the viewer's vision. Its nominal value does serve, however, as a convenient basis for comparing the performances of the conventional and HDTV services. The computation of the optimal viewing ratio depends on the degree of detail offered in the vertical dimension of the image, without reference to its picture content. The discernible detail is limited by the number of scanning lines presented to the eye and by the ability of these lines to present the details separately. The smallest detail that can

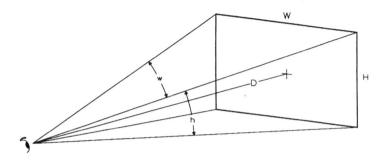

Figure 3.1 The geometry of the field of view occupied by a television image. (*From* [1]. *Used with permission.*)

be reproduced in the image is known as a picture element (*pixel*). Ideally, each detail of the scene would be reproduced by one picture element; that is, each scanning line would be available for one picture element along any vertical line in the image. In practice, however, some of the details in the scene inevitably fall between scanning lines, so that two lines are required for such picture elements. As a result, some vertical resolution is lost. Measurements of this effect show that only about 70 percent of the vertical detail is presented by the scanning lines. This ratio, known as the *Kell factor,* applies irrespective of the manner of scanning, whether the lines follow each other sequentially (progressive scan) or alternately (interlaced scan) [3]. When interlaced scanning is used, as in all the conventional systems, the 70 percent figure applies only if the image is fully stationary and the line of sight from the viewer does not move vertically by even a small amount. In practice, these conditions are seldom met, so an additional loss of vertical resolution—the *interlace factor*—occurs under typical viewing conditions. This additional loss depends on many aspects of the subject matter and viewer attention, so there is a wide range of opinion on its extent. Under favorable conditions, the interlace factor reduces the effective value of vertical resolution to not more than 50 percent; that is, no more than half the scanning lines display the vertical detail of an interlaced image. Under unfavorable conditions, a larger loss can occur. The effective loss also increases with image brightness, as the scanning beam becomes larger.

Because interlacing imposes this additional detail loss (as well as associated degradations due to flicker along horizontal edges—known as *shimmer*—and among objects aligned horizontally—known as *glitter*), it was decided by some system designers early in the development of HDTV to abandon interlaced scanning for image display. A variety of practical technical considerations, however, have required that interlace remain, at least in some transmission modes. Interlace is, of course, an integral part of the Grand Alliance DTV system, as discussed in later chapters.

Table 3.1 compares the optimal viewing ratios and associated image characteristics of the basic 1125/60 HDTV system developed by NHK with conventional television systems for both interlaced- and progressive-scan modes. The basis for the figures presented for the optimal viewing ratio is an extensive study by Fujio and associates at NHK [4]. They concluded that the preferred distance for viewing the 1125-line images of their system has a

Table 3.1 Spatial Characteristics of HDTV (NHK) and Conventional Television Systems, Based on Luminance or Equivalent Y Signal) (*After* [1].)

System	Total Lines	Active Lines	Vertical Resolution	Optimal Viewing Distance	Aspect Ratio	Horizontal Resolution	Total Picture Elements	Field of View (V[1])	Field of View, (H[1])
HDTV-I	1125	1080	540	3.3	16:9	600	575,000	17	30
NTSC-I	525	484	242	7.0	4:3	330	106,000	8	11
NTSC-P	525	484	340	5.0	4:3	330	149,000	12	16
PAL-I	625	575	290	6.0	4:3	425	165,000	10	13
PAL-P	625	575	400	4.3	4:3	425	233,000	13	18
SECAM-I	625	575	290	6.0	4:3	465	180,000	10	13
SECAM-P	625	575	400	4.3	4:3	465	248,000	13	18

Notes:
P = progressive display
I = interlaced display
[1] At optimal viewing distance, given in degrees (V = vertical, H = horizontal)
Video bandwidth (luminance): HDTV = 20 MHz, NTSC = 4.2 MHz, PAL = 5.5 MHz, SECAM = 6 MHz

median value of 3.3 times the picture height, equivalent to a vertical viewing angle of 17°. This covers about 20 percent of the total vertical visual field. The other values given in the table are based on this 3.3 figure, adjusted according to the value of vertical resolution stated for each system. The vertical resolutions in the table are 50 percent of the active scanning lines for interlaced scanning and 70 percent for progressive scanning.

3.2.3 Horizontal Detail and Picture Width

Because the fovea is approximately circular in shape, its vertical and horizontal resolutions are nearly the same. This would indicate that the horizontal resolution of a display should be equal to its vertical resolution. Such equality is the usual basis of television system design, but it is not a firm requirement [1]. A study published in 1940 by Baldwin of Bell Telephone Laboratories showed that the overall subjective sharpness of an image is greatest when the vertical and horizontal dimensions of the picture elements are the same (that is, with equal vertical and horizontal resolution) [5]. However, as shown in Figure 3.2, taken from his paper, considerable variation in the shape of the picture element produces only a minor degradation in the sharpness of the image, provided that its area is unchanged. This seminal finding led to the conclusion that sharpness depends primarily on the product of the resolutions; in other words, it depends on the total number of picture elements in the image.

The freedom to extend the horizontal dimensions of the picture elements has been taken advantage of in wide-screen movies. For example, the Fox CinemaScope system uses anamorphic optical projection to enlarge the image in the horizontal direction [6]. Because the emulsion of the film has equal vertical and horizontal resolution, this enlargement low-

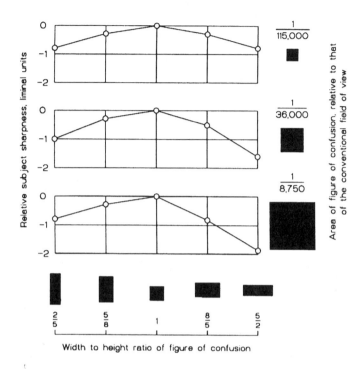

Figure 3.2 Visual sharpness as a function of the relative values of horizontal and vertical resolution. The *liminal unit* given in the figure is the least perceptible difference. (*From* [1]. *Used with permission.*)

ers the horizontal resolution of the image. A limit was reached when the image became 2.35 times as wide as it was high (aspect ratio 2.35:1). Most current motion pictures have an aspect ratio of about 1.85:1.

As discussed previously, the picture width chosen for HDTV service is 1.777 (16/9) times the picture height. Other factors being equal, the video baseband increases in direct proportion to the picture width. An aspect ratio greater than 1.33:1 was an early objective of HDTV system design. The NHK system initially was designed for a ratio of 1.666:1 (5/3), but in later work by other designers, the 16/9 ratio became the first choice. The 1.777 aspect ratio NHK system, with an optimal viewing distance of 3.3 times the picture height, offers a horizontal viewing angle of 30°—about 20 percent of the total horizontal visual field. Although this is a small percentage, it covers that portion of the field within which most visual information is conveyed. (The aspect ratios, resolutions, and fields of view of HDTV and conventional systems are given in Table 3.1.)

3.2.4   Detail Content of the Image

The vertical resolution of a video system is equal to the number of picture elements separately present in the picture height, and the number of elements in the picture width is equal to the horizontal resolution times the aspect ratio [1]. The product of the number of elements vertically and the number horizontally—the total number of picture elements in the image—is an important parameter. Because all the elements must be scanned during the time between successive frame scans, the rate of scanning (and the concomitant video bandwidth) is directly proportional to the total number of elements. Another examination of Table 3.1 provides insight into the significant bandwidth penalty that accompanies progressive scanning, relative to the interlaced mode.

3.2.5   Perception of Depth

Perception of the third spatial dimension—depth—depends in natural vision primarily on the angular separation of the images received by the eyes of the viewer [1]. There have been attempts to produce a binocular system of television, but the cost and inconvenience outweighed the benefits. Still, a considerable degree of depth perception is inferred in the flat image of television from the perspective appearance of the subject matter and from, 1) camera techniques (the choice of the focal lengths of the lenses used) and 2) changes in depth of focus. Continual adjustment of focal length by the zoom lens provides the viewer with an experience in depth perception wholly beyond the scope of natural vision.

No special steps have been taken in the design of HDTV services to offer any depth clues not previously available in conventional television systems. However, the wide field of view offered by the wide-screen HDTV display provides significant improvement in depth perception, compared with that of conventional systems.

3.2.6   Contrast and Tonal Range

The range of brightness (contrast) that can be accommodated by video displays is severely limited compared with that of natural vision, and this limitation has not been avoided in HDTV equipment [1]. Moreover, the display brightness of HDTV images is restricted by the need to spread the available light over a large area.

Within the upper and lower limits of display brightness, the quality of the image depends on the relationship between changes in brightness in the scene and the corresponding changes in the image. It is an accepted rule of image reproduction that the brightness should be directly proportional; that is, the curve relating input and output brightness should be a straight line. This continues to be the criterion for television service. Because the output-vs.-input curves of cameras and displays are, in many cases, not straight lines, intermediate adjustment (gamma correction) is needed. Gamma correction in HDTV requires careful attention to avoid excessive noise and other defects that are particularly evident at close scrutiny of the display.

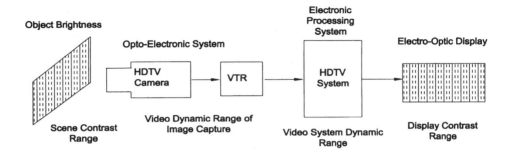

Figure 3.3 Simplified block diagram of an end-to-end video imaging system. (*After* [7].)

Video Signal Dynamic Range

As the object brightness in a scene is progressively lowered, the brightness of the television display correspondingly lowers. A point finally is encountered at which the display fails to portray to the viewer's eye any discernible lowering in the brightness, despite a continuing diminution of the scene object brightness. This point constitutes the *lower threshold* of the contrast range of that particular television system [7]. Conversely, at the upper end—as the object's brightness is progressively raised—the system reaches a point at which the display portrays no further incremental increase in apparent brightness. The *upper threshold* may occur because of overloading of the display mechanism or because of an upper boundary in the capability of the imaging device in the camera. The upper threshold divided by the lower threshold defines the *contrast handling range* of the video system.

A complete video system includes the following primary elements:

- The scene, which has a certain contrast range defined by the upper and lower levels of object brightness

- Video camera and subsequent electronic processing systems, which have a defined video dynamic range

- Final viewing display, which has its own unique contrast range

- The viewer, who possesses a finite capability in discerning brightness increments

Figure 3.3 depicts this total visual system.

The central task of visual program origination is, naturally, to capture and record images as faithfully as possible. Given the range of illumination that exists under ordinary circumstances, this objective is challenging indeed. Table 3.2 lists typical illumination ranges (in *lux*) for a variety of naturally lighted environments. Any objects illuminated within these environments reflect a proportion of this incident illumination. The object brightness is what is seen by the camera lens and what ultimately stimulates the image sensor of the device. It is apparent that the object brightness range that can be encountered outdoors is potentially enormous. On a bright, sunny day, objects of high reflectivity and dark objects in shadowed areas can create scene contrast ranges easily in excess of 20,000:1 (86 dB).

Table 3.2 Typical Day and Night Illumination Levels (*After* [7].)

Light	Illumination (Lux)
Bright Sun	50,000–100,000
Hazy Sun	25,000–50,000
Bright Cloudy	10,000–25,000
Dull Cloudy	2,000–10,000
Sunset	1–100
Full Moon	0.01–0.1
Starlight	0.0001–0.001

Table 3.3 lists typical illumination levels encountered in artificially lighted environments. Although these are considerably lower than the outdoor levels, the contrast range still may be very high within a given environment. In a studio, object brightness easily may range from greater than 2000 lux to less than 1 lux, providing scene contrast ratios in excess of 2000:1 (66 dB). Practical cameras have limitations to their dynamic range that curtail the capture of a scene with large contrast range.

3.2.7 Luminance and Chrominance

The retina contains three sets of cones that are separately sensitive to red, green, and blue light [1]. The cones are dispersed uniformly in the color-sensitive area of the retina, so that light falling on them produces a sensation of color that depends on the relative amounts of red, green, and blue light present. When the ratio of red:green:blue is approximately 30:60:10, the sensation produced is that of white light. Other ratios produce all the other color sensations. If only red cones are excited, the sensation is red. If both red and green intensities are present, the sensation is yellow; if red predominates, the sensation produced is orange.

Given this process of color vision, it would appear that color television would be most directly provided by a camera that produces three signals, proportional respectively to the relative intensities of red, green, and blue light in each portion of the scene. The three signals would be conveyed separately to the input terminals of the picture tube, so that the tube would reproduce at each point the relative intensities of the red, green, and blue discerned by the camera. This separate utilization of three component-color signals would occur at the camera and display. In the transmission between them, a different group of three would be used: luminance and two chrominance signals.

This form of transmission arose originally in the NTSC color system when it was realized that compatible operation of the millions of existing monochrome receivers required a black-and-white signal, which became known as *luminance* signal (Y) [8]. The additional information needed by color receivers was the difference between a color value and its luminance component at each point in the scene. Because the color value may have as many as three components—red, green, and blue—the chrominance signal must embrace three val-

Table 3.3 Typical Artificial Illumination Levels (*After* [7].)

Environment	Illumination (Lux)
Hospital Operating Theater	5,000–10,000
Television Studio	1,000–2,000
Business Office	200–300
Living Room	50–200
Good Street Lighting	20
Poor Street Lighting	0.1

ues: red minus luminance, green minus luminance, and blue minus luminance (R–Y, G–Y, and B–Y, respectively).

At the receiver, these color-difference signals are added individually to the luminance signal, the sums representing the individual red, green, and blue signals that are applied to the display device. The luminance component is derived from the camera output by combining, in proper proportions, the red, green, and blue signals. Thus, symbolically, Y = R + G + B. In the PAL system, the R–Y and B–Y color-difference signals are transmitted, at reduced bandwidth, with the luminance signal. The G–Y signal, equal to –(R + B) is reconstructed at the receiver. In the NTSC system, two different color-difference signals (*I* and *Q*) are used [9]. They are combinations of the R–Y and B–Y signals that serve the same purpose.

The ability of the eye to discern fine details and sharpness of outlines (*visual acuity*) is greatest for white light. Hence, the luminance signal must be designed to carry the finest detail and sharpest edges of the scene.

3.2.8 Chromatic Aspects of Vision

Two aspects of color science are of particular interest to video engineers in the implementation of HDTV services [1]. The first is the total range (*gamut*) of colors perceived by the camera and offered by the display. The second is the ability of the eye to distinguish details presented in different colors. The color gamut defines the realism of the display while attention to color acuity is essential to avoid presenting more chromatic detail than the eye can resolve. The gamut perceived by a viewer with normal vision (represented by the C.I.E. Standard Observer of 1931) is shown in the C.I.E. diagram in Figure 3.4 [10 and 11]. Color perceptions were measured by offering to viewers combinations of the three standard C.I.E. primary colors: spectrally pure red of 700 nm wavelength, green of 546.1 nm, and blue of 435.8 nm. These and all the other spectrally pure colors are located on the upper boundary of the diagram. The purple colors, which do not appear in the spectrum, appear on the line connecting the ends of the upper locus. The colors located within this outer boundary are perceived as being mixed with white light; that is, they become more pastel as the center of the diagram is approached. At the center of the diagram, the color represented is white, and in the center of the white region is a particular white (*C.I.E. Illuminant C*) that matches sunlight from the northern sky. Each point on the chromaticity diagram is identified by two numbers—its *x* and *y* coordinates—that define a particular color uniquely from all others.

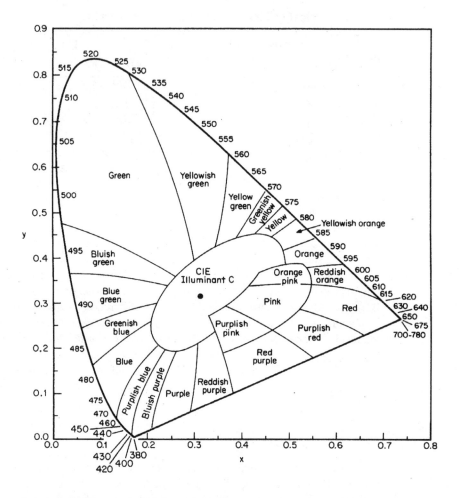

Figure 3.4 The C.I.E. Chromaticity Diagram, showing the color sensations and the triangular gamuts of the C.I.E. primaries. (*After* [11].)

Three such points identify the primary colors used in a camera, and the triangle formed by them encloses the total gamut of colors that the camera can perceive. Another triangle limits the gamut of the display.

The chromaticity diagram in Figure 3.5 compares the gamuts of interest in conventional television [12]. The largest triangle is that formed by the C.I.E. primaries. The smaller triangles are labeled to show the NTSC system and typical CRT phosphors (P22). It can be seen that the television gamuts cover only a small portion of the range of colors visible to the eye. Nevertheless, this limited range is wholly adequate for television, especially when compared with the gamut for the surface colors of the inks, dyes, and pigments used in other media [13]. These colors offer most viewers their concept of the "real" colors of material

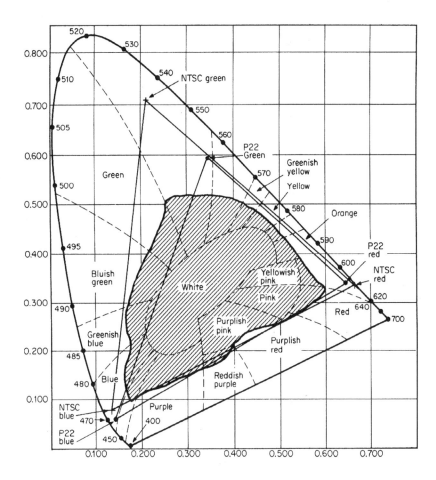

Figure 3.5 The gamuts of NTSC and common CRT phosphors (P22). (*From* [1]. *Used with permission.*)

objects, so the common acceptance of this restricted gamut is readily understood. There are slight differences in the primaries for HDTV service, but they are not intended specifically to offer a wider gamut. Rather, they were chosen based on the experiences of broadcasters and receiver manufacturers.

3.2.9 Acuity in Color Vision

The second aspect of color vision, the acuity of the eyes for the primary colors [1], was a basic factor in the development of the NTSC system. Likewise, it remained basic to the subsequent design of the PAL, SECAM, and—much later—HDTV systems. In the earliest work on color television, equal bandwidth was devoted to each primary color, although it had long been known by color scientists that the color acuity of the normal eye is greatest for green light, less for red, and much less for blue [9]. The NTSC system made use of this fact by devoting significantly less bandwidth than luminance to the red-minus-luminance channel and still less to the blue-minus-luminance. All color television services now employ this device in one form or another. Properly applied, it offers economy in the use of bandwidth without loss of resolution or color quality in the displayed images.

3.2.10 Temporal Factors in Vision

The response of the eye over time is, essentially, continuous; that is, the cones of the retina convey nerve impulses to the visual cortex of the brain in pulses recurring about 1000 times per second [1]. If television were required to follow this design, a channel several kilohertz wide would be required in HDTV for each of nearly a million picture elements, equivalent to a video baseband exceeding 1 GHz. Fortunately, television can take advantage of a temporal property of the eye known as *persistence of vision*—the retention of an image after the light is removed. The special properties of this phenomenon must be observed carefully to ensure that motion in the scene is conveyed smoothly and accurately and that the illumination of the scene is free from the effects of *flicker* in all parts of the image. A conflict encountered in HDTV design is the need to reduce the rate at which some of the spatial aspects of the image are reproduced, thus conserving bandwidth, while maintaining temporal integrity, particularly the position and shape of objects in motion. To resolve such conflicts, it has become standard practice to view the television signal and its processing simultaneously in three dimensions: width, height, and time.

3.2.11 Temporal Aspects of Illumination

Basic to television and motion pictures is the presentation of a rapid succession of slightly different still pictures (frames) [1]. Between frames, the light is cut off briefly. The visual system retains the image during the dark interval so that, under specific conditions, the image appears to be present continuously. As a result, any difference in the position of an object from one frame to the next is interpreted as motion of that object. For this process to represent visual reality, two conditions must be met:

- First, the rate of repetition of the images must be high enough for the motion to be depicted smoothly, with no sudden jumps from frame to frame.

- Second, the rate must be high enough for the persistence of vision to extend over the interval between flashes.

Here, an idiosyncrasy of natural vision occurs: the brighter the flash, the shorter the persistence of vision [13]. This is quite the opposite of what one would expect. The result is that bright pictures require rapid repetition. Otherwise, the visual system becomes aware that the light has been removed, and the illumination is perceived as unsteady. The effect is flicker, a persistent problem in television.

Continuity of Motion

Early in the development of motion pictures, it was found that motion could be depicted smoothly at any frame rate faster than about 15 frames/s [1]. This led to the establishment of 16 frames/s as a standard, a rate widely used in home movie equipment for many years. However, experience in movie production for theaters showed that very rapid motion, so prevalent in westerns, could not be shown faithfully at 16 frames/s. So, despite a 50 percent increase in the cost of film, the frame rate was changed to 24 frames/s. It remains today as a worldwide standard.

The repetition rate adopted for the European monochrome television service was 25 Hz, in accord with the 50 Hz power in that region, and this standard has persisted in the PAL and SECAM color systems. Because this rate does not match the 24-frame rate of motion pictures, it has become standard practice to telecast films in PAL and SECAM countries at 25 frames/s, thus raising the pitch of the accompanying sound by 4 percent and shortening the performance by the same amount. These discrepancies have come to be fully accepted, and only those possessing the rare sense of perfect pitch are known to object—and then usually only during musical productions.

The frame rate for the NTSC system initially was chosen to be 30 per second, not to reduce flicker, but because the 60 Hz power used would otherwise cause disturbances in scanning and signal processing. The electric and magnetic fields then causing these problems are now under such strict control that the 30/60 Hz ratio no longer is required. An illustration of this fact is that when the NTSC color service was introduced in 1953, the 30 Hz frame rate of the NTSC monochrome service was changed by 0.1 percent to 29.97 Hz, to maintain the visual-aural carrier separation at precisely 4.5 MHz [14].

The 24-frame rate of motion pictures does not solve all the problems of reproducing fast action. Prohibitive production techniques, such as limited camera angles and restricted rates of panning, must be observed when rapid motion is encountered. Proposals have been made to change the film standard to 30 frames/s to alleviate the rapid-motion problem and to ease film-scanning conversion to the NTSC and HDTV systems. This suggestion, however, has met resistance in the film community because of the heavy investment in 24-frame camera, production, and projection equipment.

3.2.12 Smear and Related Effects

The acuity of the eye to view objects in motion is impaired by the slower temporal response of the fovea, compared with that of the surrounding regions of the retina [1]. Thus, a loss of sharpness in the edges of moving objects is an inevitable aspect of natural vision, and use has been made of this fact in the design of television systems. Much greater losses of sharpness and detail, under the general term *smear*, occur whenever the image moves across the sensitive surface of the camera. Each element in the surface then receives light—not from one detail, but from a succession of them. The signal generated by the camera at that point is the sum of the passing light, and smear is the result. As in photography, this effect can be reduced through the use of a short exposure, provided there is sufficient light relative to the camera's sensitivity. Electronic shutters are used to limit the exposure to 1/1000 second or less when sufficient light is available.

Another source of smear is camera response carried over from one frame scan to the next. The retained signal elements from the previous scan then are added to those of the current scan, and any change in their relative position causes a misalignment and consequent loss of detail. Such *carryover smear* occurs when the exposure occupies the full scan time, such as under conditions of poor illumination. A similar carryover smear may occur in the display, when the light given off by one line persists long enough to be appreciably present during the successive scan of that line. Such carryover helps reduce flicker in the display, and there is room for compromise between flicker reduction and loss of detail in moving objects.

3.2.13 Flicker

The ocular peculiarity that demands rapid repetition of pictures for continuity of motion demands even faster repetition to avoid flicker [1]. As theater projectors became more powerful and the images correspondingly brighter, flicker at 24 frames/s became a serious concern. To increase the frame rate without otherwise subjecting the system to change, a 2-bladed rotating shutter was added to the projector, between the film and the screen. When this shutter was synchronized with the film advance, it shut off the light briefly while the film was stationary. In this way, the flash rate was increased to 48 per second, allowing a substantial increase in screen brightness. The appearance of moving objects was not adversely affected because the frame rate remained at 24 frames/s. In due course, wider and brighter pictures became available, and a 3-bladed shutter was incorporated to increase the flash rate to 72 Hz.

Interlaced Scanning

No such simple interruption of the image will suffice for television. To obtain two flashes for each frame, it was arranged from the beginning of public television service in 1936 to employ the technique of interlaced scanning, which divides the scanning pattern into two sets ("odd" and "even") of spaced lines that are displayed sequentially, one set fitting precisely into the spaces of the other [1]. Each set of lines is called a *field*, and the interlaced set of the two lines is a *frame*. The field rate for PAL and SECAM is 50 Hz; for NTSC it is 59.94 Hz.

This procedure, although necessary because of the technologies used to implement television in the analog era, is the source of several degradations of image quality. While the total area of the image flashes at the rate of the field scan, twice that of the frame scan, the individual lines repeat at the slower frame rate, giving rise to several degradations associated with the effect known as *interline flicker*. This effect causes small areas of the image, particularly when they are aligned horizontally, to display a shimmering or blinking visible at the usual viewing distance. A related effect is unsteadiness in extended horizontal edges of objects, because the edge is portrayed by a particular line in one field and by another line in the next. These effects become more pronounced with higher vertical resolution provided by the camera and its image-enhancement circuits.

Interlacing also introduces aberrations in the vertical and diagonal outlines of moving objects. These occur because vertically adjacent picture elements appear at different times in successive fields. An element on one line appears 1/50 second or 1/59.94 second later than the vertically adjacent element on the preceding field. If the objects in the scene are stationary, no adverse effects arise from this time difference. But if an object is in rapid motion, the time delay causes the elements of the second field to be displaced to the right of, instead of vertically or diagonally adjacent to, those of the first field. Close inspection of such moving images shows that their vertical and diagonal edges are not sharp, but display a series of stepwise serrations, usually coarser than the basic resolution of the image. Because the eye loses some acuity as it follows objects in motion, these serrations often are overlooked. Still, they are a significant impairment compared with motion picture images, for which HDTV is now available as a production tool. All the picture elements in a film frame are exposed and displayed simultaneously so that the impairments resulting from interlacing do not occur.

As previously noted, eliminating the defects of interlacing has been an important target in HDTV development. To prevent them, scanning in the camera must be progressive, using only one set of adjacent lines per frame. At the receiver, the display scan must match that at the camera. Because conventional receivers employ interlaced scanning, they cannot respond directly to a progressively scanned image. One obvious solution is to utilize a frame store to transpose the incoming signal from progressive to interlaced scanning. When this is done, much of the degradation associated with interlacing can be avoided, particularly the serrated edges of objects in motion.

The fact that the field and frame rates of PAL and SECAM (25 and 50 Hz) are different from those of NTSC (29.97 and 59.94 Hz) affected the early design efforts of HDTV service in two important aspects:

- First, their relative susceptibility to flicker. The higher rates of NTSC allow the images to be about 6 times as bright for a given flicker limit. This difference carried over into the HDTV systems. Furthermore, the wider visual field of HDTV images makes them more prone to flicker, so in planning for HDTV displays in the PAL and SECAM service areas, a field rate of 100 Hz was considered.

- Second, their impact on worldwide standardization of HDTV. Whereas standardization within the broadcast plant may be easily achieved, the prospect is less certain when it comes to transmissions to the viewing audience. In the late 1980s and early 1990s, frame-/field-rate compatibility became the predominant economic and political force affecting the choice of HDTV standards. These considerations, in fact, led to the standardization impasse detailed in Chapter 1.

3.2.14 Visual Basis of Video Bandwidth

The time consumed in scanning fields and frames for conventional television is measured in hundredths of seconds. Much less time is available to scan a picture element in HDTV. The visual basis lies in the large number of picture elements that are necessary to satisfy the eye when the image is under the close scrutiny offered by a wide-screen display [1]. The eye further requires that signals representative of each of these elements be transmitted during the short time that persistence of vision allows for flicker-free images—not more than 1/25 second. This challenging requirement is derived directly from the properties of the eye.

Table 3.4 Bandwidth Characteristics of HDTV (NHK) and Conventional Television Systems (*After* [1].)

System	Total Channel Width	Video Baseband			Scanning Rates	
		Y	R–Y	B–Y	Camera	Display
HDTV	30.0	20.0	7.0	3.0	60-I	60-I
NTSC	6.0	4.2	1.0	0.6	59.94-I	59.94-I
PAL	8.0	5.5	1.8	1.8	50-I	50-I
SECAM	8.0	6.0	2.0	2.0	50-I	50-I

Table 3.4 lists the bandwidth requirements of the systems whose spatial features are given in Table 3.1. The greater number of picture elements in HDTV, compared with the 525- and 625-line interlaced systems, dictates the use of a significantly wider video bandwidth in order to transmit the increase in horizontal resolution; it also calls for a greater number of scanning lines for a comparable increase in vertical detail. Table 3.5 tabulates the number of pixels for various television systems. If progressive rather than 2:1 interlaced scanning is used in camera-signal generation in order to prevent the introduction of artifacts with image motion, the bandwidth requirement is doubled.

These more stringent video system parameters necessitate significant improvements over the components used in 525- and 625-line systems for signal sources, signal processing, transmission and reception, and picture displays. The major requirements include:

- Image transducers with photosensitive elements having higher resolving power and significantly higher signal output to maintain an adequate S/N.

- More precise scanning registration among camera color channels, compatible with the increase in horizontal and vertical resolution.

- More precise positioning of camera and receiver vertical scanning to preserve the higher vertical resolution resulting from the increase in scanning lines.

- Picture-viewing displays with a smaller scanning-beam spot size for reproduction of the greater number of lines and horizontal picture elements.

- Wide-screen displays to accommodate the wider aspect ratio.

Although these requirements can be met with the basic system configurations found in high-quality cameras and video-processing equipment designed for 525- and 625-line signal standards, new equipment designs of the camera and other major components are required. Considerable progress continues to be made in this area.

3.2.15 Gamma

The physics of the electron gun of the CRT impose a relationship between voltage input and light output referred to as the *five-halves power law*. The intensity of light produced at the

Table 3.5 Picture Element Parameters in Various Television Systems (*After* [15].)

Parameter	CCIR Standard			1125/60, U.S. SMPTE 240M
	M, U.S. NTSC	B/G, CCIR PAL	L, France SECAM	
Information Elements per Frame	280,000	400,000	480,000	2,000,000
Picture Elements per Raster	215,000	300,000	360,000	1,600,000
Picture Elements per Line	446	520	620	1,552
Picture Element Length[1]	30	26	21	11
Picture Element Width[1]	21	17	17	10
[1] Based on raster height = 10,000 units.				

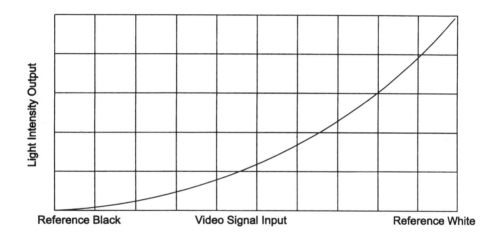

Figure 3.6 CRT light intensity output as a function of video signal input.

face of the screen is proportional to the voltage input raised to the power 5/2. The numerical value of the exponent of the power function is represented by γ (gamma). CRT monitors have voltage inputs that reflect this power function. In practice, the numerical value of gamma is close to its theoretical value of 2.5. Figure 3.6 illustrates the relationship between the signal input to a CRT and the light output of the device.

The process of precompensating for this nonlinearity by computing a voltage signal from an intensity value is referred to as *gamma correction*. The function required is approximately a 0.45-power function, whose graph is similar to that of a square root function.

In conventional video systems, gamma correction is accomplished by analog circuits at the camera. In digital video systems, including computer graphics, gamma correction may be handled by incorporating the function into a frame buffer lookup table.

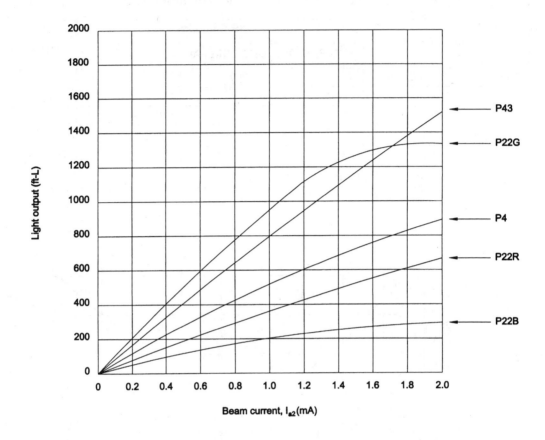

Figure 3.7 Linearity characteristics of five common phosphors. P22 is commonly used in tri-color television displays. The red, green, and blue phosphors are denoted by P22R, P22G, and P22B, respectively. (*From* [19]. *Used with permission.*)

The actual value of gamma for a particular CRT may range from about 2.3 to 2.6. By far the largest source of variation in the nonlinearity of a monitor is caused by the black level or brightness adjustment of the device. The nonlinearity in the voltage-to-intensity function of the CRT originates with the electrostatic interaction between the cathode and the grid that controls the electron-beam current. As shown in Figure 3.7, the CRT phosphors themselves are quite linear, up to a saturation point.

It is a fortunate coincidence that the CRT voltage-to-intensity function is nearly the inverse of the luminance-to-lightness relationship of human vision. The result is that coding a luminance signal into voltage, to be turned back into luminance by a CRT, is nearly the optimal coding to minimize the perceptibility of noise that is introduced into the signal as a result of transmission and reception.

In a conventional video system, gamma correction is applied at the camera for the dual purposes of coding into perceptually uniform space and precompensating for the nonlinearity of the display CRT. The first of these considerations was important in the early days of television because of the need to minimize the noise introduced by VHF over-the-air transmission. However, the same considerations of noise visibility also apply to analog videotape recording, and to minimizing the quantization noise that may be introduced at the front end of a digital system when a signal representing intensity is quantized to a limited number of bits [16].

The gamma-correction scheme has worked well for conventional television broadcasting to receivers incorporating CRT displays. During the development of HDTV, however, it was realized that a wide variety of display devices was likely to emerge in the consumer marketplace in the near future. This being the case, a new approach to gamma correction was required. It would not be possible, for example, to utilize a gamma curve that accommodated the requirements of both a CRT display and an LCD projector. Instead, the concept was advanced that program origination and processing systems should exhibit some specified *linear gamma* and that correction for specific display devices should be handled at the devices themselves.

Normalizing Transfer Characteristics

As an outgrowth of the linear gamma concept, colorimetry and transfer characteristics were examined carefully during the development of the SMPTE 240M standard. The issue was studied from the vantage point of three systems, with these goals [17]:

- The HDTV production system should provide an adequate display of picture contrast range on a reference HDTV studio monitor (direct-view CRT), implementable with today's CRT technology.

- The development should recognize the inevitability of a continuing evolution in HDTV display, both direct view and projection. This would require the capability for the display system itself to encompass the different colorimetric transformation and gamma correction appropriate to the particular display technology in question (CRT, laser, LCD, plasma, and other technologies).

- The system approach should be extended to allow the HDTV capture on videotape or other means to be subsequently transferred to 35 mm film. This, too, entailed consideration of colorimetric transformations and transfer characteristics tailored to the different film emulsions that would be encountered.

Thus, the requirement for downstream system video processing, involving different colorimetric transformations and appropriately tailored nonlinear transfer characteristics, introduced the concept of incorporating the capability to linearize the HDTV video component signals at some specific point in the postproduction process [18]. Further impetus was fueled by the anticipation of new forms of digital image manipulation that were soon to become a part of postproduction. Specifics of this work are covered in detail in Chapter 6.

3.2.16 Color Signal Transformations

As previously outlined in this chapter, the colorimetric characteristics of television component color signals are determined by three separate sets of parameters [20]:

- Color primaries and reference white, which are specified by CIE colorimetric parameters and define the relationship between scene color and the linear RGB video signals. The primaries also define the maximum gamut of color that can be transmitted with all-positive RGB signals.

- Opto-electronic transfer characteristics (gamma), used to derive the R´G´B´ gamma-corrected signals from the linear RGB values.

- The luminance equation and the color channel coding matrix derived using that luminance equation. This coding matrix defines the relationship between the gamma-corrected R´G´B´ values and the component color signals Y´PB´PR´ (analog) or Y´CB´CR´ (digital).

In current practice, the differences between the first two colorimetric characteristics are small. The luminance and color channel equations, however, are subject to some variation. In an effort to deal with these differences, the SMPTE developed Engineering Guideline 36, "Transformations Between Television Component Color Signals.". In EG 36, the analog and digital cases are treated separately because their color-difference gains differ slightly.

Background

Existing television interface standards utilize at least two different color channel coding matrices to derive their corresponding analog or digital component color signals [20]. Clearly, it is necessary to perform transformations between these color component signal sets. EG 36 describes the derivation of the transformation matrices and lists example transformations between color component signals adopting ITU-R BT.601 and ITU-R BT.709 luma/chroma equations for both digital and analog component signal sets.

Note that ITU-R BT.601 has no specification of primaries or transfer characteristics. It specifies only the third coding-matrix part of the colorimetric characteristics. The ITU-R BT.601 coding matrix is based on the color primaries and the reference white of the NTSC (1953) specification, which is practically no longer used. The equation is also used in SMPTE 170M, the EBU 625 standards, and the 1250/50 (1152 active lines) specification of ITU-R BT.709, ITU-R BT.1358, and SMPTE 293M.

In ITU-R BT.709, which includes the 1920 × 1080 *common image format* (HD-CIF), the unified colorimetric parameters are specified for HD-CIF regardless of the 1125/60 and 1250/50 formats. These unified colorimetric parameters are identical to those specified for 1125/60 and those described in ITU-R BT.1361. The equation is also used in SMPTE 274M, SMPTE 295M, and SMPTE 296M.

In EG 36, a parametric form of conversion matrix is derived for converting between signal sets with arbitrary source and target luma coefficients. Key elements of the guideline include:

- Analog signal derivation

Table 3.6 Luma Equations for Scanning Standards (*After* [20]

Television Scanning Standard	Luma Equation		
	ITU-R BT.601	**ITU-R BT.709**	**Other**
SMPTE 170M, NTSC	X		
EBU 625, PAL, SECAM	X		
SMPTE 293M, 525 / 720 × 483 / 59.94 / 1:1	X		
SMPTE 274M, 1125 / 1920 × 1080 / multiple rates / 1:1, 2:1		X	
SMPTE 295M, 1250 / 1920 × 1080 / 50 / 1:1, 2:1		X	
SMPTE 296M, 750 / 1280 × 720 / multiple rates / 1:1		X	
SMPTE 240M, 1125 / 1920 × 1035 / 60, 59.94 / 2:1			SMPTE 240M

- Digital signal representation

- Source-to-target component digital transformation matrix

- General parametric form for coding and decoding matrixes

- Numerically exact matrices for converting between ITU-R BT.601 and BT.709 (both analog and digital)

Table 3.6 lists the luma equations used by common television scanning standards. Note that SMPTE 240M uses a luma equation that is similar to ITU-R BT.709, but is not exactly the same. In many applications, the differences may be small enough to ignore, but these differences could become significant in critical applications.

3.3 Overview of the Grand Alliance System

At this point in our examination of high-definition imaging, it is appropriate to provide an overview of the Grand Alliance system, the forerunner of the ATSC standard on digital television. Details of the major components and techniques of the ATSC system will be presented in subsequent chapters.

3.3.1 Scanning Formats

Two scanning formats are provided for in the Grand Alliance HDTV system for North America [21]:

- 1080 *active* lines per frame, of 1920 pixels/line, at a frame rate of nominally 30 Hz. There are two fields/frame, and these fields are interlaced as is the present broadcast standard (NTSC).

- 720 *active* lines per picture, of 1280 pixels/line at a picture rate of nominally 60 Hz in which scanning is progressive (i.e., noninterlaced).

These image structures are, therefore: 1080 × 1920 = 2,073,600 pixels, and 720 × 1280 = 921,600 pixels each. But there are twice as many pictures in the latter format as in the former because of progressive scanning, so the pixel rates are 62,208,000 and 55,296,000 per second. Although these two pixel rates are approximately the same, the interlaced format claims higher static spatial resolution; the progressive-scan format claims higher temporal and dynamic spatial resolution and elimination of the artifacts that result from interlaced scanning, notably line flicker and aliasing of moving edges that are approximately horizontal.

3.3.2 Relative Horizontal and Vertical Spacing

The spatial distance between adjacent pixels in the horizontal axis is made *equal* to the corresponding distance in the vertical axis in both formats as 1920/16 = 1080/9 and 1280/16 = 720/9 [21]. This provides the same spatial resolution along these axes in each format. Moreover, it simplifies digital signal processing involving image rotation. The term *square pixels* generally is used to describe this format, originating from this same practice in the computer industry.

3.3.3 Luminance and Chrominance Components

Analog red, green, and blue video signals generally are quantized to 10 bits, then digitally matrixed to one luminance (Y) and two chrominance components (P_r) and (P_b) [20]. Although the luminance component requires the full pixel rate to satisfy the visual acuity needs of the observer at a distance of three picture heights, the visual acuity for chrominance components such as P_r and P_b is at least 4:1 less, which justifies a reduction in the pixel rates for both P_r and P_b by 4:1. After these chrominance components are digitally filtered in both the vertical and horizontal axes, the pixels per image number 518,400 and 230,400.

Equations for Y, P_r, and P_b for the basic HDTV system tested by the ATTC in 1995 are given by:

$$Y' = 0.701G' + 0.212R' + 0.087B' \tag{3.1}$$

$$P_b' = 0500R' - 0.384G' - 0.550B' \tag{3.2}$$

$$P_r' = 0.500R' - 0.445G' - 0.550B' \tag{3.3}$$

The use of prime values designates that the components have a nonlinear light-to-voltage transfer characteristic (gamma precorrection). The foregoing equations are provided as a matter of historical interest. The actual values adopted in the ATSC standard are slightly different for a variety of reasons. First among them was compatibility with other systems and interfaces.

The ATSC DTV standard specifies SMPTE 274M colorimetry as the default, and preferred, colorimetry. Note that SMPTE 274M colorimetry is the same as ITU-R BT.709 (1990) colorimetry. Video inputs corresponding to ITU-R Rec. 601-4 may have SMPTE 274M colorimetry or SMPTE 170M colorimetry. In generating bit streams, broadcasters

Table 3.7 System Colorimetry for SMPTE 274M (*After* [22].)

Parameter	CIE *x*	CIE *y*
Red primary	0.640	0.330
Green primary	0.300	0.600
Blue primary	0.150	0.060
White reference	0.3127	0.3290

should understand that many receivers probably will display all inputs, regardless of colorimetry, according to the default SMPTE 274M. Some receivers may include circuitry to properly display SMPTE 170M colorimetry as well, but this is not a requirement of the standard.

SMPTE 274M Colorimetry

The colorimetric analysis and optoelectronic transfer function defined in the SMPTE 274M document corresponds to ITU-R BT.709 [22]. Digital representation and treatment of wide-gamut color signals are not specified in the current edition of the international standard for HDTV colorimetry, ITU-R BT.709. In particular, coding ranges for digital primary components R', G', and B' are not specified.

Picture information is linearly represented by red, green, and blue tristimulus values (RGB), lying in the range 0 (reference black) to 1 (reference white), whose colorimetric attributes are based upon reference primaries with the chromaticity coordinates given in Table 3.7 (in conformance with ITU-R BT.709, and whose reference white conforms to CIE D65 as defined by CIE 15.2).

From the red, green, and blue tristimulus values, three nonlinear primary components R', G', and B' are computed according to the optoelectronic transfer function of ITU-R BT.709, where L denotes a tristimulus value and V' denotes a nonlinear primary signal:

$$V' = \begin{bmatrix} 4.5L & 0 \le L \le 0.018 \\ 1.099L^{0.45} - 0.099 & 0.018 \le L \le 1 \end{bmatrix} \tag{3.4}$$

To ensure the proper interchange of picture information between analog and digital representations, signal levels are completely contained in the range specified between reference black and reference white, except for overshoots and under-shoots resulting from processing.

The Y' component is computed as a weighted sum of nonlinear R' G' B' primary components, using coefficients calculated from the reference primaries according to the method given in SMPTE RP 177:

$$Y' = 0.2126 \, R' + 0.7152 \, G' + 0.0722 \, B' \tag{3.5}$$

Color-difference component signals P_b' and P_r', having the same excursion as the Y' component, are computed as follows:

$$P_b' = \frac{0.5}{1 - 0.0722}(B' - Y') \tag{3.6}$$

$$P_r' = \frac{0.5}{1 - 0.2126}(R' - Y') \tag{3.7}$$

P_b' and P_r' are filtered as specified in SMPTE 274M and may be coded as C_b' and C_r' components for digital transmission. (See [22] for additional details.)

3.3.4 Pixel Transmission Rates and Digital Modulation

The pixel rates for Y, P_r, and P_b combined are 93,312,000 per second and 82,944,000 per second for each scanning format [21]. The transmission challenge, of course, is to fit total data rates that are slightly higher than these (because of data overhead, program digital audio, and ancillary data requirements) into a 6 MHz channel while keeping the radiated field strength well below that of the existing (analog) NTSC TV signals. (Both DTV and NTSC will use only the existing spectrum allotted to broadcasting.) For example, in the UHF band (470–806 MHz), there are 55 channels. The maximum ERP permitted for NTSC service is 5 MW peak power during sync pulses (*peak-of-sync*). With this power and suitable antenna height, coverage can extend over a radius of perhaps 55 miles. The *8-state vestigial sideband modulation* (8-VSB) scheme devised for the Grand Alliance DTV system, on the other hand, was predicted to cover the same general service area with 12 dB *less* ERP.

Coverage similar to NTSC has been proved possible with radiated power lowered by 12 dB, because above the threshold *signal-to-thermal noise power* ratio, there is no further improvement in picture (or sound) quality with digital transmission techniques. This threshold is 15 dB, absent strong multipath or other interfering signals. Because of the sudden and complete failure of digital transmission at this threshold S/N, the actual coverage area is somewhat smaller unless additional ERP is provided to deal with signal fading—a well-known phenomenon, especially at UHF.

It is possible to transmit approximately 20 Mbits/s within a 6 MHz channel with the power limitations herein described. To accommodate the far higher video bit rate of HDTV, it is necessary, therefore, to compress the video signal for transmission. Video-compression ratios in the range of 50:1 are quite common within the present state of the art, even when the clock rates are as high as that required for real-time television (Y clock rate $= \approx 75$ MHz).

3.3.5 Video Compression

Real-time video compression sufficient to meet the foregoing requirement of the transmission channel has been developed through the international MPEG-2 standard, as documented by the International Standards Organization (ISO, Recommendation ITU/T H.262, 1995 E) [21].

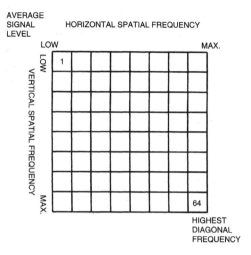

Figure 3.8 DCT 2-dimensional spectrum sorting. (*From* [21]. *Used with permission.*)

It is well known that a high degree of information redundancy exists in video signals produced by the scanning process. For example, the video signal between spatially adjacent lines in a given frame has a high degree of correlation to those parts of the picture that remain static between consecutive frames. However, those portions of a picture that are moving are not redundant, but shifted spatially frame by frame. Such movement can be detected and motion vectors transmitted to assist the receiver in reproducing the movement from the otherwise scant available data.

MPEG-II is one of many compression schemes based upon the use of the *discrete cosine transform* (DCT). In brief, the image area is divided into blocks of 8 data rows × 8 data columns. The spectrum of the data within each block is analyzed by means of *fast Fourier transform* (FFT). This results in a 2-dimensional spectrum analysis of each block. Both the vertical and horizontal frequencies are subdivided into eight ranges and stored in an array of 64 bins, as illustrated in Figure 3.8. Bin no. 1 contains the average intensity of the block, which is the video dc level of the block. Going down the left side of Figure 3.8, higher and higher vertical spatial frequencies are stored; going across the top, higher horizontal frequencies are stored. At the bottom right corner, the highest vertical and horizontal frequencies are stored. The scanning process produces high amplitudes at low spatial frequencies and, conversely, the highest spatial frequencies produce small-amplitude Fourier components. It is logical, therefore, that the low-frequency components (near the upper left) are high in amplitude and should be afforded a large portion of the available bit rate, while the high-frequency *diagonal* components in bins to the lower right generally are quite small and, thus, can be accorded fewer bits.

This disparity among different bins as to bit allocation is further enhanced by the visual acuity of the eye-brain system. It is well known that the visibility of noise is dependent, to a large extent, upon the spectrum of the noise. This is the basis of the *noise weighting func-*

Table 3.8 Typical ATSC Compression Ratios (*Data from Intel.*)

Format	Uncompressed Video Data Rate (10 bits, 4:2:2)	Compression Ratio for 1 Channel in 19.4 Mbits/s
480/60 interlaced	184 Mbits/s	9.5 to 1
480/60 progressive	368 Mbits/s	19 to 1
720/60 progressive	1106 Mbits/s	28.5 to 1
1080/60 interlaced	1244 Mbits/s	64 to 1

Table 3.9 Typical ATSC Channel Capacities as a Function of Compression Rate (*Data from Intel.*)

Format	Uncompressed Video Data Rate (10 bits, 4:2:2)	Typical Compressed Data Rate	Maximum Channels per 19.4 Mbits/s ATSC Stream
480/60 interlaced	184 Mbits/s	3 to 8 Mbits/s	4
480/60 progressive	368 Mbits/s	6 to 10 Mbits/s	3
720/60 progressive	1106 Mbits/s	14 to 16 Mbits/s	1
1080/60 interlaced	1244 Mbits/s	18 Mbits/s	1

tion (CCIR Rec. 567, Vol. XII). Therefore, high-frequency noise impairs the image less than the same noise power at lower spatial frequencies. This permits relatively coarse quantization of high frequencies in favor of finer quantization of lower frequencies and, of course, the dc level of each block must be finely quantized.

The result is a significant savings in bit rate, which translates to reduced spectrum requirements. Table 3.8 lists the typical compression ratios for a variety of formats under the DTV standard. The MPEG standards and related compression schemes are discussed in detail in Chapter 5.

One of the significant benefits of compression for the DTV system is that it permits the transmission of multiple streams of SDTV signals within the same 19.4 Mbits/s channel. Table 3.9 compares some of the possible scenarios.

3.4   References

1. Benson, K. B., and D. G. Fink: "Visual Aspects of High-Definition Images," *HDTV: Advanced Television for the 1990s*, McGraw-Hill, New York, 1990.

2. Hubel, David H.: *Eye, Brain and Vision*, Scientific American Library, New York, 1988.

3. Kell, R. D., A. V. Bedford, and M. Trainer: "Scanning Sequence and Repetition of Television Images," *Proc. IRE*, vol. 24, p. 559, April 1936.

4. Fujio, T., J. Ishida, T. Komoto, and T. Nishizawa: "High-Definition Television Systems—Signal Standards and Transmission," *SMPTE Journal*, vol. 89, pp. 579–584, August 1980.

5. Baldwin, M. W., Jr.: "The Subjective Sharpness of Simulated Television Images," *Proc. IRE*, vol. 28, p. 458, July 1940.

6. Belton, J.: "The Development of the CinemaScope by Twentieth Century Fox," *SMPTE Journal*, vol. 97, pp. 711–720, September 1988.

7. Thorpe, Laurence J.: "HDTV and Film—Issues of Video Signal Dynamic Range," Paper no. 86, 132nd SMPTE Technical Conference, SMPTE, New York, October 1990.

8. Benson, K. Blair (ed.): *Television Engineering Handbook*, McGraw-Hill, New York, pp. 21.57–21.72, 1986.

9. Fink, D. G.: "Perspectives on Television: The Role Played by the Two NTSCs in Preparing Television Service for the American Public," *Proc. IEEE,* vol. 64, pp. 1322–1331, September 1976.

10. Judd, D. B.: "The 1931 C.I.E. Standard Observer and Coordinate System for Colorimetry," *J. Opt. Soc. Am.*, vol. 23, pp. 359–374, 1933.

11. Kelly, K. L.: "Color Designation of Lights," *J. Opt. Soc. Am.*, vol. 33, pp. 627–632, 1943.

12. Pointer, R. M.: "The Gamut of Real Surface Colors," *Color Res. App*, vol. 5, pp. 145–155, 1945.

13. Porter, J. C.: *Proc. Roy. Soc.*, London, vol. 86, p. 945, 1912.

14. Fink, D. G.: *Color Television Standards*, McGraw-Hill, New York, pp. 108-111, 1955.

15. Benson, K. B., and D. G. Fink: "Picture-Signal Generation and Processing," *HDTV: Advanced Television for the 1990s*, McGraw-Hill, New York, p. 10.2, 1990.

16. Benson, K. B., and D. G. Fink: "Gamma and Its Disguises," *SMPTE Journal,* SMPTE, White Plains, N.Y., vol. 102, no. 12, pp. 1009–1108, December 1993.

17. SMPTE Standard: SMPTE 240M—1988, "Television, Signal Parameters, 1125/60 High-Definition Production System," *SMPTE Journal*, SMPTE, White Plains, N.Y., vol. 98, pp. 723–725, September 1989.

18. Internal Report of the Ad Hoc Group on Colorimetry and Transfer Characteristic (under the SMPTE Working Group on High-Definition Electronic Production, N15.04/05), SMPTE, White Plains, N.Y., October 24, 1987.

19. Whitaker, Jerry C.: *Electronic Displays: Technology, Design, and Applications*, McGraw-Hill, New York, p. 192, 1994.

20. SMPTE Engineering Guideline: EG 36 (Proposed), "Transformations Between Television Component Color Signals," SMPTE, White Plains, N.Y., 1999.

21. Rhodes, Charles W.: "Terrestrial High-Definition Television," *The Electronics Handbook*, Jerry C. Whitaker (ed.), CRC Press, Boca Raton, Fla., pp. 1599–1610, 1996.

22. SMPTE Standard: SMPTE 274-1998, *1920 × 1080 Scanning and Analog and Parallel Digital Interfaces for Multiple-Picture Rates*," SMPTE, White Plains, N.Y., 1998.

Digital Coding of Video Signals

4.1 Introduction

To fully understand digital television in general, and the ATSC DTV system in particular, it is necessary to review the basic principles of digital coding as applied to video signals.

4.2 Digital Signal Conversion

Analog-to-digital conversion (A/D) is the process of converting a continuous range of analog signals into specific digital codes. Such conversion is necessary to interface analog pickup elements and systems with digital devices and systems that process, store, interpret, transport, and manipulate the analog values. Analog-to-digital conversion is not an exact process; the comparison between the analog sample and a reference voltage is uncertain by the amount of the difference between one reference voltage and the next [1]. The uncertainty amounts to plus or minus one-half that difference. When words of 8 bits are used, this uncertainty occurs in essentially random fashion, so its effect is equivalent to the introduction of random noise (*quantization noise*). Fortunately, such noise is not prominent in the analog signal derived from the digital version. For example, in 8-bit digitization of the NTSC 4.2 MHz baseband at 13.5 megasamples per second (MS/s), the quantization noise is about 60 dB below the peak-to-peak signal level, far lower than the noise typically present in the analog signal from the camera.

4.2.1 The Nyquist Limit and Aliasing

A critical rule must be observed in sampling an analog signal if it is to be reproduced without spurious effects known as *aliasing*. The rule, first described by Nyquist in 1924 [2], states that the time between samples must be short compared with the rates of change of the analog waveform. In video terms, the sampling rate in megasamples per second must be at least twice the maximum frequency in megahertz of the analog signal. Thus, the 4.2 MHz maximum bandwidth in the luminance spectrum of the NTSC baseband requires that the NTSC signal be sampled at 8.4 MS/s or greater. Conversely, the 13.5 MS/s rate specified in the ITU-R studio digital standard can be applied to a signal having no higher frequency components than 6.75 MHz. If studio equipment exceeds this limit—and many cameras and associated amplifiers do—a low-pass filter must be inserted in the signal path before the conversion from analog to digital form takes place. A similar band limit must be met at 3.375 MHz in the chrominance channels before they are digitized in the NTSC system. If the sampling occurs at a rate lower than the Nyquist limit, the spectrum of the output analog signal contains spurious components, which are actually higher-frequency copies of the input spectrum that have been transposed so that they overlap the desired output spectrum. When this output analog signal is displayed, the spurious information shows up in a variety of forms, depending on the subject matter and its motions [1]. Moiré patterns are typical, as are distorted and randomly moving diagonal edges of objects. These aliasing effects often cover large areas and are visible at normal viewing distances.

Aliasing may occur, in fact, not only in digital sampling, but whenever any form of sampling of the image occurs. An example long familiar in motion pictures is that of vehicle wheels (usually wagon wheels) that appear to be rotating backward as the vehicle moves forward. This occurs because the image is sampled by the camera at 24 frames/s. If the rotation of the spokes of the wheel is not precisely synchronous with the film advance, another spoke takes the place of the adjacent one on the next frame, at an earlier time in its rotation. The two spokes are not separately identified by the viewer, so the spoke motion appears reversed. Many other examples of image sampling occur in television. The display similarly offers a series of samples in the vertical dimension, with results that depend not only on the time-vs.-light characteristics of the display device but also, and more important, on the time-vs.-sensation properties of the human eye. These aspects of sampling have a significant bearing on the design of HDTV systems.

4.2.2   The A/D Conversion Process

To convert a signal from the analog domain into a digital form, it is necessary to create a succession of digital words that comprise only two discrete values, 0 and 1 [1]. Figure 4.1 shows the essential elements of the analog-to-digital converter. The input analog signal must be confined to a limited spectrum to prevent spurious components in the reconverted analog output. A low-pass filter, therefore, is placed prior to the converter. The converter proper first samples the analog input, measuring its amplitude at regular, discrete intervals of time. These individual amplitudes then are matched, in the quantizer, against a large number of discrete levels of amplitude (256 levels to convert into 8-bit words). Each one of these discrete levels can be represented by a specific digital word. The process of matching each discrete amplitude with its unique word is carried out in the encoder, which, in effect, scans the list of words and picks out the one that matches the amplitude then present. The encoder passes out the series of code words in a sequence corresponding to the sequence in which the analog signal was sampled. This *bit stream* is, consequently, the digital version of the analog input.

The list of digital words corresponding to the sampled amplitudes is known as a *code*. Table 4.1 represents a simple code showing amplitude levels and their 8-bit words in three ranges: 0 to 15, 120 to 135, and 240 to 255. Signals encoded in this way are said to be *pulse-code-modulated*. Although the basic pulse-code modulation (PCM) code sometimes

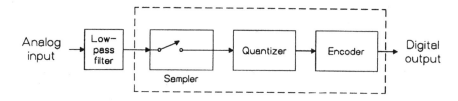

Figure 4.1 Basic elements of an analog-to-digital converter. (*From* [1]. *Used with permission.*)

Table 4.1 Binary Values of Amplitude Levels for 8-Bit Words (*From* [1]. *Used with permission.*)

Amplitude	Binary Level	Amplitude	Binary Level	Amplitude	Binary Level
0	00000000	120	01111000	240	11110000
1	00000001	121	01111001	241	11110001
2	00000010	122	01111010	242	11110010
3	00000011	123	01111011	243	11110011
4	00000100	124	01111100	244	11110100
5	00000101	125	01111101	245	11110101
6	00000110	126	01111110	246	11110110
7	00000111	127	01111111	247	11110111
8	00001000	128	10000000	248	11111000
9	00001001	129	10000001	249	11111001
10	00001010	130	10000010	250	11111010
11	00001011	131	10000011	251	11111011
12	00001100	132	10000100	252	11111100
13	00001101	133	10000101	253	11111101
14	00001110	134	10000110	254	11111110
15	00001111	135	10000111	255	11111111

is used, more elaborate codes—with many additional bits per word—generally are applied in circuits where errors may be introduced into the bit stream. Figure 4.2 shows a typical video waveform and several quantized amplitude levels based on the PCM coding scheme of Table 4.1.

Analog signals can be converted to digital codes using a number of methods, including the following [3]:

• *Integration*

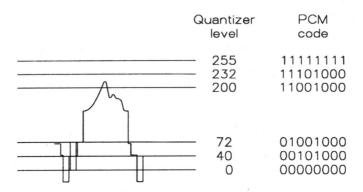

	Quantizer level	PCM code
	255	11111111
	232	11101000
	200	11001000
	72	01001000
	40	00101000
	0	00000000

Figure 4.2 Video waveform quantized into 8-bit words. (*From* [1]. *Used with permission.*)

- *Successive approximation*

- *Parallel (flash) conversion*

- *Delta modulation*

- Pulse-code modulation

- *Sigma-delta conversion*

Two of the more common A/D conversion processes are successive approximation and parallel or flash. Very high-resolution digital video systems require specialized A/D techniques that often incorporate one of these general schemes in conjunction with proprietary technology.

Successive Approximation

Successive approximation A/D conversion is a technique commonly used in medium- to high-speed data-acquisition applications. One of the fastest A/D conversion techniques, it requires a minimum amount of circuitry. The conversion times for successive approximation A/D conversion typically range from 10 to 300 μs for 8-bit systems.

The successive approximation A/D converter can approximate the analog signal to form an *n*-bit digital code in *n* steps. The *successive approximation register* (SAR) individually compares an analog input voltage with the midpoint of one of *n* ranges to determine the value of 1 bit. This process is repeated a total of *n* times, using *n* ranges, to determine the *n* bits in the code. The comparison is accomplished as follows:

- The SAR determines whether the analog input is above or below the midpoint and sets the bit of the digital code accordingly.

- The SAR assigns the bits beginning with the most significant bit.

- The bit is set to a 1 if the analog input is greater than the midpoint voltage; it is set to a 0 if the input is less than the midpoint voltage.

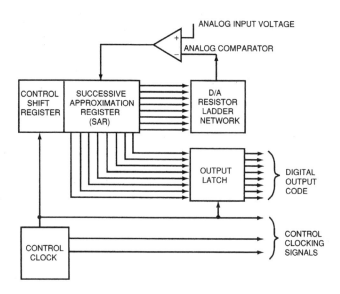

Figure 4.3 Successive approximation A/D converter block diagram. (*After* [4].)

- The SAR then moves to the next bit and sets it to a 1 or a 0 based on the results of comparing the analog input with the midpoint of the next allowed range.

Because the SAR must perform one approximation for each bit in the digital code, an n-bit code requires n approximations. A successive approximation A/D converter consists of four main functional blocks, as shown in Figure 4.3. These blocks are the SAR, the analog comparator, a D/A (digital-to-analog) converter, and a clock.

Parallel/Flash

Parallel or flash A/D conversion is used in high-speed applications such as video signal processing, medical imaging, and radar detection systems. A flash A/D converter simultaneously compares the input analog voltage with $2^n - 1$ threshold voltages to produce an n-bit digital code representing the analog voltage. Typical flash A/D converters with 8-bit resolution operate at 20 to 100 MHz and above.

The functional blocks of a flash A/D converter are shown in Figure 4.4. The circuitry consists of a precision resistor ladder network, $2^n - 1$ analog comparators, and a digital priority encoder. The resistor network establishes threshold voltages for each allowed quantization level. The analog comparators indicate whether the input analog voltage is above or below the threshold at each level. The output of the analog comparators is input to the digital priority encoder. The priority encoder produces the final digital output code, which is stored in an output latch.

An 8-bit flash A/D converter requires 255 comparators. The cost of high-resolution A/D comparators escalates as the circuit complexity increases and the number of analog convert-

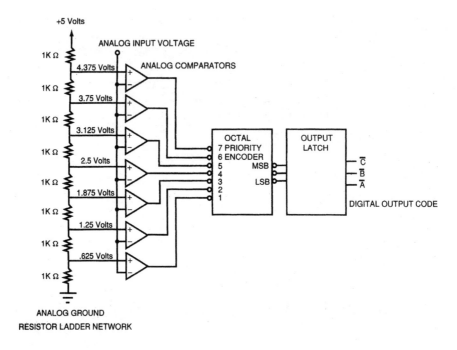

Figure 4.4 Block diagram of a flash A/D converter. (*After* [5].)

ers rises by $2^n - 1$. As a low-cost alternative, some manufacturers produce modified flash converters that perform the A/D conversion in two steps, to reduce the amount of circuitry required. These modified flash converters also are referred to as *half-flash* A/D converters because they perform only half of the conversion simultaneously.

4.2.3 The D/A Conversion Process

To convert digital codes to analog voltages, a voltage weight typically is assigned to each bit in the digital code, and the voltage weights of the entire code are summed [3]. A general-purpose D/A converter consists of a network of precision resistors, input switches, and level shifters to activate the switches to convert the input digital code to an analog current or voltage output. A D/A device that produces an analog current output usually has a faster settling time and better linearity than one that produces a voltage output.

D/A converters commonly have a fixed or variable reference level. The reference level determines the switching threshold of the precision switches that form a controlled impedance network, which in turn controls the value of the output signal. *Fixed-reference* D/A converters produce an output signal that is proportional to the digital input. In contrast, *multiplying* D/A converters produce an output signal that is proportional to the product of a varying reference level times a digital code.

D/A converters can produce bipolar, positive, or negative polarity signals. A *four-quadrant multiplying* D/A converter allows both the reference signal and the value of the binary code to have a positive or negative polarity.

4.2.4 Converter Performance Criteria

The major factors that determine the quality of performance of A/D and D/A converters are *resolution, sampling rate, speed,* and *linearity* [3]. The resolution of a D/A circuit is the smallest possible change in the output analog signal. In an A/D system, the resolution is the smallest change in voltage that can be detected by the system and produce a change in the digital code. The resolution determines the total number of digital codes, or quantization levels, that will be recognized or produced by the circuit.

The resolution of a D/A or A/D device usually is specified in terms of the bits in the digital code, or in terms of the *least significant bit* (LSB) of the system. An *n*-bit code allows for 2^n quantization levels, or $2^n - 1$ steps between quantization levels. As the number of bits increases, the step size between quantization levels decreases, therefore increasing the accuracy of the system when a conversion is made between an analog and digital signal. The system resolution also can be specified as the voltage step size between quantization levels.

The speed of a D/A or A/D converter is determined by the amount of time it takes to perform the conversion process. For D/A converters, the speed is specified as the *settling time.* For A/D converters, the speed is specified as the *conversion time.* The settling time for a D/A converter varies with supply voltage and transition in the digital code; it is specified in the data sheet with the appropriate conditions stated.

A/D converters have a maximum sampling rate that limits the speed at which they can perform continuous conversions. The sampling rate is the number of times per second that the analog signal can be sampled and converted into a digital code. For proper A/D conversion, the minimum sampling rate must be at least 2 times the highest frequency of the analog signal being sampled to satisfy the Nyquist criterion. The conversion speed and other timing factors must be taken into consideration to determine the maximum sampling rate of an A/D converter. Nyquist A/D converters use a sampling rate that is slightly greater than twice the highest frequency in the analog signal. *Oversampling* A/D converters use sampling rates of *N* times rate, where *N* typically ranges from 2 to 64.

Both D/A and A/D converters require a voltage reference to achieve absolute conversion accuracy. Some conversion devices have internal voltage references, whereas others accept external voltage references. For high-performance systems, an external precision reference is required to ensure long-term stability, load regulation, and control over temperature fluctuations.

Measurement accuracy is specified by the converter's linearity. *Integral linearity* is a measure of linearity over the entire conversion range. It often is defined as the deviation from a straight line drawn between the endpoints and through zero (or the *offset value*) of the conversion range. Integral linearity also is referred to as *relative accuracy.* The offset value is the reference level required to establish the zero or midpoint of the conversion range. *Differential linearity,* the linearity between code transitions, is a measure of the *monotonicity* of the converter. A converter is said to be monotonic if increasing input values result in increasing output values.

The accuracy and linearity values of a converter are specified in units of the LSB of the code. The linearity may vary with temperature, so the values often are specified at +25°C as well as over the entire temperature range of the device.

4.3 Space and Time Components of Video Signals

The images picked up by a television camera are sampled in space and in time [1]. The line structure represents sampling in the vertical axis. In digital processing, each line is sampled by the analog-to-digital converter, thereby producing samples in the horizontal axis of the image. These perpendicular sampling processes provide the spatial analysis. The field and frame repetitions provide the temporal sampling of the image. The signal content of all three dimensions must be exploited by the system designer to take full advantage of the channel's capacity to carry information. This approach, *spatio-temporal analysis*, has become an important tool in advanced television system development.

Two of the dimensions are divided into specific intervals: the vertical distance between the centers of the scanning lines and the time occupied by each field scan. The third dimension, in the analog operation of the system, is not divided this way because scanning the picture elements is then a continuous process. The intervals in the vertical and field-time dimensions result from the sampling of distance and time, respectively. The Nyquist limit applicable to sampled quantities imposes limits on their rates of change, and these limits must be observed in processing the signals resulting from the repetition of the scanned lines and fields.

The 3-dimensional nature of the video signal requires that it be represented as a solid figure, known as a *Nyquist volume*. Examples of this figure are given in Figure 4.5a for progressive scanning and Figure 4.5b for interlaced scanning of the luminance signal. The cross section of this volume is bounded by the maximum frequencies at which the vertical and time information can be transmitted, and its long dimension is bounded by the maximum frequency at which the picture elements can be scanned. Negative as well as positive frequencies (arising from the mathematical analysis of the signal space) are shown. They may be thought of as arising from the choice of the origin of the axis in each dimension.

Within the Nyquist volume reside all the permitted frequencies in the luminance channel. Study of the volume's contents reveals which frequencies are occupied by a given signal format, where additional signal information might be added, the extent of the interference (crosstalk) thereby created, and where empty space may be created—by filtering—to reduce or eliminate the crosstalk.

4.3.1 Dimensions of the Nyquist Volume

To determine the specific dimensions and contents of the Nyquist volume, it is appropriate first to examine its cross section [1]. This can be approached in two steps, shown in Figures 4.6 to 4.8. Figure 4.6 represents the intervals between samples for progressive scanning, and Figure 4.7 details the intervals between samples for interlaced scanning. Figure 4.8 represents the corresponding frequency limits.

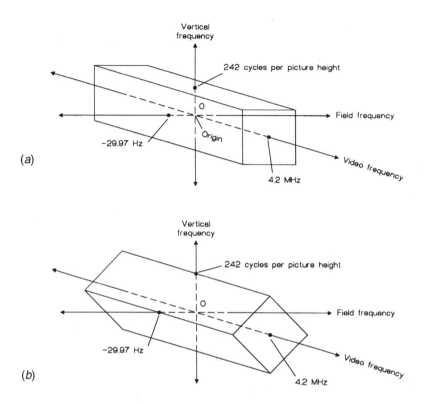

Figure 4.5 The Nyquist volume of the luminance signal: (*a*) progressive scanning, (*b*) interlaced scanning. (*From* [1]. *Used with permission.*)

The intervals in Figures 4.6 and 4.7 are the center-to-center separations between lines in the vertical dimension of the image, shown on the vertical axis, and the field-to-field time intervals, on the horizontal axis. Figure 4.6 relates to progressive scanning, in which the lines are adjacent to one another. For example, taking the number of active lines as 484 (NTSC system), the line spacing is found to be 1/484 of the picture height. The NTSC field-to-field interval is 1/59.94 s.

Positive and negative intervals are shown on both axes. Hence, the axes represent the indefinite progression of the line and field scans, with one position and one time selected for the origin. The plot, therefore, is covered by an indefinitely large number of separate points where a distance and a time coincide. These are marked on the figure as dots. It will be noted that successive intervals of 1/484 of the picture height and 1/59.94 s enclose the rectangular figure shown, and that the area of the plot is covered without gaps by such rectangles.

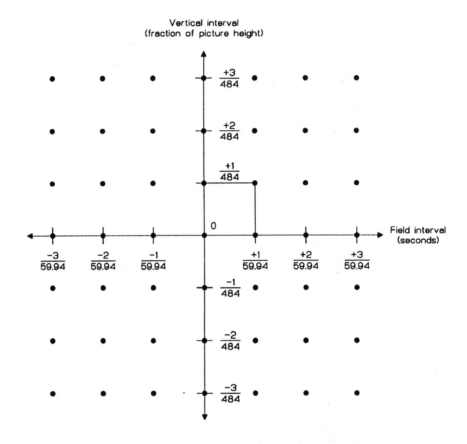

Figure 4.6 Sampled intervals in the vertical-picture and field-time axes for progressive scanning. (*From* [1]. *Used with permission.*)

Note that the mathematical analysis of the plots given in Figures 4.6 to 4.8 involves Fourier series and matrix algebra. The simplified presentation given here arrives at the conclusions of the analysis in a form appropriate for this book.

When the scanning is interlaced (Figure 4.7), a different arrangement of the line intervals occurs. Vertically above each field interval, only half the line intervals are occupied. (In interlaced scanning, every other line is omitted.) Above the next field time to the right, the same omissions exist, except that the occupied intervals lie opposite the omitted intervals to the left; that is, the line intervals are interlaced. In this plot, the figure connecting adjacent intervals is diamond-shaped. This figure is replicated throughout the area of the plot.

To convert the distance-vs.-time intervals to the corresponding sampling rates (Figure 4.8), the axis numbers are inverted. Thus, for our NTSC example case, 1/484 of the picture height becomes 484 cycles per picture height (c/ph), and 1/59.94 s becomes 59.94 Hz. Fig-

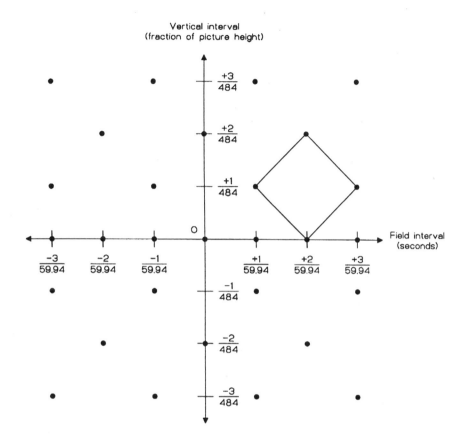

Figure 4.7 Sampled intervals in the vertical-picture and field-time axes for interlaced scanning. (*From* [1]. *Used with permission.*)

ure 4.8*a* shows inverted values over the rectangular region bounded by the values ±484 c/ph and ±59.94 Hz. These are the sampling rates for progressive scanning. The maximum frequencies for lines and fields, under the Nyquist limit, are one-half these values—242 c/ph and 29.97 Hz, respectively. These limits bound an inner rectangle, shown shaded, in which all the frequencies of the sampled quantities must reside if aliasing is to be avoided. This rectangle is the cross section of the Nyquist volume in Figure 4.5*a*. The corresponding plot for interlaced scanning is shown in Figure 4.8*b*. The area bounded by ±484 c/ph and ±59.94 Hz in this case is a diamond-shaped figure. When the Nyquist limit is applied to these values, the central diamond figure (shaded) is the area contained within ±242 c/ph and ±29.97 Hz. This area bounds the frequency range allowable in interlaced scanning, and it is the cross section of the Nyquist volume shown in Figure 4.5*b*.

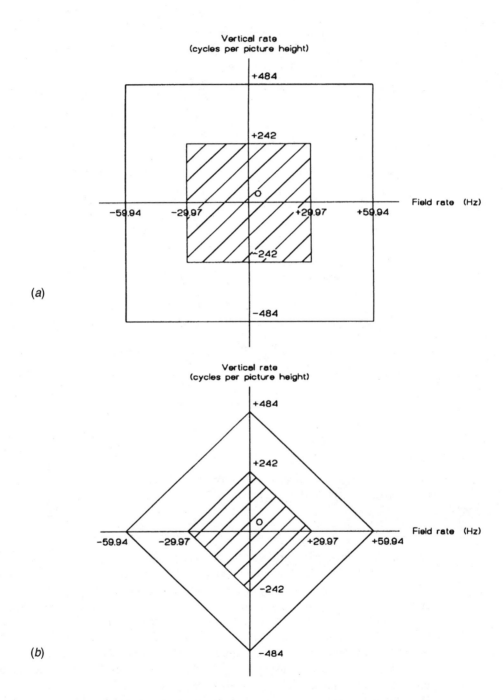

Figure 4.8 Sampling rates corresponding to the intervals in Figure 4.6 and Figure 4.7: (*a*) progressive scanning, (*b*) interlaced scanning. (*From* [1]. *Used with permission.*)

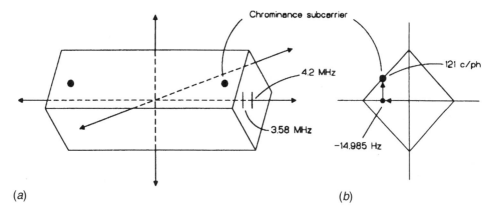

Figure 4.9 The Nyquist volume of the NTSC signal space: (*a*) location of the chrominance subcarrier, (*b*) cross section showing field- and line-frequency components. (*From* [1]. *Used with permission.*)

4.3.2   Significant Frequencies in the Nyquist Volume

On and within the Nyquist volume, a number of subcarrier frequencies and their sideband regions can be identified [1]. Figure 4.9 shows examples in the NTSC system. The horizontal dimension extends to the limit of the video baseband, ± 4.2 MHz (Figure 4.9*a*). The color subcarrier occupies a slice in Figure 4.9*b* through the volume, intersecting the horizontal axis in Figure 4.9*a* at ± 3.58 MHz. Because the subcarrier repeats its full cycle only after four fields, its picture-repetition rate is 14.985 Hz, and its line-repetition rate is reduced to 121 c/ph. When these values are laid out within the Nyquist volume (Figure 4.9*b*), the subcarrier frequency appears on its surface, at the positions marked by the dots. About these dots as origins, Nyquist volumes are erected to represent the chrominance frequency space.

The spaces occupied in common by the luminance and chrominance frequency volumes are shown in Figure 4.10. Within this common space, crosstalk between the luminance and chrominance signals occurs. This crosstalk can be reduced or eliminated if the common space is removed from the luminance volume by filtering the luminance signal at the source and again prior to the display [6].

The 3.58 MHz subcarrier frequency is one of a set of frequencies that lie on the surface of the Nyquist volume for NTSC. They are the odd multiples of half the line-scan frequency, subcarriers of alternating phase on successive scans of each line. One such multiple, 455, sets the color subcarrier in the NTSC system. The locus of these subcarrier frequencies is a horizontal line along the center of one of the Nyquist volume surfaces, marked SC-1 in Figure 4.11. Another such locus of subcarrier frequencies, SC-2, lies along the center of the adjacent face of the volume. Such a subcarrier can be formed by inverting the phase of an SC-1 type subcarrier on alternate lines. The locations of the SC-1 and SC-2 subcarriers, shown in cross section (Figure 4.11*b*), demonstrate that they can coexist in the signal space. So an additional channel, available for auxiliary information, is provided by the SC-2 subcarrier. These are known as *phase-alternate-frequency* (PAF) carriers. Such a

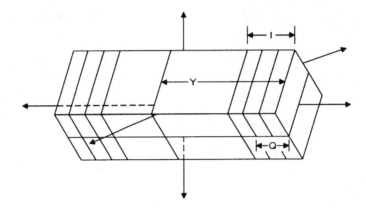

Figure 4.10 Nyquist volumes of the chrominance space positioned within the luminance volume. (*From* [1]. *Used with permission.*)

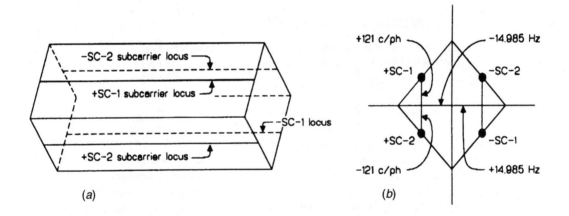

Figure 4.11 Subcarrier locations on the faces of the Nyquist volume: (*a*) relative positions, (*b*) cross-sectional view. (*From* [1]. *Used with permission.*)

subcarrier was explored as a method of transmitting auxiliary information in one of the original EDTV system proposals, ACTV-1 [7, 8].

4.3.3 Signal Distinctions in the Nyquist Volume

The Nyquist volume is indicative only of the frequency content of a signal. It does not reveal how signals may otherwise be distinguished. To separate or protect signals sharing the frequency space, several characteristics or processes may be used. These processes rely on the following:

- The relative amplitude (energy content) of conflicting signals

- Changes with the time of signal occupancy, determined by the modulation used

- The distribution of the frequency space by imposing barriers to the signal spectra by filtering

Distinction by Energy Level

The energy distribution among signal frequencies in television depends on the subject matter. This includes the following scene characteristics:

- Colors

- Luminances

- Sharpness along horizontal, vertical, and diagonal edges

- The shapes and motions of extended objects

Despite this variety in televised images, some general properties of the energy concentrations in the luminance and chrominance signals are known and can be used to define the relative importance of the frequency regions they occupy in the Nyquist volume [1].

The energy concentration in luminance is greatest at the low frequencies, by which extended shapes are defined. The high frequencies serve to define edges; above 2 MHz, most of the signal energy is of low amplitude. The energy in the chrominance signal is also greatest at the low frequencies, and it is largely removed in the Q component, to meet the NTSC standard, above 0.5 MHz. The I component extends to about 1.5 MHz. When the luminance and chrominance signals are combined in the NTSC signal spectrum, the high-energy chrominance surrounds the 3.58 MHz subcarrier, which is superposed on the low-energy region of the luminance spectrum. As a result, the crosstalk is decreased, but it still is severe enough to adversely affect the performance of most receivers. Reduction of crosstalk between luminance and chrominance has been guided by the identification of these signals in the frequency space of the Nyquist volume

Distinction by Modulation Method

Quadrature modulation is a method widely used to permit two distinct signals to occupy the same frequency space [1]. It is used in the NTSC and PAL systems to impose the two color-difference signals on the chrominance subcarrier. The Nyquist volumes of the I and Q sidebands of the NTSC system are shown in Figure 4.10 as they are located within the luminance volume. Although crosstalk occurs between luminance and chrominance, none is present between the I and Q components. This is accomplished by the use of quadrature modulation. The signals are separated by combining two color subcarriers, whose oscillations are out of phase by 90°, separately modulated by the I and Q signals. In the Nyquist volume, each point in the frequency space occupied by the subcarrier sidebands is filled first by the I sideband and next by the Q component, but in stationary images neither occupies the space at the same time.

In the NTSC and PAL systems, the subcarrier is suppressed, and only the sidebands are transmitted. To recover the I and Q sideband components at the receiver, the color burst synchronization signals are used to recreate the subcarrier, which is combined with the *I* and *Q* sidebands.

Distinction by Filtering

To make room for chrominance and other auxiliary signals, some of the frequency space in the Nyquist volume must be emptied by filtering the luminance signal [1]. The unoccupied regions are known as *channels*. Filtering may be applied in all three dimensions of the frequency space. One-dimensional filters, operating along the horizontal axis, limit the baseband of the video spectrum. Such filters are of the continuous-frequency (*wave filter*) type. Two-dimensional filtering also includes the field-repetition axis, with a periodicity of 59.94 Hz (again, for the NTSC case). This function is provided by a periodic filter of the *line-comb* type. Three-dimensional filtering adds the third axis, the vertical dimension in cycles per picture height. This function also employs a periodic filter that covers the picture scanning time, that is, a frame store (or two field stores having capacity for 262 and 263 lines, respectively, for NTSC).

The periodic filters typically used for creating channels in the Nyquist space are of the nonrecursive type with taps taken at each junction between delay cells. Consequently, 2- and 3-dimensional filtering involves many interconnections among a large number of cells. Although integrated circuits can readily meet this requirement, multidimensional filtering must meet the demands of intricate design formulas, and a number of compromises are involved (e.g., between sharpness of frequency cutoff and the introduction of signal overshoot and ringing).

An example of 1-dimensional filtering is provided by the typical conventional television receiver. No filtering between luminance and chrominance is provided at the transmitter, so crosstalk between them is inherent in the broadcast signal. To separate them at the receiver, the luminance response is cut off at about 2.5 MHz, and the chrominance sidebands are limited to ±0.5 MHz. The Nyquist frequency space of the received signal output from the second detector then appears as shown in Figure 4.12. The horizontal axis of the luminance space is cut off at 2.5 MHz, and the chrominance sidebands extend on that axis from 3.08 to 4.08 MHz, beyond the luminance limit. This cutoff of luminance bandwidth limits the horizontal resolution, but the loss has been accepted in the interest of reducing crosstalk.

When a comb filter is used, adjacent *wells* are introduced in the frequency space, into which the luminance and chrominance frequencies fall separately, appreciably reducing crosstalk. These examples apply only to receiver processing of the frequency space. A more modern approach, *cooperative processing*, involves filtering the space within the Nyquist volume at the transmitter (*prefiltering*) and the receiver (*postfiltering*). This arrangement deepens and sharpens the wells into which the luminance and chrominance fit, making the crosstalk substantially invisible in stationary images.

Considerations Regarding the Nyquist Volume

The foregoing examples serve to show the options available in the frequency space within the Nyquist volume. When the NTSC system of compatible color television was developed in the early 1950s, the mathematical and physical bases of the Nyquist volume were known to communications theorists, but not to many television engineers. Few of the NTSC mem-

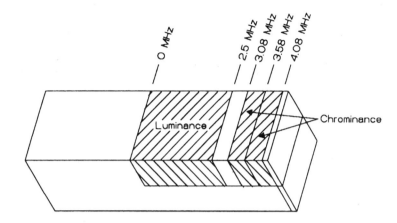

Figure 4.12 One-dimensional filtering between luminance and chrominance, as provided by a typical NTSC receiver. (*From* [1]. *Used with permission.*)

bers had a clear understanding of the unused resources in the spectrum of the compatible color signal. Nor, for that matter, could those engineers then envision the compromises that have since been found acceptable concerning the limits of luminance and chrominance spectra, and of the tradeoff between high resolution in a stationary image and low resolution in a moving one.

4.4   Digital Modulation

Digital modulation is necessary before digital data can be transmitted through a channel, be it a satellite link or HDTV [9]. *Modulation* is the process of varying some attribute of a carrier waveform as a function of the input intelligence to be transmitted. Attributes that can be varied include amplitude, frequency, and phase. In the case of digital modulation, the message sequence is a stream of digits, typically of binary value. In the simplest case, parameter variation is on a symbol-by-symbol basis; no memory is involved. Carrier parameters that can be varied under this scenario include the following:

- Amplitude, resulting in *amplitude-shift keying* (ASK)

- Frequency, resulting in *frequency-shift keying* (FSK)

- Phase, resulting in *phase-shift keying* (PSK)

So-called higher-order modulation schemes impose memory over several symbol periods. Such modulation techniques can be classified as *binary* or *M*-ary, depending on whether one of two possible signals or $M > 2$ signals per signaling interval can be sent. (Binary signaling may be defined as any signaling scheme in which the number of possible signals sent during any given signaling interval is two. *M*-ary signaling, on the other hand, is a signaling

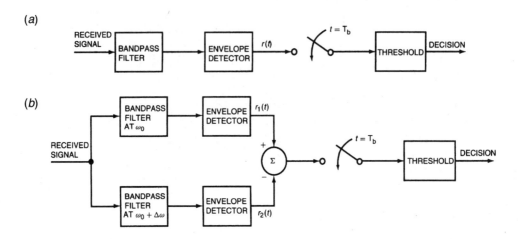

Figure 4.13 Receiver systems for noncoherent detection of binary signals: (*a*) ASK, (*b*) FSK. (*From* [9]. *Used with permission.*)

system in which the number of possible signals sent during any given signaling interval is *M*.) For the case of *M*-ary modulation when the source digits are binary, it is clear that several bits must be grouped to make up an *M*-ary word.

Another classification for digital modulation is *coherent* vs. *noncoherent*, depending upon whether a reference carrier at the receiver coherent with the received carrier is required for demodulation (the coherent case) or not (the noncoherent case). For situations in which it is difficult to maintain phase stability—for example, in channels subject to fading—it is useful to employ a modulation technique that does not require the acquisition of a reference signal at the receiver that is phase-coherent with the received carrier. ASK and FSK are two modulation techniques that lend themselves well to noncoherent detection. Receivers for detection of ASK and FSK noncoherently are shown in Figure 4.13.

One other binary modulation technique is, in a sense, noncoherent: *differentially coherent PSK* (DPSK). With DPSK, the phase of the preceding bit interval is used as a reference for the current bit interval. This technique depends on the channel being sufficiently stable so that phase changes resulting from channel perturbations from a given bit interval to the succeeding one are inconsequential. It also depends on there being a known phase relationship from one bit interval to the next. This requirement is ensured by differentially encoding the bits before phase modulation at the transmitter. Differential encoding is illustrated in Table 4.2. An arbitrary reference bit is chosen to start the process. In the table a *1* has been chosen. For each bit of the encoded sequence, the present bit is used as the reference for the following bit in the sequence. A *0* in the message sequence is encoded as a transition from the state of the reference bit to the opposite state in the encoded message sequence. A *1* is encoded as no change of state. Using these rules, the result is the encoded sequence shown in the table.

Table 4.2 Example of the Differential Encoding Process (*After* [9].**)**

Message Sequence		1	0	0	1	1	1	0
Encoded Sequence	1	1	0	1	1	1	1	0
Transmitted Phase Radians	0	0	π	0	0	0	0	π

4.4.1 QPSK

Consider the case of an MPSK signal where $M = 4$, commonly referred to as *quadriphase-shift keying* (QPSK) [9]. This common modulation technique utilizes four signals in the signal set distinguished by four phases 90° apart. For the case of an MASK signal where $M = 4$, a *quadrature-amplitude-shift keying* (QASK) condition results. With QASK, both the phase and amplitude of the carrier take on a set of values in one-to-one correspondence with the digital data to be transmitted.

Several variations of QPSK have been developed to meet specific operational requirements. One such scheme is referred to as *offset* QPSK (OQPSK) [10]. This format is produced by allowing only ±90° phase changes in a QPSK format. Furthermore, the phase changes can take place at multiples of a half-symbol interval, or a bit period. The reason for limiting phase changes to ±90° is to prevent the large envelope deviations that occur when QPSK is filtered to restrict sidelobe power, and regrowth of the sidelobes after amplitude limiting is used to produce a constant-envelope signal. This condition often is encountered in satellite systems in which, because of power-efficiency considerations, hard limiting repeaters are used in the communications system.

Another modulation technique closely related to QPSK and OQPSK is *minimum shift keying* (MSK) [10]. MSK is produced from OQPSK by weighting the in-phase and quadrature components of the baseband OQPSK signal with half-sinusoids. The phase changes linearly over a bit interval. As with OQPSK, the goal of MSK is to produce a modulated signal with a spectrum of reduced sidelobe power, one that behaves well when filtered and limited. Many different forms of MSK have been proposed and investigated over the years.

A modulation scheme related to 8-PSK is $\pi/4$-*differential QPSK* ($\pi/4$-DQPSK) [11]. This technique is essentially an 8-PSK format with differential encoding where, from a given phase state, only specified phase shifts of $\pm \pi/4$ or $\pm 3\pi/4$ are allowed.

Continuous phase modulation (CPM) [12] comprises a host of modulation schemes. These formats employ continuous phase trajectories over one or more symbols to get from one phase to the next in response to input changes. CPM schemes are employed in an attempt to simultaneously improve power and bandwidth efficiency.

4.4.2 Signal Analysis

The ideal choice of a modulation technique depends on many factors. Two of the most basic are the *bandwidth efficiency* and *power efficiency*. These parameters are defined as follows:

- Bandwidth efficiency is the ratio of the bit rate to the bandwidth occupied for a digital modulation scheme. Technically, it is dimensionless, but for convenience it is usually given the dimensions of bits/second/hertz.

- Power efficiency is the energy per bit over the noise power spectral density (E_b/N_o) required to provide a given probability of bit error for a digital modulation scheme.

Computation of these parameters is beyond the scope of this chapter. Interested readers are directed to [9] for a detailed discussion of performance parameters.

4.4.3 Digital Coding

Two issues are fundamental in assessing the performance of a digital communication system [13]:

- The reliability of the system in terms of accurately transmitting information from one point to another

- The rate at which information can be transmitted with an acceptable level of reliability

In an ideal communication system, information would be transmitted at an infinite rate with an infinite level of reliability. In reality, however, fundamental limitations affect the performance of any communication system. No physical system is capable of instantaneous response to changes, and the range of frequencies that a system can reliably handle is limited. These real-world considerations lead to the concept of *bandwidth*. In addition, random noise affects any signal being transmitted through any communication medium. Finite bandwidth and additive random noise are two fundamental limitations that prevent designers from achieving an infinite rate of transmission with infinite reliability. Clearly, a compromise is necessary. What makes the situation even more challenging is that the reliability and the rate of information transmission usually work against each other. For a given system, a higher rate of transmission normally means a lower degree of reliability, and vice versa. To favorably affect this balance, it is necessary to improve the efficiency and the robustness of the communication system. *Source coding* and *channel coding* are the means for accomplishing this task.

Source Coding

Most information sources generate signals that contain redundancies [13]. For example, consider a picture that is made up of pixels, each of which represents one of 256 grayness levels. If a fixed coding scheme is used that assigns 8 binary digits to each pixel, a 100 × 100 picture of random patterns and a 100 × 100 picture that consists of only white pixels would both be coded into the same number of binary digits, although the white-pixel version would have significantly less information than the random-pattern version.

One simple method of source encoding is the *Huffman* coding technique, which is based on the idea of assigning a code word to each symbol of the source alphabet such that the length of each code word is approximately equal to the amount of information conveyed by that symbol. As a result, symbols with lower probabilities get longer code words. Huffman coding is achieved through the following process:

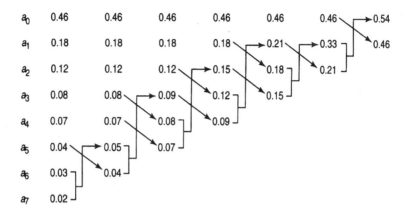

Figure 4.14 The Huffman coding algorithm. (*From* [13]. *Used with permission.*)

- List the source symbols in descending order of probabilities.

- Assign a binary 0 and a binary 1, respectively, to the last two symbols in the list.

- Combine the last two symbols in the list into a new symbol with its probability equal to the sum of two symbol probabilities.

- Reorder the list, and continue in this manner until only one symbol is left.

- Trace the binary assignments in reverse order to obtain the code word for each symbol.

A tree diagram for decoding a coded sequence of symbols is shown in Figure 4.14. It can easily be verified that the *entropy* of the source under consideration is 2.3382 bits/symbol, and the average code-word length using Huffman coding is 2.37 bits/symbol.

At this point it is appropriate to define *entropy*. In a general sense, entropy is a measure of the disorder or randomness in a closed system. With regard to digital communications, it is defined as a measure of the number of bits necessary to transmit a message as a function of the probability that the message will consist of a specific set of symbols.

Channel Coding

The previous section identified the need for removing redundancies from the message signal to increase efficiency in transmission [13]. From an efficiency point of view, the ideal scenario would be to obtain an average word length that is numerically equal to the entropy of the source. From a practical perspective, however, this would make it impossible to detect or correct errors that may occur during transmission. Some redundancy must be added to the signal in a controlled manner to facilitate detection and correction of transmission errors. This process is referred to as channel coding.

A variety of techniques exist for detection and correction of errors. For the purposes of this chapter, however, it is sufficient to understand that error-correction coding is important to reliable digital transmission and that it adds to the total bit rate of a given information stream. For closed systems, where retransmission of garbled data is possible, a minimum of error-correction overhead is practical. The error-checking *parity* system is a familiar technique. However, for transmission channels where 2-way communication is not possible, or the channel restrictions do not permit retransmission of specific packets of data, robust error correction is a requirement. More information on the basic principles of error correction can be found in [13].

4.4.4 Error-Correction Coding

Digital modulation schemes in their basic form have dependency between signaling elements over only one signaling division [9]. There are advantages, however, to providing memory over several signaling elements from the standpoint of error correction. Historically, this has been accomplished by adding redundant symbols for error correction to the encoded data, and then using the encoded symbol stream to modulate the carrier. The ratio of information symbols to total encoded symbols is referred to as the *code rate*. At the receiver, demodulation is accomplished followed by decoding.

The drawback to this approach is that redundant symbols are added, requiring a larger transmission bandwidth, assuming the same data throughput. However, the resulting signal is more immune to channel-induced errors resulting from, among other things, a marginal S/N for the channel. The end result for the system is a *coding gain*, defined as the ratio of the signal-to-noise ratios without and with coding.

There are two widely used coding methods:

- *Block coding*, a scheme that encodes the information symbols block-by-block by adding a fixed number of error-correction symbols to a fixed block length of information symbols.

- *Convolutional coding*, a scheme that encodes a sliding window of information symbols by means of a shift register and two or more modulo-2 adders for the bits in the shift register that are sampled to produce the encoded output.

Although an examination of these coding methods is beyond the scope of this chapter, note that coding used in conjunction with modulation always expands the required transmission bandwidth by the inverse of the code rate, assuming the overall bit rate is held constant. In other words, the power efficiency goes up, but the bandwidth efficiency goes down with the use of a well-designed code. Certain techniques have been developed to overcome this limitation, including *trellis-coded modulation* (TCM), which is designed to simultaneously conserve power and bandwidth [14].

4.4.5 8-VSB Modulation System

The ATSC digital terrestrial TV broadcasting system employs a single-carrier high-data-rate amplitude-modulated suppressed-carrier vestigial sideband signal known as VSB [15]. The VSB system actually provides the basis for a family of transmission systems that are employed for terrestrial broadcasts and cable. These systems share the same pilot frequency, symbol rate, data frame structure, interleaving, Reed-Solomon coding, and synchronization pulses. The VSB system has two modes:

- An 8-level VSB terrestrial broadcast mode known as 8-VSB

- A 16-level VSB high-data-rate cable mode (16-VSB)

The terrestrial broadcast mode supports one DTV signal in a single 6 MHz channel. The high-data-rate cable mode, which trades some robustness for twice the data rate, supports two DTV signals in one 6 MHz channel.

The VSB modulator accepts a 10.76 MHz coded data signal with the synchronization information added. This system employs a dc pilot carrier to allow better signal acquisition under noisy conditions. The dc pilot is created by adding a value of 1.25 to the nominal mapped signal levels. The channel response is nominally flat across the band except for the transition regions. To best accommodate the nature of the vestigial sideband signal, the selectivity of the transition regions is not the same. The baseband signals can be converted to analog form and modulated on quadrature *intermediate frequency* (IF) carriers to create the vestigial sideband IF signal by sideband cancellation.

The characteristics of practical terrestrial transmission channels require that error-correction coding be included within the data stream, which reduces the effective *data payload* capability. The terrestrial-mode 8-VSB carrier must deal with all of the primary channel impairments, including multipath reception, co-channel interference from existing analog TV services, adjacent-channel interference from existing analog TV services, and impulse interference.

To maximize coverage, the 8-VSB terrestrial broadcast mode incorporates an NTSC rejection filter in the receiver and in the trellis coding system. *Precoding* at the transmitter is incorporated into the trellis code. When the NTSC rejection filter is activated in the receiver, the trellis decoder is switched to a trellis code corresponding to the encoder trellis code concatenated with the filter.

Specifics of the 8-VSB system are discussed in Chapter 8.

4.5   Digital Filters

Digital filtering is concerned with the manipulation of discrete data sequences to remove noise, extract information, change the sample rate, and/or modify the input information in some form or context [16]. Although an infinite number of numerical manipulations can be applied to discrete data (e.g., finding the mean value, forming a histogram), the objective of digital filtering is to form a discrete output sequence $y(n)$ from a discrete input sequence $x(n)$. In some manner, each output sample is computed from the input sequence—not just from any one sample, but from many, possibly all, of the input samples. Those filters that compute their output from the present input and a finite number of past inputs are termed *finite impulse response* (FIR) filters; those that use all past inputs are termed *infinite impulse response* (IIR) filters.

4.5.1 FIR Filters

An FIR filter is a linear discrete-time system that forms its output as the weighted sum of the most recent, and a finite number of past, inputs [16]. A *time-invariant* FIR filter has finite memory, and its impulse response (its response to a discrete-time input that is unity at the first sample and otherwise zero) matches the fixed weighting coefficients of the filter. *Time-variant* FIR filters, on the other hand, may operate at various sampling rates and/or have weighting coefficients that adapt in sympathy with some statistical property of the environment in which they are applied.

Perhaps the simplest example of an FIR filter is the *moving average* operation described by the following linear constant-coefficient difference equation:

$$y[n] = \sum_{k=0}^{M} b_k x[n-k] \qquad\qquad b_k = \frac{1}{M+1} \qquad (4.1)$$

Where:
$y[n]$ = output of the filter at integer sample index n
$x[n]$ = input to the filter at integer sample index n
b_k = filter weighting coefficients, $k = 0,1,...,M$
M = filter order

In a practical application, the input and output discrete-time signals will be sampled at some regular sampling time interval, T seconds, denoted $x[nT]$ and $y[nT]$, which is related to the sampling frequency by $f_s = 1/T$, samples per second. However, for generality, it is more convenient to assume that T is unity, so that the effective sampling frequency also is unity and the Nyquist frequency is one-half. It is, then, straightforward to scale, by multiplication, this normalized frequency range, i.e. [1/2, 1], to any other sampling frequency.

The output of the simple moving average filter is the average of the $M + 1$ most recent values of $x[n]$. Intuitively, this corresponds to a smoothed version of the input, but its operation is more appropriately described by calculating the frequency response of the filter. First, however, the z-domain representation of the filter is introduced in analogy to the s- (or *Laplace*) domain representation of analog filters. The z-transform of a causal discrete-time signal $x[n]$ is defined by:

$$X(z) = \sum_{n=0}^{\infty} x[n]z^{-n} \qquad (4.2)$$

Where:
$X(z)$ = z-transform of $x[n]$
z = complex variable

The z-transform of a delayed version of $x[n]$, namely $x[n-k]$ with k a positive integer, is found to be given by $z^{-k}X(z)$. This result can be used to relate the z-transform of the output, $y[n]$, of the simple moving average filter to its input:

$$Y(z) = \sum_{k=0}^{M} b_k z^{-k} X(z) \qquad\qquad b_k = \frac{1}{M+1} \qquad (4.3)$$

The z-domain transfer function, namely the ratio of the output to input transform, becomes:

$$H(z) = \frac{Y(z)}{X(z)} = \sum_{k=0}^{M} b_k z^{-k} \qquad\qquad b_k = \frac{1}{M+1} \qquad (4.4)$$

Notice the transfer function, $H(z)$, is entirely defined by the values of the weighting coefficients, b_k, $k = 0,1,...,M$, which are identical to the discrete impulse response of the filter, and the complex variable z. The finite length of the discrete impulse response means that the transient response of the filter will last for only $M + 1$ samples, after which a steady state will be reached. The frequency-domain transfer function for the filter is found by setting

$$z = e^{j2\pi f} \qquad (4.5)$$

Where $j = \sqrt{-1}$ and can be written as:

$$H(e^{j2\pi f}) = \frac{1}{M+1} \sum_{k=0}^{M} e^{-j2\pi fk} = \frac{1}{M+1} e^{-j\pi fM} \frac{\sin[\pi f(M+1)]}{\sin(\pi f)} \qquad (4.6)$$

The magnitude and phase response of the simple moving average filter, with $M = 7$, are calculated from $H(e^{j2\pi f})$ and shown in Figure 4.15. The filter is seen clearly to act as a crude low-pass smoothing filter with a linear phase response. The sampling frequency periodicity in the magnitude and phase response is a property of discrete-time systems. The linear phase response is due to the $e^{-j\pi fM}$ term in $H(e^{j2\pi f})$ and corresponds to a constant $M/2$ group delay through the filter. A phase discontinuity of $\pm 180°$ is introduced each time the magnitude term changes sign. FIR filters that have center symmetry in their weighting coefficients have this constant frequency-independent group-delay property that is desirable in applications in which time dispersion is to be avoided, such as in pulse transmission, where it is important to preserve pulse shapes [17].

Design Techniques

Linear-phase FIR filters can be designed to meet various filter specifications, such as low-pass, high-pass, bandpass, and band-stop filtering [16]. For a low-pass filter, two frequencies are required. One is the maximum frequency of the passband below which the magnitude response of the filter is approximately unity, denoted the *passband corner frequency* f_p. The other is the minimum frequency of the stop-band above which the magnitude response of the filter must be less than some prescribed level, named the *stop-band corner frequency* f_s. The difference between the passband and stop-band corner frequencies is the *transition bandwidth*. Generally, the order of an FIR filter, M, required to meet some design specifica-

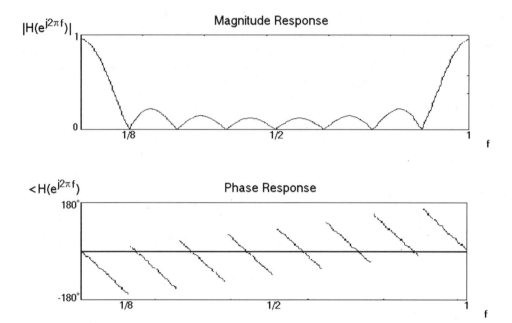

Figure 4.15 The magnitude and phase response of the simple moving average filter with $M = 7$. (*From* [16]. *Used with permission.*)

tion will increase with a reduction in the width of the transition band. There are three established techniques for coefficient design:

- *Windowing.* A design method that calculates the weighting coefficients by sampling the ideal impulse response of an analog filter and multiplying these values by a smoothing window to improve the overall frequency-domain response of the filter.

- *Frequency sampling.* A technique that samples the ideal frequency-domain specification of the filter and calculates the weighting coefficients by inverse-transforming these values.

- Optimal approximations.

The best results generally can be obtained with the optimal approximations method. With the increasing availability of desktop and portable computers with fast microprocessors, large quantities of memory, and sophisticated software packages, optimal approximations is the preferred method for weighting coefficient design. The impulse response and magnitude response for a 40th-order optimal half-band FIR low-pass filter designed with the *Parks-McClellan* algorithm [18] are shown in Figure 4.16, together with the ideal frequency-domain design specification. Notice the zeros in the impulse response. This algorithm minimizes the peak deviation of the magnitude response of the design filter from the ideal magnitude response. The magnitude response of the design filter alternates about the desired

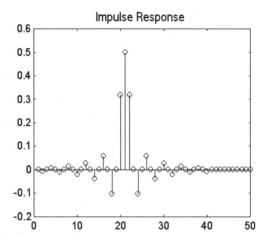

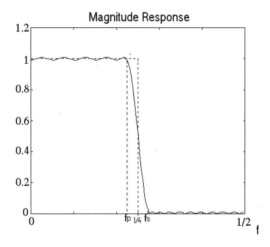

Figure 4.16 The impulse and magnitude response of an optimal 40th-order half-band FIR filter. (*From* [16]. *Used with permission.*)

specification within the passband and above the specification in the stop-band. The maximum deviation from the desired specification is equalized across the passband and stop-band; this is characteristic of an *optimal solution*.

Applications

In general, digitally implemented FIR filters exhibit the following attributes [16]:

- Absence of drift

- Reproducibility

- Multirate realizations

- Ability to adapt to time-varying environments

These features have led to the widespread use of FIR filters in a variety of applications, particularly in telecommunications. The primary advantage of the fixed-coefficient FIR filter is its unconditional stability because of the lack of feedback within its structure and its exact linear phase characteristics. Nonetheless, for applications that require sharp, selective, filtering—in standard form—they do require relatively large orders. For some applications, this may be prohibitive; therefore, recursive IIR filters are a valuable alternative.

Finite Wordlength Effects

Practical digital filters must be implemented with finite precision numbers and arithmetic [16]. As a result, both the filter coefficients and the filter input and output signals are in discrete form. This leads to four types of finite wordlength effects:

- *Discretization* (quantization) of the filter coefficients has the effect of perturbing the location of the filter poles and zeroes. As a result, the actual filter response differs slightly from the ideal response. This deterministic frequency response error is referred to as *coefficient quantization error*.

- The use of finite precision arithmetic makes it necessary to quantize filter calculations by rounding or truncation. *Roundoff noise* is that error in the filter output that results from rounding or truncating calculations within the filter. As the name implies, this error looks like low-level noise at the filter output.

- Quantization of the filter calculations also renders the filter slightly nonlinear. For large signals this nonlinearity is negligible, and roundoff noise is the major concern. However, for recursive filters with a zero or constant input, this nonlinearity can cause spurious oscillations called *limit cycles*.

- With fixed-point arithmetic it is possible for filter calculations to overflow. The term *overflow oscillation* refers to a high-level oscillation that can exist in an otherwise stable filter because of the nonlinearity associated with the overflow of internal filter calculations. Another term for this is *adder overflow limit cycle*.

4.5.2 Infinite Impulse Response Filters

A digital filter with impulse response having infinite length is known as an *infinite impulse response* filter [16]. Compared to an FIR filter, an IIR filter requires a much lower order to achieve the same requirement of the magnitude response. However, whereas an FIR filter is always stable, an IIR filter may be unstable if the coefficients are not chosen properly. Because the phase of a stable causal IIR filter cannot be made linear, FIR filters are preferable to IIR filters in applications for which linear phase is essential.

Practical *direct form* realizations of IIR filters are shown in Figure 4.17. The realization shown in Figure 4.17*a* is known as *direct form I*. Rearranging the structure results in *direct*

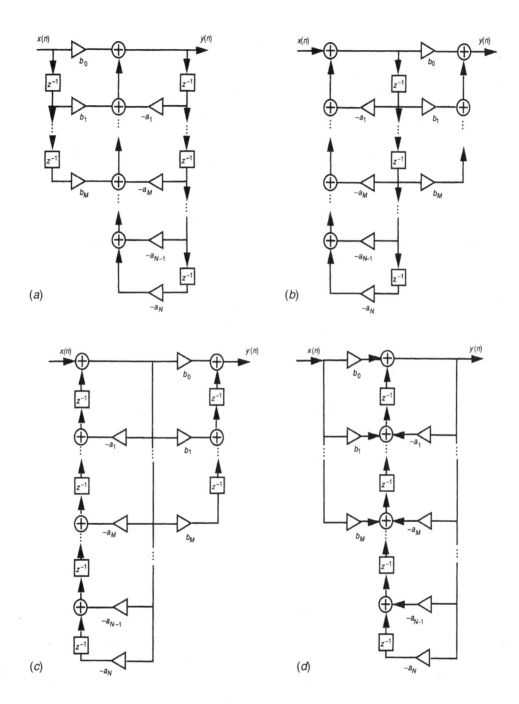

Figure 4.17 Direct form realizations of IIR filters: (*a*) direct form I, (*b*) direct form II, (*c*) transposed direct form I, (*d*) transposed direct form II. (*From* [16]. *Used with permission.*)

form II, as shown in Figure 4.17*b*. The results of transposition are *transposed direct form I* and *transposed direct form II*, as shown in Figures 4.17*c* and 4.17*d*, respectively. Other realizations for IIR filters are *state-space structure*, *wave structure*, and *lattice structure*. In some situations, it is more convenient or suitable to use software realizations that are implemented by programming a general-purpose microprocessor or a digital signal processor. (See [16] for details on IIR filter implementations.)

Designing an IIR filter involves choosing the coefficients to satisfy a given specification, usually a magnitude response parameter. There are various IIR filter design techniques, including:

- Design using an analog prototype filter, in which an analog filter is designed to meet the (analog) specification and the analog filter transfer function is transformed to a digital system function.

- Design using digital frequency transformation, which assumes that a given digital low-pass filter is available, and the desired digital filter is then obtained from the digital low-pass filter by a digital frequency transformation.

- Computer-aided design (CAD), which involves the execution of algorithms that choose the coefficients so that the response is as close as possible to the desired filter.

The first two methods are easily accomplished; they are suitable for designing standard filters (low-pass, high-pass, bandpass, and band-stop). The CAD approach, however, can be used to design both standard and nonstandard filters.

4.6 Digital Signal Processing and Distribution

Digital signal processing (DSP) techniques are being applied to the implementation of various stages of video capture, processing, storage, and distribution systems for a number of reasons, including:

- Improved cost-performance considerations

- Future product-enhancement capabilities

- Greatly reduced alignment and testing requirements

A wide variety of video circuits and systems can be readily implemented using various degrees of embedded DSP. The most important parameters are signal bandwidth and S/N, which define, respectively, the required sampling rate and the effective number of bits required for the conversion. Additional design considerations include the stability of the sampling clock, quadrature channel matching, aperture uncertainty, and the cutoff frequency of the quantizer networks.

DSP devices differ from microprocessors in a number of ways. For one thing, microprocessors typically are built for a range of general-purpose functions and normally run large blocks of software. Also, microprocessors usually are not called upon to perform real-time computation. Typically, they are at liberty to shuffle workloads and to select an action branch, such as completing a printing job before responding to a new input command. The

DSP, on the other hand, is dedicated to a single task or small group of related tasks. In a sophisticated video system, one or more DSPs may be employed as attached processors, assisting a general-purpose host microprocessor that manages the front-panel controls or other key functions of the unit.

One convenient way to classify DSP devices and applications is by their *dynamic range*. In this context, the dynamic range is the spread of numbers that must be processed in the course of an application. It takes a certain range of values, for example, to describe a particular signal, and that range often becomes even wider as calculations are performed on the input data. The DSP must have the capability to handle such data without overflow.

The processor capacity is a function of its data width, i. e., the number of bits it manipulates and the type of arithmetic that it performs (fixed or *floating point*). Floating point processing manipulates numbers in a form similar to scientific notation, enabling the device to accommodate an enormous breadth of data. Fixed arithmetic processing, as the name implies, restricts the processing capability of the device to a predefined value.

Recent advancements in very large scale integration (VLSI) technologies in general, and DSP in particular, have permitted the integration of many video system functional blocks into a single device. Such designs typically offer excellent performance because of the elimination of the traditional interfaces required by discrete designs. This high level of integration also decreases the total parts count of the system, thereby increasing the overall reliability of the system.

4.6.1 Video Sampling

The sampling of video signals is typically specified in the *x:y:z* nomenclature, which refers to the ratio of the sampling rates used in the sample-and-hold circuits in the A/D devices of the system [19]. Because satisfying the Nyquist limit requires that the sample rate must be at least twice the highest frequency of interest, this nomenclature also relates to the bandwidths of the signals. The most common form of this nomenclature is 4:2:2, which specifies the sampling rates for luminance and two chrominance difference signals. The term *4* dates from the time when multiples of the color subcarrier frequency were being considered for the sampling rate. For NTSC, this frequency would be 4× the 3.58 MHz color subcarrier, or approximately 14.3 MHz.

This approach resulted in different sampling rates for the different television systems used worldwide. Fortunately, international organizations subsequently agreed to a single sampling rate that was related to both the 50 and 60 Hz frame rates (and their related line rates.) The *4* term now refers to 13.5 MHz, which is not that different from the 14.3 MHz sampling frequency still used in NTSC composite digital systems. Thus, the 4:2:2 standard nomenclature refers to the luminance signal (Y) being sampled at 13.5 MHz, and the two color-difference signals (B − Y and R − Y) each being sampled at 6.75 MHz. The descriptions "ITU-R Rec. 601" and "4:2:2" often are used interchangeably. This practice, however, is not technically correct. ITU-R Rec. 601 refers to a video production standard; 4:2:2 simply describes a method of sampling video signals.

A variation of 4:2:2 sampling is the 4:2:2:4 scheme, which is identical to 4:2:2 except that a *key signal* (alpha channel) is included as the fourth component, sampled at 13.5 MHz.

Modern broadcast cameras originate video signals as three equal bandwidth red, green, and blue primary colors. If these color primaries are all converted to their digital equivalents, they should all use the same sampling rate. It is well known, however, that full bandwidth chrominance signals are not required for most video work. Furthermore, bandwidth economies are afforded by matrixinng the R, G, and B signals into luminance and chrominance elements. These R, G, B primaries can be mixed (or matrixed) in the analog domain into luminance and two color-difference components, as described in Chapter 3. This is essentially a lossless process that can maintain the full bandwidth of the original primaries, if desired. Unfortunately, the matrix equations are not universal, and different coefficients are used in different television systems.

Equal-sampling systems are available, such as 4:4:4. Indeed, there are a number of production devices and computer systems that work with video signals that have been sampled in this manner. The 4:4:4 sampling ratios can apply to the luminance and color difference signals, or to the R, G, and B signals themselves. RGB 4:4:4 is commonly used in computer-based equipment. An enhancement to 4:4:4 is the 4:4:4:4 scheme, which is identical to 4:4:4, but adds a key signal that is sampled at 13.5 MHz.

Acknowledging the lack of color acuity of the human visual system, the ITU-R in Rec. 601 recommends that the color-difference signals be filtered and sampled at one-half the sampling rate of the luminance signal. This results in the 4:2:2 sampling scheme that forms the basis for most digital video systems. It is generally accepted that ITU-R Rec. 601 video represents the quality reference point for standard-definition production systems. Even so, systems with greater resolution exist, ranging from 4:4:4 systems described previously, up to 8:8:8 systems for film transfer work. There is also a 4:2:2-sampling system for widescreen standard definition that uses an 18 MHz luminance sampling rate (with the color-difference sampling rate similarly enhanced.) For applications focusing on acquiring information that will be subjected to further post-production, it is important to acquire and record as much information (data) as possible. Information that is lost in the sampling process is impossible to reconstruct accurately at a later date.

While it is always best to capture as much color information as possible, there are applications where the additional information serves no direct purpose. Put another way, if the application will not benefit from sampling additional chrominance data, then there is little point in doing so. This reasoning is the basis for the development and implementation of video recording and transmission systems based on 4:1:1 and 4:2:0 sampling.

The 4:2:0 sampling scheme digitizes the luminance and color difference components for a limited range of applications, such as acquisition and play-to-air. As with 4:2:2, the first digit (4) represents the luminance sampling at 13.5 MHz, while the R – Y and B–Y components are sampled at 6.75 MHz, effectively between every other line only. In other words, one line is sampled at 4:0:0—luminance only—and the next is sampled at 4:2:2. This technique reduces the required data capacity by 25 percent compared to 4:2:2 sampling. This approach is well-suited to the capture and transmission of television signals. For post-production applications, where many generations and/or layers are required to yield the finished product, 4:2:2 sampling is preferred.

The sampling ratios of 4:1:1 are also used in professional video equipment, with luminance sampled at 13.5 MHz and chrominance sampled at 3.75 MHz. This scheme offers the same types of benefits previously outlined for 4:2:0.

4.6.2 Serial Digital Interface

Parallel connection of digital video equipment is practical only for relatively small installations. There is, then, a clear need to transmit data over a single coaxial or fiber line [20]. To reliably move large amounts of data from one location to another, it is necessary to modify the serial signal prior to transmission to ensure that there are sufficient edges (data transitions) for reliable clock recovery, to minimize the low frequency content of the transmitted signal, and to spread the transmitted energy spectrum so that radio frequency emission problems are minimized.

In the early 1980s, a serial interface for Rec. 601 signals was recommended by the EBU. This interface used 8/9 block coding and resulted in a bit rate of 243 Mbits/s. The interface did not support ten-bit precision signals, and there were some difficulties in producing reliable, cost effective integrated circuits for the protocol. The block coding-based interface was abandoned and replaced by an interface with a channel coding scheme that utilized scrambling and conversion to NRZI (*non return to zero inverted*). This serial interface was standardized as SMPTE 259M and EBU Tech. 3267, and is defined for both component and composite conventional video signals, including embedded digital audio.

Conceptually, the serial digital interface is much like a carrier system for studio applications. Baseband audio and video signals are digitized and combined on the serial digital "carrier." (SDI is not strictly a carrier system in that it is a baseband digital signal, not a signal modulated on a carrier wave.) The bit rate (carrier frequency) is determined by the clock rate of the digital data: 143 Mbits/s for NTSC, 177 Mbits/s for PAL, and 270 Mbits/s for Rec. 601 component digital. The widescreen (16×9) component system defined in SMPTE 267 will produce a bit rate of 360 Mbits/s. This serial interface may be used with normal video coaxial cable or fiber optic cable, with the appropriate interface adapters.

Following serialization of the video information, the data stream is scrambled by a mathematical algorithm and then encoded. At the receiver, an inverse algorithm is used in the deserializer to recover the data. In the serial digital transmission system, the clock is contained in the data, as opposed to the parallel system where there is a separate clock line. By scrambling the data, an abundance of transitions is assured, which is required for reliable clock recovery.

Figure 4.18 shows the SDI bitstream for 270 Mbits/s and 360 Mbits/s operation. The EAV and SAV elements of the bitstream are reserved word sequences that indicate the start and end of a video line, respectively. For the 270 Mbits/s case, each line contains 1440 10 bit 4:2:2 video samples. The horizontal interval (HANC, *horizontal ancillary data*) contains ancillary data, error detection and control, embedded audio, and other information. Vertical ancillary data (VANC) also may be used.

Embedded Audio

One of the important features of the serial digital video interface is the facility to embed (multiplex) several channels of AES/EBU digital audio in the video bitstream. SDI with embedded audio is particularly helpful in large systems where a strict link between the video and its associated audio is an important feature. In smaller systems, such as a post-production suite, it is generally preferable to maintain a separate audio path.

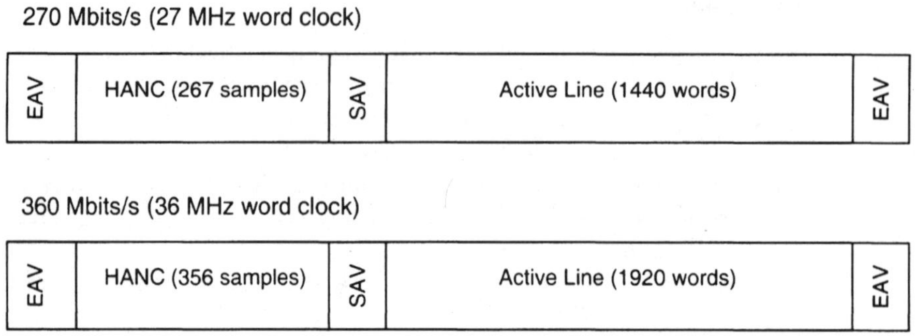

270 Mbits/s (27 MHz word clock)

EAV	HANC (267 samples)	SAV	Active Line (1440 words)	EAV

360 Mbits/s (36 MHz word clock)

EAV	HANC (356 samples)	SAV	Active Line (1920 words)	EAV

Figure 4.18 The basic SDI bitstream.

Table 4.3 SMPTE 272M Mode Definitions

A (default)	Synchronous 48 kHz, 20 bit audio, 48 sample buffer
B	Synchronous 48 kHz, composite video only, 64 sample buffer to receive 20 bits from 24 bit audio data
C	Synchronous 48 kHz, 24-bit audio and extended data packets
D	Asynchronous (48 kHz implied, other rates if so indicated)
E	44.1 kHz audio
F	32 kHz audio
G	32-48 kHz variable sampling rate audio
H	Audio frame sequence (inherent in 29.97 frame/s video systems, except 48 kHz synchronous audio—default A mode)
I	Time delay tracking
J	Non-coincident channel status Z bits in a pair.

SMPTE 272M defines the mapping of digital audio data, auxiliary data, and associated control information into the ancillary data space of the serial digital video format. Several modes of operation are defined and letter suffixes are used to help identify interoperability between equipment with differing capabilities. These descriptions are given in Table 4.3. (Note that modes *B* through *J* shown in the table require a special audio control packet.)

Some examples will help explain how Table 4.3 is used. A transmitter that can only accept 20 bit 48 kHz synchronous audio is said to conform to SMPTE 272M-A. A transmitter that supports 20 bit and 24 bit 48 kHz synchronous audio conforms to SMPTE 272M-ABC. A receiver that only uses the 20 bit data but can accept the level B sample distribution would conform to SMPTE 272M-AB because it can handle either sample distribution.

Error Detection and Handling

SMPTE Recommended Practice RP 165-1994 describes the generation of error detection checkwords and related status flags to be used optionally in conjunction with the serial digital interface [21]. Although the RP on *error detection and handling* (EDH) recommends that the specified error-checking method be used in all serial transmitters and receivers, it is not required.

Two checkwords are defined: one based on a field of active picture samples and the other on a full field of samples. This two-word approach provides continuing error detection for the active picture when the digital signal has passed through processing equipment that has changed data outside the active picture area without re-calculating the full-field checkword.

Three sets of *flags* are provided to feed-forward information regarding detected errors to help facilitate identification of faulty equipment, and the type of fault. One set of flags is associated with each of the two field-related checkwords. A third set of flags is used to provide similar information based on evaluating all of the ancillary data checksums within a field. The checkwords and flags are combined in an *error detection data packet* that is included as ancillary data in the serial digital signal. At the receiver, a recalculation of check-words can be compared to the error detection data packet information to determine if a transmission error has occurred.

All error flags indicate only the status of the previous field; that is, each flag is set or cleared on a field-by-field basis. A logical *1* is the set state and a logical *0* is the unset state. The flags are defined as follows:

- EDH, *error detected here*: Signifies that a serial transmission data error was detected. In the case of ancillary data, this means that one or more ANC data blocks did not match its check-sum.

- EDA, *error detected already*: Signifies that a serial transmission data error has been detected somewhere upstream. If device *B* receives a signal from device *A* and device *A* has set the EDH flag, when device *B* retransmits the data to device *C*, the EDA flag will be set and the EDH flag will be unset (if there is no further error in the data).

- IDH, *internal error detected here*: Signifies that a hardware error unrelated to serial transmission has been detected within a device. This feature is provided specifically for devices that have internal data error-checking facilities.

- IDA, *internal error detected already*: Signifies that an IDH flag was received and there was a hardware device failure somewhere upstream.

- UES, *unknown error status*: Signifies that a serial signal was received from equipment not supporting the RP 165 error-detection practice.

Individual error status flags (or all error status flags) may not be supported by all equipment.

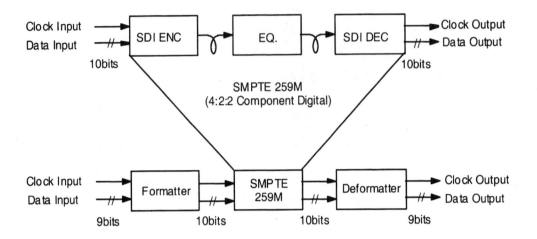

Figure 4.19 SMPTE 305M system block diagram. (*After* [22].)

SMPTE 305M

The SMPTE 305M standard specifies a data stream used to transport packetized data within a studio/production center environment [22]. The data packets and synchronizing signals are compatible with SMPTE 259M, as illustrated in Figure 4.19. Other parameters of the protocol also are compatible with the 4:2:2 component SDI format, discussed previously.

The data stream is intended to transport any packetized data signal over the active lines that have a maximum data rate up to (approximately) 200 Mbits/s for a 270 Mbits/s system or (approximately) 270 Mbits/s for 360 Mbits/s system. The maximum data rate can be increased through use of a defined extended data space.

The SMPTE 305M standard describes the assembly of a stream of 10-bit words. The resulting word stream is serialized, scrambled, coded, and interfaced according to SMPTE 259M and ITU-R BT.656. The timing reference signals (EAV and SAV) occur on every line. The signal levels, specifications, and preferred connector type are as described in SMPTE 259M.

The SMPTE 305M serial digital transport interface is described in detail in Chapter 6.

Optical Interconnect

SMPTE 297M-1997 defines an optical fiber system for transmitting bit-serial digital signals that conform to the SMPTE 259M serial digital format (143 through 360 Mbits/s). The standard's optical interface specifications and end-to-end system performance parameters are otherwise compatible with SMPTE 292M, which covers transmission rates of 1.3 through 1.5 Gbits/s.

During the Summer of 2000, a major revision was undertaken of SMPTE 297M. The document—proposed at this writing—updated applicable references and included new information on important elements covered by the standard. Among the new information were four appendices examining the following subjects:

- Maximum transmission distance range

- Minimum transmission distances

- Computing damage thresholds

- A comprehensive glossary of fiber optic terms

A discussion of SDI as applied to high-definition applications (SMPTE-292M) is given in Chapter 6.

4.6.3 IEEE 1394

IEEE 1394 is a serial bus interconnection and networking technology, and the set of protocols defining the communications methods used on the network [24]. IEEE 1394 development began in the 1980s at Apple Computer and was trademarked under the name *FireWire*. Sony has since trademarked its implementation as *iLINK*. The formal name for the standard is IEEE 1394-1995.

IEEE 1394 is widely supported by hardware, software, and semiconductor companies where it is implemented on computers, workstations, videotape recorders, cameras, professional audio, digital televisions and set top boxes, and consumer A/V equipment. 1394 enables transfers of video and data between devices without image degradation. Protocols for the transport of video, digital audio, and IP data are in place or under development at this writing. The EBU/SMPTE Task Force for Harmonized Standards for the Exchange of Program Material as Bit Streams has recognized IEEE 1394 as a recommended transport system for content.

Like other networking technologies, 1394 connects devices and transports information or data among the devices on a network. When IEEE 1394 was developed, the need to transport real-time media streams (video and audio, for example) was recognized. These signal types require consistent delivery of the data with a known latency, i.e., *isochronous* (constant or same time) delivery.

Operating Modes

When transmitting time sensitive material, such as real-time motion video, a slight impairment (a defective pixel, for example) is not as important as the delivery of the stream of pictures making up the program [24]. The IEEE 1394 designers recognized these requirements and defined 1394 from the outset with capabilities for isochronous delivery of data. This isochronous mode is a major differentiator of IEEE 1394 when compared with other networking technologies, such as Ethernet, Fibre Channel, or ATM.

IEEE 1394 divides the network bandwidth into 64 discreet channels per bus, including a special *broadcast channel* meant to transmit data to all users. This allows a single IEEE 1394 buss to carry up to 64 different independent isochronous streams. Each stream can carry a video stream, audio stream, or other types of data streams simultaneously. A net-

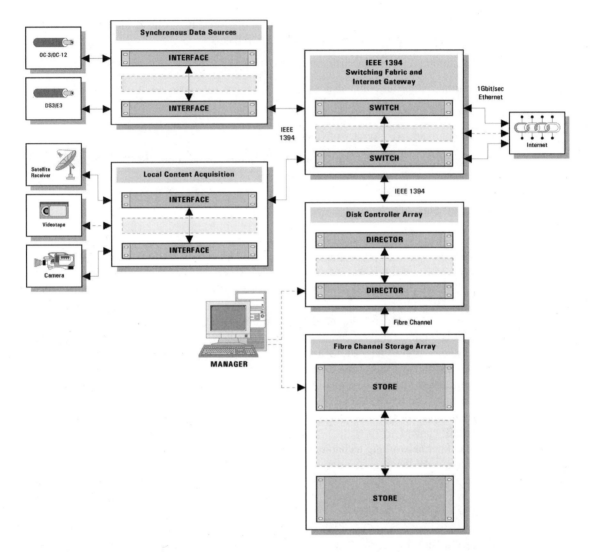

Figure 4.20 The application of IEEE 1394 in a large scale internet broadcast environment. (*From* [25]. *Used with permission.*)

work resource manager allocates bandwidth and transmission channels on the network to guarantcc a fixcd bandwidth channcl for cach strcam.

IEEE 1394 also supports asynchronous transmissions. The asynchronous channels are used to transmit data that cannot suffer loss of information. This dual transmission scheme, supporting both isochronous and asynchronous modes, makes IEEE 1394 useful for a range of applications.

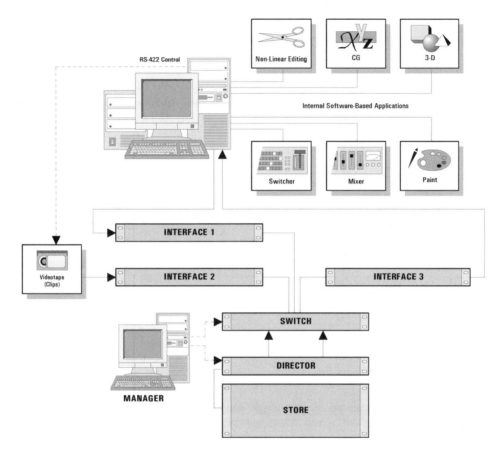

Figure 4.21 The application of IEEE 1394 in a nonlinear postproduction environment. (*From* [26]. *Used with permission.*)

Data Rates and Cable Lengths

IEEE 1394 defines a set of parameters to assure reliable system operation. The initial standards called for data rates of 100 and 200 (nominal) Mbits/s. IEEE 1394a increased the data rate to 400 Mbits/s. IEEE 1394b increases data rates to 800 Mbits/s, 1.6 Gbits/s, and 3.2 Gbits/s. At 400 Mbits/s, the isochronous payload size is 4096 bytes—considerably larger than other technologies, such as ATM. As the clock speed increases, the size of the data packets increase linearly.

Typical applications of IEEE 1394 are shown in Figures 4.20 and 4.21.

Isochronous and Asynchronous Transmissions

At first glance, the terms *isochronous* and *synchronous*, as applied to network transmissions, describe essentially the same process. Upon further investigation, however, small but significant differences emerge [26, 27].

An *isochronous* channel has a constant time interval (or integer multiples thereof) between similar "significant instants." This basic definition means that each isochronous packet in a stream begins a constant time interval after the previous packet in that stream. In IEEE 1394, isochronous packets begin 125 μs apart (at least on average). This period is controlled by a clock packet, transmitted by the IEEE 1394 bus master, which signals devices on the network that it is time to send the next isochronous packet. Another example of an isochronous channel is the 270 Mbits/s serial digital video interface, where a line of video begins every 64 μs (625/50) or 63.56 μs (525/60).

The opposite of isochronous is *anisochronous*. An anisochronous channel does not maintain constant time intervals between significant instants. A good example of this is Ethernet, a technology that does not synchronize data transmission from nodes on the network, but instead relies on signaling when conflicts occur.

The term *synchronous* refers to a timing relationship between two different channels. If two "significant instants" (e.g., packet starts) on two different channels occur at the same time every time, the channels are synchronous. A good example of synchronous channels would be a 4:2:2 4:4 SDI signal carried on two coaxial cables, where lines of video are completely time aligned on the two cables.

The opposite of synchronous is *asynchronous*. If "significant instants" on two channels do not consistently line up, then the channels are asynchronous.

4.6.4 Fibre Channel

Fibre Channel is a 1 Gbits/s data transfer interface technology that maps several common transport protocols, including IP and SCSI, allowing it to merge high-speed I/O and networking functionality in a single connectivity technology [24]. Fibre Channel is an open standard, as defined by ANSI and OSI standards, and operates over copper and fiber optic cabling at distances of up to 10 km. It is unique in its support of multiple inter-operable topologies—including point-to-point, arbitrated-loop, and switching—and it offers several qualities of service for network optimization.

Fibre Channel first appeared in enterprise networking applications and point-to-point RAID and mass storage subsystems. It has expanded to include video graphics networks, video editing, and visual imaging systems.

Fibre Channel Arbitrated Loop (FC-AL) was developed with peripheral connectivity in mind. It natively maps SCSI (as SCSI FCP), making it a powerful technology for high-speed I/O connectivity. Native FC-AL disk drives allow storage applications to take full advantage of Fibre Channel's gigabaud bandwidth, passing SCSI data directly onto the channel with access to multiple servers or nodes. FC-AL supports 127-node addressability and 10 km cabling ranges between nodes. (See Table 4.4.)

The current specification for FC-AL allows for 1, 2 and 4 Gbits/s speeds. At this writing, practical applications of the technology are at the 2 Gbits/s rate.

Fibre Channel is designed with many high-availability features, including dual ported disk drives, dual loop connectivity, and loop resiliency circuits. Full redundancy in Fibre Channel systems is achieved by cabling two fully independent, redundant loops. This cabling scheme provides two independent paths for data with fully redundant hardware.

Table 4.4 Fibre Channel General Performance Specifications (*After* [24].)

Media	Speed		Distance
Electrical Characteristics			
Coax/twinax	ECL	1.0625 Gigabits/s	24 Meters
	ECL	266 Megabits/s	47 Meters
Optical Characteristics			
9 micrometer single mode fiber	Longwave laser	1.0625 Gigabits/s	10 Kilometers
50 micrometer multi-mode fiber	Shortwave laser	1.0625 Gigabits/s	300 Meters
	Shortwave Laser	266 Megabits/s	2 Kilometer
62.5 micrometer multi-mode fiber	Longwave LED	266 Megabits/s	1 Kilometer
	Longwave LED	132 Megabits/s	500 Meters

1   *Note*: In FC-AL configurations, the distance numbers represents the distance between nodes, not the total distance around the loop.
2   *Note*: In fabric configurations, the distance numbers represent the distance from the fabric to a node, not the distance between nodes.

Table 4.5 Basic Specifications of Ethernet Performance (*After* [29].)

Parameter	10 Mbits/s		100 Mbits/s		1000 Mbits/s	
Frame size	Minimum	Maximum	Minimum	Maximum	Minimum	Maximum
Frames/s	14.8 k	812	148 k	8.1 k	1.48 M	81 k
Data rate	5.5 Mbits/s	9.8 Mbits/s	55 Mbits/s	98 Mbits/s	550 Mbits/s	980 Mbits/s
Frame interval	67 μs	1.2 ms	6.7 μs	120 μs	0.7 μs	12 μs

Most disk drives and disk arrays targeted for high-availability environments have dual ports specifically for this purpose.

4.6.5   Gigabit Ethernet

The Ethernet protocol is the dominant networking technology for data processing applications. Gigabit Ethernet adds a 1 Gbit/s variant to the existing 10 and 100 Mbits/s Ethernet family while retaining the Ethernet standard frame format, collision detection system, and flow control. For reasons of compatibility with traditional data processing systems and the wide bandwidth necessary for digital video, Gigabit Ethernet has emerged as a viable method of transporting video within a facility [28].

The basic Ethernet performance specifications are given in Table 4.5 for the minimum 46 data byte and the maximum 1500 data byte frames [29]. Gigabit Ethernet was standardized for fiber optic interconnection in July 1998, and a copper specification using 4 pairs of category 5 unshielded twisted pair (UTP) was released about a year later.

Although Gigabit Ethernet networks can be built using shared media in a *broadcast architecture*, early implementations are typically *full duplex switched* (a point-to-point architecture). In a switched network, the full capacity of the network medium is available at each device. For such a configuration, the two key switch specifications are the *aggregate capacity*, which determines the total amount of throughput for the switch, and the number of frames/s that can be handled. Ethernet has a variable frame rate and so both parameters are important. Under best-case conditions with large frames, the aggregate capacity defines performance; under-worst case conditions with small frames, the frame throughput is the limiting factor. In practice, these switches are specified to achieve *wire speed* on all their ports simultaneously. For example, an eight port switch will have 8 Gbits/s aggregate capacity. Many Gigabit Ethernet switches offer backwards compatibility with existing Ethernet installations, allowing Fast Ethernet or even 10 BASE networks to be connected.

Network Bottlenecks

In many cases the bandwidth available from a switched Gigabit network removes the network itself as a bottleneck [28]. In fact, the bottleneck moves to the devices themselves that must handle the Gigabit Ethernet data. It can be seen from Table 4.5 that even under best-case conditions, frames arrive only 12 μs apart. In this interval, the device must determine if the packet is addressed to it, verify the checksum, and move the data contents of the frame into memory. Modern network interfaces use dedicated hardware to take the load off the processor, maximizing the time it has for user applications. The network interface can verify addresses and checksums, only interrupting the processor with valid frames. It can write data to discontiguous areas of memory, removing the need for the host to move different parts of messages around in memory. The network interface can also dynamically manage interrupts so that when traffic is high, several frames will be handled with only a single interrupt, thus rninimizing processor environment switching. When traffic is low, the processor will be interrupted for a single frame to minimize latency.

These measures allow high, real data rates over Gigabit Ethernet of up to 960 Mbits/s. For systems that are CPU bound, increasing the Ethernet frame size can raise throughput by reducing the processor overhead.

A Network Solution

Ethernet on its own does not provide complete network solutions; it simply provides the lower two layers of the Open Systems Interconnect (OSI) model, specifically [28]:

- Layer 7, Application

- Layer 6, Presentation

- Layer 5, Session

- Layer 4, Transport

- Layer 3, Network

- **Layer 2, Data Link**: *logical link*—framing and flow control; *media access*—controls access to the medium

- **Layer 1, Physical**: the cable and/or fiber

The most widely used protocols for the Transport and Network layers are the *Transmission Control Protocol* (TCP) and *Internet Protocol* (IP), more commonly referred to as TCP/IP. These layers provide for reliable transmission of messages between a given source and destination over the network. Using TCP/IP on Ethernet is sufficiently common that most network hardware, for example a *network interface card* (NIC) or switch, typically has built-in support for key aspects of layers three and four.

How messages are interfaced to user applications is the function of the higher OSI layers. Here again, there are many choices depending upon the application. The *Network File System* (NFS) is a collection of protocols (developed by Sun Microsystems) with multiplatform support that presents devices on the network as disk drives. The advantage of this approach is that applications do not need to be specially coded to take advantage of the network. After a network device is *mounted*, network access looks to the application exactly the same as accessing a local drive. Familiar techniques such as "drag and drop" can continue to be used, only now working with media data over a network.

Quality of Service

A guaranteed *quality of service* (QoS) transfers allocate network bandwidth in advance and maintains it for the duration of a session [28]. After set up, such transfers are *deterministic* in that the time taken to transfer a given amount of data can be predicted. They can also be wasteful because the bandwidth is reserved, even if it is not being used, preventing it from being used by others. In some cases, a QoS transfer to a device can lock out other nodes communicating to the same device.

For established television practitioners, QoS brings familiarity to the world of data networking, however, its implications for complete system design may not be fully appreciated. One key issue is that of the capabilities of devices connected to the network, many of which have video and network interfaces. The allocation of device bandwidth among these different interfaces is an important consideration. Video transfers must be real-time and so bandwidth must be able to service them when needed. However, QoS transfers that are also deterministic need guaranteed device attention. Resolving this conflict is not a trivial matter, and different operational scenarios require different—sometimes creative—solutions.

4.7 References

1. Benson, K. B., and D. G. Fink: "Digital Operations in Video Systems," *HDTV: Advanced Television for the 1990s*, McGraw-Hill, New York, pp. 4.1–4.8, 1990.

2. Nyquist, H.: "Certain Factors Affecting Telegraph Speed," *Bell System Tech. J.*, vol. 3, pp. 324–346, March 1924.

3. Garrod, Susan A. R.: "D/A and A/D Converters," *The Electronics Handbook*, Jerry C. Whitaker (ed.), CRC Press, Boca Raton, Fla., pp. 723–730, 1996.

4. Garrod, S., and R. Borns: *Digital Logic: Analysis, Application, and Design*, Saunders College Publishing, Philadelphia, p. 919, 1991.

5. Garrod, S., and R. Borns: *Digital Logic: Analysis, Application, and Design*, Saunders College Publishing, Philadelphia, p. 928, 1991.

6. Dubois, E., and W. F. Schreiber: "Improvements to NTSC by Multidimensional Filtering," *SMPTE J.*, SMPTE, White Plains, N.Y., vol. 97, pp. 446–463, July 1988.

7. Isnardi, M. A.: "Exploring and Exploiting Subchannels in the NTSC Spectrum," *SMPTE J.*, SMPTE, White Plains, N.Y., vol. 97, pp. 526–532, July 1988.

8. Isnardi, M. A.: "Multidimensional Interpretation of NTSC Encoding and Decoding," *IEEE Transactions on Consumer Electronics*, vol. 34, pp. 179–193, February 1988.

9. Ziemer, Rodger E.: "Digital Modulation," *The Electronics Handbook*, Jerry C. Whitaker (ed.), CRC Press, Boca Raton, Fla., pp. 1213–1236, 1996.

10. Ziemer, R., and W. Tranter: *Principles of Communications: Systems, Modulation, and Noise*, 4th ed., Wiley, New York, 1995.

11. Peterson, R., R. Ziemer, and D. Borth: *Introduction to Spread Spectrum Communications*, Prentice-Hall, Englewood Cliffs, N. J., 1995.

12. Sklar, B.: *Digital Communications: Fundamentals and Applications*, Prentice-Hall, Englewood Cliffs, N. J., 1988.

13. Alkin, Oktay: "Digital Coding Schemes," *The Electronics Handbook*, Jerry C. Whitaker (ed.), CRC Press, Boca Raton, Fla., pp. 1252–1258, 1996.

14. Ungerboeck, G.: "Trellis-Coded Modulation with Redundant Signal Sets," parts I and II, *IEEE Comm. Mag.*, vol. 25 (Feb.), pp. 5-11 and 12-21, 1987.

15. Libin, Louis: "The 8-VSB Modulation System," *Broadcast Engineering*, Intertec Publishing, Overland Park, Kan., p. 22, December 1995.

16. Chambers, J. A., S. Tantaratana, and B. W. Bomar: "Digital Filters," *The Electronics Handbook*, Jerry C. Whitaker (ed.), CRC Press, Boca Raton, Fla., pp. 749–772, 1996.

17. Lee, E. A., and D. G. Messerschmitt: *Digital Communications*, 2nd ed., Kluwer, Norell, Mass., 1994.

18. Parks, T. W., and J. H. McClellan: "A Program for the Design of Linear Phase Infinite Impulse Response Filters," *IEEE Trans. Audio Electroacoustics*, AU-20(3), pp. 195–199, 1972.

19. Hunold, Kenneth: "4:2:2 or 4:1:1—What are the Differences?," *Broadcast Engineering*, Intertec Publishing, Overland Park, Kan., pp. 62–74, October 1997.

20. Fibush, David, *A Guide to Digital Television Systems and Measurement*, Tektronix, Beaverton, OR, 1994.

21. SMPTE RP 165-1994: *SMPTE Recommended Practice—Error Detection Checkwords and Status Flags for Use in Bit-Serial Digital Interfaces for Television*, SMPTE, White Plains, N.Y., 1994.

22. SMPTE Standard: SMPTE 305M-1998, *Serial Data Transport Interface*, SMPTE, White Plains, N.Y., 1998.

23. SMPTE Standard: SMPTE 297M-1997, *Serial Digital Fiber Transmission System for ANSI/SMPTE 259M Signals*, SMPTE White Plains, N.Y., 1997.

24. "Technology Brief—Networking and Storage Strategies," Omneon Video Networks, Campbell, Calif., 1999.

25. "Networking and Internet Broadcasting," Omneon Video Networks, Campbell, Calif, 1999.

26. "Networking and Production," Omneon Video Networks, Campbell, Calif., 1999.

27. Craig, Donald: "Network Architectures: What does Isochronous Mean?," *IBC Daily News*, IBC, Amsterdam, September 1999.

28. Owen, Peter: "Gigabit Ethernet for Broadcast and Beyond," *Proceedings of DTV99*, Intertec Publishing, Overland Park, Kan., November 1999.

29. Gallo and Hancock: *Networking Explained*, Digital Press, pp. 191–235, 1999.

4.8 Bibliography

Pank, Bob (ed.): *The Digital Fact Book*, 9th ed., Quantel Ltd, Newbury, England, 1998.

Video and Audio Compression

5.1 Introduction

Virtually all applications of video and visual communication deal with an enormous amount of data. Because of this, compression is an integral part of most modern digital video applications. In fact, compression is essential to the ATSC DTV system [1].

A number of existing and proposed video-compression systems employ a combination of processing techniques. Any scheme that becomes widely adopted can enjoy economies of scale and reduced market confusion. Timing, however, is critical to market acceptance of any standard. If a standard is selected well ahead of market demand, more cost-effective or higher-performance approaches may become available before the market takes off. On the other hand, a standard may be merely academic if it is established after alternative schemes already have become well entrenched in the marketplace.

These forces are shaping the video technology of the future. Any number of scenarios have been postulated as to the hardware and software that will drive the digital video facility of the future. One thing is certain, however: It will revolve around compressed video and audio signals.

5.2 Transform Coding

In technical literature, countless versions of different coding techniques can be found [2]. Despite the large number of techniques available, one that comes up regularly (in a variety of flavors) during discussions about transmission standards is *transform coding* (TC).

Transform coding is a universal bit-rate-reduction method that is well suited for both large and small bit rates. Furthermore, because of several possibilities that TC offers for exploiting the visual inadequacies of the human eye, the subjective impression given by the resulting picture is frequently better than with other methods. If the intended bit rate turns out to be insufficient, the effect is seen as a lack of sharpness, which is less disturbing (subjectively) than coding errors such as frayed edges or noise with a structure. Only at very low bit rates does TC produce a particularly noticeable artifact: the *blocking effect*.

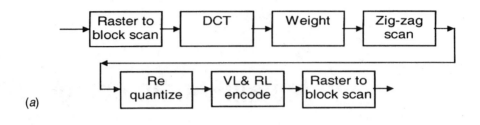

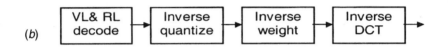

(a)

(b)

Figure 5.1 Block diagram of a sequential DCT codec: (a) encoder, (b) decoder. (*From* [2]. *Used with permission.*)

Because all pictures do not have the same statistical characteristics, the optimum transform is not constant, but depends on the momentary picture content that has to be coded. It is possible, for example, to recalculate the optimum transform matrix for every new frame to be transmitted, as is performed in the *Karhunen-Loeve transform* (KLT). Although the KLT is efficient in terms of ultimate performance, it is not typically used in practice because investigating each new picture to find the best transform matrix is usually too demanding. Furthermore, the matrix must be indicated to the receiver for each frame, because it must be used in decoding of the relevant inverse transform. A practical compromise is the *discrete cosine transform* (DCT). This transform matrix is constant and is suitable for a variety of images; it is sometimes referred to as "quick KLT."

The DCT is a near relative of the *discrete Fourier transform* (DFT), which is widely used in signal analysis. Similar to DFT techniques, DCT offers a reliable algorithm for quick execution of matrix multiplication.

The main advantage of DCT is that it *decorrelates* the pixels efficiently; put another way, it efficiently converts statistically dependent pixel values into independent coefficients. In so doing, DCT packs the signal energy of the image block onto a small number of coefficients. Another significant advantage of DCT is that it makes available a number of fast implementations. A block diagram of a DCT-based coder is shown in Figure 5.1.

In addition to DCT, other transforms are practical for data compression, such as the *Slant transform* and the *Hadamard transform* [3].

5.2.1 Planar Transform

The similarities of neighboring pixels in a video image are not only line- or column-oriented, but also area-oriented [2]. To make use of these *neighborhood relationships*, it is

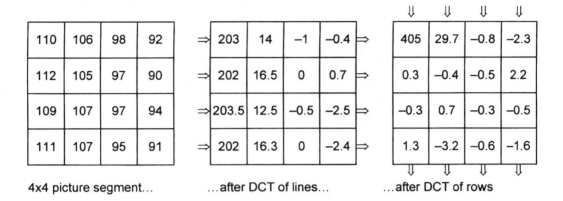

Figure 5.2 A simplified search of a best-matched block. (*From* [2]. *Used with permission.*)

desirable to transform not only in lines and columns, but also in areas. This can be achieved by a *planar transform*. In practice, *separable transforms* are used almost exclusively. A separable planar transform is nothing more than the repeated application of a simple transform. It is almost always applied to square picture segments of size $N \times N$, and it progresses in two steps, as illustrated in Figure 5.2. First, all lines of the picture segments are transformed in succession, then all rows of the segments calculated in the first step are transformed.

In textbooks, the planar transform frequently is called a *2D transform*. The transform is, in principle, possible for any segment forms—not just square ones [4]. Consequently, for a segment of the size $N \times N$, $2N$ transforms are used. The coefficients now are no longer arranged as vectors, but as a matrix. The coefficients of the i lines and the j columns are called c_{ij} ($i, j = 1 \dots N$). Each of these coefficients no longer represents a basic vector, but a *basic picture*. In this way, each $N \times N$ picture segment is composed of $N \times N$ different basic pictures, in which each coefficient gives the weighting of a particular basic picture. Figure 5.3 shows the basic pictures of the coefficients c_{11} and c_{23} for a planar 4×4 DCT. Because c_{11} represents the dc part, it is called the *dc coefficient*; the others are appropriately called the *ac coefficients*.

The planar transform of television pictures in the interlaced format is somewhat problematic. In moving regions of the picture, depending on the speed of motion, the similarities of vertically neighboring pixels of a frame are lost because changes have occurred between samplings of the two halves of the picture. Consequently, interlaced scanning may cause the performance of the system (or *output concentration*) to be greatly weakened, compared with progressive scanning. Well-tuned algorithms, therefore, try to detect stronger movements and switch to a transform in one picture half (i.e., field) for these picture regions [5]. However, the coding in one-half of the picture is less efficient because the correlation of vertically neighboring pixels is weaker than in the full picture of a static scene. Simply stated, if the picture sequences are interlaced, the picture quality may be influenced by the motion content of the scene to be coded.

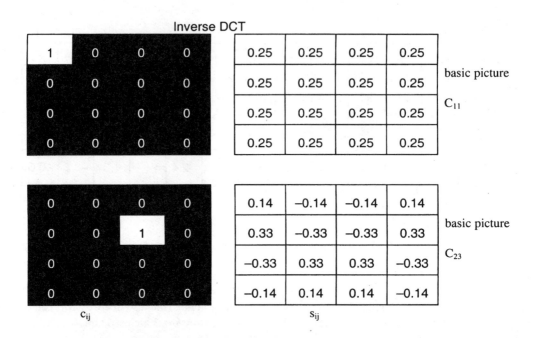

Figure 5.3 The mechanics of motion-compensated prediction. Shown are the pictures for a planar 4 × 4 DCT. Element C_{11} is located at row 1, column 1; element C_{23} is located at row 2, column 3. Note that picture C_{11} values are constant, referred to as dc coefficients. The changing values shown in picture C_{23} are known as ac coefficients. (*From* [2]. *Used with permission.*)

5.2.2 Interframe Transform Coding

With common algorithms, compression factors of approximately 8 can be achieved while maintaining good picture quality [2]. To achieve higher factors, the similarities between successive frames must be exploited. The nearest approach to this goal is the extension of the DCT in the time dimension. A drawback of such *cubic* transforms is the increase in calculation effort, but the greatest disadvantage is the higher memory requirement: for an 8 × 8 × 8 DCT, at least seven frame memories would be needed. Much simpler is the *hybrid DCT*, which also efficiently codes pictures with moving objects. This method comprises, almost exclusively, a motion-compensated *difference pulse-code-modulation* (DPCM) technique; instead of each picture being transferred individually, the motion-compensated difference of two successive frames is coded.

DPCM is, in essence, predictive coding of sample differences. DPCM can be applied for both *interframe coding*, which exploits the temporal redundancy of the input image, and *intraframe coding*, which exploits the spatial redundancy of the image. In the intraframe mode, the difference is calculated using the values of two neighboring pixels of the same frame. In the interframe mode, the difference is calculated using the value of the same pixel on two consecutive frames. In either mode of operation, the value of the target pixel is pre-

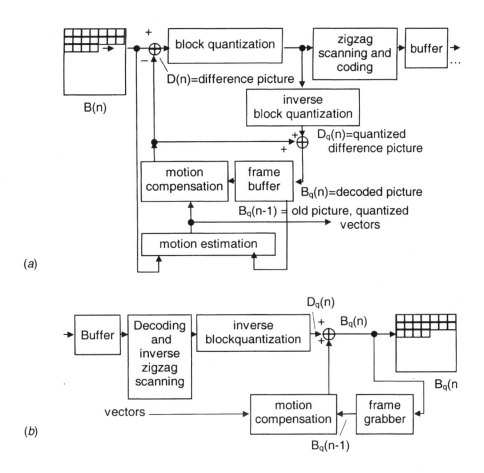

Figure 5.4 Overall block diagram of a DPCM system: (*a*) encoder, (*b*) decoder. (*From* [2]. *Used with permission.*)

dicted using the reconstructed values of the previously coded neighboring pixels. This value is then subtracted from the original value to form the differential image value. The differential image is then quantized and encoded. Figure 5.4 illustrates an end-to-end DPCM system.

5.3 The JPEG Standard

The JPEG (Joint Photographic Experts Group) standard is enjoying commercial use today in a wide variety of applications. Because JPEG is the product of a committee, it is not surprising that it includes more than one fixed encoding/decoding scheme. It can be thought of as a family of related compression techniques from which designers can choose, based upon suitability for the application under consideration. The four primary JPEG family members are [2]:

- Sequential DCT-based

- Progressive DCT-based

- Sequential lossless

- Hierarchical

As JPEG has been adapted to other environments, additional JPEG schemes have come into practice. JPEG is designed for still images and offers reduction ratios of 10:1 to 50:1. The algorithm is symmetrical—the time required for encoding and decoding is essentially the same. There is no need for motion compensation, and there are no provisions for audio in the basic standard.

The JPEG specification, like MPEG-1 and MPEG-2, often is described as a "tool kit" of compression techniques. Before looking at specifics, it will be useful to examine some of the basics.

5.3.1 Compression Techniques

As discussed briefly in previous sections, a compression system reduces the volume of data by exploiting spatial and temporal redundancies and by eliminating the data that cannot be displayed suitably by the associated display or imaging devices. The main objective of compression is to retain as little data as possible, just sufficient to reproduce the original images without causing unacceptable distortion of the images [1]. A compression system consists of the following components:

- *Digitization, sampling, and segmentation*: Steps that convert analog signals on a specified grid of picture elements into digital representations and then divide the video input—first into frames, then into blocks.

- *Redundancy reduction*: The decorrelation of data into fewer useful data bits using certain invertible transformation techniques.

- *Entropy reduction*: The representation of digital data using fewer bits by dropping less significant information. This component causes distortion; it is the main contributor in *lossy* compression.

- *Entropy coding*: The assignment of code words (bit strings) of shorter length to more likely image symbols and code words of longer length to less likely symbols. This minimizes the average number of bits needed to code an image.

Key terms important to the understanding of this topic include the following:

- *Motion compensation*: The coding of video segments with consideration to their displacements in successive frames.

- *Spatial correlation*: The correlation of elements within a still image or a video frame for the purpose of bit-rate reduction.

- *Spectral correlation*: The correlation of different color components of image elements for the purpose of bit-rate reduction.

- *Temporal correlation*: The correlation between successive frames of a video file for the purpose of bit-rate reduction.

- *Quantization compression*: The dropping of the less significant bits of image values to achieve higher compression.

- *Intraframe coding*: The encoding of a video frame by exploiting spatial redundancy within the frame.

- *Interframe coding*: The encoding of a frame by predicting its elements from elements of the previous frame.

The removal of spatial and temporal redundancies that exist in natural video imagery is essentially a lossless process. Given the correct techniques, an exact replica of the image can be reproduced at the viewing end of the system. Such lossless techniques are important for medical imaging applications and other demanding uses. These methods, however, may realize only low compression efficiency (on the order of approximately 2:1). For video, a much higher compression ratio is required. Exploiting the inherent limitations of the *human visual system* (HVS) can result in compression ratios of 50:1 or higher [6]. (See Section 3.2.) These limitations include the following:

- Limited luminance response and very limited color response

- Reduced sensitivity to noise in high frequencies, such as at the edges of objects

- Reduced sensitivity to noise in brighter areas of the image

The goal of compression, then, is to discard all information in the image that is not absolutely necessary from the standpoint of what the HVS is capable of resolving. Such a system can be described as *psychovisually lossless*.

5.3.2 DCT and JPEG

DCT is one of the building blocks of the JPEG standard. All JPEG DCT-based coders start by portioning the input image into nonoverlapping blocks of 8 × 8 picture elements. The 8-bit samples are then level-shifted so that the values range from −128 to +127. A fast Fourier transform then is applied to shift the elements into the frequency domain. Huffman coding is mandatory in a baseline system; other arithmetic techniques can be used for entropy coding in other JPEG modes. The JPEG specification is independent of color or gray scale. A color image typically is encoded and decoded in the *YUV* color space with four pixels of *Y* for each *U, V* pair.

In the *sequential DCT*-based mode, processing components are transmitted or stored as they are calculated in one single pass. Figure 5.5 provides a simplified block diagram of the coding system.

The *progressive DCT*-based mode can be convenient when it takes a perceptibly long time to send and decode the image. With progressive DCT-based coding, the picture first will appear blocky, and the details will subsequently appear. A viewer may linger on an

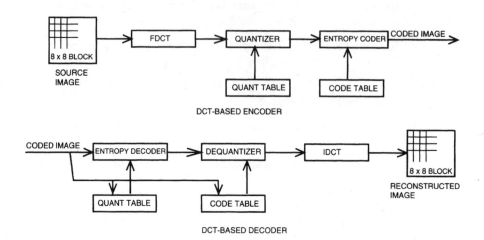

Figure 5.5 Block diagram of a DCT-based image-compression system. Note how the 8×8 source image is processed through a *forward-DCT* (FDCT) encoder and related systems to the *inverse-DCT* (IDCT) decoder and reconstructed into an 8×8 image. (*From* [1]. *Used with permission.*)

interesting picture and watch the details come into view or move onto something else, making this scheme well suited, for example, to the Internet.

In the *lossless* mode, the decoder reproduces an exact copy of the digitized input image. The compression ratio, naturally, varies with picture content. The varying compression ratio is not a problem for sending still photos, but presents significant challenges for sequential images that must be viewed in real time.

The efficiency of JPEG coding for still images led to the development of *motion* JPEG (M-JPEG) for video applications, primarily studio use. Motion JPEG uses intraframe compression, where each frame is treated as an individual signal; a series of frames is basically a stream of JPEG signals. The benefit of this construction is easy editing, making the technique a good choice for nonlinear editing applications. Also, any individual frame is self-supporting and can be accessed as a stand-alone image. The intraframe system is based, again, on DCT. Because a picture with high-frequency detail will generate more data than a picture with low detail, the data stream will vary. This is problematic for most real-time systems, which would prefer to see a constant data rate at the expense of varying levels of quality. The symmetry in complexity of decoders and encoders is another consideration in this regard.

The major disadvantage of motion JPEG is bandwidth and storage requirements. Because stand-alone frames are coded, there is no opportunity to code only the differences between frames (to remove redundancies).

M-JPEG, in its basic form, addresses only the video—not the audio—component. Many of the early problems experienced by users concerning portability of M-JPEG streams stemmed from the methods used to include audio in the data stream. Because the location of the audio may vary from one unit to the next, some decoder problems were experienced [7].

5.4   The MPEG Standard

The Moving Picture Experts Group (MPEG) was founded in 1988 with the objective of specifying an audio/visual decompression system, composed of three basic elements, which the sponsoring organization (the *International Standards Organization,* or ISO) calls "parts." They are as follows:

• *Part 1—Systems*: Describes the audio/video synchronization, multiplexing, and other system-related elements

• *Part 2—Video*: Contains the coded representation of video data and the decoding process

• *Part 3—Audio*: Contains the coded representation of audio data and the decoding process

The basic MPEG system, finalized in 1992, was designated MPEG-1. Shortly thereafter, work began on MPEG-2. The first three stages (systems, video, and audio) of the MPEG-2 standard were agreed to in November 1992. Table 5.1 lists the companies and organizations participating in the early MPEG work. Because of their combined efforts, the MPEG standards have achieved broad market acceptance.

As might be expected, the techniques of MPEG-1 and MPEG-2 are similar, and their syntax is rather extensible.

5.4.1   Basic Provisions

When trying to settle on a specification, it is always important to have a target application in mind [1]. The definition of MPEG-1 (also known as ISO/IEC 11172) was driven by the desire to encode audio and video onto a compact disc. A CD is defined to have a constant bit rate of 1.5 Mbits/s. With this constrained bandwidth, the target video specifications were:

• Horizontal resolution of 360 pixels

• Vertical resolution of 240 for NTSC, and 288 for PAL and SECAM

• Frame rate of 30 Hz for NTSC, 25 for PAL and SECAM, and 24 for film

A detailed block diagram of an MPEG-1 codec (coder-decoder) is shown in Figure 5.6.

MPEG uses the JPEG standard for intraframe coding by first dividing each frame of the image into 8×8 blocks, then compressing each block independently using DCT-based techniques. Interframe coding is based on *motion compensation* (MC) prediction that allows bidirectional temporal prediction. A block-matching algorithm is used to find the best-matched block, which may belong to either the past frame (*forward prediction*) or the future frame (*backward prediction*). The best-matched block may—in fact—be the average of two blocks, one from the previous and the other from the next frame of the target frame (*interpolation*). In any case, the placement of the best-matched block(s) is used to determine the

Table 5.1 Participants in Early MPEG Proceedings (*After* [2].)

Computer Manufacturers	IC Manufacturers
Apple	Brooktree
DEC	C-Cube
Hewlett-Packard	Cypress
IBM	Inmos
NEC	Intel
Olivetti	IIT
Sun	LSI Logic
	Motorola
Software Suppliers	National Semiconductor
Microsoft	Rockwell
Fluent Machines	SGS-Thomson
Prism	Texas Instruments
	Zoran
Audio/Visual Equipment Manufacturers	
Dolby	**Universities/Research**
JVC	Columbia University
Matsushita	Massachusetts Institute of Technology
Philips	DLR
Sony	University of Berlin
Thomson Consumer Electronics	Fraunhofer Gesellschaft
	University of Hannover

motion vector(s); blocks predicted on the basis of interpolation have two motion vectors. Frames that are bidirectionally predicted never are used themselves as reference frames.

5.4.2 Motion Compensation

At this point, it is appropriate to take a closer look at MC prediction [1]. For motion-compensated interframe coding, the target frame is divided into nonoverlapping fixed-size blocks, and each block is compared with blocks of the same size in some reference frame to find the best match. To limit the search, a small *neighborhood* is selected in the reference frame, and the search is performed by *stepwise translation* of the target block.

To reduce mathematical complexity, a simple block-matching criterion, such as the mean of the absolute difference of pixels, is used to find a best-matched block. The position of the best-matched block determines the displacement of the target block, and its location is denoted by a (motion) vector.

Block matching is computationally expensive; therefore, a number of variations on the basic theme have been developed. A simple method known as *OTS* (one-at-a-time search) is shown in Figure 5.7. First, the target block is moved along in one direction and the best match found, then it is moved along perpendicularly to find the best match in that direction. Figure 5.7 portrays the target frame in terms of the best-matched blocks in the reference frame.

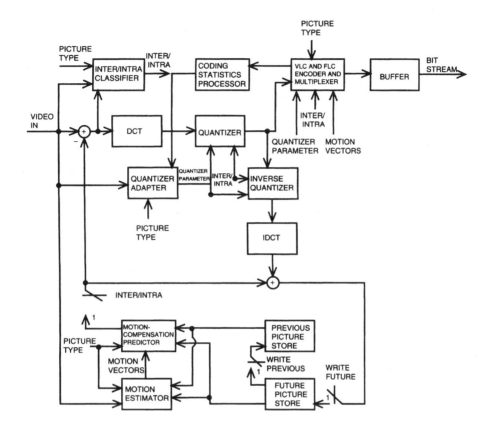

(a)

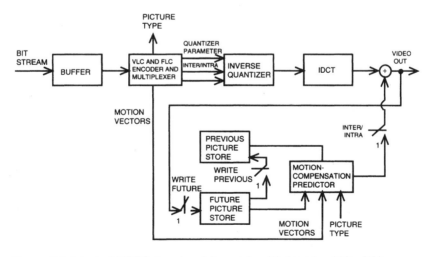

(b)

Figure 5.6 A typical MPEG-1 codec: (a) encoder, (b) decoder. (*After* [8].)

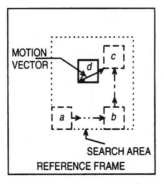

a: INITIAL POSITION OF TARGET BLOCK
   DURING THE SEARCH

b: BEST MATCH ALONG x-AXIS

c: BEST MATCH ALONG y-AXIS GIVEN b

d: LOCATION OF TARGET BLOCK IN THE
   TARGET FRAME

OTS SEARCH FOR THE BEST MATCHED BLOCK

Figure 5.7 A simplified search of a best-matched block. (*From* [1]. *Used with permission.*)

5.4.3 Putting It All Together

MPEG is a standard built upon many elements [1]. Figure 5.8 shows a *group of pictures* (GOP) of 14 frames with two different orderings. Pictures marked *I* are intraframe-coded. A *P*-picture is predicted using the most recently encoded *P*- or *I*-picture in the sequence. A macroblock in a *P*-picture can be coded using either the intraframe or the forward-predicted method. A *B*-picture macroblock can be predicted using either or both of the previous or the next *I*- and/or *P*-pictures. To meet this requirement, the transmission order and display order of frames are different. The two orders also are shown in Figure 5.8.

The MPEG-coded bit stream is divided into several layers, listed in Table 5.2. The three primary layers are:

- *Video sequence*, the outermost layer, which contains basic global information such as the size of frames, bit rate, and frame rate.

- *GOP layer*, which contains information on fast search and random access of the video data. The length of a GOP is arbitrary.

- *Picture layer*, which contains a coded frame. Its header defines the type (*I, P, B*) and the position of the frame in the GOP.

Several of the major differences between MPEG and other compression schemes (such as JPEG) include the following:

- MPEG focuses on video. The basic format uses a single color space (Y, C_r, C_b), a limited range of resolutions and compression ratios, and has built-in mechanisms for handling audio.

- MPEG takes advantage of the high degree of commonality between pictures in a video stream and the typically predictable nature of movement (*inter-picture encoding*).

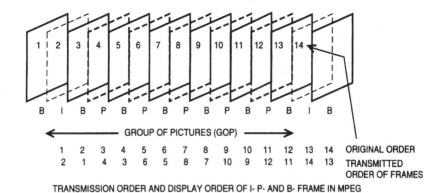

Figure 5.8 Illustration of *I*-frames, *P*-frames, and *B*-frames. (*From* [1]. *Used with permission.*)

Table 5.2 Layers of the MPEG-2 Video Bit-Stream Syntax (*After* [2].)

Syntax layer	Functionality
Video sequence layer	Context unit
Group of pictures (GOP) layer	Random access unit: video coding
Picture layer	Primary coding unit
Slice layer	Resynchronization unit
Macroblock layer	Motion-compensation unit
Block layer	DCT unit

- MPEG provides for a constant bit rate through adjustable variables, making the format predictable with regard to bandwidth requirements.

MPEG specifies the syntax for storing and transmitting compressed data and defines the decoding process. The standard does not, however, specify how encoding should be performed. Such implementation considerations are left to the manufacturers of encoding systems. Still, all conforming encoders must produce valid MPEG bit streams that can be decompressed by any MPEG decoder. This approach is, in fact, one of the strengths of the MPEG standard; because encoders are allowed to use proprietary but compliant algorithms, a variety of implementations is possible and, indeed, encouraged.

As mentioned previously, MPEG is actually a collection of standards, each suited to a particular application or group of applications, including:

- **MPEG-1**, the original implementation, targeted at multimedia uses. The MPEG-1 algorithm is intended basically for compact disc bit rates of approximately 1.5–2.0 Mbits/s.

MPEG-1 supports 525- and 625-type signal structures in progressive form with 204/288 lines per frame, sequential-scan frame rates of 29.97 and 25 per second, and 352 pixels per line. The coding of high-motion signals does not produce particularly good results, however. As might be expected, as the bit rate is reduced (compression increased), the output video quality gradually declines. The overall bit-rate reduction ratios achievable are about 6:1 with a bit rate of 6 Mbits/s and 200:1 at 1.5 Mbits/s. The MPEG-1 system is not symmetrical; the compression side is more complex and expensive than the decompression process, making the system ideal for broadcast-type applications in which there are far more decoders than encoders.

- **MPEG-2**, which offers full ITU-R Rec. 601 resolution for professional and broadcast uses, and is the chosen standard for the ATSC DTV system and the European DVB suite of applications.

- **MPEG-3**, originally targeted at high-definition imaging applications. Subsequent to development of the standard, however, key specifications of MPEG-3 were absorbed into MPEG-2. Thus, MPEG-3 is no longer in use.

- **MPEG-4**, a standard that uses very low bit rates for teleconferencing and related applications requiring high bit efficiency. Like MPEG-2, MPEG-4 is a collection of tools that can be grouped into profiles and levels for different video applications. The MPEG-4 video coding structure ranges from a *very low bit rate video* (VLBV) level, which includes algorithms and tools for data rates between 5 kbits/s and 64 kbits/s, to ITU-R. Rec. 601 quality video at 2 Mbits/s. MPEG-4 does not concern itself directly with the error protection required for specific channels, such as cellular ratio, but it has made improvements in the way payload bits are arranged so that recovery is more robust.

- **MPEG-7**, not really a compression scheme at all, but rather a "multimedia content description interface." MPEG-7 is an attempt to provide a standard means of describing multimedia content.

5.4.4 Profiles and Levels

Six *profiles* and four *levels* describe the organization of the basic MPEG-2 standard. A *profile* is a subset of the MPEG-2 bit-stream syntax with restrictions on the parts of the MPEG algorithm used. Profiles are analogous to features, describing the available characteristics. A *level* constrains general parameters such as image size, data rate, and decoder buffer size. Levels describe, in essence, the upper bounds for a given feature and are analogous to performance specifications.

By far the most popular element of the MPEG-2 standard for professional video applications is the *Main Profile* in conjunction with the *Main Level* (described in the jargon of MPEG as Main Profile/Main Level), which gives an image size of 720 × 576, a data rate of 15 Mbits/s, and a frame rate of 30 frames/s. All higher profiles are capable of decoding Main Profile/Main Level streams.

Table 5.3 lists the basic MPEG-2 classifications. With regard to the table, the following generalizations can be made:

Table 5.3 Common MPEG Profiles and Levels in Simplified Form (*After* [2] *and* [12].)

Profile	General Specifications	Parameter	Level			
			Low	Main (ITU 601)	High 1440 (HD, 4:3)	High (HD, 16:9)
Simple	Pictures: *I, P* Chroma: 4:2:0	Image size[1]		720×576		
		Image frequency[2]		30		
		Bit rate[3]		15		
Main	Pictures: *I, P, B* Chroma: 4:2:0	Image size	325×288	720×576	1440×1152	1920×1152
		Image frequency	30	30	60	60
		Bit rate	4	15	100	80
SNR-Scalable	Pictures: *I, P, B* Chroma: 4:2:0	Image size	325×288	720×576		
		Image frequency	30	30		
		Bit rate	3, 4[4]	15		
Spatially-Scalable	Pictures: *I, P, B* Chroma: 4:2:0	Image size			720×576	
		Image frequency			30	
		Bit rate			15	
	Enhancement Layer[5]	Image size			1440×1152	
		Image frequency			60	
		Bit rate			40, 60[6]	
High[7]	Pictures: *I, P, B* Chroma: 4:2:2	Image size		720×576	1440×1152	1920×1152
		Image frequency		30	60	60
		Bit rate		20	80	100
Studio	Pictures: *I, P, B* Chroma: 4:2:2	Image size		720×608		
		Image frequency		30		
		Bit rate		50		

Notes:
[1] Image size specified as samples/line × lines/frame
[2] Image frequency in frames/s
[3] Bit rate in Mbits/s
[4] For *Enhancement Layer 1*
[5] For *Enhancement Layer 1*, except as noted by [6] for *Enhancement Layer 2*
[7] For simplicity, *Enhancement Layers* not specified individually

- The three key flavors of MPEG-2 are Main Profile/Low Level (source input format, or SIF), Main Profile/Main Level (Main), and Studio Profile/Main Level (Studio).

- The SIF Main Profile/Low Level offers the best picture quality for bit rates below about 5 Mbits/s. This provides generally acceptable quality for interactive and multimedia applications. The SIF profile has replaced MPEG-1 in some applications.

- The Main Profile/Main Level grade offers the best picture quality for conventional video systems at rates from about 5 to 15 Mbits/s. This provides good quality for broadcast applications such as play-to-air, where four generations or fewer typically are required.

- The Studio Profile offers high quality for multiple-generation conventional video applications, such as post-production.

- The High Profile targets HDTV applications.

5.4.5   Studio Profile

Despite the many attributes of MPEG-2, the Main Profile/Main Level remains a less-than-ideal choice for conventional video production because the larger GOP structure makes individual frames hard to access. For this reason, the 4:2:2 *Studio Profile* was developed. The Studio Profile expands upon the 4:2:0 sampling scheme of MPEG-1 and MPEG-2. In essence, "standard MPEG" samples the full luminance signal, but ignores half of the chrominance information, specifically the color coordinate on one axis of the color grid. Studio Profile MPEG increases the chrominance sampling to 4:2:2, thereby accounting for both axes on the color grid by sampling every other element. This enhancement provides better replication of the original signal.

The Studio Profile is intended principally for editing applications, where multiple iterations of a given video signal are required or where the signal will be compressed, decompressed, and recompressed several times before it is finally transmitted or otherwise finally displayed.

SMPTE 308M

SMPTE standard 308M is intended for use in high-definition television production, contribution, and distribution applications [9]. It defines bit-streams, including their syntax and semantics, together with the requirements for a compliant decoder for 4:2:2 Studio Profile at High Level. As with the other MPEG standards, 308M does not specify any particular encoder operating parameters.

The MPEG-2 4:2:2 Studio Profile is defined in ISO/IEC 13818-2, and in SMPTE 308M, only those additional parameters necessary to define the 4:2:2 Studio Profile at High Level are specified. The primary differences are: 1) the upper bounds for sampling density are increased to 1920 samples/line, 1088 lines/frame, and 60 frames/s; 2) the upper bounds for the luminance sample rate is set at 62,668,800 samples/s; and 3) the upper bounds for bit rates is set at 300 Mbits/s.

A more detailed discussion of SMPTE 308M is given in Section 6.4.

5.5   MPEG-2 Features of Importance for DTV

The primary application of interest when the MPEG-2 standard was first defined was "true" television broadcast resolution, as specified by ITU-R Rec. 601. This is roughly four times more picture information than the MPEG-1 standard provides. MPEG-2 is a superset, or extension, of MPEG-1. As such, an MPEG-2 decoder also should be able to decode an

MPEG-1 stream. This broadcast version adds to the MPEG-1 toolbox provisions for dealing with interlace, graceful degradation, and hierarchical coding.

Although MPEG-1 and MPEG-2 each were specified with a particular range of applications and resolutions in mind, the committee's specifications form a set of techniques that support multiple coding options, including picture types and macroblock types. Many variations exist with regard to picture size and bit rates. Also, although MPEG-1 can run at high bit rates and at full ITU-R Rec. 601 resolution, it processes frames, not fields. This fact limits the attainable quality, even at data rates approaching 5 Mbits/s.

The MPEG specifications apply only to decoding, not encoding. The ramifications of this approach are:

- Owners of existing decoding software can benefit from future breakthroughs in encoding processing. Furthermore, the suppliers of encoding equipment can differentiate their products by cost, features, encoding quality, and other factors.

- Different schemes can be used in different situations. For example, although *Monday Night Football* must be encoded in real time, a film can be encoded in non-real time, allowing for fine-tuning of the parameters via computer or even a human operator.

5.5.1 MPEG-2 Layer Structure

To allow for a simple yet upgradable system, MPEG-2 defines only the functional elements—syntax and semantics—of coded streams. Using the same system of *I*-, *P*-, and *B*-frames developed for MPEG-1, MPEG-2 employs a 6-layer hierarchical structure that breaks the data into simplified units of information, as listed in Table 5.2.

The top *sequence layer* defines the decoder constraints by specifying the context of the video sequence. The sequence-layer data header contains information on picture format and application-specific details. The second level allows for random access to the decoding process by having a periodic series of pictures; it is fundamentally this GOP layer that provides the bidirectional frame prediction. Intraframe-coded (*I*) frames are the entry-point frames, which require no data from other frames in order to reconstruct. Between the *I*-frames lie the predictive (*P*) frames, which are derived from analyzing previous frames and performing motion estimation (as outlined in Section 5.4.2). These *P*-frames require about one-third as many bits per frame as *I*-frames. *B*-frames, which lie between two *I*-frames or *P*-frames, are bidirectionally encoded, making use of past and future frames. The *B*-frames require only about one-ninth of the data per frame, compared with *I*-frames.

These different compression ratios for the frames lead to different data rates, so that buffers are required at both the encoder output and the decoder input to ensure that the sustained data rate is constant. One difference between MPEG-1 and MPEG-2 is that MPEG-2 allows for a variety of data-buffer sizes, to accommodate different picture dimensions and to prevent buffer under- and overflows.

The data required to decode a single picture is embedded in the *picture layer*, which consists of a number of horizontal *slice layers*, each containing several macroblocks. Each *macroblock layer*, in turn, is made up of a number of individual blocks. The picture undergoes DCT processing, with the slice layer providing a means of synchronization, holding the precise position of the slice within the image frame.

MPEG-2 places the motion vectors into the coded macroblocks for *P*- and *B*- frames; these are used to improve the reconstruction of predicted pictures. MPEG-2 supports both field- and frame-based prediction, thus accommodating interlaced signals.

The last layer of MPEG-2's video structure is the *block layer*, which provides the DCT coefficients of either the transformed image information for *I*-frames or the residual prediction error of *B*- and *P*- frames.

5.5.2 Slices

Two or more contiguous macroblocks within the same row are grouped together to form *slices* [10]. The order of the macroblocks within a slice is the same as the conventional television raster scan, being from left to right.

Slices provide a convenient mechanism for limiting the propagation of errors. Because the coded bit stream consists mostly of variable-length code words, any uncorrected transmission errors will cause a decoder to lose its sense of code word alignment. Each slice begins with a slice start code. Because the MPEG code word assignment guarantees that no legal combination of code words can emulate a start code, the slice start code can be used to regain the sense of code-word alignment after an error. Therefore, when an error occurs in the data stream, the decoder can skip to the start of the next slice and resume correct decoding.

The number of slices affects the compression efficiency; partitioning the data stream to have more slices provides for better error recovery, but claims bits that could otherwise be used to improve picture quality.

In the DTV system, the initial macroblock of every horizontal row of macroblocks is also the beginning of a slice, with a possibility of several slices across the row.

5.5.3 Pictures, Groups of Pictures, and Sequences

The primary coding unit of a video sequence is the individual video frame or picture [10]. A video picture consists of the collection of slices, constituting the *active picture area.*

A *video sequence* consists of a collection of two or more consecutive pictures. A video sequence commences with a sequence header and is terminated by an end-of-sequence code in the data stream. A video sequence may contain additional sequence headers. Any video-sequence header can serve as an *entry point*. An entry point is a point in the coded video bit stream after which a decoder can become properly initialized and correctly parse the bitstream syntax.

Two or more pictures (frames) in sequence may be combined into a GOP to provide boundaries for interframe picture coding and registration of time code. GOPs are optional within both MPEG-2 and the ATSC DTV system. Figure 5.9 illustrates a typical time sequence of video frames.

I-Frames

Some elements of the compression process exploit only the spatial redundancy within a single picture (frame or field) [10]. These processes constitute intraframe coding, and do not take advantage of the temporal correlation addressed by temporal prediction (interframe)

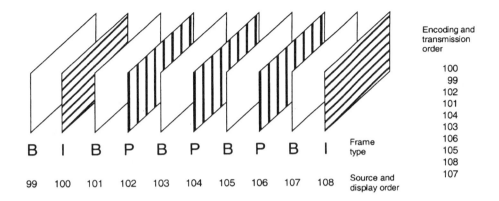

Figure 5.9 Sequence of video frames for the MPEG-2/ATSC DTV system. (*From* [10]. *Used with permission.*)

coding. Frames that do not use any interframe coding are referred to as *I*-frames (where "I" denotes *intraframe*-coded). The ATSC video-compression system utilizes both intraframe and interframe coding.

The use of periodic *I*-frames facilitates receiver initializations and channel acquisition (for example, when the receiver is turned on or the channel is changed). The decoder also can take advantage of the intraframe coding mode when noncorrectable channel errors occur. With motion-compensated prediction, an initial frame must be available at the decoder to start the prediction loop. Therefore, a mechanism must be built into the system so that if the decoder loses synchronization for any reason, it can rapidly reacquire tracking.

The frequency of occurrence of *I*-pictures may vary and is selected at the encoder. This allows consideration to be given to the need for random access and the location of scene cuts in the video sequence.

P-Frames

P-frames, where the temporal prediction is in the forward direction only, allow the exploitation of interframe coding techniques to improve the overall compression efficiency and picture quality [10]. *P*-frames may include portions that are only intraframe-coded. Each macroblock within a *P*-frame can be either forward-predicted or intraframe-coded.

B-Frames

The *B*-frame is a picture type within the coded video sequence that includes prediction from a future frame as well as from a previous frame [10]. The referenced future or previous frames, sometimes called *anchor frames*, are in all cases either *I*- or *P*-frames.

The basis of the *B*-frame prediction is that a video frame is correlated with frames that occur in the past as well as those that occur in the future. Consequently, if a future frame is available to the decoder, a superior prediction can be formed, thus saving bits and improving performance. Some of the consequences of using future frames in the prediction are:

- The B-frame cannot be used for predicting future frames.

- The transmission order of frames is different from the displayed order of frames.

- The encoder and decoder must reorder the video frames, thereby increasing the total latency.

In the example illustrated in Figure 5.9, there is one B-frame between each pair of I- and P-frames. Each frame is labeled with both its display order and transmission order. The I and P frames are transmitted out of sequence, so the video decoder has both anchor frames decoded and available for prediction.

B-frames are used for increasing the compression efficiency and perceived picture quality when encoding latency is not an important factor. The use of B-frames increases coding efficiency for both interlaced- and progressive-scanned material. B-frames are included in the DTV system because the increase in compression efficiency is significant, especially with progressive scanning. The choice of the number of bidirectional pictures between any pair of reference (I or P) frames can be determined at the encoder.

Motion Estimation

The efficiency of the compression algorithm depends on, first, the creation of an estimate of the image being compressed and, second, subtraction of the pixel values of the estimate or prediction from the image to be compressed [10]. If the estimate is good, the subtraction will leave a very small residue to be transmitted. In fact, if the estimate or prediction were perfect, the difference would be zero for all the pixels in the frame of differences, and no new information would need to be sent; this condition can be approached for still images.

If the estimate is not close to zero for some pixels or many pixels, those differences represent information that needs to be transmitted so that the decoder can reconstruct a correct image. The kinds of image sequences that cause large prediction differences include severe motion and/or sharp details. This aspect of the MPEG-2 standard is addressed in more detail in Section 5.5.8.

5.5.4 Vector Search Algorithm

The video-coding system uses motion-compensated prediction as part of the data-compression process [10]. Thus, macroblocks in the current frame of interest are predicted by macroblock-sized regions in previously transmitted frames. Motion compensation refers to the fact that the locations of the macroblock-sized regions in the reference frame can be offset to account for local motions. The macroblock offsets are known as *motion vectors*.

The DTV standard does not specify how encoders should determine motion vectors. One possible approach is to perform an exhaustive search to identify the vertical and horizontal offsets that minimize the total difference between the offset region in the reference frame and the macroblock in the frame to be coded.

5.5.5 Motion-Vector Precision

The estimation of interframe displacement is calculated with half-pixel precision, in both vertical and horizontal dimensions [10]. As a result, the displaced macroblock from the previous frame can be displaced by noninteger displacements and will require interpolation to compute the values of displaced picture elements at locations not in the original array of samples. Estimates for half-pixel locations are computed by averages of adjacent sample values.

5.5.6 Motion-Vector Coding

Motion vectors within a slice are differenced, so that the first value for a motion vector is transmitted directly, and the following sequence of motion-vector differences is sent using *variable-length codes* (VLC) [10]. Motion vectors are constrained so that all pixels from the motion-compensated prediction region in the reference picture fall within the picture boundaries.

5.5.7 Encoder Prediction Loop

The encoder prediction loop, shown in the simplified block diagram of Figure 5.10, is the heart of the video-compression system for DTV [10]. The prediction loop contains a prediction function that estimates the picture values of the next picture to be encoded in the sequence of successive pictures that constitute the TV program. This prediction is based on previous information that is available within the loop, derived from earlier pictures. The transmission of the predicted compressed information works because the same information used to make the prediction also is available at the receiving decoder (barring transmission errors, which are usually infrequent within the primary coverage area).

The subtraction of the predicted picture values from the new picture to be coded is at the core of predictive coding. The goal is to do such a good job of predicting the new values that the result of the subtraction function at the beginning of the prediction loop is zero or close to zero most of the time.

The prediction differences are computed separately for the luminance and two chrominance components before further processing. As explained in previous discussion of *I*-frames, there are times when prediction is not used, for part of a frame or for an entire frame.

Spatial Transform Block—DCT

The image prediction differences (sometimes referred to as *prediction errors*) are organized into 8×8 blocks, and a spatial transform is applied to the blocks of difference values [10]. In the intraframe case, the spatial transform is applied to the raw, undifferenced picture data. The luminance and two chrominance components are transformed separately. Because the chrominance data is subsampled vertically and horizontally, each 8×8 block of chrominance (C_b or C_r) data corresponds to a 16×16 macroblock of luminance data, which is not subsampled.

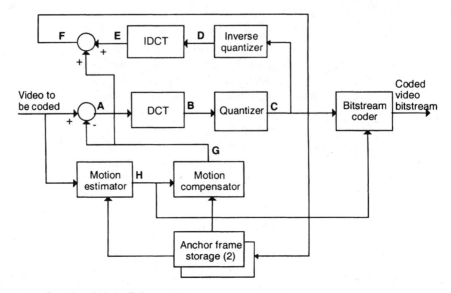

A Pixel-by-pixel prediction errors
B Transformed blocks of prediction errors (DCT coefficients)
C Prediction error DCT coefficients in quantized form
D Quantized prediction error DCT coefficients in standard form
E Pixel-by-pixel prediction errors, degraded by quantization
F Reconstructed pixel values, degraded by quantization
G Motion compensated predicted pixel values
H Motion vectors

Figure 5.10 Simplified encoder prediction loop. (*From* [10]. *Used with permission.*)

The spatial transform used is the discrete cosine transform. The formula for transforming the data is given by:

$$F(u, v) = \frac{1}{4} C(u)C(v) \sum_{x=0}^{7} \sum_{y=0}^{7} f(x, y) \cos\left[\frac{(2x+1)u\pi}{16}\right] \cos\left[\frac{(2y+1)v\pi}{16}\right] \tag{5.1}$$

where x and y are pixel indices within an 8×8 block, u and v are DCT coefficient indices within an 8×8 block, and:

$$C(w) = \frac{1}{\sqrt{2}} \qquad\qquad \text{for } w = 0 \tag{5.2}$$

$$C(w) = 1 \qquad\qquad \text{for } w = 1, 2, ..., 7 \tag{5.3}$$

Thus, an 8×8 array of numbers $f(x, y)$ is the input to a mathematical formula, and the output is an 8×8 array of different numbers, $F(u, v)$. The inverse transform is given by:

$$f(x, y) = \frac{1}{4} \sum_{u = 0}^{7} \sum_{v = 0}^{7} C(u)C(v)F(u, v)\cos\frac{(2x + 1)u\pi}{16}\cos\frac{(2y + 1)v\pi}{16} \qquad (5.4)$$

It should be noted that for the DTV implementation, the *inverse discrete cosine transform* (IDCT) must conform to the specifications noted in [11].

In principle, applying the IDCT to the transformed array would yield exactly the same array as the original. In that sense, transforming the data does not modify the data, but merely represents it in a different form.

The decoder uses the inverse transformation to approximately reconstruct the arrays that were transformed at the encoder, as part of the process of decoding the received compressed data. The approximation in that reconstruction is controlled in advance during the encoding process for the purpose of minimizing the visual effects of coefficient inaccuracies while reducing the quantity of data that needs to be transmitted.

Quantizer

The process of transforming the original data organizes the information in a way that exposes the spatial frequency components of the images or image differences [10]. Using information about the response of the human visual system to different spatial frequencies, the encoder can selectively adjust the precision of transform coefficient representation. The goal is to include as much information about a particular spatial frequency as necessary— and as possible, given the constraints on data transmission—while not using more precision than is needed, based upon visual perception criteria.

For example, in a portion of a picture that is "busy" with a great deal of detail, imprecision in reconstructing spatial high-frequency components in a small region might be masked by the picture's local "busyness." On the other hand, highly precise representation and reconstruction of the average value or dc term of the DCT block would be important in a smooth area of sky. The dc $F(0,0)$ term of the transformed coefficients represents the average of the original 64 coefficients.

As stated previously, the DCT of each 8×8 block of pixel values produces an 8×8 array of DCT coefficients. The relative precision accorded to each of the 64 DCT coefficients can be selected according to its relative importance in human visual perception. The relative coefficient precision information is represented by a *quantizer matrix*, which is an 8×8 array of values. Each value in the quantizer matrix represents the coarseness of quantization of the related DCT coefficient.

Two types of quantizer matrices are supported:

- A matrix used for macroblocks that are intraframe-coded

- A matrix used for non-intraframe-coded macroblocks

The video-coding system defines default values for both the intraframe-quantizer and the non-intraframe-quantizer matrices. Either or both of the quantizer matrices can be overridden at the picture level by transmission of appropriate arrays of 64 values. Any quantizer matrix overrides stay in effect until the following sequence start code.

The transform coefficients, which represent the bulk of the actual coded video information, are quantized to various degrees of coarseness. As indicated previously, some portions of the picture will be more affected in appearance than others by the loss of precision through coefficient quantization. This phenomenon is exploited by the availability of the quantizer scale factor, which allows the overall level of quantization to vary for each macroblock. Consequently, entire macroblocks that are deemed to be visually less important can be quantized more coarsely, resulting in fewer bits being needed to represent the picture.

For each coefficient other than the dc coefficient of intraframe-coded blocks, the quantizer scale factor is multiplied by the corresponding value in the appropriate quantizer matrix to form the quantizer step size. Quantization of the dc coefficients of intraframe-coded blocks is unaffected by the quantizer scale factor and is governed only by the (0, 0) element of the intraframe-quantizer matrix, which always is set to be 8 (ISO/IEC 13818-2).

Entropy Coder

An important effect of the quantization of transform coefficients is that many coefficients will be rounded to zero after quantization [10]. In fact, a primary method of controlling the encoded data rate is the control of quantization coarseness, because a coarser quantization leads to an increase in the number of zero-value quantized coefficients.

Inverse Quantizer

At the decoder, the coded coefficients are decoded, and an 8×8 block of quantized coefficients is reconstructed [10]. Each of these 64 coefficients is *inverse-quantized* according to the prevailing quantizer matrix, quantizer scale, and frame type. The result of inverse quantization is a block of 64 DCT coefficients.

Inverse Spatial Transform Block—IDCT

The decoded and inverse-quantized coefficients are organized as 8×8 blocks of DCT coefficients, and the inverse discrete cosine transform is applied to each block [10]. This results in a new array of pixel values, or pixel difference values that correspond to the output of the subtraction at the beginning of the prediction loop. If the prediction loop was in the interframe mode, the values will be pixel differences. If the loop was in the intraframe mode, the inverse transform will produce pixel values directly.

Motion Compensator

If a portion of the image has not moved, then it is easy to see that a subtraction of the old portion from the new portion of the image will produce zero or nearly zero pixel differences, which is the goal of the prediction [10]. If there has been movement in the portion of the image under consideration, however, the direct pixel-by-pixel differences generally will not be zero, and might be statistically very large. The motion in most natural scenes is organized, however, and can be approximately represented locally as a translation in most cases. For this reason, the video-coding system allows for *motion-compensated* prediction, whereby macroblock-sized regions in the reference frame may be translated vertically and horizontally with respect to the macroblock being predicted, to compensate for local motion.

The pixel-by-pixel differences between the current macroblock and the motion-compensated prediction are transformed by the DCT and quantized using the composition of the non-intraframe-quantizer matrix and the quantizer scale factor. The quantized coefficients then are coded.

5.5.8 Dual Prime Prediction Mode

The dual prime prediction mode is an alternative "special" prediction mode that is built on field-based motion prediction but requires fewer transmitted motion vectors than conventional field-based prediction [10]. This mode of prediction is available only for interlaced material and only when the encoder configuration does not use *B*-frames. This mode of prediction can be particularly useful for improving encoder efficiency for low-delay applications.

The basis of dual prime prediction is that field-based predictions of both fields in a macroblock are obtained by averaging two separate predictions, which are predicted from the two nearest decoded fields in time. Each of the macroblock fields is predicted separately, although the four vectors (one pair per field) used for prediction all are derived from a single transmitted field-based motion vector. In addition to the single field-based motion vector, a small *differential vector* (limited to vertical and horizontal component values of +1, 0, and −1) also is transmitted for each macroblock. Together, these vectors are used to calculate the pairs of motion vectors for each macroblock. The first prediction in the pair is simply the transmitted field-based motion vector. The second prediction vector is obtained by combining the differential vector with a scaled version of the first vector. After both predictions are obtained, a single prediction for each macroblock field is calculated by averaging each pixel in the two original predictions. The final averaged prediction then is subtracted from the macroblock field being encoded.

5.5.9 Adaptive Field/Frame Prediction Mode

Interlaced pictures may be coded in one of two ways: either as two separate fields or as a single frame [10]. When the picture is coded as separate fields, all of the codes for the first field are transmitted as a unit before the codes for the second field. When the picture is coded as a frame, information for both fields is coded for each macroblock.

When frame-based coding is used with interlaced pictures, each macroblock may be selectively coded using either field prediction or frame prediction. When frame prediction is used, a motion vector is applied to a picture region that is made up of both parity fields interleaved together. When field prediction is used, a motion vector is applied to a region made up of scan lines from a single field. Field prediction allows the selection of either parity field to be used as a reference for the field being predicted.

5.5.10 Image Refresh

As discussed previously, a given picture may be sent by describing the differences between it and one or two previously transmitted pictures [10]. For this scheme to work, there must be some way for decoders to become initialized with a valid picture upon tuning into a new channel, or to become reinitialized with a valid picture after experiencing transmission errors. Additionally, it is necessary to limit the number of consecutive predictions that can be performed in a decoder to control the buildup of errors resulting from *IDCT mismatch*.

IDCT mismatch occurs because the video-coding system, by design, does not completely specify the results of the IDCT operation. MPEG did not fully specify the results of the IDCT to allow for evolutionary improvements in implementations of this computationally intensive operation. As a result, it is possible for the reconstructed pictures in a decoder to "drift" from those in the encoder if many successive predictions are used, even in the absence of transmission errors. To control the amount of drift, each macroblock is required to be coded without prediction (intraframe-coded) at least once in any 132 consecutive frames.

The process whereby a decoder becomes initialized or reinitialized with valid picture data—without reference to previously transmitted picture information—is termed *image refresh*. Image refresh is accomplished by the use of intraframe-coded macroblocks. The two general classes of image refresh, which can be used either independently or jointly, are:

- Periodic transmission of *I*-frames

- Progressive refresh

Periodic Transmission of I-Frames

One simple approach to image refresh is to periodically code an entire frame using only intraframe coding [10]. In this case, the intra-coded frame is typically an *I*-frame. Although prediction is used within the frame, no reference is made to previously transmitted frames. The period between successive intracoded frames may be constant, or it may vary. When a receiver tunes into a new channel where *I*-frame coding is used for image refresh, it may perform the following steps:

- Ignore all data until receipt of the first sequence header

- Decode the sequence header, and configure circuits based on sequence parameters

- Ignore all data until the next received *I*-frame

- Commence picture decoding and presentation

When a receiver processes data that contains uncorrectable errors in an *I*- or *P*-frame, there typically will be a propagation of picture errors as a result of predictive coding. Pictures received after the error may be decoded incorrectly until an error-free *I*-frame is received.

Progressive Refresh

An alternative method for accomplishing image refresh is to encode only a portion of each picture using the intraframe mode [10]. In this case, the intraframe-coded regions of each picture should be chosen in such a way that, over the course of a reasonable number of frames, all macroblocks are coded intraframe at least once. In addition, constraints might be placed on motion-vector values to avoid possible contamination of refreshed regions through predictions using unrefreshed regions in an uninitialized decoder.

5.5.11 Discrete Cosine Transform

Predictive coding in the MPEG-2 compression algorithm exploits the temporal correlation in the sequence of image frames [10]. Motion compensation is a refinement of that temporal prediction, which allows the coder to account for apparent motions in the image that can be estimated. Aside from temporal prediction, another source of correlation that represents redundancy in the image data is the spatial correlation within an image frame or field. This spatial correlation of images, including parts of images that contain apparent motion, can be accounted for by a spatial transform of the prediction differences. In the intraframe-coding case, where there is by definition no attempt at prediction, the spatial transform applies to the actual picture data. The effect of the spatial transform is to concentrate a large fraction of the signal energy in a few transform coefficients.

To exploit spatial correlation in intraframe and predicted portions of the image, the image-prediction residual pixels are represented by their DCT coefficients. For typical images, a large fraction of the energy is concentrated in a few of these coefficients. This makes it possible to code only a few coefficients without seriously affecting the picture quality. The DCT is used because it has good energy-compaction properties and results in real coefficients. Furthermore, numerous fast computational algorithms exist for implementation of DCT.

Blocks of 8×8 Pixels

Theoretically, a large DCT will outperform a small DCT in terms of coefficient decorrelation and block energy compaction [10]. Better overall performance can be achieved, however, by subdividing the frame into many smaller regions, each of which is individually processed.

If the DCT of the entire frame is computed, the whole frame is treated equally. For a typical image, some regions contain a large amount of detail, and other regions contain very little. Exploiting the changing characteristics of different images and different portions of the same image can result in significant improvements in performance. To take advantage of the varying characteristics of the frame over its spatial extent, the frame is partitioned into blocks of 8×8 pixels. The blocks then are independently transformed and adaptively processed based on their local characteristics. Partitioning the frame into small blocks before taking the transform not only allows spatially adaptive processing, but also reduces the computational and memory requirements. The partitioning of the signal into small blocks before computing the DCT is referred to as the *block DCT*.

An additional advantage of using the DCT domain representation is that the DCT coefficients contain information about the spatial frequency content of the block. By utilizing the spatial frequency characteristics of the human visual system, the precision with which the DCT coefficients are transmitted can be in accordance with their perceptual importance. This is achieved through the quantization of these coefficients, as explained in the following sections.

Adaptive Field/Frame DCT

As noted previously, the DCT makes it possible to take advantage of the typically high degree of spatial correlation in natural scenes [10]. When interlaced pictures are coded on a frame basis, however, it is possible that significant amounts of motion result in relatively low spatial correlation in some regions. This situation is accommodated by allowing the DCTs to be computed either on a field basis or on a frame basis. The decision to use field- or frame-based DCT is made individually for each macroblock.

Adaptive Quantization

The goal of video compression is to maximize the video quality at a given bit rate, and this requires a careful distribution of the limited number of available bits [10]. By exploiting the perceptual irrelevancy and statistical redundancy within the DCT domain representation, an appropriate bit allocation can yield significant improvements in performance. Quantization is performed to reduce the precision of the DCT coefficient values, and through quantization and code word assignment, the actual bit-rate compression is achieved. The quantization process is the source of virtually all the loss of information in the compression algorithm. This is an important point, as it simplifies the design process and facilitates fine-tuning of the system.

The degree of subjective picture degradation caused by coefficient quantization tends to depend on the nature of the scenery being coded. Within a given picture, distortions of some regions may be less apparent than in others. The video-coding system allows for the level of quantization to be adjusted for each macroblock in order to save bits, where possible, through coarse quantization.

Perceptual Weighting

The human visual system is not uniformly sensitive to coefficient quantization error [10]. Perceptual weighting of each source of coefficient quantization error is used to increase quantization coarseness, thereby lowering the bit rate. The amount of visible distortion resulting from quantization error for a given coefficient depends on the coefficient number, or frequency, the local brightness in the original image, and the duration of the temporal characteristic of the error. The dc coefficient error results in mean value distortion for the corresponding block of pixels, which can expose block boundaries. This is more visible than higher-frequency coefficient error, which appears as noise or texture.

Displays and the HVS exhibit nonuniform sensitivity to detail as a function of local average brightness. Loss of detail in dark areas of the picture is not as visible as it is in brighter areas. Another opportunity for bit savings is presented in textured areas of the picture, where high-frequency coefficient error is much less visible than in relatively flat areas.

Brightness and texture weighting require analysis of the original image because these areas may be well predicted. Additionally, distortion can be easily masked by limiting its duration to one or two frames. This effect is most profitably used after scene changes, where the first frame or two can be greatly distorted without perceptible artifacts at normal speed.

When transform coefficients are being quantized, the differing levels of perceptual importance of the various coefficients can be exploited by "allocating the bits" to shape the quantization noise into the perceptually less important areas. This can be accomplished by varying the relative step sizes of the quantizers for the different coefficients. The perceptually important coefficients may be quantized with a finer step size than the others. For example, low spatial frequency coefficients may be quantized finely, and the less important high-frequency coefficients may be quantized more coarsely. A simple method to achieve different step sizes is to normalize or weight each coefficient based on its visual importance. All of the normalized coefficients may then be quantized in the same manner, such as rounding to the nearest integer (uniform quantization). Normalization or weighting effectively scales the quantizer from one coefficient to another. The MPEG-2 video-compression system utilizes perceptual weighting, where the different DCT coefficients are weighted according to a perceptual criterion prior to uniform quantization. The perceptual weighting is determined by quantizer matrices. The compression system allows for modifying the quantizer matrices before each picture.

5.5.12 Entropy Coding of Video Data

Quantization creates an efficient, discrete representation for the data to be transmitted [10]. Code word assignment takes the quantized values and produces a digital bit stream for transmission. Hypothetically, the quantized values could be simply represented using uniform- or fixed-length code words. Under this approach, every quantized value would be represented with the same number of bits. As outlined in general terms in Section 5.3.1, greater efficiency—in terms of bit rate—can be achieved with entropy coding.

Entropy coding attempts to exploit the statistical properties of the signal to be encoded. A signal, whether it is a pixel value or a transform coefficient, has a certain amount of information, or entropy, based on the probability of the different possible values or events occurring. For example, an event that occurs infrequently conveys much more new information than one that occurs often. The fact that some events occur more frequently than others can be used to reduce the average bit rate.

Huffman Coding

Huffman coding, which is utilized in the ATSC DTV video-compression system, is one of the most common entropy-coding schemes [10]. In Huffman coding, a code book is generated that can approach the minimum average description length (in bits) of events, given the probability distribution of all the events. Events that are more likely to occur are assigned shorter-length code words, and those less likely to occur are assigned longer-length code words.

Run Length Coding

In video compression, most of the transform coefficients frequently are quantized to zero [10]. There may be a few non-zero low-frequency coefficients and a sparse scattering of non-zero high-frequency coefficients, but most of the coefficients typically have been quantized to zero. To exploit this phenomenon, the 2-dimensional array of transform coefficients is reformatted and prioritized into a 1-dimensional sequence through either a zigzag- or alternate-scanning process. This results in most of the important non-zero coefficients (in terms of energy and visual perception) being grouped together early in the sequence. They will be followed by long runs of coefficients that are quantized to zero. These zero-value coefficients can be efficiently represented through *run length encoding*.

In run length encoding, the number (run) of consecutive zero coefficients before a non-zero coefficient is encoded, followed by the non-zero coefficient value. The run length and the coefficient value can be entropy-coded, either separately or jointly. The scanning separates most of the zero and the non-zero coefficients into groups, thereby enhancing the efficiency of the run length encoding process. Also, a special *end-of-block* (EOB) marker is used to signify when all of the remaining coefficients in the sequence are equal to zero. This approach can be extremely efficient, yielding a significant degree of compression.

In the alternate-/zigzag-scan technique, the array of 64 DCT coefficients is arranged in a 1-dimensional vector before run length/amplitude code word assignment. Two different 1-dimensional arrangements, or *scan types*, are allowed, generally referred to as *zigzag scan* (shown in Figure 5.11*a*) and *alternate scan* (shown in Figure 5.11*b*). The scan type is specified before coding each picture and is permitted to vary from picture to picture.

Channel Buffer

Whenever entropy coding is employed, the bit rate produced by the encoder is variable and is a function of the video statistics [10]. Because the bit rate permitted by the transmission system is less than the peak bit rate that may be produced by the variable-length coder, a *channel buffer* is necessary at the decoder. This buffering system must be carefully designed. The buffer controller must allow efficient allocation of bits to encode the video and also ensure that no overflow or underflow occurs.

Buffer control typically involves a feedback mechanism to the compression algorithm whereby the amplitude resolution (quantization) and/or spatial, temporal, and color resolution may be varied in accordance with the instantaneous bit-rate requirements. If the bit rate decreases significantly, a finer quantization can be performed to increase it.

The ATSC DTV standard specifies a channel buffer size of 8 Mbits. The *model buffer* is defined in the DTV video-coding system as a reference for manufacturers of both encoders and decoders to ensure interoperability. To prevent overflow or underflow of the model buffer, an encoder may maintain measures of buffer occupancy and scene complexity. When the encoder nccds to reduce the number of bits produced, it can do so by increasing the general value of the quantizer scale, which will increase picture degradation. When it is able to produce more bits, it can decrease the quantizer scale, thereby decreasing picture degradation.

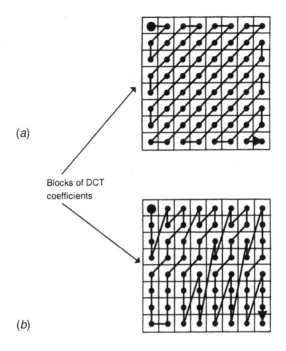

(a)

Blocks of DCT
coefficients

(b)

Figure 5.11 Scanning of coefficient blocks: (a) alternate scanning of coefficients, (b) zigzag scanning of coefficients. (*From* [10]. *Used with permission.*)

Decoder Block Diagram

As shown in Figure 5.12, the ATSC DTV video decoder contains elements that invert, or undo, the processing performed in the encoder [10]. The incoming coded video bit stream is placed in the channel buffer, and bits are removed by a *variable length decoder* (VLD).

The VLD reconstructs 8×8 arrays of quantized DCT coefficients by decoding run length/amplitude codes and appropriately distributing the coefficients according to the scan type used. These coefficients are dequantized and transformed by the IDCT to obtain pixel values or prediction errors.

In the case of interframe prediction, the decoder uses the received motion vectors to perform the same prediction operation that took place in the encoder. The prediction errors are summed with the results of motion-compensated prediction to produce pixel values.

5.5.13 Spatial and S/N Scalability

Because MPEG-2 was designed in anticipation of the need for handling different picture sizes and resolutions, including standard definition television and high-definition television, provisions were made for a hierarchical split of the picture information into a base layer and two enhancement layers [10]. In this way, SDTV decoders would not be burdened with the cost of decoding an HDTV signal.

An encoder for this scenario could work as follows. The HDTV signal would be used as the starting point. It would be spatially filtered and subsampled to create a standard resolution image, which then would be MPEG-encoded. The higher-definition information could be included in an enhancement layer.

Another use of a hierarchical split would be to provide different picture quality without changing the spatial resolution. An encoder quantizer block could realize both coarse and fine filtering levels. Better error correction could be provided for the more coarse data, so that as signal strength weakened, a step-by-step reduction in the picture signal-to-noise ratio would occur in a way similar to that experienced in broadcast analog signals today. Viewers with poor reception, therefore, would experience a more graceful degradation in picture quality instead of a sudden dropout.

5.6 Concatenation

The production of a video segment or program is a serial process: multiple modifications must be made to the original material to yield a finished product. This serial process demands many steps where compression and decompression could take place. Compression and decompression within the same format is not normally considered concatenation. Rather, concatenation involves changing the values of the data, forcing the compression technology to once again compress the signal.

Compressing video is not, generally speaking, a completely lossless process; lossless bit-rate reduction is practical only at the lowest compression ratios. It should be understood, however, that lossless compression is possible—in fact, it is used for critical applications such as medical imaging. Such systems, however, are inefficient in terms of bit usage.

For common video applications, concatenation results in artifacts and coding problems when different compression schemes are cascaded and/or when recompression is required. Multiple generations of coding and decoding are practical, but not particularly desirable. In general, the fewer generations, the better.

Using the same compression algorithm repeatedly (MPEG-2, for example) within a chain—multiple generations, if you will—should not present problems, as long as the pictures are not manipulated (which would force the signal to be recompressed). If, on the other hand, different compression algorithms are cascaded, all bets are off. A detailed mathematical analysis will reveal that such concatenation can result in artifacts ranging from insignificant and unnoticeable to considerable and objectionable, depending on a number of variables, including the following:

- The types of compression systems used

- The compression ratios of the individual systems

- The order or sequence of the compression schemes

- The number of successive coding/decoding steps

- The input signals themselves

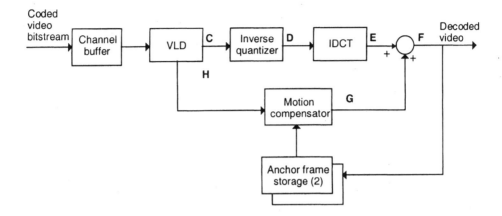

C   Prediction error DCT coefficients in quantized form

D   Quantized prediction error DCT coefficients in standard form

E   Pixel-by-pixel prediction errors, degraded by quantization

F   Reconstructed pixel values, degraded by quantization

G   Motion compensated predicted pixel values

H   Motion vectors

Figure 5.12 ATSC DTV video system decoder functional block diagram. (*From* [10]. *Used with permission.*)

The last point merits some additional discussion. Artifacts from concatenation are most likely during viewing of scenes that are difficult to code in the first place, such as those containing rapid movement of objects or noisy signals. Many video engineers are familiar with test tapes containing scenes that are intended specifically to point out the weaknesses of a given compression scheme or a particular implementation of that scheme. To the extent that such scenes represent real-world conditions, these "compression-killer" images represent a real threat to picture quality when subjected to concatenation of systems.

5.6.1 Compression Artifacts

For any video-compression system, the skill of the mathematicians writing the algorithms lies in making the best compromises between preserving the perceived original scene detail and reducing the amount of data actually transmitted. At the limits of these compromises lie artifacts, which vary depending upon the compression scheme employed. Quantifying the degradation is difficult because the errors are subjective: what is obvious to a trained observer may go unnoticed by a typical viewer or by a trained observer under less-than-ideal conditions. Furthermore, such degradation tends to be transient, whereas analog system degradations tend to be constant.

To maintain image quality in a digital system, bottlenecks must be eliminated throughout the signal path. In any system, the signal path is only as good as its weakest element or its worst compression algorithm. It is a logical assumption that the lower the compression ratio, the better the image quality. In fact, however, there is a point of diminishing return, with the increased data simply eating up bandwidth with no apparent quality improvement. These tradeoffs must be made carefully because once picture elements are lost, they cannot be fully recovered

Typical MPEG Artifacts

Although each type of program sequence consists of a unique set of video parameters, certain generalizations can be made with regard to the artifacts that may be expected with MPEG-based compression systems [13]. The artifacts are determined in large part by the algorithm implementations used by specific MPEG encoding vendors. Possible artifacts include the following:

- *Block effects*. These may be seen when the eye tracks a fast-moving, detailed object across the screen. The blocky grid appears to remain fixed while the object moves beneath it. This effect also may be seen during dissolves and fades. It typically is caused by poor motion estimation and/or insufficient allocation of bits in the coder.

- *Mosquito noise*. This artifact may be seen at the edges of text, logos, and other sharply defined objects. The sharp edges cause high-frequency DCT terms, which are coarsely quantized and spread spatially when transformed back into the pixel domain.

- *Dirty window*. This condition appears as streaking or noise that remains stationary while objects move beneath it. In this case, the encoder may not be sending sufficient bits to code the residual (prediction) error in *P*- and *B*-frames.

- *Wavy noise*. This artifact often is seen during slow pans across highly detailed scenes, such as a crowd in a stadium. The coarsely quantized high-frequency terms resulting from such images can cause reconstruction errors to modulate spatially as details shift within the DCT blocks.

It follows, then, that certain types of motion do not fit the MPEG linear translation model particularly well and are, therefore, problematic. These types of motions include:

- Zooms

- Rotations

- Transparent and/or translucent moving objects

- Dissolves containing moving objects

Furthermore, certain types of image elements cannot be predicted well. These image elements include:

- Shadows

- Changes in brightness resulting from fade-ins and fade-outs

- Highly detailed regions

- Noise effects

- Additive noise

Efforts continue to minimize coding artifacts. Success lies in the skill of the system designers in adjusting the many operating parameters of a video encoder. One of the strengths of the MPEG standard is that it allows—and even encourages—diversity and innovation in encoder design.

5.7 Video Encoding Process

The function of any video compression device or system is to provide for efficient storage and/or transmission of information from one location or device to another. The encoding process, naturally, is the beginning point of this chain. Like any chain, video encoding represents not just a single link but many interconnected and interdependent links. The bottom line in video and audio encoding is to ensure that the compressed signal or data stream represents the information required for recording and/or transmission, and *only* that information. If there is additional information of any nature remaining in the data stream, it will take bits to store and/or transmit, which will result in fewer bits being available for the required data. Surplus information is irrelevant because the intended recipient(s) do not require it and can make no use of it.

Surplus information can take many forms. For example, it can be information in the original signal or data stream that exceeds the capabilities of the receiving device to process and display. There is little point in transmitting more resolution than the receiving device can use. Noise is another form of surplus information. Noise is—by nature—random or nearly so, and this makes it essentially incompressible. Many other types of artifacts exist, ranging from filter ringing to film scratches. Some may seem trivial, but in the field of compression they can be very important. Compression relies on order and consistency for best performance, and such artifacts can compromise the final displayed images or at least lower the achievable bit rate reduction. Generally speaking, compression systems are designed for particular tasks, and make use of certain basic assumptions about the nature of the data being compressed.

5.7.1 Encoding Tools

In the migration to digital video technologies, the encoding process has taken on a new and pivotal role. Like any technical advance, however, encoding presents both challenges and rewards. The challenge involves assimilating new tools and new skills. The quality of the final compressed video is dependent upon the compression system used to perform the encoding, the tools provided by the system, and the skill of the person operating the system.

Beyond the automated procedures of encoding lies an interactive process that can considerably enhance the finished video output. These "human-assisted" procedures can make the difference between high-quality images an mediocre ones, and the difference between the efficient use of media and wasted bandwidth.

The goal of intelligent encoding is to minimize the impact of encoding artifacts, rendering them inconspicuous or even invisible. Success is in the eye of the viewer and, thus, the process involves many subjective visual and aesthetic judgments. It is reasonable to conclude, then, that automatic encoding can go only so far. It cannot substitute for the trained eye of the video professional.

In this sense, human-assisted encoding is analogous to the telecine process. In the telecine application, a skilled professional—the *colorist*—uses techniques such as color correction, filtering, and noise reduction to ensure that the video version of a motion picture or other film-based material is as true to the original as technically possible. This work requires a combination of technical expertise and video artistry. Like the telecine, human-assisted encoding is an iterative process. The operator (*compressionist*, if you will) sets the encoding parameters, views the impact of the settings on the scene, and then further modifies the parameters of the encoder until the desired visual result is achieved for the scene or segment.

5.7.2 Signal Conditioning

Correctly used, signal conditioning can provide a remarkable increase in coding efficiency and ultimate picture quality. Encoding equipment available today incorporates many different types of filters targeted at different types of artifacts. The benefits of appropriate conditioning are twofold:

- Because the artifacts are unwanted, there is a clear advantage in avoiding the allocation of bits to transmit them.

- Because the artifacts do not "belong," they generally violate the rules or assumptions of the compression system. For this reason, artifacts do not compress well and use a disproportionately high number of bits to transmit and/or store.

Filtering prior to encoding can be used to selectively screen out image information that might otherwise result in unwanted artifacts. Spatial filtering applies within a particular frame, and can be used to screen out higher frequencies, removing fine texture noise and softening sharp edges. The resulting picture may have a softer appearance, but this is often preferable to a blocking or ringing artifact. Similarly, temporal (recursive) filtering, applied from frame to frame, can be employed to remove temporal noise caused—for example—by grain-based irregularities in film.

Color correction can be used in much the same manner as filtering. Color correction can smooth out uneven areas of color, reducing the amount of data the compression algorithm will have to contend with, thus eliminating artifacts. Likewise, adjustments in contrast and brightness can serve to mask artifacts, achieving some image quality enhancements without noticeably altering the video content.

For decades, the phrase *garbage-in, garbage-out* has been the watchword of the data processing industry. If the input to some process is flawed, the output will invariably be flawed. Such is the case for video compression. Unless proper attention is paid to the entire encoding process, degradations will occur. In general, consider the following encoding checklist:

- Begin with the best. If the source material is film, use the highest-quality print or negative available. If the source is video, use the highest quality, fewest-generation tape.

- Clean up the source material before attempting to compress it. Perform whatever noise reduction, color correction, scratch removal, and other artifact-elimination steps that are possible before attempting to send the signal to the encoder. There are some defects that the encoding process may hide. Noise, scratches, and color errors are not among them. Encoding will only make them worse.

- Decide on an aspect ratio conversion strategy (if necessary). Keep in mind that once information is discarded, it cannot be reclaimed.

- Treat the encoding process like a high-quality telecine transfer. Start with the default compression settings and adjust as needed to achieve the desired result. Document the settings with an *encoding decision list* (EDL) so that the choices made can be reproduced at a later date, if necessary.

The encoding process is much more of an artistic exercise than it is a technical one. In the area of video encoding, there is no substitute for training and experience.

5.7.3 SMPTE RP 202

SMPTE Recommended Practice 202 (proposed at this writing) is an important step in the world of digital video production. Equipment conforming to this practice will minimize artifacts in multiple generations of encoding and de-coding by optimizing macroblock alignment [14]. As MPEG-2 becomes pervasive in emission, contribution, and distribution of video content, multiple compression and decompression (codec) cycles will be required. Concatenation of codecs may be needed for production, post-production, transcoding, or format conversion. Any time video transitions to or from the coefficient domain of MPEG-2 are performed, care must be exercised in alignment of the video, both horizontally and vertically, as it is coded from the raster format or decoded and placed in the raster format.

The first problem is shifting the video horizontally and vertically. Over multiple compression and decompression cycles, this could substantially distort the image. Less obvious, but just as important, is the need for macroblock alignment to reduce artifacts between encoders and decoders from various equipment vendors. If concatenated encoders do not share common macroblock boundaries, then additional quantization noise, motion estimation errors, and poor mode decisions may result. Likewise, encoding decisions that may be carried through the production and post-production process with recoding data present, will rely upon macroblock alignment. Decoders must also exercise caution in placement of the active video in the scanning format so that the downstream encoder does not receive an offset image.

With these issues in mind, RP 202 specifies the spatial alignment for MPEG-2 video encoders and decoders. Both standard definition and high-definition video formats for production, distribution, and emission systems are addressed. Table 5.4 gives the recommended coding ranges for MPEG-2 encoders and decoders.

Table 5.4 Recommended MPEG-2 Coding Ranges for Various Video Formats (*After* [14].)

Format	Resolution Pels x Lines	Coded Pels	Coded Lines			MPEG-2 Profile and Level
			Field 1	Field 2	Frame	
480I	720 × 480	0–719	23–262	286–525		MP@ML
480P	720 × 480	0–719			46–525	MP@HL
512I	720 × 512	0–719	7–262	270–525		422P@ML
512P	720 × 512	0–719			14–525	422P@HL
576I	720 × 576	0–719	23–310	336–623		MP@ML
608I	720 × 608	0–719	7–310	320–623		422P@ML
720P	1280 × 720	0–1279			26–745	MP@HL
720P	1280 × 720	0–1279			26–745	422P@HL
1080I	1920 × 1088[1]	0–1919	21–560	584–1123		MP@HL
1080I	1920 × 1088[1]	0–1919	21–560	584–1123		422P@HL
1080P	1920 × 1088[1]	0–1919			42–1121	MP@HL
1080P	1920 × 1088[1]	0–1919			42–1121	422P@HL

1 The active image only occupies the first 1080 lines.

5.8   Digital Audio Data Compression

As with video, high on the list of priorities for the professional audio industry is to refine and extend the range of digital equipment capable of the capture, storage, post production, exchange, distribution, and transmission of high-quality audio, be it mono, stereo, or 5.1 channel AC-3 [15]. This demand being driven by end-users, broadcasters, film makers, and the recording industry alike, who are moving rapidly towards a "tapeless" environment. Over the last two decades, there have been continuing advances in DSP technology, which have supported research engineers in their endeavors to produce the necessary hardware, particularly in the field of digital audio data compression or—as it is often referred to—*bit-rate reduction*. There exist a number of real-time or—in reality—near instantaneous compression coding algorithms. These can significantly lower the circuit bandwidth and storage requirements for the transmission, distribution, and exchange of high-quality audio.

The introduction in 1983 of the compact disc (CD) digital audio format set a quality benchmark that the manufacturers of subsequent professional audio equipment strive to match or improve. The discerning consumer now expects the same quality from radio and television receivers. This leaves the broadcaster with an enormous challenge.

5.8.1   PCM Versus Compression

It can be an expensive and complex technical exercise to fully implement a linear *pulse code modulation* (PCM) infrastructure, except over very short distances and within studio areas [15]. To demonstrate the advantages of distributing compressed digital audio over wireless or wired systems and networks, consider again the CD format as a reference. The CD is a 16 bit linear PCM process, but has one major handicap: the amount of circuit bandwidth the digital signal occupies in a transmission system. A stereo CD transfers information (data) at 1.411 Mbits/s, which would require a circuit with a bandwidth of approximately 700 kHz to avoid distortion of the digital signal. In practice, additional bits are added to the signal for channel coding, synchronization, and error correction; this increases the bandwidth demands yet again. 1.5 MHz is the commonly quoted bandwidth figure for a circuit capable of carrying a CD or similarly coded linear PCM digital stereo signal. This can be compared with the 20 kHz needed for each of two circuits to distribute the same stereo audio in the analog format, a 75-fold increase in bandwidth requirements.

5.8.2 Audio Bit-Rate Reduction

In general, analog audio transmission requires fixed input and output bandwidths [16]. This condition implies that in a real-time compression system, the quality, bandwidth, and distortion/noise level of both the original and the decoded output sound should not be *subjectively* different, thus giving the appearance of a lossless and real-time process.

In a technical sense, all practical real-time bit-rate-reduction systems can be referred to as "lossy." In other words, the digital audio signal at the output is not identical to the input signal data stream. However, some compression algorithms are, for all intents and purposes, lossless; they lose as little as 2 percent of the original signal. Others remove approximately 80 percent of the original signal.

Redundancy and Irrelevancy

A complex audio signal contains a great deal of information, some of which, because the human ear cannot hear it, is deemed irrelevant. [16]. The same signal, depending on its complexity, also contains information that is highly predictable and, therefore, can be made redundant.

Redundancy, measurable and quantifiable, can be removed in the coder and replaced in the decoder; this process often is referred to as *statistical compression. Irrelevancy*, on the other hand, referred to as *perceptual coding*, once removed from the signal cannot be replaced and is lost, irretrievably. This is entirely a subjective process, with each proprietary algorithm using a different psychoacoustic model.

Critically perceived signals, such as pure tones, are high in redundancy and low in irrelevancy. They compress quite easily, almost totally a statistical compression process. Conversely, noncritically perceived signals, such as complex audio or noisy signals, are low in redundancy and high in irrelevancy. These compress easily in the perceptual coder, but with the total loss of all the irrelevancy content.

Human Auditory System

The sensitivity of the human ear is biased toward the lower end of the audible frequency spectrum, around 3 kHz [16]. At 50 Hz, the bottom end of the spectrum, and 17 kHz at the top end, the sensitivity of the ear is down by approximately 50 dB relative to its sensitivity at 3 kHz (Figure 5.13). Additionally, very few audio signals—music- or speech-based—carry

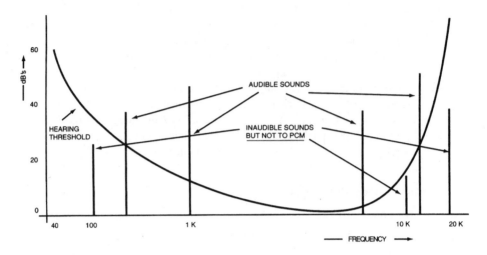

Figure 5.13 Generalized frequency response of the human ear. Note how the PCM process captures signals that the ear cannot distinguish. (*From* [16]. *Used with permission.*)

fundamental frequencies above 4 kHz. Taking advantage of these characteristics of the ear, the structure of audible sounds, and the redundancy content of the PCM signal is the basis used by the designers of the *predictive* range of compression algorithms.

Another well-known feature of the hearing process is that loud sounds mask out quieter sounds at a similar or nearby frequency. This compares with the action of an automatic gain control, turning the gain down when subjected to loud sounds, thus making quieter sounds less likely to be heard. For example, as illustrated in Figure 5.14, if we assume a 1 kHz tone at a level of 70 dBu, levels of greater than 40 dBu at 750 Hz and 2 kHz would be required for those frequencies to be heard. The ear also exercises a degree of temporal masking, being exceptionally tolerant of sharp transient sounds.

It is by mimicking these additional psychoacoustic features of the human ear and identifying the irrelevancy content of the input signal that the *transform* range of low bit-rate algorithms operate, adopting the principle that if the ear is unable to hear the sound then there is no point in transmitting it in the first place.

Quantization

Quantization is the process of converting an analog signal to its representative digital format or, as in the case with compression, the requantizing of an already converted signal [16]. This process is the limiting of a finite level measurement of a signal sample to a specific preset integer value. This means that the *actual* level of the sample may be greater or smaller than the preset *reference* level it is being compared with. The difference between these two levels, called the *quantization error*, is compounded in the decoded signal as *quantization noise*.

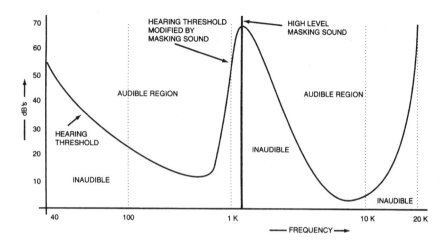

Figure 5.14 Example of the masking effect of a high-level sound. (*From* [16]. *Used with permission.*)

Quantization noise, therefore, will be injected into the audio signal after each A/D and D/A conversion, the level of that noise being governed by the bit allocation associated with the coding process (i.e., the number of bits allocated to represent the level of each sample taken of the analog signal). For linear PCM, the bit allocation is commonly 16. The level of each audio sample, therefore, will be compared with one of 2^{16} or 65,536 discrete levels or steps.

Compression or bit-rate reduction of the PCM signal leads to the requantizing of an already quantized signal, which will unavoidably inject further quantization noise. It always has been good operating practice to restrict the number of A/D and D/A conversions in an audio chain. Nothing has changed in this regard, and now the number of compression stages also should be kept to a minimum. Additionally, the bit rates of these stages should be set as high as practical; put another way, the compression ratio should be as low as possible.

Sooner or later—after a finite number of A/D, D/A conversions and passes of compression coding, of whatever type—the accumulation of quantization noise and other unpredictable signal degradations eventually will break through the noise/signal threshold, be interpreted as part of the audio signal, be processed as such, and be heard by the listener.

Sampling Frequency and Bit Rate

The bit rate of a digital signal is defined by:

sampling frequency × bit resolution × number of audio channels

The rules regarding the selection of a sampling frequency are based on Nyquist's theorem [16]. This ensures that, in particular, the lower sideband of the sampling frequency does not encroach into the baseband audio. Objectionable and audible aliasing effects would occur if the two bands were to overlap. In practice, the sampling rate is set slightly above twice the highest audible frequency, which makes the filter designs less complex and less expensive.

In the case of a stereo CD with the audio signal having been sampled at 44.1 kHz, this sampling rate produces audio bandwidths of approximately 20 kHz for each channel. The resulting audio bit rate = 44.1 kHz × 16 × 2 = 1.411 Mbits/s, as discussed previously.

5.8.3 Prediction and Transform Algorithms

Most audio-compression systems are based upon one of two basic technologies [16]:

- Predictive or *adaptive differential* PCM (ADPCM) time-domain coding

- Transform or *adaptive* PCM (APCM) frequency-domain coding

It is in their approaches to dealing with the redundancy and irrelevancy of the PCM signal that these techniques differ.

The time domain or *prediction* approach includes G.722, which has been a universal standard since the mid-70s, and was joined in 1989 by a proprietary algorithm, apt-X100. Both these algorithms deal mainly with redundancy.

The frequency domain or *transform* method adopted by a number of algorithms deal in irrelevancy, adopting psychoacoustic masking techniques to identify and remove those unwanted sounds. This range of algorithms include the industry standards ISO/MPEG-1 Layers 1, 2, and 3; apt-Q; MUSICAM; Dolby AC-2 and AC3; and others.

Subband Coding

Without exception, all of the algorithms mentioned in the previous section process the PCM signal by splitting it into a number of frequency subbands, in one case as few as two (G.722) or as many as 1024 (apt-Q) [15]. MPEG-1 Layer 1, with 4:1 compression, has 32 frequency subbands and is the system found in the Digital Compact Cassette (DCC). The MiniDisc ATRAC proprietary algorithm at 5:1 has a more flexible multisubband approach, which is dependent on the complexity of the audio signal.

Subband coding enables the frequency domain redundancies within the audio signals to be exploited. This permits a reduction in the coded bit rate, compared to PCM, for a given signal fidelity. Spectral redundancies are also present as a result of the signal energies in the various frequency bands being unequal at any instant in time. By altering the bit allocation for each subband, either by dynamically adapting it according to the energy of the contained signal or by fixing it for each subband, the quantization noise can be reduced across all bands. This process compares favorably with the noise characteristics of a PCM coder performing at the same overall bit rate.

Subband Gain

On its own, subband coding, incorporating PCM in each band, is capable of providing a performance improvement or *gain* compared with that of full band PCM coding, both being fed with the same complex, constant level input signal [15]. The improvement is defined as *subband gain* and is the ratio of the variations in quantization errors generated in each case while both are operating at the same transmission rate. The gain increases as the number of

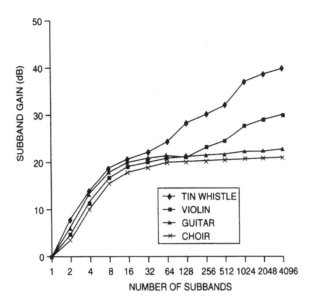

Figure 5.15 Variation of subband gain as a function of the number of subbands. (*From* [16]. *Used with permission.*)

subbands increase, and with the complexity of the input signal. However, the implementation of the algorithm also becomes more difficult and complex.

Quantization noise generated during the coding process is constrained within each subband and cannot interfere with any other band. The advantage of this approach is that the masking by each of the subband dominant signals is much more effective because of the reduction in the noise bandwidth. Figure 5.15 charts subband gain as a function of the number of subbands for four essentially stationary, but differing, complex audio signals.

In practical implementations of compression codecs, several factors tend to limit the number of subbands employed. The primary considerations include:

- The level variation of normal audio signals leading to an averaging of the energy across bands and a subsequent reduction in the coding gain

- The coding or processing delay introduced by additional subbands

- The overall computational complexity of the system

The two key issues in the analysis of a subband framework are:

- Determining the likely improvement associated with additional subbands

- Determining the relationships between subband gain, the number of subbands, and the response of the filter bank used to create those subbands

APCM Coding

The APCM processor acts in a similar fashion to an automatic gain control system, continually making adjustments in response to the dynamics—at all frequencies—of the incoming audio signal [15]. Transform coding takes a time block of signal, analyzes it for frequency and energy, and identifies irrelevant content. Again, to exploit the spectral response of the ear, the frequency spectrum of the signal is divided into a number of subbands, and the most important criteria are coded with a bias toward the more sensitive low frequencies. At the same time, through the use of psychoacoustic masking techniques, those frequencies which it is assumed will be masked by the ear are also identified and removed. The data generated, therefore, describes the frequency content and the energy level at those frequencies, with more bits being allocated to the higher-energy frequencies than those with lower energy.

The larger the time block of signal being analyzed, the better the frequency resolution and the greater the amount of irrelevancy identified. The penalty, however, is an increase in coding delay and a decrease in temporal resolution. A balance has been struck with advances in perceptual coding techniques and psychoacoustic modeling leading to increased efficiency. It is reported in [16] that, with this approach to compression, some 80 percent of the input audio can be removed with acceptable results.

This hybrid arrangement of working with time-domain subbands and simultaneously carrying out a spectral analysis can be achieved by using a *dynamic bit allocation* process for each subband. This subband APCM approach is found in the popular range of software-based MUSICAM, Dolby AC-2, and ISO/MPEG-1 Layers 1 and 2 algorithms. Layer 3—a more complex method of coding and operating at much lower bit rates—is, in essence, a combination of the best functions of MUSICAM and ASPEC, another adaptive transform algorithm. Table 5.5 lists the primary operational parameters for these systems.

Additionally, some of these systems exploit the significant redundancy between stereo channels by using a technique known as *joint stereo coding*. After the common information between left and right channels of a stereo signal has been identified, it is coded only once, thus reducing the bit-rate demands yet again.

Each of the subbands has its own defined *masking threshold*. The output data from each of the filtered subbands is requantized with just enough bit resolution to maintain adequate headroom between the quantization noise and the masking threshold for each band. In more complex coders (e.g., ISO/MPEG-1 Layer 3), any spare bit capacity is utilized by those subbands with the greater need for increased masking threshold separation. The maintenance of these signal-to-masking threshold ratios is crucial if further compression is contemplated for any postproduction or transmission process.

5.8.4 Processing and Propagation Delay

As noted previously, the current range of popular compression algorithms operate—for all intents and purposes—in real time [15]. However, this process does of necessity introduce some measurable delay into the audio chain. All algorithms take a finite time to analyze the incoming signal, which can range from a few milliseconds to tens and even hundreds of milliseconds. The amount of processing delay will be crucial if the equipment is to be used in any interactive or two-way application. As a rule of thumb, any more than 20 ms of delay in a two-way audio exchange is problematic. Propagation delay in satellite and long terrestrial

Table 5.5 Operational Parameters of Subband APCM Algorithm (*After* [16].)

Coding System	Compression Ratio	Subbands	Bit Rate, kbits/s	A to A Delay, ms[1]	Audio Bandwidth, kHz
Dolby AC-2	6:1	256	256	45	20
ISO Layer 1	4:1	32	384	19	20
ISO Layer 2	Variable	32	192–256	>40	20
IOS Layer 3	12:1	576	128	>80	20
MUSICAM	Variable	32	128–384	>35	20

[1] The total system delay (encoder-to-decoder) of the coding system.

circuits is a fact of life. A two-way hook up over a 1000 km, full duplex, telecom digital link has a propagation delay of 3 ms in each direction. This is comparable to having a conversation with someone standing 1 m away. It is obvious that even over a very short distance, the use of a codec with a long processing delay characteristic will have a dramatic effect on operation.

5.8.5   Bit Rate and Compression Ratio

The ITU has recommend the following bit rates when incorporating data compression in an audio chain [15]:

- 128 kbits/s per mono channel (256 kbits/s for stereo) as the minimum bit rate for any stage if further compression is anticipated or required.

- 192 kbits/s per mono channel (384 kbits/s for stereo) as the minimum bit rate for the first stage of compression in a complex audio chain.

These markers place a 4:1 compression ratio at the "safe" end in the scale. However, more aggressive compression ratios, currently up to a nominal 20:1, are available. Keep in mind, though, that low bit rate, high-level compression can lead to problems if any further stages of compression are required or anticipated.

With successive stages of compression, either or both the noise floor and the audio bandwidth will be set by the stage operating at the lowest bit rate. It is, therefore, worth emphasizing that after these platforms have been set by a low bit rate stage, they cannot be subsequently improved by using a following stage operating at a higher bit rate.

Bit Rate Mismatch

A stage of compression may well be followed in the audio chain by another digital stage, either of compression or linear, but—more importantly—operating at a different sampling frequency [15]. If a D/A conversion is to be avoided, a sample rate converter must be used. This can be a stand alone unit or it may already be installed as a module in existing equipment. Where a following stage of compression is operating at the same sampling frequency but a different compression ratio, the bit resolution will change by default.

If the stages have the same sampling frequencies, a direct PCM or AES/EBU digital link can be made, thus avoiding the conversion to the analog domain.

5.8.6 Editing Compressed Data

The linear PCM waveform associated with standard audio workstations is only useful if decoded [15]. The resolution of the compressed data may or may not be adequate to allow direct editing of the audio signal. The minimum audio sample that can be removed or edited from a transform-coded signal will be determined by the size of the time block of the PCM signal being analyzed. The larger the time block, the more difficult the editing of the compressed data becomes.

5.8.7 Common Compression Techniques

Subband APCM coding has found numerous applications in the professional audio industry, including [16]:

- The digital compact cassette (DCC)—uses the simplest implementation of subband APCM with the PASC/ISO/MPEG-1 Layer 1 algorithm incorporating 32 subbands offering 4:1 compression and producing a bit rate of 384 kbits/s.

- The MiniDisc with the proprietary ATRAC algorithm—produces 5:1 compression and 292 kbits/s bit rate. This algorithm uses a *modified discrete cosine transform* (MDCT) technique ensuring greater signal analysis by processing time blocks of the signal in non-uniform frequency divisions, with fewer divisions being allocated to the least sensitive higher frequencies.

- ISO/MPEG-1 Layer 2 (MUSICAM by another name)—a software-based algorithm that can be implemented to produce a range of bit rates and compression ratios commencing at 4:1.

- The ATSC DTV system—uses the subband APCM algorithm in Dolby AC-3 for the audio surround system associated with the ATSC DTV standard. AC-3 delivers five audio channels plus a bass-only effects channel in less bandwidth than that required for one stereo CD channel. This configuration is referred to as 5.1 channels.

For the purposes of illustration, two commonly used audio compression systems will be examined in some detail:

- apt-X100

- ISO/MPEG-1 Layer 2

The specifics of the Dolby AC-3 system are described in Chapter 7.

apt-X100

apt-X100 is a four subband prediction (ADPCM) algorithm [15]. Differential coding reduces the bit rate by coding and transmitting or storing only the difference between a predicted level for a PCM audio sample and the absolute level of that sample, thus exploiting the redundancy contained in the PCM signal.

Audio exhibits relatively slowly varying energy fluctuations with respect to time. Adaptive differential coding, which is dependent on the energy of the input signal, dynamically alters the step size for each quantizing interval to reflect these fluctuations. In apt-X100, this equates to the *backwards adaptation process* and involves the analysis of 122 previous samples. Being a continuous process, this provides an almost constant and optimal signal-to-quantization noise ratio across the operating range of the quantizer.

Time domain subband algorithms implicitly model the hearing process and indirectly exploit a degree of irrelevancy by accepting that the human ear is more sensitive at lower frequencies. This is achieved in the four subband derivative by allocating more bits to the lower frequency bands. This is the only application of psychoacoustics exercised in apt-X100. All the information contained in the PCM signal is processed, audible or not (i.e., no attempt is made to remove irrelevant information). It is the unique fixed allocation of bits to each of the four subbands, coupled with the filtering characteristics of each individual listeners' hearing system, that achieves the satisfactory audible end result.

The user-defined output bit rates range from 56 to 384 kbits/s, achieved by using various sampling frequencies from 16 kHz to 48 kHz, which produce audio bandwidths from 7.5 kHz mono to 22 kHz stereo.

Auxiliary data up to 9.6 kbits/s can also be imbedded into the data stream without incurring a bit overhead penalty. When this function is enabled, an audio bit in one of the higher frequency subbands is replaced by an auxiliary data bit, again with no audible effect.

An important feature of this algorithm is its inherent robustness to random bit errors. No audible distortion is apparent for normal program material at a *bit error rate* (BER) of 1:10,000, while speech is still intelligible down to a BER of 1:10.

Distortions introduced by bit errors are constrained within each subband and their impact on the decoder subband predictors and quantizers is proportional to the magnitude of the differential signal being decoded at that instant. Thus, if the signal is small—which will be the case for a low level input signal or for a resonant, highly predictable input signal—any bit error will have minimal effect on either the predictor or quantizer.

The 16 bit linear PCM signal is processed in time blocks of four samples at a time. These are filtered into four equal-width frequency subbands; for 20 kHz, this would be 0–5 kHz, 5–10 kHz, and so on. The four outputs from the *quadrature mirror filter* (QMF) tree are still in the 16-bit linear PCM format, but are now frequency-limited.

As shown in Figure 5.16, the compression process can be mapped by taking, for example, the first and lowest frequency subband. The first step is to create the difference signal. After the system has settled down on initiation, there will be a reconstructed 16 bit difference signal at the output of the inverse quantizer. This passes into a prediction loop that, having analyzed 122 previous samples, will make a prediction for the level of the next full level sample arriving from the filter tree. This prediction is then compared with the actual level.

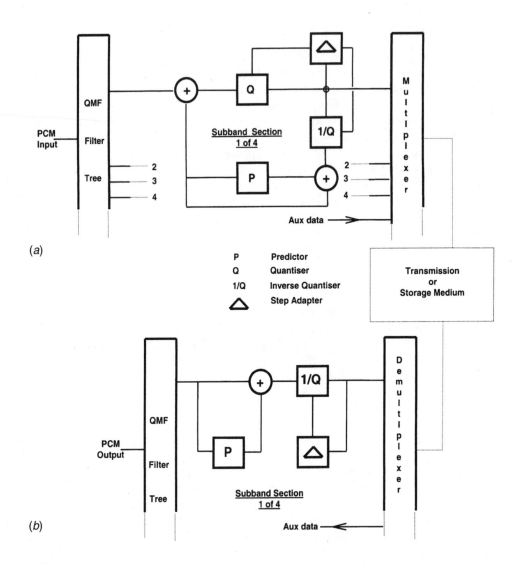

Figure 5.16 apt-X100 audio coding system: (*a*) encoder block diagram, (*b*) decoder block diagram. (*Courtesy of Audio Processing Technology.*)

The output of the comparator is the resulting 16-bit difference signal. This is requantized to a new 7-bit format, which in turn is inverse quantized back to 16 bits again to enable the prediction loop.

The output from the inverse quantizer is also analyzed for energy content, again for the same 122 previous samples. This information is compared with on-board look up tables and a decision is made to dynamically adjust, up or down as required, the level of each step of the 1024 intervals in the 7-bit quantizer. This ensures that the quantizer will always have adequate range to deal with the varying energy levels of the audio signal. Therefore, the input to the multiplexer will be a 7-bit word but the range of those bits will be varying in relation to the signal energy.

The three other subbands will go through the same process, but the number of bits allocated to the quantizers are much less than for the first subband.

The output of the multiplexer or bit stream formatter is a new 16-bit word that represents four input PCM samples and is, therefore, one quarter of the input rate; a reduction of 4:1.

The decoding process is the complete opposite of the coding procedure. The incoming 16-bit compressed data word is demultiplexed and used to control the operation of four subband decoder sections, each with similar predictor and quantizer step adjusters. A QMF filter tree finally reconstructs a linear PCM signal and separates any auxiliary data that may be present.

ISO/MPEG-1 Layer 2

This algorithm differs from Layer 1 by adopting more accurate quantizing procedures and by additionally removing redundancy and irrelevancy on the generated scale factors [15]. The ISO/MPEG-1 Layer 2 scheme operates on a block of 1152 PCM samples, which at 48 kHz sampling represents a 24 ms time block of the input audio signal. Simplified block diagrams of the encoding/decoding systems are given in Figure 5.17.

The incoming linear PCM signal block is divided into 32 equally spaced subbands using a polyphase analysis filter bank (Figure 5.17a). At 48 kHz sampling, this equates to the bandwidth of each subband being 750 Hz. The bit allocation for the requantizing of these subband samples is then dynamically controlled by information derived from analyzing the audio signal, measured against a preset psychoacoustic model.

The filter bank, which displays manageable delay and minimal complexity, optimally adapts each block of audio to balance between the effects of temporal masking and inaudible pre-echoes.

The PCM signal is also fed to a *fast Fourier transform* (FFT) running in parallel with the filter bank. The aural sensitivities of the human auditory system are exploited by using this FFT process to detect the differences between the wanted and unwanted sounds and the quantization noise already present in the signal, and then to adjust the signal-to-mask thresholds, conforming to a preset perceptual model.

This psychoacoustic model is only found in the coder, thus making the decoder less complex and permitting the freedom to exploit future improvements in coder design. The actual number of levels for each quantizer is determined by the bit allocation. This is arrived at by setting the *signal-to-mask ratio* (SMR) parameter, defined as the difference between the minimum masking threshold and the maximum signal level. This minimum masking threshold is calculated using the psychoacoustic model and provides a reference noise level of "just noticeable" noise for each subband.

In the decoder, after demultiplexing and deciphering of the audio and side information data, a dual synthesis filter bank reconstructs the linear PCM signal in blocks of 32 output samples (Figure 5.17b).

A scale factor is determined for each 12 subband sample block. The maximum of the absolute values of these 12 samples generates a *scale factor* word consisting of 6 bits, a

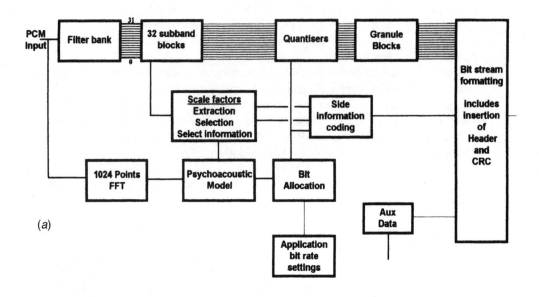

(a)

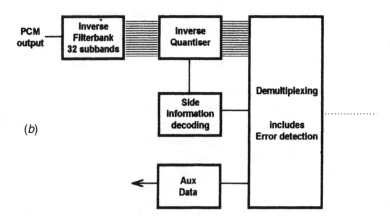

(b)

Figure 5.17 ISO/MPEG-1 Layer 2 system: (a) encoder block diagram, (b) decoder block dia-gram. (*After* [15].)

range of 63 different levels. Because each frame of audio data in Layer 2 corresponds to 36 subband samples, this process will generate 3 scale factors per frame. However, the trans-mitted data rate for these scale factors can be reduced by exploiting some redundancy in the data. Three successive subband scale factors are analyzed and a pattern is determined. This pattern, which is obviously related to the nature of the audio signal, will decide whether one, two or all three scale factors are required. The decision will be communicated by the insertion of an additional *scale factor select information* data word of 2 bits (SCFSI).

In the case of a fairly stationary tonal-type sound, there will be very little change in the scale factors and only the largest one of the three is transmitted; the corresponding data rate will be (6 + 2) or 8 bits. However, in a complex sound with rapid changes in content, the transmission of two or even three scale factors may be required, producing a maximum bit rate demand of (6 + 6 + 6 + 2) or 20 bits. Compared with Layer 1, this method of coding the scale factors reduces the allocation of data bits required for them by half.

The number of data bits allocated to the overall bit pool is limited or fixed by the data rate parameters. These parameters are set out by a combination of sampling frequency, compression ratio, and—where applicable—the transmission medium. In the case of 20 kHz stereo being transmitted over ISDN, for example, the maximum data rate is 384 kbits/s, sampling at 48kHz, with a compression ratio of 4:1.

After the number of side information bits required for scale factors, bit allocation codes, CRC, and other functions have been determined, the remaining bits left in the pool are used in the re-coding of the audio subband samples. The allocation of bits for the audio is determined by calculating the SMR, via the FFT, for each of the 12 subband sample blocks. The bit allocation algorithm then selects one of 15 available quantizers with a range such that the overall bit rate limitations are met and the quantization noise is masked as far as possible. If the composition of the audio signal is such that there are not enough bits in the pool to adequately code the subband samples, then the quantizers are adjusted down to a best-fit solution with (hopefully) minimum damage to the decoded audio at the output.

If the signal block being processed lies in the lower one third of the 32 frequency subbands, a 4-bit code word is simultaneously generated to identify the selected quantixer; this word is, again, carried as side information in the main data frame. A 3-bit word would be generated for processing in the mid frequency subbands and a 2-bit word for the higher frequency subbands. When the audio analysis demands it, this allows for *at least* 15, 7, and 3 quantization levels, respectively, in each of the three spectrum groupings. However, each quantizer can, if required, cover from 3 to 65,535 levels and additionally, if no signal is detected then no quantization takes place.

As with the scale factor data, some further redundancy can be exploited, which increases the efficiency of the quantising process. For the lowest quantizer ranges (i.e., 3, 5, and 9 levels), three successive subband sample blocks are grouped into a "granule" and this—in turn—is defined by only one code word. This is particularly effective in the higher frequency subbands where the quantizer ranges are invariably set at the lower end of the scale.

Error detection information can be relayed to the decoder by inserting a 16 bit CRC word in each data frame. This parity check word allows for the detection of up to three single bit errors or a group of errors up to 16 bits in length. A codec incorporating an error concealment regime can either mute the signal in the presence of errors or replace the impaired data with a previous, error free, data frame. The typical data frame structure for ISO/MPEG-1 Layer 2 audio is given in Figure 5.18

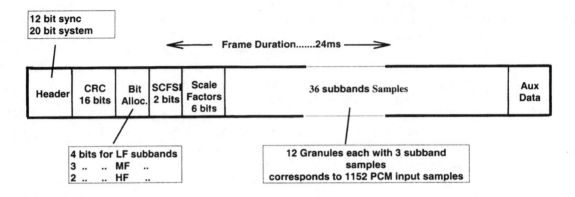

Figure 5.18 ISO/MPEG-1 Layer 2 data frame structure. (*After* [15].)

MPEG-2 AAC

Also of note is MPEG-2 *advanced audio coding* (AAC), a highly advanced perceptual code, used initially for digital radio applications. The AAC code improves on previous techniques to increase coding efficiency. For example, an AAC system operating at 96 kbits/s produces the same sound quality as ISO/MPEG-1 Layer 2 operating at 192 kbits/s—a 2:1 reduction in bit rate. There are three modes (Profiles) in the AAC standard:

- *Main*—used when processing power, and especially memory, are readily available.

- *Low complexity* (LC)—used when processing cycles and memory use are constrained.

- *Scaleable sampling rate* (SSR)—appropriate when a *scalable decoder* is required. A scalable decoder can be designed to support different levels of audio quality from a common bit stream; for example, having both high- and low-cost implementations to support higher and lower audio qualities, respectively.

Different Profiles trade off encoding complexity for audio quality at a given bit rate. For example, at 128 kbits/s, the Main Profile AAC code has a more complex encoder structure than the LC AAC code at the same bit rate, but provides better audio quality as a result.

A block diagram of the AAC system general structure is given in Figure 5.19. The blocks in the drawing are referred to as "tools" that the coding alogrithm uses to compress the digital audio signal. While many of these tools exist in most audio perceptual codes, two are unique to AAC—the *temporal noise shaper* (TNS) and the *filterbank* tool. The TNS uses a backward adaptive prediction process to remove redundancy between the frequency channels that are created by the filterbank tool.

MPEG-2 AAC provides the capability of up to 48 main audio channels, 16 low frequency effects channels, 16 overdub/multilingual channels, and 10 data streams. By comparison, ISO/MPEG-1 Layer 1 provides two channels and Layer 2 provides 5.1 channels (maximum). AAC is not backward compatible with the Layer 1 and Layer 2 codes.

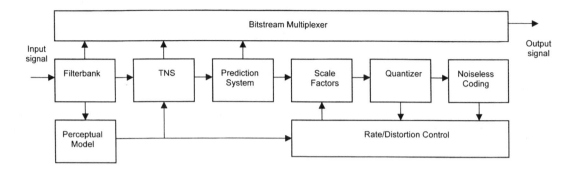

Figure 5.19 Functional block diagram of the MPEG-2 AAC coding system.

MPEG-4

MPEG-4, as with the MPEG-1 and MPEG-2 efforts, is not concerned solely with the development of audio coding standards, but also encompasses video coding and data transmission elements (as discussed previously in this chapter). In addition to building upon the audio coding standards developed for MPEG-2, MPEG-4 includes a revolutionary new element—synthesized sound. Tools are provided within MPEG-4 for coding of both natural sounds (speech and music) and for synthesizing sounds based on structured descriptions. The representations used for synthesizing sounds can be formed by text or by instrument descriptions, and by coding other parameters to provide for effects, such as reverberation and spatialization.

Natural audio coding is supported within MPEG-4 at bit rates ranging from 2–64 kbits/s, and includes the MPEG-2 AAC standard (among others) to provide for general compression of audio in the upper bit rate range (8–64 kbits/s), the range of most interest to broadcasters. Other types of coders, primarily voice coders (or *vocoders*) are used to support coding down to the 2 kbits/s rate.

For synthesized sounds, decoders are available that operate based on so-called *structured inputs*, that is, input signals based on descriptions of sounds and not the sounds themselves. Text files are one example of a structured input. In MPEG-4, text can be converted to speech in a *text-to-speech* (TTS) decoder. Synthetic music is another example, and may be delivered at extremely low bit rates while still describing an exact sound signal. The standard's *structured audio decoder* uses a language to define an orchestra made up of instruments, which can be downloaded in the bit stream, not fixed in the decoder.

TTS support is provided in MPEG-4 for unembellished text, or text with prosodic (pitch contour, phoneme duration, etc.) parameters, as an input to generate intelligible synthetic speech. It includes the following functionalities:

- Speech synthesis using the prosody of the original speech

- Facial animation control with phoneme information (important for multimedia applications)

- *Trick mode* functionality: pause, resume, jump forward, jump backward

- International language support for text

- International symbol support for phonemes

- Support for specifying the age, gender, language, and dialect of the speaker

MPEG-4 does not standardize a method of synthesis, but rather a method of describing synthesis.

5.8.8 Dolby E Coding System

Dolby E coding was developed to expand the capacity of existing two channel AES/EBU digital audio infrastructures to make them capable of carrying up to eight channels of audio plus the metadata required by the Dolby Digital coders used in the ATSC transmission system [17]. This allows existing digital videotape recorders, routing switchers, and other video plant equipment, as well as satellite and telco facilities, to be used in program contribution and distribution systems that handle multichannel audio. The coding system was designed to provide broadcast quality output even when decoded and re-encoded many times, and to provide clean transitions when switching between programs.

Dolby E encodes up to eight audio channels plus the necessary metadata and inserts this information into the payload space of a single AES digital audio pair. Because the AES protocol is used as the transport mechanism for the Dolby E encoded signal, digital VTRs, routing switchers, DAs, and all other existing digital audio equipment in a typical video facility can handle multichannel programming. It is possible to do insert or assemble edits on tape or to make audio-follow-video cuts between programs because the Dolby E data is synchronized with the accompanying video. The metadata is multiplexed into the compressed audio, so it is switched with and stays in sync with the audio.

The main challenge in designing a bit-rate reduction system for multiple generations is to prevent coding artifacts from appearing in the recovered audio after several generations. The coding artifacts are caused by a buildup of noise during successive encoding and decoding cycles, so the key to good multigeneration performance is to manage the noise optimally.

This noise is caused by the rate reduction process itself. Digitizing (quantizing) a signal leads to an error that appears in the recovered signal as a broadband noise. The smaller the quantizer steps (i.e., the more resolution or bits used to quantize the signal), the lower the noise will be. This quantizing noise is related to the signal, but becomes "whiter" as the quantizer resolution rises. With resolutions less than about 5 or 6 bits and no dither, the quantizing noise is clearly related to the program material.

Bit rate reduction systems try to squeeze the data rates down to the equivalent of a few bits (or less) per sample and, thus, tend to create quantizing noise in quite prodigious quantities. The key to recovering signals that are subjectively indistinguishable from the original signals, or in which the quantizing noise is inaudible, is in allocating the available bits to the program signal components in a way that takes advantage of the ear's natural ability to mask low level signals with higher level ones.

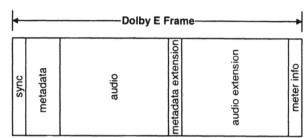

Figure 5.20 Basic frame structure of the Dolby E coding system. (*After* [18].)

The rate reduction encoder sends information about the frequency spectrum of the program signal to the decoder. A set of reconstruction filters in the decoder confines the quantizing noise produced by the bit allocation process in the encoder to the bandwidth of those filters. This allows the system designer to keep the noise (ideally) below the masking thresholds produced by the program signal. The whole process of allocating different numbers of bits to different program signal components (or of quantizing them at different resolutions) creates a noise floor that is related to the program signal and to the rate reduction algorithm used. The key to doing this is to have an accurate model of the masking characteristics of the ear, and in allocating the available bits to each signal component so that the masking threshold is not exceeded.

When a program is decoded and then re-encoded, the re-encoding process (and any subsequent ones) adds its noise to the noise already present. Eventually, the noise present in some part of the spectrum will build up to the point where it becomes audible, or exceeds the allowable *coding margin*. A codec designed for minimum data rate has to use lower coding margins (or more aggressive bit allocation strategies) than one intended to produce high quality signals after many generations

The design strategy for a multigeneration rate reduction system, such as one used for Dolby E, is therefore quite different than that of a minimum data rate codec intended for program transmission applications.

Dolby E signals are carried in the AES3 interface using a packetized structure [18]. The packets are based on the coded Dolby E frame, which is illustrated in Figure 5.20. Each Dolby E frame consists of a *synchronization field*, *metadata field*, *coded audio field*, and a *meter field*. The metadata field contains a complete set of parameters so that each Dolby E frame can be decoded independently. The Dolby E frames are embedded into the AES3 interface by mapping the Dolby E data into the audio sample word bits of the AES3 frames utilizing both channels within the signal. (See Figure 5.21.) The data can be packed to utilize 16, 20, or 24 bits in each AES3 sub-frame. The advantage of utilizing more bits per subframe is that a higher data rate is available for carrying the coded information. With a 48 kHz AES3 signal, the 16 bit mode allows a data rate of up to 1.536 Mbits/s for the Dolby E signal, while the 20 bit mode allows 1.92 Mbits/s. Higher data rate allows more generations and/or more channels of audio to be supported. However, some AES3 data paths may be restricted in data rate (e.g., some storage devices will only record 16 or 20 bits). Dolby E therefore allows the user to choose the optimal data rate for a given application.

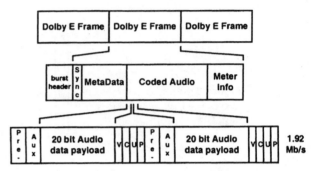

Figure 5.21 Overall coding scheme of Dolby E. (*After* [18].)

5.8.9 Objective Quality Measurements

Perceptual audio coding has revolutionized the processing and distribution of digital audio signals. One aspect of this technology, not often emphasized, is the difficulty of determining, *objectively*, the quality of perceptually coded signals. Audio professionals could greatly benefit from an objective approach to signal characterization because it would offer a simple but accurate approach for verification of good audio quality within a given facility.

Most of the discussions regarding this topic involve reference to the results of subjective evaluations of audio quality, where for example, groups of listeners compare reference audio material to coded audio material and then judge the *level of impairment* caused by the coding process. A procedure for this process has been standardized in ITU-R Rec. BS.1116, and makes use of the ITU-R five grade impairment scale:

- 5.0—Imperceptible

- 4.0—Perceptible but not annoying

- 3.0—Slightly annoying

- 2.0—Annoying

- 10—Very annoying

Quality measurements made with properly executed subjective evaluations are widely accepted and have been used for a variety of purposes, from determining which of a group of perceptual coders performs best, to assessing the overall performance of an audio broadcasting system.

The problem with subjective evaluations is that, while accurate, they are time consuming and expensive to undertake. Traditional objective benchmarks of audio performance, such as signal-to-noise ratio or total harmonic distortion, are not reliable measures of perceived audio quality, especially when perceptually coded signals are being considered.

To remedy this situation, ITU-R established Task Group 10-4 to develop a method of objectively assessing perceived audio quality. Conceptually, the result of this effort would be a device having two inputs—a reference and the audio signal to be evaluated—and would generate an audio quality estimate based on these sources.

Table 5.6 Target Applications for ITU-R Rec. BS.1116 PEAQ

Category	Application	Version
Diagnostic	Assessment of implementations	Both
	Equipment or connection status	Advanced
	Codec identification	Both
Operational	Perceptual quality line-up	Basic
	On-line monitoring	Basic
Development	Codec development	Both
	Network planning	Both
	Aid to subjective assessment	Advanced

Six organizations proposed models for accomplishing this objective, and over the course of several years these models were evaluated for effectiveness, in part by using source material from previously documented subjective evaluations. Ultimately, the task group decided that none of the models by themselves fully met the stated requirements. The group decided, instead, to use the best parts of the different models to create another model that would meet the sought-after requirements.

This approach resulted in an objective measurement method known as *Perceptual Evaluation of Audio Quality* (PEAQ). The method contains two versions—a basic version designed to support real-time implementations, and an advanced version optimized for the highest accuracy but not necessarily implementable in real-time. The primary applications for PEAQ are summarized in Table 5.6.

5.8.10 Perspective on Audio Compression

A balance must be struck between the degree of compression available and the level of distortion that can be tolerated, whether the result of a single coding pass or the result of a number of passes, as would be experienced in a complex audio chain or network [15]. There have been many outstanding successes for digital audio data compression in communications and storage, and as long as the limitations of the various compression systems are fully understood, successful implementations will continue to grow in number.

Compression is a tradeoff and in the end you get what you pay for. Quality must be measured against the coding algorithm being used, the compression ratio, bit rate, and coding delay resulting from the process.

There is continued progress in expanding the arithmetical capabilities of digital signal processors, and the supporting hardware developments would seem to be following a parallel course. It is possible to obtain a single chip containing both encoder and decoder elements, including stereo capabilities. In every five year period, it is not unreasonable to expect a tenfold increase in the processing capabilities of a single DSP chip, thus, increasing flexibility and processing power. Speculation could point to an eventual position when a completely lossless algorithm with an extremely high compression ratio would become available. In any event, the art of compressing audio data streams into narrower and narrower digital pipes will undoubtedly continue.

5.9 References

1. Lakhani, Gopal: "Video Compression Techniques and Standards," *The Electronics Handbook*, Jerry C. Whitaker (ed.), CRC Press, Boca Raton, Fla., pp. 1273–1282, 1996.

2. Solari, Steve. J.: *Digital Video and Audio Compression*, McGraw-Hill, New York, 1997.

3. Netravali, A. N., and B. G. Haskell: *Digital Pictures, Representation, and Compression*, Plenum Press, 1988.

4. Gilge, M.: "Region-Oriented Transform Coding in Picture Communication," *VDI-Verlag, Advancement Report, Series 10*, 1990.

5. DeWith, P. H. N.: "Motion-Adaptive Intraframe Transform Coding of Video Signals," *Philips J. Res.*, vol. 44, pp. 345–364, 1989.

6. Isnardi, M., and T. Smith: "MPEG Tutorial," *Proceedings of the Advanced Television Summit*, Intertec Publishing, Overland Park, Kan., 1996.

7. Nelson, Lee J.: "Video Compression," *Broadcast Engineering*, Intertec Publishing, Overland Park, Kan., p. 42, October 1995.

8. Arvind, R., et al.: "Images and Video Coding Standards," *AT&T Technical J.*, p. 86, 1993.

9. SMPTE 308M, "MPEG-2 4:2:2 Profile at High Level," SMPTE, White Plains, N.Y., 1998.

10. ATSC, "Guide to the Use of the ATSC Digital Television Standard," Advanced Television Systems Committee, Washington, D.C., doc. A/54, Oct. 4, 1995.

11. "IEEE Standard Specifications for the Implementation of 8×8 Inverse Discrete Cosine Transform," std. 1180-1990, Dec. 6, 1990.

12. Nelson, Lee J.: "Video Compression," *Broadcast Engineering*, Intertec Publishing, Overland Park, Kan., pp. 42–46, October 1995.

13. Smith, Terry: "MPEG-2 Systems: A Tutorial Overview," Transition to Digital Conference, *Broadcast Engineering*, Overland Park, Kan., Nov. 21, 1996.

14. SMPTE Recommended Practice: RP 202 (Proposed), "Video Alignment for MPEG-2 Coding," SMPTE, White Plains, N.Y., 1999.

15. Wylie, Fred: "Audio Compression Technologies," in *NAB Engineering Handbook*, 9th ed., Jerry C. Whitaker (ed.), National Association of Broadcasters, Washington, D.C., 1998.

16. Wylie, Fred: "Audio Compression Techniques," *The Electronics Handbook*, Jerry C. Whitaker (ed.), CRC Press, Boca Raton, Fla., pp. 1260–1272, 1996.

17. Lyman, Stephen, "A Multichannel Audio Infrastructure Based on Dolby E Coding," *Proceedings of the NAB Broadcast Engineering Conference*, National Association of Broadcasters, Washington, D.C., 1999.

18. Terry, K. B., and S. B. Lyman: "Dolby E—A New Audio Distribution Format for Digital Broadcast Applications," *International Broadcasting Convention Proceedings*, IBC, London, England, pp. 204–209, September 1999.

5.10 Bibliography

Bennett, Christopher: "Three MPEG Myths," *Proceedings of the 1996 NAB Broadcast Engineering Conference*, National Association of Broadcasters, Washington, D.C., pp. 129–136, 1996.

Bonomi, Mauro: "The Art and Science of Digital Video Compression," *NAB Broadcast Engineering Conference Proceedings*, National Association of Broadcasters, Washington, D.C., pp. 7–14, 1995.

Brandenburg, K., and Gerhard Stoll: "ISO-MPEG-1 Audio: A Generic Standard for Coding of High Quality Digital Audio," *92nd AES Convention Proceedings*, Audio Engineering Society, New York, N.Y., 1992, revised 1994.

Dare, Peter: "The Future of Networking," *Broadcast Engineering*, Intertec Publishing, Overland Park, Kan., p. 36, April 1996.

Fibush, David K.: "Testing MPEG-Compressed Signals," *Broadcast Engineering*, Overland Park, Kan., pp. 76–86, February 1996.

Freed, Ken: "Video Compression," *Broadcast Engineering*, Overland Park, Kan., pp. 46–77, January 1997.

IEEE Standard Dictionary of Electrical and Electronics Terms, ANSI/IEEE Standard 100-1984, Institute of Electrical and Electronics Engineers, New York, 1984.

Jones, Ken: "The Television LAN," *Proceedings of the 1995 NAB Engineering Conference*, National Association of Broadcasters, Washington, D.C., p. 168, April 1995.

Smyth, Stephen: "Digital Audio Data Compression," *Broadcast Engineering*, Intertec Publishing, Overland Park, Kan., February 1992.

Stallings, William: *ISDN and Broadband ISDN*, 2nd Ed., MacMillan, New York.

Taylor, P.: "Broadcast Quality and Compression," *Broadcast Engineering*, Intertec Publishing, Overland Park, Kan., p. 46, October 1995.

Whitaker, Jerry C., and Harold Winard (eds.): *The Information Age Dictionary*, Intertec Publishing/Bellcore, Overland Park, Kan., 1992.

6

High-Definition Production Systems

6.1   Introduction

The distinction between a production system for HDTV and a system for the transmission of high-definition images is an important one. Because of their closed-loop characteristics, production systems can be of any practical design. The process of developing a production system can focus simply on those who will directly use the system. It is not necessary to consider the larger issues of compatibility and public policy, which drive the design and implementation of over-the-air broadcast systems.

Although the foregoing is certainly correct, in the abstract, it is obvious that the economies of scale argue in favor of the development of a production system—even if only closed-loop—that meets multiple applications. The benefits of expanded markets and interoperability between systems are well documented. It was into this environment that production systems intended for HDTV applications were born.

6.2   The 1125/60 System

The 1125-line HDTV system is the grandfather of advanced television systems [1]. The 1125/60 format has been under development for more than 20 years. During that time, a full range of equipment for program origination, distribution, transmission, and display has been designed and offered commercially. A satellite-transmitted version, MUSE, began broadcasting to viewers in Japan in 1990.

The emerging "information era" increasingly demands an all-inclusive, global electronic network of services built upon a practical television production standard. Responding to a mandate set forth by the broadcasters of the world in 1983, the technical community responded with 1125/60, a working HDTV production format that satisfied the desires that had been unanimously voiced. Occasional suggestions that 1125/60 was merely a "Japanese" standard ignore historical reality. Although it is true that NHK can be credited with the pioneering efforts in HDTV, the 1125/60 production standard was a significant improvement over the originally proposed NHK system. The work of international committees—including the CCIR, SMPTE, and ATSC—resulted in numerous modifications and enhancements, especially in the areas of bandwidth, digitization, colorimetry, aspect ratio, and system interconnectability.

Any proposed standard will raise regional concerns, but the concept of a unified HDTV standard for studio origination and program exchange remains as valid today as it was in 1983, when broadcasters resolved to pursue that goal.

6.2.1 Fundamental Principles of 1125/60

It is understandable to question why North America would adopt an 1125/60-based signal format as a production standard within an environment founded upon a 525/59.94 (NTSC) scanning system [1]. The simple answer is that in seeking a future all-electronic production format, North America was cognizant of the international aspects of a new production standard. The precedent of 24-frame 35 mm film demonstrated the enormous value of a single universal standard that could be used by all program producers worldwide. The ease with which original program material could be exchanged around the globe, and subsequently transformed into high-quality versions of any of the world's broadcast television transmission systems, served as a powerful model for the long-term future of video. An HDTV production medium sought no less.

The original extensive psychophysical research underlying the 1125/60 system never claimed a magic set of scanning numbers that would define a unique, exceptional-performance HDTV system. Rather, this work produced broad guidelines for an electronic imaging system that would compare favorably with projected large-screen 35 mm film. These guidelines established important minimum requirements for such imagery, including:

- A number of scanning lines in excess of 1000 for the active picture display

- Wider aspect ratio in the general neighborhood of 16:9

- Picture field refresh rate of about 60 Hz

- Displayed picture vertical and horizontal resolution in excess of 700 TVL/ph (television lines per picture height)

- Substantial increase of color resolution compared with 525-line NTSC and 625-line PAL and SECAM systems

- Elimination of NTSC-type encoding artifacts

The aggregate of all these attributes would result in a subjective picture quality that would meet the needs of high-quality program origination commensurate with that offered by 35 mm film.

The next task was to identify specific parameters that would define a television system to form the basis of a future HDTV standard. With these criteria as a starting point, the next step was to attempt to relate the HDTV scanning parameters as optimally as possible to the disparate broadcast television standards already in place around the world. This was a definitive effort designed to facilitate convenient electronic downconversion to each of the existing television distribution systems, in the interest of establishing a broad program-distribution mechanism, an issue of considerable importance to program producers.

The 1125 format emerged as a horizontal line rate that bore a simple integer ratio relationship with the two primary line formats—525 and 625. The lowest pragmatic number above the criteria of 1000 TV lines was met by the choice of 1125. When only the active television lines of these systems (that is, excluding the vertical blanking lines, which do not contribute to picture portrayal) are considered, the relationships still hold true:

- 1035 active lines in the 1125 system

- $1035 = (15/7) \times 483$ for the 525 system

- $1035 = (9/5) \times 575$ for the 625 system

Thus, 1125 emerged as a practical and sufficient line number that would meet the basic criteria for HDTV vertical resolution and allow relatively easy downconversion to all existing television broadcast standards. The universality of future HDTV program production would be protected; 1125 represented a suitable common number that allowed line conversion to either international system via relatively simple integral ratios.

Obviously, a choice of 1050 would have been particularly favorable to the NTSC 525 system, producing a conversion ratio of 2:1 instead of 15:7. However, the 9:5 simplicity for downconversion to 625 would have been severely compromised, replaced by a 42:25 ratio that would have been far more cumbersome in digital filter implementations. Therefore, although 1125 might have seemed to represent a slightly "less friendly" choice for North America, it was preferable in the larger interests of achieving a worldwide HDTV production standard. In fact, 1125 is a better choice for North America because downconversion to 525 requires interpolation regardless of the HDTV line number. The higher overhead in 1125 (relative to 1050) allows for better interpolation, hence better 525 video.

6.2.2 The Standardization of 1125/60 HDTV

In February 1985 the SMPTE working group on high-definition electronic production (WG-HDEP) reached a first significant milestone following two years of intensive work [1]. The panel voted unanimously to recommend the parameters of an 1125/60 scanning format, having an aspect ratio of 16:9 and 2:1 interlace, to be the recommended submission by the United States for a single worldwide standard for studio origination and international program exchange. The efforts of many individuals and organizations from the broadcast, video, and film production communities of North America had led to an important consensus on an HDTV production system suitable for global applications.

This recommendation was sent to the Advanced Television Systems Committee, which—following the carefully established due process of review and voting by the HDTV T3 technology group, the executive committee, and the general membership—also voted affirmatively. In late 1985, the ATSC submitted this recommendation to the U.S. State Department delegates to the CCIR who, in turn, officially presented it to the CCIR.

Meanwhile, worldwide HDTV activities sped up. The SMPTE WG-HDEP continued its studies on details such as transfer characteristic, colorimetry, test materials, and film/video interchange. Program producers in the United States, Japan, Canada, and Europe put the first 1125/60 HDTV systems into service. By 1986, a wide variety of serious international program productions were under way, including two feature-length movies.

The first production hardware quickly revealed the need for a detailed system definition and a more unified standard that would ensure a proper interface with the 1125/60 equipment being produced by international manufacturers. In late 1986 SMPTE formalized a vigorous effort to accomplish this work. A special ad hoc subgroup was formed to produce documentation of all of the disparate 1125/60 systems and was charged with the task of seeking a single set of parameters that would closely define a unified consensus and form the basis for a standard 1125/60 system.

The Broadcast Television Association (BTA) of Japan—representing some 15 manufacturers and many broadcasters—simultaneously initiated a parallel effort. Formal liaison between SMPTE and BTA was established, and intensive activity ensued. By the fall of 1987, the work was essentially complete. SMPTE and BTA produced draft standards for 1125/60 that were identical; these were subsequently endorsed by all participating manufacturers. The proposed standard included specific definition of the following:

- Scanning format

- System colorimetry

- Electro-optical transfer characteristic

- Video component sets

- Video synchronizing waveforms and timing

- Bandwidths

Groundwork also was laid for subsequent SMPTE work on a recommended interface covering the following parameters:

- Video signal levels

- Impedance

- Clamping and signal dc content

- Connectors

- Cable

Paralleling the efforts of the standards subgroup were those of another subgroup charged with the task of defining a total systems approach to colorimetry and transfer characteristics. Driven by the early recognition of the importance of HDTV within the total future program production scenario, the need for high-quality film/HDTV interchange (in both directions) assumed special significance. An alliance between the North American broadcast and film communities produced a precise definition of colorimetry/gamma from the camera to the display, with a wide spectral taking characteristic that would afford excellent film/video interchange. The studies also anticipated new display technologies in the future; the system was tailored to ensure a defined display colorimetry, regardless of display type (direct view, projector, or flat panel). BTA endorsed the SMPTE work and incorporated the same parameters within its document.

In August 1987 the WG-HDEP voted unanimously on the proposed detailed standard for 1125/60 and passed it on to the technology committee of SMPTE and to the ATSC as a first step in due process toward formal standardization. By February 1988, SMPTE had processed the document and agreed to proceed to ANSI. The general membership of ATSC also voted (by the requisite two-thirds majority) to endorse the same parameters, thus defining 1125/60 as a voluntary national ATSC standard for HDTV studio origination.

6.2.3 HDTV Production Standard

No other single topic in television has preoccupied the attention of standardization committees worldwide like the quest to develop a high-quality HDTV studio origination standard. The effort was complicated by the desire for international unanimity on a single worldwide HDTV system. Securing a new global flexibility in high-performance television production, postproduction, and international program exchange was high on the agendas of participants within the television industry the world over.

From all this work, the parameters of a durable origination standard have been structured to endow the HDTV studio format with more electronic information than is required to satisfy, for example, the psychophysical aspects of large-screen viewing of the picture. This additional information translates into the technical overhead that can sustain complex picture processing in postproduction, multigeneration recording, downconversion, transfer to film, and other operations.

Nevertheless, the establishment of a strong production standard was essential to the definition of some of the more fundamental parameters of a broad-based HDTV system. The sheer breadth of the television industry today demands flexibility in the deployment of an HDTV format. The concept of a system hierarchy has been discussed and, indeed, a carefully structured hierarchy offers an ideal methodology for tailoring a basic HDTV system to a wide variety of applications while meeting specific performance requirements, system needs, and budgets.

Within business and industrial (B&I) applications—and other nonentertainment video uses—there exists a wide diversity of imaging requirements. The picture quality and system facilities needed may be unique to each application. The spearhead of a B&I HDTV evolution, however, was recognized to be dependent on a lower-cost method for storing and distributing HDTV images. Central to the realization of such systems is the challenge of reducing the bandwidth of the HDTV studio signal without compromising the essence of the high-definition image.

6.2.4 Technical Aspects of SMPTE 240M

SMPTE formalized the technical parameters of the 1125/60 high-definition production standard in April 1987, with the SMPTE document (240M-1988) issued shortly thereafter. The basic elements of the HDTV production standard are as follows:
Scanning

- 1125 total lines per frame

- 1035 total active lines per frame

- 16:9 aspect ratio

- 2:1 interlace-scanning structure

- Duration of active horizontal picture period = 25.86 μs (total line time = 29.63 μs)

- Duration of horizontal blanking period = 3.77 μs

- 60 Hz field rate

Bandwidth and resolution

- Luminance = 30 MHz

- Two color-difference signals = 15 MHz

- Horizontal luminance resolution = 872.775 (872) TVL/ph

- Vertical luminance resolution = 750 TVL/ph

During its consideration of the technical parameters of the HDTV production format, the WG-HDEP modified the horizontal blanking interval to accommodate an improved sync waveform. The final decision involved a highly complex relationship among five factors:

- Aspect ratio

- Digital coding and sampling frequency

- Preservation of geometric compatibility with existing aspect ratio software

- Practical limitations in camera and monitor scanning circuits at the time

- Characteristics of the new sync waveform

SMPTE and the ATSC, working with the BTA in Japan, established a horizontal blanking interval that satisfied the requirements outlined.

Sync Waveform

Both the ATSC and SMPTE identified a potential problem for implementing early HDTV 1125/60 systems: different synchronizing waveforms from various 1125/60 equipment manufacturers [1]. The ATSC and SMPTE searched for a single sync waveform standard that would ensure system compatibility. Other objectives were precise synchronization and relative timing of the three component video signals, and a sync structure sufficiently robust to survive multigeneration recording and other noisy environments.

The SMPTE/ATSC/BTA standard sync agreed upon was a trilevel bipolar waveform with a large horizontal timing edge occupying the center of the video signal dynamic range. To permit the implementation of a precise, minimum-jitter sync separator system, the timing edge had a defined midpoint centered on the video blanking level. The sync system was tested extensively. Timing accuracy was maintained even in noisy environments.

While refining the basic parameters of the scanning format, SMPTE also examined electro-optical transfer and system colorimetry.

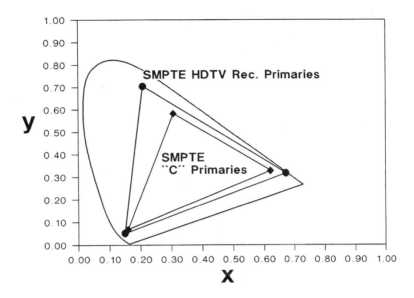

Figure 6.1 Color gamut curve for SMPTE 240M. (*From* [1]. *Courtesy of Sony.*)

HDTV Colorimetry

In NTSC, the highly nonlinear transfer characteristic of studio monitors and home TV receivers forces the use of precorrection with electronic *gamma correction* in the TV camera [1]. Although this system works as intended, it would not be able to accommodate the different gamma characteristics of future HDTV displays using LCD, laser, plasma, and other technologies yet to be invented.

The SMPTE working group considered the case for improved colorimetry: to ensure high-quality film/HDTV interchange, to provide enhanced TV display, and to meet the requirements for digital implementation. These demands led the working group to specify the camera transfer characteristic with high mathematical precision, a guideline intended to lead HDTV camera designers toward a predictable and unified specification. The highly precise specification makes it possible to perform precise linearization of the signal, thus permitting digital processing on linear signals when required. The working group's comprehensive definition of a total system of television colorimetry and transfer characteristic may be its greatest accomplishment. The specification began with a substantially broader spectral-taking characteristic that promised to revolutionize film/tape interchange. Colorimetry and gamma were precisely specified throughout the reproduction chain. Figure 6.1 shows the curve for reference primaries.

NTSC color television always has had a less expressive color palette than color photography. More than just a philosophical issue, this constraint becomes a practical concern whenever the broad color gamut of film is transferred onto the narrow color gamut of television. A more extensive color gamut, long thought desirable for broadcasting, is absolutely essential for a better fit with the capabilities of computer graphics, tape-to-film transfer, and print

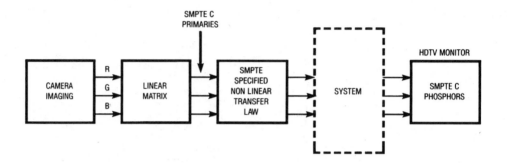

Figure 6.2 Colorimetry process under the initial (step one) guidelines of SMPTE 240M. (*From* [1]. *Used with permission.*)

media. Because of these requirements, SMPTE was determined to establish improved colorimetry for HDTV 1125/60.

The difficulties in fully describing television colorimetry are well known. They had impeded the practical emergence of a standard for colorimetry within the present 525-line system. However, through their combined efforts, the film and video industries successfully achieved a standard for HDTV production. The secret to this success was the active dialogue within SMPTE and BTA, as well as between the program production community and the manufacturers of cameras and displays. Candid revelations as to the practical constraints of current camera and display technology led to a pragmatic—but quite ingenious—2-step approach to building hardware.

Step One: The Near Term

The SMPTE standard took a significant step toward coordinating HDTV cameras and monitors [1]. This effort included the following recommendations:

- All HDTV monitor manufacturers were urged to conform to a single phosphor set: the SMPTE C phosphors that are the de facto standard in North America for 525-line studio monitors.

- All camera manufacturers were urged to conform HDTV camera colorimetry to a SMPTE C reference primary and to a D65 reference white.

- All camera manufacturers were urged to incorporate a nonlinear transfer circuit of precise mathematical definition.

Through official liaison with SMPTE, Japan's BTA examined the U.S. recommendations and fully accepted them. The SMPTE standards subsequently were written into the BTA studio standard document. The step-one process is illustrated in Figure 6.2.

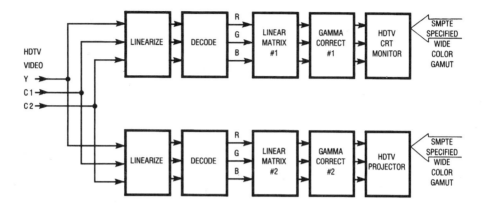

Figure 6.3 Colorimetry process under the secondary (step two) guidelines of SMPTE 240M. (*From* [1]. *Used with permission.*)

Step Two: Into the Future

The second step makes a more dramatic leap toward wide color gamut colorimetry, defined from camera origination through to final display [1]. In addition, step two ensures accurate and equal reproduction on displays employing vastly different technologies.

The standard called for the unprecedented use of nonlinear electronic processing within the display. First, the source HDTV component signal would be linearized (which can now be accurately implemented because of the precisely specified camera nonlinear transfer characteristic), then decoded to RGB, if necessary. The signal would next be tailored to the specific electro-optical characteristics of the display via linear matrix and gamma correction circuits. As a result, the full range of current and future displays could portray identical colorimetry. This process is illustrated in Figure 6.3.

6.2.5 Bandwidth and Resolution Considerations

The collective parameters that contribute to any scanning TV system are contained within a well-known expression that relates the *minimum bandwidth* required to reproduce all of the information within a given television standard to the scanning parameters themselves [1]. Specifically:

$$F_{min} = \frac{I}{2} \times K \times A \times N^2 \times f_F \frac{(1 + 1/K_h)}{(1 + 1/K_v)} \tag{6.1}$$

Where:
I = interlace ratio
K = Kell factor, generally assumed to have an approximate value of 0.7
A = aspect ratio

N = total number of scanning lines in a single interlaced frame
f_F = frame repetition rate
K_n = horizontal retrace ratio
K_v = vertical retrace ratio

This basic relationship permits the effects of individual scanning parameters of a TV system to be examined.

Bandwidth Considerations

Equation 6.1 demonstrates the dependence of bandwidth on the number of scanning lines (a square power law) [1]. It is worthwhile to examine HDTV from the viewpoint of bandwidth requirements. If, for example, values for the 1125/60 system are entered:

$$F_{min} = 1 \times 0.7 \times 1.77 \times 1125^2 \times 30 \frac{(1 + 0.13)}{(1 + 0.06)} = 25 \ \text{MHz} \tag{6.2}$$

The foregoing demonstrates that a minimum bandwidth of 25 MHz is required to properly reproduce this television standard. Furthermore:

$$H_{res} = \frac{2 \times T_A}{A} \times f_b \tag{6.3}$$

Where:
H_{res} = horizontal resolution in lines of picture height
T_A = active picture period
A = aspect ratio
f_b = bandwidth (MHz)

Considering the 525-line NTSC system with its 4.2 MHz maximum bandwidth, it follows that:

$$H_{res} = \frac{2 \times (52.6 \times 10^{-6})}{1.33} \times (4.2 \times 10^{-6}) = 332 \ \text{TVL/ph} \tag{6.4}$$

Vertical resolution is determined by the number of active TV lines, which for the 525 NTSC system is given by:

$$525 - (2 \times 21) = 483 \ \text{TVL} \tag{6.5}$$

In practice, this equation is modified by the Kell and interlace factors, generally assumed to have a value of approximately 0.7. Therefore, effective vertical resolution for 525 NTSC is:

$$483 \times 0.7 = 338 \ \text{TVL/ph} \tag{6.6}$$

If this approach is applied to the 1125/60 HDTV system, the result is:

$$H_{res} = \frac{2 \times T_A}{A} \times f_b \tag{6.7}$$

$$H_{res} = \frac{2 \times (25.86 \times 10^{-6})}{1.77} \times (25 \times 10^6) = 730 \ \text{TVL/ph} \tag{6.8}$$

Again, the vertical resolution is given by the number of active lines, which, in this case, has been defined to be 1035. The Kell and interlace factors again come into play, reducing the effective vertical definition to:

$$1035 \times 0.7 = 725 \ \text{TVL} \tag{6.9}$$

Comparing the 1125/60 HDTV system with the 525 NTSC system:
Horizontal resolution (TVL/ph)

- 332 lines for NTSC

- 730 lines for HDTV

Vertical resolution (TVL)

- 338 lines for NTSC

- 725 lines for HDTV

It can be seen that the 1125/60 HDTV system more than meets the established criterion of at least doubling both horizontal and vertical resolution of NTSC.

HDTV Video Component Considerations

The original NHK HDTV proposal had recommended a set of primary signals that were composed of a luminance signal (Y) and two separate color-difference components (C_W and C_N) [1]. The bandwidths proposed by NHK were as follows:

- $Y = 20$ MHz

- $C_W = 7.5$ MHz

- $C_N = 5.5$ MHz

In the United States, however, the ATSC focused on the first step—namely, choosing video components that would form the basis for the highest-quality HDTV studio origination. Basic to some of the choices was an understanding of the importance of facilitating the conversion process from 1125/60 to the current 525/60 and 625/50 systems, based upon ITU-R Rec. 601 (not just NTSC, PAL, and SECAM). The first key premise established was that the horizontal resolution be twice that of the 4:2:2 code specified in ITU-R Rec. 601, not twice that of conventional 525 NTSC. In the 4:2:2 code, the luminance video signal has 720 active samples per line. Hence, for HDTV the number of active samples should be:

$$720 \times \frac{5.33}{4} \times 2 = 1920 \ \text{active samples} \tag{6.10}$$

The 4:2:2 code in ITU-R Rec. 601 refers to the digital sampling pattern and the resolution characteristics of the studio standard for the video components of luminance and two color-difference frequencies. A primary sampling frequency of 13.5 MHz/4 is specified in ITU-R Rec. 601. This produces (for ITU-R Rec. 601) the following sampling frequencies:

$$Y = \frac{13.5}{4} \times 4 = 13.5 \text{ MHz} \tag{6.11}$$

$$R - Y = \frac{13.5}{4} \times 2 = 6.75 \text{ MHz} \tag{6.12}$$

$$B - Y = \frac{13.5}{4} \times 2 = 6.75 \text{ MHz} \tag{6.13}$$

It is desirable for conversion purposes to preserve simple clocking relationships between HDTV and the 4:2:2 digital codes. The ATSC and SMPTE, after some consideration, adopted a code of 22:11:11. This code produced for the 1125/60 system the following parameters:

$$Y = \frac{13.5}{4} \times 22 = 74.25 \text{ MHz} \tag{6.14}$$

$$R - Y = \frac{13.5}{4} \times 11 = 37.625 \text{ MHz} \tag{6.15}$$

$$B - Y = \frac{13.5}{4} \times 11 = 37.625 \text{ MHz} \tag{6.16}$$

Assuming the same order filters in the D/A converters as ITU-R Rec. 601, then the bandwidths of the analog HDTV video components become:

$$Y = \frac{74.25}{2.45} = 30 \text{ MHz} \tag{6.17}$$

$$R - Y = \frac{37.625}{2.45} = 15 \text{ MHz} \tag{6.18}$$

$$B - Y = \frac{37.625}{2.45} = 15 \text{ MHz} \tag{6.19}$$

This produces a horizontal luminance resolution of:

$$H_{res} = \frac{2 \times T_A}{A} \times f_b \tag{6.20}$$

$$H_{res} = \frac{2 \times (25.86 \times 10^{-6})}{1.77} \times (30 \times 10^{6}) = 876 \text{ TVL/ph} \tag{6.21}$$

This figure represents performance beyond what was originally assumed. Hence the SMPTE 240M standard includes significant headroom insofar as the minimum horizontal resolution is concerned.

6.2.6 Interlace and Progressive Scanning

During the many deliberations on HDTV, various potential weaknesses of the 1125/60 approach were postulated, and alternative approaches were offered [1]. Prominent among these discussions were issues relating to:

- Frame rate and motion portrayal

- Impact of interlace scanning on motion and on aliasing artifacts

- Impact of interlace scanning on digital processing techniques

Of the various alternative systems proposed in the late 1980s, that of RCA/NBC probably drew the most interest, because it rested squarely on the premise that a superior new TV system should avoid all of the historic limitations of interlace and, instead, should be based upon a progressive-scanning system. To simply take the 1125/60 system and make it progressive scanning would, unfortunately, dramatically elevate the bandwidth according to:

$$F_{min} = \frac{0.7 \times 1.77 \times 1125^2 \times 60}{2} \times \frac{1.13}{1.06} = 50 \ \text{MHz} \tag{6.22}$$

Few disputed that this would represent a magnificent HDTV signal, but because of the enormous bandwidth, practical implementation would await radically new technologies. RCA/NBC, accordingly, carefully structured the system to utilize approximately the same bandwidth as the 1125/60 system, but traded line numbers for the elevated frame rate of progressive scanning. They suggested a 750-line system, which produced:

$$F_{min} = \frac{0.7 \times 1.77 \times 750^2 \times 60}{2} \times 1.1 = 23 \ \text{MHz} \tag{6.22}$$

A prototype system was assembled and demonstrated, side by side with the 1125/60 interlaced system. With both HDTV cameras viewing the same scene and directly feeding their respective monitors, it was generally agreed that, for normal scene material, the two systems produced subjective results that were very close. But the inherent strength of the 1125/60 system clearly emerged when the two systems were subjected to tests that embraced the broader aspects of real-world applications. Notable among these was the transfer of HDTV original material to 35 mm film.

Video-to-film transfer is a highly complex process involving the tandem connection of many elements of a total system. The weaknesses of each element cascade. The inherent information—the actual number of horizontal and vertical samples—in the originating picture source becomes crucial to the survival of the signal through this complex transformation.

The debate over interlace vs. progressive scanning has, of course, been going on for years. The controversy intensified during the work on the Grand Alliance DTV standard and the SMPTE 1125-line production standard before it, but progressive scanning had the historical advantage, interlace having been hailed as a great invention when it was developed in the 1930s [2]. Interlace provided, at that time, a badly needed increase in large-area flicker frequency with no increase in bandwidth, or—for a given flicker frequency and bandwidth—an increase in resolution. Interlace has its price, however [3]. The interlace artifacts that have been a part of the "video look" for more than 5 decades finally are being eliminated with the implementation of DTV and HDTV.

6.2.7 Digital Representation of SMPTE 240M

Following the emergence of the SMPTE 240M standard, the need for hardware and software tools for the digital capture, storage, transmission, and manipulation of HDTV images in the 1125/60 format created a sense of urgency within the SMPTE WG-HDEP to complete a digital characterization of 1125/60 HDTV signal parameters [1]. The WG-HDEP created an ad hoc group on the digital representation of 1125/60 with a charter to study and document the digital parameters of the basic system, as defined within the body of SMPTE 240M. The unified digital description of the 1125/60 HDTV signal was expected to stimulate the development of all-digital equipment and to enhance the development of universal interfaces for the interconnection of digital HDTV equipment from various manufacturers. Indeed, as SMPTE worked toward internal consensus, some manufacturers were committing to the original ATSC digital recommendations—even before the standardization process was complete.

To fulfill its task, the ad hoc group brought together a cross section of industry experts, including:

- Technical representatives of international manufacturers of HDTV equipment

- Designers of digital video-processing and computer graphics equipment

- Current users of 4:2:2 digital 525-/625-line equipment

- Motion picture engineers, who sought to ensure the highest standards of image quality for motion-picture-related HDTV imaging

- Technical members of various broadcast and research organizations

Numerous studies and recommendations were brought into focus. A document for the digital representation of and the design of a bit-parallel digital interface for the 1125/60 studio HDTV standard was agreed to in 1992 (SMPTE 260M-1992). Specific areas addressed by SMPTE 260M included the following:

- Digital encoding parameters of the 1125/60 HDTV signal

- Dynamic range considerations

- Transient regions

- Filtering characteristics

- Design of the bit-parallel digital interface

The process of converting analog signals into their digital counterparts was characterized by the following parameters:

- Specification of signal component sets

- Number of bits per component sample

- Correspondence between digital and analog video values (assignment of quantization levels)

- Sampling frequency

- Sampling structure

Signal Component Sets

The specification of the analog characteristics of the 1125/60 HDTV signal, as documented in the SMPTE 240M standard, established two sets of HDTV components [1]:

- A set consisting of three full-bandwidth signals, G', B', R', each characterized by a bandwidth of 30 MHz.

- A set of luminance, Y', and color-difference components (P_R' and P_B') with band-widths of 30 MHz and 15 MHz, respectively.

It should be noted that the primed G', B', R', Y', P_R', and P_B' components result when linear signals pass through the nonlinear optoelectronic transfer characteristic of the HDTV camera. According to SMPTE 240M, the luminance signal, Y', is defined by the following linear combination of G', B', and R' signals:

$$Y' = 0.701G' + 0.087B' + 0.212R' \tag{6.23}$$

The color-difference component, P_R', is amplitude-scaled ($R' - Y'$), according to the relation ($R' - Y'$)/1.576, or in other terms:

$$P_{R'} = (-0.445G') - 0.055B' + 0.500R' \tag{6.24}$$

In the same manner, the color-difference component, P_B', is amplitude-scaled ($B' - Y'$) according to ($B' - Y'$)/1.826, or in other terms:

$$P_{B'} = 0.384G' + 0.500B' - 0.11R' \tag{6.25}$$

It should be noted that these baseband encoding equations differ from those for NTSC (or ITU-R Rec. 601) because they relate to a specified SMPTE 240M colorimetry and white point color temperature (D65).

6.3 SMPTE 260M

The SMPTE 260M-1992 standard specifies the digital representation of the signal parameters of the 1125/60 high-definition production system as given in their analog form by SMPTE 240M-1988 [4]. The standard, subsequently revised in 1999 as SMPTE 260M-1999, also specifies the signal format and the mechanical and electrical characteristics of the bit-parallel digital interface for the interconnection of digital television equipment operating in the 1125/60 HDTV domain.

6.3.1 Sampling and Encoding

The use of 8- or 10-bit quantization has become common in digital recording of conventional component and composite video signals. Today, 10-bit linear quantization per sample is the practical limit. This limit is determined by both technical and economic constraints. In any event, experience has demonstrated that 8-bit quantization is quite adequate for conventional video systems. For HDTV applications, particularly those involving eventual transfer to—or intercutting with—35 mm film, 10-bit quantization provides greater resolution, flexibility, and operating "headroom."

Pulse-code modulation (PCM) typically is used to convert the 1125/60 HDTV signals into their digital form [1]. An A/D converter uses a linear quantization law with a coding precision of 8 or 10 bits per sample of the luminance signal and for each color-difference signal. The encoding characteristics of the 1125/60 HDTV signal follow those specified in ITU-R Rec. 601 (encoding parameters for 525-/625-line digital TV equipment) for use with 8- and 10-bit systems. The experience gained over the past 15 years or so with 4:2:2 hardware has shown these to be an equally optimal set of numbers for HDTV.

For an 8-bit system, luminance (Y') is coded into 220 quantization levels with the black level corresponding to level 16 and the peak white level corresponding to level 235. The color-difference signals (P_R', P_B') are coded into 225 quantization levels symmetrically distributed about level 128, corresponding to the zero signal.

For a 10-bit system, luminance (Y') is coded into 887 quantization levels with the black level corresponding to level 64 and the peak white level corresponding to level 940. The color-difference signals (P_R', P_B') are coded into 897 quantization levels symmetrically distributed about level 512, corresponding to the zero signal.

The A/D quantizing levels have a direct bearing on the ultimate S/N performance of a digital video system. Over the linear video A/D output range of a camera from black to the nominally exposed reference white level, the video S/N is given by S/N = $10.8 + (6 \times N)$, where N = number of bits assigned to the range. Table 6.1 summarizes the quantization S/N levels of various common digital amplitude sampling systems.

Quantization-Level Assignment

For the 8-bit system, 254 of the 256 levels (quantization levels 1 through 254) of the 8-bit word are used to express quantized values [4]. Data levels 0 and 255 are used to indicate timing references. For the 10-bit system, 1016 of the 1024 levels (digital levels 4 through 1019) of the 10-bit word are used to express quantized values. Data levels 0 to 3 and 1020 to 1023 are for indication of timing references.

Sampling Frequency

In the world of 4:2:2 digital video signals (as established by ITU-R Rec. 601 for 525-/625-line TV systems), the frequency values of 13.5 and 6.75 MHz have been selected for sampling of the luminance and color-difference components, respectively. It is interesting to note that 13.5 MHz is an integer multiple of 2.25 MHz; more precisely, 6×2.25 MHz = 13.5 MHz).

The importance of the 2.25 MHz frequency lies in the fact that 2.25 MHz represents the minimum frequency found to be a common multiple of the scanning frequencies of 525-

Table 6.1 Quantizing S/N Associated with Various Quantization Levels

Number of Bits	Quantization S/N Levels
8 bits	S/N = 10.8 + (6 × 8) = 58.8 dB
9 bits	S/N = 10.8 + (6 × 9) = 64.8 dB
10 bits	S/N = 10.8 + (6 × 10) = 70.8 dB
11 bits	S/N = 10.8 + (6 × 11) = 76.8 dB
12 bits	S/N = 10.8 + (6 × 12) = 82.8 dB

and 625-line systems. Hence, by establishing sampling based on an integer multiple of 2.25 MHz (in this case, 6 × 2.25 MHz = 13.5 MHz), an integer number of samples is guaranteed for the entire duration of the horizontal line in the digital representation of 525-/625-line component signals (i.e., 858 for the 525-line system and 864 for the 625-line system). More important, however, is the fact that a common number of 720 pixels defines the active picture time of both TV systems. Also, the sampling frequencies of 13.5 MHz for the luminance component and 6.75 MHz for each of the color-difference signals permits the specification of a *digital hierarchy* for various classes of signals used in the digital TV infrastructure. For example, the studio-level video signal was identified by the nomenclature 4:2:2 (indicating a ratio of the sampling structures for the component signals), while processing of three full-bandwidth signals such as G', B', and R' were denoted by 4:4:4.

In the early 1980s, numerous international studies were conducted with the purpose of defining basic picture attributes of high-definition television systems. One of those picture parameters related to the requirement of twice the resolution provided by 4:2:2 studio signals scaled by the difference in picture aspect ratios (that is, between the conventional 4:3 picture aspect ratio and the 16:9 aspect ratio). The CCIR recommended the number of pixels for the active portion of the scanning line to be 1920. In other words:

$$720 \times 2 \times \frac{16/9}{4/3} = 1920 \tag{6.26}$$

The desire to maintain a simple relationship between the sampling frequencies of the 1125/60 HDTV signals and the already established digital world of 4:2:2 components led to the selection of a sampling frequency that was an integer multiple of 2.25 MHz. The sampling frequency value of 74.25 MHz is 33 times 2.25 MHz. When considering the total horizontal line-time of the 1125/60 HDTV signal of 29.63 µs, the sampling rate gives rise to a total number of 2200 pixels. This number conveniently accommodates the 1920 pixels already agreed upon by the international television community as the required number of active pixels for HDTV signals.

Other sampling frequencies are possible. Among others, the values of 72 and 81 MHz (also being integer multiples of 2.25 MHz) have been examined. However, higher values of the sampling frequency result in narrow horizontal retrace intervals for the 1125/60 HDTV signal, if 1920 pixels are assigned to the active part of the picture. On the other hand, the

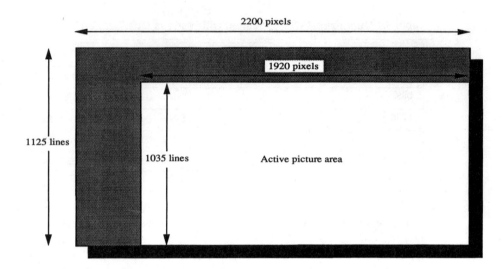

Figure 6.4 Overall pixel count for the digital representation of the 1125/60 HDTV production standard. (*From* [1]. *Used with permission.*)

sampling frequency of 74.25 MHz allows the practical implementation of a horizontal retrace interval (horizontal blanking time) of 3.77 μs.

Another important characteristic of the 74.25 MHz sampling frequency is that none of its harmonics interfere with the values of international distress frequencies (121.5 and 243 MHz).

For the case of sampling and color-difference components, one-half of the value of the sampling frequency for the luminance signal would be used (37.125 MHz). This gives rise to a total of 960 pixels for each of the color-difference components during the active period of the horizontal line and 1100 for the entire line.

Overall, 74.25 MHz emerged as the sampling frequency of choice in 1125/60 HDTV signal set because it provided the optimum compromise among many related parameters, including:

- Practical blanking intervals

- Total data rates for digital HDTV VTRs

- Compatibility with signals of the ITU-R Rec. 601 digital hierarchy

- Manageable signal-processing speeds

The favored set of numbers for the 1125/60 HDTV scanning line exhibits the following number of pixels (see Figure 6.4):

- G', B', R', Y' (luminance) 2200 total pixels; 1920 active pixels

- P_R', P_B' (color-difference) 1100 total pixels; 960 active pixels

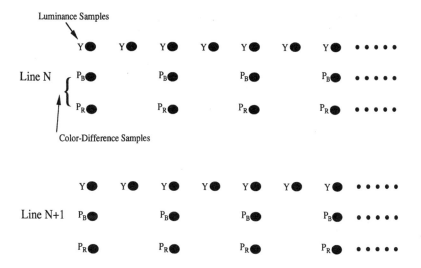

Figure 6.5 Basic sampling structure for the digital representation of the 1125/60 HDTV production standard. (*From* [1]. *Used with permission.*)

Sampling Structure

The fact that the full-bandwidth components G', B', R', and Y' are sampled using the same sampling frequency of 74.25 MHz results in identical sampling structures (locations of the pixels on the image raster) for these signals [1]. Furthermore, because of the integer number of samples per total line (2200), the sampling pattern aligns itself vertically, forming a rectangular grid of samples, as illustrated in Figure 6.5. This pattern is known as an *orthogonal sampling structure* that is line-, field-, and frame-repetitive. This type of sampling structure facilitates the decomposition of most 2-D and 3-D processing algorithms into simpler operations that can be carried out independently in the horizontal, vertical, and temporal directions, thereby enabling the use of less complex modular hardware and software systems. Also, the relationship between the sampling positions of the luminance and color-difference signals is such that P_B' and P_R' samples are co-sited with odd (1st, 3rd, 5th, 7th) samples of the luminance component in each line.

6.3.2 Principal Operating Parameters

SMPTE 260M-1992 (and SMPTE 260M-1999) digital coding is based on the use of one luminance signal (E_Y') and two color-difference signals (E_{CB}' and E_{CR}') or on the use of three primary color signals (E_G', E_B', and E_R') [4]. E_Y', E_{CB}', and E_{CR}' are transmitted at the 22:11:11 level of the CCIR digital hierarchy for digital television signals, with a nominal sampling frequency of 74.25 MHz for the luminance signal and 37.125 MHz for each of the color-difference signals. (See Table 6.2.) E_G', E_B', and E_R' signals are transmitted at the 22:22:22 level of the CCIR digital hierarchy with a nominal sampling frequency of

Table 6.2 Encoding Parameter Values for SMPTE-260M (*After* [4].)

Parameter		Value
Matrix formulas	E_Y', E_{CB}', E_{CR}' E_G', E_B', E_R'	E_Y', E_{CB}', E_{CR}' are derived from gamma-corrected values of E_G', E_B', E_R' as defined by the linear matrix specified in SMPTE-240M
Number of samples/line	Video components	$E_Y' = 2200$ $E_G' = 2200$ $E_{CB}' = 1100$ $E_B' = 2200$ $E_{CR}' = 1100$ $E_R' = 2200$
	Auxiliary channel	2200
Sampling structure	E_G', E_B', E_R', Luminance E_Y', Auxiliary channel	Identical sampling structures: orthogonal sampling, line, field, and frame repetitive.
	Color difference signals (E_{CB}', E_{CR}')	Samples are co-sited with odd (1st, 3rd, 5th, ...) E_Y samples in each line.
Sampling frequency (tolerance ±10 ppm)	Video components	$E_Y' = 74.25$ MHz $E_G' = 74.25$ MHz $E_{CB}' = 37.125$ MHz $E_B' = 74.25$ MHz $E_{CR}' = 37.125$ MHz $E_R' = 74.25$ MHz
	Auxiliary channel	74.25 MHz
Form of encoding		Uniformly quantized, PCM 8- or 10-bits/sample for each of the video component signals and the auxiliary channel.
Active number of samples/line	Video components	$E_Y' = 1920$ $E_G' = 1920$ $E_{CB}' = 960$ $E_B' = 1920$ $E_{CR}' = 960$ $E_R' = 1920$
	Auxiliary channel	1920

74.25 MHz. Provisions exist within the standard for the coding of SMPTE 240M signals to a precision of 8 or 10 bits.

The SMPTE 260M standard describes the bit-parallel digital interface only. The complete specification of the bit-serial interface is contained in another document, SMPTE 292M-1996. However, when the digital representation of the 1125/60 production standard was defined, consideration was given to making the signal format equally applicable to the bit-serial interface.

The parallel interface consists of one transmitter and one receiver in a point-to-point connection. The bits of the digital code words that describe the video signal are transmitted in parallel using 10-conductor cables (shielded twisted pairs) for each of the component signals. Each pair carries bits at a nominal sample rate of 74.25 Mwords/s. For the transmission of E_Y', E_{CB}', and E_{CR}', the color-difference components are time-multiplexed into a single signal E_{CB}'/E_{CR}' of 74.25 Mwords/s. The digital bit-parallel interface uses a 93 multitipin connector for transmission of E_Y', E_{CB}', and E_{CR}' or for transmission of E_G', E_B', and E_R' with 8- or 10-bit precision.

Table 6.2 Encoding Parameter Values for SMPTE-260M (*After* [4].)

Parameter		Value
Timing relationship between video data and the analog synchronizing waveform		The time duration between the *end of active video* (EAV) timing reference code and the reference point 0H of the horizontal sync waveform = 88 clock intervals.
Correspondence between video signal levels and quantization levels[1]	8-bit system: E_G', E_B', E_R', Luminance E_Y', Auxiliary channel	220 quantization levels with the black level corresponding to level 16 and the peak white level corresponding to level 235.
	Each color difference signal (E_{CB}', E_{CR}')	225 quantization levels symmetrically distributed about level 128, which corresponds to the zero signal.
	10-bit system: E_G', E_B', E_R', Luminance E_Y', Auxiliary channel	877 quantization levels with the black level corresponding to level 64 and the peak white level corresponding to level 940.
	Each color difference signal (E_{CB}', E_{CR}')	877 quantization levels symmetrically distributed about level 512, which corresponds to the zero signal.
Quantization level assignment[1]	8-bit system	254 of the 256 levels (digital levels 1 through 254) of the 8-bit word used to express quantized values. Data levels 0 and 255 are reserved to indicate timing references.
	10-bit system	1016 of the 1024 levels (digital levels 4 through 1019) of the 10-bit word used to express quantized values. Data levels 0–3 and 1020–1023 are reserved to indicate timing references.

[1] These values refer to precise nominal video signal levels. Signal processing may occasionally cause the signal level to deviate outside this range.

Figures 6.6 and 6.7 illustrate the line-numbering scheme of SMPTE 260M. Lines are numbered from 1 through 1125, as shown in the drawings.

6.3.3 Production Aperture

SMPTE 240M precisely defines a picture aspect ratio of 16:9 with 1920 pixels per active line and 1035 active lines. However, digital processing and associated spatial filtering can produce various forms of *transient effects* at picture blanking edges and within the adjacent active video that should be taken into account to allow practical implementation of the standard [4]. A number of factors can contribute to such effects, including the following:

- Bandwidth limitations of component analog signals (most noticeably, the ringing of color-difference signals)

- Analog filter implementation

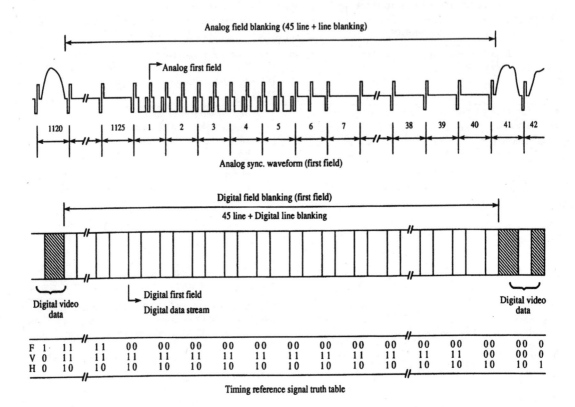

Figure 6.6 Line numbering scheme, first digital field (*F* = 0). (*From* [4]. *Used with permission.*)

- Amplitude clipping of analog signals as a result of the finite dynamic range imposed by the quantization process

- Use of digital blanking in repeated analog-digital-analog conversions

- Tolerance in analog blanking

To accommodate realistic tolerances for analog and digital processes during postproduction operations, the SMPTE-260M standard includes a list of recommendations with regard to production aperture.

Clean Aperture

It is generally unrealistic to impose specific limits on the number of cascaded digital processes that might be encountered in a practical postproduction system [4]. In light of those picture-edge transient effects, therefore, the definition of a system design guideline was introduced in the form of a subjectively artifact-free area called the *clean aperture*.

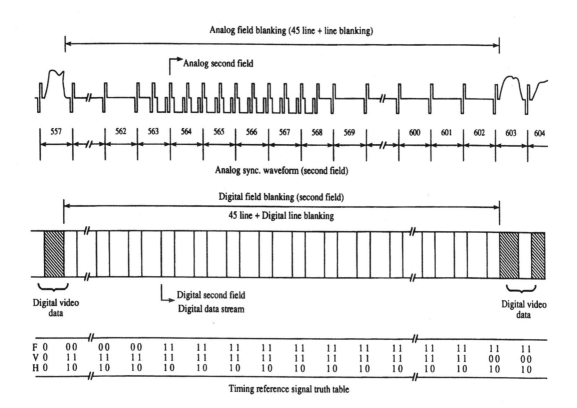

Figure 6.7 Line numbering scheme, second digital field ($F = 1$). (*From* [4]. *Used with permission.*)

The concept of a clean aperture defines an inner picture area within which the picture information is subjectively uncontaminated by all edge transient distortions. This clean aperture should be as wide as needed to accommodate cascaded digital manipulations of the picture. Computer simulations have shown that a transient-effect area defined by 16 samples on each side and nine lines at both the top and bottom within the digital production aperture affords an acceptable (and practical) worst-case level of protection in allowing 2-dimensional transient ringing to settle below a subjectively acceptable level.

This gives rise to a possible picture area—the clean aperture—of 1888 horizontal active pixels by 1017 active lines within the digital production aperture whose quality is guaranteed for final release. This concept is illustrated in Figure 6.8.

6.3.4 SMPTE 240M-1995

This standard, issued 7 years after the first HDTV production standard was published, defines the basic characteristics of the analog video signals associated with origination

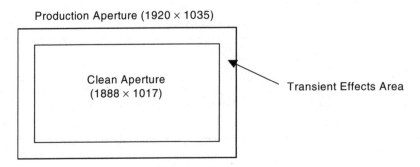

Figure 6.8 Production and clean aperture recommendations in SMPTE-260M. (*After* [4].)

equipment for 1125-line high-definition television production systems [5]. SMPTE 240M-1995 describes systems operating at 60 and 59.94 Hz field rates. As stated previously, the digital representation of the signals described in this standard are given in SMPTE 260M. Between them, these two documents define both digital and analog implementations of 1125-line HDTV production systems.

The differences between SMPTE 240M-1988 and the later standard are minimal, except for the inclusion of the 59.94 Hz field rate. The video signals of the standard represent a scanned raster with the characteristics shown in Table 6.3. For the sake of completeness, the principal operating parameters of the standard will be outlined in this section. Because much of the technical background on the 1125-line HDTV system was discussed in detail previously in this chapter, those technical arguments will not be repeated here.

Operational Parameters

The HDTV production system specification is intended to create a *metameric reproduction* (visual color match) of the original scene as if lit by CIE illuminant D65 [5]. To this end, the combination of the camera optical spectral analysis and linear signal matrixing is set to match the CIE color-matching functions (1931) of the reference primaries. Furthermore, the combination of a linear matrixing reproducer and reproducing primaries is equivalent to the reference primaries. The reference red, green, and blue primary x,y chromaticities, thus, are (CIE 1931):

- Red: $x = 0.630, y = 0.340$

- Green: $x = 0.310, y = 0.595$

- Blue: $x = 0.155, y = 0.070$

The system reference white is an illuminant that causes equal primary signals to be produced by the reference camera; it is produced by the reference reproducer when driven by equal primary signals. For this system, the reference white is specified in terms of its 1931 CIE chromaticity coordinates, which have been chosen to match those of CIE illuminant D65, namely: $x = 0.3127, y = 0.3290$.

Table 6.3 Scanned Raster Characteristics of SMPTE-240M-1995 (*After* [5].)

Parameter	1125/60 System	1125/59.94 System
Total scan lines/frame	1125	1125
Active lines/frame	1035	1035
Scanning format	Interlaced 2:1	Interlaced 2:1
Aspect ratio	16:9	16:9
Field repetition rate	60.00 Hz ± 10 ppm	59.94 Hz[1] ± 10 ppm
Line repetition rate (derived)	33750.00 Hz	33716.28 Hz[2]

[1] The 59.94 Hz notation denotes an approximate value. The exact value = 60/1.001.

[2] The 33716.28 Hz notation denotes an approximate value. The exact value = (60 × 1125)/(2 × 1.001).

The captured image is represented by three parallel, time-coincident video signals. Each incorporates a synchronizing waveform. The signals may be either of the following sets:

- Primary color set: E_G' = green, E_B' = blue, E_R' = red

- Difference color set: E_Y' = luminance, E_{PB}' = blue color difference, E_{PR}' = red color difference

Where E_G', E_B', and E_R' are the signals appropriate to directly drive the primaries of the reference reproducer and E_Y', E_{PB}', and E_{PR}' can be derived from E_G', E_B', and E_R' as follows:

$$E_{Y'} = (0.701 \times E_{G'}) + (0.087 \times E_{B'}) + (0.212 \times E_{R'})$$ (6.27)

E_{PB}' is amplitude-scaled $(E_B' - E_Y')$, according to:

$$E_{PB'} = \frac{(E_{B'} - E_{Y'})}{1.826}$$ (6.28)

E_{PR}' is amplitude-scaled $(E_R' - E_Y')$, according to:

$$E_{PR'} = \frac{(E_{R'} - E_{Y'})}{1.576}$$ (6.29)

Where the scaling factors are derived from the signal levels given in Table 6.4. The agreement of these values with SMPTE-240M-1988 is evident (see Section 6.2.7).

The color primary set E_G', E_B', and E_R' comprises three equal-bandwidth signals whose nominal bandwidth is 30 MHz. The color-difference set E_Y', E_{PB}', and E_{PR}' comprises a luminance signal E_Y' whose nominal bandwidth is 30 MHz, and color-difference signals E_{PB}' and E_{PR}' whose nominal bandwidth is 30 MHz for analog originating equipment, and 15 MHz for digital originating equipment.

Table 6.4 Analog Video Signal Levels (*After* [5].)

E_Y', E_G', E_B', E_R' Signals	
Reference black level	0 mV
Reference white level	700 mV
Synchronizing level	± 300 mV
E_{PB}', E_{PR}' Signals	
Reference zero signal level	0 mV
Reference peak levels	± 350 mV
Synchronizing level	± 300 mV
All Signals	
Sync pulse amplitude	300 ± 6 mV
Amplitude difference between positive- and nega-tive-going sync pulses	< 6 mV

1999 Revision

A revision of SMPTE 240M-1995 was undertaken in 1999. The document included essentially no technical changes, but did update certain parameter values. SMPTE 240M-1995 references ITU-R BT.709-2, "Parameter Values for the HDTV Standards for Production and International Programme Exchange." The 1999 revision updates that reference to ITU-R BT.709-3, which was released in February 1998.

6.4   DTV-Related Raster-Scanning Standards

In the years after the ATSC standard for digital television was approved for implementation, the SMPTE coordinated considerable efforts to define a family of scanning formats and interfaces for the multiple picture rates accommodated by DTV. The results of this work—some still ongoing as this book went to press—are outlined in the following sections.

6.4.1   1920 x 1080 Scanning Standard

SMPTE 274M defines a family of raster-scanning systems for the representation of images sampled temporally at a constant frame rate and within the following parameters [6]:

- An image format of 1920×1080 samples (pixels) inside a total raster of 1125 lines

- An aspect ratio of 16:9

The standard specifies the following elements:

- R', G', B' color encoding

- R', G', B' analog and digital interfaces

Table 6.5 Scanning System Parameters for SMPTE-274M (*After* [6].)

System Description	Samples per Active Line (S/AL)	Active Lines per Frame	Frame Rate (Hz)	Scanning Format	Interlace Sampling Frequency f_s (MHz)	Samples per Total Line (S/TL)	Total Lines per Frame
1: 1920 × 1080/ 60/1:1	1920	1080	60	Progressive	148.5	2200	1125
2: 1920 × 1080/ 59.94/1:1	1920	1080	60/1.001	Progressive	148.5/1.001	2200	1125
3. 1920 × 1080/ 50/1:1	1920	1080	50	Progressive	148.5	2640	1125
4: 1920 × 1080/ 60/2:1	1920	1080	30	2:1 Interlace	74.25	2200	1125
5: 1920 × 1080/ 59.94/2:1	1920	1080	30/1.001	2:1 Interlace	74.25/10.0	2200	1125
6. 1920 × 1080/ 50/2:1	1920	1080	25	2:1 Interlace	74.25	2540	1125
7: 1920 × 1080/ 30/1:1	1920	1080	30	Progressive	74.25	2200	1125
8: 1920 × 1080/ 29.97/1:1	1920	1080	30/1.001	Progressive	74.25/1.001	2200	1125
9: 1920 × 1080/ 25/1:1	1920	1080	25	Progressive	74.25	2640	1125
10: 1920 × 1080/24/1:1	1920	1080	24	Progressive	74.25	2750	1125
11: 1920 × 1080/23.98/1:1	1920	1080	24/1.001	Progressive	74.25/1.001	2750	1125

- Y', P_B', P_R' color encoding and analog interface

- Y', C_B', C_R' color encoding and digital interface

An auxiliary component A may optionally accompany Y', C_B', C_R', denoted Y', C_B', C_R', A.

It should be noted that, in this standard, references to signals represented by a single letter, that is, R', G', and B', are equivalent to the nomenclature in earlier documents of the form E_R', E_G', and E_B', which, in turn, refer to signals to which specified transfer characteristics have been applied. Such signals commonly are described as having been gamma-corrected.

The system parameters for SMPTE 274M are listed in Table 6.5. Analog interface timing details are given in Figure 6.9 and digital interface vertical timing details are shown in Figure 6.10.

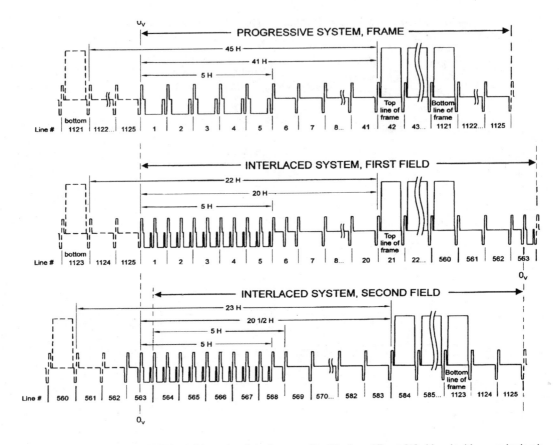

Figure 6.9 SMPTE 274M analog interface vertical timing. (*From* [6]. *Used with permission.*)

SMPTE 295M

SMPTE 295M-1997 defines a family of systems similar to SMPTE 274M, but for 50 Hz line rates [7]. The principal operating system parameters are given in Table 6.6.

6.4.2 1280 x 720 Scanning Standard

SMPTE 296M-1997 defines a family of raster scanning systems having an image format of 1280 × 720 samples and an aspect ratio of 16:9 [8]. The standard specifies the following parameters:

- R', G', B' color encoding
- R', G', B' analog and digital representation
- Y', P_B', P_R' color encoding, analog representation and analog interface
- Y', C_B', C_R' color encoding and digital representation

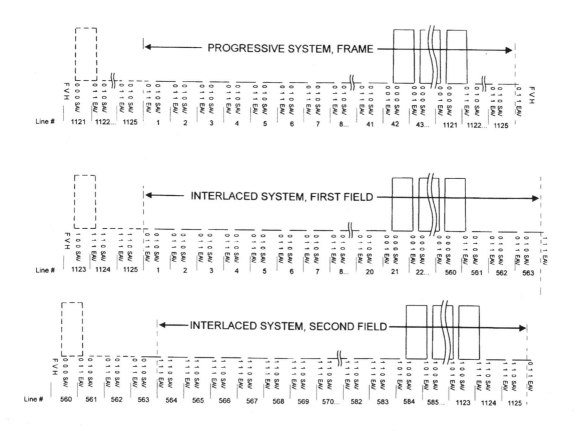

Figure 6.10 SMPTE 274M digital interface vertical timing. (*From* [6]. *Used with permission.*)

Table 6.6 Scanning Systems Parameters for SMPTE-295M-1997 (*After* [7].)

System Designation	Samples per Active Line (S/AL)	Active Lines per Frame	Frame Rate (Hz)	Scanning Format	Interlace Sampling Frequency f_s (MHz)	Samples per Total Line (S/TL)	Total Lines per Frame
1: 1920 × 1080/ 50/1:1	1920	1080	50	Progressive	148.5	2376	1250
2: 1920 × 1080/ 50/2:1	1920	1080	25	2:1 interlace	74.25	2376	1250

An auxiliary component A may optionally accompany Y', C_B', C_R'; this representation is denoted Y', C_B', C_R', A. A bit-parallel digital interface also is specified in the standard.

Table 6.7 Scanning System Parameters for SMPTE 296M-1997 (*After* [8].)

System Type	Samples per Active Line (S/AL)	Active Lines per Frame (AL/F)	Frame Rate (Hz)	Scanning Format	Reference Clock (f_s) MHz	Samples per Total Line (S/TL)	Total Lines per Frame
1. 1280 × 720/60/1:1	1280	720	60	Progressive	74.25	1650	750
2. 1280 × 720/59.94/ 1:1	1280	720	60/1.001	Progressive	74.25/1.001	1650	750

Table 6.8 Scanning System Parameters for SMPTE 293M-1996 (*After* [9].)

System Type	Samples per Digital Active Line (S/AL)	Lines per Active Image	Frame Rate	Sampling Frequency f_s (MHz)	Samples per Total Line	Total Lines per Frame	Colorimetry
720 × 483/ 59.94	720	483	60/1.001	27.0	858	525	ANSI/SMPTE 170M

The basic format parameters are listed in Table 6.7.

6.4.3 720 × 483 Scanning Standard

SMPTE 293M-1996 defines the digital representation of stationary or moving two-dimensional images of sampling size 720 × 483 at an aspect ratio of 16:9 [9]. The scanning format details are given in Table 6.8. This standard includes both R', G', B' and Y', C_B', C_R' expressions for the signal representations.

The principal application of this standard is for the production of content for EDTV-II, which employs an NTSC letterbox encoding scheme, compatible with ANSI/SMPTE 170M.

Bit-Serial Interface

SMPTE 294M-1997 defines two alternatives for bit-serial interfaces for the 720 × 483 active line at 59.94 Hz progressive scan digital signal for production, defined in SMPTE 293M [10]. Interfaces for coaxial cable are defined, each having a high degree of commonality with interfaces operating in accordance with SMPTE 259M. The two basic system modes are as follows:

Table 6.9 SMPTE 294M-1997 Interface Parameters (*After* [10].)

System (Total Serial Data Rate)	4:2:2 P (2 x 270 Mbits/s) Dual Link	4:2:0 P (360 Mbits/s) Single Link
Frame rate	60/1.001 Hz	60/1.001 Hz
Word length	10 bits	10 bits
Parallel and multiplexed word rate: channels Y', Y'' and C_B'/C_R', C_B''/C_R'' + SAV, EAV and auxiliary data	2×27 Mwords/s	36 Mword/s
Active lines per frame	483	483
Words per active line (channels Y' and Y'')	720 and 720	720 and 720
Words per active line (channels C_B' and C_R')	$2 \times (360$ and $360)$	360 and 360
Words per horizontal blanking area (SAV/EAV and auxiliary data)	2×276 (Total: $2 \times 483 \times 276$ $= 2 \times 133,308$/frame)	128 (Total: 483×128 $= 61,824$/frame)
Words in the active picture area	$2 \times (1440 \times 483) = 2$ x 695,520	$2160 \times 483 = 1,043,280$
Words in the vertical blanking interval (SAV/EAV and auxiliary data)	$2 \times (1716 \times 42) = 2$ x 72,072	$2288 \times 42 = 96,096$

- *Dual-link interface* (4:2:2 P): Each link operates at 270 Mbits/s; the active data in the Y', C_B', C_R' format (equivalent to 8:4:4) are transparently divided, line sequentially, into two data streams, each equivalent to the 4:2:2 component signal of SMPTE 259M.

- *Single-link interface* (4:2:0 P): Operates at 360 Mbits/s; the active data representing the color-difference components in the Y', C_B', C_R' format (equivalent to 8:4:4) are quincunx down-converted by a factor of two, prior to reformatting with the full luminance data, into a single data stream equivalent to the component signal of SMPTE 259M (but at a higher rate, conceptually 8:4:0).

Basic system parameters are listed in Table 6.9.

SMPTE 170M-1999

This standard, a revision of SMPTE 170M-1994, describes the composite analog color video signal for studio applications of NTSC: 525 lines, 59.94 fields, 2:1 interlace, with an aspect ratio of 4:3 [11]. The standard specifies the interface for analog interconnection and serves as the basis for the digital coding necessary for digital interconnection of NTSC equipment.

The composite color video signal contains an electrical representation of the brightness and color of a scene being analyzed (the active picture area) along defined paths (scan lines). The signal also includes synchronizing and color reference signals that allow the geometric and colorimetric aspects of the original scene to be correctly reconstituted at the display. The synchronizing and color reference signals are placed in parts of the composite color video signal that are not visible on a correctly adjusted display. Certain portions of the composite color video signal that do not contain active picture information are blanked in order to allow the retrace of scanning beams in some types of cameras and display devices.

The video signal representing the active picture area consist of:

- A wideband luminance component with setup, and no upper bandwidth limitation for studio applications

- A pair of simultaneous chrominance components, amplitude modulated on a pair of suppressed subcarriers of identical frequency (f_{sc} = 3.579545... MHz) in quadrature.

The reference reproducer for this system is representative of cathode ray tube displays, which have an inherently nonlinear electro-optical transfer characteristic. To achieve an overall system transfer characteristic that is linear, SMPTE 170M specifies compensating nonlinearity at the signal source (gamma correction). For purposes of precision, particularly in digital signal processing applications, exactly inverse characteristics are specified in the document for the reference camera and reproducer.

6.4.4 MPEG-2 4:2:2 Profile at High Level

ISO/IEC 13818-2, commonly known as MPEG-2 video, includes specification of the MPEG-2 4:2:2 profile [12]. Based on ISO/IEC 13818-2, the SMPTE 308M standard (proposed) provides additional specification for the MPEG-2 4:2:2 profile at high level. This standard is intended for use in high-definition television production, contribution, and distribution applications. As in ISO/IEC 13818-2, SMPTE 308M defines bit streams, including their syntax and semantics, together with the requirements for a compliant decoder for 4:2:2 profile at high level, but does not specify particular encoder operating parameters.

Because the MPEG-2 4:2:2 profile at main level is defined in ISO/IEC 13818-2, only those additional parameters necessary to define the 4:2:2 profile at high level are specified in SMPTE 308M. The 4:2:2 high profile does not have a hierarchical relationship to other profiles. Syntactic constraints for the 4:2:2 profile are specified in ISO/IEC 13818-2. No new constraints are specified for the 4:2:2 profile at high level.

The parameter constraints for the 4:2:2 profile at high level are the same as those for the main profile at the main level, except as follows:

- The upper bounds for sampling density is 1920 samples/line, 1088 lines/frame, 60 frames/s

- Upper bounds for luminance sample rate is 62,668,800 samples/s

- Upper bounds for bit rates is 300 Mbits/s

Additional constraints include that at bit rates of 175 to 230 Mbits/s, no two consecutive frames shall be coded as *nonintracoded pictures*. This constraint describes a bit-stream that a compliant decoder is required to properly decode. It is understood that bit-stream splicing might result in consecutive nonintracoded pictures, but that operation of the decoder is not ensured in such a case.

At bit rates of 230 to 300 Mbits/s, only intracoded pictures can be used for interlaced scan images and no two consecutive frames can be coded as nonintracoded pictures for progressive scan images.

6.5   High-Definition Serial Digital Interface

Based upon the experience gained from the application of digital technology to the 525/625 broadcast and production environments, it is evident that wide acceptance of new professional video equipment takes place only after an efficient means of interconnection is available [13]. For studios today, the means of interconnection is typically the serial digital interface (SDI). In an effort to address the facility infrastructure requirements of HDTV, the SMPTE and BTA developed a standard for digital serial transmission of studio HDTV signals.

The overall transmission rate for transporting a digital studio HDTV signal (1125-line, 2:1 interlace, with 10-bit component sampling) is approximately 1.5 Gbits/s. The active payload is on the order of 1.2 Gbits/s (for 1035/1080 active lines). The transmission of video signals at these bit rates represents a far more difficult technical challenge than serial distribution at 270 Mbits/s used for conventional television signals.

The introduction of the serial digital interface for conventional video in 1986 (SMPTE 259M) was well received by the television industry and has become the backbone of digital audio/video networking for broadcast and post production installations around the world. SDI is ideally suited to the task of transporting uncompressed component/composite digital video and multichannel audio signals over a single coaxial cable. To emulate the same level of operational usability and system integration of conventional television equipment in the HDTV world, the implementation of a *high-definition serial digital interface* (HD-SDI) system—based on an extension of SMPTE 259M—was essential.

Work on the HD-SDI system began in 1992 under the auspices of SMPTE and BTA. The end result of these efforts was the BTA document BTA S-004 (May 1995), followed closely by SMPTE 292M-1996, both with similar technical content.

The source formats of SMPTE 292M adhere to those signal characteristics specified in SMPTE 260M and 274M. In particular, the field frequencies of 59.94 Hz and 60.00 Hz, and active line numbers of 1035/1080 are used by HD-SDI. Table 6.10 lists the basic parameters of the input source formats.

Subsequently, a revision of SMPTE 292M was undertaken, resulting in SMPTE 292M-1998 [14]. The revised source format parameters are given in Table 6.11. Note that the total data rate is either 1.485 Gbits/s or 1.485/1.001 Gbits/s. In the table, the former is indicated by a rate of "1" and the later by a rate of "M," which is equal to 1.001.

6.5.1   A Practical Implementation

To better understand the operation of the HD-SDI system, it is instructive to consider a practical example. A set of HD-SDI transmitter/receiver modules was developed (Sony) to provide the desired interconnection capabilities for HDTV [13].

Table 6.10 SDI Reference Source Format Parameters (*After* [13].)

Reference Document	SMPTE-260M	SMPTE-274M	SMPTE-274M
Parallel word rate (each channel Y, C_R/C_B)	74.25 Mword/s	74.25 Mword/s	74.25/1.001 Mword/s
Lines per frame	1125	1125	1125
Words per active line (each channel Y, C_R/C_B)	1920	1920	1920
Total active line	1035	1080	1080
Words per total line (each channel Y, C_R/C_B)	2200	2200	2200
Frame rate	30 Hz	30 Hz	30/1.001 Hz
Total Fields per frame	2	2	2
Total data rate	1.485Gbits/s	1.485Gbits/s	1.485/1.001 Gbits/s
Field 1 EAV V = 1	Line 1121	Line 1124	Line 1124
Field 1 EAV V = 0	Line 41	Line 21	Line 21
Field 2 EAV V = 0	Line 558	Line 561	Line 561
Field 2 EAV V = 0	Line 603	Line 584	Line 584
EAV F = 0	Line 1	Line 1	Line 1
EAV F = 1	Line 564	Line 563	Line 563

Each module of the system makes use of a *coprocessor* IC that implements the data structure (protocol) specified in SMPTE 292M. This device has both transmitting and receiving functions, which makes it possible to transmit and receive video, audio, and ancillary data, as well as EAV/SAV, line number, and other parameters.

The transmission module consists of two main ICs. (A block diagram of the system is shown in Figure 6.11.) The first is the coprocessor, which is used to embed EAV/SAV, line number, and CRC information in the input digital video signal. This device also serves to multiplex the audio data and audio channel status information in the HANC area of the chrominance channel. Conventional ancillary data is multiplexed in the ancillary space (HANC and/or VANC of Y, and P_B/P_R signals).

The second element of the transport system is the *P/S converter* IC, which converts the two channels of parallel data (luminance and chrominance) into a single serial bit stream. This device also performs the encoding operation of scrambled NRZI, which is the channel coding technique stipulated in SMPTE 292M/BTA S-004. At the output of the P/S converter IC, the serial data rate is 1.485 Gbits/s. The input video signal is represented by 10 bits in parallel for each of the Y and P_B/P_R samples. External timing signals (EXTF/EXTH) are provided as additional inputs to the IC for cases when EAV/SAV information is not present in the input parallel video data.

The video clock frequency is 74.25 MHz (or 74.25/1.001 MHz), which is synchronous with the video data. The audio data packet is multiplexed in the HANC area of the chromi-

Table 6.11 Source Format Parameters for SMPTE 292M (*After*[14].)

Reference SMPTE Standard	260M		295M	274M								296M	
Format	A	B	C	D	E	F	G	H	I	J	K	L	M
Lines per frame	1125	1125	1250	1125	1125	1125	1125	1125	1125	1125	1125	750	750
Words per active line (each channel $Y, C_B/C_R$)	1920	1920	1920	1920	1920	1920	1920	1920	1920	1920	1920	1280	1280
Total active lines	1035	1035	1080	1080	1080	1080	1080	1080	1080	1080	1080	720	720
Words per total line (each channel $Y, C_B/C_R$)	2200	2200	2376	2200	2200	2640	2200	2200	2640	2750	2750	1650	1650
Frame rate (Hz)	30	30/M	25	30	30/M	25	30	30/M	25	24	24/M	60	60/M
Fields per frame	2	2	2	2	2	2	1	1	1	1	1	1	1
Data rate divisor	1	M	1	1	M	1	1	M	1	1	M	1	M

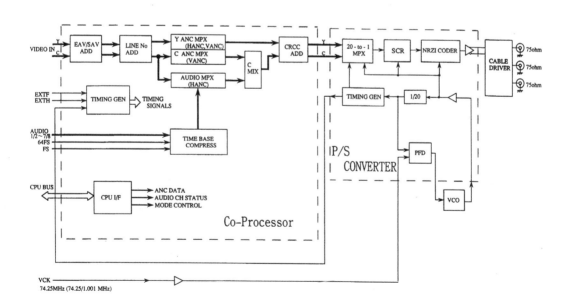

Figure 6.11 Block diagram of an HD-SDI transmission system. (*From* [13]. *Courtesy of Sony.*)

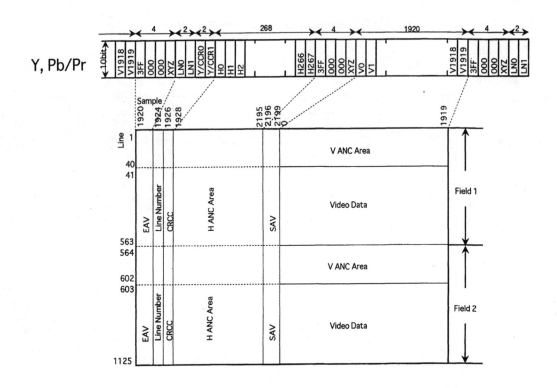

Figure 6.12 The HD-SDI signal format from the coprocessor IC to the P/S device. (*From* [13]. *Courtesy of Sony.*)

nance channel, with the exception of lines 8 and 570. All of the video and ancillary data present in the input digital signal, and the embedded data (line numbers, EAV/SAV, CRC, and other data) are transmitted to the P/S device without alteration, where the data is converted into a serial form. Figure 6.12 shows the data footprint for the system.

As mentioned previously, the interface between the coprocessor and P/S converter consists of two channels of I/O bit parallel data at 74.25 Mwords/s. The P/S IC converts these two parallel channels into a serial data stream at 1.485 Gbits/s by means of a 20-to-1 multiplexer. This serial data is then encoded using scrambled NRZI. The generator polynomial for the NRZI scrambler is [13]:

$$G_{(x)} = (x^9 + x^4 + 1) \cdot (x + 1) \tag{6.30}$$

From the P/S IC, the encoded serial data is output in ECL form and distributed by 3 channels of coaxial outputs. The signal amplitude of the coaxial output is 800 mV p-p into 75 Ω.

Figure 6.13 shows a block diagram of the receiver module. Cable equalizer circuitry compensates for high-frequency losses of the coaxial cable, and the clock recovery circuit

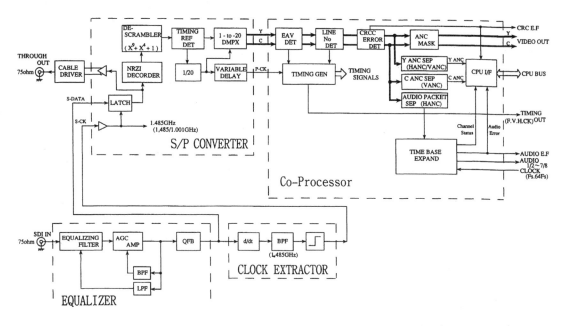

Figure 6.13 Block diagram of an HD-SDI receiver system. (*From* [13]. *Courtesy of Sony.*)

extracts the 1.485 GHz serial clock. The serial-to-parallel converter section reconstructs the parallel data from the serial bitstream, while the coprocessor IC separates the video, audio, and ancillary data. The cable equalizer is an automatic system that compensates for the frequency-dependent attenuation of coaxial cable. The attenuation characteristic of a 5C-2V cable (75 Ω, comparable to Belden 8281) is shown in Figure 6.14.

By using a PIN diode, the compensation curve of the receiving system can be controlled while satisfying the condition of linear phase. The cable length is detected by a band-pass filter, and the equalizing filter is controlled to keep the amplitude constant. A dc restorer also is used. After a decision stage, the threshold-controlled signal is fed back through a low-pass filter, permitting recovery of the dc component. The data given in Figure 6.15 show the eye-patterns for two cable lengths. The bit error rate for 1.485 Gbits/s transmission over these lengths is less than 1 error in 10^{10}.

The clock recovery circuit is a clock frequency filter that works on the serial data recovered from the cable equalizer. The transition edges of the serial data contain the 1.485 GHz clock frequency that is selected by the filter. The HDTV source formats make use of two field frequencies: 60.00 Hz and 59.94 Hz. Hence, the HD-SDI system provides two values for the serial clock frequency, that is, 1.485 GHz and 1.4835 GHz (1.485 GHz/1.001). Figures 6.16*a* and 6.16*b* show the recovered clock frequencies of 1.485 GHz and 1.4835 GHz, respectively.

The serial-to-parallel converter reconstructs the parallel data from the serial bit stream. The serial data recovered from the cable equalizer is reclocked by the serial clock frequency

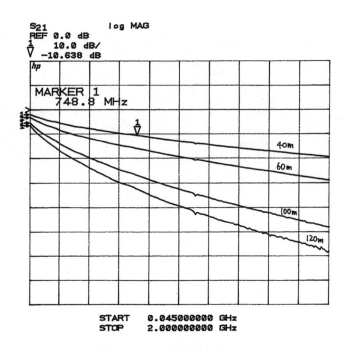

S21   log MAG
REF 0.0 dB
1   10.0 dB/
▽   -10.638 dB

hp

MARKER 1
748.8 MHz

1
▽

40m

60m

100m

120m

START   0.045000000 GHz
STOP   2.000000000 GHz

Figure 6.14 Attenuation characteristics of 5C-2V coaxial cable. (*From* [13]. *Courtesy of Sony.*)

that is produced at the output of the clock recovery circuit. At this point, the channel coding scrambled NRZI is decoded and the unique pattern of the timing reference signal EAV/SAV is detected. In order to generate the 74.25 MHz clock frequency for the parallel data words, the 1.485 GHz serial clock is divided by 20 and synchronized using detection of the timing reference signals. The serial data is next latched by the parallel clock waveform to generate the 20 bits of parallel data. The serial data, latched by the serial high-frequency clock, also is provided as an active loop-through output.

The coprocessor IC separates video, audio, and ancillary data from the 20-bit parallel digital signal. EAV/SAV information and line number data are detected from the input signal, permitting the regeneration of F/V/H video timing waveforms. Transmission errors are detected by means of CRC coding.

The embedded audio packet is extracted from the HANC space of the P_B/P_R channel and the audio data are written into memory. The audio data are then read out by an external audio clock frequency F_s, enabling the reconstruction of the received audio information. The coprocessor IC can receive up to 8 channels of embedded audio data.

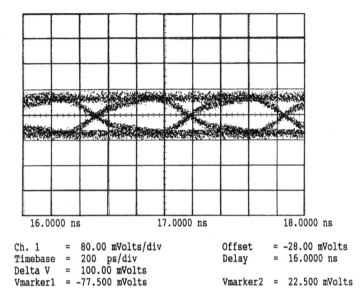

(a)

Ch. 1	= 80.00 mVolts/div	Offset	= -28.00 mVolts
Timebase	= 200 ps/div	Delay	= 16.0000 ns
Delta V	= 100.00 mVolts		
Vmarker1	= -77.500 mVolts	Vmarker2	= 22.500 mVolts

Trigger on External at Pos. Edge at -1.902 Volts

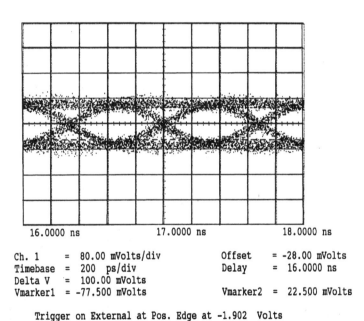

(b)

Ch. 1	= 80.00 mVolts/div	Offset	= -28.00 mVolts
Timebase	= 200 ps/div	Delay	= 16.0000 ns
Delta V	= 100.00 mVolts		
Vmarker1	= -77.500 mVolts	Vmarker2	= 22.500 mVolts

Trigger on External at Pos. Edge at -1.902 Volts

Figure 6.15 Eye diagram displays of transmission over 5C-2V coaxial cable: (*a*) 3 m length, (*b*) 100 m length. (*After* [13]. *Courtesy of Sony.*)

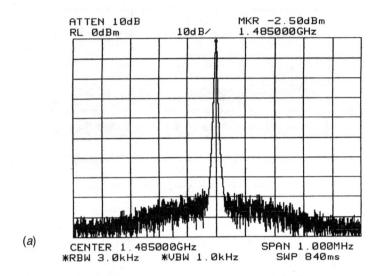

(a)

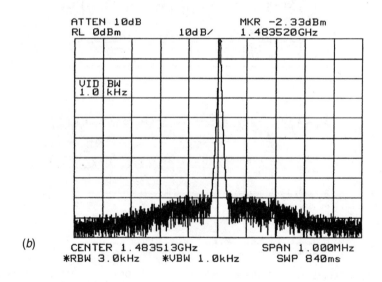

(b)

Figure 6.16 Recovered signal clocks: (a) 1.485 GHz 60 Hz system, (b) 1.4835 GHz 59.94 Hz system. (*After* [13]. *Courtesy of Sony.*)

6.5.2 Audio Interface Provisions

SMPTE 299M-1997 defines the mapping of 24-bit AES digital audio data and associated control information into the ancillary data space of a serial digital video stream conforming to SMPTE 292M [15]. The audio data are derived from ANSI S4.40, more commonly referred to as AES audio.

An audio signal, sampled at a clock frequency of 48 kHz locked (synchronous) to video, is the preferred implementation for intrastudio applications. As an option, this standard supports AES audio at synchronous or asynchronous sampling rates from 32 kHz to 48 kHz. The number of transmitted audio channels ranges from a minimum of two to a maximum of 16. Audio channels are transmitted in pairs, and where appropriate, in groups of four. Each group is identified by a unique ancillary data ID.

Audio data packets are multiplexed into the horizontal ancillary data space of the C_B/C_R parallel data stream, and audio control packets are multiplexed into the horizontal ancillary data space of the Y parallel data stream.

MPEG-2 Audio Transport

SMPTE 302M specifies the transport of AES3 data in an MPEG-2 transport stream for television applications at a sampling rate of 48 ksamples/s [16]. The MPEG audio standard itself defines compressed audio carriage, but does not define uncompressed audio for carriage in an MPEG-2 transport system. SMPTE 302M augments the MPEG standard to address the requirement to carry AES3 streams, which may consist of linear PCM audio or other data. MPEG-2 transport streams convey one or more programs of coded data, and may be constructed from one or more elementary coded data streams, program streams, or other transport streams.

The specifications are described in terms of a model that starts with AES3 data, constructs *elementary streams* (ES) from the AES3 data, then constructs *packetized elementary streams* (PES) from the elementary streams, and finally constructs *MPEG-2 transport streams* (MTS) from the packetized elementary streams. Although this model is used to describe the transport of AES3 streams in MPEG-2 transport streams, the model is not mandatory. MPEG-2 transport streams may be constructed by any method that results in a valid stream.

The SMPTE audio data elementary streams consists of audio sample words, which may be derived from AES3 digital audio subframes, together with validity, user, and channel status (*V, U, C*) bits and a framing (*F*) bit. There may be 2, 4, 6, or 8 channels of audio data conveyed in a single audio elementary stream and corresponding packetized elementary stream. Multiple packetized elementary streams may be used in applications requiring more channels.

6.5.3 Data Services

SMPTE 334M (proposed at this writing) defines a method of coding that allows data services to be carried in the vertical ancillary data space of a bit-serial component television signal conforming with SMPTE 292M or ANSI/SMPTE 259M [17]. This includes data broadcast services intended for the public as well as broadcaster internal control and communications. Despite the reference to the bit-serial interface, nothing in the specification precludes its use in a parallel digital interface for component digital video signals. The data described in the standard can also be transported in K-L-V format according to SMPTE 336M, or via other means.

The data packets are located in the active line portion of one or more lines in the vertical ancillary space. Data can be located in any lines in the area from the second line after the line specified for switching to the last line before active video, inclusively. Individual data services are not assigned to any specific data lines; receiving equipment should identify and select services on the basis of their ANC DID and SDID fields.

Because ANC data may be located in the lines immediately preceding active video, manufacturers of video compression equipment must ensure that these data bits are not included in video compression calculations.

The chrominance (C_b/C_r) and luminance (Y) data are carried in two separate streams within the 292M signal, complete with their own ANC data flags and CRCs. Defined data services are carried in the Y stream. Other data services can be inserted into either one of these streams without restrictions.

In the 259M/125M signal, the chrominance and luminance data are carried in a single stream. In this case, all data services are carried in this stream with a single ANC data flag and CRC.

There is no specific provision in SMPTE 334M for ensuring that the relative timing between the video and its embedded VANC data is correct. The only timing relationship that exists is created when the data are embedded in the video. Once that relationship is established, the deterministic nature of 292M or 259M transport ensures that the relationship is preserved.

6.5.4 Time Division Multiplexing on SMPTE 292M

Given the wide variety of commercially accepted standards for both standard and high-definition television systems used in studio production, post-production, and distribution facilities, a large number of incompatible interfaces exists for the interconnection and routing of the various 4:2:2I, 4:4:4I, 4:2:2:4I, 4:4:4:4I, and 4:2:2P component and composite video signals throughout a given plant [18]. In addition, high-definition video signals used in a number of television facilities operate at a bandwidth of 1.485 Gbits/s.

A higher bandwidth interface such as that used to carry high-definition component video could carry several lower-bandwidth component or composite video signals over a single physical link, reducing the number of cable or fiber routing signals throughout the plant and simplifying the overall routing requirements. In fact, the lower-bandwidth signals could be formatted into digital active line areas of existing high-definition interfaces in such a way that the resulting signal would appear to most existing pieces of equipment as a regular high-definition video signal.

In addition, the low-bandwidth signals could be any generic data stream, including compressed standard- and high-definition video, not just component video signals. The obvious benefit of this multiplexing scheme is that existing equipment for serializing and deserializing high-definition bit-parallel data signals for distribution throughout the plant could also be used to serialize and deserialize the high-bandwidth multiplexed data stream without any additional hardware.

All that is needed to implement this system is hardware for multiplexing and demultiplexing between bit-parallel high- and low-bandwidth formats.

With these real-world needs in mind, the SMPTE developed an innovative solution aimed at bridging the video hardware of the present with that of the future. SMPTE 346M (proposed at this writing) defines the *time division multiplexing* (TDM) of various standard-definition digital video and generic 8-bit data signals over the high-definition serial digital interface specified in SMPTE 292M. The objective of this high-definition multiplexing interface is to use a single physical link to transmit, distribute, route, and switch a complete family of existing 10-bit video formats and various data formats.

Active video and vertical blanking areas in the HD SDI stream are time divided into 19 interleaved channels. The word in the first channel is used to indicate the data validation of the remaining 18 channels. A single video or data stream is multiplexed into one or multiple channels of the total 18 data channels. A control packet is multiplexed into the horizontal ancillary data space of the luminance parallel data area after the switching point of the HD stream for each video or data stream. The control packet indicates how and which channels are used for this video or data stream. It also contains *stream clock reference* information for clock recovery of the original clock signal.

Multiple standard-definition video or data streams can be multiplexed in and demultiplexed from a single HD SDI stream with a total delay of a fraction of a horizontal line. By dividing the payload into segments, which are subdivided into channels, time-division multiplexing can be applied to several data sources in such a way that very low latency is generated during the demultiplexing process. A fixed number of channels in each segment provides an efficient implementation of different data by reducing the complexity of the multiplexing and demultiplexing processes.

SMPTE 346M specifies the format for multiplexing multiple asynchronous standard-definition video streams and generic data into the high-definition system interfaces as defined in SMPTE 274M and ANSI/SMPTE 296M in the bit-parallel source format for the bit-serial interface defined in SMPTE 292M. The major standard-definition system interfaces are defined by ANSI/SMPTE 259M and ITU-R BT.601-5. They are the 4:2:2 digital component video signal interfaces defined in ANSI/SMPTE 125M, ANSI/SMPTE 267M, and ANSI/SMPTE 293M. These standards specify interfaces for the 270-, 360-, and 540-Mbits/s bit-parallel formats. The standard also covers other serial video standards such as 143 Mbits/s 525-line and 177 Mbits/s 625-line composite digital signals.

SMPTE 346M can be used for a 270-Mbits/s SDI interface to carry multiple streams of data and a single 8-bit or 10-bit composite digital video signal. In a similar manner, a 540-Mbits/s SDI system can carry multiple streams of video and data. SMPTE 346M can also be extended to future higher bit rate serial digital standards.

The standard definition video formats referred to in SMPTE 346M are listed in Table 6.12.

6.5.5 Packet Transport

SMPTE 348M (proposed at this writing) provides the mechanisms necessary to facilitate the transport of packetized data over a synchronous data carrier [19]. The HD-SDTI data packets and synchronizing signals provide a data transport interface that is compatible with SMPTE 292M such that it can be readily used by the infrastructure provided by the standard. SMPTE 348M uses a dual-channel technique where each line carries two data chan-

Table 6.12 Summary of SD Video Formats Referenced in SMPTE 346M (After [18]

SD System/Sampling Structure	525 × 60 or 625 × 50, 4 × 3 13.5 MHz	525 × 59.94 or 625 × 50, 16 × 9 18 MHz
4:0:0I	135 Mbits/s	180 Mbits/s
4:2:2I	270 Mbits/s ANSI/SMPTE 267M	360 Mbits/s ANSI/SMPTE 267M
4:2:2:4I	360 Mbits/s	540 Mbits/s
4:4:4I	360 Mbits/s ITU-R BT.601-5	540 Mb/s ITU-R BT.601-5
4:2:0P	360 Mbits/s	540 Mbits/s
4:4:4:4I	540 Mbits/s SMPTE RP 174	720 Mbits/s
8:4:4I	540 Mbits/s	720 Mbits/s
4:2:2P	540 Mbits/s ANSI/SMPTE 293M	720 Mbits/s

nels, each forming an independent HD-SDTI data transport mechanism. The two channels are word-multiplexed onto the HD-SDI stream such that one line-channel occupies the C data space and the other line-channel occupies the Y data space.

The standard provides for a baseline operation that supports a constant payload length per line-channel having a maximum payload data rate up to approximately 1.0 Gbits/s. It further provides for an extended operation mode that supports a variable payload length through the advancement of the SAV sequence to ensure a constant payload data rate regardless of the HD-SDI frame rate. The HD-SDTI protocol is compatible with SMPTE 305M.

SMPTE 348M describes the assembly of two channels of 10-bit words multiplexed into one HD-SDI line for the purpose of transporting the data streams in a structured framework. The HD-SDTI data blocks and synchronizing signals provide a data transport protocol that can readily be added to the infrastructure provided by SMPTE 292M.

SMPTE 292M requires a sequence of 10-bit words that define a television horizontal line comprising five areas in the following sequence (the first two areas are often described together):

- **EAV**: a four-word unique timing sequence defining the end of active video (of the previous line)

- **LN/CRC**: two words defining the line number followed by a two-word CRC error detection code

- **Digital line blanking**

- **SAV**: a four-word unique timing sequence defining the start of active video

- **Digital active line**

An associated television source format standard defines the rate of television horizontal lines by specifying the following parameters:

- The number of words per line

- The number of words in the digital active line (and hence the number of words in the digital line blanking period)

- The number of lines per frame

- The number of frames per second

SMPTE 292M currently defines four source format standards (1152, 1035, 1080, and 720 active lines per frame). SMPTE 125M describes the meaning of the EAV and SAV word sequences that can be applied to all relevant source formats.

A decoder operating under the SMPTE 348M standard is not be required to decode all the source formats available to SMPTE 292M. The source formats that must be supported by the decoder are specified in the application document.

6.5.6   540 Mbits/s Interface

SMPTE 344M (proposed at this writing) specifies a serial digital interface that operates at a nominal rate of 540 Mbits/s [20]. The standard is intended for applications in television studios over specified lengths of coaxial cable. Separate SMPTE documents specify the mapping of source image formats onto the 540 Mbits/s serial interface.

The connector has mechanical characteristics conforming to the 50-Ω BNC type. Mechanical dimensions of the connector may produce either a nominal 50- or 75-Ω impedance and are usable at frequencies up to 540 MHz, assuming a return loss that is greater than 15 dB. However, the electrical characteristics of the connector and its associated interface circuitry must provide a resistive impedance of 75 Ω. Where a 75-Ω connector is used, its mechanical characteristics must reliably interface with the nominal 50-Ω BNC type defined by IEC 60169-8.

The application of SMPTE 344M does not require a particular type of coax. It is necessary, however, for the coax to be a 75-Ω type and for the frequency response of the coax, in dB, to be approximately proportional to $1/\sqrt{f}$ from 1 MHz to 540 MHz to ensure correct operation of automatic cable equalizers over moderate to maximum lengths.

The channel coding specified in SMPTE 344M is scrambled NRZI. The LSB of any data word is transmitted first. To maintain synchronization and word alignment at the serial receiver, EAV and SAV timing references are inserted into the data stream.

6.6   Serial Data Transport Interface

The serial data transport interface (SDTI) is a standard for transporting packetized audio, video, and data between cameras, VTRs, editing/compositing systems, video servers, and transmitters in professional and broadcast video environments [21]. SDTI builds on the familiar SDI standard that is now widely used in studios and production centers to transfer uncompressed digital video between video devices. SDTI provides for faster-than-real-time video transfers and a reduction in the number of decompression/compression generations required during the video production process, while utilizing the existing SDI infrastructure.

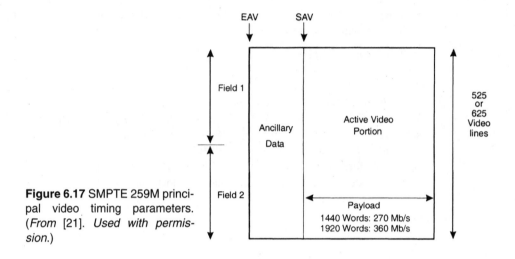

Figure 6.17 SMPTE 259M principal video timing parameters. (*From* [21]. *Used with permission.*)

The SMPTE 305M SDTI specification evolved from a collaborative effort on the part of equipment vendors and interested parties, under the auspices of the SMPTE PT20.04 Workgroup on Packetized Television Interconnections, to define a common interchange interface for compressed audio and video.

Because SDTI is built upon the SMPTE 259M SDI specification, it shares the same mechanical, electrical, and transport mechanisms. BNC connectors and coaxial cables establish the mechanical link.

SDI transports uncompressed digital video using 10-bit words in the 4:2:2 *Y, U, V* component mode for 525- and 625-line applications. Words are serialized, scrambled, and coded into a 270-Mbits/s or 360-Mbits/s serial stream. In order to synchronize video timing between the transmitter and the receiver, SDI defines words in the bitstream called *end of active video* (EAV) and *start of active video* (SAV), as illustrated in Figure 6.17. At 270 Mbits/s, the active portion of each video line is 1440 words and at 360 Mbits/s, the active portion is 1920 words. The area between EAV and SAV can be used to transmit ancillary data such as digital audio and time code. The ancillary data space is defined by SMPTE 291M-1998.

SDI and SDTI can co-exist in a facility using the same cabling, distribution amplifiers, and routers. Cable lengths of more than 300 meters are supported. SDI repeaters can be used to reach longer distances. A versatile studio configuration that supports all the required point-to-point connections can be established using an SDI router.

6.6.1 SDTI Data Structure

SDTI uses the ancillary data space in SDI to identify that a specific video line carries SDTI information [21]. The packetized video is transported within the active video area, providing 200 Mbits/s of payload capacity on 270-Mbits/s links and 270 Mbits/s of payload capacity on 360-Mbits/s links.

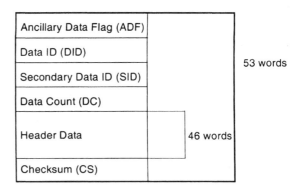

Figure 6.18 Header data packet structure. (*From* [21]. *Used with permission.*)

A 53-word header data packet is inserted in the ancillary data space. The rest of the ancillary data space is left free to carry other ancillary data. The 53-word SDTI header data structure is in accordance with the SMPTE 291M ancillary data specification, shown in Figure 6.18. The specification contains an ancillary data flag (ADF), a data ID (DID) specified as code 140h for SDTI, secondary data ID (SDID) specified as code 101h for SDTI, and 46 words for header data. A checksum for data integrity checking also is included.

The 46 words of header data define source and destination addresses and the formatting of the payload (Figure 6.19). Line number and line number CRC bits are used to ensure data continuity. The code identifies the size of the payload to be either 1440 words or 1920 words long. An authorized address identifier (AAI) defines the addressing method utilized.

Currently, the Internet Protocol (IP) addressing scheme is defined for SDTI. The source and destination addresses are 128 bits long, allowing essentially a limitless number of addressable devices to be supported.

The bock type identifies whether data is arranged in fixed-sized data blocks—with or without error correction code (ECC)—or variable-sized data blocks. In the case of fixed-sized blocks, the block type also defines the number of blocks that are transported in one video line. The block type depends on the data type of the payload.

Between SAV and EAV, the payload itself is inserted. The payload can be of any valid data type registered with SMPTE. The data structure of the payload includes a data type code preceding the data block, shown in Figure 6.20*a*. In addition, separator, word count, and end code are required for data types that feature variable-sized blocks (Figure 6.20*b*).

SMPTE 305M does not specify the data structure inside the data block, which is left to the registrant of the particular data type.

A significant revision of SMPTE 305M was undertaken in March 2000. The document, proposed at this writing, updated normative references and provided additional detail on the operation elements of the SDTI system. The revision was intended to keep the standard in conformance with related documents and to address certain implementation issues.

Header Data Packet (53 words)

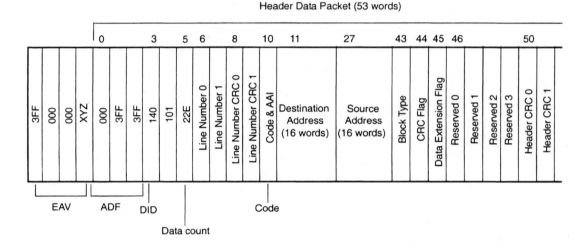

Figure 6.19 SDTI header data structure. (*From* [21]. *Used with permission.*)

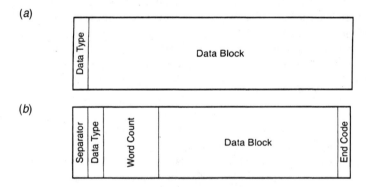

Figure 6.20 SDTI data structure: (*a*) fixed block size, (*b*) variable block size. (*From* [21]. *Used with permission.*)

6.6.2   SDTI in Computer-Based Systems

Transferring material to and from computer-based nonlinear editing (NLE)/compositing systems and video servers is one of the primary uses of SDTI [21]. Computers can interface with other SDTI devices through the use of an adapter. Typically, an SDTI-enabled computer-based NLE system will contain the following components:

• A CPU motherboard

• A/V storage and a controller to store the digital audio and compressed video material

• Editing board with audio/video I/O, frame buffer, digital video codecs, and processors

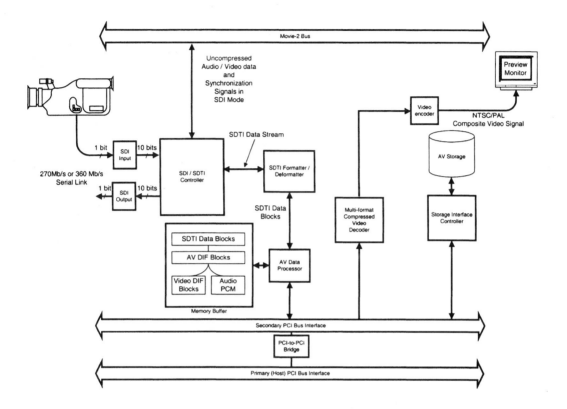

Figure 6.21 SDI/SDTI implementation for a computer-based editing/compositing system. (*From* [21]. *Used with permission.*)

- Network and console display adapters

- SDI/SDTI adapter

As an example of a typical SDI/SDTI hardware adapter, consider the PCI-bus implementation illustrated in Figure 6.21. With the PCI adapter installed, the computer can share and move SDI and SDTI streams with and among other devices and terminals. Such a configuration also can be used to transcode from SDI to SDTI, and vice-versa.

6.6.3 SDTI Content Package Format

The SDTI *content package format* (SDTI-CP) is an extension of the basic SDTI system that allows for package-based delivery of data signals. The SMPTE developed a collection of standards that define the technology and protocol of SDTI-CP. The applicable documents include the following:

- SMPTE 326M: SDTI Content Package Format

- SMPTE 331M: Element and Metadata Definitions for the SDTI-CP

- SMPTE 332M: Encapsulation of Data Packet Streams over SDTI

- Recommended Practice RP 204: SDTI-CP MPEG Decoder Templates (proposed at this writing)

The general parameters of these documents are described in the following sections.

SMPTE 326M

The SDTI-CP standard specifies the format for the transport of *content packages* (CP) over the serial digital transport interface [22]. Known as SDTI-CP, this format is a packaging structure for the assembly of system, picture, audio, and auxiliary data items in a specified manner. SMPTE 326M defines the structure of the content package mapped onto the SDTI transport; element and metadata formats are defined by SMPTE 331M [23].

The baseline operation of the SDTI-CP standard is defined by the transport of content packages locked to the SDTI transport frame rate. The standard additionally defines format extension capabilities that include:

- Content package transfers at higher and lower than the specified rate through isochronous and asynchronous transfer modes

- Provision of a timing mode to reduce delay and provision for two content packages in each SDTI transport frame

- Carriage of content packages in a low-latency mode

- Multiplexing of content packages from different sources onto one SDTI transport

The SMPTE 326M standard is limited to SDTI systems operating at a bit rate of 270 Mbits/s and 360 Mbits/s as defined by SMPTE 305M.

An SDTI-CP compliant receiver must be capable of receiving and parsing the structure of the SDTI-CP format. An SDTI-CP compliant decoder is defined by the ability to both receive and decode a defined set of elements and metadata according to an associated decoder template document. The MPEG decoder template is detailed in RP 204 [24]. Other decoder template recommended practices can be defined as required for other applications of the SDTI-CP.

Figure 6.22 shows the basic layered structure of a *content package*. It is constructed of up to four *items*, where each item is constructed of one or more *elements*, which include:

- *System item*—carries content package metadata and may contain a control element. The system item also carries metadata that is related to elements in the other items.

- *Picture item*—can consist of up to 255 picture stream elements.

- *Audio item*—can consist of up to 255 audio stream elements.

- *Auxiliary item*—can consist of up to 255 auxiliary data elements.

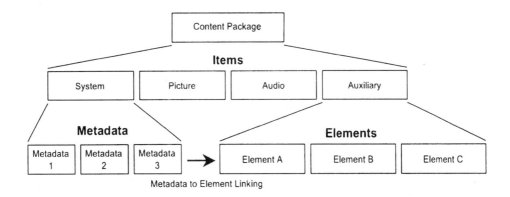

Figure 6.22 The basic content package structure of SMPTE 326M. (*From* [22]. *Used with permission.*)

A content package contains the associated contents of one content package frame period starting with a system item and optionally containing picture, audio, and auxiliary items.

Data Packet Encapsulation

SMPTE 332M specifies an open framework for encapsulating data packet streams and associated control metadata over the SDTI transport (SMPTE 305M) [25]. Encapsulating data packet streams on SDTI allows them to be routed through conventional SDI (SMPTE 259M) equipment. The standard specifies a range of packet types that can be carried over SDTI, which may be expanded as requirements develop.

The standard does not attempt to specify the payload contained in any packet. It offers options to add specific information to each individual packet including localized user data space, forward error correction (FEC), and a mechanism for accurate packet retiming at the decoder.

The standard also offers a limited capability for metadata to be added, providing packet control information to aid the successful transfer of packets. The specification of the metadata follows the K-L-V approach of the SMPTE dynamic metadata dictionary and provides extensibility for future requirements.

Timing Issues

Most packet streams do not have critical timing requirements and a decoder can output packets in the order in which they were encoded, but with increased *packet jitter* resulting from the buffering of packets onto SDTI lines [25]. The result of the SDTI-PF packet encapsulation process is to introduce both delay and jitter to the packet stream. However, MPEG-2 transport stream (MPEG-2 TS) packets are one case where a relatively small packet jitter specification is required to ensure minimal impact on MPEG-2 transport stream clock recovery and buffering circuits. SMPTE 332M contains provisions to allow the packet jitter to be reduced to insignificant levels; the delay is an issue addressed by the method of packet buffering at the encoder. As a benchmark, the specification is defined so that a low packet jitter source can be carried through the SDTI-PF and be decoded to create an output with negligible packet jitter.

Although MPEG-2 TS packets are the most critical packet type for decoder timing accuracy, this standard also allows for other kinds of packets to be carried over the SDTI, with or without buffering, to reduce packet jitter. Such packets may be ATM cells and packets based on the *unidirectional Internet protocol* (Uni-IP).

MPEG Decoder Templates

SMPTE Recommended Practice RP 204 defines decoder templates for the encoding of SDTI content packages with MPEG coded picture streams [24]. The purpose of RP 204 is to provide appropriate limits to the requirements for a receiver/decoder in order to allow practical working devices to be supplied to meet the needs of defined operations. Additional MPEG templates are expected to be added to the practice as the SDTI-CP standard matures. The SMPTE document recommends that each new template be a superset of previous templates so that any decoder defined by a template in the document can operate with both the defined template and all subsets.

6.7 Camera Systems

The imaging device is starting point for any video system. The implementation of HDTV programming has imposed a number of new technical challenges for video engineers. There are a number of key requisites for acquisition systems supporting HDTV program origination, including the following [26]:

- **Aspect ratio management.** The need to service both the standard 4:3 aspect ratio and the new 16:9 widescreen image format introduces a difficult operational challenge the program originator.

- **Highest picture quality.** With the arrival of HDTV, an entirely new yardstick of picture quality is emerging. This trend is being propelled by a plethora of new digital delivery media that bring MPEG-2 digital component video directly into the home.

- **Highest signal/noise performance.** Noise is the enemy of compression. Video program masters will be subjected to increasingly frequent digital compression in order to service distribution via DTV broadcasting, digital satellite and cable, and digital packaged media ranging from CD-ROM to DVD. A formerly benign noise interference (in the analog NTSC context) can, in an era of heavy digital compression, easily be translated into new and disturbing picture artifacts.

The implications of each of these issues are complex, and need to be carefully evaluated against the contemporary technologies available to camera equipment manufacturers.

With the foregoing issues in mind, the overall performance of a television camera can be divided into two distinct categories [26]:

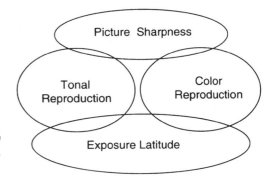

Figure 6.23 The elements that combine to define the quality of a video camera. (*Courtesy of Sony.*)

- Those separate imaging attributes that collectively contribute to overall picture quality (that is, the aesthetics and beauty of the picture).

- Those separate artifacts of the camera system that collectively detract from the overall picture quality.

The name of the game in high-end video camera design is to optimize *all* of the picture quality factors, while at the same time, minimizing *all* of the picture artifacts. In describing the overall aesthetics of the HDTV picture, it is necessary to examine the multi-dimensional aspect of image quality and to assign priorities to the contribution of each of those picture dimensions.

For the purpose of overall camera performance analysis, there are four core attributes of picture quality, as illustrated in Figure 6.23. They can be separately considered (and separately specified) as the key contributing dimensions of picture quality:

- *Picture sharpness*—the overall resolution of the image.

- *Tonal reproduction*—the accuracy of reproduction of the gray scale.

- *Color reproduction*—the total color gamut that can be captured, and the accuracy of reproduction of the luminance, hue, and saturation of each color.

- *Exposure latitude*—the total camera dynamic range, or the ability of the camera to simultaneously capture picture detail in deep shadows and in areas of the scene that are overexposed.

The overall performance of a camera is largely determined by the front-end imaging system, namely the combination of optics and imager. These elements predetermine the core attributes of a video picture. The image quality must be fully retained and—where possible—enhanced within the complex RGB video processing system that follows the imaging elements.

6.7.1   The Optical System

The optical system of a camera is used to form a precise image of the original scene on the surface of the imaging devices. The optical system consists of [27]:

- A lens to capture the image.

- Optical filters to condition the image.

- A color separation system to split the incoming light into the three primary color components.

With the exception of the highest levels of program production, where fixed focal length (also called *prime lenses*) are sometimes used, the zoom lens is the universal lens used with virtually all video cameras. Zoom lenses are available at a wide range of prices and performance levels, up to and including performance levels required for use with high-definition television systems.

At first look, the requirements for a lens intended for use with a video camera appear to be quite similar to lenses intended for use with a film camera. Unfortunately, lenses appropriate for a high-quality video work differ in one critical parameter from lenses designed for film cameras: the *back focal distance* (i.e., the distance from the end of the lens to the plane where the image is formed) is increased significantly compared to film lenses to allow the insertion of the prism-type color separation system between the lens and the CCD imagers.

Several types of optical filters are used to achieve the high level of performance found in modern cameras, including the following:

- Color correction and neutral density filters

- Infrared filter

- Quarter-wavelength filter

- Anti-aliasing filter

Each device serves a specific purpose, or solves a specific shooting problem.

Figure 6.24 illustrates the optical system for one channel (green) of a modern HDTV camera.

6.7.2   Digital Signal Processing

The standard imager for modern cameras is the CCD, a thoroughly analog device despite the fact that the information from the imager is read out in discrete packets [27]. The dynamic range of a CCD is quite large. It is not uncommon for the early processing stages of an analog camera to faithfully process video signals as high as 600 percent of nominal exposure. Digital processing with an inordinately high number of bits per sample (greater than 12 bit A/D) is required to handle this large dynamic range while retaining the ability to resolve fine shades of luminance difference.

Although the front-end of the camera is required to process highlights of up to 600 percent, it is acceptable to subsequently compress these highlights to a more reasonable range. An analog *pre-knee* circuit has been found to execute this function quite efficiently, limiting

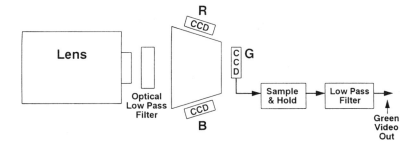

Figure 6.24 Optical/sampling path for a high-performance HDTV camera. (*After* [26]. *Courtesy of Sony.*)

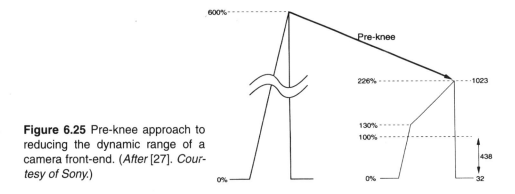

Figure 6.25 Pre-knee approach to reducing the dynamic range of a camera front-end. (*After* [27]. *Courtesy of Sony.*)

the digital signal processing circuitry to a dynamic range of perhaps 225 percent. Until the current analog imagers are displaced by true digital imagers, studio and portable cameras will remain hybrid analog/digital devices. Figure 6.25 illustrates the analog pre-knee technique with 10-bit linear A/D bit assignment.

The benefits of digital signal processing for video cameras include the following:

• **High stability and reliability**. However careful its design, the circuitry in analog cameras is inevitably subject to drifting that requires manual correction. Subsequent readjustments are then subject to operator interpretation, and so the actual set-up of a camera at any one time is difficult to define. With digital processing, parameters can be held permanently in memory. Potentiometers, the least reliable component of any camera, are reduced from about 150 for analog processing to less than 6 for cameras with digital processing. As a result, the need for operator adjustment is dramatically reduced. A further advantage of digital processing is that it is much easier to implement digital circuits in ICs and LSIs than analog circuits, allowing the size and weight of cameras to be reduced.

• **Precise alignment**. The accuracy of a camera set-up can be defined with great precision by digital processing. Moreover, variations between cameras, which are difficult to avoid with analog processing, can be reduced to a minimum with digital techniques by simply choosing the same setup parameters.

- **Flexible signal processing and parameter setting**. A significant advantage of digital signal processing is that it can provide very flexible operation. Many camera parameters can be controlled and each setting can be adjusted over a wide range of values. It has been well known that different camera adjustments can dramatically improve the ability of the device to capture difficult scene material. With analog cameras, such custom adjustment is difficult and time consuming to implement, as well as to restore. With digital processing cameras, control adjustments can be manipulated quickly and easily, stored, and then recalled with great accuracy.

Figure 6.26 compares the signal flow for analog- and digital-based cameras. Before the widespread use of DSP in cameras, the functional blocks of a tube-based and CCD-based system were not all that different—the imaging devices, preamplifiers, and power supply notwithstanding, of course. The extensive use of DSP has now brought the individual signal processing blocks of the analog system into a single package, or chip set (as illustrated in the figure). The digital signal path for a high-definition video camera is shown in Figure 6.27.

6.7.3 Camera Specifications

A video camera performs the complex task of creating an electronic image of a real scene, ranging from scenes with extreme highlights—scenes with large dynamic range that must be compressed to fit within the capability of the video system—to scenes with marginal illumination. Defining the performance of a camera in a complete but concise set of specifications is a difficult exercise [27]. It is not unusual to find a low cost camera with virtually the same published specifications as a camera costing significantly more. Actual day-to-day performance, on the other hand, will probably show the more expensive camera to be far superior in handling difficult lighting situations. For this reason the published camera specifications are no more than a basic guide for which cameras to consider for actual evaluation.

It is usually unnecessary to limit the choice of camera to the one with the best picture quality because virtually all HDTV cameras make high-quality images. Such factors as ease of use, cost, and operational features are frequently the deciding factors in choosing one camera over another for a specific application.

6.7.4 Camera-Related Standards

SMPTE 304M defines a connector primarily intended for use in television broadcasting and video equipment, such as camera head to camera control-unit connections [28]. The standard defines hybrid connectors, which contain a combination of electrical contacts and fiber-optic contacts for single-mode fibers. The document also contains dimensional tolerances that ensure functional operability of the electrical interface.

SMPTE 311M describes the minimum performance for a hybrid cable containing single-mode optical fibers and electrical conductors to convey signal and control information in a

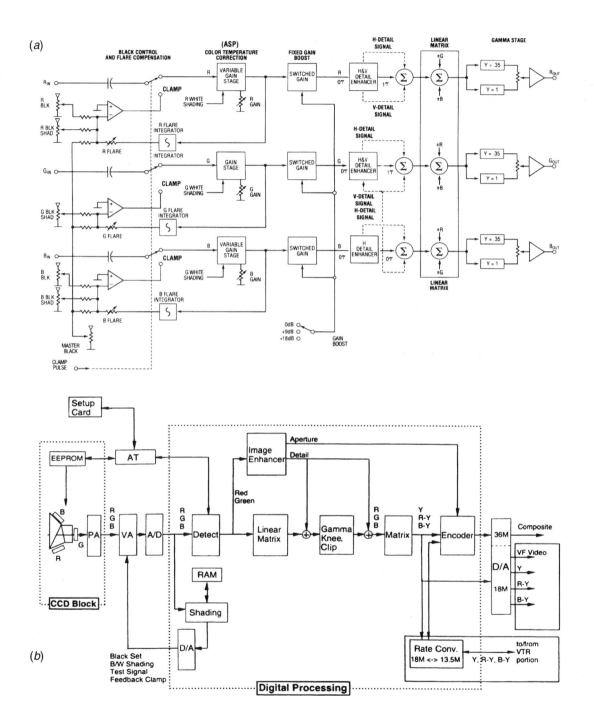

Figure 6.26 Video camera signal flow: (*a*) analog system, (*b*) digital system. (*From* [27]. *Used with permission. Courtesy of Sony.*)

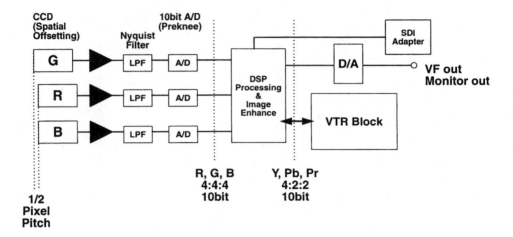

Figure 6.27 Block diagram of the digital processing elements of an HDTV camera. (*After* [26]. *Courtesy of Sony.*)

variety of environments where moisture, weather, and ozone resistance are required [29]. The cable described in the standard is intended to be used to interconnect cameras and base stations in conjunction with the connector interface standard.

SMPTE 315M provides a method for multiplexing camera positioning information into the ancillary data space described in SMPTE 291M [30]. Applications of the standard include the 525-line, 625-line, component or composite, and high-definition digital television interfaces that provide 10-bit ancillary data space. Two types of camera positioning information are defined in the standard: binary and ASCII.

Camera positioning information, as described in the document, includes the following parameters:

- Camera relative position

- Camera pan

- Camera tilt

- Camera roll

- Origin of world coordinate longitude

- Origin of world coordinate latitude

- Origin of world coordinate altitude

- Vertical angle of view

- Focus distance

- Lens opening (iris or f-value)

- Time address information

- Object relative position

Data for each parameter can be obtained from several kinds of pickup devices, such as rotary encoders. These data are formatted as an ancillary data packet and multiplexed into the ancillary data space of serial digital video and conveyed to the receiving end.

SMPTE 315M defines the packet structure, word structure, coordinate, range, and accuracy of each parameter, and the method of multiplexing packets.

6.7.5 HDTV Camera/Film Production Issues

The psychophysical sensation of image depth, which can impart an almost 3-dimensional quality to the large-screen HDTV image, is very much a function of the displayed contrast range [31]. Contrast range is one of the vital "multidimensions" of the displayed HDTV image and, therefore, it is an important element in the creation, by the image, of a new sense of reality [32]. However, the issue of digitization of the high-definition video signal is a significant determinant in what ultimately can be achieved.

Quite apart from the issues surrounding the final portrayal of HDTV are the more important implications of what happens when images from the two mediums are brought together in the production/postproduction environment [33]. In this context, the primary concern is about *matching* the two images to achieve a seamless intercut or, possibly, a blue-screen composite. This integration of images, originated from two separate mediums (video and film), can take place in one of two domains:

- *Film domain*—Image processing and integration follows the transfer of HDTV to film by an electronic-beam recorder or laser recorder.

- *Electronic domain*—Image processing and integration follows the transfer by telecine of the original film imagery to HDTV.

Electronic integration is perhaps the more likely option because of the numerous advantages of all-electronic postproduction and image manipulation. Indeed, the concept of electronic intermediate film postproduction has been a topic of considerable interest since the original adoption of SMPTE 240M in 1988.

During evaluation of any new electronic imaging system, discussions tend to focus on horizontal pixel counts and scanning line numbers, to the exclusion of other important parameters. In the case of HDTV, the image is, in fact, multidimensional. The aggregate aesthetic quality of the picture is a complex combination of the following elements:

- Aspect ratio

- Horizontal resolution

- Vertical resolution

- Colorimetry

- Gray scale

Table 6.13 Different Terminology Used in the Video and Film Industries for Comparable Imaging parameters (*After* [30].)

Video	Film
Sensitivity	Exposure index (EI) Speed
Resolution	Resolving power
Colorimetry	Sharpness Modulation transfer curves
Gray scale	Color reproduction
Dynamic range	Exposure latitude
Noise	Diffuse rms granularity

- Total dynamic range for still images, and other "dimensions," such as temporal resolution and lag, for moving images

The historic difficulty in finding a common ground for discussing high-definition video and film imaging lies as much in the disparate terminology used as it does in the task of properly quantifying some of the technical parameters. (See Table 6.13.) The particular mood or ambience that a program director might want to achieve in a given scene often is created in real time by artificial manipulation of one or more variables. This process applies equally to both HDTV and film origination, specifically:

- *Spatial resolution*—Modified by employing fog filters on the lens, or through the use of electronic image enhancement

- *Colorimetry*—Modified by using color filters, or through the application of electronic painting

- *Gray scale*—Modified by using a specific film process, or through manipulating the electronic camera transfer characteristics

When images, originated either on film or video via an HDTV camera, are to be brought together in the electronic domain for possible integration (a *composite* or an *intercut*), it is desirable that they match each other as closely as possible in all of the imaging dimensions discussed previously. The seamless operation of an electronic intermediate film system relies upon the success of such matching. Depending on scene content, a disparity in any one of the dimensions could easily compromise the realism of the composite image.

A good match between the characteristics of separate images originated on film and HDTV is dependent upon the specific transfer characteristics of each, and upon the exercise of certain operational discretionary practices during their separate shooting [34]. Some fundamental disparities traditionally have existed between film and video that worked against such an ultimate matching of gray scale and other parameters. However, a number of advancements have been made to improve the degree of match between the overall operational transfer characteristics of the two mediums, including [35–37]:

- Substantial improvements in video camera pickup devices

- A better understanding of the respective media transfer characteristics

- Innovations in manipulation of the video camera transfer characteristics

- An increasing interest in HDTV possibilities within the program creative community

The marriage of film and HDTV is an important, ongoing effort. Regardless of which high-definition imaging system is used, it must offer interoperability with film. In spite of its many variations in format and aspect ratio, film has served—and will continue to serve—as a major worldwide exchange standard, unencumbered by the need for standards conversion or transcoding [36]. Thus, the image quality achievable from 35 mm film has served as a guide for the development of HDTV.

Debate will no doubt continue over the "film look" vs. the "video look." However, with the significant progress that has been made in HDTV resolution, gamma, chromaticity, and dynamic range embodied in SMPTE 240M-1995 and SMPTE 260M, the ultimate performance of film and video have never been closer.

6.8   References

1. Thorpe, Laurence J.: "High-Definition Production Systems," *Television Engineering Handbook*, rev. ed., K. B. Benson and Jerry C. Whitaker (eds.), McGraw-Hill, New York, pp. 24.1–24.28, 1992.

2. Ballard, Randall C.: U.S. Patent 2,152,234, filed July 19,1932.

3. Powers, Kerns H.: "High Definition Production Standards—Interlace or Progressive?," *Implementing HDTV: Television and Film Applications*, S. Baron (ed.), SMPTE, White Plains, N.Y., pp. 27–35, 1996.

4. "SMPTE Standard for Television—Digital Representation and Bit-Parallel Interface—1125/60 High-Definition Production System," SMPTE 260M-1992, SMPTE, White Plains, N.Y., 1992.

5. "SMPTE Standard for Television—Signal Parameters—1125-Line High-Definition Production Systems," SMPTE 240M-1995, SMPTE, White Plains, N.Y., 1995.

6. "SMPTE Standard for Television—1920 × 1080 Scanning and Analog and Parallel Digital Interfaces for Multiple-Picture Rates," SMPTE 274-1998, SMPTE, White Plains, N.Y., 1998.

7. "SMPTE Standard for Television—1920 × 1080 50 Hz Scanning and Interfaces," SMPTE 295M-1997, SMPTE, White Plains, N.Y., 1997.

8. "SMPTE Standard for Television—1280 × 720 Scanning, Analog and Digital Representation and Analog Interface," SMPTE 296M-1997, SMPTE, White Plains, N.Y., 1997.

9. "SMPTE Standard for Television—720 × 483 Active Line at 59.94 Hz Progressive Scan Production—Digital Representation," SMPTE 293M-1996, SMPTE, White Plains, N.Y., 1996.

10. "SMPTE Standard for Television—720 × 483 Active Line at 59.94-Hz Progressive Scan Production Bit-Serial Interfaces," SMPTE 294M-1997, SMPTE, White Plains, N.Y., 1997.

11. "SMPTE Standard for Television—Composite Analog Video Signal NTSC for Studio Applications," SMPTE 170M-1999, SMPTE, White Plains, N.Y., 1999.

12. "SMPTE Standard for Television—MPEG-2 4:2:2 Profile at High Level," SMPTE 308M-1998, SMPTE, White Plains, N.Y., 1998.

13. Gaggioni, H., M. Ueda, F. Saga, K. Tomita, and N. Kobayashi, "The Development of a High-Definition Serial Digital Interface," Sony Technical Paper, Sony Broadcast Group, San Jose, Calif., 1998.

14. "SMPTE Standard for Television—Bit-Serial Digital Interface for High-Definition Television Systems," SMPTE 292M-1998, SMPTE, White Plains, N.Y., 1998.

15. "SMPTE Standard for Television—24-Bit Digital Audio Format for HDTV Bit-Serial Interface," SMPTE 299M-1997, SMPTE, White Plains, N.Y., 1997.

16. "SMPTE Standard for Television—Mapping of AES3 Data into MPEG-2 Transport Stream," SMPTE 302M-1998, SMPTE, White Plains, N.Y., 1998.

17. "SMPTE Standard for Television (Proposed)—Vertical Ancillary Data Mapping for Bit-Serial Interface," SMPTE 334M, SMPTE, White Plains, N.Y., 2000.

18. "SMPTE Standard for Television (Proposed)—Signals and Generic Data over High-Definition Interfaces," SMPTE 346M, SMPTE, White Plains, N.Y., 2000.

19. "SMPTE Standard for Television (Proposed)—High Data-Rate Serial Data Transport Interface (HD-SDTI)," SMPTE 348M, SMPTE, White Plains, N.Y., 2000.

20. SMPTE 344M (Proposed), "540 Mb/s Serial Digital Interface," SMPTE, White Plains, N.Y., 2000.

21. Legault, Alain, and Janet Matey: "Interconnectivity in the DTV Era—The Emergence of SDTI," *Proceedings of Digital Television '98*, Intertec Publishing, Overland Park, Kan., 1998.

22. "SMPTE Standard for Television—SDTI Content Package Format (SDTI-CP)," SMPTE 326M, SMPTE, White Plains, N.Y., 2000.

23. "SMPTE Standard for Television—Element and Metadata Definitions for the SDTI-CP," SMPTE 331M, SMPTE, White Plains, N.Y., 2000.

24. "SMPTE Recommended Practice—SDTI-CP MPEG Decoder Templates," RP 204 (Proposed), SMPTE, White Plains, N.Y., 1999.

25. "SMPTE Standard for Television—Encapsulation of Data Packet Streams over SDTI (SDTI-PF)," SMPTE 332M, SMPTE, White Plains, N.Y., 2000.

26. Thorpe, Laurence J.: "The HDTV Camcorder and the March to Marketplace Reality," *SMPTE Journal*, SMPTE, White Plains, N.Y., pp. 164–177, March 1998.

27. Gloeggler, Peter: "Video Pickup Devices and Systems," in *NAB Engineering Handbook*, 9th Ed., Jerry C. Whitaker (ed.), National Association of Broadcasters, Washington, D.C., 1999.

28. "SMPTE Standard for Television—Broadcast Cameras: Hybrid Electrical and Fiber-Optic Connector," SMPTE 304M-1998, SMPTE, White Plains, N.Y., 1998.

29. "SMPTE Standard for Television—Hybrid Electrical and Fiber-Optic Camera Cable," SMPTE 311M-1998, SMPTE, White Plains, N.Y., 1998.

30. "SMPTE Standard for Television—Camera Positioning Information Conveyed by Ancillary Data Packets," SMPTE 315M-1999, SMPTE, White Plains, N.Y., 1999.

31. Thorpe, Laurence J.: "HDTV and Film—Digitization and Extended Dynamic Range," 133rd SMPTE Technical Conference, Paper no. 133-100, SMPTE, White Plains, N.Y., October 1991.

32. Ogomo, M., T. Yamada, K. Ando, and E. Yamazaki: "Considerations on Required Property for HDTV Displays," *Proc. of HDTV 90 Colloquium*, vols. 1, 2B, 1990.

33. Tanaka, H., and L. J. Thorpe: "The Sony PCL HDVS Production Facility," *SMPTE J.*, SMPTE, White Plains, N.Y., vol. 100, pp. 404–415, June 1991.

34. Mathias, H.: "Gamma and Dynamic Range Needs for an HDTV Electronic Cinematography System," *SMPTE J.*, SMPTE, White Plains, N.Y., vol. 96, pp. 840–845, September 1987.

35. Thorpe, L. J., E. Tamura, and T. Iwasaki: "New Advances in CCD Imaging," *SMPTE J.*, SMPTE, White Plains, N.Y., vol. 97, pp. 378–387, May 1988.

36. Thorpe, L., et. al.: "New High Resolution CCD Imager," *NAB Engineering Conference Proceedings*, National Association of Broadcasters, Washington, D.C., pp. 334–345, 1988.

37. Favreau, M., S. Soca, J. Bajon, and M. Cattoen: "Adaptive Contrast Corrector Using Real-Time Histogram Modification," *SMPTE J.*, SMPTE, White Plains, N.Y., vol. 93, pp. 488–491, May 1984.

6.9 Bibliography

ATSC Standard A/53 (1995), "ATSC Digital Television Standard," Advanced Television Systems Committee, Washington, D.C., 1995.

Benson, K. B., and D. G. Fink: *HDTV: Advanced Television for the 1990s*, McGraw-Hill, New York, 1990.

"SMPTE Standard for Television—10-Bit 4:2:2 Component and $4f_{sc}$ Composite Digital Signals—Serial Digital Interface," SMPTE 259M-1997, SMPTE, White Plains, N.Y., 1997.

"SMPTE Standard for Television—Ancillary Data Packet and Space Formatting," SMPTE 291M-1998, SMPTE, White Plains, N.Y., 1998.

"SMPTE Standard for Television—Bit-Serial Digital Interface for High-Definition Television Systems," SMPTE 292M-1996, SMPTE, White Plains, N.Y., 1996.

"SMPTE Standard for Television—Digital Representation and Bit-Parallel Interface— 1125/60 High-Definition Production System," SMPTE 260M-1999, SMPTE, White Plains, N.Y., 1999.

"SMPTE Standard for Television—Serial Data Transport Interface," SMPTE 305M-1998, SMPTE, White Plains, N.Y., 1998.

Turow, Dan: "SDTI and the Evolution of Studio Interconnect," *International Broadcasting Convention Proceedings,* IBC, Amsterdam, September 1998.

Wilkinson, J. H., H. Sakamoto, and P. Horne: "SDDI as a Video Data Network Solution," *International Broadcasting Convention Proceedings,* IBC, Amsterdam, September 1997.

7

DTV Audio Encoding and Decoding

7.1 Introduction

Monophonic sound is the simplest form of aural communication. A wide range of acceptable listening positions are practical, although it is obvious from most positions that the sound is originating from one source rather than occurring in the presence of the listener. Consumers have accepted this limitation without much thought in the past because it was all that was available. However, monophonic sound creates a poor illusion of the sound field that the program producer might want to create.

Two channel stereo improves the illusion that the sound is originating in the immediate area of the reproducing system. Still, there is a smaller acceptable listening area. It is difficult to keep the sound image centered between the left and right speakers, so that the sound and the action stay together as the listener moves in the room.

The AC-3 surround sound system is said to have 5.1 channels because there is a left, right, center, left surround, and right surround, which make up the 5 channels. A sixth channel is reserved for the lower frequencies and consumes only 120 Hz of the bandwidth; it is referred to as the 0.1 or *low-frequency effects* (LFE) channel. The center channel restores the variety of listening positions possible with monophonic sound.

The AC-3 system is effective in providing either an enveloping (ambient) sound field or allowing precise placement and movement of special effects because of the channel separation afforded by the multiple speakers in the system.

For efficient and reliable interconnection of audio devices, standardization of the interface parameters is of critical importance. The primary interconnection scheme for professional digital audio systems is AES Audio.

7.2 AES Audio

AES audio is a standard defined by the Audio Engineering Society and the European Broadcasting Union. Each AES stream carries two audio channels, which can be either a stereo pair or two independent feeds. The signals are pulse code modulated (PCM) data streams carrying digitized audio. Each sample is quantized to 20 or 24 bits, creating an audio *sam-*

Table 7.1 Theoretical S/N as a Function of the Number of Sampling Bits

Number of Sampling Bits	Resolution (number of quantizing steps)	Maximum Theoretical S/N
18	262,144	110 dB
20	1,048,576	122 dB
24	16,777,216	146 dB

ple word. Each word is then formatted to form a *subframe*, which is multiplexed with other subframes to form the AES digital audio stream. The AES stream can then be serialized and transmitted over coaxial or twisted-pair cable. The sampling rates supported range from 32 to 50 kHz. Common rates and applications include the following:

- 32 kHz—used for radio broadcast links

- 44.1 kHz—used for CD players

- 48 kHz—used for professional recording and production

Although 18-bit sampling was commonly used in the past, 20 bits has become prevalent today.

At 24 bits/sample, the S/N is 146 dB. This level of performance is generally reserved for high-end applications such as film recording and CD mastering. Table 7.1 lists the theoretical S/N ratios as a function of sampling bits for audio A/D conversion.

Of particular importance is that the AES format is designed to be independent of the audio conversion sample rate. The net data rate is exactly 64 times the sample rate, which is generally 48 kHz for professional applications. Thus, the most frequently encountered bit rate for AES3 data is 3.072 Mbits/s.

The AES3-1992 standard document precisely defines the AES3 twisted pair interconnection scheme. The signal, which is transmitted on twisted pair copper cable in a balanced format, is *bi-phase coded*. Primary signal parameters include the following:

- Output level can range from 2–10 V p-p

- Source impedance 110 Ω

- Receiver sensitivity 200 mV minimum

- Input impedance is recommended to be 110 Ω

- Interconnecting cable characteristic impedance 110 Ω

Electrical interface guidelines also have been set by the SMPTE and AES3 committees to permit transmission of AES3 data on coaxial cable. This single-ended interface is known as AES3-ID. The signal level, when terminated with 75 Ω, is 1 V p-p, ±20 percent. The source impedance is 75 Ω .

AES3 is inherently synchronous. A master local digital audio reference is normally used so that all audio equipment will be frequency- and phase-locked. The master reference can

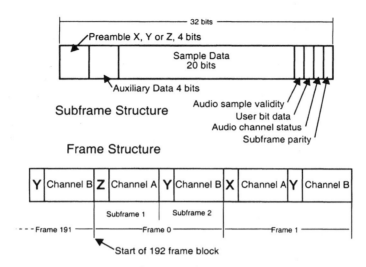

Figure 7.1 AES audio data format structure. (*From* [1]. *Used with permission.*)

originate from the digital audio equipment in a single room or an external master system providing a reference signal for larger facilities.

7.2.1 AES3 Data Format

The basic format structure of the AES data frames is shown in Figure 7.1. Each sample is carried by a subframe containing the following elements [1]:

- 20 bits of sample data

- 4 bits of auxiliary data, which may be used to extend the sample to 24 bits

- 4 additional bits of data

- A preamble

Two subframes make up a frame, which contains one sample from each of the two channels. Frames are further grouped into 192-frame blocks, which define the limits of *user data* and *channel status data* blocks. A special preamble indicates the channel identity for each sample (X or Y preamble) and the start of a 192-frame block (Z preamble). To minimize the direct-current (dc) component on the transmission line, facilitate clock recovery, and make the interface polarity insensitive, the data is channel coded in the biphase-mark mode.

The preambles specifically violate the biphase-mark rules for easy recognition and to ensure synchronization. When digital audio is embedded in the serial digital video data stream, the start of the 192-frame block is indicated by the Z bit, which corresponds to the occurrence of the Z-type preamble.

The *validity bit* indicates whether the audio sample bits in the subframe are suitable for conversion to an analog audio signal. User data is provided to carry other information, such as time code. Channel status data contains information associated with each audio channel.

There are three levels of implementation of the channel status data: minimum, standard, and enhanced. The standard implementation is recommended for use in professional video applications; the channel status data typically contains information about signal emphasis, sampling frequency, channel mode (stereo, mono, etc.), use of auxiliary bits (extend audio data to 24 bits or other use), and a CRC for error checking of the total channel status block.

7.2.2 SMPTE 324M

SMPTE 324M (proposed at this writing) defines a synchronous, self-clocking serial interface for up to 12 channels of linearly encoded audio and auxiliary data [2]. The interface is designed to allow multiplexing of six two-channel streams compliant with AES3. Audio sampled at 48 kHz and clock-locked to video is the preferred implementation for studio applications. However, the 324M interlace supports any frequency of operation supported by AES3, provided that all the audio channels are sampled by a common clock. Ideally, all the channels should be *audio synchronous* for guaranteed audio phase coherence. An audio channel is defined as being synchronous with another when the two channels are running from the same clock and the analog inputs are concurrently sampled.

The 324M standard is intended to provide a reliable method of distributing multiple cophased channels of digital audio around the studio without losing the initial relative sample-phase relationship. A mechanism is provided to allow more than one 12-channel stream to be realigned after a relative misalignment of up to ±8 samples.

The interface, intended to be compatible with the complete range of digital television scanning standards and standard film rates, may be used for distribution of multiple channels of audio in either a pre-mix or post-mix situation. In the post-mix case, channel assignment is defined in SMPTE 320M.

7.3   Audio Compression

Efficient recording and/or transmission of digital audio signals demands a reduction in the amount of information required to represent the aural signal [3]. The amount of digital information needed to accurately reproduce the original PCM samples taken of an analog input may be reduced by applying a digital compression algorithm, resulting in a digitally compressed representation of the original signal. (In this context, the term *compression* applies to the digital information that must be stored or recorded, not to the dynamic range of the audio signal.) The goal of any digital compression algorithm is to produce a digital representation of an audio signal which, when decoded and reproduced, sounds the same as the original signal, while using a minimum amount of digital information (bit rate) for the compressed (or encoded) representation. The AC-3 digital compression algorithm specified in the ATSC DTV system can encode from 1 to 5.1 channels of source audio from a PCM representation into a serial bit stream at data rates ranging from 32 to 640 kbits/s.

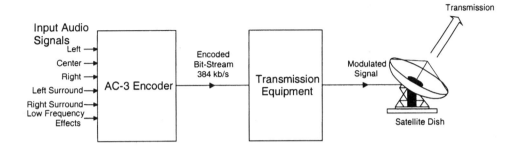

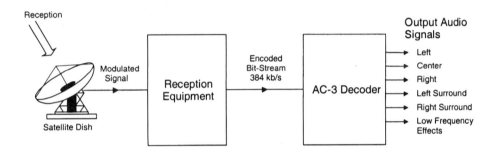

Figure 7.2 Example application of the AC-3 audio subsystem for satellite audio transmission. (*From* [3]. *Used with permission.*)

A typical application of the bit-reduction algorithm is shown in Figure 7.2. In this example, a 5.1 channel audio program is converted from a PCM representation requiring more than 5 Mbits/s (6 channels × 48 kHz × 18 bits = 5.184 Mbits/s) into a 384 kbits/s serial bit stream by the AC-3 encoder. Radio frequency (RF) transmission equipment converts this bit stream into a modulated waveform that is applied to a satellite transponder. The amount of bandwidth and power thus required by the transmission has been reduced by more than a factor of 13 by the AC-3 digital compression system. The received signal is demodulated back into the 384 kbits/s serial bit stream, and decoded by the AC-3 decoder. The result is the original 5.1 channel audio program.

Digital compression of audio is useful wherever there is an economic benefit to be obtained by reducing the amount of digital information required to represent the audio signal. Typical applications include the following:

• Terrestrial audio broadcasting

• Delivery of audio over metallic or optical cables, or over RF links

• Storage of audio on magnetic, optical, semiconductor, or other storage media

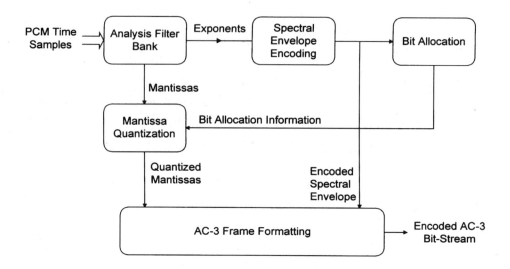

Figure 7.3 Overview of the AC-3 audio-compression system encoder. (*From* [3]. *Used with permission.*)

7.3.1 Encoding

As discussed briefly in Chapter 5, the AC-3 encoder accepts PCM audio and produces the encoded bit stream for the ATSC DTV standard [3]. The AC-3 algorithm achieves high *coding gain* (the ratio of the input bit rate to the output bit rate) by coarsely quantizing a frequency-domain representation of the audio signal. A block diagram of this process is given in Figure 7.3. The first step in the encoding chain is to transform the representation of audio from a sequence of PCM time samples into a sequence of blocks of frequency coefficients. This is done in the *analysis filterbank*. Overlapping blocks of 512 time samples are multiplied by a time window and transformed into the frequency domain. Because of the overlapping blocks, each PCM input sample is represented in two sequential transformed blocks. The frequency-domain representation then may be decimated by a factor of 2, so that each block contains 256 frequency coefficients. The individual frequency coefficients are represented in binary exponential notation as a *binary exponent* and a *mantissa*. The set of exponents is encoded into a coarse representation of the signal spectrum, referred to as the *spectral envelope*. This spectral envelope is used by the core bit-allocation routine, which determines how many bits should be used to encode each individual mantissa. The spectral envelope and the coarsely quantized mantissas for six audio blocks (1536 audio samples) are formatted into an AC-3 *frame*. The AC-3 bit stream is a sequence of AC-3 frames.

The actual AC-3 encoder is more complex than shown in the simplified system of Figure 7.3. The following functions also are included:

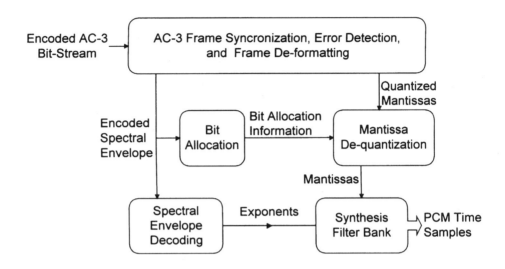

Figure 7.4 Overview of the AC-3 audio-compression system decoder. (*From* [3]. *Used with permission.*)

- A frame header is attached, containing information (bit rate, sample rate, number of encoded channels, and other data) required to synchronize to and decode the encoded bit stream.

- Error-detection codes are inserted to allow the decoder to verify that a received frame of data is error-free.

- The analysis filterbank spectral resolution may be dynamically altered to better match the time/frequency characteristic of each audio block.

- The spectral envelope may be encoded with variable time/frequency resolution.

- A more complex bit-allocation may be performed, and parameters of the core bit-allocation routine may be modified to produce a more optimum bit allocation.

- The channels may be coupled at high frequencies to achieve higher coding gain for operation at lower bit rates.

- In the 2-channel mode, a rematrixing process may be selectively performed to provide additional coding gain, and to allow improved results to be obtained in the event that the 2-channel signal is decoded with a matrix surround decoder.

7.3.2 Decoding

The decoding process is, essentially, the inverse of the encoding process [3]. The basic decoder, shown in Figure 7.4, must synchronize to the encoded bit stream, check for errors,

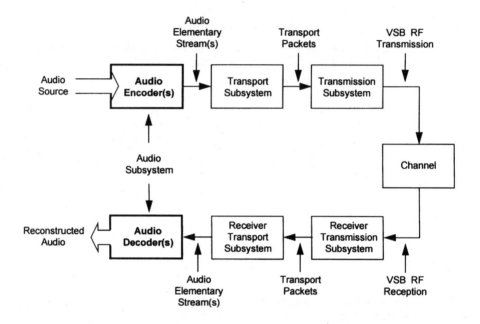

Figure 7.5 The audio subsystem in the DTV standard. (*From* [4]. *Used with permission.*)

and deformat the various types of data (i.e., the encoded spectral envelope and the quantized mantissas). The bit-allocation routine is run, and the results are used to unpack and dequantize the mantissas. The spectral envelope is decoded to produce the exponents. The exponents and mantissas are transformed back into the time domain to produce the decoded PCM time samples. Additional steps in the audio decoding process include the following:

- Error concealment or muting may be applied in the event a data error is detected.

- Channels that have had their high-frequency content coupled must be decoupled.

- Dematrixing must be applied (in the 2-channel mode) whenever the channels have been *rematrixed*. (See section 7.3.4.)

- The synthesis filterbank resolution must be dynamically altered in the same manner as the encoder analysis filterbank was altered during the encoding process.

7.4 Implementation of the AC-3 System

As illustrated in Figure 7.5, the audio subsystem of the ATSC DTV standard comprises the audio-encoding/decoding function and resides between the audio inputs/outputs and the transport subsystem [4]. The audio encoder is responsible for generating the *audio elementary stream*, which is an encoded representation of the baseband audio input signals. (Note that more than one audio encoder may be used in a system.) The flexibility of the transport system allows multiple audio elementary streams to be delivered to the receiver. At the receiver, the transport subsystem is responsible for selecting which audio streams to deliver to the audio subsystem. The audio subsystem is then responsible for decoding the audio elementary stream back into baseband audio.

An audio program source is encoded by a *digital television audio encoder*. The output of the audio encoder is a string of bits that represent the audio source (the audio elementary stream). The transport subsystem packetizes the audio data into PES (*packetized elementary system*) packets, which are then further packetized into *transport packets*. The transmission subsystem converts the transport packets into a modulated RF signal for transmission to the receiver. At the receiver, the signal is demodulated by the receiver transmission subsystem. The receiver transport subsystem converts the received audio packets back into an audio elementary stream, which is decoded by the digital television audio decoder.

The partitioning shown in Figure 7.5 is conceptual, and practical implementations may differ. For example, the transport processing may be broken into two blocks; the first would perform PES packetization, and the second would perform transport packetization. Or, some of the transport functionality may be included in either the audio coder or the transmission subsystem.

7.4.1   Audio-Encoder Interface

The audio system accepts baseband inputs with up to six channels per audio program bit stream in a channelization scheme consistent with ITU-R Rec. BS-775 [5]. The six audio channels are:

- Left

- Center

- Right

- Left surround

- Right surround

- Low-frequency enhancement (LFE)

Multiple audio elementary bit streams may be conveyed by the transport system.

The bandwidth of the LFE channel is limited to 120 Hz. The bandwidth of the other (main) channels is limited to 20 kHz. Low-frequency response may extend to dc, but it is more typically limited to approximately 3 Hz (−3 dB) by a dc-blocking high-pass filter. Audio-coding efficiency (and thus audio quality) is improved by removing dc offset from audio signals before they are encoded. The input audio signals may be in analog or digital form.

For analog input signals, the input connector and signal level are not specified [3]. Conventional broadcast practice may be followed. One commonly used input connector is the 3-pin XLR female (the incoming audio cable uses the male connector) with pin 1 ground, pin 2 hot or positive, and pin 3 neutral or negative.

Likewise, for digital input signals, the input connector and signal format are not specified. Commonly used formats such as the AES3-1992 2-channel interface are suggested. When multiple 2-channel inputs are used, the preferred channel assignment is:

- Pair 1: Left, Right

- Pair 2: Center, LFE

- Pair 3: Left surround, Right surround

Sampling Parameters

The AC-3 system conveys digital audio sampled at a frequency of 48 kHz, locked to the 27 MHz system clock [3]. If analog signal inputs are employed, the A/D converters should sample at 48 kHz. If digital inputs are employed, the input sampling rate should be 48 kHz, or the audio encoder should contain sampling rate converters that translate the sampling rate to 48 kHz. The sampling rate at the input to the audio encoder must be locked to the video clock for proper operation of the audio subsystem.

In general, input signals should be quantized to at least 16-bit resolution. The audio-compression system can convey audio signals with up to 24-bit resolution.

7.4.2 Output Signal Specification

Conceptually, the output of the audio encoder is an elementary stream that is formed into PES packets within the transport subsystem [3]. It is possible that digital television systems will be implemented wherein the formation of audio PES packets takes place within the audio encoder. In this case, the output of the audio encoder would be PES packets. Physical interfaces for these outputs (elementary streams and/or PES packets) may be defined as voluntary industry standards by SMPTE or other organizations; they are not, however, specified in the core ATSC standard.

7.5 Operational Details of the AC-3 Standard

The AC-3 audio-compression system consists of three basic operations, as illustrated in Figure 7.6 [6]. In the first stage, the representation of the audio signal is changed from the time domain to the frequency domain, which is a more efficient domain in which to perform psychoacoustically based audio compression. The resulting frequency-domain coefficients are then encoded. The frequency-domain coefficients may be coarsely quantized because the resulting quantizing noise will be at the same frequency as the audio signal, and relatively low S/N ratios are acceptable because of the phenomenon of psychoacoustic masking. Based on a psychoacoustic model of human hearing, a bit-allocation operation determines the actual S/N acceptable for each individual frequency coefficient. Finally, the frequency coefficients are coarsely quantized to the necessary precision and formatted into the audio elementary stream.

The basic unit of encoded audio is the AC-3 *sync frame*, which represents 1536 audio samples. Each sync frame of audio is a completely independent encoded entity. The elemen-

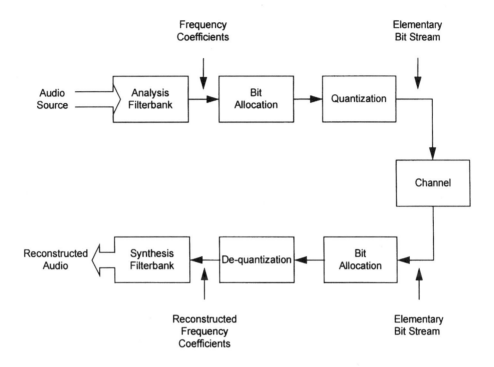

Figure 7.6 Overview of the AC-3 audio-compression system. (*From* [6]. *Used with permission.*)

tary bit stream contains the information necessary to allow the audio decoder to perform the identical (to the encoder) bit allocation. This permits the decoder to unpack and dequantize the elementary bit stream frequency coefficients, resulting in the reconstructed frequency coefficients. The synthesis filterbank is the inverse of the analysis filterbank, and it converts the reconstructed frequency coefficients back into a time-domain signal.

7.5.1 Transform Filterbank

The process of converting the audio from the time domain to the frequency domain requires that the audio be blocked into overlapping blocks of 512 samples [6]. For every 256 new audio samples, a 512-sample block is formed from the 256 new samples and the 256 previous samples. Each audio sample is represented in two audio blocks, so the number of samples to be processed initially is doubled. The overlapping of blocks is necessary to prevent audible blocking artifacts. New audio blocks are formed every 5.33 ms. A group of six blocks is coded into one AC-3 sync frame.

Window Function

Prior to being transformed into the frequency domain, the block of 512 time samples is *windowed* [6]. The windowing operation involves a vector multiplication of the 512-point block with a 512-point window function. The window function has a value of 1.0 in its center, tapering down to almost zero at the ends. The shape of the window function is such that the overlap/add processing at the decoder will result in a reconstruction free of blocking artifacts. The window function shape also determines the shape of each individual filterbank filter.

Time-Division Aliasing Cancellation Transform

The analysis filterbank is based on the fast Fourier transform [6]. The particular transformation employed is the oddly stacked *time-domain aliasing cancellation* (TDAC) transform. This particular transformation is advantageous because it allows removal of the 100 percent redundancy that was introduced in the blocking process. The input to the TDAC transform is 512 windowed time-domain points, and the output is 256 frequency-domain coefficients.

Transient Handling

When extreme time-domain transients exist (an impulse, such as a castanets click), there is a possibility that quantization error—incurred by coarsely quantizing the frequency coefficients of the transient—will become audible as a result of *time smearing* [6]. The quantization error within a coded audio block is reproduced throughout the block. It is possible for the portion of the quantization error that is reproduced prior to the impulse to be audible. Time smearing of quantization noise may be reduced by altering the length of the transform that is performed. Instead of a single 512-point transform, a pair of 256-point transforms may be performed—one on the first 256 windowed samples, and one on the last 256 windowed samples. A transient detector in the encoder determines when to alter the transform length. The reduction in transform length prevents quantization error from spreading more than a few milliseconds in time, which is adequate to prevent audibility.

7.5.2 Coded Audio Representation

The frequency coefficients that result from the transformation are converted to a binary floating point notation [6]. The scaling of the transform is such that all values are smaller than 1.0. An example value in binary notation (base 2) with 16-bit precision would be:

0.0000 0000 1010 11002

The number of leading zeros in the coefficient, 8 in this example, becomes the *raw exponent*. The value is left-shifted by the exponent, and the value to the right of the decimal point (1010 1100) becomes the *normalized mantissa* to be coarsely quantized. The exponents and the coarsely quantized mantissas are encoded into the bit stream.

Exponent Coding

A certain amount of processing is applied to the raw exponents to reduce the amount of data required to encode them [6]. First, the raw exponents of the six blocks to be included in a single AC-3 sync frame are examined for block-to-block differences. If the differences are small, a single exponent set is generated that is usable by all six blocks, thus reducing the amount of data to be encoded by a factor of 6. If the exponents undergo significant changes within the frame, exponent sets are formed over blocks where the changes are not significant. Because of the frequency response of the individual filters in the analysis filterbank, exponents for adjacent frequencies rarely differ by more than ±2. To take advantage of this fact, exponents are encoded differentially in frequency. The first exponent is encoded as an absolute, and the difference between the current exponent and the following exponent then is encoded. This reduces the exponent data rate by a factor of 2. Finally, where the spectrum is relatively flat, or an exponent set only covers 1 or 2 blocks, differential exponents may be shared across 2 or 4 frequency coefficients, for an additional savings of a factor of 2 or 4.

The final coding efficiency for AC-3 exponents is typically 0.39 bits/exponent (or 0.39 bits/sample, because there is an exponent for each audio sample). Exponents are coded only up to the frequency needed for the perception of full frequency response. Typically, the highest audio frequency component in the signal that is audible is at a frequency lower than 20 kHz. In the case that signal components above 15 kHz are inaudible, only the first 75 percent of the exponent values are encoded, reducing the exponent data rate to less than 0.3 bits/sample.

The exponent processing changes the exponent values from their original values. The encoder generates a local representation of the exponents that is identical to the decoded representation that will be used by the decoder. The decoded representation then is used to shift the original frequency coefficients to generate the normalized mantissas that are subsequently quantized.

Mantissas

The frequency coefficients produced by the analysis filterbank have a useful precision that is dependent upon the word length of the input PCM audio samples as well as the precision of the transform computation [6]. Typically, this precision is on the order of 16 to18 bits, but may be as high as 24 bits. Each normalized mantissa is quantized to a precision from 0 to16 bits. Because the goal of audio compression is to maximize the audio quality at a given bit rate, an optimum (or near-optimum) allocation of the available bits to the individual mantissas is required.

7.5.3 Bit Allocation

The number of bits allocated to each individual mantissa value is determined by the bit-allocation routine [6]. The identical core routine is run in both the encoder and the decoder, so that each generates an identical bit allocation.

The core bit-allocation algorithm is considered *backward adaptive*, in that some of the encoded audio information within the bit stream (fed back into the encoder) is used to compute the final bit allocation. The primary input to the core allocation routine is the decoded exponent values, which give a general picture of the signal spectrum. From this version of the signal spectrum, a *masking curve* is calculated. The calculation of the masking model is based on a model of the human auditory system. The masking curve indicates, as a function of frequency, the level of quantizing error that may be tolerated. Subtraction (in the log power domain) of the masking curve from the signal spectrum yields the required S/N as a function of frequency. The required S/N values are mapped into a set of *bit-allocation pointers* (BAPs) that indicate which quantizer to apply to each mantissa.

Forward Adaptive

The AC-3 encoder may employ a more sophisticated psychoacoustic model than that used by the decoder [6]. The core allocation routine used by both the encoder and the decoder makes use of a number of adjustable parameters. If the encoder employs a more sophisticated psychoacoustic model than that of the core routine, the encoder may adjust these parameters so that the core routine produces a better result. The parameters are subsequently inserted into the bit stream by the encoder and fed forward to the decoder.

In the event that the available bit-allocation parameters do not allow the ideal allocation to be generated, the encoder can insert explicit codes into the bit stream to alter the computed masking curve, hence the final bit allocation. The inserted codes indicate changes to the base allocation and are referred to as *delta bit-allocation codes*.

7.5.4 Rematrixing

When the AC-3 encoder is operating in a 2-channel stereo mode, an additional processing step is inserted to enhance interoperability with Dolby Surround 4-2-4 matrix encoded programs [6]. This extra step is referred to as *rematrixing*.

The signal spectrum is broken into four distinct rematrixing frequency bands. Within each band, the energy of the left, right, sum, and difference signals are determined. If the largest signal energy is in the left and right channels, the band is encoded normally. If the dominant signal energy is in the sum and difference channels, then those channels are encoded instead of the left and right channels. The decision as to whether to encode left and right or sum and difference is made on a band-by-band basis and is signaled to the decoder in the encoded bit stream.

7.5.5 Coupling

In the event that the number of bits required to transparently encode the audio signals exceeds the number of bits that are available, the encoder may invoke *coupling* [6]. Coupling involves combining the high-frequency content of individual channels and sending the individual channel signal envelopes along with the combined coupling channel. The psychoacoustic basis for coupling is that within narrow frequency bands, the human ear detects high-frequency localization based on the signal envelope rather than on the detailed signal waveform.

The frequency above which coupling is invoked, and the channels that participate in the process, are determined by the AC-3 encoder. The encoder also determines the frequency banding structure used by the coupling process. For each coupled channel and each cou-

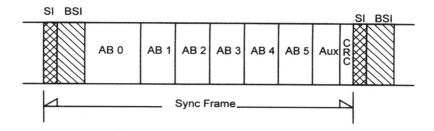

Figure 7.7 The AC-3 synchronization frame. (*From* [6]. *Used with permission.*)

pling band, the encoder creates a sequence of *coupling coordinates*. The coupling coordinates for a particular channel indicate what fraction of the common coupling channel should be reproduced out of that particular channel output. The coupling coordinates represent the individual signal envelopes for the channels. The encoder determines the frequency with which coupling coordinates are transmitted. If the signal envelope is steady, the coupling coordinates do not need to be sent every block, but can be reused by the decoder until new coordinates are sent. The encoder determines how often to send new coordinates, and it can send them as often as every block (every 5.3 ms).

7.5.6   Bit Stream Elements and Syntax

An AC-3 serial-coded audio bit stream is made up of a sequence of *synchronization frames*, as illustrated in Figure 7.7 [6]. Each synchronization frame contains six coded audio blocks, each of which represent 256 new audio samples. A *synchronization information* (SI) header at the beginning of each frame contains information needed to acquire and maintain synchronization. A *bit stream information* (BSI) header follows each SI, containing parameters describing the coded audio service. The coded audio blocks may be followed by an auxiliary data (Aux) field. At the end of each frame is an error-check field that includes a CRC word for error detection. An additional CRC word, the use of which is optional, is located in the SI header.

A number of bit stream elements have values that may be transmitted, but whose meaning has been reserved. If a decoder receives a bit stream that contains reserved values, the decoder may or may not be able to decode and produce audio.

Splicing and Insertion

The ideal place to splice encoded audio bit streams is at the boundary of a sync frame [6]. If a bit stream splice is performed at the boundary of the sync frame, the audio decoding will proceed without interruption. If a bit stream splice is performed randomly, there will be an audio interruption. The frame that is incomplete will not pass the decoder's error-detection test, and this will cause the decoder to mute. The decoder will not find sync in its proper place in the next frame, and it will enter a sync search mode. After the sync code of the new bit stream is found, synchronization will be achieved, and audio reproduction will resume.

This type of outage will be on the order of two frames, or about 64 ms. Because of the windowing process of the filterbank, when the audio goes to mute, there will be a gentle fade-down over a period of 2.6 ms. When the audio is recovered, it will fade up over a period of 2.6 ms. Except for the approximately 64 ms of time during which the audio is muted, the effect of a random splice of an AC-3 elementary stream is relatively benign.

Error-Detection Codes

Each AC-3 sync frame ends with a 16-bit CRC error-check code [6]. The decoder may use this code to determine whether a frame of audio has been damaged or is incomplete. Additionally, the decoder may make use of error flags provided by the transport system. In the case of detected errors, the decoder may try to perform error concealment, or it may simply mute.

7.5.7 Loudness and Dynamic Range

It is important for the digital television system to provide uniform subjective loudness for all audio programs [6]. Consumers often find it annoying when audio levels fluctuate between broadcast channels (observed when channel hopping) or between program segments on a particular channel (such as commercials being much louder than entertainment programs). One element found in most audio programming is the human voice. Achieving an approximate level match for dialogue (spoken in a normal voice, without shouting or whispering) in all audio programming is a desirable goal. The AC-3 audio system provides syntactical elements that make this goal achievable.

Because the digital audio-coding system can provide more than 100 dB of dynamic range, there is no technical reason for dialogue to be encoded anywhere near 100 percent, as it commonly is in NTSC television. However, there is no assurance that all program channels, or all programs or program segments on a given channel, will have dialogue encoded at the same (or even a similar) level. Without a uniform coding level for dialogue (which would imply a uniform headroom available for all programs), there would be inevitable audio-level fluctuations between program channels or even between program segments. These issues are addressed by the DTV audio standard and are described in Section 7.4.

Dynamic Range Compression

It is common practice for high-quality programming to be produced with wide dynamic range audio, suitable for the highest-quality audio reproduction environment [6]. Because they serve audiences with a wide range of receiver capabilities, however, broadcasters typically process audio to reduce its dynamic range. This processed audio is more suitable for most of the audience, which does not have an audio reproduction environment that matches the original audio production studio. In the case of NTSC, all viewers receive the same audio with the same dynamic range; it is impossible for any viewer to enjoy the original wide dynamic range of the audio production.

For DTV, the audio-coding system provides an embedded dynamic range control scheme that allows a common encoded bit stream to deliver programming with a dynamic range appropriate for each individual listener. A *dynamic range control value* (DynRng) is provided in each audio block (every 5 ms). These values are used by the audio decoder to alter the level of the reproduced sound for each audio block. Level variations of up to ±24 dB can be indicated.

7.5.8 Encoding the AC-3 Bit Stream

Because the ATSC DTV standard AC-3 audio system is specified by the syntax and decoder processing, the encoder itself is not precisely specified [3]. The only normative requirement on the encoder is that the output elementary bit stream follow the AC-3 syntax. Therefore, encoders of varying levels of sophistication may be produced. More sophisticated encoders may offer superior audio performance, and they may make operation at lower bit rates acceptable. Encoders are expected to improve over time, and all decoders will benefit from encoder improvements. The encoder described in this section, although basic in operation, provides good performance and offers a starting point for future designs. A flow chart diagram of the encoding process is given in Figure 7.8.

Input Word Length/Sample Rate

The AC-3 encoder accepts audio in the form of PCM words [3]. The internal dynamic range of AC-3 allows input word lengths of up to 24 bits to be useful.

The input sample rate must be locked to the output bit rate so that each AC-3 sync frame contains 1536 samples of audio. If the input audio is available in a PCM format at a different sample rate than that required, sample rate conversion must be performed to conform the sample rate.

Individual input channels may be high-pass filtered. Removal of dc components of the input signals can allow more efficient coding because the available data rate then is not used to encode dc. However, there is the risk that signals that do not reach 100 percent PCM level before high-pass filtering will exceed the 100 percent level after filtering, and thus be clipped. A typical encoder would high-pass filter the input signals with a single pole filter at 3 Hz.

The LFE channel normally is low-pass-filtered at 120 Hz. A typical encoder would filter the LFE channel with an 8th-order elliptic filter whose cutoff frequency is 120 Hz.

Transients are detected in the full-bandwidth channels to decide when to switch to short-length audio blocks to improve pre-echo performance. High-pass filtered versions of the signals are examined for an increase in energy from one subblock time segment to the next. Subblocks are examined at different time scales. If a transient is detected in the second half of an audio block in a channel, that channel switches to a short block.

The transient detector is used to determine when to switch from a *long transform block* (length 512) to a *short transform block* (length 256). It operates on 512 samples for every audio block. This is done in two passes, with each pass processing 256 samples. Transient detection is broken down into four steps:

- High-pass filtering

- Segmentation of the block into submultiples

- Peak amplitude detection within each subblock segment

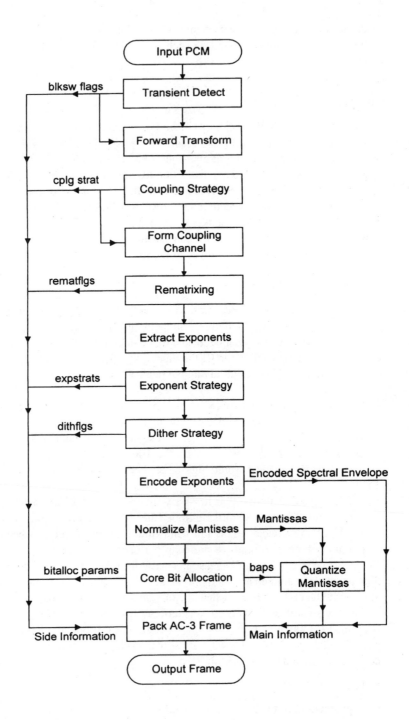

Figure 7.8 Generalized flow diagram of the AC-3 encoding process. (*From* [6]. *Used with permission.*)

- Threshold comparison

7.5.9   AC-3/MPEG Bit Stream

The AC-3 elementary bit stream is included in an MPEG-2 multiplex bit stream in much the same way an MPEG-1 audio stream would be included, with the AC-3 bit stream packetized into PES packets [7]. An MPEG-2 multiplex bit stream containing AC-3 elementary streams must meet all audio constraints described in the MPEG model. It is necessary to unambiguously indicate that an AC-3 stream is, in fact, an AC-3 stream, and not an MPEG audio stream. The MPEG-2 standard does not explicitly state codes to be used to indicate an AC-3 stream. Also, the MPEG-2 standard does not have an audio descriptor adequate to describe the contents of the AC-3 bit stream in its internal tables. The solution to this problem is beyond the scope of this chapter; interested readers should consult [7] for additional information on the subject.

7.5.10   Decoding the AC-3 Bit Stream

An overview of AC-3 decoding is diagrammed in Figure 7.9, where the decoding process flow is shown as a sequence of blocks down the center of the illustration, and some of the key information flow is indicated by arrowed lines at the sides [3]. This decoder should be considered only as an example; other methods certainly exist to implement decoders, and those other methods may have advantages in certain areas (such as instruction count, memory requirements, number of transforms required, and other parameters). The input bit stream typically will come from a transmission or storage system. The interface between the source of AC-3 data and the AC-3 decoder is not specified in the ATSC DTV standard.

Continuous or Burst Input

The encoded AC-3 data may be input to the decoder as a continuous data stream at the nominal bit rate, or chunks of data may be burst into the decoder at a high rate with a low duty cycle [3]. For burst-mode operation, either the data source or the decoder may be the master controlling the burst timing. The AC-3 decoder input buffer may be smaller if the decoder can request bursts of data on an as-needed basis, but the external buffer memory may need to be larger.

Most applications of the standard will convey the elementary AC-3 bit stream with byte or (16-bit) word alignment. The *sync frame* is always an integral number of words in length. The decoder may receive data as a continuous serial stream of bits without any alignment, or the data may be input to the decoder with either byte or word alignment. Byte or word alignment of the input data may allow some simplification of the decoder. Alignment does reduce the probability of false detection of the sync word.

Synchronization and Error Detection

The AC-3 bit steam format allows for rapid synchronization [3]. The 16-bit sync word has a low probability of false detection. With no input stream alignment, the probability of false detection of the sync word is 0.0015 percent per input stream bit position. For a bit rate of

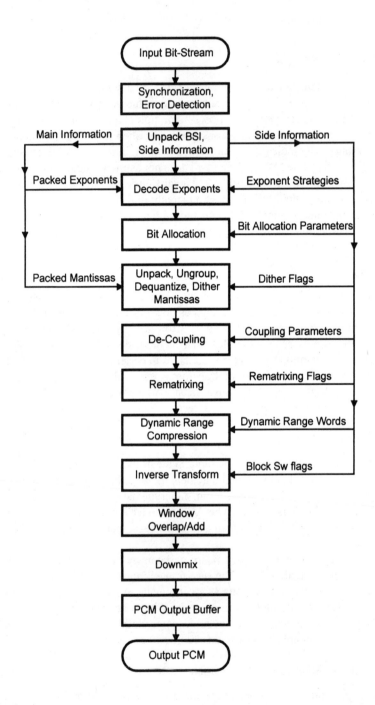

Figure 7.9 Generalized flow diagram of the AC-3 decoding process. (*From* [3]. *Used with permission.*)

384 kbits/s, the probability of false sync word detection is 19 percent per frame. Byte alignment of the input stream drops this probability to 2.5 percent, and word alignment drops it to 1.2 percent.

When a sync pattern is detected, the decoder may be estimated to be in sync, and one of the CRC words (CRC1 or CRC2) may be checked. Because CRC1 comes first and covers the first five-eighths of the frame, the result of a CRC1 check may be available after only five-eighths of the frame has been received. Or, the entire frame size can be received and CRC2 checked. If either CRC word checks, the decoder may safely be presumed to be in sync, and decoding and reproduction of audio may proceed. The chance of false sync in this case would be the concatenation of the probabilities of a false sync word detection and a CRC misdetection of error. The CRC check is reliable to 0.0015 percent. This probability, concatenated with the probability of a false sync detection in a byte-aligned input bit stream, yields a probability of false synchronization of 0.000035 percent (or about once in 3 million synchronization attempts).

If this small probability of false sync is too large for a specific application, several methods may be used to reduce it. The decoder may only presume correct sync in the case that both CRC words check properly. The decoder also may require multiple sync words to be received with the proper alignment. If the data transmission or storage system is aware that data is in error, this information may be made known to the decoder.

Inherent to the decoding process is the *unpacking* (demultiplexing) of the various types of information included in the bit stream. Among the options for distribution of this bit stream information are:

- Selected data may be copied from the input buffer to dedicated registers.

- Data from the input buffer may be copied to specific working memory locations.

- The data may simply be located in the input buffer, with pointers to the data saved to another location for use when the information is required.

Decoding Components

The audio-compression system exponents are delivered in the bit stream in an encoded form [3]. To unpack and decode the exponents, two types of "side information" are required:

- The number of exponents must be known.

- The exponent "strategy" in use by each channel must be known.

The *bit-allocation computation* reveals how many bits are used for each mantissa. The inputs to the bit-allocation computation are the decoded exponents and the bit-allocation side information. The outputs of the bit-allocation computation are a set of *bit-allocation pointers* (BAPs), one BAP for each coded mantissa. The BAP indicates the quantizer used for the mantissa, and how many bits in the bit stream were used for each mantissa.

The coarsely quantized mantissas make up the bulk of the AC-3 data stream. Each mantissa is quantized to a level of precision indicated by the corresponding BAP. To pack the mantissa data more efficiently, some mantissas are grouped together into a single transmitted value. For instance, two 11-level quantized values are conveyed in a single 7-bit code (3.5 bits/value) in the bit stream.

The mantissa data is unpacked by peeling off groups of bits as indicated by the BAPs. Grouped mantissas must be ungrouped. The individual coded mantissa values are converted into a dequantized value. Mantissas that are indicated as having zero bits may be reproduced as either zero or by a random dither value (under control of a dither flag).

Other steps in the decoding process include the following:

- *Decoupling.* When *coupling* is in use, the channels that are coupled must be decoupled. Decoupling involves reconstructing the high-frequency section (exponents and mantissas) of each coupled channel, from the common coupling channel and the coupling coordinates for the individual channel. Within each coupling band, the coupling-channel coefficients (exponent and mantissa) are multiplied by the individual channel coupling coordinates.

- *Rematrixing.* In the 2/0 audio-coding mode, rematrixing may be employed as indicated by a *rematrix flag.* When the flag indicates that a band is rematrixed, the coefficients encoded in the bit stream are sum and difference values, instead of left and right values.

- *Dynamic range compression.* For each block of audio, a dynamic range control value may be included in the bit stream. The decoder, by default, will use this value to alter the magnitude of the coefficient (exponent and mantissa) as required to properly process the data.

- *Inverse transform.* The decoding steps described in this section will result in a set of frequency coefficients for each encoded channel. The inverse transform converts these blocks of frequency coefficients into blocks of time samples.

- *Window, overlap/add.* The individual blocks of time samples must be windowed, and adjacent blocks are overlapped and added together to reconstruct the final continuous-time-output PCM audio signal.

- *Downmixing.* If the number of channels required at the decoder output is smaller than the number of channels that are encoded in the bit stream, then downmixing is required. Downmixing in the time domain is shown in the example decoder of Figure 7.9. Because the inverse transform is a linear operation, it also is possible to downmix in the frequency domain prior to transformation.

- *PCM output buffer.* Typical decoders will provide PCM output samples at the PCM sampling rate. Because blocks of samples result from the decoding process, an output buffer typically is required.

- *Output PCM.* The output PCM samples are delivered in a form suitable for interconnection to a digital-to-analog converter (D/A), or in some other form required by the receiver.

7.5.11 Algorithmic Details

The actual audio information conveyed by the AC-3 bit stream consists of the quantized frequency coefficients [3]. The coefficients, delivered in floating point form, are 5-bit values that indicate the number of leading zeros in the binary representation of a frequency coefficient. The exponent acts as a scale factor for each mantissa, equal to 2^{-exp}. Exponent values are allowed to range from 0 (for the largest-value coefficients with no leading zeros) to 24. Exponents for coefficients that have more than 24 leading zeros are fixed at 24, and the corresponding mantissas are allowed to have leading zeros. Exponents require 5 bits to represent all allowed values.

AC-3 bit streams contain coded exponents for all independent channels, all coupled channels, and for the coupling and low-frequency effects channels (when they are enabled). Because audio information is not shared across frames, block 0 of every frame will include new exponents for every channel. Exponent information may be shared across blocks within a frame, so blocks 1 through 5 may reuse exponents from previous blocks.

AC-3 exponent transmission employs *differential coding*, in which the exponents for a channel are differentially coded across frequency. These differential exponents are combined into groups in the audio block. This grouping is done by one of three methods, which are referred to as *exponent strategies*. The number of grouped differential exponents placed in the audio block for a particular channel depends on the exponent strategy and on the frequency bandwidth information for that channel. The number of exponents in each group depends only on the exponent strategy.

An AC-3 audio block contains two types of fields with exponent information. The first type defines the exponent coding strategy for each channel, and the second type contains the actual coded exponents for channels requiring new exponents. For independent channels, frequency bandwidth information is included along with the exponent strategy fields. For coupled channels, and the coupling channel, the frequency information is found in the coupling strategy fields.

7.5.12 Bit Allocation

The bit allocation routine analyzes the spectral envelope of the audio signal being coded with respect to masking effects to determine the number of bits to assign to each transform coefficient mantissa [3]. In the encoder, the bit allocation is performed globally on the ensemble of channels as an entity, from a common bit pool. Because there are no preassigned exponent or mantissa bits, the routine is allowed to flexibly allocate bits across channels, frequencies, and audio blocks in accordance with signal demand.

The bit allocation contains a parametric model of human hearing for estimating a noise-level threshold, expressed as a function of frequency, which separates audible from inaudible spectral components. Various parameters of the hearing model can be adjusted by the encoder depending upon signal characteristics. For example, a prototype masking curve is defined in terms of two piecewise continuous line segments, each with its own slope and *y*-axis intercept. One of several possible slopes and intercepts is selected by the encoder for each line segment. The encoder may iterate on one or more such parameters until an optimal result is obtained. When all parameters used to estimate the noise-level threshold have been selected by the encoder, the final bit allocation is computed. The model parameters are conveyed to the decoder with other side information. The decoder then executes the routine in a single pass.

The estimated noise-level threshold is computed over 50 bands of nonuniform bandwidth (an approximate 1/6-octave scale). The defined banding structure is independent of sampling frequency. The required bit allocation for each mantissa is established by performing a table lookup based upon the difference between the input signal *power spectral density* (PSD), evaluated on a fine-grain uniform frequency scale, and the estimated noise-level threshold, evaluated on the coarse-grain (*banded*) frequency scale. Therefore, the bit allocation result for a particular channel has spectral granularity corresponding to the exponent strategy employed.

7.6   Audio System Level Control

The AC-3 system provides elements that allow the encoded bit stream to satisfy listeners in many different situations. Two principal techniques are used to control the subjective loudness of the reproduced audio signals:

- Dialogue normalization

- Dynamic range compression

7.6.1   Dialogue Normalization

The *dialogue normalization* (DialNorm) element permits uniform reproduction of spoken dialogue when decoding any AC-3 bit stream [3]. When audio from different sources is reproduced, the apparent loudness often varies from source to source. Examples include the following:

- Audio elements from different program segments during a broadcast (for example, a movie vs. a commercial message)

- Different broadcast channels

- Different types of media (for example, disc vs. tape)

The AC-3 coding technology solves this problem by explicitly coding an indication of loudness into the AC-3 bit stream.

The subjective level of normal spoken dialogue is used as a reference. The 5-bit dialogue normalization word that is contained in the bit stream, DialNorm, is an indication of the subjective loudness of normal spoken dialogue compared with digital 100 percent. The 5-bit value is interpreted as an unsigned integer (most significant bit transmitted first) with a range of possible values from 1 to 31. The unsigned integer indicates the headroom in decibels above the subjective dialogue level. This value also may be interpreted as an indication of how many decibels the subjective dialogue level is below digital 100 percent.

The DialNorm value is not directly used by the AC-3 decoder. Rather, the value is used by the section of the sound reproduction system responsible for setting the reproduction volume, such as the system volume control. The system volume control generally is set based on listener input as to the desired loudness, or *sound-pressure level* (SPL). The listener adjusts a volume control that directly adjusts the reproduction system gain. With AC-3 and the DialNorm value, the reproduction system gain becomes a function of both the listener's desired reproduction sound-pressure level for dialogue, and the DialNorm value that indicates the level of dialogue in the audio signal. In this way, the listener is able to reliably set the volume level of dialogue, and the subjective level of dialogue will remain uniform no matter which AC-3 program is decoded.

Example Situation

An example will help to illustrate the DialNorm concept [3]. The listener adjusts the volume control to 67 dB. (With AC-3 dialogue normalization, it is possible to calibrate a system volume control directly in sound-pressure level, and the indication will be accurate for any AC-3 encoded audio source). A high quality entertainment program is being received, and the AC-3 bit stream indicates that the dialogue level is 25 dB below the 100 percent digital level. The reproduction system automatically sets the reproduction system gain so that full-scale digital signals reproduce at a sound-pressure level of 92 dB. Therefore, the spoken dialogue (down 25 dB) will reproduce at 67 dB SPL.

The broadcast program cuts to a commercial message, which has dialogue level at –15 dB with respect to 100 percent digital level. The system level gain automatically drops, so that digital 100 percent is now reproduced at 82 dB SPL. The dialogue of the commercial (down 15 dB) reproduces at a 67 dB SPL, as desired.

For the dialogue normalization system to work, the DialNorm value must be communicated from the AC-3 decoder to the system gain controller so that DialNorm can interact with the listener-adjusted volume control. If the volume-control function for a system is performed as a digital multiplier inside the AC-3 decoder, then the listener-selected volume setting must be communicated into the AC-3 decoder. The listener-selected volume setting and the DialNorm value must be combined to adjust the final reproduction system gain.

Adjustment of the system volume control is not an AC-3 function. The AC-3 bit stream simply conveys useful information that allows the system volume control to be implemented in a way that automatically removes undesirable level variations between program sources.

7.6.2   Dynamic Range Compression

The *dynamic range compression* (DynRng) element allows the program provider to implement subjectively pleasing dynamic range reduction for most of the intended audience, while allowing individual members of the audience the option to experience more (or all) of the original dynamic range [3].

A consistent problem in the delivery of audio programming is that members of the audience may prefer differing amounts of dynamic range. Original high-quality programs (such as feature films) typically are mixed with quite a wide dynamic range. Using dialogue as a reference, loud sounds, such as explosions, often are at least 20 dB louder; faint sounds, such as rustling leaves, may be 50 dB quieter. In many listening situations, it is objectionable to allow the sound to become very loud, so the loudest sounds must be compressed downward in level. Similarly, in many listening situations, the very quiet sounds would be inaudible, and they must be brought upward in level to be heard. Because most of the television audience will benefit from a limited program dynamic range, motion picture soundtracks that have been mixed with a wide dynamic range generally are compressed. The dynamic range is reduced by bringing down the level of the loud sounds and bringing up the level of the quiet sounds. Although this satisfies the needs of much of the audience, some audience members may prefer to experience the original sound program in its intended form. The AC-3 audio-coding technology solves this conflict by allowing dynamic range control values to be placed into the AC-3 bit stream.

The dynamic range control values, DynRng, indicate a gain change to be applied in the decoder to implement dynamic range compression. Each DynRng value can indicate a gain change of ±24 dB. The sequence of DynRng values constitute a compression control signal. An AC-3 encoder (or a bit stream processor) will generate the sequence of DynRng values. Each value is used by the AC-3 decoder to alter the gain of one or more audio blocks. The DynRng values typically indicate gain reductions during the loudest signal passages and gain increases during the quiet passages. For the listener, it is often desirable to bring the loudest sounds down in level, toward dialogue level, and bring the quiet sounds up in level, again toward dialogue level. Sounds that are at the same loudness as normal spoken dialogue typically will not have their gain changed.

The compression actually is applied to the audio in the AC-3 decoder. The encoded audio has full dynamic range. It is permissible for the AC-3 decoder to (optionally, under listener control) ignore the DynRng values in the bit stream. This will result in reproduction of the full dynamic range of the audio. It also is permissible (again under listener control) for the decoder to use some fraction of the DynRng control value and to use a different fraction of positive or negative values. Therefore, the AC-3 decoder can reproduce sounds according to one of the following parameters:

- Fully compressed audio (as intended by the compression control circuit in the AC-3 encoder)

- Full dynamic range audio

- Audio with partially compressed dynamic range, with different amounts of compression for high-level and low-level signals.

Example Situation

A feature film soundtrack is encoded into AC-3 [3]. The original program mix has dialogue level at −25 dB. Explosions reach a full-scale peak level of 0 dB. Some quiet sounds that are intended to be heard by all listeners are 50 dB below dialogue level (−75 dB). A compression control signal (a sequence of DynRng values) is generated by the AC-3 encoder. During those portions of the audio program when the audio level is higher than dialogue level, the DynRng values indicate negative gain, or gain reduction. For full-scale 0 dB signals (the loudest explosions), a gain reduction of −15 dB is encoded into DynRng. For very quiet signals, a gain increase of 20 dB is encoded into DynRng.

A listener wishes to reproduce this soundtrack quietly so as not to disturb anyone, but wishes to hear all of the intended program content. The AC-3 decoder is allowed to reproduce the default, which is full compression. The listener adjusts dialogue level to 60 dB SPL. The explosions will go only as loud as 70 dB (they are 25 dB louder than dialogue but receive −15 dB applied gain), and the quiet sounds will reproduce at 30 dB SPL (20 dB of gain is applied to their original level of 50 dB below dialogue level). The reproduced dynamic range, therefore, will be 70 dB − 30 dB = 40 dB.

The listening situation changes, and the listener now wishes to raise the reproduction level of dialogue to 70 dB SPL, but still wishes to limit the loudness of the program. Quiet sounds may be allowed to play as quietly as before. The listener instructs the AC-3 decoder to continue to use the DynRng values that indicate gain reduction, but to attenuate the values that indicate gain increases by a factor of 1/2. The explosions still will reproduce 10 dB above dialogue level, which is now 80 dB SPL. The quiet sounds now are increased in level by 20 dB/2 = 10 dB. They now will be reproduced 40 dB below dialogue level, at 30 dB SPL. The reproduced dynamic range is now 80 dB – 30 dB = 50 dB.

Another listener prefers the full original dynamic range of the audio. This listener adjusts the reproduced dialogue level to 75 dB SPL and instructs the AC-3 decoder to ignore the dynamic range control signal. For this listener, the quiet sounds reproduce at 25 dB SPL, and the explosions hit 100 dB SPL. The reproduced dynamic range is 100 dB – 25 dB = 75 dB. This reproduction is exactly as intended by the original program producer.

For this dynamic range control method to be effective, it must be used by all program providers. Because all broadcasters wish to supply programming in the form that is most usable by their audiences, nearly all will apply dynamic range compression to any audio program that has a wide dynamic range. This compression is not reversible unless it is implemented by the technique embedded in AC-3. If broadcasters make use of the embedded AC-3 dynamic range control system, listeners can have significant control over the reproduced dynamic range at their receivers. Broadcasters must be confident that the compression characteristic that they introduce into AC-3 will, by default, be heard by the listeners. Therefore, the AC-3 decoder must, by default, implement the compression characteristic indicated by the DynRng values in the data stream. AC-3 decoders may optionally allow listener control over the use of the DynRng values, so that the listener may select full or partial dynamic range reproduction.

7.6.3   Heavy Compression; COMPR, COMPR2

The *compression* (COMPR) element allows the program provider (or broadcaster) to implement a large dynamic range reduction (heavy compression) in a way that ensures that a monophonic downmix will not exceed a certain peak level [3]. The heavily compressed audio program may be desirable for certain listening situations, such as movie delivery to a hotel room or to an airline seat. The peak level limitation is useful when, for example, a monophonic downmix will feed an RF modulator, and overmodulation must be avoided.

Some products that decode the AC-3 bit stream will need to deliver the resulting audio via a link with very restricted dynamic range. One example is the case of a television signal decoder that must modulate the received picture and sound onto an RF channel to deliver a signal usable by a low-cost television receiver. In this situation, it is necessary to restrict the maximum peak output level to a known value—with respect to dialogue level—to prevent overmodulation. Most of the time, the dynamic range control signal, DynRng, will produce adequate gain reduction so that the absolute peak level will be constrained. However, because the dynamic range control system is intended to implement a subjectively pleasing reduction in the range of perceived loudness, there is no assurance that it will control instantaneous signal peaks adequately to prevent overmodulation.

To allow the decoded AC-3 signal to be constrained in peak level, a second control signal, COMPR, (COMPR2 for channel 2 in 1+1 mode) may be included in the AC-3 data stream. This control signal should be present in all bit streams that are intended to be received by, for example, a television set-top decoder. The COMPR control signal is similar to the DynRng control signal in that it is used by the decoder to alter the reproduced audio level. The COMPR control signal has twice the control range as DynRng (±48 dB compared with ±24 dB) with half the resolution (0.5 vs. 0.25 dB).

7.7   Audio System Features

The audio subsystem offers a host of services and features to meet varied applications and audiences [3]. An AC-3 elementary stream contains the encoded representation of a single audio service. Multiple audio services are provided by multiple elementary streams. Each elementary stream is conveyed by the transport multiplex with a unique *program ID* (PID). A number of audio service types may be coded (individually) into each elementary stream; each AC-3 elementary stream is tagged as to its service type. There are two types of *main service* and six types of *associated service*. Each associated service may be tagged (in the AC-3 audio descriptor) as being associated with one or more main audio services. Each AC-3 elementary stream also may be tagged with a language code.

Associated services may contain complete program mixes or only a single program element. Associated services that are complete mixes may be decoded and used "as is." Associated services that contain only a single program element are intended to be combined with the program elements from a main audio service.

In general, a complete audio program (what is presented to the listener over the set of loudspeakers) may consist of a main audio service, an associated audio service that is a complete mix, or a main audio service combined with an associated audio service. The capability to simultaneously decode one main service and one associated service is required in order to form a complete audio program in certain service combinations. This capability may not exist in some receivers.

7.7.1   Complete Main Audio Service (CM)

The CM type of main audio service contains a complete audio program (complete with dialogue, music, and effects) [3]. This is the type of audio service normally provided. The CM service may contain from 1 to 5.1 audio channels, and it may be further enhanced by means of the VI, HI, C, E, or VO associated services described in the following sections. Audio in multiple languages may be provided by supplying multiple CM services, each in a different language.

7.7.2   Main Audio Service, Music and Effects (ME)

The ME type of main audio service contains the music and effects of an audio program, but not the dialogue for the program [3]. The ME service may contain from 1 to 5.1 audio channels. The primary program dialogue is missing and (if any exists) is supplied by simultaneously encoding a D associated service. Multiple D associated services in different languages may be associated with a single ME service.

7.7.3   Visually Impaired (VI)

The VI associated service typically contains a narrative description of the visual program content [3]. In this case, the VI service is a single audio channel. The simultaneous reproduction of both the VI associated service and the CM main audio service allows the visually impaired user to enjoy the main multichannel audio program, as well as to follow (by ear) the on-screen activity.

The dynamic range control signal in this type of service is intended to be used by the audio decoder to modify the level of the main audio program. Thus, the level of the main audio service will be under the control of the VI service provider, and the provider may signal the decoder (by altering the dynamic range control words embedded in the VI audio elementary stream) to reduce the level of the main audio service by up to 24 dB to ensure that the narrative description is intelligible.

Besides being provided as a single narrative channel, the VI service may be provided as a complete program mix containing music, effects, dialogue, and the narration. In this case, the service may be coded using any number of channels (up to 5.1), and the dynamic range control signal would apply only to this service.

7.7.4   Hearing Impaired (HI)

The HI associated service typically contains only dialogue that is intended to be reproduced simultaneously with the CM service [3]. In this case, the HI service is a single audio channel. This dialogue may have been processed for improved intelligibility by hearing-impaired users. Simultaneous reproduction of both the CM and HI services allows the hearing-impaired users to hear a mix of the CM and HI services in order to emphasize the dialogue while still providing some music and effects.

Besides being available as a single dialogue channel, the HI service may be provided as a complete program mix containing music, effects, and dialogue with enhanced intelligibility. In this case, the service may be coded using any number of channels (up to 5.1).

7.7.5   Dialogue (D)

The D associated service contains program dialogue intended for use with an ME main audio service [3]. The language of the D service is indicated in the AC-3 bit stream and in the audio descriptor. A complete audio program is formed by simultaneously decoding the D service and the ME service, then mixing the D service into the center channel of the ME main service (with which it is associated).

If the ME main audio service contains more than two audio channels, the D service is monophonic (1/0 mode). If the main audio service contains two channels, the D service may also contain two channels (2/0 mode). In this case, a complete audio program is formed by simultaneously decoding the D service and the ME service, mixing the left channel of the ME service with the left channel of the D service, and mixing the right channel of the ME service with the right channel of the D service. The result will be a 2-channel stereo signal containing music, effects, and dialogue.

Audio in multiple languages may be provided by supplying multiple D services (each in a different language) along with a single ME service. This is more efficient than providing multiple CM services, but, in the case of more than two audio channels in the ME service, requires that dialogue be restricted to the center channel.

Some receivers may not have the capability to simultaneously decode an ME and a D service.

7.7.6 Commentary (C)

The commentary associated service is similar to the D service, except that instead of conveying essential program dialogue, the C service conveys optional program commentary [3]. The C service may be a single audio channel containing only the commentary content. In this case, simultaneous reproduction of a C service and a CM service will allow the listener to hear the added program commentary.

The dynamic range control signal in the single-channel C service is intended to be used by the audio decoder to modify the level of the main audio program. Thus, the level of the main audio service will be under the control of the C service provider; the provider may signal the decoder (by altering the dynamic range control words embedded in the C audio elementary stream) to reduce the level of the main audio service by up to 24 dB to ensure that the commentary is intelligible.

Besides providing the C service as a single commentary channel, the C service may be provided as a complete program mix containing music, effects, dialogue, and the commentary. In this case, the service may be provided using any number of channels (up to 5.1).

7.7.7 Emergency (E)

The E associated service is intended to allow the insertion of emergency or high priority announcements [3]. The E service is always a single audio channel. An E service is given priority in transport and in audio decoding. Whenever the E service is present, it will be delivered to the audio decoder. Whenever the audio decoder receives an E-type associated service, it will stop reproducing any main service being received and reproduce only the E service out of the center channel (or left and right channels if a center loudspeaker does not exist). The E service also may be used for nonemergency applications. It may be used whenever the broadcaster wishes to force all decoders to quit reproducing the main audio program and reproduce a higher priority single audio channel.

7.7.8 Voice-Over (VO)

The VO associated service is a single-channel service intended to be reproduced along with the main audio service in the receiver [3]. It allows typical voice-overs to be added to an already encoded audio elementary stream without requiring the audio to be decoded back to baseband and then reencoded. The VO service is always a single audio channel and has second priority; only the E service has higher priority. It is intended to be simultaneously decoded and mixed into the center channel of the main audio service. The dynamic range control signal in the VO service is intended to be used by the audio decoder to modify the level of the main audio program. Thus, the level of the main audio service may be controlled by the broadcaster, and the broadcaster may signal the decoder (by altering the dynamic range control words embedded in the VO audio elementary stream) to reduce the level of the main audio service by up to 24 dB during the voice-over.

Some receivers may not have the capability to simultaneously decode and reproduce a voice-over service along with a program audio service.

7.7.9 Multilingual Services

Each audio bit stream may be in any language [3]. Table 7.2 lists the language codes for the ATSC DTV system. To provide audio services in multiple languages, a number of main audio services may be provided, each in a different language. This is the (artistically) preferred method, because it allows unrestricted placement of dialogue along with the dialogue reverberation. The disadvantage of this method is that as much as 384 kbits/s is needed to provide a full 5.1-channel service for each language. One way to reduce the required bit rate is to reduce the number of audio channels provided for languages with a limited audience. For instance, alternate language versions could be provided in 2-channel stereo with a bit rate of 128 kbits/s. Or, a mono version could be supplied at a bit rate of approximately 64 to 96 kbits/s.

Another way to offer service in multiple languages is to provide a main multichannel audio service (ME) that does not contain dialogue. Multiple single-channel dialogue associated services (D) can then be provided, each at a bit rate in the range of 64 to 96 kbits/s. Formation of a complete audio program requires that the appropriate language D service be simultaneously decoded and mixed into the ME service. This method allows a large number of languages to be efficiently provided, but at the expense of artistic limitations. The single channel of dialogue would be mixed into the center reproduction channel, and could not be panned. Also, reverberation would be confined to the center channel, which is not optimum. Nevertheless, for some types of programming (sports and news, for example), this method is very attractive because of the savings in bit rate that it offers. Some receivers may not have the capability to simultaneously decode an ME and a D service.

Stereo (2-channel) service without artistic limitation can be provided in multiple languages with added efficiency by transmitting a stereo ME main service along with stereo D services. The D and appropriate-language ME services are combined in the receiver into a complete stereo program. Dialogue may be panned, and reverberation may be included in both channels. A stereo ME service can be sent with high quality at 192 kbits/s, and the stereo D services (voice only) can make use of lower bit rates, such as 128 or 96 kbits/s per language. Some receivers may not have the capability to simultaneously decode an ME and a D service.

Note that during those times when dialogue is not present, the D services can be momentarily removed, and the data capacity can be used for other purposes. Table 7.3 lists the typical bit rates for various types of service.

Table 7.2 Language Code Table for AC-3 (*After* [4].)

Code	Language	Code	Language	Code	Language	Code	Language
0x00	unknown/not applicable	0x20	Polish	0x40	background sound/ clean feed	0x60	Moldavian
0x01	Albanian	0x21	Portuguese	0x41		0x61	Malaysian
0x02	Breton	0x22	Romanian	0x42		0x62	Malagasay
0x03	Catalan	0x23	Romansh	0x43		0x63	Macedonian
0x04	Croatian	0x24	Serbian	0x44		0x64	Laotian
0x05	Welsh	0x25	Slovak	0x45	Zulu	0x65	Korean
0x06	Czech	0x26	Slovene	0x46	Vietnamese	0x66	Khmer
0x07	Danish	0x27	Finnish	0x47	Uzbek	0x67	Kazakh
0x08	German	0x28	Swedish	0x48	Urdu	0x68	Kannada
0x09	English	0x29	Turkish	0x49	Ukrainian	0x69	Japanese
0x0A	Spanish	0x2A	Flemish	0x4A	Thai	0x6A	Indonesian
0x0B	Esperanto	0x2B	Walloon	0x4B	Telugu	0x6B	Hindi
0x0C	Estonian	0x2C		0x4C	Tatar	0x6C	Hebrew
0x0D	Basque	0x2D		0x4D	Tamil	0x6D	Hausa
0x0E	Faroese	0x2E		0x4E	Tadzhik	0x6E	Gurani
0x0F	French	0x2F		0x4F	Swahili	0x6F	Gujurati
0x10	Frisian	0x30	reserved for national assignment	0x50	Sranan Tongo	0x70	Greek
0x11	Irish	0x31	"	0x51	Somali	0x71	Georgian
0x12	Gaelic	0x32	"	0x52	Sinhalese	0x72	Fulani
0x13	Galician	0x33	"	0x53	Shona	0x73	Dari
0x14	Icelandic	0x34	"	0x54	Serbo-Croat	0x74	Churash
0x15	Italian	0x35	"	0x55	Ruthenian	0x75	Chinese
0x16	Lappish	0x36	"	0x56	Russian	0x76	Burmese
0x17	Latin	0x37	"	0x57	Quechua	0x77	Bulgarian
0x18	Latvian	0x38	"	0x58	Pustu	0x78	Bengali
0x19	Luxembourgian	0x39	"	0x59	Punjabi	0x79	Belorussian
0x1A	Lithuanian	0x3A	"	0x5A	Persian	0x7A	Bambora
0x1B	Hungarian	0x3B	"	0x5B	Papamiento	0x7B	Azerbijani
0x1C	Maltese	0x3C	"	0x5C	Oriya	0x7C	Assamese
0x1D	Dutch	0x3D	"	0x5D	Nepali	0x7D	Armenian
0x1E	Norwegian	0x3E	"	0x5E	Ndebele	0x7E	Arabic
0x1F	Occitan	0x3F	"	0x5F	Marathi	0x7F	Amharic

Table 7.3 Typical Bit Rates for Various Services (*After* [4].)

Type of Service	Number of Channels	Typical Bit Rates
CM, ME, or associated audio service containing all necessary program elements	5	320–384 kbits/s
CM, ME, or associated audio service containing all necessary program elements	4	256–384 kbits/s
CM, ME, or associated audio service containing all necessary program elements	3	192–320 kbits/s
CM, ME, or associated audio service containing all necessary program elements	2	128–256 kbits/s
VI, narrative only	1	48–128 kbits/s
HI, narrative only	1	48–96 kbits/s
D	1	64–128 kbits/s
D	2	96–192 kbits/s
C, commentary only	1	32–128 kbits/s
E	1	32–128 kbits/s
VO	1	64–128 kbits/s

7.8   Channel Assignments and Levels

To facilitate the reliable exchange of programs, the SMPTE developed a standard for channel assignments and levels on multichannel audio media. The standard, SMPTE 320M, provides specifications for the placement of a 5.1 channel audio program onto multitrack audio media [8]. As specified in ITU-R BS.775-1, the internationally recognized multichannel sound system consists of left, center, right, left surround, right surround, and low-frequency effects (LFE) channels. SMPTE RP 173 specifies the locations and relative level calibration of the loudspeakers intended to reproduce these channels. SMPTE 320M specifies a mapping between the audio signals intended to feed loudspeakers, and a sequence of audio tracks on multitrack audio storage media. The standard also specifies the relative levels of the audio signals. Media prepared according to the standard should play properly on a loudspeaker system calibrated according to RP 173.

In consumer audio systems, the LFE channel is considered optional in reproduction. Media that conform to SMPTE 320M should be prepared so that they sound satisfactory even if the LFE channel is not reproduced. When an audio program originally produced as a feature film for theatrical release is transferred to consumer media, the LFE channel is often derived from the dedicated theatrical subwoofer channel. In the cinema, the dedicated subwoofer channel is always reproduced, and thus film mixes may use the subwoofer channel to convey important low frequency program content. Therefore, when transferring programs originally produced for the cinema over to television media, it may be necessary to remix some of the content of the subwoofer channel into the main full bandwidth channels.

7.9   References

1. Fibush, David K., *A Guide to Digital Television Systems and Measurements*, Tektronix, Beaverton, Ore., 1997.

2. SMPTE Standard for Television: "12-Channel Serial Interface for Digital Audio and Auxiliary Data," SMPTE 324M (Proposed), SMPTE, White Plains, N.Y., 1999.

3. ATSC, "Digital Audio Compression Standard (AC-3)," Advanced Television Systems Committee, Washington, D.C., Doc. A/52, Dec. 20, 1995.

4. ATSC, "Digital Television Standard," Advanced Television Systems Committee, Washington, D.C., Doc. A/53, Sep.16, 1995.

5. ITU-R Recommendation BS-775, "Multi-channel Stereophonic Sound System with and Without Accompanying Picture."

6. ATSC, "Guide to the Use of the Digital Television Standard," Advanced Television Systems Committee, Washington, D.C., Doc. A/54, Oct. 4, 1995.

7. ATSC, "Digital Audio Compression Standard (AC-3), Annex A: AC-3 Elementary Streams in an MPEG-2 Multiplex," Advanced Television Systems Committee, Washington, D.C., Doc. A/52, Dec. 20, 1995.

8. SMPTE Standard for Television: "Channel Assignments and Levels on Multichannel Audio Media," SMPTE 320M-1999, SMPTE, White Plains, N.Y., 1999.

7.10   Bibliography

Ehmer, R. H.: "Masking of Tones Vs. Noise Bands," *J. Acoust. Soc. Am.*, vol. 31, pp. 1253–1256, September 1959.

Ehmer, R. H.: "Masking Patterns of Tones," *J. Acoust. Soc. Am.*, vol. 31, pp. 1115–1120, August 1959.

Moore, B. C. J., and B. R. Glasberg: "Formulae Describing Frequency Selectivity as a Function of Frequency and Level, and Their Use in Calculating Excitation Patterns," *Hearing Research*, vol. 28, pp. 209–225, 1987.

Todd, C., et. al.: "AC-3: Flexible Perceptual Coding for Audio Transmission and Storage," AES 96th Convention, Preprint 3796, Audio Engineering Society, New York, February 1994.

Zwicker, E.: "Subdivision of the Audible Frequency Range Into Critical Bands (Frequenzgruppen)," *J. Acoust. Soc. of Am.*, vol. 33, p. 248, February 1961.

The ATSC DTV System

8.1 Introduction

The ATSC DTV standard describes a system designed to transmit high-quality video, audio, and ancillary data over a single 6 MHz channel [1]. The system can reliably deliver about 19 Mbits/s of throughput in a 6 MHz terrestrial broadcasting channel and about 38 Mbits/s of throughput in a 6 MHz cable television channel. This means that encoding a video source whose resolution can be as high as 5 times that of conventional television (NTSC) requires bit-rate reduction by a factor of 50 or higher. To achieve this bit-rate reduction, the system is designed to be efficient in utilizing available channel capacity by exploiting advanced video- and audio-compression technologies, as outlined in previous chapters.

The objective of the system designers was to maximize the information passed through the data channel by minimizing the amount of data required to represent the video image sequence and its associated audio, while preserving the level of quality required for the given application.

8.1.1 System Overview

A basic block diagram representation of the DTV system is shown in Figure 8.1 [1]. This representation is based on a model adopted by the International Telecommunication Union, Radiocommunication Sector (ITU-R), Task Group 11/3 (Digital Terrestrial Television Broadcasting). According to this model, the digital television system can be seen to consist of three subsystems [2].

- Source coding and compression

- Service multiplex and transport

- RF/transmission

Source coding and compression refers to the bit-rate reduction methods (data compression) appropriate for application to the video, audio, and ancillary digital data streams. The term *ancillary data* encompasses the following functions:

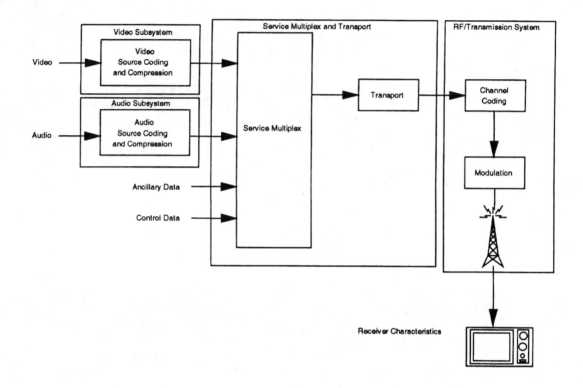

Figure 8.1 ITU-R digital terrestrial television broadcasting model. (*From* [1]. *Used with permission.*)

- Control data

- Conditional-access control data

- Data associated with the program audio and video services, such as closed captioning

Ancillary data also can refer to independent program services. The purpose of the coder is to minimize the number of bits needed to represent the audio and video information.

Service multiplex and transport refers to the means of dividing the digital data stream into *packets* of information, the means of uniquely identifying each packet or packet type, and the appropriate methods of multiplexing video data-stream packets, audio data-stream packets, and ancillary data-stream packets into a single data stream. In developing the transport mechanism, interoperability among digital media—such as terrestrial broadcasting, cable distribution, satellite distribution, recording media, and computer interfaces—was a prime consideration. The DTV system employs the MPEG-2 transport-stream syntax for the packetization and multiplexing of video, audio, and data signals for digital broadcasting systems [3]. The MPEG-2 transport-stream syntax was developed for applications where channel bandwidth or recording media capacity is limited, and the requirement for an effi-

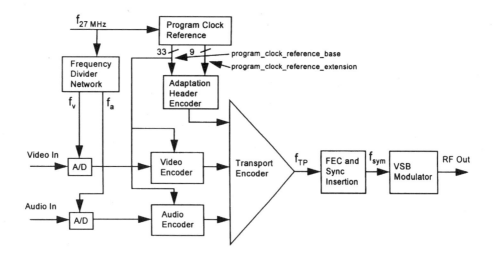

Figure 8.2 High-level view of the DTV encoding system. (*From* [4]. *Used with permission.*)

cient transport mechanism is paramount. The MPEG-2 transport stream also was designed to facilitate interoperability with the *asynchronous transfer mode* (ATM) transport stream.

RF/transmission refers to channel coding and modulation. The channel coder takes the data bit stream and adds additional information that can be used by the receiver to reconstruct the data from the received signal which, because of transmission impairments, may not accurately represent the transmitted signal. The modulation (or *physical layer*) uses the digital data-stream information to modulate the transmitted signal. The modulation subsystem offers two modes:

- Terrestrial broadcast mode (8-VSB)

- High-data-rate mode (16-VSB).

Figure 8.2 gives a high-level view of the encoding equipment. This view is not intended to be complete, but is used to illustrate the relationship of various clock frequencies within the encoder. There are two domains within the encoder where a set of frequencies are related: the source-coding domain and the channel-coding domain.

The source-coding domain, represented schematically by the video, audio, and transport encoders, uses a family of frequencies that are based on a 27 MHz clock (f_{27MHz}). This clock is used to generate a 42-bit sample of the frequency, which is partitioned into two elements defined by the MPEG-2 specification:

- The 33-bit *program clock reference base*

- The 9-bit *program clock reference extension*

The 33-bit program clock reference base is equivalent to a sample of a 90 kHz clock that is locked in frequency to the 27 MHz clock, and is used by the audio and video source encoders when encoding the *presentation time stamp* (PTS) and the *decode time stamp* (DTS).

The audio and video sampling clocks, f_a and f_v, respectively, must be frequency-locked to the 27 MHz clock. This condition can be expressed as the requirement that there exist two pairs of integers, (n_a, m_a) and (n_V, m_V), such that:

$$f_a = \left(\frac{n_a}{m_a}\right) \times 27 \text{ MHz} \tag{8.1}$$

and

$$f_v = \left(\frac{n_v}{m_v}\right) \times 27 \text{ MHz} \tag{8.2}$$

The channel-coding domain is represented by the forward error correction/sync insertion subsystem and the vestigial sideband (VSB) modulator. The relevant frequencies in this domain are the VSB symbol frequency (f_{sym}) and the frequency of the transport stream (f_{TP}), which is the frequency of transmission of the encoded transport stream. These two frequencies must be locked, having the relation:

$$f_{TP} = 2 \times \left(\frac{188}{208}\right)\left(\frac{312}{313}\right)f_{sym} \tag{8.3}$$

The signals in the two domains are not required to be frequency-locked to each other and, in many implementations, will operate asynchronously. In such systems, the frequency drift can necessitate the occasional insertion or deletion of a NULL packet from within the transport stream, thereby accommodating the frequency disparity.

8.1.2 Video Systems Characteristics

Table 8.1 lists the television production standards that define video formats relating to compression techniques applicable to the ATSC DTV standard. These picture formats may be derived from one or more appropriate video input formats. It is anticipated that additional video production standards will be developed in the future that extend the number of possible input formats.

As discussed in Chapter 5, the DTV video-compression algorithm conforms to the Main Profile syntax of ISO/IEC 13818-2 (MPEG-2). The allowable parameters are bounded by the upper limits specified for the Main Profile/High Level. Table 8.2 lists the allowed compression formats under the ATSC DTV standard.

8.1.3 Transport System Characteristics

The transport format and protocol for the DTV standard is a compatible subset of the MPEG-2 system specification (defined in ISO/IEC 13818-1) [4]. It is based on a fixed-length packet transport stream approach that has been defined and optimized for digital television delivery applications.

Table 8.1 Standardized Video Input Formats

Video Standard	Active Lines	Active Samples/Line
SMPTE 274M-1995 SMPTE 295M-1997 (50 Hz)	1080	1920
SMPTE 296M-1997	720	1280
ITU-R Rec. 601-4 SMPTE 293M-1996 (59.94, P) SMPTE 294M-1997 (59.94, P)	483	720

Table 8.2 ATSC DTV Compression Format Constraints (*After* [4].)

Vertical Size Value	Horizontal Size Value	Aspect Ratio Information	Frame-Rate Code	Progressive Sequence
1080[1]	1920	16:9, square pixels	1,2,4,5	Progressive
			4,5	Interlaced
720	1280	16:9, square pixels	1,2,4,5,7,8	Progressive
480	704	4:3, 16:9	1,2,4,5,7,8	Progressive
			4,5	Interlaced
	640	4:3, square pixels	1,2,4,5,7,8	Progressive
			4,5	Interlaced

Frame-rate code: 1 = 23.976 Hz, 2 = 24 Hz, 4 = 29.97 Hz, 5 = 30 Hz, 7 = 59.94 Hz, 8 = 60 Hz

[1] Note that 1088 lines actually are coded in order to satisfy the MPEG-2 requirement that the coded vertical size be a multiple of 16 (progressive scan) or 32 (interlaced scan).

As illustrated in Figure 8.3, the transport function resides between the application (e.g., audio or video) encoding and decoding functions and the transmission subsystem. The encoder transport subsystem is responsible for formatting the coded elementary streams and multiplexing the different components of the program for transmission. At the receiver, it is responsible for recovering the elementary streams for the individual application decoders and for the corresponding error signaling. The transport subsystem also incorporates other higher-protocol-layer functionality related to synchronization of the receiver.

The overall system multiplexing approach can be thought of as a combination of multiplexing at two different layers. In the first layer, single-program transport bit streams are formed by multiplexing transport packets from one or more *packetized elementary stream* (PES) sources. In the second layer, many single-program transport bit streams are combined to form a system of programs. The *program-specific information* (PSI) streams contain the information relating to the identification of programs and the components of each program.

Not shown explicitly in Figure 8.3, but nonetheless essential to the practical implementation of the standard, is a control system that manages the transfer and processing of the ele-

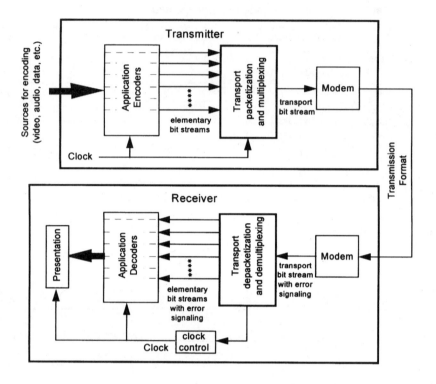

Figure 8.3 Sample organization of functionality in a transmitter-receiver pair for a single DTV program. (*From* [4]. *Used with permission.*)

mentary streams from the application encoders. The rules followed by this control system are not a part of the DTV standard, but must be established as recommended practices by the users of the standard. The control system implementation must adhere to the requirements of the MPEG-2 transport system as specified in ISO/IEC 13818-1 and with additional constraints specified in the DTV standard.

8.2 Overview of Video Compression and Decompression

The need for compression in a digital HDTV system is apparent from the fact that the bit rate required to represent an HDTV signal in uncompressed form is about 1 Gbit/s, and the bit rate that can reliably be transmitted within a standard 6 MHz television channel is about 20 Mbits/s [1]. This implies a need for a compression ratio of about 50:1 or greater.

The DTV standard specifies video compression using a combination of compression techniques, and for reasons of compatibility, these compression algorithms have been selected to conform to the specifications of MPEG-2.

Table 8.3 Picture Formats Under the DTV Standard (*After* [1].)

Vertical Lines	Pixels	Aspect Ratio	Picture Rate
1080	1920	16:9	60I, 30P, 24P
720	1280	16:9	60P, 30P, 24P
480	704	16:9 and 4:3	60P, 60I, 30P, 24P
480	640	4:3	60P, 60I, 30P, 24P

8.2.1 MPEG-2 Levels and Profiles

The DTV standard is based on the MPEG-2 Main Profile, which includes three types of frames for prediction (*I*-frames, *P*-frames, and *B*-frames), and an organization of luminance and chrominance samples (designated 4:2:0) within the frame [1]. The Main Profile does not include a *scalable algorithm*, where scalability implies that a subset of the compressed data can be decoded without decoding the entire data stream. The High Level includes formats with up to 1152 active lines and up to 1920 samples per active line, and for the Main Profile is limited to a compressed data rate of no more than 80 Mbits/s. The parameters specified by the DTV standard represent specific choices within these general constraints.

8.2.2 Overview of the DTV Video System

The DTV video-compression system takes in an analog video source signal and outputs a compressed digital signal that contains information that can be decoded to produce an approximate version of the original image sequence [1]. The goal is for the reconstructed approximation to be imperceptibly different from the original for most viewers, for most images, for most of the time. To approach such fidelity, the algorithms are flexible, allowing for frequent adaptive changes in the algorithm depending on scene content, history of the processing, estimates of image complexity, and perceptibility of distortions introduced by the compression.

Table 8.3 lists the picture formats allowed in the DTV standard. The parameters given include the following:

- *Vertical lines*. The number of active lines in the picture

- *Pixels*. The number of pixels during the active line

- *Aspect ratio*. The picture aspect ratio

- *Picture rate*. The number of frames or fields per second (*P* refers to progressive scanning; *I* refers to interlaced scanning)

Note that both 60 Hz and 59.94 (60 × 1000/1001) Hz picture rates are allowed. Dual rates also are allowed at the picture rates of 30 and 24 Hz.

The sampling rates for the DTV system are as follows:

- 1080-line format (1125 total lines/frame and 2200 total samples/line): sampling frequency = 74.25 MHz for the 30 frames/s rate.

- 720-line format (750 total lines/frame and 1650 total samples/line): sampling frequency = 74.25 MHz for the 60 frames/s rate.

- 480-line format (704 pixels, with 525 total lines/frame and 858 total samples/line): sampling frequency = 13.5 MHz for the 59.94 Hz field rate.

Note that both 59.94 and 60 are acceptable as frame or field rates for the system.

For the 480-line format, there may be 704 or 640 pixels in the active line. If the input is based on ITU-R Rec. 601, it will have 483 active lines with 720 pixels in the active line. Only 480 of the 483 active lines are used for encoding. Only 704 of the 720 pixels are used for encoding; the first eight and the last eight are dropped. The 480-line 640-pixel picture format is not related to any current video production format. It does, however, correspond to the IBM VGA graphics format and may be used with ITU-R Rec. 601-4 sources by using appropriate resampling techniques.

The DTV standard specifies SMPTE 274M colorimetry as the default, and preferred, colorimetry. Note that SMPTE 274M colorimetry is the same as ITU-R BT.709 (1990) colorimetry. Video inputs corresponding to ITU-R Rec. 601-4 may have SMPTE 274M colorimetry or SMPTE 170M colorimetry. In generating bit streams, broadcasters should understand that many receivers probably will display all inputs, regardless of colorimetry, according to the default SMPTE 274M. Some receivers may include circuitry to properly display SMPTE 170M colorimetry as well, but this is not a requirement of the standard.

Video preprocessing converts the analog input signals into digital samples in the form required for subsequent compression. The analog inputs are typically in the R, G, B form. Samples normally are obtained using A/D converters of 8-bit precision. After preprocessing, the various luminance and chrominance samples typically are represented using 8 bits per sample of each component.

8.2.3 Color Component Separation and Processing

The input video source to the DTV compression system is in the form of R, G, B (RGB) components matrixed into luminance (Y) and chrominance (C_b and C_r) components using a linear transformation (3×3 matrix, specified in the standard) [1]. As with NTSC, the luminance component represents the intensity, or black-and-white picture, and the chrominance components contain color information. The original RGB components are highly correlated with each other; the resulting Y, C_b, and C_r signals have less correlation, so they are easier to code efficiently. The luminance and chrominance components correspond to functioning of the biological vision system; that is, the human visual system responds differently to the luminance and chrominance components.

The coding process also may take advantage of the differences in the ways that humans perceive luminance and chrominance. In the Y, C_b, C_r color space, most of the high frequencies are concentrated in the Y component; the human visual system is less sensitive to high frequencies in the chrominance components than to high frequencies in the luminance component. To exploit these characteristics, the chrominance components are low-pass-filtered in the DTV video-compression system and subsampled by a factor of 2 along both the horizontal and vertical dimensions, producing chrominance components that are one-fourth the spatial resolution of the luminance component.

The Y, C_b, and C_r components are applied to appropriate low-pass filters that shape the frequency response of each of the three elements. Prior to horizontal and vertical subsampling of the two chrominance components, they may be processed by half-band filters to prevent aliasing [5].

8.2.4 Number of Lines Encoded

The video-coding system requires that the number of lines in the coded picture area is a multiple of 32 for an interlaced format and a multiple of 16 for a noninterlaced format [1]. This means that for encoding the 1080-line format, a coder must actually deal with 1088 lines ($1088 = 32 \times 34$). The extra eight lines are, in effect, "dummy" lines with no content, and the coder designers will choose dummy data that simplifies the implementation. The extra eight lines are always the last eight lines of the encoded image. These dummy lines do not carry useful information, but add little to the data required for transmission.

8.2.5 Film Mode

When a large fraction of pixels do not change from one frame in the image sequence to the next, a video encoder may automatically recognize that the input was film with an underlying frame rate of less than 60 frames/s [1]. In the case of 24 frames/s film material that is sent at 60 Hz using a 3:2 pulldown operation, the processor may detect the sequences of three nearly identical pictures followed by two nearly identical pictures, and encode only the 24 unique pictures per second that existed in the original film sequence. When 24 frames/s film is detected by observation of the 3:2 pulldown pattern, the input signal is converted back to a progressively scanned sequence of 24 frames/s prior to compression. This prevents the sending of redundant information and allows the encoder to provide an improved quality of compression. The encoder indicates to the decoder that the *film mode* is active.

In the case of 30 frames/s film material that is sent at 60 Hz, the processor may detect the sequences of two nearly identical pictures followed by two nearly identical pictures. In that case, the input signal is converted back to a progressively scanned sequence of 30 frames/s.

8.2.6 Pixels

The analog video signals are sampled in a sequence that corresponds to the scanning raster of the television format (that is, from left to right within a line, and in lines from top to bottom) [1]. The collection of samples in a single frame, or in a single field for interlaced images, is treated together, as if they all corresponded to a single point in time. (In the case of the film mode, they do, in fact, correspond to a single time or exposure interval.) The individual samples of image data are referred to as picture elements (*pixels*, or *pels*). A single frame or field, then, can be thought of as a rectangular array of pixels.

When the ratio of active pixels/line to active lines/frame is the same as the display aspect ratio, which is 16:9 in the case of the DTV standard, the format is said to have "square" pixels. This term refers to the spacing of samples and does not refer to the shape of the pixel, which might ideally be a point with zero area from a mathematical sampling point of view.

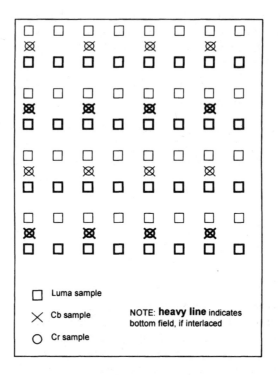

Figure 8.4. Placement of luma/chroma samples for 4:2:0 sampling. (*From* [1]. *Used with permission.*)

As described previously, the chrominance component samples are subsampled by a factor of 2 in both horizontal and vertical directions. This means the chrominance samples are spaced twice as far apart as the luminance samples, and it is necessary to specify the location of chrominance samples relative to the luminance samples.

Figure 8.4 illustrates the spatial relationship between chrominance and luminance samples. For every four luminance samples, there are one each of the C_b and C_r chroma samples. The C_b and C_r chroma samples are located in the same place. Note that the vertical spatial location of chrominance samples does not correspond to an original sample point, but lies halfway between samples on two successive lines. Therefore, the 4:2:0 sampling structure requires the C_b and C_r samples to be interpolated. For progressively scanned source pictures, the processor may simply average the two adjacent (upper and lower) values to compute the subsampled values.

In the case of interlaced pictures, it can be seen in Figure 8.4 that the vertical positions of the chrominance samples in a field are not halfway between the luminance samples of the same field. This is done so that the spatial locations of the chrominance samples in the frame are the same for both interlaced and progressive sources.

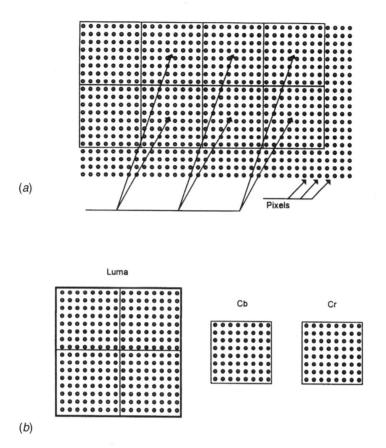

(a)

Pixels

Luma

Cb   Cr

(b)

Figure 8.5 DTV system block structure: (a) blocks, (b) macroblocks. (*From* [1]. *Used with permission.*)

Pixels are organized into *blocks* for the purpose of further processing. A block consists of an array of pixel values or an array that is some transform of pixel values. A block for the DTV system is defined as an array of 8 × 8 values representing either luminance or chrominance information. (See Figure 8.5a.)

Next, blocks of information are organized into *macroblocks*. A macroblock consists of four blocks of luminance (or a 16-pixel × 16-line region of values) and two chroma (C_b and C_r) blocks. The term *macroblock* may be used to refer directly to pixel data or to the transformed and coded representation of pixel data. As shown in Figure 8.5b, this yields 256 luminance samples, 64 C_b samples, and 64 C_r samples (a total of 384 samples) per macroblock. Organized by format, the macroblock structure is as follows:

- 1080-line format (1920 samples/line): Structured as 68 rows of macroblocks (including the last row that adds eight dummy lines to create the 1088 lines for coding), with 120 macroblocks per row.

- 720-line format (1280 samples/line): Structured as 45 rows of macroblocks, with 80 macroblocks per row.

- 480-line format (704 samples/line): Structured as 30 rows of macroblocks, with 44 macroblocks per row.

- 480-line format (640 samples/line): Structured as 30 rows of macroblocks, with 40 macroblocks per row.

8.2.7 Transport Encoder Interfaces and Bit Rates

The MPEG-2 standard specifies the inputs to the transport system as MPEG-2 elementary streams [4]. It also is possible that systems will be implemented wherein the process of forming PES packets takes place within the video, audio, or other data encoders. In such cases, the inputs to the transport system would be PES packets. Physical interfaces for these inputs (elementary streams and/or PES packets) may be defined as voluntary industry standards by SMPTE or other standardizing organizations.

8.2.8 Concatenated Sequences

The MPEG-2 standard, which underlies the ATSC DTV standard, clearly specifies the behavior of a compliant video decoder when processing a single video sequence [1]. A coded video sequence commences with a sequence header, may contain some repeated sequence headers and one or more coded pictures, and is terminated by an end-of-sequence code. A number of parameters specified in the sequence header are required to remain constant throughout the duration of the sequence. The sequence-level parameters include, but are not limited to:

- Horizontal and vertical resolution

- Frame rate

- Aspect ratio

- Chroma format

- Profile and level

- All-progressive indicator

- *Video buffering verifier* (VBV) size

- Maximum bit rate

It is a common requirement for coded bit streams to be spliced for editing, insertion of commercial advertisements, and other purposes in the video production and distribution chain. If one or more of the sequence-level parameters differ between the two bit streams to be spliced, then an end-of-sequence code must be inserted to terminate the first bit stream, and a new sequence header must exist at the start of the second bit stream. Thus, the situation of concatenated video sequences arises.

Although the MPEG-2 standard specifies the behavior of video decoders for the processing of a single sequence, it does not place any requirements on the handling of concatenated sequences. Specification of the decoding behavior in the former case is feasible because the MPEG-2 standard places constraints on the construction and coding of individual sequences. These constraints prohibit channel buffer overflow and coding of the same field parity for two consecutive fields. The MPEG-2 standard does not prohibit these situations at the junction between two coded sequences; likewise, it does not specify the behavior of decoders in this case.

Although it is recommended, the DTV standard does not require the production of *well-constrained* concatenated sequences. Well-constrained concatenated sequences are defined as having the following characteristics:

- The extended decoder buffer never overflows and may underflow only in the case of low-delay bit streams. Here, "extended decoder buffer" refers to the natural extension of the MPEG-2 decoder buffer model to the case of continuous decoding of concatenated sequences.

- When field parity is specified in two coded sequences that are concatenated, the parity of the first field in the second sequence is opposite that of the last field in the first sequence.

- Whenever a progressive sequence is inserted between two interlaced sequences, the exact number of progressive frames is such that the parity of the interlaced sequences is preserved as if no concatenation had occurred.

8.2.9 Guidelines for Refreshing

Although the DTV standard does not require refreshing at less than the intraframe-coded macroblock refresh rate (defined in IEC/ISO 13818-2), the following general guidelines are recommended [1]:

- In a system that uses periodic transmission of *I*-frames for refreshing, the frequency of occurrence of *I*-frames will determine the channel-change time performance of the system. In this case, it is recommended that *I*-frames be sent at least once every 0.5 second for acceptable channel-change performance. It also is recommended that sequence-layer information be sent before every *I*-frame.

- To spatially localize errors resulting from transmission, intraframe-coded slices should contain fewer macroblocks than the maximum number allowed by the standard. It is recommended that there be four to eight slices in a horizontal row of intraframe-coded macroblocks for the intraframe-coded slices in the *I*-frame refresh case, as well as for the intraframe-coded regions in the progressive refresh case. Nonintraframe-coded slices can be larger than intraframe-coded slices.

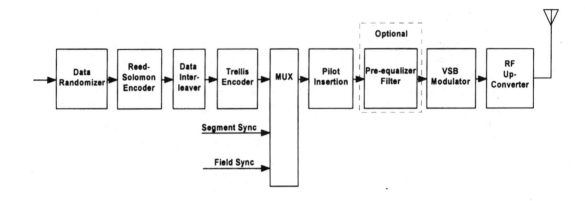

Figure 8.6 Simplified block diagram of an example VSB transmitter. (*From* [4]. *Used with permission.*)

8.3   Transmission Characteristics for Terrestrial Broadcast

The terrestrial broadcast mode (8-VSB) will support a payload data rate of approximately 19.28 Mbits/s in a 6 MHz channel [4]. (In the interest of simplicity, the bit stream data rate is rounded to two decimal points.) A functional block diagram of a representative 8-VSB terrestrial broadcast transmitter is shown in Figure 8.6. The input to the transmission subsystem from the transport subsystem is a serial data stream composed of 188-byte MPEG-compatible data packets (including a sync byte and 187 bytes of data, which represent a payload data rate of 19.28 Mbits/s).

The incoming data is randomized, then processed for *forward error correction* (FEC) in the form of *Reed-Solomon* (RS) coding (whereby 20 RS parity bytes are added to each packet). This is followed by 1/6-data-field interleaving and 2/3-rate trellis coding. The randomization and FEC processes are not applied to the sync byte of the transport packet, which is represented in transmission by a *data segment sync* signal. Following randomization and FEC processing, the data packets are formatted into *data frames* for transmission, and data segment sync and *data field sync* are added.

Figure 8.7 shows how the data is organized for transmission. Each data frame consists of two *data fields*, each containing 313 *data segments*. The first data segment of each data field is a unique synchronizing signal (data field sync) and includes the *training sequence* used by the equalizer in the receiver. The remaining 312 data segments each carry the equivalent of the data from one 188-byte transport packet plus its associated FEC overhead. The actual data in each data segment comes from several transport packets because of data interleaving. Each data segment consists of 832 *symbols*. The first four symbols are transmitted in binary form and provide segment synchronization. This data segment sync signal also represents the *sync byte* of the 188-byte MPEG-compatible transport packet. The remaining 828 symbols of each data segment carry data equivalent to the remaining 187 bytes of a transport packet and its associated FEC overhead. These 828 symbols are transmitted as 8-

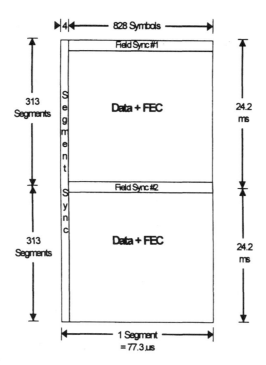

Figure 8.7 The basic VSB data frame. (*From* [4]. *Used with permission.*)

level signals and, therefore, carry 3 bits per symbol. Thus, 2484 (828 × 3) bits of data are carried in each data segment, which exactly matches the requirement to send a protected transport packet:

- 187 data bytes + 20 RS parity bytes = 207 bytes

- 207 bytes × 8 bits/byte = 1656 bits

- 2/3-rate trellis coding requires 3/2 × 1656 bits = 2484 bits.

The exact symbol rate S_r is given by the equation:

$$S_r = \frac{4.5}{286} \times 684 = 10.76 \quad \text{MHz} \tag{8.4}$$

The frequency of a data segment f_{seg} is given in the equation:

$$f_{seg} = \frac{S_r}{832} = 12.94\ldots \times 10^3 \text{ data segments/s} \tag{8.5}$$

The data frame rate f_{frame} is given by:

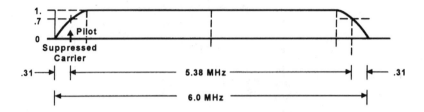

Figure 8.8 Nominal VSB channel occupancy. (*From* [4]. *Used with permission.*)

$$f_{frame} = \frac{f_{seg}}{626} = 20.66...\text{frames/s} \tag{8.6}$$

The symbol rate S_r and the transport rate T_r are locked to each other in frequency.

The 8-level symbols combined with the binary data segment sync and data field sync signals are used to suppressed-carrier-modulate a single carrier. Before transmission, however, most of the lower sideband is removed. The resulting spectrum is flat, except for the band edges where a nominal square-root raised-cosine response results in 620 kHz transition regions. The nominal VSB transmission spectrum is shown in Figure 8.8. It includes a small pilot signal at the suppressed-carrier frequency, 310 kHz from the lower band edge.

8.3.1 Channel Error Protection and Synchronization

All payload data is carried with the same priority [4]. A data randomizer is used on all input data to randomize the data payload (except for the data field sync, data segment sync, and RS parity bytes). The data randomizer *exclusive-ORs* (XORs) all the incoming data bytes with a 16-bit maximum-length *pseudorandom binary sequence* (PRBS), which is initialized at the beginning of the data field. The PRBS is generated in a 16-bit shift register that has nine feedback taps. Eight of the shift register outputs are selected as the fixed randomizing byte, where each bit from this byte is used to individually XOR the corresponding input data bit. The randomizer-generator polynomial and initialization are shown in Figure 8.9.

Although a thorough knowledge of the channel error protection and synchronization system is not required for typical end-users, a familiarity with the basic principles of operation—as outlined in the following sections—is useful in understanding the important functions performed.

Reed-Solomon Encoder

The Reed-Solomon (RS) code used in the VSB transmission subsystem is a $t = 10$ (207,187) code [4]. The RS data block size is 187 bytes, with 20 RS parity bytes added for error correction. A total RS block size of 207 bytes is transmitted per data segment. In creating bytes from the serial bit stream, the most significant bit (MSB) is the first serial bit. The 20 RS parity bytes are sent at the end of the data segment. The parity-generator polynomial and the

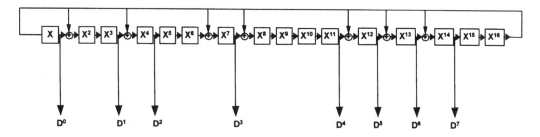

The generator is shifted with the Byte Clock and one 8 bit Byte
of data is extracted per cycle.

Figure 8.9 Randomizer polynomial for the DTV transmission subsystem. (*From* [4]. *Used with permission.*)

primitive-field-generator polynomial (with the fundamental supporting equations) are shown in Figure 8.10.

Reed-Solomon encoding/decoding is expressed as the total number of bytes in a transmitted packet to the actual application payload bytes, where the overhead is the RS bytes used (i.e., 207,187).

Interleaving

The interleaver employed in the VSB transmission system is a 52-data-segment (*intersegment*) convolutional byte interleaver [4]. Interleaving is provided to a depth of about one-sixth of a data field (4 ms deep). Only data bytes are interleaved. The interleaver is synchronized to the first data byte of the data field. Intrasegment interleaving also is performed for the benefit of the trellis coding process. The convolutional interleave stage is shown in Figure 8.11.

Trellis Coding

The 8-VSB transmission subsystem employs a 2/3-rate ($R = 2/3$) trellis code (with one unencoded bit that is precoded). Put another way, one input bit is encoded into two output bits using a 1/2-rate convolutional code while the other input bit is precoded [4]. The signaling waveform used with the trellis code is an 8-level (3-bit) 1-dimensional constellation. Trellis code intrasegment interleaving is used. This requires 12 identical trellis encoders and precoders operating on interleaved data symbols. The code interleaving is accomplished by

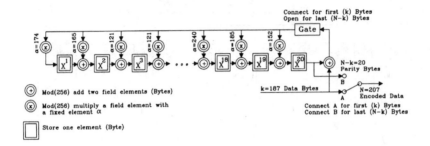

$$\prod_{i=0}^{i=2t-1}(X+\alpha^i) = X^{20}+X^{19}\alpha^{17}+X^{18}\alpha^{60}+X^{17}\alpha^{79}+X^{16}\alpha^{50}+X^{15}\alpha^{81}+X^{14}\alpha^{162}+X^{13}\alpha^{28}+X^{12}\alpha^{187}+X^{11}\alpha^{202}+X^{10}\alpha^{162}+X^{9}\alpha^{221}+X^{8}\alpha^{225}+X^{7}\alpha^{63}+X^{6}\alpha^{239}+X^{5}\alpha^{156}+X^{4}\alpha^{164}+X^{3}\alpha^{212}+X^{2}\alpha^{212}+X^{1}\alpha^{188}+\alpha^{190}$$

$$= X^{20}+152X^{19}+185X^{18}+240X^{17}+5X^{16}+111X^{15}+99X^{14}+6X^{13}+220X^{12}+112X^{11}+150X^{10}+69X^{9}+36X^{8}+187X^{7}+22X^{6}+228X^{5}+198X^{4}+121X^{3}+121X^{2}+165X^{1}+174$$

Figure 8.10 Reed-Solomon (207,187) $t = 10$ parity-generator polynomial. (*From* [4]. *Used with permission.*)

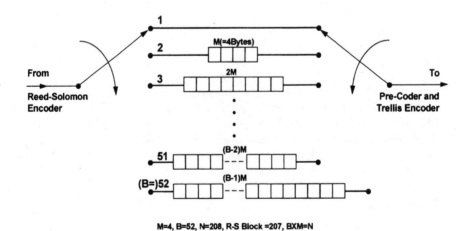

M=4, B=52, N=208, R-S Block =207, BXM=N

Figure 8.11 Convolutional interleaving scheme (byte shift register illustration). (*From* [4]. *Used with permission.*)

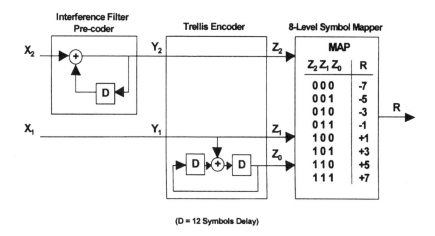

Figure 8.12 An 8-VSB trellis encoder, precoder, and symbol mapper. (*From* [4]. *Used with permission.*)

encoding symbols (0, 12, 24, 36 ...) as one group, symbols (1, 13, 25, 37 ...) as a second group, symbols (2, 14, 26, 38 ...) as a third group, and so on for a total of 12 groups.

In creating serial bits from parallel bytes, the MSB is sent out first: (7, 6, 5, 4, 3, 2, 1, 0). The MSB is precoded (7, 5, 3, 1) and the LSB (least significant bit) is feedback-convolutional encoded (6, 4, 2, 0). Standard 4-state optimal Ungerboeck codes are used for the encoding. The trellis code utilizes the 4-state feedback encoder shown in Figure 8.12. Also shown in the figure is the precoder and the symbol mapper. The trellis code and precoder intrasegment interleaver, which feed the mapper shown in Figure 8.12, are illustrated in Figure 8.13. As shown in Figure 8.13, data bytes are fed from the byte interleaver to the trellis coder and precoder; then they are processed as whole bytes by each of the 12 encoders. Each byte produces four symbols from a single encoder.

The output multiplexer shown in Figure 8.13 advances by four symbols on each segment boundary. However, the state of the trellis encoder is not advanced. The data coming out of the multiplexer follows normal ordering from encoders 0 through 11 for the first segment of the frame; on the second segment, the order changes, and symbols are read from encoders 4 through 11, and then 0 through 3. The third segment reads from encoder 8 through 11 and then 0 through 7. This 3-segment pattern repeats through the 312 data segments of the frame. Table 8.4 shows the interleaving sequence for the first three data segments of the frame.

After the data segment sync is inserted, the ordering of the data symbols is such that symbols from each encoder occur at a spacing of 12 symbols.

A complete conversion of parallel bytes to serial bits needs 828 bytes to produce 6624 bits. Data symbols are created from 2 bits sent in MSB order, so a complete conversion operation yields 3312 data symbols, which corresponds to four segments of 828 data sym-

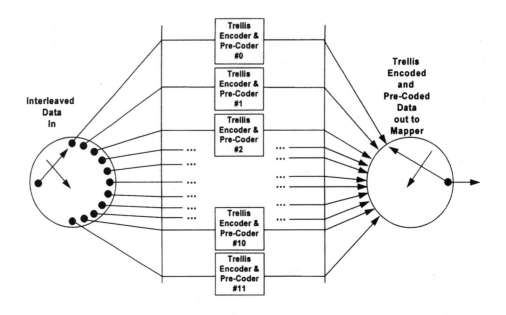

Figure 8.13 Trellis code interleaver. (*From* [4]. *Used with permission.*)

Table 8.4 Partial Trellis Coding Interleaving Sequence (*After* [4].)

Segment	Block 0				Block 1				...	Block 68			
0	D0	D1	D2 ...	D11	D0	D1	D2 ...	D11	...	D0	D1	D2 ...	D11
1	D4	D5	D6 ...	D3	D4	D5	D6 ...	D3	...	D4	D5	D6 ...	D3
2	D8	D9	D10 ...	D7	D8	D9	D10 ...	D7	...	D8	D9	D10 ...	D7

bols. A total of 3312 data symbols divided by 12 trellis encoders gives 276 symbols per trellis encoder, and 276 symbols divided by 4 symbols per byte gives 69 bytes per trellis encoder.

The conversion starts with the first segment of the field and proceeds with groups of four segments until the end of the field. A total of 312 segments per field divided by 4 gives 78 conversion operations per field.

During segment sync, the input to four encoders is skipped, and the encoders cycle with no input. The input is held until the next multiplex cycle, then is fed to the correct encoder.

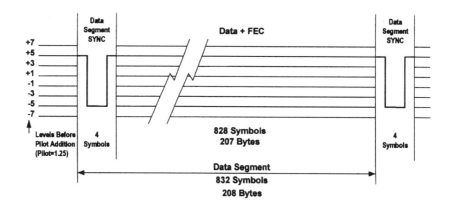

Figure 8.14 The 8-VSB data segment. (*From* [4]. *Used with permission.*)

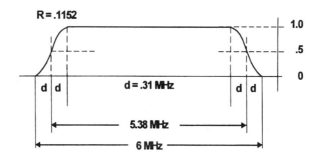

Figure 8.15 Nominal VSB system channel response (linear-phase raised-cosine Nyquist filter). (*From* [4]. *Used with permission.*)

8.3.2 Modulation

You will recall that in Figure 8.12 the mapping of the outputs of the trellis decoder to the nominal signal levels (–7, –5, –3, –1, 1, 3, 5, 7) was shown. As detailed in Figure 8.14, the nominal levels of data segment sync and data field sync are –5 and +5. The value of 1.25 is added to all these nominal levels after the bit-to-symbol mapping function for the purpose of creating a small pilot carrier [4]. The frequency of the pilot is the same as the sup-pressed-carrier frequency (as shown in Figure 8.8). The in-phase pilot is 11.3 dB below the average data signal power.

The VSB modulator receives the 10.76 Msymbols/s, 8-level trellis-encoded composite data signal with pilot and sync added. The DTV system performance is based on a *linear-phase raised-cosine* Nyquist filter response in the concatenated transmitter and receiver, as illustrated in Figure 8.15. The system filter response is essentially flat across the entire band, except for the transition regions at each end of the band. Nominally, the rolloff in the transmitter has the response of a *linear-phase root raised-cosine* filter.

8.3.3 Service Multiplex and Transport Systems

The transport system employs the fixed-length transport-stream packetization approach defined by MPEG [1]. This approach to the transport layer is well suited to the needs of terrestrial broadcast and cable transmission of digital television. The use of moderately long, fixed-length packets matches well with the needs and techniques for error protection in both terrestrial broadcast and cable television distribution environments. At the same time, it provides great flexibility to accommodate the initial needs of the service to multiplex video, audio, and data while providing a well-defined path to add services in the future in a fully backward compatible manner. By basing the transport layer on MPEG-2, maximum interoperability with other media and standards is maintained.

A transport layer based on a fixed-length packetization approach offers a great deal of flexibility and some significant advantages for multiplexing data related to several applications on a single bit stream. These benefits are discussed in subsequent sections.

Dynamic Capacity Allocation

Digital systems generally are described as flexible, but the use of fixed-length packets offers complete flexibility to allocate channel capacity among video, audio, and auxiliary data services [1]. The use of a *packet identifier* (PID) in the packet header as a means of bit stream identification makes it possible to have a mix of video, audio, and auxiliary data that is flexible and that does not need to be specified in advance. The entire channel capacity can be reallocated in bursts for data delivery. This capability could be used to distribute decryption keys to a large audience of receivers during the seconds preceding a popular pay-per-view program or to download program-related computer software to a "smart receiver."

Scalability

The transport format is scalable in the sense that the availability of a larger bandwidth channel also may be exploited by adding more elementary bit streams at the input of the multiplexer, or even by multiplexing these elementary bit streams at the second multiplexing stage with the original bit stream [1]. This is a valuable feature for network distribution, and it also facilitates interoperability with a cable plant's capability to deliver a higher data rate within a 6 MHz channel.

Extensibility

Because there will be demands for future services that could not have been anticipated when the DTV standard was developed, it is important that the transport architecture provide open-ended extensibility of services [1]. New elementary bit streams could be handled at the transport layer without hardware modification by assigning new packet IDs at the transmitter and filtering on these new PIDs in the bit stream at the receiver. Backward compatibility is assured when new bit streams are introduced into the transport system because existing decoders automatically will ignore the new PIDs. This capability could be used to compatibly introduce new display formats by sending augmentation data along with the basic signal.

Robustness

Another fundamental advantage of the fixed-length packetization approach is that the fixed-length packet can form the basis for handling errors that occur during transmission [1]. Error-correction and detection processing (which precedes packet demultiplexing in the receiver subsystem) may be synchronized to the packet structure so that the system deals at the decoder with units of packets when handling data loss resulting from transmission impairments. Essentially, after detecting errors during transmission, the system recovers the data bit stream from the first good packet. Recovery of synchronization within each application also is aided by the transport packet header information. Without this approach, recovery of synchronization in the bit streams would be completely dependent on the properties of each elementary bit stream.

Cost-Effective Receiver Implementations

A transport system based on fixed-length packets enables simple decoder bit stream demultiplex architectures, suitable for high-speed implementations [1]. The decoder does not need detailed knowledge of the multiplexing strategy or the source bit-rate characteristics to extract individual elementary bit streams at the demultiplexer. All the receiver needs to know is the identity of the packet, which is transmitted in each packet header at fixed and known locations in the bit stream. The only important timing information is for bit-level and packet-level synchronization.

MPEG-2 Compatibility

Although the MPEG-2 system layer has been designed to support many different transmission and storage scenarios, care has been taken to limit the burden of protocol inefficiencies caused by this generality in definition [1].

An additional advantage of MPEG-2 compatibility is interoperability with other MPEG-2 applications. The MPEG-2 format is likely to be used for a number of other applications, including storage of compressed bit streams, computer networking, and non-HDTV delivery systems. MPEG-2 transport system compatibility implies that digital television transport bit streams may be handled directly in these other applications (ignoring for the moment the issues of bandwidth and processing speed).

The transport format conforms to the MPEG-2 system standard, but it will not exercise all the capabilities defined in the MPEG-2 standard. Therefore, a digital television decoder need not be fully MPEG-2 system-compliant, because it will not need to decode any arbitrary MPEG-2 bit stream. However, all MPEG-2 decoders should be able to decode the digital television bit stream syntax at the transport system level.

8.3.4   Overview of the Transport Subsystem

Figure 8.16 illustrates the organization of a digital television transmitter-receiver pair and the location of the transport subsystem in the overall system [1]. The transport resides

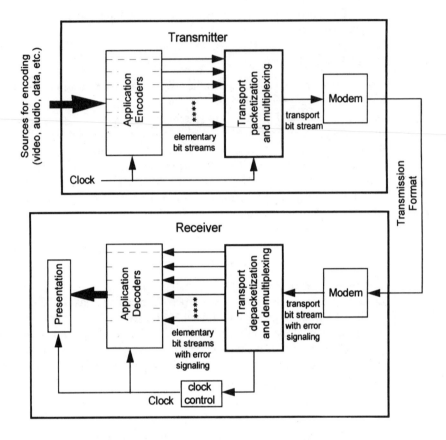

Figure 8.16. Sample organization of a transmitter-receiver system for a single digital television program. (*From* [4]. *Used with permission.*)

between the application (for example, audio or video) encoding/decoding function and the transmission subsystem. At its lowest layer, the encoder transport subsystem is responsible for formatting the encoded bits and multiplexing the different components of the program for transmission. At the receiver, it is responsible for recovering the bit streams for the individual application decoders and for the corresponding error signaling. (At a higher layer, multiplexing and demultiplexing of multiple programs within a single bit stream can be achieved with an additional system-level multiplexing or demultiplexing stage before the modem in the transmitter and after the modem in the receiver.) The transport subsystem also incorporates other higher-level functionality related to identification of applications and synchronization of the receiver.

As described previously, the data transport mechanism is based on the use of fixed-length packets that are identified by headers. Each header identifies a particular application bit stream (elementary bit stream) that forms the payload of the packet. Applications supported include video, audio, data, program, and system-control information. The elementary bit streams for video and audio are themselves wrapped in a variable-length packet structure (the packetized elementary stream) before transport processing. The PES layer provides functionality for identification and synchronization of decoding as well as presentation of the individual application.

Moving up one level in the description of the general organization of the bit streams, elementary bit streams sharing a common time base are multiplexed, along with a control data stream, into *programs*. Note that a program in the digital television system is analogous to a channel in the NTSC system in that it contains all of the video, audio, and other information required to make up a complete television program. These programs and an overall system-control data stream then are asynchronously multiplexed to form a multiplexed *system*. At this level, the transport is quite flexible in two primary aspects:

- It permits programs to be defined as any combination of elementary bit streams; specifically, the same elementary bit stream can be present in more than one program (e.g., two different video bit streams with the same audio bit stream); a program can be formed by combining a basic elementary bit stream and a supplementary elementary bit stream (e.g., bit streams for scalable decoders); and programs can be tailored for specific needs (e.g., regional selection of language for broadcast of secondary audio).

- Flexibility at the systems layer allows different programs to be multiplexed into the system as desired, and allows the system to be reconfigured easily when required. The procedure for extraction of separate programs from within a system also is simple and well defined.

The transport system provides other features that are useful for normal decoder operation and for the special features required in broadcast and cable applications. These include:

- Decoder synchronization

- Conditional access

- Local program insertion

The transport bit stream definition directly addresses issues relating to the storage and playback of programs. Although this is not directly related to the transmission of digital television programs, it is a fundamental requirement for creating programs in advance, storing them, and playing them back at the desired time. The programs are stored in the same format in which they are transmitted, that is, as transport bit streams. The bit-stream format also contains the *hooks* needed to support the design of consumer digital products based on recording and playback of these bit streams.

General Bit Stream Interoperability Issues

Bit stream interoperability at the transport level is an important feature of the digital television system [1]. Two aspects of interoperability should be considered:

- Whether the transport bit stream can be carried on other communication systems

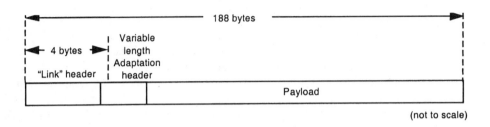

Figure 8.17 Basic transport packet format for the DTV standard. (*From* [4]. *Used with permission.*)

- The ability of the system to carry bit streams generated by other communication systems

In general, there is nothing that prevents the transmission of a bit stream as the payload of a different transmission system. It may be simpler to achieve this functionality in certain systems (e.g., cable television, DBS, and ATM) than in others (e.g., computer networks based on protocols such as FDDI and IEEE 802.6), but it is always possible. Because ATM is expected to form the basis of future broadband communications networks, the issue of bit stream interoperability with ATM networks is especially important. ATM interoperability has been specifically addressed in the design of the protocol, as discussed in Section 8.3.7.

The second aspect of interoperability is the transmission of other nontelevision bit streams within the digital television system. This makes more sense for bit streams linked to television broadcast applications, such as cable television and DBS, but also is possible for other "private" bit streams. This function is achieved by transmitting these other bit streams as the payload of identifiable transport packets. The only requirement is to have the general nature of these bit streams recognized within the system context. Note that a certain minimum system-level processing requirement defined by the DTV standard must be implemented to extract all (even private) bit streams. Furthermore, it is important to remember that this is essentially a broadcast system, hence any private transmissions that are based on a 2-way communications protocol will not be directly supported, unless this functionality is provided external to the system definition.

The Packetization Approach

The transport bit stream consists of fixed-length packets with a fixed and a variable component to the header field, as illustrated in Figure 8.17 [1]. Each packet consists of 188 bytes and is constructed in accordance with the MPEG-2 transport syntax and semantics. The choice of this packet size was motivated by several factors. The packets need to be large enough that the overhead resulting from the transport headers does not become a significant portion of the total data carried. They should not, however, be so large that the probability of packet error becomes significant under standard operating conditions (because of inefficient error correction). It also is desirable to have packet lengths consistent with the block sizes of typical block-oriented error-correction methods, so that packets may be synchronized to error-correction blocks, and the physical layer of the system can aid the packet-level synchronization process in the decoder. Another reason for this particular packet-length selection is interoperability with the ATM format.

The contents of each packet are identified by the packet headers. The packet header structure is layered and may be described as a combination of a fixed-length *link layer* and a variable-length *adaptation layer*. Each layer serves a different function, similar to the link- and transport-layer functions in the OSI layered model of a communications system. In the digital television system, this link and adaptation-level functionality are used directly for the terrestrial broadcast link on which the MPEG-2 transport bit stream is transmitted. However, in a different communications system—ATM, for example—the MPEG-2 headers would not play a role in implementing a protocol layer in the overall transmission scheme. The MPEG-2 headers would be carried as part of the payload in such a case and would continue to serve as identifiers for the contents of the data stream.

Random Entry Into the Compressed Bit Stream

Random entry into the application bit streams (typically video and audio) is necessary to support functions such as program acquisition and program switching [1]. Random entry into an application is possible only if the coding for the elementary bit stream for the application supports this functionality directly. For example, the video bit stream supports random entry through the concept of intraframes (*I*-frames) that are coded without any prediction and, therefore, can be decoded without any prior information. The beginning of the video sequence header information preceding data for an *I*-frame could serve as a random entry point into a video elementary bit stream. In general, random entry points also should coincide with the start of PES packets where they are used. The support for random entry at the transport layer comes from a flag in the adaptation header of the packet that indicates whether the packet contains a random access point for the elementary bit stream. In addition, the data payload of packets that are random access points starts with the data that forms the random access point of entry into the elementary bit stream itself. This approach allows packets to be discarded directly at the transport layer when switching channels and searching for a resynchronization point in the transport bit stream; it also simplifies the search for the random access point in the elementary bit stream after transport-level resynchronization is achieved.

A general objective in the DTV standard is to have random entry points into the programs as frequently as possible, to enable rapid channel switching.

Local Program Insertion

The transport system supports insertion of local programs and commercials through the use of flags and features dedicated to this purpose in the transport packet *adaptation header* [1]. The syntax allows local program insertion to be supported and its performance to improve as techniques and equipment are developed around these syntax tools. These syntax elements must be used within some imposed constraints to ensure proper operation of the video decoders. There also may need to be constraints on some current common broadcast practices, imposed not by the transport, but rather by virtue of the compressed digital data format.

The functionality of program segment insertion and switching of channels at a broadcast headend are quite similar, the differences being:

- The time constants involved in the splicing process.

- The fact that, in program segment insertion, the bit stream is switched back to the original program at the end of the inserted segment; in the channel-switching case, a cut is most likely made to yet another program at the end of the splice.

Other detailed issues related to the hardware implementation may differ for these two cases, including input source devices and buffering requirements. For example, if local program insertion is to take place on a bit stream obtained directly from a network feed, and if the network feed does not include placeholders for local program insertion, the input program transport stream will need to be buffered for the duration of the inserted program segment. If the program is obtained from a local source, such as a video server or a tape machine, it may be possible to pause the input process for the duration of the inserted program segment.

Two layers of processing functionality must be addressed for local program insertion. The lower-layer functionality is related to splicing transport bit streams for the individual elements of the program. The higher-level functionality is related to coordination of this process between the different elementary bit streams that make up the program transport stream. Figure 8.18 illustrates the approach recommended by the ATSC DTV standard to implement program insertion.

The first step for program insertion is to extract (by demultiplexing) the packets—identified by the PIDs—of the individual elementary bit streams that make up the program that is to be replaced, including the bit stream carrying the *program map table*. After these packets have been extracted, as illustrated in Figure 8.18, program insertion can take place on an individual PID basis. If applicable, some packets may be passed through without modification. There is also the flexibility to add and drop elementary bit streams.

8.3.5 Higher-Level Multiplexing Functionality

As described previously, the overall multiplexing approach can be described as a combination of multiplexing at two different layers [1]. In the first layer, program transport streams are formed by multiplexing one or more elementary bit streams at the transport layer. In the second layer, the program transport streams are combined (using asynchronous packet multiplexing) to form the overall system. The functional layer in the system that contains both this program- and system-level information is the *program-specific information* (PSI) layer.

A program transport stream is formed by multiplexing individual elementary bit streams (with or without PES packetization) that share a common time base. (Note that this terminology can be confusing. The term *program* is analogous to a channel in NTSC, as discussed in Section 8.3.4. The term *program stream* refers to a particular bit stream format defined by MPEG but not used in the DTV standard. *Program transport stream* is the term used to describe a transport bit stream generated for a particular program.) As the elementary streams are multiplexed, they are formed into *transport packets*, and a control bit stream that describes the program (also formed into transport packets) is added. The elementary bit streams and the control bit stream—also called the *elementary stream map*—are identified by their unique PIDs in the link header field. The organization of this multi-

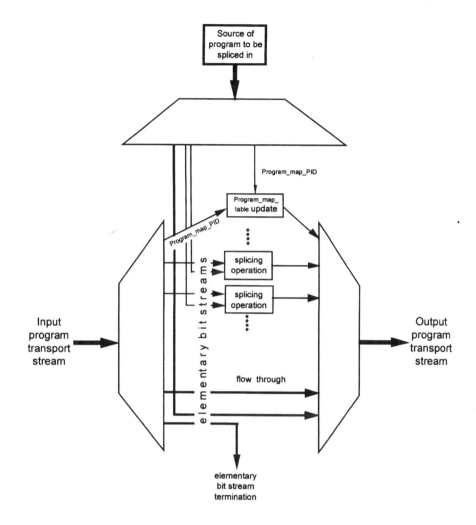

Figure 8.18 Example program insertion architecture. (*From* [4]. *Used with permission.*)

plex function is illustrated in Figure 8.19. The control bit stream contains the *program map table* that describes the elementary stream map. The program map table also includes information about the PIDs of the transport streams that make up the program, the identification of the applications (audio, video, data) that are being transmitted on these bit streams, the relationship between the bit streams, and other parameters.

The transport syntax allows a program to be composed of a large number of elementary bit streams, with no restriction on the types of applications required within a program. For example, a program transport stream does not need to contain a single video or audio bit stream; it could be a data "program." On the other hand, a program transport stream could

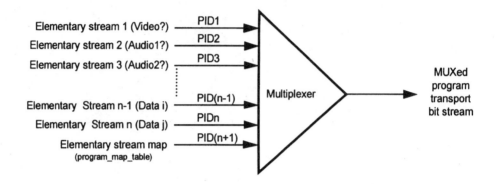

Figure 8.19 Illustration of the multiplex function in the formation of a program transport stream. (*From* [4]. *Used with permission.*)

contain multiple related video and audio bit streams, as long as they share a common time base.

Note that, for the different elementary bit streams that make up a program, the link-level functions are carried out independently without program-level coordination. This includes functions such as PID manipulation, bit stream filtering, scrambling and descrambling, and definition of random entry packets. The coordination between the elements of a program is primarily controlled at the presentation (display) stage, based on the use of the common time base. This time base is imposed by having all elementary bit streams in a program derive timing information from a single clock, then transmitting this timing information on one of the elementary bit streams that constitute the program. The data for timing of the presentation is contained within the elementary bit stream for each individual application.

System Multiplex

The *system multiplex* defines the process of multiplexing different program transport streams [4]. In addition to the transport bit streams (with the corresponding PIDs) that define the individual programs, a *system-level control bit stream* with PID = 0 is defined. This bit stream carries the *program association table* that maps program identities to their program transport streams; the program identity is represented by a number in the table. A "program" in this context corresponds to what has traditionally been called a "channel" (e.g., PBS, C-SPAN, CNN). The map indicates the PID of the bit stream containing the program map table for a specific program. Thus, the process of identifying a program and its contents takes place in two stages:

- First, the program association table in the PID = 0 bit stream identifies the PID of the bit stream carrying the program map table for the program.

- Second, the PIDs of the elementary bit streams that make up the program are obtained from the appropriate program map table.

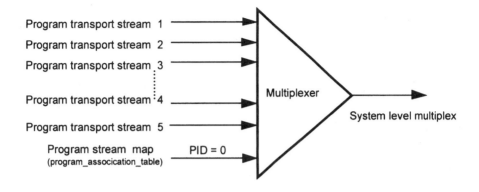

Figure 8.20 The system-level bit stream multiplex function. (*From* [4]. *Used with permission.*)

After these steps are completed, the filters at a demultiplexer can be set to receive the transport bit streams that correspond to the program of interest.

The system layer of multiplexing is illustrated in Figure 8.20. Note that during the process of system-level multiplexing, there is the possibility of PIDs on different program streams being identical at the input. This poses a problem because PIDs for different bit streams must be unique. A solution to this problem lies at the multiplexing stage, where some of the PIDs are modified just before the multiplex operation. The changes have to be recorded in both the program association table and the program map table. Hardware implementation of the PID reassignment function in real time is aided by the fact that this process is synchronous at the packet clock rate. Another approach is to make certain, up front, that the PIDs being used in the programs that make up the system are unique. This is not always possible, however, with stored bit streams.

The architecture of the bit stream is scalable. Multiple system-level bit streams can be multiplexed together on a higher bandwidth channel by extracting the program association tables from each system multiplexed bit stream and reconstructing a new PID = 0 bit stream. Note again that PIDs may have to be reassigned in this case.

Note also that in all descriptions of the higher-level multiplexing functionality, no mention is made of the functioning of the multiplexer and multiplexing policy that should be used. This function is not a part of the DTV standard and is left to the discretion of individual system designers. Because its basic function is one of filtering, the transport demultiplexer will function on any digital television bit stream regardless of the multiplexing algorithm used.

Figure 8.21 illustrates the entire process of extracting the elementary bit streams of a program at the receiver. It also serves as one possible implementation approach, although perhaps not the most efficient. In practice, the same demultiplexer hardware could be used to extract both the program association table and the program map table control bit streams.

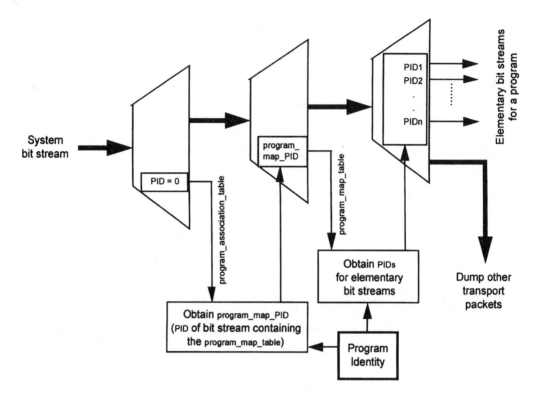

Figure 8.21 Overview of the transport demultiplexing process for a program. (*From* [4]. *Used with permission.*)

This also represents the minimum functionality required at the transport layer to extract any application bit stream (including those that may be private).

It should be noted that once the packets are obtained for each elementary bit stream in the program, further processing to obtain the random entry points for each component bit stream—to achieve decoder system clock synchronization or to obtain presentation (or decoding) synchronization—must take place before the receiver decoding process reaches normal operating conditions for receiving a program.

It is important to clarify here that the layered approach to defining the multiplexing function does not necessarily imply that program and system multiplexing always should be implemented in separate stages. A hardware implementation that includes both the program and system-level multiplexing within a single multiplexer stage is allowed, as long as the multiplexed output bit stream has the correct properties, as described in the ATSC DTV standard.

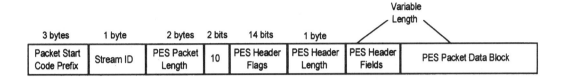

Figure 8.22 Structural overview of a PES packet. (*From* [4]. *Used with permission.*)

8.3.6   The PES Packet Format

The PES packet may be generated by either the application encoder or the transport encoder; however, for the purposes of explanation, the PES encoding is assumed here to be a transport function [1]. As described previously, some elementary bit streams—including the compressed video and compressed audio streams—will go through a PES-layer packetization process prior to transport-layer packetization. The PES header carries various rate, timing, and data descriptive information, as set by the source encoder. The PES packetization interval is application-dependent. The resulting PES packets are of variable length, with a maximum size of 2^{16} bytes, when the PES packet-length field is set to its maximum value. This value is set to zero for the video stream, indicating that the packet size is unconstrained and that the header information cannot be used to skip over the particular PES packet. Note also that the PES packet format has been defined to also be of use as an input bit stream for *digital storage media* (DSM) applications. Although the capability to handle input bit streams in the DSM format is not essential for a receiver, it may be useful for VCR applications.

Note that the format for carrying the PES packet within the transport layer is a subset of the general definition in MPEG-2. These choices were made to simplify the implementation of the digital television receiver and also to assist in error recovery.

A PES packet consists of a PES *packet start code*, PES header flags, PES packet header fields, and a payload (or data block), as illustrated in Figure 8.22. The payload is created by the application encoder. The packet payload is a stream of contiguous bytes of a single elementary stream. For video and audio packets, the payload is a sequence of access units from the encoder. These access units correspond to the video pictures and audio frames.

Each elementary stream is identified by a unique *stream ID*. The PES packets formed from elementary streams supplied by each encoder carry the corresponding stream ID. PES packets carrying various types of elementary streams can be multiplexed to form a program or transport stream in accordance with the MPEG-2 system standard.

Figure 8.23 provides a detailed look at the PES header flags. These flags are a combination of indicators of the properties of the bit stream, as well as indicators of the existence of additional fields in the PES header. The PES header fields are organized according to Figure 8.24 for the PES packets for video elementary streams. Most fields require *marker bits* to be inserted to avoid the occurrence of long strings of zeros, which could resemble a start code.

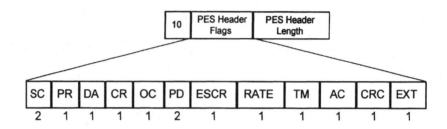

Figure 8.23 PES header flags in their relative positions (all sizes in bits). (*From* [4]. *Used with permission.*)

8.3.7 High Data-Rate Mode

The high data-rate mode trades off transmission robustness (28.3 dB S/N threshold) for payload data rate (38.57 Mbits/s) [4]. Most parts of the high data-rate mode VSB system are identical, or at least similar, to the terrestrial system. A pilot, data segment sync, and data field sync all are used to provide robust operation. The pilot in the high data-rate mode is 11.3 dB below the data signal power; and the symbol, segment, and field signals and rates all are the same as well, allowing either receiver type to lock up on the other's transmitted signal. Also, the data frame definitions are identical. The primary difference is the number of transmitted levels (8 vs.16) and the use of trellis coding and NTSC interference-rejection filtering in the terrestrial system.

The RF spectrum of the high data-rate modem transmitter looks identical to the terrestrial system. Figure 8.25 illustrates a typical data segment, where the number of data levels is seen to be 16 as a result of the doubled data rate. Each portion of 828 data symbols represents 187 data bytes and 20 Reed-Solomon bytes, followed by a second group of 187 data bytes and 20 Reed-Solomon bytes (before convolutional interleaving).

Figure 8.26 shows a functional block diagram of the high data-rate transmitter. It is identical to the terrestrial VSB system, except that the trellis coding is replaced with a mapper that converts data to multilevel symbols. The interleaver is a 26-data-segment intersegment convolutional byte interleaver. Interleaving is provided to a depth of about one-twelfth of a data field (2 ms deep). Only data bytes are interleaved. Figure 8.27 shows the mapping of the outputs of the interleaver to the nominal signal levels (−15, −13, −11, ..., 11, 13, 15). As shown in Figure 8.25, the nominal levels of data segment sync and data field sync are −9 and +9. The value of 2.5 is added to all these nominal levels after the bit-to-symbol mapping for the purpose of creating a small pilot carrier. The frequency of the in-phase pilot is the same as the suppressed-carrier frequency. The modulation method of the high data-rate mode is identical to the terrestrial mode except that the number of transmitted levels is 16 instead of 8.

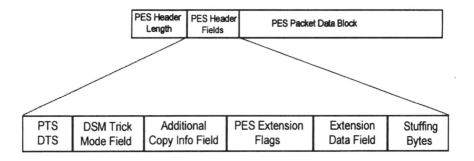

Figure 8.24 Organization of the PES header. (*From* [4]. *Used with permission.*)

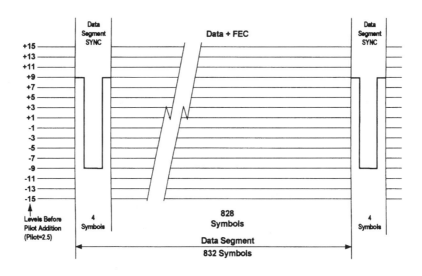

Figure 8.25 Typical data segment for the 16-VSB mode. (*From* [1]. *Used with permission.*)

8.3.8 Compatibility With Other Transport Systems

The DTV transport system interoperates with two of the most important alternative transport schemes [1]:

- It is identical in syntax with the MPEG-2 transport stream definition; the DTV standard is, in fact, a subset of the MPEG-2 specification.

- The DTV transport system has a high degree of interoperability with the ATM definition for broadband ISDN.

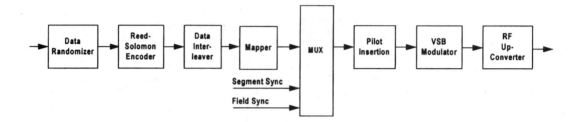

Figure 8.26 Functional block diagram of the 16-VSB transmitter. (*From* [1]. *Used with permission.*)

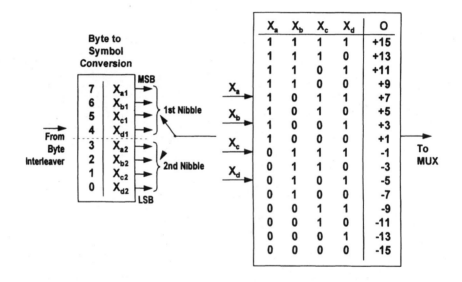

Figure 8.27 A 16-VSB mapping table. (*From* [1]. *Used with permission.*)

Furthermore, because several of the cable television and direct broadcast satellite (DBS) systems currently in use employ MPEG-2 transport-layer syntax, the degree of interoperability with such deployed systems is quite high.

Interoperability has two aspects. The first is syntactic and refers only to the coded representation of the digital television information. The second relates to the delivery of the bit stream in real time. This aspect of interoperability is beyond the scope of this chapter, but it should be noted that to guarantee interoperability with a digital television receiver conforming to the DTV standard, the output bit stream of the alternative transport system must have the proper real-time characteristics.

The DTV system also supports bit streams and services beyond the compressed video and audio services, such as text-based services, emergency messages, and other future ancillary services. A means of identifying such bit streams is necessary, but is not part of the MPEG-2 definition.

The MPEG-2 transport packet size is such that it can be easily partitioned for transfer in a link layer that supports ATM transmission. The MPEG-2 transport layer and the ATM layer serve different functions in a video delivery application: the MPEG-2 transport layer solves MPEG-2 presentation problems and performs the multimedia multiplexing function, and the ATM layer solves switching and network-adaptation problems.

Architecture of ATM

Asynchronous transfer mode is a technology based on high-speed packet switching. It is an ideal protocol for supporting professional video/audio and other complex multimedia applications. ATM is capable of data rates of up to 622 Mbits/s.

ATM was developed in the early 1980s by Bell Labs as a backbone switching and transportation protocol. It is a high-speed integrated multiplexing and switching technology that transmits information using fixed-length cells in a connection-oriented manner. Physical interfaces for the *user-network interface* (UNI) of 155.52 Mbits/s and 622.08 Mbits/s provide integrated support for high-speed information transfers and various communications modes—such as *circuit* and *packet* modes—and constant, variable, or burst bit-rate communications. These capabilities lead to four basic types of service classes of interest to video users [6]:

- *Constant bit rate* (CBR), which emulates a leased line service, with fixed network delay

- *Variable bit rate* (VBR), which allows for bursts of data up to a predefined peak cell rate

- *Available bit rate* (ABR), in which capacity is negotiated with the network to fill capacity gaps

- *Unspecified bit rate* (UBR), which provides unnegotiated use of available network capacity

These tiers of service are designed to maximize the traffic capabilities of the network. The CBR data streams are fixed and constant with time; the VBR and ABR systems vary. The bandwidth of the UBR class of service is a function of whatever network capacity is left over after all other users have claimed their stake to the bandwidth. Not surprisingly, CBR is usually the most expensive class of service, and UBR is the least expensive. Figure 8.28 shows the typical packing of an ATM trunk.

One of the reasons ATM is attractive for video applications is that the transport of video and audio fits nicely into the established ATM service classes. For example, consider the following applications:

- Real-time video—which demands real-time transmission for scene capture, storage, processing, and relay—fits well into the CBR service class.

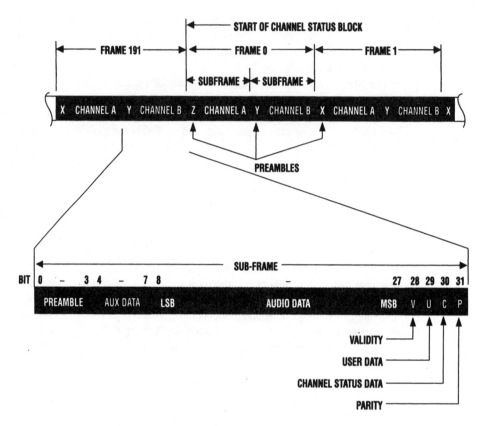

Figure 8.28 The typical packing of an internodal ATM trunk. (*After* [6].)

- Non-real-time video—such as recording and editing from servers, distributing edited masters, and other operations that can be considered essentially off-line—can use the ABR service.

- Machine control and file transfer—such as sending still clips from one facility to another—find the VBR service attractive.

ATM is growing and maturing rapidly. It already has been implemented in many industries, deployed by customers who anticipate such advantages as:

- Enabling high-bandwidth applications including desktop video, digital libraries, and real-time image transfer

- Coexistence of different types of traffic on a single network platform to reduce both the transport and operations costs

- Long-term network scalability and architectural stability

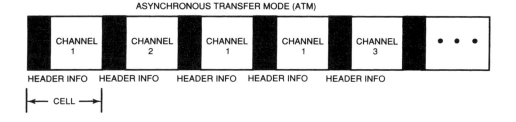

Figure 8.29 The ATM cell format. (*After* [7].)

In addition, ATM has been used in both local- and wide-area networks. It can support a variety of high-layer protocols and will cope with future network speeds of gigabits per second.

ATM Cell Structure

It is worthwhile to explore the ATM channel format in some detail because its features are the key to the usefulness of ATM for video applications. ATM channels are represented by a set of fixed-size cells and are identified through the channel indicator in the *cell header* [7]. The ATM cell has two basic parts: the header (5 bytes) and the payload (48 bytes). This structure is shown in Figure 8.29. ATM switching is performed on a cell-by-cell basis, based on the routing information contained in the cell header.

Because the main function of the ATM layer is to provide fast multiplexing and routing for data transfer based on information included in the header, this element of the protocol includes not only information for routing, but also fields to indicate the type of information contained in the cell payload. Other data is included in the header to perform the following support functions:

- Assist in controlling the flow of traffic at the UNI

- Establish priority for the cell

- Facilitate header error-control and cell-delineation functions

One key feature of ATM is that the cells can be independently labeled and transmitted on demand. This allows facility bandwidth to be allocated as needed, without the fixed hierarchical channel rates required by other network protocols. Because the connections supported are either permanent or semipermanent and do not require call control, real-time bandwidth management, or processing capabilities, ATM has the flexibility for video/multimedia applications.

The ATM cell header, the key to the use of this technology for networking purposes, consists of the fields shown in Table 8.5. Figure 8.30 illustrates the differences between the format of an ATM cell and the format of the MPEG-2 transport packet.

Table 8.5 ATM Cell Header Fields (*After* [1].)

GFC	A 4-bit *generic flow control* field: used to manage the movement of traffic across the user network interface (UNI).
VPI	An 8-bit network *virtual path identifier.*
VCI	A 16-bit network *virtual circuit identifier.*
PT	A 3-bit *payload type* (i.e., user information type ID).
CLP	A 1-bit *cell loss priority* flag (eligibility of the cell for discard by the network under congested conditions).
HEC	An 8-bit *header error control* field for ATM header error correction.
AAL	ATM *adaptation-layer* bytes (user-specific header).

8.4 Program and System Information Protocol

The *program and system information protocol* (PSIP) is a collection of tables designed to operate within every transport stream for terrestrial broadcast of digital television [8]. The purpose of the protocol, described in ATSC document A/65, is to describe the information at the system and event levels for all virtual channels carried in a particular transport stream. Additionally, information for analog channels—as well as digital channels from other transport streams—may be incorporated.

As outlined previously, the typical 6 MHz channel used for analog broadcast supports about 19 Mbits/s throughput. Because program signals with standard resolution can be compressed using MPEG-2 to sustainable rates of approximately 6 Mbits/s, three or four SD digital television channels can be safely supported within a single physical channel. Moreover, enough bandwidth remains within the same transport stream to provide several additional low bandwidth non-conventional services such as:

- Weather reports
- Stock reports
- Headline news
- Software download (for games or enhanced applications)
- Image-driven classified ads
- Home shopping
- Pay-per-view information

It is, therefore, practical to anticipate that in the future, the list of services (*virtual channels*) carried in a physical transmission channel (6 MHz of bandwidth for the U.S.) may easily reach ten or more. Furthermore, the number and types of services may also change continuously, thus becoming a dynamic medium for entertainment, information, and commerce.

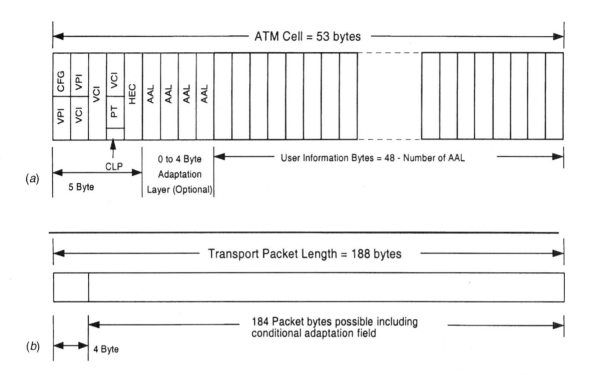

Figure 8.30 Comparison of the ATM cell structure and the MPEG-2 transport packet structure: (*a*) structure of the ATM cell, (*b*) structure of the transport packet. (*After* [1].)

An important feature of terrestrial broadcasting is that sources follow a distributed information model rather than a centralized one. Unlike cable or satellite, terrestrial service providers are geographically distributed and have no interaction with respect to data unification or even synchronization. It is, therefore, necessary to develop a protocol for describing *system information* and *event descriptions* that are followed by every organization in charge of a physical transmission channel. System information allows navigation of and access to each of the channels within the transport stream, whereas event descriptions give the user content information for browsing and selection.

8.4.1 Elements of PSIP

PSIP is a collection of hierarchically associated tables, each of which describes particular elements of typical digital television services [8]. Figure 8.31 shows the primary components and the notation used to describe them. The packets of the base tables are all labeled with a *base PID* (base_PID). The base tables are:

- *System time table* (STT)

- *Rating region table* (RRT)

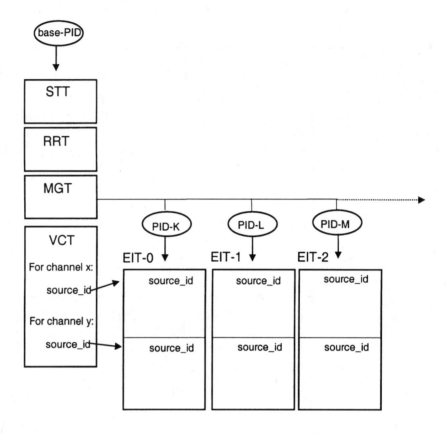

Figure 8.31 Overall structure for the PSIP tables. (*From* [8]. *Used with permission.*)

- *Master guide table* (MGT)

- *Virtual channel table* (VCT)

The *event information tables* (EIT) are a second set of tables whose packet identifiers are defined in the MGT. The *extended text tables* (ETT) are a third set of tables, and similarly, their PIDs are defined in the MGT.

The system time table is a small data structure that fits in one packet and serves as a reference for time-of-day functions. Receivers can use this table to manage various operations and scheduled events.

Transmission syntax for the U.S. voluntary program rating system is included in the ATSC standard. The rating region table has been designed to transmit the rating standard in use for each country using the system. Provisions also have been made for multicountry regions.

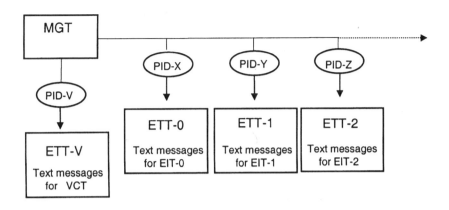

Figure 8.32 Extended text tables in the PSIP hierarchy. (*From* [8]. *Used with permission.*)

The master guide table provides general information about all of the other tables that comprise the PSIP standard. It defines table sizes necessary for memory allocation during decoding, defines version numbers to identify those tables that need to be updated, and generates the packet identifiers that label the tables.

The virtual channel table, also referred to as the *terrestrial VCT* (TVCT), contains a list of all the channels that are or will be on-line, plus their attributes. Among the attributes given are the channel name, navigation identifiers, and stream components and types.

As part of PSIP, there are several event information tables, each of which describes the events or television programs associated with each of the virtual channels listed in the VCT. Each EIT is valid for a time interval of 3 hours. Because the total number of EITs is 128, up to 16 days of programming may be advertised in advance. At minimum, the first four EITs must always be present in every transport stream.

As illustrated in Figure 8.32, there can be several extended text tables, each defined in the MGT. As the name implies, the purpose of the extended text table is to carry text messages. For example, for channels in the VCT, the messages can describe channel information, cost, coming attractions, and other related data. Similarly, for an event such as a movie listed in the EIT, the typical message would be a short paragraph that describes the movie itself. Extended text tables are optional in the ATSC system.

8.4.2 A/65 Technical Corrigendum and Amendment

During the summer of 1999, the ATSC membership approved a Corrigendum and an Amendment to the PSIP Standard (A/65). The *Technical Corrigendum No.1 to ATSC Standard A/65* documented changes for clarification, added an informative annex describing how PSIP could be used over cable, and made editorial corrections. The *Amendment No.1 to ATSC Standard A/65* made a technical change by defining a previously reserved bit to enable the unambiguous communication of virtual channel identification for channels that are not currently being used.

PSIP for Cable Applications

As outlined previously in this chapter, certain data specified in the Program and System Information Protocol (PSIP) forms a mandatory part of every ATSC-compliant digital multiplex signal delivered via terrestrial broadcast [9]. ATSC Document A/66, Annex G, provides an overview of the use of PSIP for digital cable applications.

PSIP was designed, as much as possible, to be independent of the physical system used to deliver the MPEG-2 multiplex. Therefore, the system time table, master guide table, virtual channel table (VCT), and event information tables and extended text tables are generally applicable equally as well to cable as to terrestrial broadcast delivery methods. The differences can be summarized as follows:

- For cable, the *cable virtual channel table* (CVCT) provides the VCT function, while the terrestrial virtual channel table applies for terrestrial broadcast. The cable VCT includes two parameters not applicable to the terrestrial broadcast case, and the syntax of several parameters in the table is slightly different for cable compared with the terrestrial broadcast case.

- Use of the program guide portion of PSIP (EIT and ETT) for cable is considered optional, while it is mandatory when PSIP is used for terrestrial broadcasting. Cable operators are free to *not* provide any program guide data at all if they so choose, or provide the data in a format other than PSIP if they do support an EPG.

While the syntax of the cable and terrestrial VCTs are nearly identical, the cable VCT has two parameters not present in the terrestrial VCT: a "path select" bit, and a bit that can indicate that a given virtual channel is transported *out-of-band* (OOB). Also, the semantics of the major and minor channel number fields and the source ID differ for the cable VCT compared with its terrestrial broadcast counterpart.

Channel Numbers

When PSIP is used for terrestrial broadcast, care must be taken in the assignment of major and minor channel numbers to avoid conflicts [9]. For example, the PSIP standard indicates that for the U.S. and its possessions, a terrestrial broadcaster with an existing NTSC license must use a major channel number for digital services that corresponds to the NTSC RF channel number in present use for the analog signal. For cable, such restrictions are technically unnecessary. For terrestrial broadcast, the major channel number is limited to the range 1 to 99 for ATSC digital television or audio services. For cable, major channel numbers can range from 1 to 999.

PSIP Data on Cable

PSIP data carried on cable in-band is analogous to PSIP included in the terrestrial digital broadcast multiplex: a receiver can discover the structure of digital services carried on that multiplex by collecting the current VCT from it [9]. A cable-ready digital TV can visit each digital signal on the cable, in sequence, and record from each a portion of the full cable VCT. This is exactly the same process a terrestrial digital broadcast receiver performs to build the terrestrial channel map.

Re-Multiplexing Issues

A cable operator may wish to take incoming digital transport streams from various sources (terrestrial broadcast, satellite, or locally generated), add or delete services or elementary streams, and then re-combine them into output transport streams [9]. If the incoming transport streams carry PSIP data, care must be taken to properly process this data in the re-multiplexer. Specifically, the re-multiplexer needs to account for any MPEG or PSIP fields or variables that are scoped to be unique within the transport stream. Such fields include PID values, MPEG program numbers, certain source ID tags, and event ID fields.

Enhancements to the PSIP Standard

In May 2000, the ATSC revised the PSIP standard to include an amendment that provides functionality known as *Directed Channel Change* (DCC), and also clarified existing aspects of the standard. The new feature allows broadcasters to tailor programming or advertising based upon viewer demographics. For example, viewers who enter location information such as their zip code into a DCC-equipped receiver can receive commercials that provide specific information about retail stores in their neighborhood. Segments of newscasts, such as weather reports can also be customized based upon this location information.

A channel change may also be based upon the subject matter of the content of the program. Nearly 140 categories of subject matter have been tabulated that can be assigned to describe the content of a program. A broadcaster can use this category of DCC request switching to direct a viewer to a program based upon the viewer's desire to receive content of that subject matter.

8.4.3 Conditional Access System

One of the important capabilities of the DTV standard is support for delivering pay services through *conditional access* (CA) [10]. The ATSC, in document A/70, specified a CA system for terrestrial broadcast that defines a middle protocol layer in the ATSC DTV system. The standard does not describe precisely all the techniques and methods to provide CA, nor the physical interconnection between the CA device (typically a "smart card" of some type) and its host (the DTV receiver or set-top box). Instead, it provides the data envelopes and transport functions that allow several CA systems of different types to operate simultaneously. In other words, a broadcaster may offer pay-TV services by means of one or more CA systems, each of which may have different transaction mechanisms and different security strategies.

Such services generally fall into one of five major categories:

- *Periodic subscription*, where the subscriber purchases entitlements, typically valid for one month.

- *Order-ahead pay-per-view* (OPPV), where the subscriber pre-pays for a special event.

- *Pay-per-view* (PPV) and *impulse PPV* (IPPV), distinguished by the consumer deciding to pay close (or very close) to the time of occurrence of the event.

- *Near video on demand* (NVOD), where the subscriber purchases an event that is being transmitted with multiple start times. The subscriber is connected to the next showing, and may be able to pick up a later transmission of the same program after a specified pause.

All of these services can be implemented without requiring a return channel in the DTV system by storing a "balance" in the CA card and then "deducting" charges from that balance as programs are purchased. While video is implied in each of these services, the CA system can also be used for paid data delivery.

The ATSC standard uses the SimulCrypt technique that allows the delivery of one program to a number of different decoder populations that contain different CA systems and also for the transition between different CA systems in any decoder population. *Scrambling* is defined as a method of continuously changing the form of a data signal so that without suitable access rights and an electronic descrambling key, the signal is unintelligible. *Encryption*, on the other hand, is defined as a method of processing keys needed for descrambling, so that they can be conveyed to authorized users. The information elements that are exchanged to descramble the material are the *encrypted keys*. The ATSC key is a 168-bit long data element and while details about how it is used are secret, essential rules to enable coexistence of different CA systems in the same receiver are in the ATSC standard.

System Elements

The basic elements of a conditional access system for terrestrial DTV are the headend broadcast equipment, the conditional access resources, a DTV host, and the security module(s) [11]. Figure 8.33 illustrates these basic elements and some possible interactions among them. The headend broadcast equipment generates the scrambled programs for over-the-air transmission to the constellation of receivers. The DTV host demodulates the transmitted signals and passes the resulting transport stream to the security module for possible descrambling.

Security modules are distributed by CA providers in any of a number of ways. For example, either directly, through consumer electronics manufacturers, through broadcasters, or through their agents. Security modules typically contain information describing the status of the subscriber. Every time the security module receives from the host a TS with some of its program components scrambled, the security module will decide, based on its own information and information in the TS, if the subscriber is allowed access to one or more of those scrambled programs. When the subscriber is allowed access then the security module starts its most intensive task, the descrambling of the selected program.

The packets of the selected program are descrambled one by one in real time by the security module, and the resulting TS is passed back to the DTV host for decoding and display. According to the ATSC A/70 standard, two types of security module technologies are acceptable: NRSS-A and NRSS-B. A DTV host with conditional access support needs to include hardware and/or software to process either A or B, or both. The standard does not define a communication protocol between the host and the security module. Instead, it man-

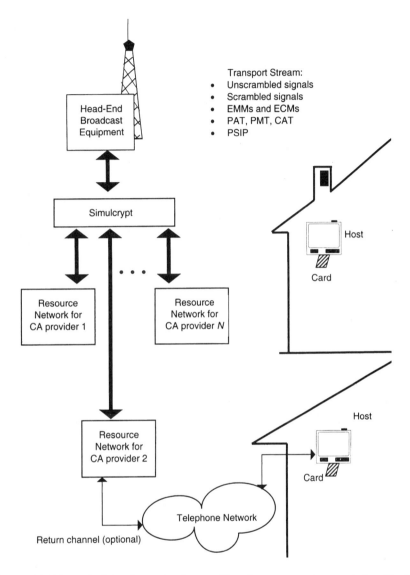

Transport Stream:
- Unscrambled signals
- Scrambled signals
- EMMs and ECMs
- PAT, PMT, CAT
- PSIP

Figure 8.33 The principle elements of a conditional access system as defined in ATSC A/70. (*After* 11].)

dates the use of NRSS. Similarly, for copy protection of the interface between the host and the security module, the standard relies on NRSS specifications.

Besides the scrambled programs, the digital multiplex carries streams dedicated to the transport of *entitlement control messages* (ECM) and *entitlement management messages* (EMM). ECMs are data units that mainly carry the key for descrambling the signals. EMMs provide general information to subscribers and most likely contain information about the status of the subscription itself. Broadcasters interested in providing conditionally accessed services through one or more CA providers need to transmit ECMs and EMMs for each of those CA providers. The A/70 standard defines only the envelope for carrying ECMs and EMMs. A security module is capable of understanding the content of EMMs and ECMs privately defined by one (or more) CA provider.

The digital multiplex for terrestrial broadcast carries a program guide according to the specifications defined in ATSC document A/65. The program guide contains detailed information about present and future events that may be useful for the implementation of a CA system. For this reason, the A/70 standard defines a descriptor that can be placed in either the channel or event tables of A/65. Similar to the definitions of ECMs and EMMs, only the generic descriptor structure is defined while its content is private. The host is required to parse the program guide tables in search of this descriptor and pass it to the security module. Because of the private nature of the content, it is the security module that ultimately processes the information. Note that although A/65 PSIP does not require use of conditional access, the conditional access standard (A/70) requires the use of A/65 PSIP.

Most of the communications between the CA network and a subscriber receiver can be performed automatically through broadcast streams using EMMs. EMMs are likely to be addressed to a specific security module (or receiver, possibly groups of security modules or receivers). Security modules will receive their EMMs by monitoring the stream of EMMs in the multiplex, or by searching for addressed EMMs by *homing*. Homing is the method for searching EMM streams while the receiver is in stand-by mode. Homing is initiated by the host according to schedules and directives as defined in NRSS specifications for the security module. As Figure 8.33 shows, the security module can communicate with the CA network using a telephone modem integrated into the host. This return channel is optional according to the A/70 standard, but if it exists, the host and the security module must adhere to the communication resource specifications of NRSS. A combination of homing and return channel EMM delivery can be used at the discretion of the CA provider.

Figure 8.33 shows that the interconnection between CA networks and the broadcast headend equipment requires Simulcrypt. Simulcrypt is a DVB protocol defining equipment and methods for adequate information exchange and synchronization. The most important information elements exchanged are the scrambling keys. According to Simulcrypt procedures, a new key can be generated by the headend equipment after a certain time interval that ranges from a fraction of a second to almost two hours. Before the encoder activates the new key, it is transferred to each CA system for encapsulation using their own protocols. Encapsulated keys in the form of ECMs are transferred back to the transmission equipment and are broadcast to all receivers. Shortly after the encoder has transmitted the new ECMs, the encoder uses the new key to scramble the content.

8.4.4 Transport Stream Identification

One of the many elements of configuration information needed to transmit a DTV bit stream that complies with the ATSC standard is a unique identification number [12]. The Transport Stream Identifier (TSID) was defined by the MPEG committee in ISO/IEC 13818-1, which is the Systems volume of the MPEG-2 standard. It also is required by ATSC

Standard A/65 Program and System Information Protocol for Terrestrial Broadcast and Cable (PSIP).

The TSID is a critical number that provides DTV sets the ability to identify and tune a channel even if its frequency of operation is changed. It is a key link to the creation of *electronic program guides* (EPGs) to present the available program choices to the consumer. If the TSID is not a unique number assignment, then a receiver can fail to correctly associate all programs with the correct broadcaster or cable programmer. When present and unique, this value can not only enable EPGs, but can even help a DTV set find a broadcaster's DTV signal when it is carried on cable. When the DTV signal is put on a cable as an 8VSB signal on a different RF channel (relative to the over-the-air broadcast), the TSID provides the link to the channel identity of the broadcaster. The TSID lets the receiver remap the RF frequency and select the program based on the major/minor channel number in the PSIP data stream. A DTV transport stream that is delivered via QAM retains the broadcaster's PSIP identification because the TSID in the cable system's Program Association Table (PAT) could be different. The TSID also permits use of DTV translators that appear to the consumer as if they had not been frequency shifted (i.e., they look like the main transmission, branded with the NTSC channel number).

To accomplish this functionality, the ATSC started with the ISO/IEC 13818-1 standard, which defined the TSID. It is a 16-bit number that must be placed in the PAT. According to MPEG-2, it serves as a label to identify the Transport Stream from any other multiplex within a network. MPEG-2 left selection of its value to the user.

As an element of the PSIP Standard (A/65), the ATSC has defined additional functions for the TSID and determined that its value for terrestrial broadcasts must be unique for a "network" consisting of geographically contiguous areas. The first such network is North America. The TSID is carried in the Virtual Channel Table (both cable and terrestrial versions) in PSIP.

The ATSC also established an identification number for existing analog television stations that is paired with the DTV TSID (differing in the least significant bit only). This Transmission Signal ID number is carried in the XDS packets associated with the closed captioning system on VBI line 21. The Consumer Electronics Manufacturers Association (CEMA) formalized how to carry this optional identifier in EIA 752. This number can provide a precise linkage between the NTSC service and the DTV service, Also, if the NTSC channel is RF shifted and contains the complementary TSID, it then can be located by DTV sets and labeled with the original RF channel number.

All DTV stations need PSIP generation software, which generally is presented as an option by the multiplexer or encoder vendor. It is important to realize there is a negative impact on operation of receivers (and thus consumers) if a station does not incorporate PSIP. or if it fails to comply with the ATSC PSIP Standard.

PSIP established rules to make the information presented to receivers uniform. For broadcasters with existing NTSC licenses, the *major channel* number must be set to the current NTSC RF channel number. For example, assume a broadcaster who has an NTSC broadcast license for RF channel 13 is assigned RF channel 39 for the DTV broadcast. That broadcaster should use major channel number 13 for identification of the digital transport it is controlling on RF channel 39 as well as the analog NTSC signal on RF channel 13. For a future DTV broadcaster without a NTSC license, the rule is that the major channel number shall be set to the FCC-assigned RF channel number for the DTV station.

This rule assigns major channel number values 2 through 69 uniquely to broadcasters with licenses to transmit NTSC and/or digital ATSC signals in a way that avoids conflicts and the need for coordination among broadcasters. However, other higher values also are possible. Values for major channel numbers from 70 to 99 may be used to identity groups of digital services carried in an ATSC multiplex that the broadcaster wishes to be identified by a different major channel number.

The other part of the PSIP service identification is the *minor channel* number. The minor channel 0 is reserved for identification and announcement of the programming on the NTSC service via the DTV service. Although there are no requirements for assigning the first DTV program as minor channel 1, that is a good value to use. Some stations may want to reserve 1 for any HD material and use 2, 3, 4, and so on for times when they chose to send multiple SD programs.

In March 1998, the ATSC asked the FCC to assign and maintain TSID numbers for DTV broadcasters. At this writing, the FCC had not taken formal action on the request. When consumer DTV receivers were initially deployed, it was discovered that some models incorrectly identified channels because the broadcasters had not coordinated TSID assignments and, therefore, were transmitting the default TSID set by the multiplexer manufacturers (usually 0X0000 or 0X0001). The problem was brought to the attention of the Technical Committee of the HDTV Model Station Project, which then created a list of proposed TSIDs for U.S. DTV broadcasters. It was hoped that the FCC would use this list as the starting point for their maintenance of the assignments and coordination of assignments for Canada and Mexico.

8.5   Closed Captioning

On July 31, 2000, the FCC issued a Report and Order (R&O) in ET Docket No. 99-254 regarding DTV Closed Captions (DTVCC) [13]. The R&O amended Part 15 of the FCCs Rules, adopting technical standards for the display of closed captions on digital television receivers. It also amended Part 79 to require all captions to be passed through program distribution facilities and reflect the changes in Part 15. The R&O also clarified the compliance date for including closed captions in digital programming.

In 1990, Congress passed the Telecommunications Decoder Circuitry Act (TDCA), which required television receivers with picture screen diagonals of 13-in. or larger to contain built-in closed caption decoders and have the ability to display closed captioned television transmissions. The Act also required the FCC to take appropriate action to ensure that closed captioning services continue to be available to consumers as new technology was developed. In 1991, the FCC amended its rules to include standards for the display of closed captioned text on analog NTSC TV receivers. The FCC said that with the advent of DTV broadcasting, it would again update its rules to fulfill its obligations under the TDCA.

The R&O adopted Section 9 of EIA-708-B, which specifies the methods for encoding, delivery, and display of DTVCC. Section 9 recommends a minimum set of display and performance standards for DTVCC decoders. However, based on comments filed by numerous consumer advocacy groups, the FCC decided to require DTV receivers to support display features beyond those contained in Section 9. In addition, the FCC incorporated by reference the remaining sections of EIA-708-B into its rules for informational purposes only.

Manufacturers must begin to include DTVCC functionality, in accordance with the rules adopted in the R&O, in DTV devices manufactured as of July 1, 2002. Specifically:

- All digital television receivers with picture screens in the 4:3 aspect ratio measuring at least 13-in. diagonally.

- Digital television receivers with picture screens in the 16:9 aspect ratio measuring 7.8-in. or larger vertically (this size corresponds to the vertical height of an analog receiver with a 13-in. diagonal screen).

- All DTV tuners shipped in interstate commerce or manufactured in the U.S. The rules apply to DTV tuners whether or not they are marketed with display screens.

The R&O further stated that programming prepared or formatted for display on digital television receivers before the July 1, 2002, date that digital television decoders are required to be included in digital television devices is considered "pre-rule" programming (as defined in the FCC's existing the closed captioning rules). Therefore, programming prepared or formatted for display on digital television after that date will be considered *new programming*. The existing rules require an increasing amount of captioned new programming over an eight-year transition period with 100 percent of all new non-exempt programming required to be captioned by January 1, 2006. This means that as of July 1, 2002, DTV services have the same hourly captioning requirement as NTSC services. The average amount required per day in 2002 is nearly 10 hrs (900 hrs/quarter). Those stations operating for part of a quarter are expected to meet the prorated or average daily amount.

There are three ways that the stations can originate DTVCC:

- If the DTV captions arrive already formatted and embedded in an MPEG-2 video stream, then the broadcaster is required to insure they are passed through and transmitted to receivers.

- If the DTV program is being up-converted from an NTSC source, then the caption data in that NTSC program must, at a minimum, be encapsulated into EIA-708-B format captions (using CC types 00 or 01) and broadcast with the DTV program.

- If the program is locally originated (and not exempt) and captions are being locally created but are not in DTVCC. then again—at a minimum—the caption information must be encapsulated into EIA-708-B format.

In the R&O, the FCC also stated that in order for cable providers to meet their closed captioned obligations, they must pass through closed captions they receive to digital television sets. Also, they must transmit those captions in a format that will be understandable to DTVCC decoders.

Regarding DTV set-top boxes, converter boxes, and standalone tuners, if they have outputs that are intended to be used to display digital programming on analog receivers, then the device must deliver the encoded "analog" caption information on those outputs to the attached analog receiver.

8.5.1 SMPTE 333M

To facilitate the implementation of closed-captioning for DTV facilities, the SMPTE developed SMPTE 333M, which defines rules for interoperation of DTVCC data server devices and video encoders. The caption data server devices provide partially-formatted EIA 708-A data to the DTV video encoder using a *request/response* protocol and interface, as defined in the document. The video encoder completes the formatting and includes the EIA 708-A data in the video elementary stream.

8.6 Data Broadcasting

It has long been felt that data broadcasting will hold one of the keys to profitability for DTV stations, at least in the early years of implementation. The ATSC Specialist Group on Data Broadcasting, under the direction of the Technology Group on Distribution (T3), was charged with investigating the transport protocol alternatives to add data to the suite of ATSC digital television standards [14]. The Specialist Group subsequently prepared a standard to address issues relating to data broadcasting using the ATSC DTV system.

The foundation for data broadcasting is the same as for video, audio, and PSIP—the MPEG-2 standard for transport streams (ISO/IEC 13818-1). Related work includes the following:

- ISO standardization of the Digital Storage Media Command Control framework in ISO/IEC 13818-6.

- The Internet Engineering Task Force standardization of the Internet Protocol (IP) in RFC 791.

- The ATSC specification of the data download protocol, addressable section encapsulation, data streaming, and data piping protocols.

The service-specific areas and the applications are not standardized. The DTV data broadcasting standard, in conjunction with the other referenced standards, defines how data can be transported using four different methods:

- *Data piping*, which describes how to put data into MPEG-2 transport stream packets. This approach supports private data transfer methods to devices that have service-specific hardware and/or software.

- *Data streaming*, which provides additional functionality, especially related to timing issues. The standard is designed to support synchronous data broadcast, where the data is sent only once (much as the video or audio is sent once). The standard is based on PES packets as defined by MPEC-2.

- *Addressable section encapsulation*, built using the DSM-CC framework. The ATSC added specific information to customize the framework for the ATSC environment, especially in conjunction with the PSIP standard, while retaining maximum commonality with the DVB standards. These methods enable repeated transmission of the same data elements, thus enabling better availability or reliability of the data.

- Data download

Because receivers will have different capabilities as technology evolves, methods to enable the receivers to determine if they could support data services also were developed. This type of "data about the data" is referred to as *control data* or *metadata*. The control information describes where and when the data service is being transmitted and provides linkage information, The draft standard uses and builds upon the PSIP standard. A *data information table* (DIT), which is structured in a manner similar to an *event information table* (EIT), transmits the information for data-only services. For data that is closely related to audio or video, the EIT can contain announcement information as well. Each data service is announced with key information about its data rate profile and receiver buffering requirements. Both opportunistic and fixed data rate allocations are defined in four profiles. These data facilitate receivers only presenting services to the consumer that the receiver can actually deliver. Each data service has its own minor channel number, which must have a value greater than 100.

Because data services can be quite complex, and related data might be provided to the receiver via different paths than the broadcast channel, an additional structure to standardize the linkage methods was developed. The structure consists of two tables—the *data service table* (DST) and the *network resource table* (NRT)—that together are called the *service description framework* (SDF). The DST contains one of thirteen protocol encapsulations and the linkage information for related data that is in the same MPEG TS. The NRT contains information to associate data streams that are not in the TS.

8.6.1 PSIP and SDF for Data Broadcasting

The PSIP and SDF are integral elements of the data broadcasting system. These structures provide two main functions [15]:

- Announcement of the available services

- Detailed instructions for assembling all the components of the data services as they are delivered

Generally, data of any protocol type is divided and subdivided into packets before transmission. PES packets are up to 64 kB long; sections are up to 4 kB, and transport stream (TS) packets are 188 bytes. Protocol standards document the rules for this orderly subdivision (and reassembly). All information is transmitted in 188 byte TS packets. The receiver sees the packets, and by using the packet identifier (PID) in the header of each packet, routes each packet to the appropriate location within the receiver so that the information can be recovered by reversing the subdivision process.

A data service can optionally be announced in either an EIT or a DIT. in conjunction with additional entries in the virtual channel table (VCT). These tables use the MPEG section structure. Single data services associated with a program are announced in the EIT. The DIT is used to support direct announcement of data services that are associated with a video program, but start and stop at different times within that program. Like the EIT, there are four required DITs that are used for separately announced data services. DITs contain the start time and duration of each event, the data ID number, and a data broadcast descriptor.

This descriptor contains information about the type of service profile, the necessary buffer sizes, and synchronization information.

There are three profiles for services that need constant or guaranteed delivery rates:

• G1—up to 384 kbits/s

• G2—up to 3.84 Mbits/s

• G3—up to the full 19.4 Mbits/s transport stream

For services that are delivered opportunistically, at up to the full transport data rate, the profile is called A1. Also present are other data that enable the receiver to determine if it has the capability to support a service being broadcast.

The VCT contains the virtual channel for each data service. The ATSC PSIP data service is intended to facilitate human interaction through a program guide that contains a linkage to the SDF, which provides the actual road map for reassembly of the fragmented information. The process of following this roadmap is known as *discovery and binding*. The SDF information is part of the bandwidth of the data service, not part of the broadcaster's overhead bandwidth (PSIP is in the overhead).

As mentioned previously, the SDF contains two distinct structures, the DST and the NRT (each use MPEG-2 private sections). The concepts for application discovery and binding rely upon standard mechanisms defined in ISO/IEC-13818-6 (MPEG-2 Digital Storage Media Command and Control, DSM-CC). The MPEG-2 transport stream packets conveying the DST and the NRT for each data service are referenced by the same PID, which is different from the PID value used for the SDF of any other data service. The DST must be sent at least once during the delivery of the data service. Key elements in the DST include the following:

• A descriptor defining signal compatibility and the requirements of an application

• The name of the application

• Method of data encapsulation

• List of author-created associations and application data structure parameters

Some services will not need the NRT because it contains information to link to data services outside the TS.

8.6.2 Opportunistic Data

The SMPTE 325M standard defines the flow control protocol to be used between an emission multiplexer and data server for *opportunistic data broadcast* [16]. Opportunistic data broadcast inserts data packets into the output multiplex to fill any available free bandwidth. The emission multiplexer maintains a buffer from which it draws data to be inserted. The multiplexer requests additional MPEG-2 transport packets from the data server as its buffer becomes depleted. The number of packets requested depends upon the implementation, with the most stringent requirement being the request of a single MPEG-2 transport packet where the request and delivery can occur in less than the emission time of an MPEG-2 transport packet from the multiplexer.

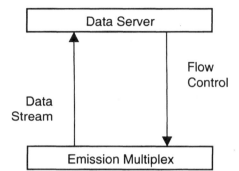

Figure 8.34 The opportunistic flow control environment of SMPTE 325M. (*After* [16].)

This protocol is designed to be extensible and provide the basis for low latency, real-time backchannel communications from the emission multiplexer. Encapsulated in MPEG-2 transport packets, the messages of the flow control protocol are transmitted via MPEG-2 DSM-CC sections, following the message format defined in ISO/IEC 13818-6, Chapter 2. Such sections provide the capability to support error correction or error detection (or to ignore either).

The environment for this standard is illustrated in Figure 8.34. Within an emission station, one or more data servers provide broadcast data (contained within MPEG-2 transport packets, with appropriate protocol encapsulations) to an emission multiplexer. A real-time control path is available from the emission multiplexer to the data server for flow control messages. Opportunistic data broadcast will attempt to fill any bandwidth available in the emission multiplex with broadcast data on a nearly instantaneous basis. The emission multiplexer is in control of the opportunistic broadcast because it is aware of the instantaneous gaps in the multiplex.

The operational model of SMPTE 325M is that the emission multiplexer will maintain an internal buffer from which it can draw MPEG-2 data packets to insert into the emission multiplex as opportunity permits. As the buffer is emptied, the mux requests a number of packets from the data server to maintain buffer fullness over the control path. These data packets are delivered over the data path. To avoid buffer overflow problems (should the data server be delayed in servicing the packet request), the following conventions are recommended:

• The data server should not queue requests for a given service (that is, a new request will displace one that has not been acted upon).

• The emission multiplexer should request no more than half of its buffer size at a time.

Support for multiple opportunistic streams (multiple data servers and multiple opportunistic broadcasts from a single server) is provided by utilizing the MPEG-2 transport header PID as a session identifier.

8.6.3 RP 203

SMPTE Recommended Practice 203 (proposed at this writing) defines the means of implementing opportunistic data flow control in a DTV MPEG-2 transport broadcast according to flow control messages defined in SMPTE 325M [17]. An emissions multiplexer requests opportunistic data packets as the need for them arises and a data server responds by forwarding data already inserted into MPEG-2 transport stream packets. The control protocol that allows this transfer of asynchronous data is extensible in a backward-compatible manner to allow for more advanced control as may be necessary in the future. Control messages are transmitted over a dedicated data link of sufficient quality to ensure reliable real-time interaction between the multiplexer and the data server.

8.6.4 ATSC A/90 Standard

With the foregoing efforts as a backdrop, the ATSC undertook the task of preparing a standard to define in specific terms data broadcast features, functions, and formats. The ATSC data broadcast system, described in document A/90 and released in August 2000, is illustrated in diagram form in Figure 8.35. The standard covers the delivery of data from the last part of the distribution chain (emission transmitter) to a receiver. While they are significant, issues related to delivery of data from originating points to this "last transmitter" using transport mechanisms other than ATSC transmission (such as disks, tapes, various types of network connections, and so on) are not described in the document.

Receivers are assumed to vary greatly in the number of services they are capable of presenting and their ability to store data or process it in some meaningful way [18]. Some may decode and present several audio/video broadcasts along with multiple data services. Others may be designed to perform a single function (such as delivering a stock ticker) as inexpensively as possible.

The A/90 standard defines the carriage of data using the *non-flow controlled scenario* and the *data carousel scenario* of the DSM-CC user-to-network download protocol. The ATSC use of the DSM-CC download protocol supports the transmission of the following:

- Asynchronous data modules

- Asynchronous data streaming

- Non-streaming synchronized data

Data carried by the download protocol may be error protected, since the DSM-CC sections used include a checksum field for that purpose.

The standard defines transmission of *datagrams* in the payload of MPEG-2 TS packets by encapsulating the datagrams in DSM-CC addressable sections. This mechanism is used for the asynchronous delivery of datagrams having the following characteristics:

- No MPEG-2 systems timing is associated with the delivery of data

- The smoothing buffer can go empty for indeterminate periods of time

- The data is carried in DSM-CC sections or DSM-CC addressable sections

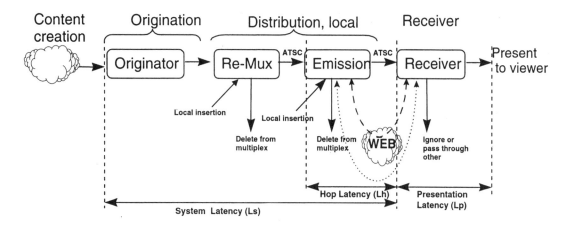

Figure 8.35 ATSC data broadcast system diagram. (*After* [18].)

The A/90 standard supports synchronous and synchronized data streaming using PES. *Synchronous data streaming* is defined as the streaming of data with timing requirements in the sense that the data and clock can be regenerated at the receiver into a synchronous data stream. Synchronous data streams have no strong timing association with other data streams and are carried in PES packets.

Synchronized data streaming implies a strong timing association between PES streams referenced by different PIDs. Synchronized streaming data are carried in PES packets. An example is application data associated with a video stream.

The standard defines *data piping* as a mechanism for delivery of arbitrary user defined data inside an MPEG-2 TS. Data are inserted directly into the payload of MPEG-2 TS packets. No methods are specified in the standard for fragmentation or re-assembly of data sent in this manner.

A *data service*, as defined in document A/90, is a collection of one or more data broadcast types. For example, a data service may include streaming synchronized data and asynchronous multiprotocol encapsulated data.

The tables, codes, and commands necessary for making this system work are described in [18].

8.7   References

1. ATSC, "Guide to the Use of the Digital Television Standard," Advanced Television Systems Committee, Washington, D.C., Doc. A/54, Oct. 4, 1995.

2. ITU-R Document TG11/3-2, "Outline of Work for Task Group 11/3, Digital Terrestrial Television Broadcasting," June 30, 1992.

3. Chairman, ITU-R Task Group 11/3, "Report of the Second Meeting of ITU-R Task Group 11/3, Geneva, Oct. 13-19, 1993," p. 40, Jan. 5, 1994.

4. ATSC, "ATSC Digital Television Standard," Advanced Television Systems Committee, Washington, D.C., Doc. A/53, Sep.16, 1995.

5. Cadzow, James A.: *Discrete Time Systems*, Prentice-Hall, Inc., Englewood Cliffs, N.J., 1973.

6. Piercy, John: "ATM Networked Video: Moving From Leased-Lines to Packetized Transmission," *Proceedings of the Transition to Digital Conference*, Intertec Publishing, Overland Park, Kan., 1996.

7. Wu, Tsong-Ho: "Network Switching Concepts," *The Electronics Handbook*, Jerry C. Whitaker (ed.), CRC Press, Boca Raton, Fla., p. 1513, 1996.

8. ATSC: "Program and System Information Protocol for Terrestrial Broadcast and Cable," Advanced Television Systems Committee, Washington, D.C., Doc. A/65, February 1998.

9. ATSC: "Technical Corrigendum No.1 to ATSC Standard: Program and System Information Protocol for Terrestrial Broadcast and Cable," Doc. A/66, ATSC, Washington, D.C., December 17, 1999.

10. NAB: "Pay TV Services for DTV," *NAB TV TechCheck*, National Association of Broadcasters, Washington, D.C., October 4, 1999.

11. ATSC: "Conditional Access System for Terrestrial Broadcast," Advanced Television Systems Committee, Washington, D.C., Doc. A/70, July 1999.

12. *NAB TV TechCheck*: National Association of Broadcasters, Washington, D.C., January 4, 1999 and June 7, 1999.

13. NAB: "Digital TV Closed Captions," *NAB TV TechCheck*, National Association of Broadcasters, Washington, D.C., August 7, 2000.

14. "An Introduction to DTV Data Broadcasting," *NAB TV TechCheck*, National Association of Broadcasters, Washington, D.C., August 2, 1999.

15. "Navigation of DTV Data Broadcasting Services," *NAB TV TechCheck*, National Association of Broadcasters, Washington, D.C., November 1, 1999.

16. SMPTE Standard: SMPTE 325M-1999, "Opportunistic Data Broadcast Flow Control," SMPTE, White Plains, N.Y., 1999.

17. SMPTE Recommended Practice RP 203 (Proposed): "Real Time Opportunistic Data Flow Control in an MPEG-2 Transport Emission Multiplex," SMPTE, White Plains, N.Y., 1999.

18. ATSC: "ATSC Data Broadcast Standard," Advanced Television Systems Committee, Washington, D.C., Doc. A/90, July 26, 2000.

8.8   Bibliography

ATSC: "Amendment No. 1 to ATSC Standard: Program and System Information Protocol for Terrestrial Broadcast and Cable," Doc. A/67, ATSC, Washington, D.C, December 17, 1999.

ATSC: "Implementation of Data Broadcasting in a DTV Station," Advanced Television Systems Committee, Washington, D.C., Doc. IS/151, November 1999.

FCC Report and Order: "Closed Captioning Requirements for Digital Television Receivers," Federal Communications Commission, Washington, D.C., ET Docket 99-254 and MM Docket 95-176, adopted July 21, 2000.

SMPTE Standard: SMPTE 333M, "DTV Closed-Caption Server to Encoder Interface," SMPTE, White Plains, N.Y., 1999.

9

DTV Transmission Issues

9.1   Introduction

As detailed in previous chapters, the VSB system offers two modes: a simulcast terrestrial broadcast mode and a high-data-rate mode. The two modes share the same pilot, symbol rate, data frame structure, interleaving, Reed-Solomon coding, and synchronization pulses [1]. The terrestrial broadcast mode is optimized for maximum service area, and it supports one ATV signal in a 6 MHz channel. The high-data-rate mode, which trades off some robustness for twice the data rate, supports two ATV signals in one 6 MHz channel.

Both modes of the VSB transmission subsystem take advantage of a pilot, segment sync, and training sequence for robust acquisition and operation. The two system modes also share identical carrier, sync, and clock recovery circuits, as well as phase correctors and equalizers. Additionally, both modes use the same Reed-Solomon (RS) code for forward error correction (FEC).

To maximize service area, the terrestrial broadcast mode incorporates both an NTSC rejection filter (in the receiver) and trellis coding. Precoding at the transmitter is incorporated in the trellis code. When the NTSC rejection filter is activated in the receiver, the trellis decoder is switched to a trellis code corresponding to the encoder trellis code concatenated with the filter.

The high-data-rate mode, on the other hand, does not experience an environment as severe as that of the terrestrial system. Therefore, a higher data rate is transmitted in the form of more data levels (bits/symbol). No trellis coding or NTSC interference-rejection filters are employed.

VSB transmission inherently requires processing only the in-phase (I) channel signal, sampled at the symbol rate, thus optimizing the receiver for low-cost implementation. The decoder requires only one A/D converter and a real (not complex) equalizer operating at the symbol rate of 10.76 Msamples/s.

The parameters for the two VSB transmission modes are shown in Table 9.1.

Table 9.1 Parameters for VSB Transmission Modes (*From* [1]. *Used with permission.*)

Parameter	Terrestrial Mode	High-Data-Rate Mode
Channel bandwidth	6 MHz	6 MHz
Excess bandwidth	11.5 percent	11.5 percent
Symbol rate	10.76 Msymbols/s	10.76 Msymbols/s
Bits per symbol	3	4
Trellis FEC	2/3 rate	None
Reed-Solomon FEC	T = 10 (207,187)	T = 10 (207,187)
Segment length	832 symbols	832 symbols
Segment sync	4 symbols/segment	4 symbols/segment
Frame sync	1/313 segments	1/313 segments
Payload data rate	19.28 Mbits/s	38.57 Mbits/s
NTSC co-channel rejection	NTSC rejection filter in receiver	N/A
Pilot power contribution	0.3 dB	0.3 dB
C/N threshold	14.9 dB	28.3 dB

9.1.1   Real World Conditions

In any discussion of the ATSC DTV terrestrial transmission system, it is important to keep in mind that many of the technical parameters established by the FCC are based on laboratory and best-case conditions. Results from the field tests performed as part of the Grand Alliance system development were certainly taken into account in the standard, however, many technical issues are subject to considerable variation, primary among them interference and receiver sensitivity.

9.1.2   Bit-Rate Considerations

The exact symbol rate of the transmission subsystem is given by [1]:

$$\frac{4.5}{286} \times 684 \ = \ 10.76 \text{ MHz} \tag{9.1}$$

The symbol rate must be locked in frequency to the transport rate. The transmission subsystem carries two information bits per trellis-coded symbol, so the gross payload is:

$$10.76... \times 2 \ = \ 21.52...\text{Mbits/s} \tag{9.2}$$

To find the net payload delivered to a decoder, it is necessary to adjust Equation 9.2 for the overhead of the data segment sync, data field sync, and Reed-Solomon FEC. Then, the net payload bit rate of the 8-VSB terrestrial transmission subsystem becomes:

$$21.52... \times \frac{312}{313} \times \frac{828}{832} \times \frac{187}{207} \ = \ 19.28...\text{Mbits/s} \tag{9.3}$$

The factor of 312/313 accounts for the data field sync overhead of one data segment per field. The factor of 828/832 accounts for the data segment sync overhead of four symbol intervals per data segment, and the factor of 187/207 accounts for the Reed-Solomon FEC overhead of 20 bytes per data segment.

The calculation of the net payload bit rate of the high-data-rate mode is identical, except that 16-VSB carries 4 information bits per symbol. Therefore, the net bit rate is twice that of the 8-VSB terrestrial mode:

$$19.28\ldots \times 2 \; = \; 38.57\ldots\text{Mbits/s} \tag{9.4}$$

To arrive at the net bit rate seen by a transport decoder, however, it is necessary to account for the fact that the MPEG sync bytes are removed from the data-stream input to the 8-VSB transmitter. This amounts to the removal of 1 byte per data segment. These MPEG sync bytes then are reconstituted at the output of the 8-VSB receiver. The net bit rate seen by the transport decoder is:

$$19.28\ldots \times \frac{188}{187} \; = \; 19.39\ldots\text{Mbits/s} \tag{9.5}$$

The net bit rate seen by the transport decoder for the high-data-rate mode is:

$$19.39\ldots \times 2 \; = \; 38.78\,\text{Mbits/s} \tag{9.6}$$

9.2 Performance Characteristics of the Terrestrial Broadcast Mode

The terrestrial VSB system can operate in a signal-to-additive-white-Gaussian-noise (S/N) environment of 14.9 dB [1]. The 8-VSB 4-state segment-error-probability curve in Figure 9.1 shows a segment-error probability of 1.93×10^{-4}. This is equivalent to 2.5 segment errors/s, which has been established by measurement as the threshold of visibility of errors.

The *cumulative distribution function* (CDF) of the peak-to-average power ratio, as measured on a low-power transmitted signal with no nonlinearities, is plotted in Figure 9.2. The plot shows that the transient peak power is within 6.3 dB of the average power for 99.9 percent of the time.

9.2.1 Transmitter Signal Processing

A preequalizer filter is recommended for use in over-the-air broadcasts where the high-power transmitter may have significant in-band ripple or rolloff at band edges [1]. This linear distortion can be detected by an equalizer in a reference demodulator ("ideal" receiver), located at the transmitter site, that is receiving a small sample of the antenna signal feed provided by a directional coupler, which is recommended to be located at the sending end of the antenna feed transmission line. The reference demodulator equalizer tap weights can be transferred into the transmitter preequalizer for precorrection of transmitter linear distortion.

A suitable preequalizer is an 80-tap feed-forward transversal filter. The taps are symbol-spaced (93 ns), with the main tap being approximately at the center, giving about ±3.7 μs

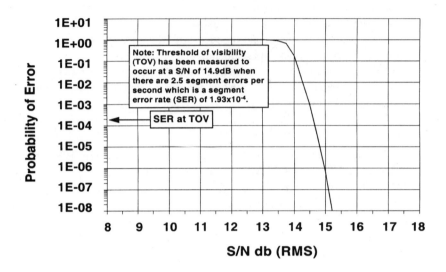

Figure 9.1 Segment-error probability for 8-VSB with 4-state trellis coding; RS (207,187). (*From* [1]. *Used with permission.*)

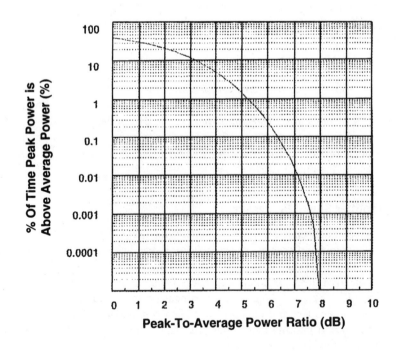

Figure 9.2 The cumulative distribution function of 8-VSB peak-to-average power ratio. (*From* [1]. *Used with permission.*)

correction range. It operates on the *I* channel data signal (there is no *Q* channel data in the transmitter), and shapes the frequency spectrum of the intermediate frequency (IF) signal so that there is a flat in-band spectrum at the output of the high-power transmitter that feeds the antenna. There is essentially no effect on the out-of-band spectrum of the transmitted signal.

The transmitter VSB filtering may be implemented by complex-filtering the baseband data signal, creating precision-filtered and stable in-phase and quadrature-phase modulation signals. This filtering process provides the root raised-cosine Nyquist filtering as well as the sin *x*/*x* compensation for the D/A converters. The orthogonal baseband signals are converted to analog form (D/A converters), then modulated on quadrature IF carriers to create the vestigial sideband IF signal by sideband cancellation (*phasing method*). Other techniques may be used as well, with varying degrees of success.

9.2.2   Upconverter and RF Carrier Frequency Offsets

Modern NTSC TV transmitters use a 2-step modulation process [1]. The first step usually is modulation of the video signal onto an IF carrier, which is the same frequency for all channels, followed by translation to the desired RF channel. The VSB transmitter applies this same 2-step modulation process. The RF upconverter translates the filtered flat IF data signal spectrum to the desired RF channel.

The frequency of the RF upconverter oscillator in DTV terrestrial broadcasts typically will be the same as that used for NTSC (except for NTSC offsets). However, in extreme co-channel situations, the DTV system is designed to take advantage of precise RF carrier frequency offsets with respect to the NTSC co-channel carrier. Because the VSB data signal sends repetitive synchronizing information (*segment syncs*), precise offset causes NTSC co-channel carrier interference into the VSB receiver to phase alternate from sync to sync. The VSB receiver circuits average successive syncs to cancel the interference and make data segment sync detection more reliable.

For DTV co-channel interference into NTSC, the interference is noiselike and does not change with precise offset. Even the DTV pilot interference into NTSC does not benefit from precise frequency offset because it is so small (11.3 dB below the data power) and falls far down the Nyquist slope (20 dB or more) of NTSC receivers.

The DTV co-channel pilot should be offset in the RF upconverter from the dominant NTSC picture carrier by an odd multiple of half the data segment rate. A consequential spectrum shift of the VSB signal into the upper adjacent channel is required. An additional offset of 0, +10, or –10 kHz is required to track the principal NTSC interferer.

For DTV-into-DTV co-channel interference, precise carrier offset prevents possible misconvergence of the adaptive equalizer. If, by chance, the two DTV data field sync signals fall within the same data segment time, the adaptive equalizer could misinterpret the interference as a ghost. To prevent this, a carrier offset of $f_{seg}/2 = 6.47$ kHz is recommended in the DTV standard for close DTV-into-DTV co-channel situations. This causes the interference to have no effect in the adaptive equalizer.

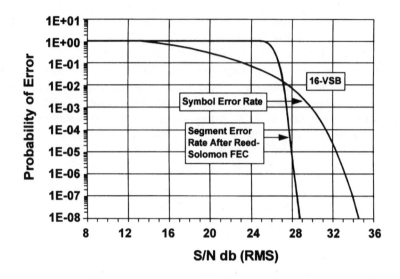

Figure 9.3 The error probability of the 16-VSB signal. (*From* [1]. *Used with permission.*)

9.2.3   Performance Characteristics of High-Data-Rate Mode

The high-data-rate mode can operate in a signal-to-white-noise environment of 28.3 dB [1]. The error-probability curve is shown in Figure 9.3.

The cumulative distribution function of the peak-to-average power ratio, as measured on a low-power transmitted signal with no nonlinearities, is plotted in Figure 9.4 and is slightly higher than that of the terrestrial mode.

9.3   Spectrum Issues

Spectrum-allocation considerations played a significant role in the development of the DTV standard. Because of the many demands on use of frequencies in the VHF and UHF bands, spectrum efficiency was paramount. Furthermore, a migration path was required from NTSC operation to DTV operation. These restraints led to the development of a spectrum-utilization plan for DTV that initially called for the migration of stations to the UHF-TV block of spectrum, and the eventual return of the VHF-TV frequencies after a specified period of simulcasting.

9.3.1   UHF Taboos

Early on in the DTV standardization process, the FCC decreed that such transmissions must be limited to a single 6 MHz channel and that the transmissions would be primarily in the UHF band (470 to 806 MHz) channels 14 to 69 [2]. Other radio services occupy spectrum

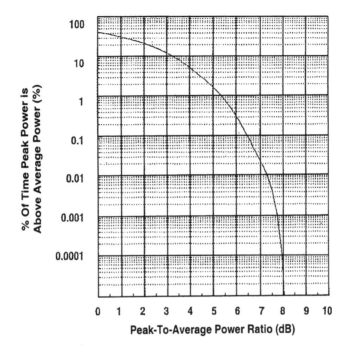

Figure 9.4 Cumulative distribution function of the 16-VSB peak-to-average power ratio. (*From* [1]. *Used with permission.*)

immediately below channel 14 and above channel 69, and channel 37 (608 to 614 MHz) is not allotted in any community (reserved for radio astronomy). Thus, during the transition from NTSC to DTV, these signals must share the available spectrum resource.

In any given community, the number of UHF channels allotted is a small fraction of the 54 UHF channels available because of interference considerations known as *UHF taboos*. If, for example, a transmitter is to be operated on channel n, no transmitter can be located within 20 miles of it on any of the following channels: $n \pm 2, 3, 4, 5, 7$, and 8. In addition, for a distance of 60 miles, channel $n + 14$ cannot be used; for 75 miles, channel $n + 15$ cannot be used. It also is common practice that VHF channels $n \pm 1$, which are called *first adjacent channels*, cannot be used because of the limited selectivity and dynamic range of consumer receivers. Strong signals on nearby channels, most obviously $n \pm 1$, can cause interference by linear and nonlinear mechanisms, while $n + 14$ and $n + 15$ represent the *image response channels* for consumer receivers having a 44 MHz intermediate frequency.

DTV transmitters could operate on any of these locally unused channels, provided that their *effective radiated power* (ERP) was substantially lower than what would be required for NTSC operation, and that their signal spectrum was designed to provide the minimum interference potential to NTSC. In effect, this requirement dictates the use of digital modulation where the spectral power density is constant within the assigned channel. (See Figure

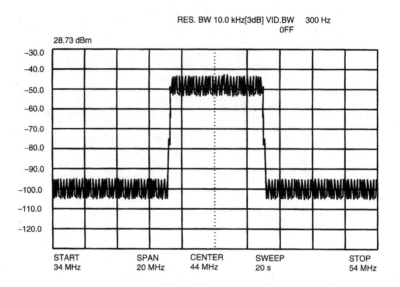

RES. BW 10.0 kHz[3dB] VID.BW 300 Hz
OFF

28.73 dBm

Figure 9.5 Typical spectrum of a digital television signal. (*From* [2]. *Used with permission.*)

9.5.) Digital signals having an equal probability of occurrence of each of their possible states—8-VSB has eight states—generate a signal whose spectrum appears to be random and, as such, can be considered similar to Gaussian noise. To ensure that the data signal looks like random noise at all times, the data is randomized at the transmitter, then derandomized at the DTV receiver. Binary data can be randomized by adding to it a pseudorandom binary sequence in an exclusive-OR gate, and at the receiver, repeating this operation with the same pseudorandom binary sequence, which must be known and synchronous to that employed at the transmitter. In this way, the spectral power density per hertz of spectrum is extremely low and independent of time and frequency, just as if the transmitter were a source of random noise.

9.3.2 Co-Channel Interference

Although frequency reuse is limited for the NTSC signal to a minimum distance between co-channel transmitters of 275 km in the UHF band, it is inevitable that much closer spacing between DTV and NTSC transmitters must be possible for each NTSC station to be assigned a second channel for DTV [2]. In fact, the minimum spacing to meet this requirement is about 112 miles, with a few exceptions well below this distance. Therefore, tests were conducted to determine the minimum NTSC-to-DTV signal levels for a subjectively acceptable level of interference that would not be annoying to the average viewer. The subjective data is plotted in Figure 9.6 [3]. The *desired/undesired signal ratio* (D/U)—at which the impairment was judged perceptible but not annoying—averaged about 33 dB at the edge of the NTSC coverage area. Typically, the NTSC signal might be –55 dBm where there is

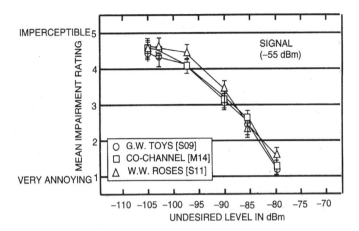

Figure 9.6 Measured co-channel interference, DTV-into-NTSC [3]. (*From* [2]. *Used with permission.*)

inevitably some thermal noise visible, which tends to mask the DTV interference. This ratio does not include whatever discrimination may be realized by means of the receiving antenna. The gain resulting from antenna directivity depends on the design of the antenna and the angle between the NTSC and DTV signals as they arrive at the receiving site.

One way to improve the D/U would be to employ vertical wave polarization for DTV, because most NTSC wave polarization is in the horizontal plane. This technique is widely used in Europe to increase protection of TV signals from other TV signals [4]. The protection provided by cross-polarization is greatest when the undesired signal comes from the same direction as the desired signal (where directionality of the receiving antenna is zero), and it is minimal when the undesired signal comes from behind the receiving antenna. The combined effects of directionality and cross-polarization may be summed to a constant 15 dB, which is the practice in the United Kingdom.

9.3.3 Adjacent-Channel Interference

Interference from DTV signals on the adjacent channels ($n \pm 1$) into NTSC receivers depends greatly upon the design of the tuner in the NTSC consumer receiver [2]. Under weak signal conditions, the RF amplifier operates at maximum gain, so the undesired signal on an adjacent channel is amplified before it reaches the mixer, where it may generate intermodulation and/or cross-modulation products (which produce artifacts). At high undesired signal levels, the undesired signal is converted to frequencies just outside of the IF bandpass and may not be adequately attenuated by the IF filter. Such adjacent-channel interference, when it reaches the second detector, contributes noise to the picture.

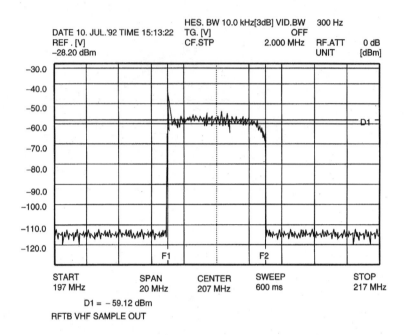

Figure 9.7 Spectrum of the 8-VSB DTV signal on channel 12. Note the pilot carrier at 204.3 MHz. (*From* [2]. *Used with permission.*)

The 8-VSB signal includes a pilot carrier, which is shown in Figure 9.7. This is the one coherent element of the 8-VSB signal, and it can cause interference to NTSC on the lower adjacent channel. Because this pilot is a coherent form of interference, it can be suppressed by means of precise carrier offset. In this case, the offset is between the DTV pilot carrier and the NTSC visual carrier on the lower adjacent channel.

Interference from a DTV signal on the upper adjacent channel to the desired NTSC signal also can cause impairment to the BTSC (Broadcast Television Systems Committee) stereo/SAP (*second audio program*) sound reception. The BTSC SAP channel has a slightly lower FM improvement factor than the BTSC stereo signal, which is less rugged than the monophonic FM sound of the NTSC signal. The aural carrier is located 0.25 MHz *below* the upper edge of the assigned channel, so a DTV signal on the upper adjacent channel may introduce what amounts to noise into the aural IF amplifier, and this noise may not be adequately rejected by the limiter that generally precedes the FM discriminator. Noise from the DTV signal also can generate, in the limiter, even wider bandwidth noise, some of which may be at the aural IF (4.500 MHz), thus being heard as either noise or distortion.

Under laboratory conditions, where nonlinearity can be held below the noise floor of its amplifiers, the 8-VSB signal has a spectrum that fits inside the 6 MHz channel. However, this is not practical for high-power amplifiers employed in actual DTV transmitters. Figure

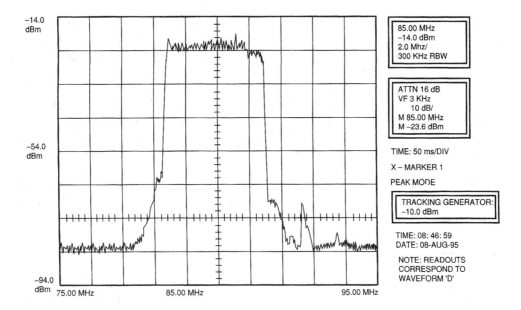

Figure 9.8 The spectrum of the 8-VSB DTV signal at the output of a VHF transmitter after bandpass filtering. Note the shoulders 38 dB down, which are intermodulation products. (*From* [2]. *Used with permission.*)

9.7 shows the 8-VSB signal on channel 12 under laboratory conditions. The noise floor of this signal in adjacent channels is about 54 dB below the spectral power density within channel 12. Figure 9.8 shows the spectrum of the 8-VSB signal being transmitted on channel 6, and Figure 9.9 shows the same for a UHF transmitter on channel 53. Nonlinearity of the high-power amplifier generates intermodulation products, some of which can be seen as "shoulders" of the signal spectrum in Figures 9.8 and 9.9, but not in Figure 9.7. The noise density 0.25 MHz below the channel edge, the aural carrier frequency of the BTSC signal on the lower adjacent channel, is much higher in Figures 9.8 and 9.9 than it is in Figure 9.7.

Sideband splatter from a DTV signal on the lower adjacent channel into NTSC will not impact the audio, but it may affect the picture, introducing low-frequency luminance noise. This can be minimized with a notch filter at the output of the DTV transmitter that is tuned to the NTSC visual carrier frequency.

Siting of HDTV Transmitters

The use of first adjacent channels in the same community has never been the practice either in AM or FM radio broadcasting or television [2]. It is only the differences in characteristics of analog NTSC and digital DTV that make this possible. However, it must be recognized that the DTV transmitter on either $n \pm 1$, relative to a locally allotted NTSC station on channel n, is possible only by co-siting these transmitters. In this context, the term co-siting is

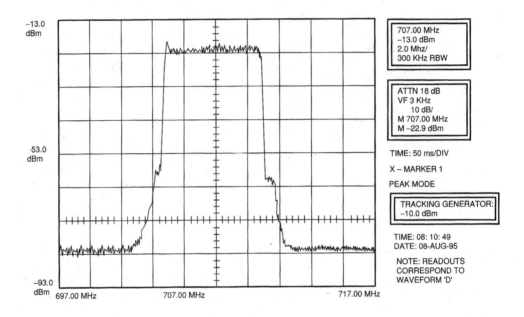

Figure 9.9 The spectrum of the 8-VSB DTV signal at the output of a UHF transmitter after bandpass filtering. Note the shoulders 35 dB down, which are intermodulation products, as in Figure 9.8. (*From* [2]. *Used with permission.*)

defined as using the same tower for the adjacent-channel transmitting antennas, but not necessarily using the same antenna. It is desirable that the radiation patterns of the two antennas are closely matched as to their respective nulls.

9.3.4 Power Ratings of NTSC and DTV

The visual transmitter for analog television signals universally employs negative modulation sense, meaning that the average power decreases with increasing scene brightness or *average picture level* (APL). Black level corresponds to a very high instantaneous power output with maximum power output during synchronizing pulses. The power output during sync pulses is constant and independent of APL. Therefore, it is the peak power output of the transmitter that is measured in analog television systems. The average power is an inverse function of the APL, a meaningless term.

Conversely, the average power of the DTV signal is constant. There are, however, transient peaks whose amplitude varies over a large range above the average power and whose amplitudes are data-dependent. Therefore, it is common practice to measure the average power of the DTV signal. Peaks at least 7 dB above average power are generated, and even higher peaks are possible. These transient peaks are the result of the well-known *Gibb's*

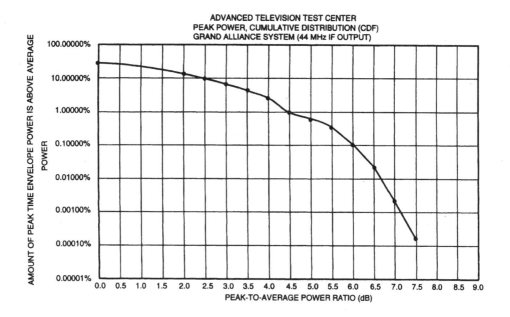

Figure 9.10 The cumulative distribution function transient peaks, relative to the average power of the 8-VSB DTV signal at the output of the exciter (IF). (*From* [2]. *Used with permission.*)

phenomena, which explains why sharp cutoff filters produce overshoot and ringing. Filters in the transmitter, generally digital in nature, are responsible for the transient peaks.

In the case of 8-VSB, transient peaks have been observed at least 7 dB above the average power at the IF output of the exciter, as shown in Figure 9.10. The use of a statistical metric is required to deal with the probabilistic nature of these transient peaks. It can be readily shown that the maximum symbol power of this signal is 4.8 dB above the average power, so the "headroom" above symbols of maximum power to these transient peaks is approximately 2.2 dB. Because the symbols must not be subjected to either AM/AM compression or AM/PM conversion, which is tantamount to *incidental carrier phase modulation* (ICPM) in NTSC, it follows that the digital transmitter must remain linear—for all intents and purposes—over its maximum power output range. Compression of transient peaks generates the out-of-channel components seen in Figures 9.8 and 9.9, which contribute to adjacent-channel interference. Extreme compression would increase the in-channel noise to the point that the combined receiver-plus-transmitter noise exceeds the receiver noise alone to a significant degree, reducing the range of the transmitter.

9.3.5 Sixth Report and Order

On February 18, 1998, the FCC released its final actions on DTV service rules and channel assignments. These actions resolved 231 petitions for reconsideration filed in response to the Fifth and Sixth Report and Order rule makings on DTV, released in April 1997. In general, the requests for reconsideration of technical issues relating to the Table of Assignments and other matters were denied, however, a total of 137 channel allotments in the Continental U.S. were changed from the April 1997 table. The DTV *core spectrum* (the frequency band for television broadcasting after the cessation of NTSC broadcasting) was set to be channels 2 to 51. There was no change to the DTV implementation schedule, which had drawn a great deal of interest and concern on the part of broadcasters. Additional flexibility was provided, however, for broadcasters to optimize their coverage through modification of their technical plants.

Although an examination of the Table of Allotments is beyond the scope of this book, it is important to point out that ERP levels were increased by more than 1 kW for 677 stations whose channel allotments did not change. Typically, these were 4 to 5 percent increases (669 were less than 5 percent). There also were 42 technical adjustments made to eliminate severe DTV-to-DTV adjacent channel interference. Testing at the Advanced Television Technology Center after the April 1997 order was released showed the original DTV-to-DTV interference criteria to be inadequate. The out-of-band emission limits were tightened to further address DTV-to-DTV adjacent channel interference problems.

Another important action was that the maximum DTV ERP allowable for UHF stations was increased to 200 kW, and in certain cases to 1000 kW.

The Commission said in its February 1998 order that it would address a number of related issues in separate rulemakings. These included:

- *Must carry* and retransmission consent rules

- Fees for pay services (separately for commercial and noncommercial broadcasters)

- Local zoning ordinance impact on the DTV rollout

On the business side, the FCC rules were modified to allow negotiated exchanges of DTV allotments, subject to not causing interference, or such interference being accepted by the impacted station.

Filing Guidelines

On August 10, 1998, the FCC released a Public Notice titled "Additional Application Processing Guidelines for Digital Television (DTV)." Following the issuance of the Memorandum Opinion and Order on Reconsideration of the Sixth Report and Order in MM Docket 87-268 (February 1998), questions arose as to how to implement some of the policies adopted in that action. The Public Notice includes detailed filing instructions in the following areas:

- Processing priorities (high to low): 1) checklist applications; 2) November 1, 1998, volunteer stations and top 10 market stations with May 1, 1999 on-air deadlines; 3) stations in markets 11–30 with November 1, 1999 on-air deadlines; and 4) all other DTV applications.

- Conditions when technical or interference studies must be included in the application.

- How to apply the 2%/10% de minimus interference criteria to applications.

- How to submit applications that include the use of antenna beam tilt to direct higher power toward close-in viewers while not exceeding predicted field strength at the FCC-defined noise-limited contour.

- How to submit applications that include negotiated agreements among affected stations for channel exchanges or changes to the technical parameters of allotments in a community.

In the Order, the Commission stated that would be necessary to limit modifications of NTSC facilities where such modifications would conflict with DTV allotments, and that it would consider the impact on DTV allotments in determining whether to grant applications for modification of NTSC facilities that were pending after April 3, 1997 (NTSC modification proposals are not permitted to cause any additional interference to DTV).

9.3.6   ATSC Standards for Satellite

In February 1997, the ATSC's Technology Group on Distribution (referred to as T3) formed a specialist group on satellite transmission, designated T3/S14 to define the parameters necessary to transmit the ATSC digital television standard (transport, video, audio, and data) over satellite. Although the ATSC had identified the particulars for terrestrial transmission of DTV (using FEC-encoded 8VSB modulation), this method would be wholly inappropriate for satellite transmission because of the many differences existing between terrestrial and satellite transmission environments. Also, the work of T3/S14 focused on the contribution and distribution application requirements of broadcasters, not the delivery of DTV signals to end-users.

The ATSC satellite transmission standard is intended to serve as a resource for satellite equipment manufacturers and broadcasters as satellite support for DTV becomes necessary. The general goals include:

- Assist manufacturers in building equipment that will be interoperable across vendors and be easily integrated into broadcast facilities.

- Help broadcasters identify suitable products and be in a position to share resources outside of their own closed network, when necessary.

The work of T3/S14 was restricted to considering the modulator and demodulator portions of a satellite link. Normally, for standards of this type, describing one end of a complementary system usually defines the other end. This was the approach T3/S14 took in preparing the document.

ATSC A/80 Standard

The ATSC, in document A/80, defined a standard for modulation and coding of data delivered over satellite for digital television applications [5]. These data can be a collection of program material including video, audio, data, multimedia, or other material generated in a digital format. They include digital multiplex bit streams constructed in accordance with ISO/IEC 13818-1 (MPEG-2 systems), but are not limited to these. The standard includes provisions for arbitrary types of data as well.

Document A/80 covers the transformation of data using error correction, signal mapping, and modulation to produce a digital carrier suitable for satellite transmission. In particular, quadrature phase shift modulation (QPSK), eight phase shift modulation (8PSK), and 16 quadrature amplitude modulation (16QAM) schemes are specified. The main distinction between QPSK, 8PSK, and 16QAM is the amount of bandwidth and power required for transmission. Generally, for the same data rate, progressively less bandwidth is consumed by QPSK, 8PSK, and 16QAM, respectively, but the improved bandwidth efficiency is accompanied by an increase in power to deliver the same level of signal quality.

A second parameter, coding, also influences the amount of bandwidth and power required for transmission. Coding, or in this instance, forward error correction (FEC) adds information to the data stream that reduces the amount of power required for transmission and improves reconstruction of the data stream received at the demodulator. While the addition of more correction bits improves the quality of the received signal, it also consumes more bandwidth in the process. So, the selection of FEC serves as another tool to balance bandwidth and power in the satellite transmission link. Other parameters exist as well, such as transmit filter shape factor (α), which have an effect on the overall bandwidth and power efficiency of the system.

System operators optimize the transmission parameters of a satellite link by carefully considering a number of trade-offs. In a typical scenario for a broadcast network, material is generated at multiple locations and requires delivery to multiple destinations by transmitting one or more carriers over satellite, as dictated by the application. Faced with various size antennas, available satellite bandwidth, satellite power, and a number of other variables, the operator will tailor the system to efficiently deliver the data payload. The important tools available to the operator for dealing with this array of system variables include the selection of the modulation, FEC, and α value for transmission.

Services and Applications

The ATSC satellite transmission standard includes provisions for two distinct types of services [5]:

- *Contribution service*—the transmission of programming/data from a programming source to a broadcast center. Examples include such material as digital satellite news gathering (DSNG), sports, and special events.

- *Distribution service*—the transmission of material (programming and/or data) from a broadcast center to its affiliate or member stations.

The A/80 document relies heavily upon previous work done by the Digital Video Broadcasting (DVB) Project of the European Broadcast Union (EBU) for satellite transmission. Where applicable, the ATSC standard sets forth requirements by reference to those standards, particularly EN 300 421 (QPSK) and prEN 301 210 (QPSK, 8PSK, and 16QAM).

The modulation and coding defined in the standard have mandatory and optional provisions. QPSK is considered mandatory as a mode of transmission, while 8PSK and 16QAM

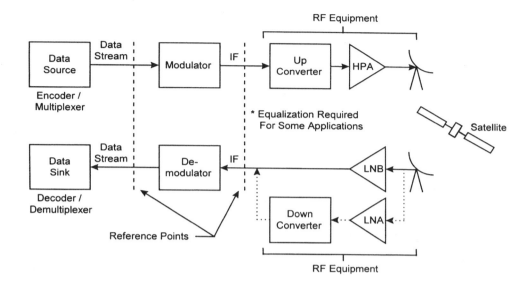

Figure 9.11 Overall system block diagram of a digital satellite system. The ATSC standard described in document A/80 covers the elements noted by the given reference points. (*From* [5]. *Used with permission.*)

are optional. Whether equipment implements optional features is a decision for the manufacturer. However, when optional features are implemented they must be in accordance with this standard in order to be compliant with it.

System Overview

A digital satellite transmission system is designed to deliver data from one location to one or more destinations. A block diagram of a simple system is shown in Figure 9.11 [5]. The drawing depicts a *data source* and *data sink*, which might represent a video encoder/multiplexer or decoder/demultiplexer for ATSC applications, but can also represent a variety of other sources that produce a digital data stream.

This particular point, the accommodation of arbitrary data streams, is a distinguishing feature between the systems supported by the ATSC standard and those supported by the DVB specifications EN 300 421 and prEN 301 210, which deal solely with MPEG transport streams. ATSC-compliant satellite transmission systems, for contribution and distribution applications, will accommodate arbitrary data streams.

The subject of this standard is the segment between the dashed lines designated by the reference points on Figure 9.11; it includes the modulator and demodulator. Only the modulation parameters are specified; the receive equipment is designed to recover the transmitted signal. The ATSC standard does not preclude combining equipment outside the dashed lines with the modulator or demodulator, but it sets a logical demarcation between functions.

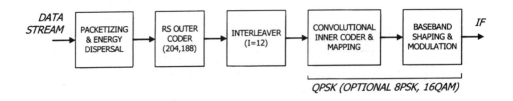

Figure 9.12 Block diagram of the baseband and modulator subsystem. (*From* [5]. *Used with permission.*)

In the figure, the modulator accepts a data stream and operates upon it to generate an intermediate frequency (IF) carrier suitable for satellite transmission. The data are acted upon by forward error correction (FEC), interleaving and mapping to QPSK, 8PSK or 16QAM, frequency conversion, and other operations to generate the IF carrier. The selection of the modulation type and FEC affects the bandwidth of the IF signal produced by the modulator. Selecting QPSK, 8PSK, or 16QAM consumes successively less bandwidth as the modulation type changes. It is possible, then, to use less bandwidth for the same data rate or to increase the data rate through the available bandwidth by altering the modulation type.

Coding or FEC has a similar impact on bandwidth. More powerful coding adds more information to the data stream and increases the occupied bandwidth of the IF signal emitted by the modulator. There are two types of coding applied in the modulator. An outer Reed-Solomon code is concatenated with an inner convolutional/trellis code to produce error correction capability exceeding the ability of either coding method used alone. The amount of coding is referred to as the *code rate*, quantified by a dimensionless fraction (k/n) where n indicates the number of bits out of the encoder given k input bits (e.g., rate 1/2 or rate 7/8). The Reed-Solomon code rate is fixed at 204,188 but the inner convolutional/trellis code rate is selectable, offering the opportunity to modify the transmitted IF bandwidth.

One consequence of selecting a more bandwidth-efficient modulation or a higher inner code rate is an increase in the amount of power required to deliver the same level of performance. The key measure of power is the E_b/N_o (energy per useful bit relative to the noise power per Hz), and the key performance parameter is the bit error rate (BER) delivered at a particular E_b/N_o. For digital video, a BER of about 10^{-10} is necessary to produce high-quality video. Thus, noting the E_b/N_o required to produce a given BER provides a way of comparing modulation and coding schemes. It also provides a relative measure of the power required from a satellite transponder, at least for linear transponder operation.

The basic processes applied to the data stream are illustrated in Figure 9.12. Specifically,

- Packetizing and energy dispersal

- Reed-Solomon outer coding

- Interleaving

Table 9.2 System Interfaces Specified in A/80 (*After* [5].)

Location	System Inputs/Outputs	Type	Connection
Transmit station	Input	MPEG-2 transport (Note 1) or arbitrary	From MPEG-2 multiplexer or other device
	Output	70/140 MHz IF, L-band IF, RF (Note 2)	To RF devices
Receive installation	Input	70/140 MHz IF, L-band IF (Note 2)	From RF devices
	Output	MPEG-2 transport (Note 1) or arbitrary	To MPEG-2 de-multiplexer or other device
1 In accordance with ISO/IEC 13838-1				
2 The IF bandwidth may impose a limitation on the maximum symbol rate.				

Table 9.3 Input Data Stream Structures (*After* [5].)

Type	Description
1	The packet structure shall be a constant rate MPEG-2 transport per ISO/IEC 13818-1 (188 or 204 bytes per packet including 0x47 sync, MSB first).
2	The input shall be a constant rate data stream that is arbitrary. In this case, the modulator takes successive 187 byte portions from this stream and prepends a 0x47 sync byte to each portion, to create a 188 byte MPEG-2 like packet. (The demodulator will remove this packetization so as to deliver the original, arbitrary stream at the demodulator output.)

- Convolutional inner coding

- Baseband shaping for modulation

- Modulation

The input to the modulator is a data stream of specified characteristics. The physical and electrical properties of the data interface, however, are outside the scope of this standard. (Work was underway in the ATSC and other industry forums to define appropriate data interfaces as this book went to press.) The output of the modulator is an IF signal that is modulated by the processed input data stream. This is the signal delivered to RF equipment for transmission to the satellite. Table 9.2 lists the primary system inputs and outputs.

The data stream is the digital input applied to the modulator. There are two types of packet structures supported by the standard, as given in Table 9.3.

9.4 Transmitter Considerations

Two parameters determine the basic design of any transmitter: the operating frequency and the power level. For DTV, the frequency is spelled out in the FCCs Table of Allotments; the power parameter, however, deserves—in fact, requires—additional consideration.

9.4.1 Operating Power

The FCC allocation table for DTV lists ERP values that are given in watts rms. The use of the term *average power* is not always technically correct. The intent was to specify the true heating power or rms watts of the total DTV signal averaged over a long period of time [6]. The specification of transmitter power is further complicated by the DTV system characteristic peak-to-average ratio, which has a significant impact on the required power output rating of the transmitter. For example, assume a given FCC UHF DTV ERP allocation of 405 kW rms, an antenna power gain of 24, and a transmission-line efficiency of 70 percent. The required DTV transmitter power T_x will equal

$$T_x = \frac{405}{24} \div 0.7 = 24.1 \text{ kW} \tag{9.7}$$

Because the DTV peak-to-average ratio is 4 (6 dB), the actual DTV transmitter power rating must be 96.4 kW (peak). This 4× factor is required to allow sufficient headroom for signal peaks (as discussed previously in this chapter). Figure 9.13 illustrates the situation. The transmitter rating is a peak value because the RF peak envelope excursions must traverse the linear operating range of the transmitter on a peak basis to avoid high levels of IMD spectral spreading [6]. In this regard, the *DTV peak envelope power* (PEP) is similar to the familiar NTSC *peak-of-sync* rating for setting the transmitter power level. Note that NTSC linearizes the PEP envelope from sync tip to zero carrier for best performance. Although many UHF transmitters use pulsed sync systems, where the major portion of envelope linearization extends only from black level to maximum white, the DTV signal has no peak repetitive portion of the signal to apply a pulsing system and, therefore, must be linearized from the PEP value to zero carrier. Many analog transmitters also linearize over the full NTSC amplitude range and, as a result, the comparison between NTSC and DTV peak RF envelope power applies for setting the transmitter power. The DTV power, however, always is stated as average (rms) because this is the only consistent parameter of an otherwise pseudorandom signal.

Figure 9.14 shows a DTV signal RF envelope operating through a transmitter *intermediate power amplifier* (IPA) stage with a spectral spread level of –43 dB. Note the large peak circled on the plot at 9 dB. This is significantly above the previously noted values. The plot also shows the average (rms) level, the 6 dB peak level, and other sporadic peaks above the 6 dB dotted line. If the modulator output were measured directly, where it can be assumed to be very linear, peak-to-average ratios as high as 11 dB could be seen.

Figure 9.15 shows another DTV RF envelope, but in this case the power has been increased to moderately compress the signal. Note that the high peaks above the 6 dB dotted line are nearly gone. The peak-to-average ratio is 6 dB.

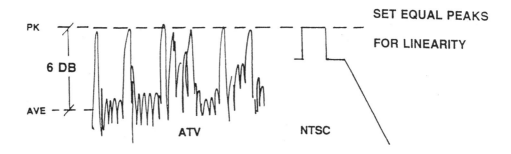

Figure 9.13 Comparison of the DTV peak transmitter power rating and NTSC peak-of-sync. (*After* [6].)

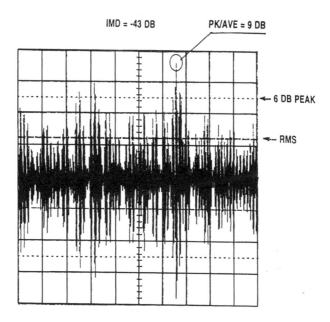

Figure 9.14 DTV RF envelope at a spectral spread of –43 dB. (*After* [6].)

9.4.2 Technology Options

With the operating power and frequency established, the fundamental architecture of the transmitter can be set. Three basic technologies are used for high-power television broadcasting today:

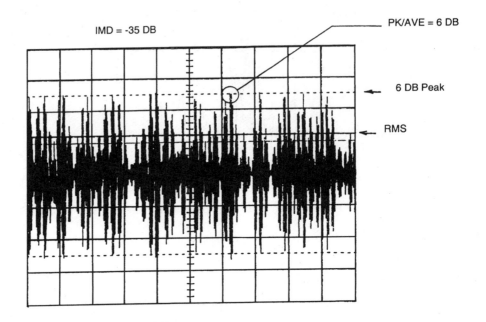

Figure 9.15 DTV RF envelope at a spectral spread of –35 dB. (*After* [6].)

- Solid-state—bipolar, *metal-oxide-semiconductor field-effect transistor* (MOSFET), LDMOS, silicon carbide, and others.

- Grid-based power vacuum tubes—tetrode, UHF tetrode, and Diacrode

- Klystron-based UHF devices—conventional klystron, MSDC (*multistage depressed collector*) klystron, and IOT (*inductive output tube*), also known as the Klystrode (Varian)

Each class of device has its strengths and weaknesses. Within each class, additional distinctions also can be made.

Solid-State Devices

Solid-state devices play an increasingly important role in the generation of RF energy. As designers find new ways to improve operating efficiency and to remove heat generated during use, the maximum operating power per device continues to increase.

Invariably, parallel amplification is used for solid-state RF systems, as illustrated in Figure 9.16. Parallel amplification is attractive from several standpoints. First, redundancy is a part of the basic design. If one device or one amplifier fails, the remainder of the system will continue to operate. Second, lower-cost devices can be used. It is often less expensive to put two 150 W transistors into a circuit than it is to put in one 300 W device. Third, troubleshooting the system is simplified because an entire RF module can be substituted to return the system to operation.

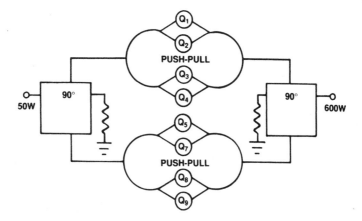

Figure 9.16 Schematic diagram of a 600 W VHF amplifier using eight FETs in a parallel device/parallel module configuration.

Solid-state systems are not, however, without their drawbacks. A high-power transmitter using vacuum tubes is much simpler in design than a comparable solid-state system. The greater the number of parts, the higher the potential for system failure. It is only fair to point out, however, that failures in a parallel fault-tolerant transmitter usually will not cause the entire system to fail. Instead, some parameter, typically peak output power, will drop when one or more amplifier modules is out of service.

This discussion assumes that the design of the solid-state system is truly fault-tolerant. For a system to provide the benefits of parallel design, the power supplies, RF divider and combiner networks, and supervisory/control systems also must be capable of independent operation. Furthermore, hot swapping of defective modules is an important attribute.

The ac-to-RF efficiency of a solid-state transmitter may or may not be any better than a tube transmitter of the same operating power and frequency. Much depends on the type of modulation used and the frequency of operation. Fortunately, the lower average power and duty cycle of the DTV signal suggests that a high-efficiency solid-state solution may be possible [7]. The relatively constant signal power of the DTV waveform eliminates one of the biggest problems in NTSC applications of class AB solid-state amplifiers: the continual level changes in the video signal vary the temperature of the class AB amplifier junctions and, thus, their bias points. This, in turn, varies all of the transistor parameters, including gain and linearity. Sophisticated adaptive bias circuits are required for reduction or elimination of this limitation to class AB operation.

Solid-state amplifiers operating class A do not suffer from such linearity problems, but the class A operation imposes a substantial efficiency penalty. Still, many designs use class A because of its simplicity and excellent linearity.

The two primary frontiers for solid-state devices are 1) power dissipation and 2) improved materials and processes [8]. With regard to power dissipation, the primary factor in determining the amount of power a given device can handle is the size of the active junctions on the chip. The same power output from a device also can be achieved through the use of several smaller chips in parallel within a single package. This approach, however, may result in unequal currents and uneven distribution of heat. At high power levels, heat management becomes a significant factor in chip design. Specialized layout geometries have been developed to ensure even current distribution throughout the device.

The second frontier—improved materials and processes—is being addressed with technologies such as LDMOS and silicon carbide (SiC). The bottom line with regard to solid-state is that, from the standpoint of power-handling capabilities, there is a point of diminishing returns for a given technology. The two basic semiconductor structures, bipolar and FET, have seen numerous fabrication and implementation enhancements over the years that have steadily increased the maximum operating power and switching speed. Power MOSFET, LDMOS, and SiC devices are the by-products of this ongoing effort.

With any new device, or class of devices, economies always come into play. Until a device has reached a stable production point, at which it can be mass-produced with few rejections, the per-device cost is usually high, limiting its real-world applications. For example, if an SiC device that can handle more than 4 times the power of a conventional silicon transistor is to be cost-effective in a transmitter, the per-device cost must be less than 4 times that of the conventional silicon product. It is fair to point out in this discussion that the costs of the support circuitry, chassis, and heat sink are equally important. If, for example, the SiC device—though still at a cost disadvantage relative to a conventional silicon transistor—requires fewer support elements, then a cost advantage still may be realized.

If increasing the maximum operating power for a given frequency is the primary challenge for solid-state devices, then using the device more efficiently ranks a close second. The only thing better than being able to dissipate more power in a transistor is not generating the waste heat in the first place. The real performance improvements in solid-state transmitter efficiency have not come as a result of simply swapping out a single tube with 200 transistors, but from using the transistors in creative ways so that higher efficiency is achieved and fewer devices are required.

This process has been illustrated dramatically in AM broadcast transmitters. Solid-state transmitters have taken over that market at all power levels, not because of their intrinsic feature set (graceful degradation capability when a module fails, no high voltages used in the system, simplified cooling requirements, and other attributes), but because they lend themselves to enormous improvements in operating efficiency as a result of the waveforms being amplified. For television, most notably UHF, the march to solid-state has been much slower. One of the promises of DTV is that clever amplifier design will lead to similar, albeit less dramatic, improvements in intrinsic operating efficiency.

Power Grid Devices

Advancements in vacuum tube technology have permitted the construction of numerous high-power UHF transmitters based on tetrodes [9, 10]. Such devices are attractive for television applications because they are inherently capable of operating in an efficient class AB mode. UHF tetrodes operating at high power levels provide essentially the same specifications, gain, and efficiency as tubes operating at lower powers. The anode power-supply voltage of the tetrode is much lower than the collector potential of a klystron- or IOT-based system (8 kV is common). Also, the tetrode does not require focusing magnets.

Efficient removal of heat is the key to making a UHF tetrode practical at high power levels. Such devices typically use water or vapor-phase cooling. Air-cooling at such levels is impractical because of the fin size that would be required. Also, the blower for the tube would have to be quite large, reducing the overall transmitter ac-to-RF efficiency.

Another drawback inherent in tetrode operation is that the output circuit of the device appears electrically in series with the input circuit and the load [7]. The parasitic reactance of the tube elements, therefore, is a part of the input and output tuned circuits. It follows, then, that any change in the operating parameters of the tube as it ages can affect tuning. More important, the series nature of the tetrode places stringent limitations on internal-element spacings and the physical size of those elements in order to minimize the electron transit time through the tube vacuum space. It is also fair to point out, however, that the tetrode's input-circuit-to-output-circuit characteristic has at least one advantage: power delivered to the input passes through the tube and contributes to the total power output of the transmitter. Because tetrodes typically exhibit low gain compared with klystron-based devices, significant power may be required at the input circuit. The pass-through effect, therefore, contributes to the overall operating efficiency of the transmitter.

The expected lifetime of a tetrode in UHF service usually is shorter than that of a klystron of the same power level. Typical lifetimes of 8000 to 15,000 hours have been reported. Intensive work, however, has led to products that offer higher output powers and extended operating lifetimes, while retaining the benefits inherent in tetrode devices. One such product is the TH563 (Thomson), which is capable of 50 kW in NTSC visual service and 25 to 30 kW in combined aural/visual service at 10 to 13 dB aural/visual ratios. With regard to DTV application possibilities, the linearity of the tetrode is excellent, a strong point for DTV consideration [11]. Minimal phase distortion and low intermodulation translate into reduced correction requirements for the amplifier.

The Diacrode (Thomson) is a promising adaptation of the high-power UHF tetrode. The operating principle of the Diacrode is basically the same as that of the tetrode. The anode current is modulated by an RF drive voltage applied between the cathode and the power grid. The main difference is in the position of the active zones of the tube in the resonant coaxial circuits, resulting in improved reactive current distribution in the electrodes of the device.

Figure 9.17 compares the conventional tetrode with the Diacrode. The Diacrode includes an electrical extension of the output circuit structure to an external cavity [11]. The small dc-blocked cavity rests on top of the tube, as illustrated in Figure 9.18.

The cavity is a quarter-wave transmission line, as measured from the top of the cavity to the vertical center of the tube. The cavity is short-circuited at the top, reflecting an open circuit (current minimum) at the vertical center of the tube and a current maximum at the base of the tube, like the conventional tetrode, and a second current maximum above the tube at the cavity short-circuit. (Figure 9.17 helps to visualize this action.)

With two current maximums, the Diacrode has an RF power capability twice that of the equivalent tetrode, while the element voltages remain the same. All other properties and aspects of the Diacrode are basically identical to those of the TH563 high-power UHF tetrode, upon which the Diacrode is patterned.

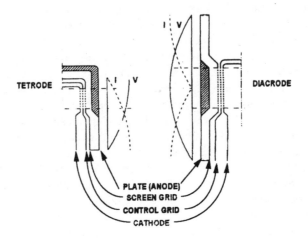

Figure 9.17 Cutaway view of the tetrode (*left*) and the Diacrode (*right*). Note that the RF current peaks above and below the Diacrode center, but the tetrode has only one peak at the bottom. (*After* [11].)

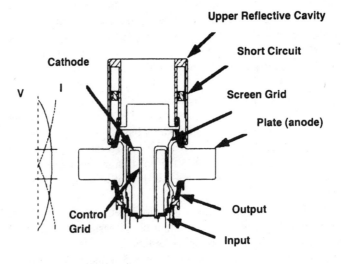

Figure 9.18 The elements of the Diacrode, including the upper cavity. Double current, and consequently, double power is achieved with the device because of the current peaks at the top and bottom of the tube, as shown. (*After* [11].)

Some of the benefits of such a device, in addition to the robust power output available, are its low high-voltage requirements (low relative to a klystron/IOT-based system, that is), small size, and simple replacement procedures. On the downside, there is little installed service lifetime data at this writing because the Diacrode is relatively new to the market.

Klystron-Based Devices

The klystron is a *linear-beam* device that overcomes the transit-time limitations of a grid-controlled vacuum tube by accelerating an electron stream to a high velocity before it is modulated [12]. Modulation is accomplished by varying the velocity of the beam, which causes the drifting of electrons into *bunches* to produce RF *space current*. One or more cavities reinforce this action at the operating frequency. The output cavity acts as a transformer to couple the high-impedance beam to a low-impedance transmission line. The frequency response of a klystron is limited by the impedance-bandwidth product of the cavities, which can be extended by stagger tuning or by the use of multiple-resonance filter-type cavities.

The klystron is one of the primary means of generating high power at UHF frequencies and above. Output powers for multicavity devices range from a few thousand watts to 10 MW or more. The klystron provides high gain and requires little external support circuitry. Mechanically, the klystron is relatively simple. It offers long life and requires a minimum of routine maintenance.

The klystron, however, is inefficient in its basic form. Efficiency improvements can be gained for television applications through the use of beam pulsing; still, a tremendous amount of energy must be dissipated as waste heat. Years of developmental research have produced two high-efficiency devices for television use: the MSDC klystron and the IOT, also known as the Klystrode.

The MSDC device is essentially identical to a standard klystron, except for the collector assembly. Beam reconditioning is achieved by including a *transition region* between the RF interaction circuit and the collector under the influence of a magnetic field. From an electrical standpoint, the more stages of a multistage depressed collector klystron, the better. Predictably, the tradeoff is increased complexity and, therefore, increased cost for the product. Each stage that is added to the depressed collector system is a step closer to the point of diminishing returns. As stages are added above four, the resulting improvements in efficiency are proportionally smaller. Because of these factors, a 4-stage device was chosen for television service. (See Figure 9.19.)

The IOT is a hybrid of a klystron and a tetrode. The high reliability and power-handling capability of the klystron is due, in part, to the fact that electron-beam dissipation takes place in the collector electrode, quite separate from the RF circuitry. The electron dissipation in a tetrode is at the anode and the screen grid, both of which are inherent parts of the RF circuit; therefore, they must be physically small at UHF frequencies. An advantage of the tetrode, on the other hand, is that modulation is produced directly at the cathode by a grid so that a long drift space is not required to produce density modulation. The IOT has a similar advantage over the klystron—high efficiency in a small package.

The IOT is shown schematically in Figure 9.20. The electron beam is formed at the cathode, density-modulated with the input RF signals by a grid, then accelerated through the anode aperture. In its bunched form, the beam drifts through a field-free region and interacts with the RF field in the output cavity. Power is extracted from the beam in the same way it is extracted from a klystron. The input circuit resembles a typical UHF power grid tube. The output circuit and collector resemble a klystron.

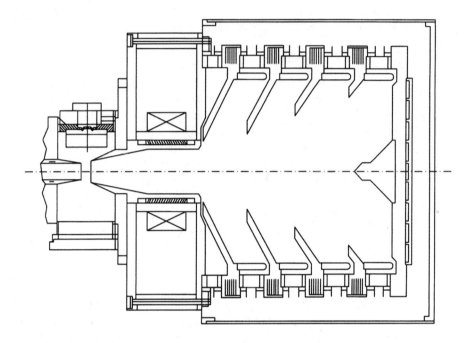

Figure 9.19 Mechanical design of the multistage depressed collector assembly. Note the "V" shape of the 4-element system.

Because the IOT provides beam power variation during sync pulses (as in a pulsed klystron) as well as over the active modulating waveform, it is capable of high efficiency. The device thus provides full-time beam modulation as a result of its inherent structure and class B operation.

For DTV service, the IOT is particularly attractive because of its good linearity characteristics. The IOT provides –60 dB or better intermodulation performance in combined 10 dB aural/visual service [7]. Tube life data varies depending upon the source, but one estimate puts the life expectancy at more than 35,000 hours [13]. Of course, tube cost, relative to the less expensive Diacrode, is also a consideration.

The maximum power output available from the standard IOT (60 kW visual-only service) had been an issue in some applications. In response, a modified tube was developed to produce 55 kW visual plus 5.5 kW aural in common amplification [14]. Early test results indicated that the device (EEV) was capable of delivering peak digital powers in excess of 100 kW and, therefore, was well suited to DTV applications. The tube also incorporated several modifications to improve performance. One change to the input cavity was shown to improve intermodulation performance of the device. Figure 9.21 shows a cross-section of the improved input cavity.

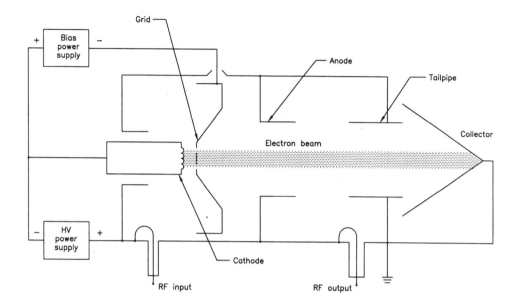

Figure 9.20 Functional schematic diagram of an IOT.

Constant Efficiency Amplifier

Because of the similarity between the spent electron beam in an IOT and that of a klystron, it is possible to consider the use of a multistage depressed collector on an IOT to improve the operating efficiency [15]. This had been considered by Priest and Shrader [16] and by Gilmore [17], but the idea was rejected because of the complexity of the multistage depressed collector assembly and because the IOT already exhibited fairly high efficiency. Subsequent development by Symons [15, 18] has led to a working device. An inductive output tube, modified by the addition of a multistage depressed collector, has the interesting property of providing linear amplification with (approximately) constant efficiency.

Figure 9.22 shows a schematic representation of the constant efficiency amplifier (CEA) [15]. The cathode, control grid, anode and output gap, and external circuitry are essentially identical with those of the IOT amplifier. Drive power introduced into the input cavity produces an electric field between the control grid and cathode, which draws current from the cathode during positive half-cycles of the input RF signal. For operation as a linear amplifier, the peak value of the current—or more accurately, the fundamental component of the current—is made (as nearly as possible) proportional to the square root of the drive power, so that the product of this current and the voltage it induces in the output cavity will be proportional to the drive power.

Following the output cavity is a multistage depressed collector in which several typical electron trajectories are shown. These are identified by the letters *a* through *e*. The collector electrodes are connected to progressively lower potentials between the anode potential and

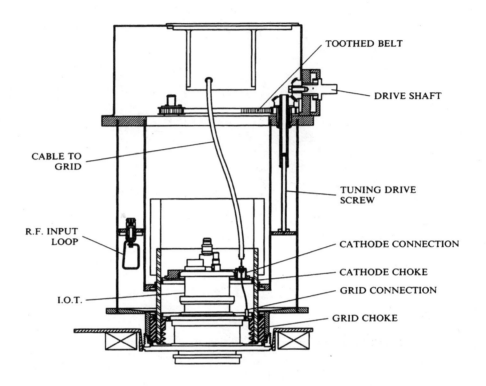

Figure 9.21 Mechanical configuration of a high-power IOT (EEV). (*After* [14].)

the cathode potential so that more energetic electrons penetrate more deeply into the collector structure and are gathered on electrodes of progressively lower potentials.

In considering the difference between an MSDC IOT and an MSDC klystron, it is important to recognize that in a class B device, no current flows during the portion of the RF cycle when the grid voltage is below cutoff and the output gap fields are accelerating. As a result, it is not necessary to have any collector electrode at a potential equal to or below cathode potential. At low output powers, when the RF output gap voltage is just equal to the difference in potential between the lowest-potential collector electrode and the cathode, all the current will flow to that electrode. Full class B efficiency is thus achieved under these conditions.

As the RF output gap voltage increases with increased drive power, some electrons will have lost enough energy to the gap fields so they cannot reach the lowest potential collector, and so current to the next-to-the-lowest potential electrode will start increasing. The efficiency will drop slightly and then start increasing again until all the current is just barely collected by the two lowest-potential collectors, and so forth.

Maximum output power is reached when the current delivered to the output gap is sufficient to build up an electric field or voltage that will just stop a few electrons. At this output

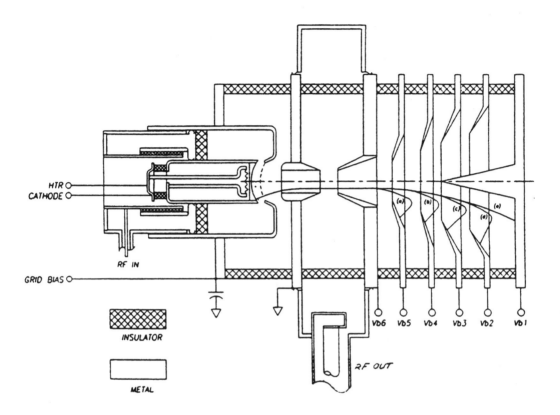

Figure 9.22 Schematic overview of the MSDC IOT or constant efficiency amplifier. (After [15].)

power, the current is divided between all of the collector electrodes and the efficiency will be somewhat higher than the efficiency of a single collector, class B amplifier. Computer simulations have demonstrated that it is possible to select the collector voltages so as to achieve very nearly constant efficiency from the MSDC IOT device over a wide range of output powers [15].

The challenge of developing a multistage depressed collector for an IOT is not quite the same as that of developing a collector for a conventional klystron [18]. It is different because the dc component of beam current rises and falls in proportion to the square root of the output power of the tube. The dc beam current is not constant as it is in a klystron. As a result, the energy spread is low because the output cavity RF voltage is low at the same time that the RF and dc beam currents are low. Thus, there will be small space-charge forces, and the beam will not spread as much as it travels deep into the collector toward electrodes having the lowest potential. For this reason, the collector must be rather long and thin when compared to the multistage depressed collector for a conventional klystron, as described previously.

Additional development on the IOT/depressed collector concept resulted in the Energy Saving Collector IOT (ESCIOT, Marconi Applied Technologies). This device, in the testing stages as this book went to press, uses a three stage collector with the following voltage divisions:

- Collector one at ground potential, water-cooled

- Collector two at 12 kV below ground potential, water-cooled

- Collector three at cathode potential (38 kV below ground potential), air-cooled

The tube was designed to serve as a plug-in replacement for a conventional IOT in DTV service. Preliminary data indicated a 23 percent energy savings when operated at an average digital power output of more than 16 kW and with performance within FCC DTV specifications [19].

9.4.3 Digital Signal Pre-Correction

Because amplitude modulation contains at least three parent signals (the carrier plus two sidebands) its amplification has always been fraught with in-band intermodulation distortion [20]. A common technique to correct for these distortion products is to pre-correct an amplifier by intentionally generating an *anti-intermodulation signal*. The following process is typically used:

- Mix the three tones together

- Adjust the amplitude of the anti-IM product to match that of the real one

- Invert the phase

- Delay the anti-IM product or the parent signals in time so that it aligns with the parent signals

- Add the anti-IM signal into the main parent signal path to cause cancellation in the output circuit of the amplifier

This is a common approach that works well in countless applications. Each intermodulation product may have its own anti-intermodulation product intentionally generated to cause targeted cancellation.

In a vacuum tube or transistor stage that is amplifying a noise-like signal, such as the 8-VSB waveform, sideband components arise around the parent signal at the output of the amplifier that also appear noise-like and are correlated with the parent signal. The amplifying device, thus, behaves precisely like two mixers not driven into switching. As with the analog pre-correction process described previously, pre-correction for the digital waveform also is possible.

The corrector illustrated in Figure 9.23 mimics the amplifying device, but contains phase shift, delay, and amplitude adjustment circuitry to properly add (subtract), causing cancellation of the intermodulation components adjacent to the parent signal at the amplifier output.

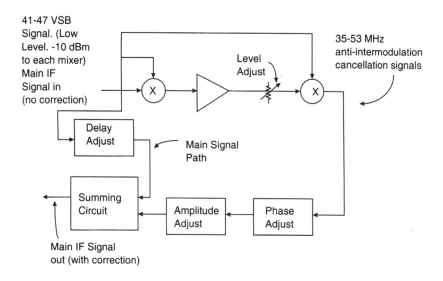

Figure 9.23 An example 8-VSB sideband intermodulation distortion corrector functional block diagram for the broadcast television IF band from 41 to 47 MHz for use with DTV television transmitters. (*From* [20]. *Used with permission.*)

Because the intentional mixing circuit is done in a controlled way, it may be used at the IF frequency before the RF conversion process in the transmitter so that an equal and opposite phase signal is generated and added to the parent signal at IF. It will then go through the same RF conversion and amplification process as the parent signal that will spawn the real intermodulation products. The result is intermodulation cancellation at the output of the amplifier.

Figure 9.24 shows the relative amplitude content of the correction signal spanning 6 MHz above and below the desired signal. Because the level of the correction signal must match that of the out-of-band signal to be suppressed, it must be about 43 dB below the in-band signal according to the example. This level of in-band addition is insignificant to the desired signal, but just enough to cause cancellation of the out-of-band signal.

Using such an approach that intentionally generates the correct anti-intermodulation component and causes it to align in time, be opposite phase, and equal in amplitude allows for cancellation of the unwanted component, at least in part. The degree of cancellation has everything to do with the precise alignment of these three attributes. In practice, it has been demonstrated [20] that only the left or right shoulder may be optimally canceled by up to about 3–5 dB. This amount may not seem to be significant, but it must be remembered that 3 dB of improvement is cutting the power in half.

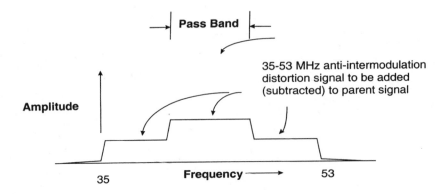

Figure 9.24 After amplification, the out-of-band signal is corrected or suppressed by an amount determined by the ability of the correction circuit to precisely match the delay, phase, and amplitude of the cancellation signal. (*From* [20]. *Used with permission.*)

9.4.4 FCC Emissions Mask

The FCC emissions mask has been the subject of many heated discussions by various industry officials ever since its issuance by the FCC on February 27, 1998. The paramount feature of the mask is its exceptionally sharp cut off characteristics and the deep level of out-of-band emissions (–110 dB down from reference power) [21]. There has never been such a demanding requirement imposed on the industry before, barring a few land mobile protection situations. The mask requirement applies to all DTV broadcasters.

Specifically, the DTV out-of-band emissions mask requires that:

- In the first 500 kHz from the authorized channel edge, transmitter emissions must be attenuated no less than 47 dB below the average transmitted power.

- At more than 6 MHz from the channel edge, emissions must be attenuated no less than 110 dB below average transmitted power.

- At any frequency between. 5 and 6 MHz from the channel edge, emissions must be attenuated no less than the value determined by the formula: attenuation in dB = $-11.5 (\Delta f + 3.6)$, where Δf = the frequency difference in MHz from the edge of the channel.

Figure 9.25 illustrates a compliant and noncompliant signal. In (*a*), IMD levels spill over the FCC attenuation limits (shown as sloping lines), which means an aggressive filter system will need to be used at the transmitter output, as illustrated in (*b*). Note that in (*b*) the IMD spectral response is contained within the FCC mask limits, and it appears that the FCC out-of-band emission issue is solved. There are, however, a number of practical issues to be considered in such a scenario, including:

- The response shown in Figure 9.25(*b*) is expensive to build

- The sharp tuned response shown calls for elaborate temperature compensation to keep the transmitted band edges from being distorted during temperature drifting

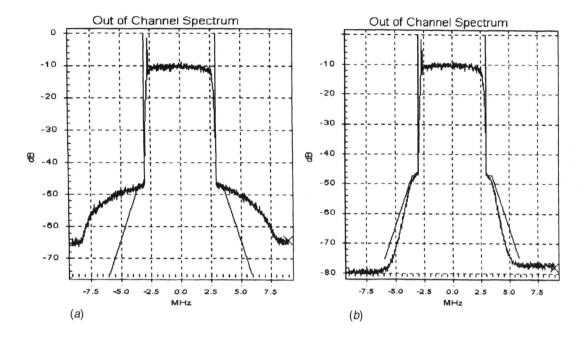

Figure 9.25 FCC linearity emissions requirements: (*a*) typical transmitter output with linearity such that the IMD shoulder levels are at –37 dB, (*b*) DTV mask filter on transmitter output to be fully mask compliant. (*After* [21].)

- The filter is best configured as a constant impedance unit, which is more expensive to build than other, more common, designs

A powerful method of easing, or even eliminating these issues is to employ digital signal pre-correction techniques, as discussed in the previous section.

9.4.5 Implementation Issues

The design and installation of a transmitter for DTV operation is a complicated process that must take into consideration a number of variables. Some of the more important issues include:

- The actual cost of the transmitter, both the purchase price and the ongoing maintenance expenses for tubes and other supplies.

- The actual ac-to-RF efficiency, which relates directly, of course, to the operating costs.

- Maintenance issues, the most important of which is the *mean time between failure* (MTBF). Also significant is the *mean time to repair* (MTTR), which relates directly to the accessibility of transmitter components and the type of amplifying devices used in the unit.

- Environmental issues, not the least of which is the occupied space needed for the transmitter. The cooling requirements also are important and may, for example, affect the ongoing maintenance costs.

- The availability of sufficient ac power and power of acceptable reliability and regulation at the site.

9.4.6 Channel-Combining Considerations

A number of techniques are practical to utilize an existing tower for both NTSC and DTV transmissions. Combining RF signals allows broadcasters to use an existing structure to transmit NTSC and DTV from a common line and antenna or, in the case of a VHF and UHF combination, to utilize the same line to feed two separate antennas.

In the transition period from NTSC to DTV, many broadcasters may choose to use the existing tower to transmit both NTSC and DTV channels. Some may choose to add a new line and DTV antenna; others may combine their DTV with NTSC and transmit from a common antenna and line; and still others may choose to consolidate to a new structure common to many local channels. For most stations, it is a matter of cost and feasibility.

Channel combiners, also known as *multiplexers* or *diplexers*, are designed for various applications. These systems can be generally classified as follows [22]:

- *Constant impedance*—designs that consist of two identical filters placed between two hybrids.

- *Starpoint*—designs that consist of single bandpass filters phased into a common output tee.

- *Resonant loop*—types that utilize two coaxial lines placed between two hybrids; the coaxial lines are of a calculated length.

- *Common line*—types that use a combination of band-stop filters matched into a common output tee.

A detailed discussion of these combiner systems is beyond the scope of this chapter. Interested readers are referred to [22] and [23] for more information on combining techniques.

The *dual-mode channel combiner* (Micro Communications) is a device that shows promise for using a single transmission line on a tower to feed two separate antennas. Dual-mode channel combining is the process by which two channels are combined within the same transmission line, but in separate orthogonal modes of propagation [24].

The device combines two different television channels from separate coaxial feedlines into a common circular waveguide. Within the circular waveguide, one channel propagates in the TE_{11} mode while the other channel propagates in the TM_{01} mode. The dual-mode channel combiner is reciprocal and, therefore, also may be used to efficiently separate two TE_{11}/TM_{01} mode-isolated channels that are propagating within the same circular waveguide into two separate coaxial lines. This provides a convenient method for combining at the transmitters and splitting at the antennas. The operating principles of the dual-mode channel combiner are described in [24].

9.4.7 Antenna Systems

The availability of suitable locations for new television transmission towers is diminishing, even in the secondary markets, and sites are practically nonexistent in major markets [25]. After the hurdles of zoning variance and suitable tower location are overcome, FAA restrictions and environmental concerns may delay the construction of a new tower for years. Not surprisingly, many broadcasters are looking at the pros and cons of using existing towers to support their new DTV antennas even though the prime tower-top spots are occupied.

For any given antenna, directional or omnidirectional, the tower will modify the as-designed antenna pattern. For optimum coverage, the as-installed pattern must be known—not just at the carrier frequency, but throughout the entire channel—before the relative position of the antenna and its azimuthal pattern orientation can be fixed. There is usually one position that will provide the optimum coverage without exceeding the structural limitations of the tower. This optimum position can be calculated (see [25]).

Coverage considerations are particularly important to DTV because all undesired energies, such as reflections, translate into a loss of coverage, whereas the undesired energies in NTSC translate primarily into a loss of picture quality.

Another transmission-optimization technique that holds promise for DTV is *circular polarization* (CP). Although mixed results were achieved with the use of CP for NTSC broadcasts, DTV applications may prove more amenable to improvement through the use of this technique. The transmission of CP has obvious drawbacks in the form of a 2× increase in required transmitter power and transmission line, as well as a more complex antenna. (The 2× increase is due to the fact that ERP is measured in just one radiation plane.) For the DTV signal, *polarization diversity* can be achieved if the vertically polarized signal is transmitted through CP. A polarization-diversity system at the receive antenna can provide missing signal level when one of the horizontal or vertical signal components experiences a deep fade. It follows that the inherent diversity attributes of CP operation could be put to good use in reducing the cliff-edge effect of the terrestrial DTV signal [26].

9.5 References

1. ATSC, "Guide to the Use of the Digital Television Standard," Advanced Television Systems Committee, Washington, D.C., Doc. A/54, Oct. 4, 1995.

2. Rhodes, Charles W.: "Terrestrial High-Definition Television," *The Electronics Handbook*, Jerry C. Whitaker (ed.), CRC Press, Boca Raton, Fla., pp. 1599–1610, 1996.

3. ACATS, "ATV System Description: ATV-into-NTSC Co-channel Test #016," Grand Alliance Advisory Committee on Advanced Television, p. I-14-10, Dec. 7, 1994.

4. CCIR Report 122-4, 1990.

5. ATSC Standard: "Modulation And Coding Requirements For Digital TV (DTV) Applications Over Satellite," Doc. A/80, ATSC, Washington, D.C., July, 17, 1999.

6. Plonka, Robert J.: "Planning Your Digital Television Transmission System," *Proceedings of the 1997 NAB Broadcast Engineering Conference*, National Association of Broadcasters, Washington, D.C., p. 89, 1997.

7. Ostroff, Nat S.: "A Unique Solution to the Design of an ATV Transmitter," *Proceedings of the 1996 NAB Broadcast Engineering Conference*, National Association of Broadcasters, Washington, D.C., p. 144, 1996.

8. Whitaker, Jerry C.: "Solid State RF Devices," *Radio Frequency Transmission Systems: Design and Operation*, McGraw-Hill, New York, p. 101, 1990.

9. Whitaker, Jerry C.: "Microwave Power Tubes," *The Electronics Handbook*, Jerry C. Whitaker (ed.), CRC Press, Boca Raton, Fla., p. 413, 1996.

10. Tardy, Michel-Pierre: "The Experience of High-Power UHF Tetrodes," *Proceedings of the 1993 NAB Broadcast Engineering Conference*, National Association of Broadcasters, Washington, D.C., p. 261, 1993.

11. Hulick, Timothy P.: "60 kW Diacrode UHF TV Transmitter Design, Performance and Field Report," *Proceedings of the 1996 NAB Broadcast Engineering Conference*, National Association of Broadcasters, Washington, D.C., p. 442, 1996.

12. Whitaker, Jerry C.: "Microwave Power Tubes," *Power Vacuum Tubes Handbook*, Van Nostrand Reinhold, New York, p. 259, 1994.

13. Ericksen, Dane E.: "A Review of IOT Performance," *Broadcast Engineering*, Intertec Publishing, Overland Park, Kan., p. 36, July 1996.

14. Aitken, S., D. Carr, G. Clayworth, R. Heppinstall, and A. Wheelhouse: "A New, Higher Power, IOT System for Analogue and Digital UHF Television Transmission," *Proceedings of the 1997 NAB Broadcast Engineering Conference*, National Association of Broadcasters, Washington, D.C., p. 531, 1997.

15. Symons, Robert S.: "The Constant Efficiency Amplifier," *Proceedings of the NAB Broadcast Engineering Conference*, National Association of Broadcasters, Washington, D.C., pp. 523–530, 1997.

16. Priest, D. H., and M. B. Shrader: "The Klystrode—An Unusual Transmitting Tube with Potential for UHF-TV," *Proc. IEEE*, vol. 70, no. 11, pp. 1318–1325, November 1982.

17. Gilmore, A. S.: *Microwave Tubes*, Artech House, Dedham, Mass., pp. 196–200, 1986.

18. Symons, R., M. Boyle, J. Cipolla, H. Schult, and R. True: "The Constant Efficiency Amplifier—A Progress Report," *Proceedings of the NAB Broadcast Engineering Conference*, National Association of Broadcasters, Washington, D.C., pp. 77–84, 1998.

19. Crompton, T., S. Aitken, S. Bardell, R. Heppinstall, M. Keelan, A. Wheelhouse, and G. Clayworth: "The ESCIOT High-Power DTV Amplifier System—A Progress Report,"

NAB Broadcast Engineering Conference Proceedings, National Association of Broadcasters, Washington, D.C., 2000.

20. Hulick, Timothy P.: "Very Simple Out-of-Band IMD Correctors for Adjacent Channel NTSC/DTV Transmitters," *Proceedings of the Digital Television '98 Conference*, Intertec Publishing, Overland Park, Kan., 1998.

21. Plonka, Robert J., "Bandpass Filter and Linearization Requirements for the New FCC Mask," *Proceedings of the NAB Broadcast Engineering Conference*, National Association of Broadcasters, Washington, D.C., pp. 387–396, 1999.

22. Heymans, Dennis: "Channel Combining in an NTSC/ATV Environment," *Proceedings of the 1996 NAB Broadcast Engineering Conference*, National Association of Broadcasters, Washington, D.C., p. 165, 1996.

23. Whitaker, Jerry C.: "RF Combiner and Diplexer Systems," *Power Vacuum Tubes Handbook*, Van Nostrand Reinhold, New York, p. 368, 1994.

24. Smith, Paul D.: "New Channel Combining Devices for DTV, *Proceedings of the 1997 NAB Broadcast Engineering Conference*, National Association of Broadcasters, Washington, D.C., p. 218, 1996.

25. Bendov, Oded: "Coverage Contour Optimization of HDTV and NTSC Antennas," *Proceedings of the 1996 NAB Broadcast Engineering Conference*, National Association of Broadcasters, Washington, D.C., p. 69, 1996.

26. Plonka, Robert J.: "Can ATV Coverage Be Improved With Circular, Elliptical, or Vertical Polarized Antennas?" *Proceedings of the 1996 NAB Broadcast Engineering Conference*, National Association of Broadcasters, Washington, D.C., p. 155, 1996.

Receiver Systems and Display Devices

10.1   Introduction

The introduction of a new television system must be viewed as a chain of elements that begins with image and sound pickup and ends with image display and sound reproduction. The DTV receiver is a vital link in the chain. By necessity, the ATSC system places considerable requirements upon the television receiver. The level of complexity of a DTV-compliant receiver is unprecedented, and that complexity is made possible only through advancements in large-scale integrated circuit design and fabrication.

The goal of any one-way broadcasting system, such as television, is to concentrate the hardware requirements at the source as much as possible and to make the receivers—which greatly outnumber the transmitters—as simple and inexpensive as possible. Despite the significant complexity of a DTV receiver, this principal has been an important design objective from the start.

10.1.1   Noise Figure

One of the more important specifications that receiver designers must consider is the *noise figure* (NF). A number of factors enter into the ultimate carrier-to-noise ratio within a given receiver. For example, the receiver planning factors applicable to UHF DTV service are shown in Table 10.1 [1]. A consumer can enhance the noise performance of the installation by improving any contributing factor; examples include installation of a *low-noise amplifier* (LNA) or a high-gain antenna.

The assumptions made in the DTV system planning process were based on *threshold-of-visibility* (TOV) measurements taken during specified tests and procedures [2]. These TOV numbers were correlated to *bit error rate* (BER) results from the same tests. When characterizing the Grand Alliance VSB modem hardware in the lab and in field tests, only BER measurements were taken. In Table 10.2, the results of these tests are expressed in equivalent TOV numbers derived from the BER measurements. It should be noted that the exact amount of adjacent-channel or co-channel interference entering the receiver terminals is a function of the overall antenna gain pattern used (not just the front/back ratio), which is also a function of frequency.

Table 10.1 Receiver Planning Factors Used by the FCC (*After* [1].)

Planning Factors	Low VHF	High VHF	UHF
Antenna impedance (ohms)	75	75	75
Bandwidth (MHz)	6	6	6
Thermal noise (dBm)	−106.2	−106.2	−106.2
Noise figure (dB)	10	10	7
Frequency (MHz)	69	194	615
Antenna factor (dBm/dBμ)	−111.7	−120.7	−130.7[1]
Line loss (dB)	1	2	4
Antenna gain (dB)	4	6	10
Antenna F/B ratio (dB)	10	12	14
1. See Appendix B of the Sixth Report and Order (MM 87-268), adopted April 3, 1997, for a discussion of the dipole factor.			

Table 10.2 DTV Interference Criteria (*After* [2].)

Co-channel DTV-into-NTSC	33.8 dB
Co-channel NTSC-into-DTV	2.07 dB
Co-channel DTV-into-DTV	15.91 dB
Upper-adjacent DTV-into-NTSC	−16.17 dB
Upper-adjacent NTSC-into-DTV	−47.05 dB
Upper-adjacent DTV-into-DTV	−42.86 dB
Lower-adjacent DTV-into-NTSC	−17.95 dB
Lower-adjacent NTSC-into-DTV	−48.09 dB
Lower-adjacent DTV-into-DTV	−42.16 dB

10.2 Receiver System Overview

At this writing, second-generation DTV-compliant consumer receivers were entering the marketplace. Because of the rapidly changing nature of the consumer electronics business, this chapter will focus on the overall nature of the DTV receiving subsystem, rather than on any specific implementation.

Figure 10.1 shows a general block diagram of the VSB terrestrial broadcast transmission system. The major circuit elements include:

- Tuner

- Channel filtering and VSB carrier recovery

- Segment sync and symbol clock recovery

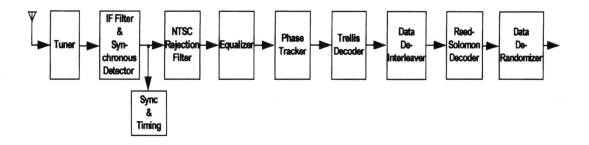

Figure 10.1 Simplified block diagram of a VSB receiver. (*From* [2]. *Used with permission.*)

- Noncoherent and coherent *automatic gain control* (AGC)
- Data field synchronization
- Interference-rejection filter
- Channel equalizer
- Phase-tracking loop
- Trellis decoder
- Data de-interleaver
- Reed-Solomon decoder
- Data derandomizer
- Receiver loop acquisition sequencing

Descriptions of the major system elements are provided in the following sections.

10.2.1 Tuner

The basic tuner, illustrated in Figure 10.2, receives the 6 MHz signal (UHF or VHF) from an external antenna [2]. The tuner is a high-side injection double-conversion type with a first IF frequency of 920 MHz. This puts the image frequencies above 1 GHz, making them easy to reject by a fixed front-end filter. This selection of first IF frequency is high enough that the input bandpass filter selectivity prevents the local oscillator (978 to 1723 MHz) from leaking out the tuner front end and interfering with other UHF channels; yet, it is low enough for the second harmonics of UHF channels (470 to 806 MHz) to fall above the first IF bandpass. Harmonics of cable channels could possibly occur in the first IF passband but are not a significant problem because of the relatively flat spectrum (within 10 dB) and small signal levels (–28 dBm or less) used in cable systems.

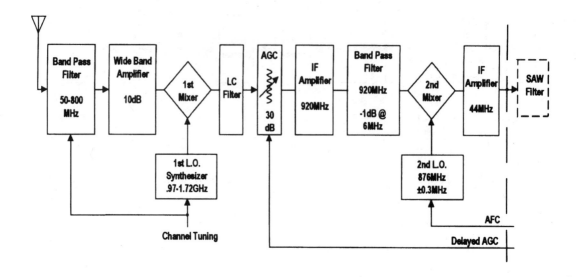

Figure 10.2 Block diagram of the tuner subsystem. (*From* [2]. *Used with permission.*)

The tuner input has a bandpass filter that limits the frequency range to 50 to 810 MHz, rejecting all other nontelevision signals that may fall within the tuner's image frequency range (beyond 920 MHz). In addition, a broadband tracking filter rejects other television signals, especially those much larger in signal power than the desired signal. This tracking filter is not narrow, nor is it critically tuned, as is the case of present-day NTSC tuners that must reject image signals only 90 MHz away from the desired channel. Minimal channel tilt, if any, exists because of this tracking filter.

At 10 dB gain, the wideband RF amplifier increases the signal level into the first mixer, and is the dominant determining factor of receiver noise figure (7 to 9 dB over the entire VHF, UHF, and cable bands). The first mixer, a highly linear double-balanced circuit designed to minimize even-harmonic generation, is driven by a synthesized low-phase-noise *local oscillator* (LO) above the first IF frequency (*high-side injection*). Both the channel tuning (first LO) and broadband tracking filters (input bandpass filter) are controlled by a microprocessor. The system is capable of tuning the entire VHF and UHF broadcast bands, as well as all standard, IRC, and HRC cable bands.

The mixer is followed by an LC filter in tandem with a narrow 920 MHz bandpass ceramic resonator filter. The LC filter provides selectivity against the harmonic and subharmonic spurious responses of the ceramic resonator. The 920 MHz ceramic resonator bandpass filter has a –1 dB bandwidth of about 6 MHz. A 920 MHz IF amplifier is placed between the two filters. Delayed AGC of the first IF signal is applied immediately following the first LC filter. The 30 dB range AGC circuit protects the remaining active stages from large signal overload.

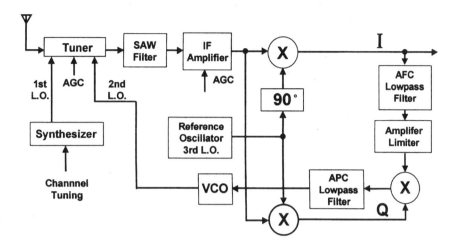

Figure 10.3 Block diagram of the tuner IF FPLL system. (*From* [2]. *Used with permission.*)

The second mixer is driven by the second LO, which is an 876 MHz voltage-controlled *surface acoustic wave* (SAW) oscillator. It is controlled by the *frequency and phase-locked loop* (FPLL) synchronous detector. The second mixer, whose output is the desired 44 MHz second IF frequency, drives a constant-gain 44 MHz amplifier. The output of the tuner feeds the IF SAW filter and synchronous detection circuitry.

10.2.2 Channel Filtering and VSB Carrier Recovery

Carrier recovery is performed on the small pilot by an FPLL circuit, illustrated in Figure 10.3 [2]. The first LO is synthesized by a PLL and controlled by a microprocessor. The third LO is a fixed-reference oscillator. Any frequency drift or deviation from nominal must be compensated in the second LO. Control for the second LO comes from the FPLL synchronous detector, which integrally contains both a frequency loop and a phase-locked loop in one circuit. The frequency loop provides a wide frequency pull-in range of ±100 kHz, while the phase-locked loop has a narrow bandwidth (less than 2 kHz).

During frequency acquisition, the frequency loop uses both the in-phase I and quadrature-phase Q pilot signals. All other data-processing circuits in the receiver use only the I channel signal. Prior to phase-lock, which is the condition after a channel change, the *automatic frequency control* (AFC) low-pass filter acts on the beat signal created by the frequency difference between the VCO and the incoming pilot. The high-frequency data (as well as noise and interference) is mostly rejected by the AFC filter, leaving only the pilot beat frequency. After limiting this pilot beat signal to a constant amplitude (±1) square wave, and using it to multiply the quadrature signal, a traditional bipolar S-curve AFC characteristic is obtained. The polarity of the S-curve error signal depends upon whether the

VCO frequency is above or below the incoming IF signal. Filtered and integrated by the *automatic phase control* (APC) low-pass filter, this dc signal adjusts the tuner's second LO to reduce the frequency difference.

When the frequency difference comes close to zero, the APC loop takes over and phase-locks the incoming IF signal to the third LO. This is a normal phase-locked loop circuit, with the exception that it is biphase-stable. The correct phase-lock polarity is determined by forcing the polarity of the pilot to be equal to the known transmitted positive polarity. Once locked, the detected pilot signal is constant, the limiter output feeding the third multiplier is at a constant +1, and only the phase-locked loop is active (the frequency loop is automatically disabled). The APC low-pass filter is wide enough to reliably allow ±100 kHz frequency pull-in, yet narrow enough to consistently reject strong white noise (including data) and NTSC co-channel interference signals. The PLL has a bandwidth that is sufficiently narrow to reject most of the AM and PM generated by the data, yet is wide enough to track out phase noise on the signal (hence, on the pilot) to about 2 kHz. Tracking out low-frequency phase noise (as well as low-frequency FM components) allows the phase-tracking loop to be more effective.

10.2.3   Segment Sync and Symbol Clock Recovery

The repetitive data segment sync signals (Figure 10.4) are detected from among the synchronously detected random data by a narrow bandwidth filter [2]. From the data segment sync, a properly phased 10.76 MHz symbol clock is created along with a coherent AGC control signal. A block diagram of this circuit is shown in Figure 10.5.

The 10.76 Msymbols/s (684 ÷ 286 × 4,500,000 Hz) I channel composite baseband data signal (sync and data) from the synchronous detector is converted by an A/D device for digital processing. Traditional analog data eyes can be viewed after synchronous detection. After conversion to a digital signal, however, the data eyes cannot be seen because of the sampling process. A PLL is used to derive a clean 10.76 MHz symbol clock for the receiver.

With the PLL free-running, the data segment sync detector—containing a 4-symbol sync correlator—looks for the two level sync signals occurring at the specified repetition rate. The repetitive segment sync is detected while the random data is not, enabling the PLL to lock on the sampled sync from the A/D converter and achieve data symbol clock synchronization. Upon reaching a predefined level of confidence (using a confidence counter) that the segment sync has been found, subsequent receiver loops are enabled.

10.2.4   Noncoherent and Coherent AGC

Prior to carrier and clock synchronization, noncoherent automatic gain control (AGC) is performed whenever any signal (locked or unlocked signal, or noise/interference) overruns the A/D converter [2]. The IF and RF gains are reduced accordingly, with the appropriate AGC *delay* applied.

When data segment sync signals are detected, coherent AGC occurs using the measured segment sync amplitudes. The amplitude of the bipolar sync signals, relative to the discrete levels of the random data, is determined in the transmitter. After the sync signals are detected in the receiver, they are compared with a reference value, with the difference

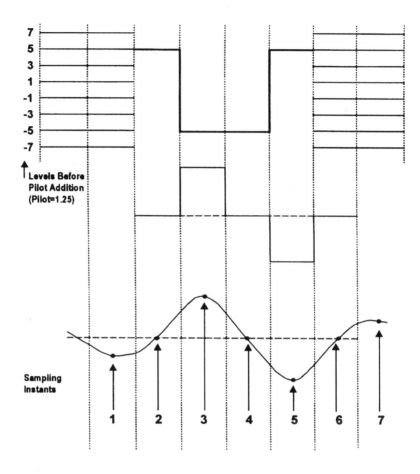

Figure 10.4 Data segment sync waveforms. (*From* [2]. *Used with permission.*)

(error) integrated. The integrator output then controls the IF and *delayed* RF gains, forcing them to whatever values provide the correct sync amplitudes.

10.2.5 Data Field Synchronization

Data field sync detection, shown in Figure 10.6, is achieved by comparing each received data segment from the A/D converter (after interference-rejection filtering to minimize co-channel interference) with ideal field 1 and field 2 reference signals in the receiver [2]. Oversampling of the field sync is not necessary, because a precision data segment and symbol clock already has been reliably created by the clock recovery circuit. Therefore, the field sync recovery circuit knows exactly where a valid field sync correlation should occur within each data segment, and needs only to perform a symbol-by-symbol difference. Upon reach-

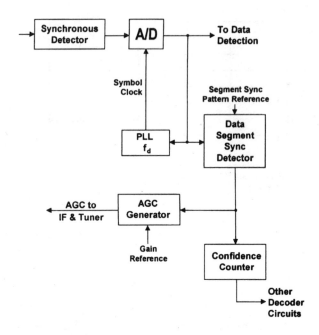

Figure 10.5 Segment sync and symbol clock recovery with AGC. (*From* [2]. *Used with permission.*)

ing a predetermined level of confidence (using a confidence counter) that field sync signals have been detected on given data segments, the data field sync signal becomes available for use by subsequent circuits. The polarity of the middle of the three alternating 63-bit *pseudorandom* (PN) sequences determines whether field 1 or field 2 is detected. This procedure makes field sync detection robust, even in heavy noise, interference, or ghost conditions.

10.2.6 Interference-Rejection Filter

The interference-rejection properties of the VSB transmission system are based on the frequency location of the principal components of the NTSC co-channel interfering signal within the 6 MHz television channel and the periodic nulls of a VSB receiver baseband comb filter [2]. Figure 10.7a shows the location and approximate magnitude of the three principal NTSC components:

- The visual carrier V, located 1.25 MHz from the lower band edge

- Chrominance subcarrier C, located 3.58 MHz higher than the visual carrier frequency

- Aural carrier A, located 4.5 MHz higher than the visual carrier frequency

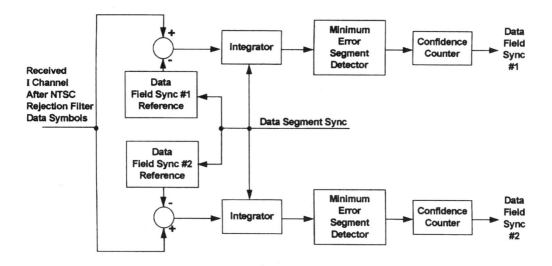

Figure 10.6 The process of data field sync recovery. (*From* [2]. *Used with permission.*)

The NTSC interference-rejection filter (a comb filter) is a 1-tap linear feed-forward device, as shown in Figure 10.8. Figure 10.7*b* illustrates the frequency response of the comb filter, which provides periodic spectral nulls spaced $57 \times f_H$ (10.762 MHz/12, or 896.85 kHz) apart. There are seven nulls within the 6 MHz channel. The NTSC visual carrier frequency falls close to the second null from the lower band edge. The sixth null from the lower band edge is correctly placed for the NTSC chrominance subcarrier, and the seventh null from the lower band edge is near the NTSC aural carrier.

Comparing Figure 10.7*a* and Figure 10.7*b* shows that the visual carrier falls 2.1 kHz below the second comb filter null, the chroma subcarrier falls near the sixth null, and the aural carrier falls 13.6 kHz above the seventh null. The NTSC aural carrier is at least 7 dB below its visual carrier.

The comb filter, while providing rejection of steady-state signals located at the null frequencies, has a finite response time of 12 symbols (1.115 μs). Thus, if the NTSC interfering signal has a sudden step in carrier level (low to high or high to low), one cycle of the zero-beat frequency (offset) between the DTV and NTSC carrier frequencies will pass through the comb filter at an amplitude proportional to the NTSC step size as instantaneous interference. Examples of such steps of NTSC carrier are the leading and trailing edge of sync (40 IRE units). If the *desired to undesired* (D/U) signal power ratio is large enough, data slicing errors will occur. Interleaving will spread the interference, however, and make it easier for the Reed-Solomon code to correct them (RS can correct up to 10 byte errors per segment).

Although the comb filter reduces the NTSC interference, the data also is modified. The seven data eyes (eight levels) are converted to 14 data eyes (15 levels). This conversion is caused by the *partial response process*, which is a special case of intersymbol interference

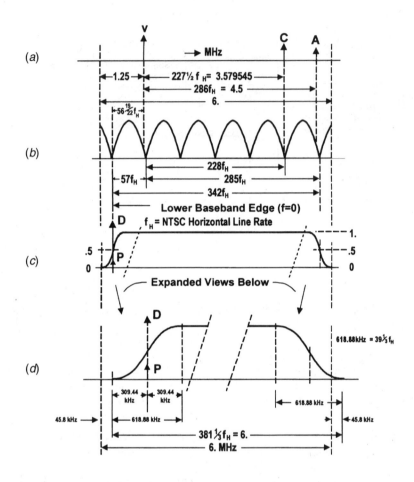

Figure 10.7 Receiver filter characteristics: (*a*) primary NTSC components, (*b*) comb filter response, (*c*) filter band edge detail, (*d*) expanded view of band edge. (*From* [2]. *Used with permission.*)

that does not close the data eye, but creates double the number of eyes of the same magnitude. The modified data signal can be properly decoded by the trellis decoder. Note that, because of time sampling, only the maximum data eye value is seen after A/D conversion.

The detail at the band edges for the overall channel is shown in Figure 10.7*c* and Figure 10.7*d*. Figure 10.7*d* shows that the frequency relationship of $56 \times 19/22 \times f_H$ between the NTSC visual carrier and the ATV carrier requires a shift in the ATV spectrum with respect to the nominal channel. The shift equals +45.8 kHz, or about +0.76 percent. This is slightly higher than currently applied channel offsets and reaches into the upper adjacent channel at a level of about –40 dB. If the upper adjacent channel is another DTV channel, its spectrum is also shifted upward

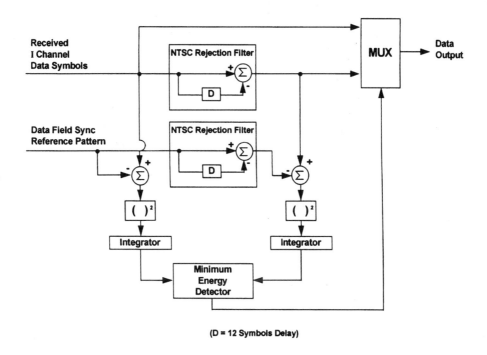

Figure 10.8 Block diagram of the NTSC interference-rejection filter. (*From* [2]. *Used with permission.*)

and, therefore, no spectral overlapping occurs. If it is an NTSC channel, the shift is below the (RF equivalent of the) Nyquist slope of an NTSC receiver where there is high attenuation, and it is slightly above the customary lower adjacent-channel sound trap. No adverse effects of this shift should be expected. An additional shift of the DTV spectrum is used to track the dominant NTSC interferer, which may be assigned an offset of –10, 0, or +10 kHz.

NTSC interference can be detected by the circuit shown in Figure 10.8, where the signal-to-interference plus noise ratio of the binary data field sync is measured at the input and output of the comb filter and compared. This is accomplished by creating two error signals. The first is generated by comparing the received signal with a stored reference of the field sync. The second is generated by comparing the rejection filter output with a combed version of the internally stored reference field sync. The errors are squared and integrated. After a predetermined level of confidence is achieved, the path with the largest signal-to-noise ratio (lowest interference energy) is switched in and out of the system automatically.

It is not advisable to leave the rejection comb filter switched in all the time. The comb filter, while providing needed co-channel interference benefits, degrades white noise performance by 3 dB. This occurs because the filter output is the subtraction of two full-gain paths, and as white noise is uncorrelated from symbol-to-symbol, the noise power doubles. There is an additional 0.3 dB degradation resulting from the 12-symbol differential coding.

If little or no NTSC interference is present, the comb filter is automatically switched out of the data path. When the NTSC service is phased out, the comb filter can be omitted from digital television receivers.

10.2.7 Channel Equalizer

The equalizer/ghost canceller compensates for linear channel distortions, such as tilt and ghosts [2]. These distortions can originate in the transmission channel or result from imperfect components within the receiver.

The equalizer uses a *least-mean-square* (LMS) algorithm and can adapt on the transmitted binary training sequence, as well as on the random data. The LMS algorithm computes how to adjust the filter taps to reduce the error present at the output of the equalizer. The system does this by generating an estimate of the error present in the output signal, which then is used to compute a cross-correlation with various delayed data signals. These correlations correspond to the adjustment that needs to be made for each tap to reduce the error at the output. The equalizer algorithm can achieve equalization in three ways:

- Adapt on the binary training sequence

- Adapt on data symbols throughout the frame when the eyes are open

- Adapt on data when the eyes are closed (*blind equalization*)

The principal difference among these three methods is how the error estimate is generated.

For adapting on the training sequence, the training signal presents a fixed data pattern in the data stream. Because the data pattern is known, the exact error is generated by subtracting the training sequence from the output. The training sequence alone, however, may not be enough to track dynamic ghosts because these conditions require tap adjustments more often than the training sequence is transmitted. Therefore, after equalization is achieved, the equalizer can switch to adapting on data symbols throughout the frame, and it can produce an accurate error estimate by slicing the data with an 8-level slicer and subtracting it from the output signal.

For fast dynamic ghosts such as airplane flutter, it is necessary to use a blind equalization mode to aid in acquisition of the signal. Blind equalization models the multilevel signal as binary data signal plus noise, and the equalizer produces the error estimate by detecting the sign of the output signal and subtracting a (scaled) binary signal from the output to generate the error estimate.

To perform the LMS algorithm, the error estimate (produced using the training sequence, 8-level slicer, or the binary slicer) is multiplied by delayed copies of the signal. The delay depends upon which tap of the filter is being updated. This multiplication produces a cross-correction between the error signal and the data signal. The size of the correlation corresponds to the amplitude of the residual ghost present at the output of the equalizer and indicates how to adjust the tap to reduce the error at the output.

A block diagram of the equalizer is shown in Figure 10.9. The dc bias of the input signal is first removed by subtraction. The dc may be caused by circuit offsets, nonlinearities, or shifts in the pilot caused by ghosts. The dc offset is tracked by measuring the dc value of the training signal.

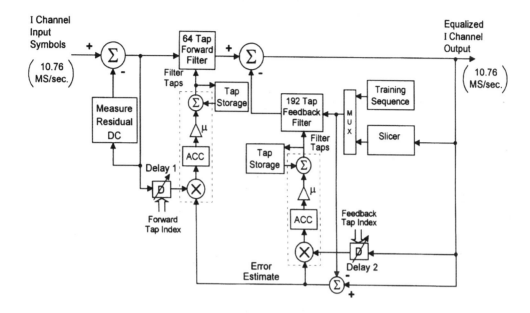

Figure 10.9 Simplified block diagram of the VSB receiver equalizer. (*From* [2]. *Used with permission.*)

The equalizer filter consists of two parts: a 64-tap feed-forward transversal filter followed by a 192-tap decision-feedback filter. The equalizer operates at the 10.762 MHz symbol rate (*T*-sampled equalizer).

The output of the forward filter and feedback filter are summed to produce the output. This output is sliced by either an 8-level slicer (15-level slicer when the comb filter is used) or a binary slicer depending upon whether the data eyes are open. (As pointed out previously, the comb filter does not close the data eyes, but creates twice as many of the same magnitude.) This sliced signal has the training signal and segment sync signals reinserted, as these are fixed patterns of the signal. The resultant signal is fed into the feedback filter and subtracted from the output signal to produce the error estimate. The error estimate is correlated with the input signal (for the forward filter) or with the output signal (for the feedback filter). This correlation is scaled by a step-size parameter and is used to adjust the value of the tap. The delay setting of the adjustable delays is controlled according to the index of the filter tap that is being adjusted.

10.2.8 Phase-Tracking Loop

The phase-tracking loop is an additional decision-feedback loop that further tracks out phase noise that has not been removed by the IF PLL operating on the pilot [2]. Thus, phase

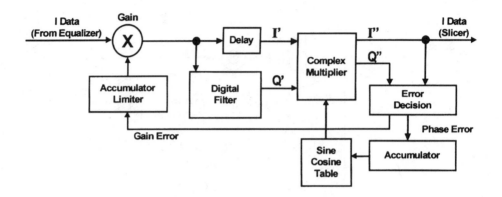

Figure 10.10 The phase-tracking loop system. (*From* [2]. *Used with permission.*)

noise is tracked out by not just one loop, but two concatenated loops. Because the system is already frequency-locked to the pilot by the IF PLL (independent of the data), the phase-tracking loop bandwidth is maximized for phase tracking by using a first-order loop. Higher-order loops, which are needed for frequency tracking, do not perform phase tracking as well as first-order loops. Therefore, they are not used in the VSB system.

A block diagram of the phase-tracking loop is shown in Figure 10.10. The output of the real equalizer operating on the *I* signal is first gain-controlled by a multiplier, then fed into a filter that recreates an approximation of the *Q* signal. This is possible because of the VSB transmission method, where the *I* and *Q* components are related by a filter function that is almost a Hilbert transform. The complexity of this filter is minor because it is a *finite impulse response* (FIR) filter with fixed antisymmetric coefficients and with every other coefficient equal to zero. In addition, many filter coefficients are related by powers of 2, thus simplifying the hardware design.

These *I* and *Q* signals then are fed into a *de-rotator* (complex multiplier), which is used to remove the phase noise. The amount of de-rotation is controlled by decision feedback of the data taken from the output of the de-rotator. Because the phase tracker is operating on the 10.76 Msymbols/s data, the bandwidth of the phase-tracking loop is fairly large, approximately 60 kHz. The gain multiplier also is controlled with decision feedback.

10.2.9 Trellis Decoder

To help protect the trellis decoder against short burst interference, such as impulse noise or NTSC co-channel interference, 12-symbol code intrasegment interleaving is employed in the transmitter [2]. As shown in Figure 10.11, the receiver uses 12 trellis decoders in parallel, where each trellis decoder sees every 12th symbol. This code interleaving has all the same burst noise benefits of a 12 symbol interleaver, but also minimizes the resulting code expansion (and hardware) when the NTSC rejection comb filter is active.

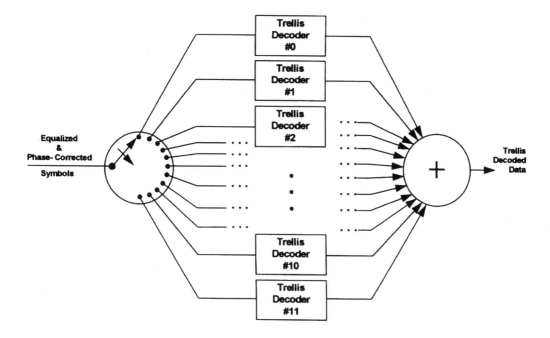

Figure 10.11 Functional diagram of the trellis code de-interleaver. (*From* [2]. *Used with permission.*)

Before the 8-VSB signal can be processed by the trellis decoder, it is necessary to suspend the segment sync. The segment sync is not trellis-encoded at the transmitter. The segment sync suspension system is illustrated in Figure 10.12.

The trellis decoder performs the task of slicing and convolutional decoding. It has two modes: one when the NTSC-rejection filter is used to minimize NTSC co-channel energy, and the other when it is not used. This is illustrated in Figure 10.13. The insertion of the NTSC-rejection filter is determined automatically (before the equalizer), with this information passed to the trellis decoder. When there is little or no NTSC co-channel interference, the NTSC-rejection filter is not used, and an optimal trellis decoder is used to decode the 4-state trellis-encoded data. Serial bits are recreated in the same order in which they were created in the encoder.

In the presence of significant NTSC co-channel interference, when the NTSC-rejection filter (12-symbol, feed-forward subtractive comb) is employed, a trellis decoder optimized for this partial response channel is used. This optimal code requires eight states. This approach is necessary because the NTSC-rejection filter, which has memory, represents another state machine seen at the input of the trellis decoder. To minimize the expansion of trellis states, two measures are taken: first, special design of the trellis code; and second, 12-to-1 interleaving of the trellis encoding. The interleaving, which corresponds exactly to the

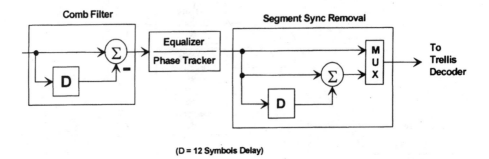

Figure 10.12 Block diagram of the 8-VSB receiver segment sync suspension system. (*From* [2]. *Used with permission.*)

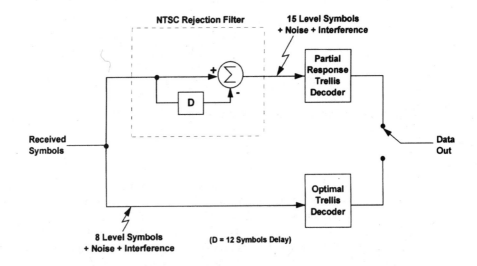

Figure 10.13 Trellis decoding with and without the NTSC rejection filter. (*From* [2]. *Used with permission.*)

12-symbol delay in the NTSC-rejection filter, ensures that each trellis decoder sees only a 1-symbol-delay NTSC-rejection filter. Minimizing the delay stages seen by each trellis decoder also minimizes the expansion of states. Only a 3.5 dB penalty in white noise performance is paid as the price for having good NTSC co-channel performance. The additional 0.5 dB beyond the 3 dB comb filter noise threshold degradation is the result of the 12-symbol differential coding.

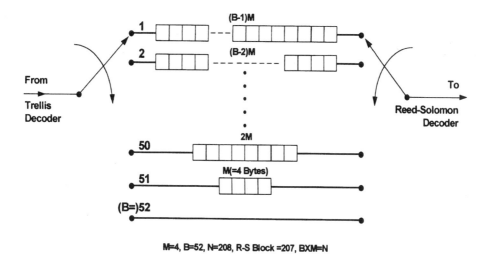

Figure 10.14 Functional diagram of the convolutional de-interleaver. (*From* [2]. *Used with permission.*)

As noted previously, after the transition period, when NTSC is no longer being transmitted, the NTSC-rejection filter and the 8-state trellis decoder can be eliminated from digital television receivers.

10.2.10 Data De-Interleaver

The convolutional de-interleaver performs the exact inverse function of the transmitter convolutional interleaver [2]. The 1/6-data-field depth and intersegment *dispersion* properties allow noise bursts lasting approximately 193 µs to be handled. Even strong NTSC co-channel signals, passing through the NTSC-rejection filter and creating short bursts due to NTSC vertical edges, are handled reliably because of the interleaving and RS coding process. The de-interleaver uses data field sync for synchronizing to the first data byte of the data field. The convolutional de-interleaver is shown in Figure 10.14.

10.2.11 Other Receiver Functional Blocks

The trellis-decoded byte data is sent to the Reed-Solomon decoder, where it uses the 20 parity bytes to perform the byte error correction on a segment-by-segment basis [2]. Up to 10 byte errors per data segment are corrected by the RS decoder. Burst errors created by impulse noise, NTSC co-channel interference, or trellis-decoding errors are greatly reduced by the combination of the interleaving and RS error correction.

As detailed in Chapter 9, the data is randomized at the transmitter by a pseudorandom sequence (PRS). The derandomizer at the receiver accepts the error-corrected data bytes from the RS decoder and applies the same PRS randomizing code to the data. The PRS code is generated identically as in the transmitter, using the same PRS generator feedback and output taps. Because the PRS is locked to the reliably recovered data field sync (and not some code word embedded within the potentially noisy data), it is exactly synchronized with the data, and it performs reliably.

The receiver incorporates a *universal reset* feature that initiates a number of "confidence counters" and "confidence flags" involved in the lockup process. A universal reset occurs, for example, when another station is being tuned or the receiver is being turned on. The various loops within the VSB receiver acquire and lock up sequentially, with "earlier" loops being independent from "later" loops. The order of loop acquisition is as follows:

- Tuner first LO synthesizer acquisition

- Noncoherent AGC reduces unlocked signal to within the A/D range

- Carrier acquisition (FPLL)

- Data segment sync and clock acquisition

- Coherent AGC of signal (IF and RF gains properly set)

- Data field sync acquisition

- NTSC-rejection filter insertion decision made

- Equalizer completes tap adjustment algorithm

- Trellis and RS data decoding begin

Most of the loops mentioned here have confidence counters associated with them to ensure proper operation, but the build-up or letdown of confidence is not designed to be equal. The confidence counters build confidence quickly for quick acquisition times, but lose confidence slowly to maintain operation in noisy environments.

High-Data-Rate Mode

The VSB digital transmission system provides the basis for a family of DTV receivers suitable for receiving data transmissions from a variety of media [2]. This family shares the same pilot, symbol rate, data frame structure, interleaving, Reed-Solomon coding, and synchronization pulses. The VSB system offers two modes: a simulcast terrestrial broadcast mode and a high-data-rate mode.

Most elements of the high-data-rate mode VSB system are similar or identical to the terrestrial system. A pilot, data segment sync, and data field sync all are used to provide robust operation. The pilot in the high-data-rate mode also adds 0.3 dB to the data power. The symbol, segment, and field signals and rates are all the same, allowing either receiver to lock up on the other's transmitted signal. Also, the data frame definitions are identical. The primary difference is the number of transmitted levels (8 vs. 16) and the use of trellis coding and NTSC interference-rejection filtering in the terrestrial system.

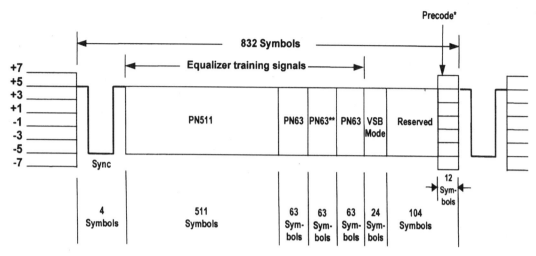

Figure 10.15 Data field sync waveform. (*From* [2]. *Used with permission.*)

The high-data-rate mode receiver is identical to the VSB terrestrial receiver, except that the trellis decoder is replaced by a slicer, which translates the multilevel symbols into data. Instead of an 8-level slicer, a 16-level slicer is used.

10.2.12 Receiver Equalization Issues

The VSB signal contains features that allow the design of receivers that reliably perform the functions of acquiring and locking onto the transmitted signal [2]. The equalization of the signal for channel frequency response and ghosts is facilitated by the inclusion of specific features in the data field sync, as illustrated in Figure 10.15. Utilization of these features is made more reliable by the inclusion of means to first acquire and synchronize to the VSB signal, particularly by the segment sync. The data field sync can then be used both to identify itself and to further perform equalization of linear transmission distortions. The VSB signal also may be equalized by data-based or *blind equalization* methods that do not use the data field sync.

Data field sync is a unique type of data segment in the VSB signal. All payload data in the VSB signal is contained in data segments, which are processed with data interleaving, Reed-Solomon error coding, and trellis coding. The data field sync (and the segment sync portion of every data segment), however, is not processed this way because its purpose is to provide direct measurement and compensation for transmission channel linear distortion.

Equalizer training signals consisting of pseudonoise sequences are major parts of the data field sync.

Equalizer Performance Using Training Signals

Theoretically, in a noise-free signal, ghosts of amplitude up to 0 dB with respect to the largest signal—and within a total window of 63 symbols—can be exactly canceled in one pass using the information in the sequence [2]. Ghosts of any delay can be canceled using the information in the sequence, with a single-pass accuracy of approximately –27 dB, which improves by averaging over multiple passes. The operation of a complete receiver system with 0 dB ghosts may not be achievable because of the failure of carrier acquisition, but operation of the equalizer itself at 0 dB is demonstrable under test by supplying an external carrier.

The number of ghosts to be canceled has, in itself, little effect on the theoretical performance of the system. Theoretical limits to cancellation depend on the amount of noise gain that occurs in compensating the frequency response that results from the particular ghosted signal.

Because the equalizer training signals recur with a period of approximately 24 ms, the receiver cannot perform equalization updates at a faster rate. Therefore, the signal provides information such that the equalization system, theoretically, can have a bandwidth of 20 Hz. Additional constraints are implied by the desire to average out the correlation between PN sequences.

The speed of convergence is not the only important criterion of performance, however. Ultimate accuracy and response to noise also are of importance. Equalization techniques generally proceed by successive approximation to the desired state, and therefore exhibit a convergence time measured as a number of frame periods. If the information in the signal is used in different ways, the speed of operation can be traded off for accuracy and noise immunity. The success of such a tradeoff may depend on nonlinear techniques, such as switching between a quick acquisition mode and slower refinement mode or using varying step sizes in a *steepest-descent* technique.

Receiver Implementation Using Blind Equalization

Blind equalization techniques are not based on a training signal reference, but they may be appropriate for use with the VSB transmission system [2]. Blind equalization techniques are particularly useful when the channel impairments vary more rapidly than the transmission of the training waveform.

As in many modern digital data communication systems, an adaptive equalizer is used in the ATSC system to compensate for changing conditions on the broadcast channel. In communication systems that use an adaptive equalizer, it is necessary to have a method of adapting the equalizer's filter response to adequately compensate for channel distortions. Several algorithms are available for adapting the filter coefficients. The most widely used is the LMS, or least-mean-square method [3].

When the equalizer is started, the tap weights usually are not set to adequately compensate for the channel distortions. To force initial convergence of the equalizer coefficients, a known training signal (that is, a signal known by both the transmitter and receiver) is used as the reference signal. The error signal is formed by subtracting a locally generated copy of the training signal from the output of the adaptive equalizer. When using the training signal, the eye diagram is typically closed. The training signal serves to open the eye. After adaptation with the training signal, the eye has opened, and the equalizer may be switched to a decision-directed mode of operation. The decision-directed mode uses the symbol values at the output of the decision device instead of the training signal.

A problem arises in this scenario when a training signal is not available. In this case, a method of acquiring initial convergence of the equalizer taps and forcing the eye open is necessary. Blind equalization has been studied extensively for QAM systems. Several methods typically are employed: the *constant modulus algorithm* (CMA) and the *reduced constellation algorithm* (RCA) are among the most popular [4, 5]. For VSB systems, however, neither of these methods are directly applicable. CMA relies on the fact that, at the decision instants, the modulus of the detected data symbols should lie on one of several circles of varying diameters. Thus, it inherently relies on the underlying signal to be 2-dimensional. Because VSB is essentially a 1-dimensional signal (at least for the data-carrying portion), CMA is not directly applicable. RCA, on the other hand, relies on forming *super constellations* within the main constellation. The data signal is forced to fit into a super constellation, then the super constellations are subdivided to include the entire constellation. Again, as typically used, RCA implies a 2-dimensional constellation.

Blind equalization can be performed on the VSB constellation, however, using a *modified reduced constellation algorithm* (MRCA). The key part of this modification is to realize the existence of a 1-dimensional version of the RCA algorithm that is appropriate for VSB signals. The MRCA consists of an algorithm to determine appropriate decision regions for a VSB decision device, so as to generate decisions that allow an adaptive equalizer to converge without the use of a training signal.

In VSB systems, the decision regions typically span one data symbol of the full constellation, and the upper and lower boundaries of each decision region are set midway between the constellation points. If these decision regions are used for initial convergence of the equalizer, the equalizer will not converge because, as a result of the presence of intersymbol interference, a significant number of the decisions from the decision device will be incorrect.

To force more correct decisions to be made, an algorithm for determining new upper and lower decision region boundaries was designed. The algorithm clusters the full VSB constellation into several sets, determines upper and lower boundaries for decision regions, and determines appropriate decision device output *symbol* values. These first sets are further divided into smaller sets until each set of symbols contains exactly one symbol, and the decision regions correspond to the standard decision regions for VSB described previously. The function of each stage is to allow for more decisions to be correct, thereby driving the equalizer toward convergence. In this way, each stage in the blind equalization process serves to further open the eye.

In general, the MRCA algorithm consists of clustering the decision regions of the decision device into finer portions of the VSB constellation. The method starts with a binary (2-level) slicer, then switches to a 4-level slicer, then an 8-level slicer, and so on. It should be noted that the MRCA algorithm is applicable to both linear equalization and decision-feedback equalization.

10.2.13 Receiver Performance in the Field

Shortly after the initiation of DTV service, receiver manufacturers began studying the performance of their first-generation designs. One widely-quoted study was an ambitious project undertaken by Zenith [6]. The objectives of the DTV field tests included the following:

- To build a large database of DTV field measurements to refine the propagation models used for DTV field strength prediction, and ultimately to develop a first order prediction of DTV service.

- Gathering performance data on the DTV RF transmission system in a variety of terrain, as well as urban, suburban, and rural environments throughout the U.S.

- Analysis of DTV service availability and system performance in all three FCC-allocated frequency bands (low VHF, high VHF, and UHF).

In the initial phase, twelve field tests had been completed—a total of 2,682 outdoor sites in 9 U.S. cities and 242 indoor sites in 5 cities. All but two of the tests were conducted on UHF channels. Each DTV field test followed a standardized data gathering methodology (described in [6]). Test results indicated that DTV service availability would approach NTSC replication in most instances. Taller transmitter antenna installations were found to have a greater beneficial affect on signal reception than greater transmitter power.

Table 10.3 contains target DTV receiver parameters for reasonable performance in the field (as identified in the field study). These values served as a starting point for receiver designers. As more field test data became available, these target values were expected to be updated.

Extensive laboratory tests were conducted by the Advanced Television Test Center [7] that compared production line receivers under various conditions. Although these tests were not intended to represent either actual reception conditions or a specification for receiver design, they did indicate that the performance of DTV receivers could be improved beyond the Grand Alliance "blue rack" reference receiver. However, the results of these tests also demonstrated the need to further improve the performance of DTV receivers beyond what was current available.

Canadian DTV Receiver Tests

The Communications Research Center of Canada (CRC) carried out laboratory and field tests to evaluate the performance of ATSC 8-VSB receivers against channel impairments such as noise, multipath, and other distortions. A number of receivers were tested in the laboratory; among them were consumer models available on the market, prototypes, and professional models. The summary report presented at the International Broadcasting Convention in September 2000 cautioned that the results were valid only for the particular units used for the tests; no attempt was made to establish if the same results would be obtained with another receiver of the same model [8].

All receivers were tested against random noise, and random noise in the presence of static multipath ensembles. As a simplified version of the multipath test, a *single echo* test was also conducted to verify the delay range of the receiver equalizer, and to observe the

Table 10.3 Target DTV Receiver Parameters (*After* [6].)

DTV Receiver Parameter	Target Value
Dynamic range: minimum level	< –80 dBm
Dynamic range: maximum level	> 0 dBm
Noise figure	< 10 dB
Synchronization white noise (lock) limit	< 3 dB
AGC speed (10 dB peak/valley fade)	> 75 Hz
White noise threshold of errors	< 15.5 dB
Phase noise threshold of errors	> –76 dBc/Hz @ 20 kHz offset
Gated white noise burst duration	> 185 μs
Co-channel N/D interference @ –45 dBm	< 3 dB, D/U
Co-channel D/D interference @ –45 dBm	< 15.5 dB, D/U
First adjacent N/D interference @ –60 dBm	< –40 dB, D/U
First Adjacent D/D Interference @ –60 dBm	< –28 dB, D/U
Inband CW Interference @ –45 dBm	< 10 dB, D/U
Out-of-band CW interference @ –45 dBm	< –40 dB, D/U
Equalizer length (pre-ghost)	< –3 μs
Equalizer length (post-ghost)	> +22 μs
Quasi-static multipath (1 μs @ < 0.2 Hz)	< +4 dB, D/U
Dynamic multipath: (1 μs @ < 5 Hz)	< +7 dB, D/U
Dynamic multipath: (1 μs @ <10 Hz)	< +10 dB, D/U

sensitivity of the receiver to the phase of an echo. Specific tests conducted as a part of the program included:

- **Susceptibility to random noise**. Measurements were intended to determine the performance of the receivers under random noise impairment. The effect of random noise on the ATSC 8-VSB receiver was evaluated subjectively at the threshold of visibility (TOV) by observing pictures on the monitor. The average power of the DTV signal was adjusted to three different levels: strong (–28 dBm), moderate (–53 dBm), and weak (–68 dBm), as recommended by FCC/ACATS [9]. For each power level of the DTV signal, random noise was added and increased until TOV was reached. The ratio between the average power of the DTV signal and the random noise (C/N) was measured; the results are summarized in Table 10.4. The value measured by the ATTC for the Grand Alliance prototype receiver [10] is included in the table for comparison purposes.

- **Multipath test**. The multipath measurements were intended to determine the robustness of ATSC 8-VSB receivers against random noise in the presence of static multipath (an ensemble of five echoes). TOV was determined for the combinations of multipath used by ATTC/ACATS [9]. Normally, the main (or desired) signal would be adjusted to the

Table 10.4 Susceptibility to Random Noise Test Results (*After* [8])

Receiver	C	E	F	G	H	Blue Rack
Desired Signal Level	C/N (dB) at TOV	C/N (dB) at TOV	C/N (dB) at TOV	C/N (dB) at TOV	C/N (dB) at TOV	C/N (dB) at TOV
Strong: –28 dBm	16.8	15.3	15.4	15.2	16.0	15.28
Moderate: –53 dBm	17.2	15.3	15.3	15.2	16.0	N/A
Weak: –68 dBm	17.8	15.3	15.8	15.0	16.7	N/A

Table 10.5 Susceptibility to Random Noise in the Presence of Static Multipath (*After* [8])

Static Multipath Ensemble	Degradation Relative to No Ghost at TOV (dB) for the Following Receiver					
	C	E	F	G	H	Blue Rack
No Ghost	18.6	16.0	18.0	15.3	16.6	15.16
A *	Not working	Δ 1.8	Δ 3.3	Δ 0.5	Δ 2.3	Δ 3.28
B *	Not working	Δ 2.2	Δ 3.8	Δ 0.9	Δ 3.5	Δ 2.40
C *	Not working	Δ 1.3	Δ 2.1	Δ 0.0	Δ 2.5	Δ 2.18
D *	Not working	Δ 1.9	Δ 3.1	Δ 0.2	Δ 4.7	Δ 2.89
E *	Not working	Δ 2.0	Δ 3.8	Δ 1.1	Δ 2.6	Δ 3.64
F *	Not working	Δ 1.2	Δ 2.7	Δ 0.9	Δ 2.5	Δ 1.20
G *	Not working	Δ 0.1	Δ 0.1	Δ 0.0	Δ 0.1	Δ 1.68
* Combinations of multipath used by ATTC/ACATS [9]						

strong signal level (–28 dBm), but because of the poor C/N ratio and insertion loss of the channel simulator, the main or desired signal level was adjusted to –43 dBm.

The reference carrier-to-noise ratio (C/N) with no ghost was measured by increasing the level of the random noise until TOV was reached. The degradation in C/N was then measured the same way for each multipath ensemble.

The phase of each echo was selected arbitrarily for each ensemble and the same values were used for each receiver tested. The C/N measurements relative to the no ghost cases for the receivers tested are presented in Table 10.5. Receivers E, F, and G obtained better results than the receiver H and the blue-rack. Receiver C could not work without any errors, with or without random noise for any combination of multipath.

- **Single echo without random noise**. A single echo was added to the main signal. The range of operation of all the equalizers under test was covered with an echo delay range of –10 to 24 μs, except for receiver G with a range that extended to 44 μs. The step size was 1 μs. For every delay, the relative attenuation of the echo was adjusted until TOV was reached.

The phase of the echo was varied from 0 degree to 360 degrees at a rotation speed of 0.1 Hz (Doppler). The goal of this test was to quickly determine how the phase of the echo is critical for each receiver or the worst case. With a phase variation limited to 0.1 Hz, the receivers were not sensitive to the Doppler effect and were affected only by the instant phase of the echo. Receivers E, F, and G performed better than receiver C.

- **Single echo with random noise**. This test was designed to verify the susceptibility of the receiver to random noise in the presence of a static single echo. The main difference from the previous test was that both the relative attenuation and phase of the single echo were kept fixed for the complete range of echo delays, and only the level of the random noise was increased until TOV was reached.

The following broad conclusions were drawn by the authors:

- In the random noise test, the difference between the performance of the worst and the best receivers (C and G respectively) is approximately 2.0 dB at the moderate signal level. For the multipath ensemble, the difference in C/N degradation, between receiver H and receiver G is approximately 0.1 to 4.5 dB. The older receiver C did not work without any errors for this test.

- In the single echo without random noise test, receiver G did work with an echo from 2 to 7 dB higher than receiver C. This is for the worst case (phase rotation) over the delay range.

- In the test with single echo in the presence of random noise, receivers F and G did work properly over most of the delay range for an echo of –4 dB, and this is 4 dB better than receiver E and at least 6 dB better than receiver C.

Further tests were conducted with these receivers for outdoor and indoor reception to determine to what extent the differences measured in the laboratory could be translated into variations in field performance.

The results of the indoor DTV reception measurements were reported at the IEEE Broadcast Symposium in September 2000 [11]. The test program focused on 46 residential homes in the Ottawa metropolitan area. The goal of the measurement program was to gain a better understanding of the following issues:

- The location availability of DTV using simple off-the-shelf indoor antennas

- The types of antennas (active or passive) needed for indoor DTV reception

- The role that low field strength and/or multipath play in DTV reception problems

The tests were conducted using an NTSC signal on channel 65 and a DTV signal on channel 67. Initially, two different DTV receivers were used for the field tests; however, because one receiver consistently outperformed the other, after the first month, the tests were performed with only the better receiver.

The following types of indoor antennas were used at each site:

- An active loop antenna with a gain of 32 dB in the UHF band and another active antenna with a gain of about 30 dB

- A passive log-periodic antenna with a gain of 5 to 7.5 dB across the UHF band

The antennas were mounted on a tripod 5 ft. above the floor, close to a window. They were then oriented for best DTV reception (that is, highest carrier-to-noise ratio and maximum spectrum flatness on channel 67).

The engineers recorded the DTV and NTSC field strengths at each measurement location and stored a "snapshot" of the DTV signal pass-band using a PC controlled spectrum analyzer. DTV reception at each site was characterized as being either:

- *Reliable* (not sensitive to people moving across the room)

- *Sensitive* to the movement of people across the room

- *Completely unreliable* (frozen and/or broken picture and concomitant loss of audio)

If the TV reception was found to be reliable, then, while the technicians were moving around in the proximity of the antenna (to generate dynamic ghosts), the DTV signal was attenuated until the first visible impairments were observed. This level of attenuation was recorded as the margin relative to the threshold of visibility (TOV). Next, the DTV signal was attenuated further until the *point of unusability* (POU) was determined. NTSC reception was also evaluated using the ITU-R picture impairment scale at each site.

The principal results and conclusions of the field tests were:

- Three sites were rejected because of technical problems.

- DTV signals could reliably be received with both types of antennas (active and passive) in 21 of the 43 sites (49 percent), and with one or the other antenna types in 27 sites (63 percent). For the former sites (21/43), NTSC pictures received on channel 65 were rated between 1.5 and 4.5 with a mean of 3.4 (slightly annoying impairments).

- DTV reception failed using the active antenna at 13/43 sites (30 percent) where the NTSC pictures were rated between 0.5 and 2 with a mean of 0.8 (very annoying impairments). A combination of low field strength and multipath was the most common cause of these failures at 11 of the 13 failed sites (85 percent).

- DTV reception failed with the passive antenna at 17/43 sites (40 percent) where the NTSC pictures were rated between 0.5 and 3 with a mean of 1.3 (very annoying impairments). A combination of low field strength and multipath was again the most common cause of these failures, being present at 13 of these 17 sites (76 percent). Only in very few cases (3/17 or 18 percent) did low field strength alone cause such failures. On the other hand, the DTV failure could be explained by excessive multipath alone in only one case (6 percent) where the corresponding NTSC picture was rated at 3 (slightly annoying impairments).

Based on these figures and a comparison of the median margins to TOV and POU for each type of antenna, the CRC concluded that the benefits of using an active indoor antenna—chiefly boosting the received signal strength—outweigh the potentially negative effects of increased levels of intermodulation distortion contained in the amplifier.

10.2.14 Compatibility Standards

As second generation DTV receivers were making their way to retail outlets, the Consumer Electronics Association (CES) released a set of technical specifications regarding compatibility requirements for DTV receivers and digital cable systems. The document defined the minimum requirements that must be met by digital cable TV systems and digital TV receivers so that the receivers and the cable systems will interoperate to support the following baseline services:

- NTSC analog television signals.

- Digital television programs.

- Signals utilizing a *point-of-deployment* (POD) security module, supplied by the cable TV system operator, to decode scrambled digital television programs that can be authorized by one-way downstream data transmission to the POD module. These services include subscription television programs and pay-per-view programs that are separately ordered by telephone.

- Event information data equivalent to that provided by terrestrial broadcasters to support certain navigation function in the receiver.

There are two fundamental parts to the specification:

- **Part A**—defines the minimum requirements that must be met by the cable TV system

- **Part B**—defines the minimum requirements that must be met by the TV receiver or other cable-compatible consumer device

These two parts go together and must be coordinated. However, each element has been written so that they can be considered as separate specifications.

The CEA requirements are minimum requirements and are not intended to constrain either the cable system operator or the consumer electronics manufacturer from offering and supporting additional services and/or features.

The document contains interface specifications for a number of parameters, including:

- The physical interface, which defines the RF interconnection—including the signal levels, modulation types, and channel plans supported.

- The video formats supported, which includes all of the ATSC Table 3 formats and adds 1440×1080, 720×480, 544×480, 528×480, and 352×480.

- How to support System Information for digital transport streams and Emergency Alert System messaging.

The specifications also detail the use of POD modules. A POD module is a replaceable security module supplied to the consumer by the cable TV system operator that can be plugged into a cable-compatible digital television receiver or other cable-compatible digital consumer device through a standardized POD interface to provide access to scrambled services. This feature is significant because it allows DTV receivers to be connected directly to a cable system without the use of a set top box.

10.2.15 Digital Receiver Advancements

Experience with first-generation consumer sets demonstrated that improvement were needed to provide reliable service to indoor locations where consumers are currently enjoying NTSC [12]. The DTV reception problems were believed to be the result of the inability of the receiver to decode signals in the presence of long-delay static and dynamic multipath conditions. To combat this problem, a number of second-generation improvements were announced by receiver manufacturers. Examples included the following:

- The Philips TDA8961 demodulator chip with feed-forward adaptive equalization to deal with ghosts from –2.3 to +22.5 µs, with a capability to reach +80 µs using external software. The equalization could use either the 8-VSB training signal, or the data itself (*blind equalization*), to handle complex urban multipath situations. The device also featured low insertion loss NTSC co-channel interference filtering.

- The Samsung KS 1402 decoder chip, which eliminated dynamic multipath impairment in the range of –5.1 to +18.6 µs with a 56-tap feed-forward filter. It was capable of decoding both 8- and 16-VSB, with automatic detection and filtering of an interfering NTSC signal.

- The NxtWave (a spin-off of the David Sarnoff Research Center) NXT2000 receiver subsystem. The ATSC-compliant device performed in either the 8-VSB mode for terrestrial broadcasting or in 64-QAM, 256-QAM or 16-VSB modes for cable. The NXT2000 was designed to cancel transmission channel impairments such as static and dynamic multipath, phase noise, adjacent or co-channel NTSC interference, and impulse noise. The NXT2000 and the complementary proprietary equalization scheme provided demodulation even when the VSB pilot was destroyed because of severe channel conditions. The equalizer range extended from –4.5 to +44.5 µs. Other features included an integrated FEC and a programmable symbol rate for the different VSB/QAM modes.

- The Motorola MCT2100 8-VSB decoder chip, the result of a partnership between Motorola and Sarnoff. Motorola reported that the MCT2100 completely eliminated multipath issues from the equation in home and pedestrian portable reception. The MCT2100 corrected static multipath with delays of up to 41 µs. It accomplished this by incorporating a full equalizer capable of blind equalization.

From these second-generation efforts, progress in receiver design continued to be made, with most attention given to improved chip sets that form the basis of the major receiver subsystems.

10.3 HDTV Display Considerations

The cathode-ray tube (CRT) has remained the primary display device for television since electronic television was developed in the 1930s [13]. It survived the conversion from monochrome to color television, but it may not survive the cessation of analog television broadcasting (at least in its present form). The CRT is fundamentally a 3-dimensional structure and, as such, is limited in the size of image available on direct-view tubes. For HDTV, the viewer cannot resolve the available detail unless viewing is at or less than 3 picture heights. Numerous studies have demonstrated that HDTV requires large screens for family or group viewing for the definition of images to be apparent. Viewers farther than 3 picture heights from the screen will see noise-free, interference-free pictures whose color rendition is sharper and more faithful than anything they have known with NTSC. Failing the availability of substantially larger screens than the CRT can reasonably provide, however, the data capacity of the digital channel will not be maximized. Although projection displays can provide extremely large images, they too are 3-dimensional boxes, which in many homes are simply unacceptably large.

It must be acknowledged, however, that the CRT is exceptionally flexible and usable over a significant range of scanning standards, including HDTV, NTSC, and any number of computer formats. This flexibility results because such displays are analog-addressed, with scanning by means of sawtooth currents flowing in their deflection yokes. Most display technology development at this writing has focused on 2-dimensional flat-panel displays, which are digitally addressed and, as a consequence, must be designed and constructed to operate on one specific number of lines per picture height and at a given number of pixels per line.

It is undeniable that great progress has been made in solid state displays of various designs over the past few years. Solid-state flat panel display devices are nearing a cost-competitive stance for computer applications utilizing small screen sizes. The ultimate goal of flat panel displays, however, remains elusive: a display flat and light enough that it can be hung on the wall like a picture, and reasonably price. While promising new products continue to be developed with each passing year, the hang-it-on-the-wall display is still (at this writing) perhaps five years away. Having said that, it is only fair to point out that such devices have been about five years away for the past thirty years.

10.3.1　Color Space Issues in Digital Video

If all TV programming were generated using only a video camera, the issues of color space and the resulting "legality" of signals would essentially disappear. Generally speaking, a set of component signals is considered *legal* if each element is contained within the specified voltage range of the format. Although this concept is rather simple, the execution can lead to problems. For example, a legal Y, $B - Y$, $R - Y$ set of signals may result in illegal R, G, B transcoded signals, as well as illegal NTSC or PAL composite signals. A more complete explanation, then, would be: A set of legal component analog signals is considered valid if it results in legal signals in the format into which it is transcoded.

Television-camera-originated signals are always valid because the camera generates R, G, B signals in the first place. These signals then may be subsequently transcoded into legal Y, $B - Y$, and $R - Y$ signals and NTSC or PAL composite signals. However, some video equipment, such as graphics systems or character generators, are capable of producing invalid signals even though they are legal in their original format. Test signal generators also are capable of generating legal but invalid signals.

The root of these concerns lies in the basic colorimetry of television.

Video Colorimetry

A video display may be regarded as a series of small visual colorimeters. In each picture element, a *colorimetric match* is made to an element of the original scene [14]. The primaries are the red, green, and blue phosphors. The mixing occurs inside the eye of the observer because the eye cannot resolve the individual phosphor dots; they are too closely spaced. The outputs (R, G, B) of the three phosphors may be regarded as *tristimulus values*. The coefficients in the related equations depend on the chromaticity coordinates of the phosphors and on the luminous outputs of each phosphor for a given unit of electrical input. Usually, the gains of each of the three channels are set so that equal electrical inputs to the three produce a standard *display white*, such as CIE illuminant $D65$.

For each picture element, then, a video camera must produce three electrical signals that are representative of the three tristimulus values (R, G, B) of the required display. To accomplish this, the system must have three optical channels with spectral sensitivities equal to the color-matching functions $\bar{r}(\lambda)$, $\bar{g}(\lambda)$, and $\bar{b}(\lambda)$, corresponding to the three primaries of the display.

Consequently, the information to be conveyed by the electronic circuits comprising the camera, transmission system, and receiver/display is the amount of each of the three primaries (phosphors) required to match the input color. This information is based on the following:

- An agreement concerning the chromaticities of the three primaries to be used

- The representation of the amounts of these three primaries by electrical signals suitably related to them

- The specification (typically) that the electrical signals shall be equal at some specified chromaticity

Given these conditions, the signal voltages are then representative of the tristimulus values of the original scene. They obey all the laws to which tristimulus values conform, including the property of being transformable to represent the amounts of primaries of chromaticities other than those for which the signals originally were composed. Such transformations can be arranged by forming three sets of linear combinations of the original signals.

Unfortunately, the simple objective of producing an exact colorimetric match between each picture element in the display and the corresponding element of the original scene is difficult to fulfill and, in any case, may not further the ultimate objective of equality of appearance between the display and the original scene. There are several reasons for this, such as:

- It may be difficult to achieve the luminance of the original scene because of the limitation of the maximum luminance that can be generated by the reproducing system.

- The adaptation of the eye may be different for the reproduction than it is for the original scene because the surrounding conditions are different.

- Ambient light complicates viewing the reproduced picture and changes its effective contrast ratio.

• The angle subtended by the reproduced picture may be different from that of the original scene.

Although it is an oversimplification, adequate reproduction often is considered to be achieved when the chromaticity is accurately reproduced and the luminance is reproduced *proportionally* to the luminance of the original scene. Even though the adequacy of this approach is somewhat questionable, it provides a starting point for system designers and enables the establishment of targets for system performance.

Gamma

So far in this discussion, a linear relationship has been assumed between corresponding electrical and optical quantities in both the camera and the receiver. In practice, the *transfer function* is normally not linear. For example, over the useful operating range of a typical color receiver, the light output of each phosphor follows a power-law relationship to the video voltage applied to the grid or cathode of the CRT. The light output L is proportional to the video-driving voltage E_v raised to the power γ:

$$L = K E_v{}^{\gamma} \tag{10.1}$$

Where γ is typically about 2.5 for a color CRT. This produces black compression and white expansion. Compensation for these three nonlinear transfer functions is accomplished by three electronic *gamma correctors* in the color camera video-processing amplifiers. Thus, the three signals that are encoded, transmitted, and decoded are not, in fact, R, G, and B; rather, they are R', G', and B', given by:

$$R' = R^{1/\gamma}, G' = G^{1/\gamma}, B' = B^{1/\gamma} \tag{10.2}$$

If the rest of the system is linear, application of these signals to the color picture tube causes light outputs that are linearly related to the R, G, and B tristimulus inputs to the color camera, so the correct reproduction is achieved.

Scene White

When the original scene is illuminated by daylight (of which $D65$ is representative), it is a clearly reasonable aim to reproduce the chromaticities of each object exactly in the final display. Many video images, however, are taken in a studio with incandescent illumination of about 3000 K. In viewing the original scene, the eye adapts to a great extent so that most objects have a similar appearance in both daylight and incandescent light. In particular, whites appear white under both types of illumination. If the chromaticities were to be reproduced exactly, however, studio whites would appear much yellower than the outdoor whites. This is because the viewer's adaptation is controlled more by the ambient viewing illuminant than by the scene illuminant and, therefore, does not correct fully for the change of scene illuminant. Because of this property, exact reproduction of chromaticities is not necessarily a good objective.

The ideal goal, instead, is that the reproduction have exactly the same *appearance* as the original scene, but not enough is known about the chromatic adaptation of the human eye to define what this means in terms of chromaticity. A simpler criterion is to reproduce objects

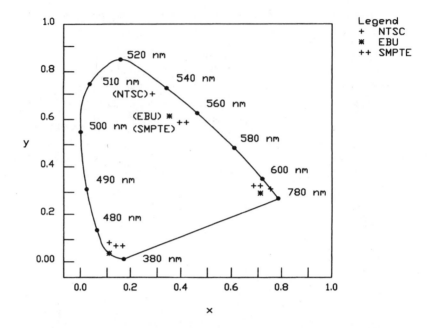

Figure 10.16 The CIE 1931 chromaticity diagram showing three sets of phosphors used in color television displays. (*From* [14]. *Used with permission.*)

with the same chromaticity that they would have if the original scene were illuminated by D65.

The phosphor chromaticities specified by the NTSC in 1953 were based on phosphors in common use for color television displays at that time. Since then, different phosphors have been introduced, mainly to increase the brightness of displays. These modern phosphors, especially the green ones, have different chromaticities so that the gamut of reproducible chromaticities has been reduced. Because of the increased brightness, however, the overall effect on color rendition has been beneficial. Figure 10.16 shows three sets of modern phosphors plotted on the CIE chromaticity diagram.

Video as Data

To process video in a computer requires some form of manipulation of the video signal [15]. Analog composite or component video obviously must be digitized; composite signals also must be decoded into R, G, B or Y, U, V elements. Even component digital video signals may require Y, C_b, C_r to R, G, B color space conversion and/or pixel aspect ratio correction. For compressed-video systems, conversion to an M-JPEG, MPEG, or similar bit stream also must be accomplished. These processes, in all likelihood, will need to be reversed when bringing the video back out of the computer realm, and this is where color space issues often arise.

Inside a computer, video is most likely to be represented by 24-bit R, G, B values. An optional 8-bit alpha channel sometimes is added to facilitate keying functions. The R, G, B analog format utilizes three separate monochrome video circuits to convey the complete signal. Sync and setup are optional on each signal. If sync is not present on any of the R, G, B signals, it must be carried as a fourth signal in the format referred to as R, G, B, S.

Component digital video (ITU-R Rec. 601) is based upon Y, C_b, C_r component signals. It should be noted that Y, C_b, C_r is the correct terminology for this luminance (Y), $R - Y$ (C_r), and $B - Y$ (C_b) component format. These values sometimes are incorrectly referred to as Y, U, V (which also are related to the unscaled and offset signals from which Y, C_b, C_r are derived). The component digital signals are developed from the standard gamma-corrected R', G', B' signals according to the following equations:

$$Y = 0.257R' + 0.504G' + 0.098B' + 16 \tag{10.3}$$

$$C_b = -0.148R' - 0.291G' + 0.439B' + 128 \tag{10.4}$$

$$C_r = 0.439R' - 0.368G' - 0.071B' + 128 \tag{10.5}$$

By definition, the Y signal has a range of 16 to 235, and the C_b/C_r signals have a range of 16 to 240 (assuming 8-bit resolution). This leaves some "digital headroom" for overshoots and undershoots of the video signal. Even so, much of the Y, C_b, C_r color space is outside of the standard R, G, B gamut, as shown in Figure 10.17. This is a result of restricting the R, G, B values to a range of 0 to 255.

To make matters worse, some R, G, B combinations translate into illegal colors when encoded to NTSC composite, primarily the result of excessive chroma levels. For this reason, levels on all signals brought in from Y, C_b, C_r space (for example, from a digital paint system) must be carefully controlled.

A quick examination of the typical video production cycle will reveal that most video work is done by transferring component video in and out of a computer. This can be accomplished in one of three ways:

- Analog component video

- Digital component video

- Data transfer

To feed component digital video into and out of a computer requires a conversion process. Serial digital video (ANSI/SMPTE 259M) most likely will be color-space-converted between Y, C_b, C_r and R, G, B. The levels also may be scaled to match the computer. For example, $Y = 16$ black would scale to $Y = 0$, or $R = G = B = 0$; $Y = 235$ white then becomes $Y = R = G = B = 255$. A related part of this process is that the color bandwidth is expanded from 4:2:2 to the computer equivalent of 4:4:4 sampling. Although these operations are mostly transparent, illegal colors and/or optional filters can cause variations in color ranging from slight to significant. Such problems could make the processed images stand out if they are edited back into the original clip or if the scenes contain elements whose colors are well known and easily recognizable.

Video frames also can be transferred simply as data files, of course, essentially removing the translation issues (or at least moving them downstream).

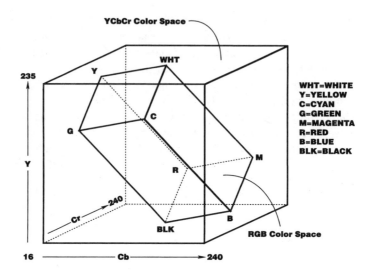

Figure 10.17 Comparison of the *Y*, *C_b*, *C_r* and *R*, *G*, *B* color spaces. Note that about one-half of the *Y*, *C_b*, *C_r* values are outside of the *R*, *G*, *B* gamut. (*From* [15]. *Used with permission.*)

Another Look at Gamut and Color Space

As noted previously, the color gamut in current television systems is limited to that of the CRT phosphors used in today's studio monitors [16]. Current television systems produce very acceptable color rendition, so gamut is clearly not a problem now and, in any event, common display devices are not capable of displaying a larger gamut.

Future television displays and receivers, however, are likely to use new technologies and may be capable of displaying a larger color gamut. Indeed, several new displays already have appeared, and some can produce a larger color gamut than current CRTs. Also, interoperability with other imaging systems, such as film, will benefit from a larger gamut, particularly in DTV service.

The question then becomes: How large a gamut is big enough? One extreme would require that the system be capable of transporting all physically realizable colors. Although that may be attractive from an aesthetic standpoint, the engineering challenges would be significant, especially considering that large parts of the signal space would rarely be used. For future television systems to be capable of capturing and transmitting a larger color gamut (relative to NTSC), they will need to abandon the rigid coupling of the camera-transmission-system colorimetry to the CRT display. Depending on the method used, some new display devices will require color transformations to correct the transmitted signal and make it correct for that specific display technology.

Just such an approach was envisioned in the development of the 1125/60 HDTV system (SMPTE 240M). The SMPTE Working Group on High-Definition Television considered the demand for improved colorimetry to ensure high-quality film/HDTV interchange, to provide enhanced TV display, and to meet the requirements for digital implementation [17].

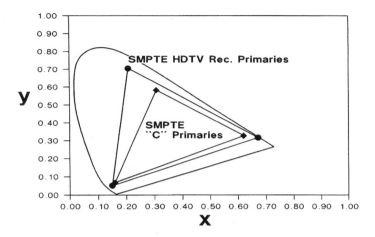

Figure 10.18 Color gamut for SMPTE 240M. (*From* [17]. *Courtesy of Sony.*)

These demands led the working group to specify the camera transfer characteristic with high mathematical precision. The guide was intended to lead HDTV camera designers toward a predictable and unified specification. Linearization of the signal can be performed precisely, thus permitting digital processing on linear signals when required. Colorimetry and gamma are specified precisely throughout the reproduction chain. Figure 10.18 shows the color gamut curve for reference primaries. Note the significant improvement over SMPTE C [18].

10.3.2 Display Technology Trends

Advanced display system design is an area of great technological interest across a broad range of industries. As a result, considerable engineering expertise is being directed toward improved displays of all types, from consumer television to specialized aeronautical applications. Key evaluation metrics for any display include the following:

- Overall luminous efficiency

- Viewability (brightness and contrast)

- Uniformity of reproduction, both large- and small-area

- Gray scale

- Color capability, gamut, and accuracy

- Life expectancy and reliability

- Cost of the display device and supporting circuitry

Important technology trends for the principal display technologies include the following:

- **Cathode Ray Tube.** In a cathode ray tube, a deflected electron beam is used to excite a cathodoluminescent phosphor. In this very mature technology, continued emphasis is being placed on achieving higher resolution, lower cost, sunlight viewability, and longer life, for both direct-view and projection devices. Improved computer modeling will lead to smaller and more intense electron beams. Additional trends include continued emphasis on achieving flatter faceplates and wider deflection angles.

- **Flat CRT.** The flat CRT is similar in principal to the conventional CRT only that it is configured in a flat (or flatter, relative to the conventional CRT) design. This type of device may or may not use a deflection system. A number of different electron sources are used. Deflected beam versions of classic flat CRT designs have allowed low-cost, portable, small-size displays. Large-area multiplexed versions offering full color and gray scale reproduction have also been produced. Improvements in LCD display technologies, however, have eroded what market existed for flat CRT designs in small-sized screens.

- **Liquid Crystal Display.** In a liquid crystal display (LCD), an electric field is applied across a material having both liquid and crystalline properties. This field is used to modulate light by controlling the amplitude, wave vector, or phase vector of the device. LCDs are likely to dominate in low-cost vector graphic applications, particularly if low power consumption and overall physical size are important. For large area information display, the future depends on continued progress of active matrix concepts. The large number of companies pursuing LCDs give this technology a significant advantage over other non-CRT display systems.

- **Plasma/Gas Discharge.** In the plasma/gas discharge display technology, an electric field is applied across a gas atmosphere, which creates an avalanche effect. Photons are emitted when the excited atoms return to the ground state. This technology can be divided logically into two basic configurations: ac and dc-based. For the *ac plasma display panel* (AC PDP), developmental work is concentrating on large, high information content applications, particularly for harsh environments. As color and gray scale performance improves, new application areas will develop. This technology shows considerable promise for HDTV applications. Circuitry and panel costs remain high at this writing, but improvements are likely with volume production. For the *dc plasma display panel* (DC PDP), efforts are primarily related to applications requiring large size, good color rendition and gray scale representation, such as conventional and advanced television. DC plasma displays face stiff competition in moderate size *alphanumeric and graphics* (A/N&G) applications from other display technologies. While DC PDP holds promise for flat panel HDTV display applications, panel complexity—and therefore cost—are greater than for the AC PDP.

- **Electroluminescent Display.** In an electroluminescent (EL) display device, an electric field applied across a polycrystalline phosphor stimulates the material and light energy is subsequently emitted. This technology can be divided into two basic classes; *high field type* (includes *ac powder*, *dc powder*, *ac thin film*, *dc thin film*, and combinations of these schemes, with and without memory), and *low field type* (LEDs—organic and inorganic).

At this writing, *ac thin film* (ACTF) is the most advanced. The future of this technology is highly dependent on the ability of developers to achieve full color, acceptable gray scale, larger display size, and lower cost. It is likely that drivers and decoding logic will be integrated on the display panel.

For each technology, a trade-off must be made between panel complexity and electronics complexity. Typically, technologies with the simplest addressing techniques have the most complex structures, and vice-versa. Technologies requiring high voltage drive (EL, PDP, and most CRTs) use relatively expensive drivers. However, many of these same technologies require fewer drivers for a given panel size. Currently, electronics cost is often viewed as a more significant problem to overcome than panel cost. The cost of drivers alone can be significantly more than the cost of an entire equivalent performance CRT monitor.

Despite significant progress in solid-state display systems, the conventional CRT remains the most common display device. The primary advantages of the CRT over competing technologies include the following:

- Low cost for high information content

- Full color available (greater than 256 colors)

- High resolution, high pixel count displays readily available

- Direct view displays of up to 40-in diagonal practical

- Devices available in high volume

- It is a mature, well-understood technology

The CRT, however, is not without its drawbacks, which include:

- High voltages required for operation

- Relatively high power consumption

- Excessive weight for large-screen tubes

- Limited brightness under high-ambient light conditions

- Conventional tubes have a long neck, making the overall display somewhat bulky

Flat-panel devices are not expected to dislodge conventional CRTs any time in the near-term future for video applications, including HDTV. The reasons for this continued dominance of conventional devices include:

- More than 95 percent of all TV sets sold in the world are 34-in diagonal or smaller. (It is fair to point out that this percentage will likely change with the appearance of HDTV sets.)

- Initial flat-panel HDTV screens probably will be projection systems using CRT sources (or perhaps *digital micromirror device* (DMD) projectors). A major barrier to the LCD flat-panel projection display is the typically low average lifetime of the light source.

- Because of years of experience in producing CRT television sets, they are far less costly and easier to manufacture than new plasma and LCD technologies.

Another consideration is that CRTs continue to improve as manufacturing methods are refined and new techniques developed [19].

Rapid progress, however, is also being made in alternative display systems for video in general, and HDTV in particular [20]. A number of promising display devices have been demonstrated, although cost continues to be an issue. Clearly, the rate of progress in developing solid-state display technologies is remarkable.

Projection systems using light-valve devices are capable of the resolution and brightness required for HDTV. High purchase prices and maintenance expenses, however, have priced such systems out of the reach of most segments of the consumer market. Projection LCD systems offer high resolution and medium brightness. Recent progress should permit such systems to eventually reach acceptable consumer performance and pricing levels.

Another contender for the HDTV projection market is the DMD, which combines electronic, mechanical, and optical functionality [21]. DMDs, along with associated optics and signal processing electronics, have been used to develop true all-digital display systems, where the video signal processing—as well as the display itself—remains entirely in the digital domain.

Whatever type of display is used, the size of the television market is staggering: approximately 100 million black-and-white and color sets are purchased worldwide each year. Of the sets produced, by far the most incorporate color displays (in excess of 75 percent). The market for monochrome devices, however, is expected to remain strong for many years to come. New applications for CRT display, usually through computer control of processes and systems, are the primary reason for the continued strength of the monochrome CRT.

In entertainment, communications, and computer systems, the display usually represents the single most expensive component. It is often the product differentiator as well. Offering a variety of attributes, *flat-panel displays* (FPDs) are becoming the platform of choice for new information systems. Flat-panel display systems are of interest to design engineers because of their favorable operational characteristics relative to the CRT. These advantages include:

- Portability

- Low occupied volume

- Low weight (in sizes below about 15-in diagonal)

- Modest power requirements

One of the significant challenges for any display device in DTV service is the variety of possible scanning rates that the display may be called upon to reproduce. Here again, cost—not necessarily technology—is the central issue. A multiscan DTV receiver is certainly possible, but it also would be expensive. A multiscan DTV set would be required to span a wide range of horizontal scan frequencies, stretching from 15.7 kHz for 480-line interlaced SDTV images to 45 kHz for 720-line progressive HDTV. One possible solution to this problem is to insert a scan converter into the DTV receiver before the display device. Although this complicates the system, it permits the display to operate at a single specified scan rate. In the case of solid-state display technologies built around a fixed pixel pattern, scan conversion is likely a necessity.

10.3.3   Color CRT Display Devices

Many types of color CRTs have been developed for video, data, and special display applications, including:

- The shadow-mask tube

- The parallel-stripe tube

- The voltage-penetration tube

Most color tubes used for consumer and professional applications fall into three size categories:

- 19-in diagonal

- 21-in diagonal

- 25-in diagonal

Within each category are four primary grades of resolution, based on the center-to-center spacing between phosphor dots of the same color (*pitch*):

- Low resolution: dot pitch 0.44 to 0.47 mm

- Medium resolution: dot pitch 0.32 to 0.43 mm

- High resolution: dot pitch 0.28 to 0.31 mm

- Ultrahigh resolution: dot pitch 0.21 mm (or less) to 0.27 mm

Shadow-Mask CRT

The shadow-mask CRT is the most common type of color display device. As illustrated in Figure 10.19, it utilizes a cluster of three electron guns in a wide neck, one gun for each of the colors—red, green, and blue. All the guns are aimed at the same point at the center of the shadow-mask, which is an iron-alloy grid with an array of perforations in triangular arrangement, generally spaced 0.025-in between centers for entertainment television. For high-resolution studio monitor or computer graphic monitor applications, color CRTs with shadow-mask aperture spacing of 0.012-in center to center or less, are readily available. This triangular arrangement of electron guns and shadow-mask apertures is known as the *delta-gun* configuration. Phosphor dots on the faceplate just beyond the shadow-mask are arranged so that after passing through the perforations, the electron beam from each gun can strike only the dots emitting one color.

All three beams are deflected simultaneously by a single large-diameter deflection yoke, which is usually permanently bonded to the CRT envelope by the tube manufacturer. The three basic phosphors together are designated P-22, individual phosphors of each color being denoted by the numbers P-22R, P-22G, and P-22B. Most modern color CRTs are constructed with rare-earth-element-activated phosphors, which offer superior color and brightness compared with phosphors previously used.

Because of the close proximity of the phosphor dots to each other and the strict dependence on angle of penetration of the electrons through the apertures, tight control over elec-

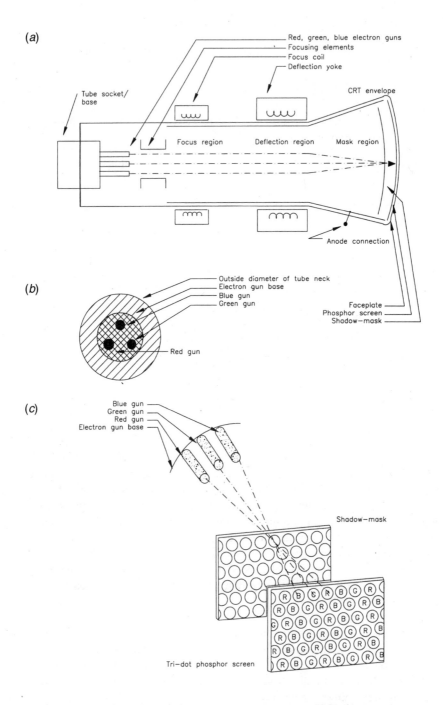

(a)

Red, green, blue electron guns
Focusing elements
Focus coil
Deflection yoke

CRT envelope

Tube socket/
base

Focus region   Deflection region   Mask region

Anode connection

(b)

Outside diameter of tube neck
Electron gun base
Blue gun
Green gun

Red gun

Faceplate
Phosphor screen
Shadow—mask

(c)

Blue gun
Green gun
Red gun
Electron gun base

Shadow—mask

Tri—dot phosphor screen

Figure 10.19 Basic concept of a shadow-mask color CRT: (a) overall mechanical configuration, (b) delta-gun arrangement on tube base, (c) shadow-mask geometry.

tron optics must be maintained. Close attention also is paid to shielding the CRT from extraneous ambient magnetic fields and to degaussing of the shield and shadow-mask (usually carried out automatically when the equipment is switched on).

Even if perfect alignment of the mask and phosphor triads is assumed, the shadow-mask CRT still is subject to certain limitations, mainly with regard to resolution and brightness. Resolution restriction results from the need to align the mask apertures and the phosphor dot triads; the mask aperture size controls the resolution that can be attained by the device.

Electron-beam efficiency in a shadow-mask tube is low, relative to a monochrome CRT. Typical beam efficiency is 10 percent; considering the three beams of the color tube, total efficiency is approximately 30 percent. By comparison, a monochrome tube may easily achieve 80 percent electron-beam efficiency. This restriction leads to a significant reduction in brightness for a given input power to the shadow-mask CRT.

Parallel-Stripe Color CRT

The parallel-stripe class of CRT, such as the popular *Trinitron (Sony)*, incorporates fine stripes of red-, green-, and blue-emitting phosphors deposited in continuous lines repetitively across the faceplate, generally in a vertical orientation. (See Figure 10.20.) This device, unlike a shadow-mask CRT, uses a single electron gun that emits three electron beams across a diameter perpendicular to the orientation of the phosphor stripes. This type of gun is called the *in-line gun*. Each beam is directed to the proper color stripe by the internal beam-aiming structure and a slitted aperture grille.

The Trinitron phosphor screen is built in parallel stripes of alternating red, green, and blue elements. A grid structure placed in front of the phosphors, relative to the CRT gun, is used to focus and deflect the beams to the appropriate color stripes. Because the grid spacing and stripe width can be made smaller than the shadow-mask apertures and phosphor dot triplets, higher resolutions may be attained with the Trinitron system.

Elimination of mask transmission loss, which reduces the electron-beam-to-luminance efficiency of the shadow-mask tube, permits the Trinitron to operate with significantly greater luminance output for a given beam input power.

The in-line gun is directed through a single lens of large diameter. The tube geometry minimizes beam focus and deflection aberrations, greatly simplifying convergence of the red, green, and blue beams on the phosphor screen.

Tube Geometry

Figure 10.21 illustrates the shadow-mask geometry for a tube at face center using in-line guns and a shadow-mask of round holes. As an alternative, the shadow-mask may consist of vertical slots, as shown in Figure 10.22. The three guns and their undeflected beams lie in the horizontal plane, and the beams are shown converged at the mask surface. The beams may overlap more than one hole, and the holes are encountered only as they happen to fall in the scan line. By convention, a beam in the figure is represented by a single straight line projected backward at the incident angle from an aperture to an apparent *center of deflection* located in the *deflection plane*. In Figure 10.21, the points B′, G′, and R′, lying in the deflection plane, represent such apparent *centers of deflection* for blue, green, and red beams striking an aperture under study. (These deflection centers move with varying deflec-

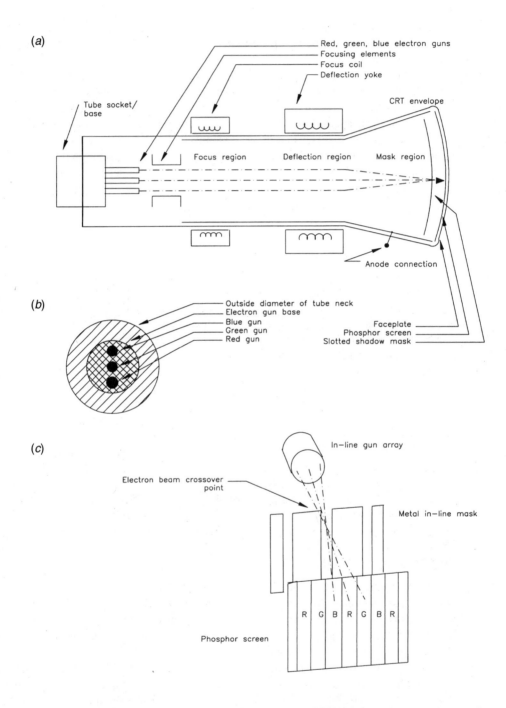

Figure 10.20 Basic concept of the Trinitron color CRT: (*a*) overall mechanical configuration, (*b*) in-line gun arrangement on tube base, (*c*) mask geometry.

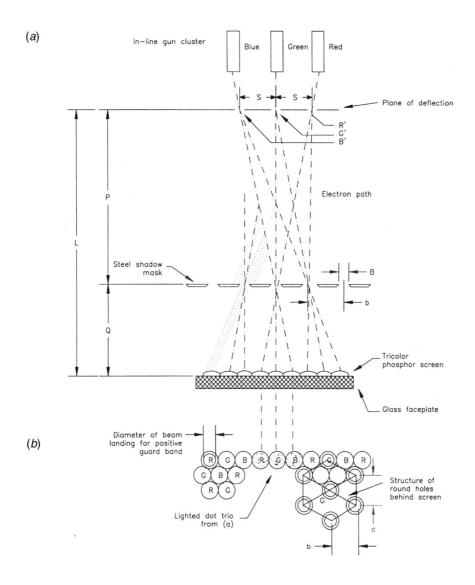

Figure 10.21 Shadow-mask CRT using in-line guns and round mask holes: (*a*) overall tube geometry, (*b*) detail of phosphor dot layout.

tion angles.) Extending the rays forward to the faceplate denotes the printing location for the respective colored dots (or stripes) of a tricolor group. Thus, centers of deflection become color centers with a spacing S in the deflection plane. The distance S projects in the ratio Q/P as the dot spacing within the trio. Figure 10.21 also shows how the mask-hole hor-

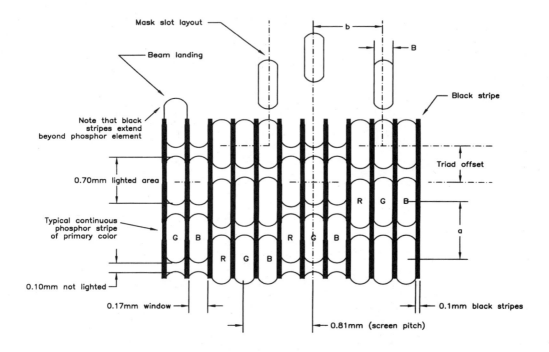

Figure 10.22 Shadow-mask CRT using in-line guns and vertical stripe mask holes.

izontal pitch b projects as screen horizontal pitch in the ratio L/P. The same ratio applies for projection of mask vertical pitch a. The Q-space (mask-to-panel spacing) is optimized to obtain the largest dots that are theoretically possible without overlap. At panel center, the ideal screen geometry is then a mosaic of equally spaced dots (or stripes).

The stripe screen shown in Figure 10.22 is used extensively in color CRTs. One variation of this stripe (or line) screen uses a cylindrical faceplate with a vertically tensioned grill shadow-mask without tie bars. Prior to the stripe screen, the standard construction was a tri-dot screen with a delta gun cluster, as shown in Figure 10.23.

Guard Band

The use of *guard bands* is a common feature for aiding purity in a CRT. The guard band, where the lighted area is smaller than the theoretical tangency condition, may be either *positive* or *negative*. In Figure 10.21, the leftmost red phosphor exemplifies a positive guard band; the lighted area is smaller than the actual phosphor segment, accomplished by mask-hole-diameter design. Figure 10.22, on the other hand, shows negative guard band (NGB) or *window-limited* construction for stripe screens. Vertical black stripes about 0.1 mm (0.004-in) wide separate the phosphor stripes, forming windows to be lighted by a beam wider than

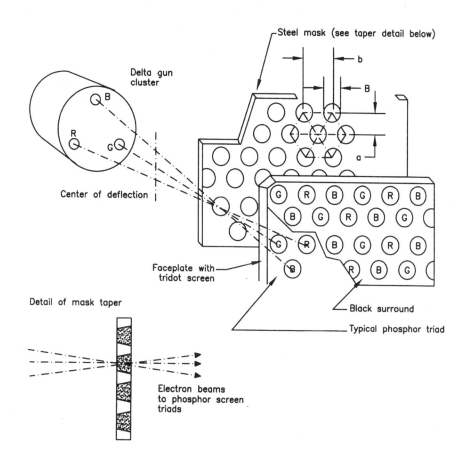

Figure 10.23 Delta-gun, round-hole mask negative guard band tri-dot screen. The taper on the mask holes is shown in the detail drawing only.

the window opening by about 0.1 mm. Figure 10.23 shows NGB construction of a tri-dot screen.

Shadow-Mask Design

The shadow-mask for a color CRT typically is constructed of 0.13 mm low-carbon sheet steel that is chemically etched to the desired pattern of apertures using photoresist techniques. The photographic masters are made by a precision laser plotter. The completed flat mask is then press-formed to a contour approximately concentric to the faceplate. Mask-to-panel distance (*Q-spacing*) is locally modified to achicve optimum nesting of screen triplets.

The array layout for a typical round-hole shadow-mask is shown in Figures 10.21 and 10.23. The round holes, numbering approximately 440,000 in a conventional consumer television display, are placed at the vertices and centers of iterated regular hexagons (long-axis vertical).

As illustrated in Figure 10.23, each aperture is tapered to present a more sharply defined limiting-aperture plane to an angled incident beam. This construction increases transmission efficiency and prevents color desaturation resulting from electrons being scattered by sidewalls of the apertures. Apertures are graded radially to smaller diameters at screen edge, because that is where the trio configuration and beam quality are less ideal, and registry is more critical. At the tube center, shadow-mask transmission is typically 16 percent for non-matrix construction and 22 percent for *black matrix* shadow-mask construction.

The black matrix screen is designed to overcome the loss in luminance and resultant brightness caused by use of a neutral density filter as the faceplate. Such a filter may be used to increase the contrast ratio of the device. This condition arises as a result of the large amount of area in the viewing surface that is not covered by any phosphor. The need to align the phosphor dots with the mask holes to improve convergence can lead to 50 percent of the surface merely reflecting ambient light (in a medium-quality consumer television display). The black matrix screen covers this area with black, thus reducing backscattered and reflected light by the same amount as the faceplate. This scheme reduces the loss in the faceplate without affecting contrast. A brighter displayed picture results.

The beam triad pattern may become distorted near the screen edges, resulting in poorer *packing factors* (*geometric nesting*) for the related phosphor dots. For delta-gun systems, the beam triad triangles compress radially at all screen edges because of foreshortening. For in-line gun dot screens, on the other hand, the beam triad represents three circles in a horizontal line. Using compass points to designate axes and areas, there is no foreshortening at N and S; at E and W, the mask-panel spacing can be sufficiently increased to restore nesting quality. Near the screen corners, however, there will be some rotation of the line trio, thereby demanding smaller holes, dots, and beam landings; the result is less efficient nesting.

For a slot mask, the relationship between a and b, horizontal and vertical pitch, can be chosen by the designer. Only the horizontal pitch will affect Q-space. The factors affecting the choice of horizontal pitch include:

- Display resolution

- Practical mask-panel spacing

- Attainable slot widths

- Ability to manufacture masks and screens with reasonable yields

The main considerations relating to the choice of vertical pitch and tie bars (or *bridges*) include:

- Avoidance of moiré pattern

- Strength of the tie bars

- Transmission efficiency

Vertical screen stripes may be regarded as aligned oblong dots that have been merged vertically. As a result, there is no color-purity or registration requirement in the vertical direction, which simplifies the design and manufacture of display devices.

Resolution and Moiré

Resolution is a measure of the definition or sharpness of detail in a displayed image. It may be measured vertically and horizontally. The layout and center-line spacings of the mask apertures are designed to provide sufficient horizontal and vertical lines in the pattern to ensure that they are not the limiting factor in resolution, compared with the number of raster lines. In addition, the number of lines running horizontally in the layout pattern is chosen to avoid moiré, fringes resulting from a "beat" with the scan lines. (This latter criterion is not applicable to the cylindrical grill structure of continuous slits.)

For round-hole masks, the effective number of pattern lines running horizontally, allowing for the staggered pattern, is typically about 2.25 times the number of picture horizontal scan lines. In a conventional television display, there are approximately 440,000 dot trios on the screen, compared with about 200,000 pixels in the raster. These criteria apply to tube sizes from 19- to 25-in (48- to 63-cm) screen diagonal. For smaller tubes, the pattern is made relatively coarser to avoid excessively small apertures and to avoid a moiré pattern in the display. The hexagonal pattern used to lay out the centers of round holes is favorable for increasing the number of pattern lines, reducing the likelihood of moiré, and for nesting of screen dots. The zigzag, or staggering, of alternate columns effectively increases the number of columns and rows, as long as the pattern is significantly finer than the raster scan.

For slot apertures, spacings between columns and the resulting number of columns are chosen for resolution and cosmetic appearance. The tie bars or bridges, which are only about 0.13 mm (0.005-in) in width, are placed according to moiré and strength considerations and are not thought to be significant for resolution in most applications.

Mask-Panel Temperature Compensation

As mentioned previously, color CRTs for consumer, computer, and industrial applications are shifting toward larger, flatter physical dimensions and higher resolution. Manufacturing such devices for good color purity and stability is a major engineering challenge.

Conventional printing of screen patterns requires that the mask assembly be removed and reinserted without loss of registry. In the most common mask-suspension system, the interior skirt of the glass faceplate has three or four protruding metal studs that engage spring clips welded to the mask frame. Because the shadow-mask may intercept perhaps 75 percent of the beam current, it is subject to thermal expansion during operation. Misregistration would occur if thermal correction were not provided. This temperature compensation is accomplished by automatically shifting the mask slightly closer to the screen surface by thermal expansion of a bimetal structure incorporated into each spring support, or by a designed lever action resulting from the transverse expansion. The exact mechanism used varies from one tube manufacturer to another.

Because the shadow-mask is one of the most critical components in determining CRT performance, the material used to form the structure is an important design parameter. *Aluminum killed* (AK) steel has been extensively employed as mask material because of its

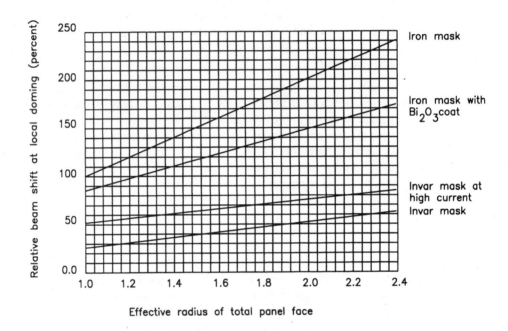

Figure 10.24 The relationship between beam shift at local doming and the effective radius of the faceplate with various mask materials. (*After* [22]).

etching characteristics, low cost, and relatively good magnetic properties. To build a wide-aspect-ratio (HDTV) CRT at a size of 30 in or larger requires mask thickness of approximately 0.2 mm [22]. The pitch of the device must be 70 to 100 percent higher than that of a conventional CRT. Because of the practical limits of etching technology, mask transmission is significantly lower than for a conventional tube of similar size. Consequently, the percentage of electron-beam bombardment onto the mask will be significantly greater, requiring higher beam current to achieve sufficient brightness. This requirement contributes to color-purity problems when the beam is mislanded onto the screen because of mask thermal expansion.

To overcome the effects of thermal expansion, an iron-nickel alloy was introduced as a substitute to AK steel. *Invar*, a *Fe-36Ni* alloy, exhibits a thermal expansion coefficient 10 times lower than that of AK steel. The effects of misregistration resulting from thermal expansion, therefore, are significantly reduced, as illustrated in Figure 10.24. Because of the relatively low modulus of elasticity of Invar (40 percent less than that of AK), however, manufacturing difficulties included mask-forming and mask-mounting to minimize spring-back effects and microphonics, respectively.

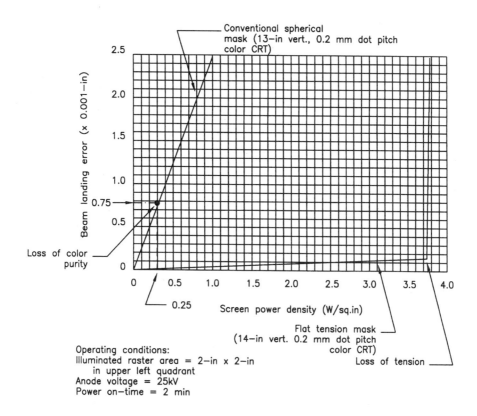

Figure 10.25 Comparison of small-area-mask doming for a conventional CRT and a flat-tension-mask CRT. (*After* [23].)

Tension Mask

Mask stability also may be improved by utilizing the *tension mask* method, in which the mask is tensioned during manufacture to a predetermined stress [23]. When the proper level of tension is applied to the mask, thermal impact on the structure reduces the tension to a certain extent, but will not cause displacement of the aperture features. Therefore, the effects on color purity resulting from thermal expansion are minimized, as shown in Figure 10.25.

The *taut shadow-mask* is a variation on the basic tension mask concept. This design departs from the conventional domed shadow-mask approach and, instead, stretches the shadow-mask into a perfectly flat contour, as illustrated in Figure 10.26 [24]. This design provides greater power-handling capability and is less susceptible to Z-axis vibration. The taut shadow-mask typically achieves a fourfold increase in maximum beam current for a specified color purity.

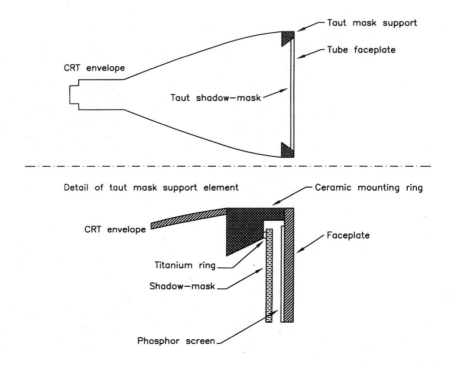

Figure 10.26 Mechanical structure of a taut shadow-mask CRT. (*After* [24].)

Magnetic Shielding

Because of the earth's magnetic effect, electron-beam deflection will change, to some extent, depending on the location and orientation of the CRT. This effect influences color purity. An external or internal magnetic metal shield is included, therefore, to compensate for the effect of the earth's magnetic field and stray magnetic fields. The low-carbon steel shadow-mask itself also contributes to the shielding. The shield does not extend into the yoke field because this would result in loading. Also, external shields must be clear of the anode button area. With the display orientation in the chosen compass direction, the shield and tube must be thoroughly degaussed to gain full purity. This treatment not only removes any residual magnetization of the shield and shadow-mask, but also induces a residual static magnetic field that bucks the ambient magnetic field.

X-Radiation

The shadow-mask color tube operates with extremely high anode voltages (typically 20 to 30 kV). The possibility of x-rays must be considered for the safety of technicians and end-users. The color-tube envelope is made from glass that has been formulated for x-ray-absorbing characteristics. By closely controlling glass thickness, the glassmaker controls

Table 10.6 Color CRT Diagonal Dimension vs. Weight (*After* [14].)

Diagonal Visible (in)	Weight (kg)	Weight (lb.)
19V	12	26
25V	23	51
30V	40	88

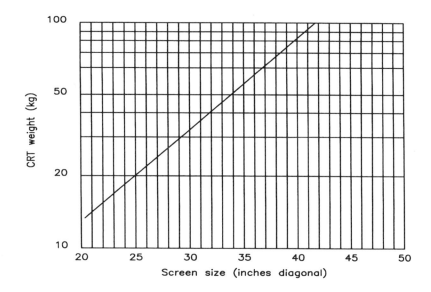

Figure 10.27 Relationship between screen size and CRT weight. (*After* [25].)

the extent of x-ray absorption. Also, the power levels at which the tube is operated must be controlled. Tube-type data sheets provide the relevant absolute maximum ratings (for voltages and currents) that will keep x-radiation within accepted safety levels.

Screen Size

Consumer demand for ever-larger picture sizes brings to the forefront serious limitations concerning not only the weight and depth of the CRT, but also the higher power and voltage requirements. These limitations are reflected in the sharply increasing costs of receiver cabinets to accommodate the larger tubes (a major share of the manufacturing costs) and in more complex circuitry for high-voltage operation.

To withstand the atmospheric pressures on the evacuated glass envelope, the CRT weight increases exponentially with the viewable diagonal. Typical figures for television receiver tubes designed for a 4:3 aspect ratio are shown in Table 10.6; Figure 10.27 charts the relationship. Nevertheless, manufacturers have continued to meet the demand for larger screen

Table 10.7 Comparative Resolution of Shadow-Mask Designs (*After* [26].)

19V (48 cm) NGB tube type	Mask Material (mm)[1]	Vertical Pitch (mm)[1]	Center Hole Diameter (mm)[1]	Screen Vertical Pitch (mm)	N_t, Trios in Screen	N_r = sq. rt. (NT/1.33)
Conventional	0.15	0.56	0.27	0.60	400,000	500 lines
Monitor	0.15	0.40	0.19	0.43	800,000	775 lines
High resolution	0.13	0.30	0.15	0.32	1,400,000	1025 lines
1 Flat shadow mask						

sizes with larger direct-view tubes. Examples include an in-line gun, 110° deflection tube with a 35-in-diagonal screen (Mitsubishi and Matsushita). In the Trinitron configuration, 37-, 38-, and 43-in-diagonal tubes have been produced (Sony).

Because of the weight of 35-in and larger tubes and the depth of the required receiver cabinets, tubes of that size are of questionable practicality for home use. Consequently, a 27-in tube is probably the largest size suitable for most home viewing situations.

CRTs designed for HDTV applications suffer an additional weight disadvantage because of the wide aspect ratio of the format (16:9). Furthermore, large-diameter electron guns and small deflection angles (90°) often are employed to improve resolution, making the tube necks fat and giving the devices greater depth and weight.

Resolution Improvement Techniques

In a shadow-mask display device, higher resolution can be achieved through attention to the following parameters:

• Finer pitch

• Smaller-diameter apertures and screen dots

• Smaller Q-space

• Thinner mask material

Etching of mask holes becomes more demanding when diameters are smaller than the material thickness. The black matrix tri-dot system has a mask-manufacturing advantage because mask-aperture diameters are larger than in comparable nonmatrix tri-dot tubes or in comparable matrix (or nonmatrix) stripe tubes.

Table 10.7 shows comparative resolution capabilities of three 19-in (48 cm) visible (19V) devices with round-hole masks as screen pitch is reduced to increase resolution. The value N_r, a comparative measure of resolution, assumes that resolution is proportional to the square root of the number of trios in a square area with sides equal to screen height. To achieve higher resolution than that of the third design shown in the table (or for smaller tube sizes), it is necessary to use thinner mask material and smaller holes. The Q-space also reduces proportionally to pitch, becoming critically small for manufacturing.

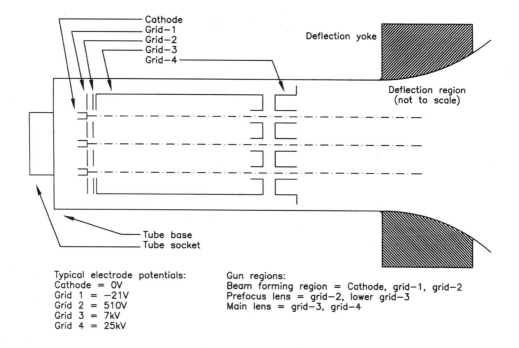

Typical electrode potentials:
Cathode = 0V
Grid 1 = −21V
Grid 2 = 510V
Grid 3 = 7kV
Grid 4 = 25kV

Gun regions:
Beam forming region = Cathode, grid−1, grid−2
Prefocus lens = grid−2, lower grid−3
Main lens = grid−3, grid−4

Figure 10.28 Simplified mechanical structure of a bipotential color electron gun.

The construction of large, wide screen glass tubes presents several design and fabrication problems. The tensile stresses on the bulb must be minimized to avoid cracks resulting from temperature imbalance during and after manufacture, and to prevent subsequent implosion. Extensive study of the stress points in the design stages through computer simulation has reduced the nominal physical stress to about the same level as in conventional tubes of 4:3 aspect ratio.

Reduction of moiré to an unnoticeable level requires an increase in the frequency of the pattern or a decrease in the amplitude. The pattern can be minimized through the use of certain specific values of triad pitch. However, the selection of the pitch is limited not only by the thinner shadow-mask for a finer pitch, but also by a sacrifice in resolution for a more coarse pitch. For a 40-in wide screen tube, a pitch of 0.46 mm provides a resolution of 1000 television lines and satisfies the other requirements of shadow-mask strength and moiré-pattern reduction.

Electron Gun

Figure 10.28 illustrates the general electrode configuration for a shadow-mask color electron gun. The device can be subdivided into three major regions:

- Beam-forming region, which consists of the cathode, grid-1, and grid-2 electrodes.

- Prefocus lens region, which consists of the grid-2 and lower grid-3 electrodes.

- Main lens region, which consists of the grid-3 and grid-4 electrodes. These elements create a focusing field for the electron beam.

In more complicated lens systems, additional elements follow grid 4.

Electron guns for color tubes can be classified according to the main lens configuration, which include:

- Unipotential

- Bipotential

- Tripotential

- Hybrid lenses

The unipotential gun is the simplest of all designs. This type of gun rarely is used for color applications, except for small screen sizes. The system suffers from a tendency to arc at high anode voltage and from relatively large low-current spots.

The bipotential lens is the most commonly used gun in shadow-mask color tubes. The main lens of the gun is formed in the gap between grids 3 and 4. When grid 3 operates at 18 to 22 percent of the grid-4 voltage, the lens is referred to as a *low-bipotential* configuration, often called *LoBi* for short. When grid 3 operates at 26 to 30 percent of the grid-4 voltage, the lens is referred to as a *high-bipotential* or *HiBi* configuration.

The LoBi configuration has the advantages of a short grid 3 and shorter overall length, with parts assembly generally less critical than the HiBi configuration. However, with its shorter object distance (grid-3 length), the lens suffers from somewhat larger spot size than the HiBi. The HiBi, on the other hand, with a longer grid-3 object distance, has improved spot size and resolution. The focus voltage supply for the LoBi also can be less expensive.

Further improvement in focus characteristics can been achieved with a tripotential lens. The lens region has more than one gap and requires two focus supplies, one typically at 40 percent and the other at 24 percent of the anode potential. With this refinement, the lens has lower spherical aberration. These features, together with a longer object distance (grids 3 to 5), yield a smaller spot size at the screen than is achieved with the bipotential design. Among the drawbacks of the tripotential gun are:

- The assembly is physically longer.

- It requires two focus supplies.

- It requires a special base to deliver the high focus voltage through the stem of the tube.

Improved performance may be realized by combining elements of the unipotential and bipotential lenses in series. The two more common configurations are known as *UniBi* and *BiUni*. The UniBi structure (sometimes referred to as *quadripotential focus*) combines the HiBi main lens gap with the unipotential type of lens structure to collimate the beam. As shown in Figure 10.29, the grid-2 voltage is tied to grid 4 (inserted in the object region of the gun), causing the beam bundle to collimate to a smaller diameter in the main lens. With

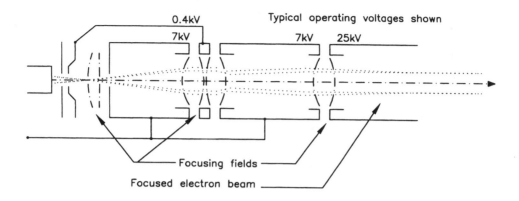

Figure 10.29 Electrode arrangement of a UniBi gun.

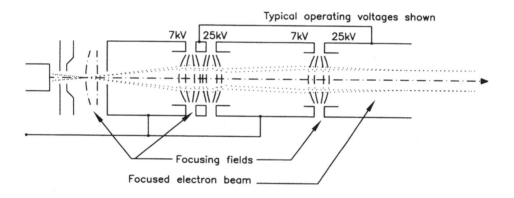

Figure 10.30 Electrode arrangement of a BiUni gun.

this added focusing, the gun is slightly shorter than a bipotential gun having an equal focus voltage.

The BiUni structure, illustrated in Figure 10.30, achieves a similar beam collimation by tying the added element to the anode, rather than to grid 2. The gun structure is shorter because of the added focusing early in the device. With three high-gradient gaps, arcing can be a problem in the BiUni configuration.

The Trinitron gun consists of three beams focused through the use of a single large main-focus lens of unipotential design. Figure 10.31 shows the three in-line beams mechanically tilted to pass through the center of the lens, then reconverged toward a common point on the screen. The gun is somewhat longer than other color guns, and the mechanical structuring of the device requires unusual care and accuracy in assembly.

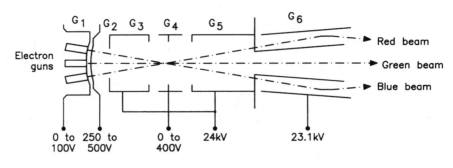

Typical operating voltages shown

Figure 10.31 Electrode arrangement of the Trinitron gun. (*After* [27].)

Guns for High-Resolution Application

Improved versions of both the delta and in-line guns previously noted are used in high-resolution display applications. In both cases, the guns are adjusted for the lower beam current and higher resolution needed in data and/or graphics display. The advantages and disadvantages of delta and in-line guns also apply here. For example, the use of a delta-type cylinder, or *barrel-type* gun, requires as many as 20 carefully tailored convergence waveforms to obtain near-perfect convergence over the full face of the tube. The in-line gun, with a self-converging yoke, avoids the need for these waveforms, but at the expense of slightly larger spots, particularly in the corners where overfocused haze tails can cause problems.

Figure 10.32 compares spot sizes (at up to 1 mA of beam current) for high-resolution designs with those of commercial receiver-type devices. Both delta and in-line 13- and 19-in vertical (13V and 19V) devices are shown. Note the marked improvement in spot size at current levels below 500 µA for the high-resolution devices.

Deflecting Multiple Electron Beams

Deflection of the electron beams in a color CRT is a difficult technical exercise. The main problems that occur when the three beams are deflected by a common deflection system are associated with spot distortions that occur in single-beam tubes. The effect is intensified, however, by the need for the three beams to cross over and combine as a spot on the shadow-mask. The two most significant effects are:

- Curvature of the field

- Astigmatism

Misalignment and misregistration of the three beams of the color CRT will lead to loss of purity for colors produced by combinations of the primary colors. Such reproduction distortions also may result in a reduction in luminance output because a smaller part of the beams are passing through the apertures. Additional errors that must be considered include:

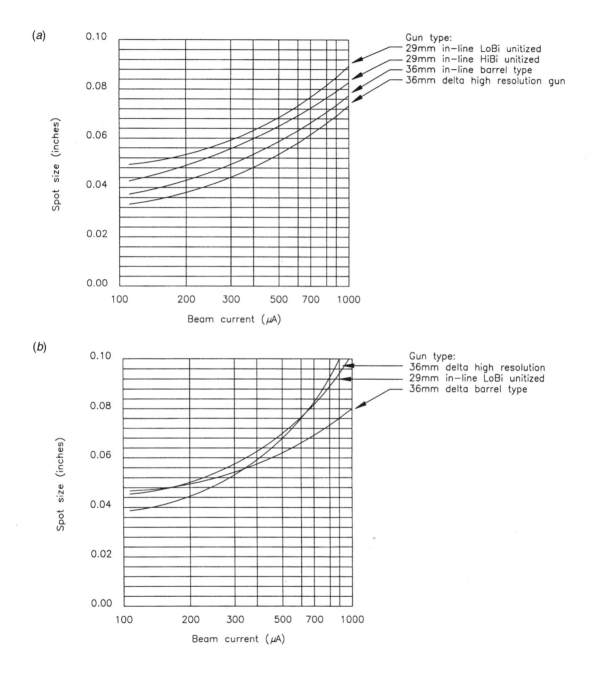

Figure 10.32 Spot-size comparison of high-resolution guns vs. commercial television guns: (a) 13-in vertical display, (b) 19-in vertical display. (*After* [28].)

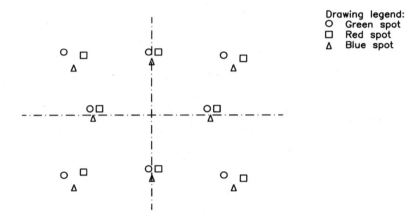

Figure 10.33 Astigmatism errors in a color CRT. (*After* [29].)

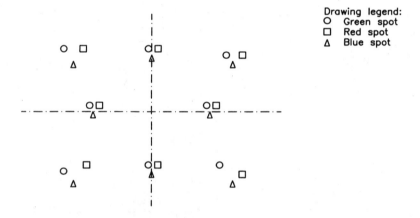

Figure 10.34 Astigmatism and coma errors. (*After* [29].)

- Deflection-angle changes in the yoke-deflection center

- Stray electromagnetic fields

Curvature of field and astigmatism distortions result in a misconvergence of the beam. Figure 10.33 illustrates distortion resulting from astigmatism (for a delta-gun tube). Figure 10.34 illustrates misconvergence of the beam resulting from astigmatism and coma. The misconvergence that occurs in the four corners of the raster is shown in Figure 10.35. The result is that color rendition will not be true, particularly at the edges of the screen. This can

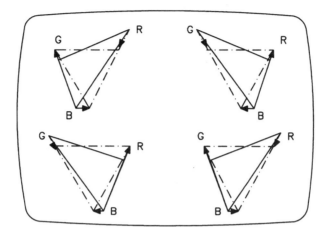

Figure 10.35 Misconvergence in the four corners of the raster in a color CRT. (*After*[29].)

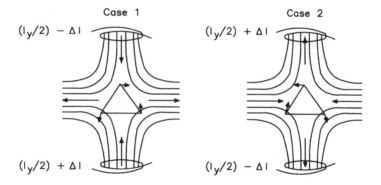

Figure 10.36 Field configurations suitable for correcting misconvergence in a color CRT. (*After*[29].)

be partially compensated by introducing quadripole fields, which cause the beams to be twisted, restoring the equilateral nature of the triangle. The shapes these fields can take are illustrated in Figure 10.36 along with the currents required to produce the fields.

The registration challenge is less severe in the case of the in-line gun; the three beams need only be converged into a vertical line, rather than the round spot required by the delta gun. For the in-line gun, a precision self-converging system is used where the yoke is designed to operate with one specific tube. This self-converging yoke causes the beams to diverge horizontally in the yoke, resulting in nonuniform fields that counteract the overcon-

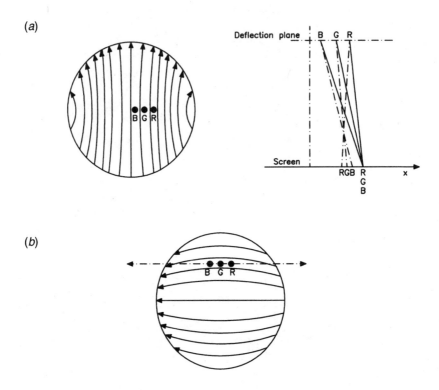

Figure 10.37 Self-converging deflection field in an in-line gun device: (*a*) horizontal deflection field, (*b*) vertical deflection field. (*After* [30].)

vergence. The shape of the fields for horizontal and vertical deflection are shown in Figure 10.37. The horizontal yoke generates a pincushion field, and the vertical yoke generates a barrel-shaped field to accomplish these ends. However, complete self-convergence without the need for compensating adjustments is possible only with narrow-angle tubes; for 110° tubes, it is necessary to add some convergence adjustments, although much less than is required for the delta gun. The drawback is that the yoke usually must be tailored to the specific tube with which it is used.

New Consumer Devices

Progress continues to be made in CRT technology. Numerous improvements have been made over the years, and considerable new efforts have accompanied the move to DTV. A case in point is the FD Trinitron Wega (Sony), the first consumer TV with a flat face CRT. The purpose of making a flat face display was to eliminate distortion, reduce reflections, and improve the picture quality [31].

Conventional curved CRT screens cause geometric distortion in the displayed images and a loss of part of the image when viewed from a side angle. Furthermore, the spherical shape of the screen causes ambient room lighting to be reflected to the viewer. A flat display screen eliminates the geometric distortion and reduces reflections.

Glass is stronger when it is curved. For this reason, detailed modeling and strength analysis were necessary to make the face of the tube capable of withstanding the applied atmospheric pressure without failing or deforming. Because of the pressure gradient on the glass, it was necessary to develop a new tempered glass based upon the principles of automobile windshield glass. For additional implosion protection, a 188 μm thick film was laminated on the panel surface. The film also serves as an anti-reflective coating to improve display contrast, a common practice in computer monitors.

To realize a flat *aperture grille* (AG) with high tension, there were two major problems to overcome: mask wrinkle and AG vibration. Mask wrinkle is caused by localized heating during the welding process. It was found that the AG wires and even the frame itself were susceptible to sympathetic vibrations from the receiver speakers. The answer was to double the springs that attach the frame to the CRT panel.

As the faceplate gets flatter, it becomes more difficult for the electron gun to make a round spot towards the edges of the screen. As a solution to this problem, the focal length of the electron gun was elongated by 20 percent compared with a conventional Trinitron gun. The result was improved focus at the edges of the flat CRT. In order to keep the overall length of the TV set the same, other parts of the gun were widened and shortened.

A new deflection yoke was developed using a horizontal coil with a large square-front-bend to eliminate the usual pincushion distortion introduced by the flat CRT. A modulation coil was added to the vertical winding that used the horizontal frequency to correct vertical misconvergence.

Additional improvements included the following:

- Dynamic focus to vary the voltage across scan lines for sharp focus from edge-to-edge

- Velocity modulation to change the speed of the beam as it scans horizontally across the screen and sharpens the focus

This device is representative of the new generation of CRT displays now finding their way to consumers. The trend to higher quality, lower priced displays will likely accelerate as DTV penetration in the home increases.

10.3.4 Projection Systems

As the need to present high-resolution video and graphics information steadily increases, the use of large-screen projection displays rapidly expands. High-definition television will require large screens (greater than 40-in diagonal) to provide effective presentation. Given the physical limitations of CRTs, some form of projection is the only practical solution. The role of HDTV and film in theaters of the future also is being explored, along with performance criteria for effective large-screen video presentations.

New projection-system hardware is taking advantage of *liquid crystal* (LC) and *thin-film transistor* (TFT) technology originally developed for direct-view flat-panel displays. Also, new deformable-membrane light valves used in Schlieren systems are being developed to update oil film light-valve projectors, which have been the mainstay of large-screen projection technology for many years.

Extensive developmental efforts are being directed toward large-screen display systems. Much of this research is aimed at advancing new technologies such as plasma, electroluminescent and LCD, lasers, and new varieties of CRTs. Currently, CRT systems lead the way for applications requiring full color and high resolution. LCD systems, which are advancing rapidly, may capture a sizable portion of the video-only marketplace. The DMD shows great promise for high-resolution applications as well.

Large-screen projectors fall into four broad classes, or grades:

- **Graphics**. Graphics projectors are the highest-quality—and generally the highest-priced—projectors. These systems are capable of the highest operating frequency and resolution. They can gen-lock to almost any computer or image source, and they offer resolutions of 2000 × 2000 pixels (or more) with horizontal sweep rates to 89 kHz and higher.

- **Data**. Data projectors are less expensive than graphics projectors and are suitable for use with common computer image generators, such as PCs equipped with VGA graphics cards. Data projectors typically offer resolutions of 1024 × 768 pixels and horizontal sweep rates of approximately 49 kHz.

- **HDTV**. HDTV projectors provide the quality level necessary to take full advantage of high-definition imaging systems. Resolution in excess of 1000 TV lines is provided at an aspect ratio of 16:9.

- **Video**. Video projectors provide resolution suitable for NTSC-level images. Display performance of 380 to 480 TV lines is typical.

Computer signal sources can follow many different—and sometimes incompatible—standards. It is not a trivial matter to connect any given projector to any given source, although multisync projectors are becoming the norm. Generally speaking, projectors are *downward compatible*. In other words, most graphics projectors can function as data projectors, HDTV projectors, and video projectors; most data projectors also can display HDTV and video, and so on.

Displays for HDTV Applications

The most significant differences between a conventional display and one designed for high-definition video are the increased resolution and wider aspect ratio of HDTV. Four technologies have emerged as practical for viewing high-definition images of 40-in diagonal screens and larger:

- **Light-valve projection display**. Capable of modulating high-power external light sources, light valves are used mainly for large-screen displays measuring greater than 200 in.

- **CRT projection displays**. Widely used for midsize-screen displays of 45 to 200 in, CRT projection systems are popular because of their relative ease of manufacturing (hence,

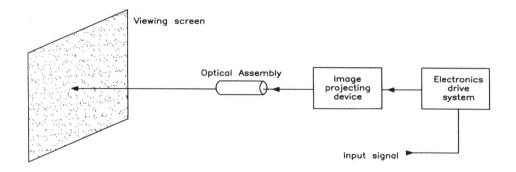

Figure 10.38 Principal elements of a video projection system.

competitive cost) and good performance. These displays may be the mainstay technology for HDTV in the near future.

- **Flat-panel plasma display panel (PDP) and LCD (liquid crystal display)**. Flat-panel PDP displays have been produced in 40-in and larger displays. Similar LCD systems also have been shown. Both hold great promise for future HDTV applications.

- **Digital micromirror device**. The DMD (Texas Instruments) is a spatial light modulator that combines electronic, mechanical, and optical functions to yield high-resolution projection images.

In addition, several hybrid technologies that are under study combine various elements from the mainstay projection techniques.

Projection-System Fundamentals

Video projection systems provide a method of displaying a much larger image than can be generated on a direct-view cathode-ray picture tube [28]. Optical magnification and other techniques are employed to throw an expanded image on a passive viewing surface that may have a diagonal dimension of 75 in or more.

The basic elements of a projection system, illustrated in Figure 10.38, include:

- Viewing screen

- Optical elements

- Image source

- Drive electronics

The major differences between projection systems and direct-view displays are embodied in the first three elements, but the electronics assembly of a typical projection system is essentially the same as that of a direct-view system.

Table 10.8 Performance Levels of Video and Theater Displays (*After* [32].)

Display System	Luminous Output (Brightness), nits (ft-L)	Contrast Ratio at Ambient Illumination (fc)	Resolution (TVL)
Television receiver	200–400, 60–120	30:1 at 5	275
Theater (film projector)	34–69, 10–20[1]	100:1 at 0.1[2]	1000 and up
[1] U. S. standard (PH-22.124-1961); see[32].			
[2] Limited by lens flare.			

To provide an acceptable image, a projection system must approach or equal the performance of a direct-view device in terms of brightness, contrast, and resolution. Whereas brightness and contrast may be compromised to some extent, large displays must excel in resolution because of the tendency of viewers to be positioned at less than the normal standard-definition-image relative distance of 4 to 8 times the picture height from the viewing surface. Table 10.8 provides performance levels achieved by direct-view video displays and conventional film theater equipment.

Evaluation of overall projection system brightness *B*, as a function of its optical components, can be calculated using the following equation:

$$B = \frac{L_G \times G \times T \times R^M \times D}{4 \times W_G (F/N)^2 \times (1+m)^2} \tag{10.6}$$

Where:
L_G = luminance of the green source (CRT or other device)
G = screen gain
T = lens transmission
R = mirror reflectance
M = number of mirrors
D = dichroic efficiency
W_G = green contribution to desired white output (percent)
f/N = lens *f*-number
m = magnification

For systems in which dichroics or mirrors are not employed, those terms drop out.

Two basic categories of viewing screens are employed for projection video displays. As illustrated in Figure 10.39, the systems are:

- Front projection, where the image is viewed from the same side of the screen onto which it is projected.

- Rear projection, where the image is viewed from the opposite side of the screen onto which it is projected.

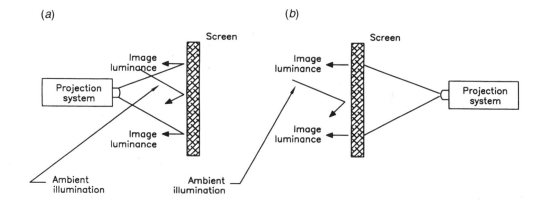

Figure 10.39 Projection-system screen characteristics: (*a*) front projection, (*b*) rear projection.

Front projection depends upon reflectivity to provide a bright image, while rear projection requires high transmission to achieve that same characteristic. In either case, screen size influences display brightness inversely as follows:

$$B = \frac{L}{A} \qquad\qquad (10.7)$$

Where:
B = apparent brightness (cd/m^2)
L = projector light output (lm)
A = screen viewing area (m^2)

Thus, for a given projector luminance output, viewed brightness varies in proportion to the reciprocal of the square of any screen linear dimension (width, height, or diagonal). An increase in screen width from the conventional aspect ratio of 4:3 (1.33) to an HDTV ratio of 16:9 (1.777) requires an increase in projector light output of approximately 33 percent for the same screen brightness.

To improve apparent brightness, directional characteristics can be designed into viewing screens. This property is termed *screen gain G*, and the previous equation becomes:

$$B = G \times \frac{L}{A} \qquad\qquad (10.8)$$

Gain is expressed as screen brightness relative to a *lambertian* surface. Table 10.9 lists some typical front-projection screens and their associated gains.

Screen contrast is a function of the manner in which ambient illumination is treated. Figure 10.39 illustrates that a highly reflective screen (used in front projection) reflects ambient illumination, as well as the projected illumination (image). Thus, the reflected light tends to dilute contrast, although highly directional screens diminish this effect. A rear-projection screen depends upon high transmission for brightness but can capitalize on low reflectance

Table 10.9 Screen Gain Characteristics for Various Materials (*After* [33].)

Screen Type	Gain
Lambertian (flat-white paint, magnesium oxide)	1.0
White semigloss	1.5
White pearlescent	1.5–2.5
Aluminized	1–12
Lenticular	1.5–2
Beaded	1.5–3
Ektalite (Kodak)	10–15
Scotch-light (3M)	Up to 200

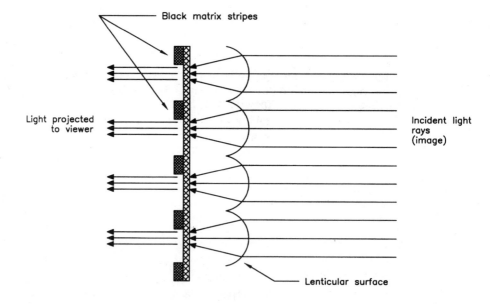

Figure 10.40 High-contrast rear-projection system.

to improve contrast. A scheme for achieving this is equivalent to the black matrix utilized in tricolor CRTs. Illustrated in Figure 10.40, the technique focuses projected light through lenticular lens segments onto strips of the viewing surface, allowing intervening areas to be coated with a black (nonreflective) material. The lenticular segments and black stripes normally are oriented in the vertical dimension to broaden the horizontal viewing angle. The overall result is a screen that transmits most of the light (typically 60 percent) incident from the rear, while absorbing a large percentage of the light (typically 90 percent) incident from the viewing side, thus providing greater contrast.

Rear-projection screens usually employ extra elements, including diffusers and directional correctors, to maximize brightness and contrast in the viewing area.

As with direct-view CRT screens, resolution can be affected by screen construction. This is not usually a problem with front-projection screens, although granularity or lenticular patterns can limit image detail. In general, any screen element (such as the matrix arrangement previously described) that quantizes the image (breaks it into discrete segments) limits resolution. For 525- and 625-line video, this factor does not provide the limiting aperture. High-resolution applications, however, require attention to this parameter.

Optical Projection Systems

Both refractive and reflective lens configurations have been used for the display of a CRT raster on screens that are 40-in diagonal or larger [26]. The first attempts merely placed a lenticular Fresnel lens, or an inefficient $f/1.6$ projection lens, in front of a shadow-mask direct view tube, as shown in Figure 10.41a. The resulting brightness of no greater than 2 or 2 ft-L was suitable for viewing only in a darkened room. Figure 10.41b shows a variation on this basic theme. Three individual CRTs are combined with cross-reflecting mirrors and focused onto the screen. The in-line projection layout is shown in Figure 10.41c using three tubes, each with its own lens. This is the most common system used for multitube displays. Typical packaging to reduce cabinet size for front or rear projection is shown in Figures 10.41d and 10.41e, respectively.

Because of off-center positioning of the outboard color channels, the optical paths differ from the center channel, and keystone scanning height modulation is necessary to correct for differences in optical throw from left to right. The problem, illustrated in Figure 10.42, is more severe for wide-screen formats.

Variables to be evaluated in choosing from the many available projection schemes include the following:

• Source luminance

• Source area

• Image magnification (screen size required)

• Optical-path transmission efficiency

• Light-collection efficiency (of the lens)

• Cost, weight, and complexity of components and corrective circuitry

The lens package is a critical factor in rendering a projection system cost-effective. The package must offer good luminance-collection efficiency (small f-number), high transmission, good *modulation transfer function* (MTF), light weight, and low cost.

The total light incident upon a projection screen is equal to the total light emerging from the projection optical system, neglecting losses in the intervening medium. Distribution of this light generally is not uniform. Its intensity is less at screen edges than at the center in

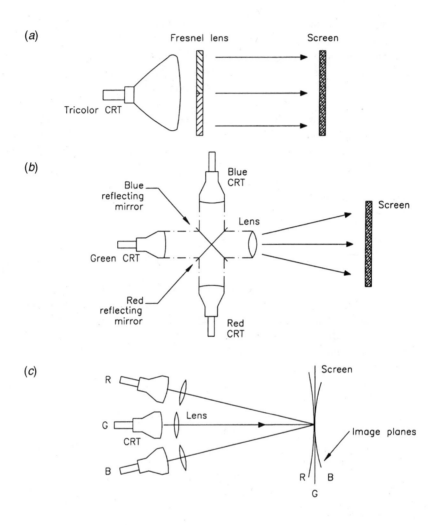

Figure 10.41 Optical projection configurations: (*a*) single-tube, single-lens rear-projection system; (*b*) crossed-mirror trinescope rear projection; (*c*) 3-tube, 3-lens rear-projection system; (*d*, next page) folded optics, front projection with three tubes in-line and a dichroic mirror; (*e*, next page) folded optics, rear projection. (*After* [26].)

most projection systems as a result of light-ray obliquity through the lens ($cos^4 \theta$ *law*) and *vignetting effects* [33].

Light output from a lens is determined by collection efficiency and transmittance, as well as source luminance. Typical figures for these characteristics are:

- Collection efficiency: 15 to 25 percent

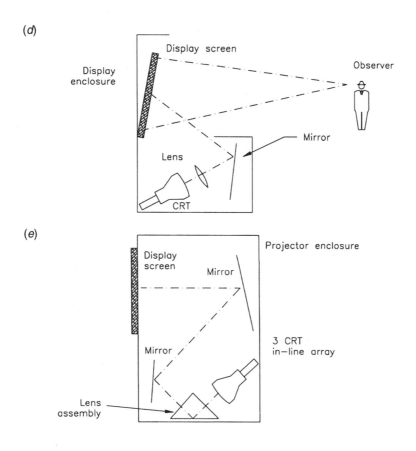

Figure 10.41 continued

- Transmittance: 75 to 90 percent

Collection efficiency is partially a function of the light source, and the figure given is typical for a lambertian source (CRT) and lens having a half-field angle of approximately 25°.

Optical Distortions

Optical distortions are important to image geometry and resolution [26]. Geometry generally is corrected electronically, for both pincushion/barrel effects and keystoning, which result from the fact that the three image sources are not coaxially disposed in the common in-line array.

Resolution, however, is affected by lens astigmatism, coma, spherical aberration, and chromatic aberration. The first three factors are dependent upon the excellence of the lens,

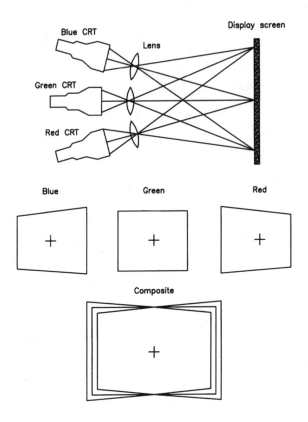

Figure 10.42 A 3-tube in-line array. The optical axis of the outboard (red and blue) tubes intersect the axis of the green tube at screen center. Red and blue rasters show trapezoidal (keystone) distortion and result in misconvergence when superimposed on the green raster, thus requiring electrical correction of scanning geometry and focus. (*After* [26].)

but chromatic aberration can be minimized by using line emitters (monochromatic) or narrowband emitters for each of the three image sources. Because a specific lens design possesses different magnification for each of the three primary colors (the index of refraction varies with wavelength), throw distance for each also must be adjusted independently to attain proper registration.

To determine the ultimate luminance performance of the device, transmission, reflectance, and scattering by additional optical elements such as dichroic filters, optical-path folding mirrors, or corrective lenses must also be accounted for. Dichroics exhibit light attenuations of 5 to 30 percent, and mirrors can reduce light transmission by as much as 5 percent each. Front-surface mirrors exhibit minimum absorption and scattering, but they are susceptible to damage during cleaning.

Contrast also is affected by the number and nature of optical elements employed in the projection system. Each optical interface generates internal reflections and scattering, which dilute contrast and reduce MTF amplitude response. Optical coatings may be utilized to minimize these effects, but their contribution must be balanced against their cost.

Image Devices

CRTs and light valves are the two most common devices for creating images to be optically projected [26]. Each is available in a multitude of variations. Projection CRTs have historically ranged in size from 1-in (2.5 cm) to 13-in (33 cm) diagonal (diameter for round envelope types). Because screen power must increase in proportion to the square of the magnification ratio, it is clear that faceplate dissipation for CRTs used in projection systems must be extremely high. Electrical-to-luminance conversion efficiency for common video phosphors is on the order of 15 percent [34]. A 50-in (1.3 m) diagonal measure screen at 60 ft-L requires a 5-in (12.7 cm) CRT to emit 6000 ft-L, exclusive of system optical losses, resulting in a faceplate dissipation of approximately 20 W in a 3-in (7.6 cm) by 4-in (10.2 cm) raster. A practical limitation for ambient air-cooled glass envelopes (to minimize thermal breakage) is 1 mW/mm^2 or 7.74 W for this size display. This incompatibility must be accommodated through either improved phosphor efficiency or reduced strain on the envelope via cooling. Because phosphor development is a mature science, maximum benefits are found in the latter course of action, with liquid-cooling assemblies employed to equalize differential strain on the CRT faceplate. Such an implementation produces an added benefit through reduction of phosphor *thermal quenching*, thereby supplying up to 25 percent more luminance output than is attainable in an uncooled device at equal screen dissipation [35].

A liquid-cooled CRT assembly, shown in Figure 10.43, depends upon a large heat sink to carry away and dissipate a substantial portion of the heat generated in the CRT. Large-screen projectors using such assemblies commonly operate CRTs at 4 to 5 times their rated thermal capacities. Economic constraints must be weighed against the added cost of cooling assemblies, however, and methods to improve phosphor conversion efficiency and optical coupling/transmission efficiencies continue to be investigated.

Concomitant to high power is high voltage and/or high beam current. Each has its benefits and penalties. Resolution, dependent on spot diameter, is improved by increased anode voltage and reduced beam current. For a 525- or 625-line display, spot diameter should be 0.006-in (0.16 mm) on the 3 × 4-in (7.6 × 10.2 cm) raster previously discussed. Higher-resolving-power displays require yet smaller spot diameters, but a practical maximum anode-voltage limit is 30 to 32 kV when x-radiation, arcing, and stray emission are considered. Exceeding 32 kV typically requires special shielding and CRT processing during assembly.

One form of the in-line array benefits from the relatively large aperture and light transmission efficiency of Schmidt reflective optics by combining the electron optics and phosphor screen with the projection optics in a single tube. The principal components of an integral Schmidt system are shown in Figure 10.44. Electrons emitted from the electron gun pass through the center opening in the spherical mirror of the reflective optical system to scan a metal-backed phosphor screen. Light from the phosphor (red, green, or blue, depending on the color channel) is reflected from the spherical mirror through an *aspheric corrector lens*, which serves as the face of the projection tube. Schmidt reflective optical systems

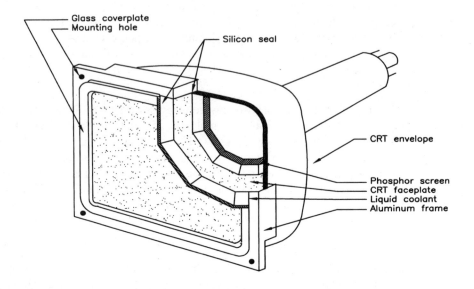

Figure 10.43 Mechanical construction of a liquid-cooled projection CRT.

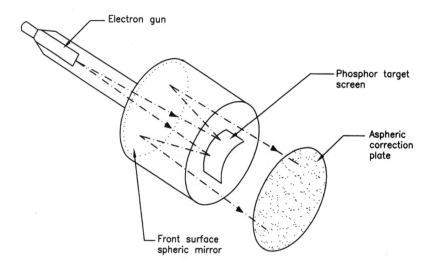

Figure 10.44 Projection CRT with integral Schmidt optics.

are significantly more efficient than refractive systems because of the lower f characteristic and the reduced attenuation by glass in the optical path.

Application Example

A CRT-based 61-in high-definition projection television conforming to the ATSC standard and intended for consumer applications has been developed (Hitachi) that provides 900 TV lines resolution and 550 Cd/m^2 peak brightness with a contrast ratio of 100. The key elements of the new display device include the following [36]:

- **Projection lens system**. The design requirements included minimum distortion, necessary because of the high-resolution displayed image, and high optical efficiency for maximum brightness. The system consists of a spherical glass lens and five aspherical plastic lenses, which are used primarily to control monochromatic aberrations.

- **Large diameter CRT gun**. A 36.5-mm diameter CRT neck was used to house the double-cylinder type *high focus voltage unipotential focusing* (Hi-UPF) electron gun. The gun has three lenses to finely control and shape the emitted beam with a minimum of distortion.

- **Deflection system**. In general, the deflection sensitivity of the deflection yoke (DY) and convergence yoke (CY) is reduced as a result of the larger neck diameter of the tube. Thus, more power must be delivered to the DY and CY devices. Coupled with this requirement was the higher horizontal scanning frequency of 33.75 kHz, rather than the conventional 15.73 kHz for a consumer television receiver. The higher frequency operation brought into play the *skin effect*, which required considerable changes in the windings to achieve the required performance.

- **Video driver circuit**. The requirements of high-definition operation demanded considerable improvements in the video driver, relative to a conventional display. To achieve high brightness, the circuit must operate up to approximately 150 V p-p at a bandwidth of 30 MHz. These performance levels are not unusual in themselves—high-end computer monitors achieve these levels and greater—but in the case of a consumer receiver, the cost also must be reasonable for commercial success.

- **Digital convergence system**. A two-mode convergence system—a *full scan mode* and a so-called *smooth wide mode*—was implemented to maintain the accuracy of beam landing across the face of the device. The convergence accuracy is 12 bits for vertical R, G, and B components. Horizontally, G resolution is 10 bits and both B and R are 9 bits. The resolution of the green channel is higher than that of the red and blue channels because the eye is more sensitive to green. Eight photo sensors are mounted on the screen frame to facilitate the auto convergence function.

The display is packaged as a rear-projection system. The 61-in diagonal picture size is set in a 16:9 aspect ratio. The effective optical resolution of the display is 900 × 1600 pixels. The total resolution in pixels is 1400 and the screen pitch is 0.72.

10.3.5 Light-Valve Systems

Light valves may be defined as devices that, like film projectors, employ a fixed light source modulated by an optical-valve intervening source and projection optics [26]. Although light-valve displays have been commercially available for some time, it is still a

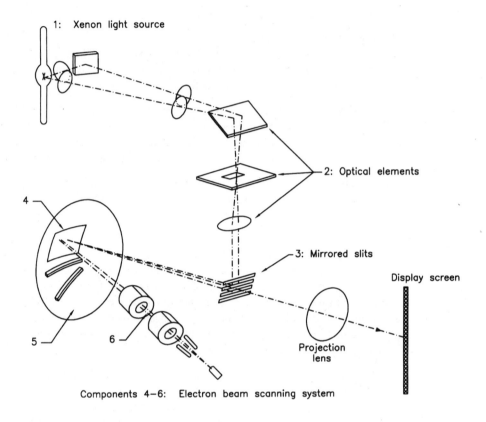

Figure 10.45 Mechanical configuration of the Eidophor projector optical system. (*After* [26].)

rapidly developing discipline. Light-valve systems offer high brightness, variable image size, and high resolution. Progress in light-valve technology for HDTV depends upon developments in two key areas:

- Materials and technologies for light control

- Integrated electronic driving circuits for addressing picture elements

Eidophor Reflective Optical System

Light-valve systems are capable of producing images of substantially higher resolution than are required for 525-/625-line systems [26]. They are ideally suited to large-screen theater displays of HDTV. The *Eidophor* system (Gretag) is in common usage.

In a manner similar to the operation of a film projector, a fixed light source is modulated by an optical valve system (Schlieren optics) located between the light source and the projection optics (see Figure 10.45). In the basic Eidophor system, collimated light—typically

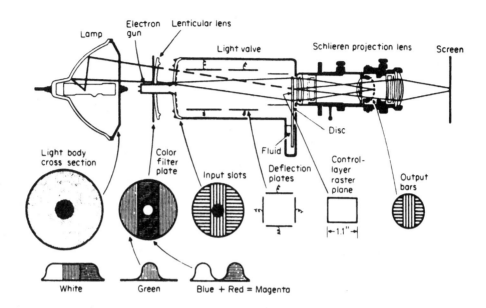

Figure 10.46 Functional operation of the General Electric single-gun light-valve system. (*After* [26].)

from a 2 kW xenon source (component 1 in the figure)—is directed by a mirror to a viscous oil surface in a vacuum by a grill of mirrored slits (component 3).

The slits are positioned relative to the oil-coated reflective surface so that when the surface is flat, no light is reflected back through the slits. An electron beam scanning the surface of the oil with a video picture raster (components 4 to 6) deforms the surface in varying amounts, depending upon the video modulation of the scanning beam. Where the oil is deformed by the modulated electron scanning beam, light rays from the mirrored slits are reflected at an angle that permits them to pass through the slits to the projection lens. The viscosity of the liquid is high enough to retain the deformation over a period slightly greater than a television field.

Projection of color signals is accomplished through the use of three units, one for each of the red, green, and blue primary colors converged on a screen.

Talaria Transmissive Color System

The *Talaria* system (General Electric) also uses the principle of deformation of an oil film to modulate light rays with video information [26]. The oil film, however, is transmissive rather than reflective. In addition, for full-color displays, only one gun is used to produce red, green, and blue colors. This is accomplished in a single light valve by the more complex Schlieren optical system shown in Figure 10.46.

Colors are created by writing *diffraction grating*, or grooves, for each pixel on the fluid by modulating the electron beam with video information. These gratings break up the transmitted light into its spectral colors, which appear at the output bars where they are spatially filtered to permit only the desired color to be projected onto the screen.

Green light is passed through the horizontal slots and is controlled by modulating the width of the raster scan lines. This is accomplished by means of a high-frequency carrier, modulated by the green information applied to the vertical deflection plates. Magenta light, composed of red and blue primaries, is passed through the vertical slots and is modulated by diffraction gratings created at right angles (*orthogonal diffraction*) to the raster lines by velocity-modulating the electron beam in the horizontal direction. This typically is achieved by applying 16 and 12 MHz carrier signals for red and blue, respectively, to the horizontal deflection plates and modulating them with the red and blue video signals. The grooves created by the 16 MHz carrier have the proper spacing to diffract the red portion of the spectrum through the output slots while the blue light is blocked. For the 12 MHz carrier, the blue light is diffracted onto the screen while the red light is blocked. The three primary colors are projected simultaneously onto the screen in register as a full-color picture.

To meet the requirements of HDTV, the basic Talaria system can be modified as shown in Figure 10.47. In the system (Talaria MLV-HDTV), one monochromatic unit with green dichroic filters produces the green spectrum. Because of the high scan rate for HDTV, the green video is modulated onto a 30 MHz carrier instead of the 12 or 15 MHz used for 525- or 625-line displays. Adequate brightness levels are produced using a 700 W xenon lamp for the green light valve and a 1.3 kW lamp for the magenta (red and blue) light valve.

A second light valve with red and blue dichroic filters produces the red and blue primary colors. The red and blue colors are separated through the use of orthogonal diffraction axes. Red is produced when the writing surface diffracts light vertically. This is accomplished by negative amplitude modulation of a 120 MHz carrier, which is applied to the vertical diffraction plates of the light valve. Blue is produced when the writing surface diffracts light horizontally. This is accomplished by modulating a 30 MHz carrier with the blue video signal and applying it to the horizontal plate, as is done in the green light valve.

The input slots and the output bar system of the conventional light valve are used, but with wider spacing of the bars. Therefore, the resolution limit is increased. The wider bar spacing is achievable because the red and blue colors do not have to be separated on the same diffraction axis as in the single light-valve system. This arrangement eliminates the cross-color artifact present with the single light-valve system, and therefore improves the overall colorimetric characteristics.

High-performance electron guns help provide the required resolution and modulation efficiency for HDTV systems. The video carriers are optimized to increase the signal bandwidth capability to approximately 30 MHz.

In the 3-element system, all three devices are monochrome light valves with red, blue, and green dichroic filters. The use of three independent light valves improves color brightness, resolution, and colorimetry. Typically, the three light valves are individually illuminated by xenon arc lamps operating at 1 kW for the green and at 1.3 kW for the red and blue light valves.

The contrast ratio is an important parameter in light-valve operation. The amount of light available from the arc lamp is basically constant; the oil film modulates the light in response

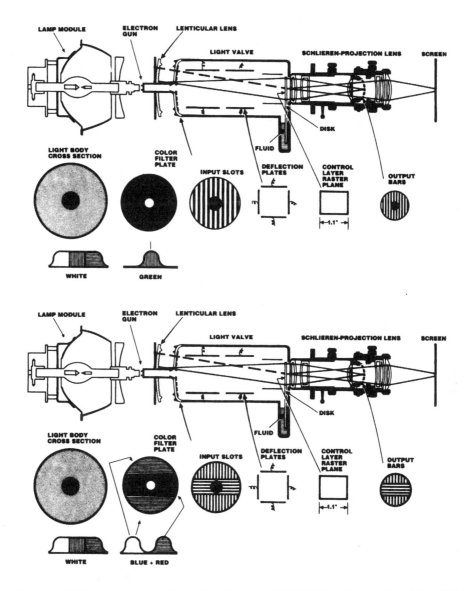

Figure 10.47 Functional operation of the 2-channel HDTV light-valve system. (*After* [26].)

to picture information. The key parameter is the amount of light that is blocked during picture conditions when totally dark scenes are being displayed (the *darkfield* performance). Another important factor is the capability of the display device to maintain a linear relationship between the original scene luminance and the reproduced picture. The amount of pre-

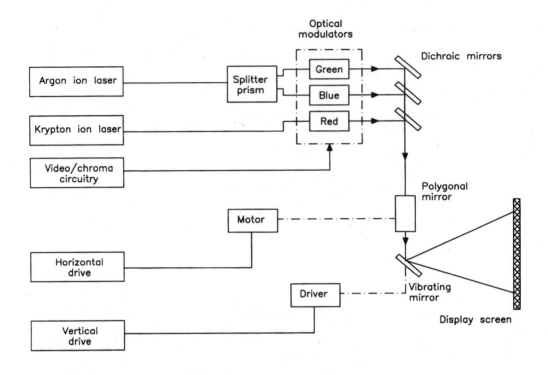

Figure 10.48 Block diagram of a laser-scanning projection display. (*After* [26].)

distortion introduced by the camera must be compensated for by an opposite amount of distortion at the display device.

Laser-Beam Projection Scanning System

Several approaches to laser projection displays have been implemented [26]. The most successful employs three optical laser light sources whose coherent beams are modulated electro-optically and deflected by electromechanical means to project a raster display on a screen. The scanning functions typically are provided by a rotating polygon mirror and a separate vibrating mirror. A block diagram of the basic system is shown in Figure 10.48.

The flying spots of light used in this approach are one scan line (or less) in height and a small number of pixels wide. This means that any part of the screen may be illuminated for only a few nanoseconds. The scanned laser light projector is capable of high contrast ratios (as high as 1000:1) in a darkened environment. A laser projector, however, may be subject to a brightness variable referred to as *speckle*. Speckle is a sparkling effect resulting from interference patterns in coherent light. The effect causes a flat, dull projection surface to look as if it has a beaded texture. This tends to increase the perception of brightness, at the expense of image quality.

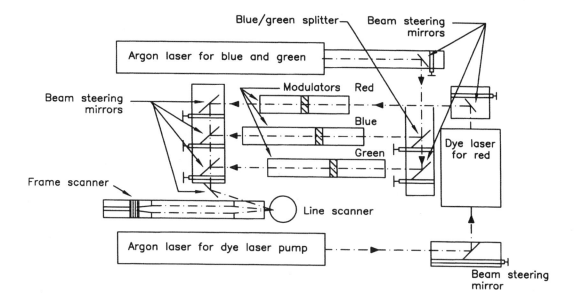

Figure 10.49 Configuration of a color laser projector. (*After* [37].)

Figure 10.49 shows the configuration of a laser projector using continuous-wave lasers and mechanical scanners. The requirements for the light wavelengths of the lasers are critical. The blue wavelength must be shorter than 477 nm, but as long as possible. The red wavelength should be longer than 595 nm, but as short as possible. Greens having wavelengths of 510, 514, and 532 nm have been used with success. Because laser projectors display intense colors, small errors in color balance can result in significant distortions in gray scale or skin tone. The requirement for several watts of continuous-wave power further limits the usable laser devices. Although several alternatives have been considered, color laser projectors typically use argon ion lasers for blue and green, and an argon ion pumped-dye laser for red.

Production of a conventional video signal requires a modulator with a minimum operating frequency of 6 MHz. Several approaches are available. The bulk acousto-optic modulator is well suited to this task in high-power laser projection. A directly modulated laser diode is used for some low-power operations. The modulator does not absorb the laser beam; instead, it deflects it as needed. The angle of deflection is determined by the frequency of the acoustic wave.

The scanner is the component of the laser projector that moves the point of modulated light across and down the image plane. Several types of scanners may be used, including mechanical, acousto-optic, and electro-optic. Two categories of scanning devices are used in the system:

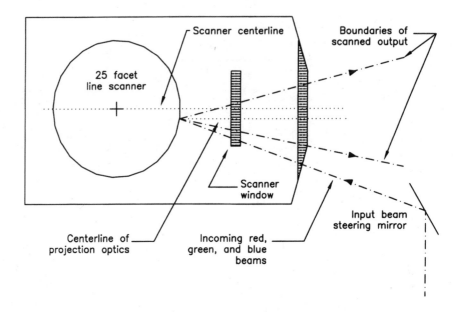

Figure 10.50 Rotating polygon line scanner for a laser projector. (*After* [37].)

- *Line scanner*, which scans the beam in horizontal lines across the screen. Lines are traced at 15,000 to 70,000 times per second. The rotating polygon scanner is commonly used, featuring 24 to 60 (or more) mirrored facets. Figure 10.50 illustrates a mechanical rotating polygon scanner.

- *Frame scanner*, which scans the beam vertically and forms frames of horizontal scan lines. The frame scanner cycles 50 to 120 (or more) times per second. A galvanometer-based scanner typically is used in conjunction with a mirrored surface. Because the mirror must fly back to the top of the screen in less than a millisecond, the device must be small and light. Optical elements are used to force the horizontally scanned beam into a small spot on the frame scanner mirror. The operating speed of the deflection components may be controlled from an internal clock, or it may be locked to an external time base.

- The laser projector is said to have *infinite focus*. To accomplish this, optics are necessary to keep the scanned laser beam thin. Such optics allow a video image to remain in focus from approximately 4 ft to infinity. For high-resolution applications, the focus is more critical.

Heat is an undesirable by-product of laser operation. Most devices are water-cooled and use a heat exchanger to dump the waste heat.

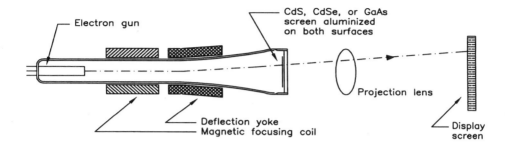

Figure 10.51 Laser-screen projection CRT. (*After* [26].)

Electron-Beam Projection System

Another approach to large-screen display employs an electron-beam pumped monocrystal-line screen in a CRT to produce a 1-in raster [26]. The raster screen image is projected by conventional optics, as shown in Figure 10.51. This technology promises three optical benefits:

- High luminance in the image plane

- Highly directional luminance output for efficient optical coupling

- Compact, lightweight, and inexpensive projection optics

Full-color operation is accomplished using three such devices, one for each primary color.

10.3.6   LCD Projection Systems

Liquid crystal displays have been widely employed in high-information content systems and high-density projection devices. The steady progress in active-matrix liquid crystal displays, in which each pixel is addressed by means of a thin-film transistor (TFT), has led to the development of full-color video TFT-addressed *liquid crystal light valve* (LCLV) projectors. Compared with conventional CRT-based projectors, LCLV systems have a number of advantages, including:

- Compact size

- Light weight

- Low cost

- Accurate color registration

Total light output (brightness) and contrast ratio, however, are issues of concern. Improvements in the transmittance of polarizer elements have helped increase display brightness. By arrangement of the direction of the polarizers, the LCD can be placed in either the *normally*

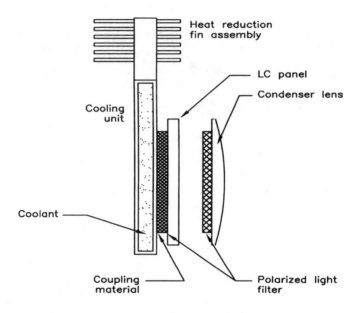

Figure 10.52 Structure of a liquid-cooled LC panel. (*After* [38].)

black (NB) or the *normally white* (NW) mode. In the NW mode, light will be transmitted through the cell at V_{off} and will be blocked at V_{on}. The opposite situation applies for the NB mode. A high-contrast display cannot be obtained without a satisfactory dark state.

Cooling the LC panels represents a significant technical challenge as the light output of projection systems increases. The contrast ratio of the displayed image will decrease as the temperature of the LC panels rises. Furthermore, long-term operation under conditions of elevated temperature will result in shortened life for the panel elements.

The conventional approach to cooling has been to circulate forced air over the LC panels. This approach is simple and works well for low- to medium-light output systems. Higher-power operation, however, may require liquid-cooling, as illustrated in Figure 10.52. The cooling unit, mounted behind the LC panel, is made of frame glass and two glass panels. The coolant is a mixture of water and ethylene glycol, which prevents the substance from freezing at low temperatures.

Figure 10.53 compares the temperature in the center of the panel for conventional air-cooling and for liquid-cooling. Because LC panel temperature is a function of the brightness distribution of the light source, the highest temperature is at the center of the panel, where most of the light usually is concentrated. As the amount of light from the lamp increases, the cooling activity of the liquid-based system accelerates. To carry waste heat away from the projector, cooling air is directed across the heat-reduction fins.

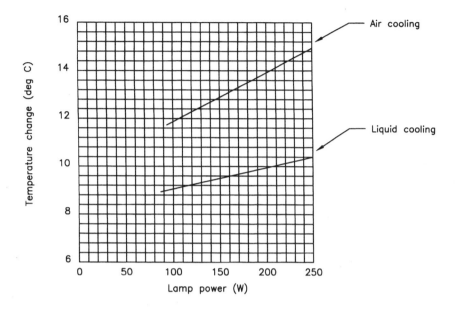

Figure 10.53 LC panel temperature dependence on lamp power in a liquid crystal projector. (*After* [38].)

In addition to the basic LCD projector, a number of variations on the fundamental technology have been developed, including:

- Homeotropic LCLV

- Laser-addressed LCLV

- Image light amplifier

- Digital micromirror device

The applications of these technologies are discussed in the following sections.

Homeotropic LCLV

The *homeotropic* (perpendicular alignment) LCLV (Hughes) is based on the optical switching element shown in Figure 10.54 [39]. Sandwiched between two transparent idium-tin oxide electrodes are a layer of cadmium sulfide photoconductor, a cadmium telluride light-blocking layer, a dielectric mirror, and a 6-µm-thick layer of liquid crystal. An ac bias voltage connects the transparent electrodes. The light-blocking layer and the dielectric mirror are thin and have high dielectric constants, so the ac field is primarily across the photoconductor and the liquid crystal layer. When there is no writing light impinging on the photo-

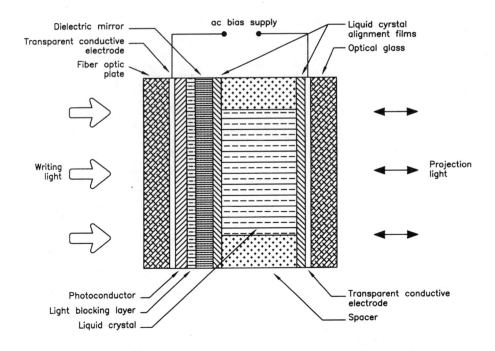

Figure 10.54 Layout of a homeotropic liquid crystal light valve. (*After* [39].)

conductor, the field or voltage drop is primarily across the photoconductor and not across the liquid crystal. When a point on the photoconductor is activated by a point of light from an external writing CRT, the impedance at that point drops and the ac field is applied to the corresponding point in the liquid crystal layer. The lateral impedances of the layers are high, so the photocarriers generated at the point of exposure do not spread to adjacent points. Thus, the image from the CRT that is exposing the photoconductor is reproduced as a voltage image across the liquid crystal. The electric field causes the liquid crystal molecules to rotate the plane of polarization of the projected light that passes through the liquid crystal layer.

The homeotropic LCLV provides a perpendicular alignment in the *off* state (no voltage across the liquid crystal) and a pure optical birefringence effect of the liquid crystal in the *on* state (voltage applied across the liquid crystal). To implement this electro-optical effect, the liquid crystal layer is fabricated in a perpendicular alignment configuration; the liquid crystal molecules at the electrodes are aligned with their long axes perpendicular to the electrode surfaces. In addition, they are aligned with their long axes parallel to each other along a preferred direction that is fabricated into the surface of the electrodes.

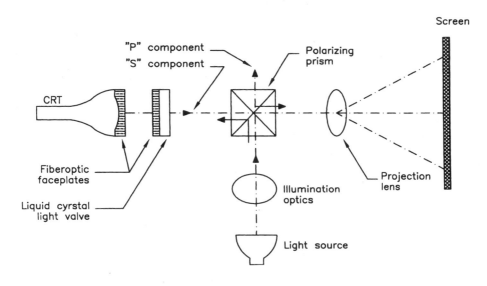

Figure 10.55 Block diagram of an LCLV system. (*After* [39].)

A conceptual drawing of an LCLV projector using this technology is shown in Figure 10.55. Light from a xenon arc lamp is polarized by a polarizing beam-splitting prism. The prism is designed to reflect *S* polarized light (the polarization axis, the E-field vector, is perpendicular to the plane of incidence) and transmit *P* polarized light (the polarization axis, the E-field vector, is parallel to the plane of incidence). The *S* polarization component of the light is reflected to the light valve. When the light valve is activated by an image from the CRT, the reflecting polarized light is rotated 90° and becomes *P* polarized. The *P* polarization component transmits through the prism to the projection lens. In this type of system, the prism functions as both a polarizer and an analyzer. Both monochrome and full-color versions of this design have been implemented. Figure 10.56 shows a block diagram of the color system.

The light valve is coupled to the CRT assembly by a fiber optic backplate on the light valve and a fiber optic faceplate on the CRT.

The heart of this system is the fluid-filled prism assembly. This device contains all of the filters and beam-splitter elements that separate the light into different polarizations and colors (for the full-color model). Figure 10.57 shows the position of the beam-splitter elements inside the prism assembly.

The light that exits the arc lamp housing enters the prism through the ultraviolet (UV) filter. The UV filter is a combination of a UV reflectance dichroic on a UV absorption filter. The UV energy below 400 nm can be destructive to the life of the LCLV. Two separate beam-splitters *prepolarize* the light: the green prepolarizer and the red/blue prepolarizer. Prepolarizers are used to increase the extinction ratio of the two polarization states. The

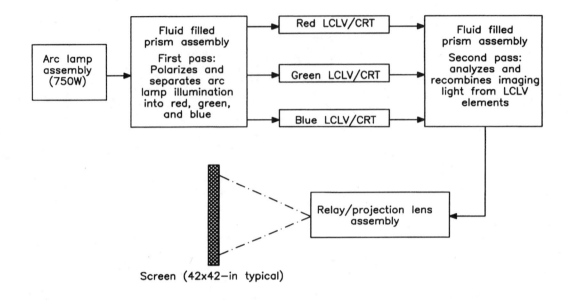

Figure 10.56 Functional optical system diagram of an LCLV projector. (*After* [39].)

green prepolarizer is designed to reflect only the *S* polarization component of green light. The *P* polarization component of green and both the *S* and *P* polarizations of red and blue are transmitted through the green prepolarizer. The green prepolarizer is oriented 90° with respect to the other beam-splitters in the prism. This change of orientation means that the transmitted *P* component of the green light appears to be *S* polarized relative to the other beam-splitters in the prism assembly. The red/blue prepolarizer is designed to reflect the *S* polarization component of red and blue. The remaining green that is now *S* polarized and the red and blue *P* polarized components are transmitted to the main beam-splitter.

The main beam-splitter is a broadband polarizer designed to reflect all visible *S* polarized light and transmit all visible *P* polarized light. The *S* polarized green light is reflected by the main beam-splitter to the green mirror. The green mirror reflects the light to the green LCLV. The *P* polarized component of red and blue transmits through the main polarizer. The red/blue separator reflects the blue light and transmits the red to the respective LCLVs.

Image light is rotated by the light valve and reflected to the prism. The rotation of the light or phase change (*P* polarization to *S* polarization) allows the image light from the red and blue LCLVs to reflect off the main beam-splitter toward the projection lens. When the green LCLV is activated, the green image light is transmitted through the main beam-splitter.

The fluid-filled prism assembly is constructed with thin-film coatings deposited on quartz plates and immersed in an index-matching fluid. Because polarized light is used, it is necessary to eliminate stress birefringence in the prism, which, if present, would cause con-

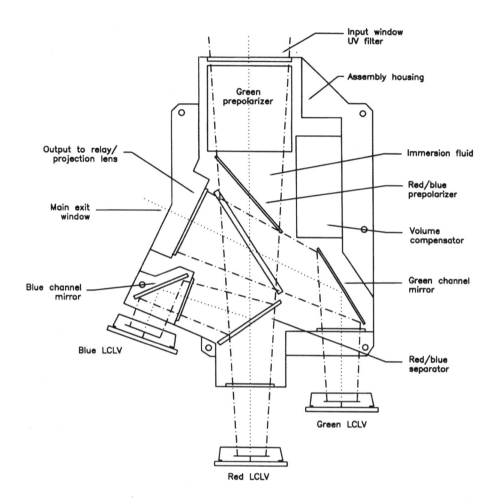

Figure 10.57 Layout of the fluid-filled prism assembly for an LCLV projector. (*After* [39].)

trast degradation and background nonuniformity in the projected image. Using a fluid-filled prism effectively eliminates mechanically and thermally induced stress birefringence, as well as residual stress birefringence. Expansion of the fluid with temperature is compensated for by the use of a bellows arrangement.

Laser-Addressed LCLV

The laser-addressed liquid crystal light valve is a variation on the CRT-addressed LCLV that was discussed in the previous section. Instead of using a CRT to write the video information

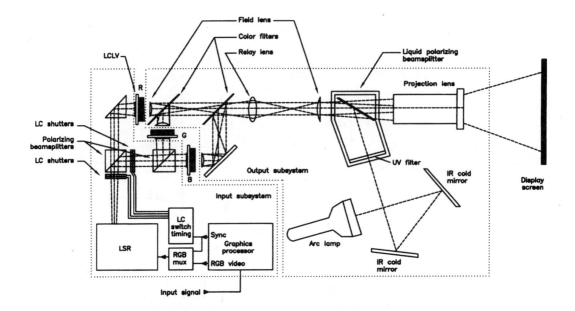

Figure 10.58 Block diagram of a color laser-addressed light-valve display system. (*After* [40].)

on the LC element, however, a laser is used [40]. The principal advantage of this approach is the small spot size possible with laser technology, a factor that increases the resolution of the displayed image. In addition, a single beam from a laser scanner can be directed to three separate RGB light valves; the use of only one source for the image reduces the potential for convergence problems. A block diagram of the scheme is shown in Figure 10.58. Major components of the system include:

- Laser
- Video raster scanner
- Polarizing multiplexing switches
- LCLV assemblies
- Projection optics

Projector operation can be divided into two basic subsystems:

- Input subsystem, which includes the laser, scanning mechanism, and polarizing beam-splitters
- Output subsystem, which includes the LCLV assemblies, projection optics, and light source

The output subsystem is similar in nature to the homeotropic LCLV discussed previously.

The heart of the input subsystem is the *laser raster scanner* (LRS). The X-Y deflection system incorporated into the LRS uses four acousto-optic devices to achieve all solid-state modulation and scanning. A diode-pumped Nd:YAG, doubled to 535 nm (18 mW at 40:1 polarization ratio), is used as the laser source for the basic raster scan system.

Because only a single-channel scanner is used to drive the LCLV assemblies, a method is required for sequencing the RGB video fields to their respective light valves. This sequencing is accomplished through the use of LC polarization switches. These devices act in a manner similar to the LCLV in that they change the polarization state of incident light. In this case, the intent is to rotate the polarization by $0°$ or $90°$ and exploit the beam-steering capabilities of polarizing beam-splitters to direct the image field to the proper light valve at the proper time.

As shown in Figure 10.58, each switch location actually consists of a pair of individual switches. This feature was incorporated to achieve the necessary fall time required by the LRS to avoid overlapping of the RGB fields. Although these devices have a short rise time (approximately 100 μs), their decay time is relatively long (approximately 1 ms). By using two switches in tandem, compensation for the relatively slow relaxation time is accomplished by biasing the second switch to relax in an equal, but opposite, polarization sense. This causes the intermediate polarization state that is present during the switch-off cycle to be nulled out, allowing light to pass through the LC switch during this phase of the switching operation.

Image Light Amplifier (ILA)

The Image Light Amplifier (Hughes, JVC) combines the advantages of a CRT-type projector with the high-brightness capability of a xenon arc lamp projection source. The enabling technology of the projector is the ILA modulator, which accepts a low-intensity image from a CRT and replicates the image on a high-intensity white xenon arc lamp beam. The ILA uses solid-state thin-film and liquid crystal technologies to produce a system with up to 12,000 lm output, at a contrast ratio of 1000:1. Resolutions of 2800×1500 TV lines (and higher) have been produced [41]. The ILA projector incorporates three modulators for the RGB channels of a color system.

The ILA light valve (ILA-LV) is a spatial light modulator that accepts a low-intensity input image and converts it, in real time, into an output image with light from another source. [42]. In the system, the image light sources are high-resolution CRTs, and the output light source is a xenon arc lamp. The ILA-LV is designed to operate in a reflective mode so that the input CRT and output xenon light beam are incident on opposite faces of the device. A cross-sectional diagram of the ILA-LV is shown in Figure 10.59. Note the similarities to the homeotropic LCLV shown in Figure 10.54.

The basic ILA projector optical channel is shown in Figure 10.60. This is the building block for the full-color system, which is shown in Figure 10.61. The input image, provided by the high-resolution CRT, is imaged on the ILA-LV through a relay lens. The xenon arc lamp and condensing optics provide the output projection light beam, which is linearly polarized by a McNeille-type *polarizing beam-splitter* (PBS) before reaching the ILA-LV. The PBS polarizes the light to a high extinction ratio without absorption. The projected

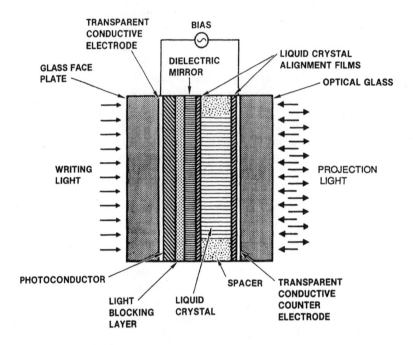

Figure 10.59 Cross-sectional diagram of the Image Light Amplifier light valve. (*After* [41]. *Courtesy of Hughes-JVC Technology.*)

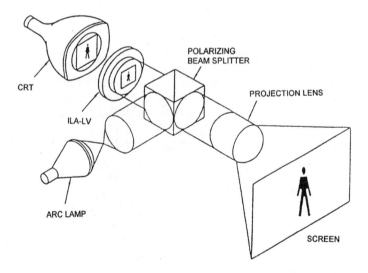

Figure 10.60 Optical channel of the ILA projector. (*After* [36]. *Courtesy of Hughes-JVC Technology.*)

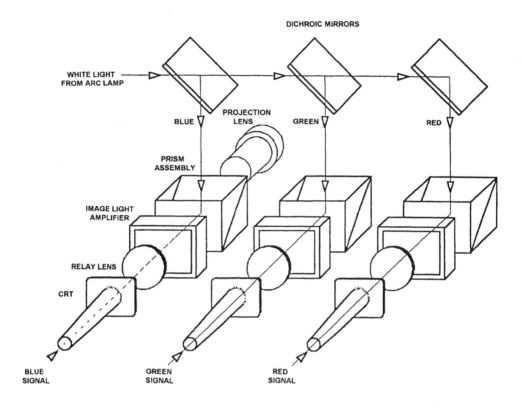

Figure 10.61 Full-color optical system for the ILA projector (Series 300/400). (*After* [41]. *Courtesy of Hughes-JVC Technology.*)

beam passes through the liquid crystal, reflects from the dielectric mirror, and passes through the liquid crystal again before returning to the polarizing beam-splitter.

As the beam passes through the liquid crystal, the direction of the linearly polarized light is rotated in direct response to the level of input image modulation of the liquid crystal birefringence. The PBS then operates on the output image from the ILA-LV, passing rotated polarized light to the projection lens and returning nonrotated light toward the lamp. Finally, the projection lens focuses and magnifies the ILA-LV image onto a screen.

Among the benefits of this system are:

- High brightness

- Undefined pixel structure (which permits a variety of input signal types and formats)

- High resolution

- Good contrast ratio

- Excellent color rendition, as illustrated in Figure 10.62

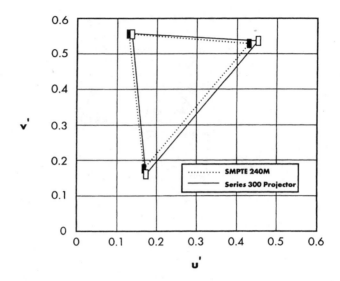

Figure 10.62 Color gamut of the ILA projector (Series 300/400) compared with SMPTE 240M. (*After* [41]. *Courtesy of Hughes-JVC Technology.*)

Digital Micromirror Device

DMD is a semiconductor-based array of fast, reflective digital light switches that precisely control a light source using a binary pulse-width modulation (PWM) technique [42]. Individual DMD elements can be combined with image processing, memory, a light source, and optics to form a *Digital Light Processing* (DLP, Texas Instruments) system capable of projecting high-resolution color images. DLP-based projection displays are well suited to high-brightness and high-resolution applications. Attributes of the technology include the following:

- The digital light switch is reflective and has a high *fill factor*, resulting in high optical efficiency at the pixel level and low pixelation effects in the projected image.

- As the resolution and size of the DMD array increases, the overall system optical efficiency grows because of higher lamp-coupling efficiency.

- The DMD operates with conventional CMOS voltage levels (5 V), so integrated row and column drivers are readily employed to minimize the complexity and cost of scaling to higher resolutions.

- Because the DMD is a reflective technology, the DMD chip can be effectively cooled through the chip substrate, thus facilitating the use of high-power projection lamps without thermal degradation of the DMD.

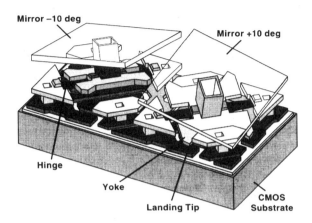

Mirror −10 deg

Mirror +10 deg

Hinge

Yoke

Landing Tip

CMOS
Substrate

Figure 10.63 A pair of DMD pixels with one mirror shown at −10° and the other at +10°. (*After* [42]. *Courtesy of Texas Instruments.*)

The DMD light switch, shown in Figure 10.63, is a member of a class of devices known as *microelectromechanical systems* (MEMS). Other MEMS devices include pressure sensors, accelerometers, and microactuators. The DMD is monolithically fabricated by CMOS-like processes over a CMOS memory element. Each light switch has a 16-μm-square aluminum mirror that can reflect light in one of two directions, depending on the state of the underlying memory cell. Rotation of the mirror is accomplished through electrostatic attraction produced by voltage differences developed between the mirror and the underlying memory. With the memory cell in the *on* (1) state, the mirror rotates to +10°; with the memory cell in the *off* (0) state, the mirror rotates to −10°.

When the DMD is combined with a suitable light source and projection optics, as illustrated in Figure 10.64, the mirror reflects incident light either into or out of the pupil of the projection lens by a simple beam-steering process. As a result, the (1) state of the mirror appears bright, and the (0) state of the mirror appears dark. Compared with diffraction-based light switches, the beam-steering action of the DMD light switch provides a favorable tradeoff between contrast ratio and the overall brightness efficiency of the system.

Image gray scale is achieved through pulse-width modulation of the incident light. Color is achieved by using color filters, either stationary or rotating, in combination with one, two, or three DMD chips. A detailed photo of a DMD element is shown in Figure 10.65.

The simultaneous update of all mirrors produces an inherently low flicker display; there is no line-to-line temporal phase shift. Furthermore, a bit-splitting PWM algorithm produces short-duration light pulses that are distributed uniformly throughout the video field time, eliminating a temporal decay in brightness.

DLP optical systems have been designed in a variety of configurations, distinguished by the number of DMD chip arrays in the system [43]. The 1- and 2-chip schemes rely on a rotating color disk to time-multiplex the colors. The 1-chip configuration is used for lower

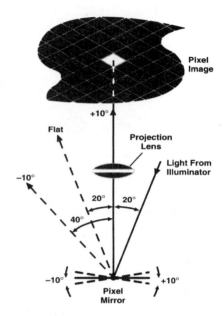

Figure 10.64 The basics of DMD optical switching. (*After* [42]. *Courtesy of Texas Instruments.*)

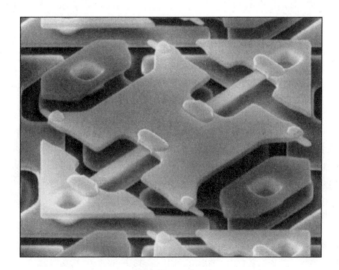

Figure 10.65 A scanning electron microscope view of the DMD yoke and spring tips (the mirror has been removed). (*After* [42]. *Courtesy of Texas Instruments.*)

brightness applications and is the most compact. The 2-chip systems yield higher brightness performance, but are primarily intended to compensate for the color deficiencies resulting

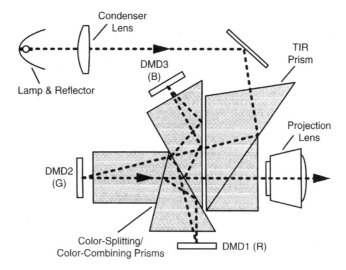

Figure 10.66 Optical system of the DLP 3-chip implementation. (*After* [42]. *Courtesy of Texas Instruments.*)

from spectrally imbalanced lamps (e.g., the red deficiency in many metal halide lamps). For applications requiring the highest brightness, 3-chip systems are used.

A 3-chip DLP optical system is shown in Figure 10.66. Because the DMD is a simple array of reflective light switches, no polarizers are required. Light from a metal halide or xenon lamp is collected by a condenser lens. For proper operation of the DMD light switch, this light must be directed at 20° relative to the normal of the DMD chip. To accomplish this in a method that eliminates mechanical interference between the illuminating and projecting optics, a *total internal reflection* (TIR) prism is interposed between the projection lens and the DMD color-splitting/color-combining prisms.

The color-splitting/color-combining prisms use dichroic interference filters deposited on their surfaces to split the light by reflection and transmission into red, green, and blue components. The red and blue prisms require an additional reflection from a TIR surface of the prism in order to direct the light at the correct angle to the red and blue DMDs. Light reflected from the on-state mirrors of the three DMDs is directed back through the prisms, and the color components are recombined. The combined light then passes through the TIR prism and into the projection lens because its angle has been reduced to below the critical angle for total internal reflection in the prism air gap.

As the DMD resolution is increased, the pixel pitch is held constant and the chip diagonal is allowed to increase, as detailed in Figure 10.67. This approach to display design has several advantages:

• The high optical efficiency and contrast ratio of the pixel are maintained at all resolutions.

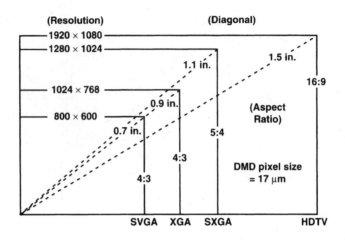

Figure 10.67 DMD display resolution as a function of chip diagonal. (*After* [42]. *Courtesy of Texas Instruments.*)

- Pixel timing is common to all designs, and high address margins are maintained.

- The chip diagonal increases with resolution, which improves the DMD system optical efficiency.

DLP projection systems have been demonstrated at a variety of resolutions (and aspect ratios), including a 16:9 aspect ratio high-definition (1920 × 1080) system [44, 45, 46].

10.3.7 Projection Requirements for Cinema Applications

Screen brightness is a critical element in providing an acceptable HDTV large-screen display. Without adequate brightness, the impact on the audience is reduced. Theaters have historically used front-projection systems for 35 mm film. This arrangement provides for efficient and flexible theater seating. Large-screen HDTV is likely to maintain the same arrangement. Typical motion picture theater projection-system specifications are as follows:

- Screen width: 30 ft

- Aspect ratio: 2.35:1

- Contrast ratio: 300:1

- Screen luminance: 15 ft-L

- Center-to-edge brightness uniformity: 85 percent

These specifications meet the expectations of motion picture viewers. It follows that for HDTV projectors to be competitive in large-screen theatrical display applications, they must

Table 10.10 Minimum Projector Specifications for Consumer and Theater Displays (*After* [47].)

Parameter	Consumer Display	Theater Display
Resolution	Greater than 750 TV lines	Greater than 750 TV lines
Light output	1000 lm	10,000 lm
Cost	Less than $2000	Less than $50,000
Response time	Less than 10 ms	Less than 10 ms
Power consumption	Less than 300 W	Less than 3000 W
Small area uniformity	± 0.25 percent	± 0.25 percent
Contrast ratio	Greater than 90:1	Greater than 90:1
Flicker	Undetectable	Undetectable

meet similar specifications. The major hurdle for making the electronic cinema a commercial reality is the availability—actually, the lack—of high-performance HDTV projectors with theater brightness at a reasonable cost [47]. Illuminating a 20- × 36-ft theater screen with 20 ft-L (assuming a screen gain of 1.5) requires 9600 lm output from the projector. Table 10.10 compares key performance requirements for theater HDTV systems and consumer displays.

Operational Considerations

Accurate convergence is critical to display resolution. Figure 10.68 shows the degradation in resolution resulting from convergence error. It is necessary to keep convergence errors to less than half the distance between the scanning lines to hold resolution loss below 3 dB. Errors in convergence also result in color contours in a displayed image. Estimates have put the detectable threshold of color contours at 0.75 to 0.5 minutes of arc. This figure also indicates that convergence error must be held under 0.5 scanning lines.

Raster stability influences the short-term stability of the display. The signal-to-noise ratio (S/N) and raster stability relationships for deflection circuits are shown in Table 10.11. The S/N equivalent of one-fifth of a scanning line is necessary to obtain sufficient raster stability. In HDTV applications, this translates to approximately 80 dB S/N. Other important factors are high speed and improved efficiency of the deflection circuit.

Some manufacturers have begun to incorporate automatic convergence systems into their products. These systems usually take the form of a CCD camera sensor that scans various portions of the screen as test patterns are displayed.

10.4 Consumer Computer and Networking Issues

One of the promises of DTV is a marriage of video and computer technologies in a wide range of consumer devices. To accomplish this goal, several fundamental technologies and

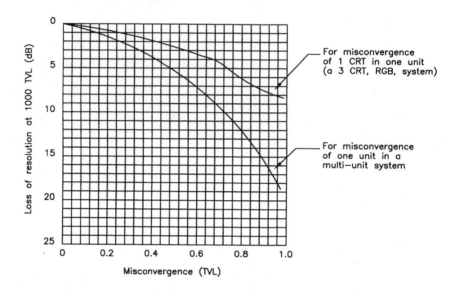

Figure 10.68 The relationship between misconvergence and display resolution. (*After* [48].)

Table 10.11 Appearance of Scan Line Irregularities in a Projection Display (*After* [48].)

Condition of Adjacent Lines	S/N (p-p)
Clearly overlapped	Less than 69 dB
Just before overlapped	75 dB
No irregularities	Greater than 86 dB

standards must be in place, and certain agreements must be reached among the many potential information-providers.

10.4.1 IEEE 1394

As discussed in Chapter 4, IEEE 1394 is an international standard, low-cost digital interface designed to integrate entertainment, communication, and computing electronics into consumer multimedia products [49]. Originated by Apple Computer as a desktop LAN and developed by the IEEE 1394 working group, IEEE 1394 has the following attributes:

- High data rate capabilities—the hardware and software standard can transport data at 100, 200, or 400 Mbits/s

- Physically small—the serial cable can replace larger and more expensive interfaces

- Ease of use—there is no need for terminators, device IDs, or elaborate setup procedures

- *Hot plug* capable—users can add or remove 1394 devices with the bus active

- Scaleable architecture—users can mix 100, 200, and 400 Mbits/s devices on a single bus

- Flexible topology—support is provided for daisy chaining and branching, facilitating true peer-to-peer communication

- Non-proprietary—there are no significant licensing issues for implementation in products or systems

Serial bus management provides overall configuration control of the 1394 bus in the form of optimizing arbitration timing, guarantee of adequate electrical power for all devices on the bus, assignment of which compliant device is the *cycle master*, assignment of the isochronous channel ID, and notification of errors. Bus management is built upon the IEEE 1212 standard register architecture.

There are two types of IEEE 1394 data transfer:

- **Asynchronous**: the traditional computer memory-mapped, load-and-store interface. Data requests are sent to a specific address and an acknowledgment is returned.

- **Isochronous**: data channels provide guaranteed data transport at a pre-determined rate. This is especially important for time-critical multimedia data where just-in-time delivery eliminates the need for costly buffering.

Much like LANs and WANs, IEEE 1394 is defined by the high level application interfaces that use it, not a single physical implementation. Therefore, as new silicon technologies allow higher speeds, longer distances, and alternate media, IEEE 1394 can scale to enable new applications.

Perhaps most important for use as a digital interface for consumer electronics is that IEEE 1394 is a peer-to-peer interface. This allows not only dubbing from one camcorder to another without a computer, for example, but also allows multiple computers to share a given camcorder without any special support in the camcorders or computers. All of these features of IEEE 1394 are key reasons why it is a preferred audio/video digital interface.

Enhancements

A number of extensions to IEEE 1394 were under consideration at this writing by the 1394 Trade Association and the IEEE 1394.1 Study Group. These extensions include the following [50]:

- Gigabit speeds for cables

- 100 Mbits/s for backplane implementations

- Longer-distance cables using copper wire and plastic fiber

- Audio/video command and control protocols

- 1394 to 1394 bus bridges

- IEEE 1394 gateways to communication interfaces, such as ATM

It is believed that ATM and IEEE 1394 will have beneficial effects on each other's markets. It has been predicted that ATM will become the world-wide standard for voice/video/data public switched networks. However, ATM is considered too expensive for devices such as disk drives, cameras, and desktop computers. IEEE 1394, therefore, is being positioned by proponents as a complementary technology, serving as the device interface for ATM systems.

10.4.2 Digital Home Network

It is expected that in the near-term future, data for audio, video, telephony, printing, and control functions are all likely to be transported through the home over a digital network [50]. This network will allow the connection of devices such as computers, digital TVs, digital VCRs, digital telephones, printers, stereo systems, and remotely controlled appliances. To enable this scope of interoperability of home network devices, standards for physical layers, network information, and control protocols need to be generally agreed upon and accepted. While it would be preferable to have a single stack of technology layers, no one selection is likely to satisfy all cost, bandwidth, and mobility requirements for in-home devices.

From a broadcasting perspective, the ability to provide unrestricted entertainment services to consumer devices in the home is a key point of interest. Also, ancillary data services directed to various devices offer significant marketplace promise. Control and protocol standards to enable the delivery of selected programming from cable set-top boxes to DTV sets using IEEE 1394 have been approved by the Consumer Electronics Manufacturers Association (CEMA) and the Society of Cable and Telecommunications Engineers (SCTE). The CEMA standard is EIA-775. The SCTE has approved two complementary standards, DVS 194 and DVS 195 (with copy protection—the *SC* system—and without copy protection).

The standards from these organizations differ in some respects, and the SCTE version has some troublesome aspects for broadcasters. For example, the SCTE standards require the cable set-top box to be the program selection device and only transfer data related to one selected program at a time to the 1394 bus. Accessing data from a broadcast stream that is unrelated to the current program selection on the cable box is also not defined in the SCTE standards.

More generally, the principal physical layer interconnections at the DTV set are expected to be RF (NTSC, VSB, QAM), baseband component (RGB, Y Pr Pb), and digital (1394). Composite video and S-video will be around a long time as well. All of these but 1394 are one-way paths.

Other appliances (or sensors) in the home may use RF on the power lines, dedicated coax, and dedicated twisted pair for control functions. From this plethora of physical layer choices, the signaling defined in the set of 1394 standards appears to be emerging as the choice for high-speed (at the 196 Mbits/s level) local connections and as the backbone for passing information around the home. This physical layer comes in several versions. The most mature technology is the IEEE 1394-1995 standard for communicating over 4.5 m of unshielded twisted pair (UTP). In addition, 60- and 100-m versions have been demonstrated to work over fiber. Commercial products for 1394 over plastic fiber are now available.

On top of the physical layer, network and control layers are needed. This is the part of the communications stack that was defined for the DTV interface (i.e., EIA-775), and was being defined at this writing for long distance 1394 by the Video Electronics Standards Association (VESA) Home Network group (VHN). Their objective was to define a network layer approach to allow seamless operation across different physical layers using Internet protocol (IP). HAVI (Home Audio/Video Interoperability) is a different network layer solution for home networking that has the same unifying objective as VHN. Optimized for the 1394 physical layer, HAVI is focused on audio/video applications and requirements.

On top of the network layer, widely differing approaches exist to the application interfaces, operating system, rendering engines, browsers, and the degree of linkage with the Internet.

10.4.3 Advanced Television Enhancement Forum

The Advanced Television Enhancement Forum (ATVEF) is a cross-industry group formed to specify a single public standard for delivering interactive television experiences that can be authored once—using a variety of tools—and deployed to a variety of television, set-top, and PC-based receivers [51]. The ATVEF Version 1.1 document is a foundation specification, defining the fundamentals necessary to enable creation of HTML-enhanced television content so that it can be reliably broadcast across any network to any compliant receiver.

The ATVEF specification for enhanced television programming uses existing Internet technologies. It delivers enhanced TV programming over both analog and digital video systems using terrestrial, cable, satellite, and Internet networks. The specification can be used with both one-way broadcast and two-way video systems, and is designed to be compatible with all international standards for both analog and digital video systems.

The ATVEF specification consists of three principal parts:

- Content specifications to establish minimum requirements for receivers

- Delivery specifications for transport of enhanced TV content

- A set of specific bindings

A central design point of the ATVEF document was to use existing standards wherever possible and to minimize the creation of new specifications. The content creators in the group determined that existing web standards, with only minimal extensions for television integration, provide a rich set of capabilities for building enhanced TV content in today's marketplace. The ATVEF specification references full existing specifications for HTML, ECMAScript, DOM, CSS, and media types as the basis of the content specification. The guidelines are not a limit on what content can be sent, but rather are intended to provide a common set of capabilities so that content developers can author content once and reproduce it on a wide variety of players.

Another key design goal was to provide a single solution that would work on a wide variety of networks. ATVEF is capable of running on both analog and digital video systems as well as networks with no video capabilities at all. The specification further supports transmission across terrestrial broadcast, cable, and satellite systems, and the Internet. In addition, it will bridge between networks; for example, in a compliant system, data on an analog terrestrial broadcast will bridge to a digital cable system. This design goal was achieved through the definition of a transport-independent content format and the use of IP as the reference binding. Because IP bindings already exist for each of these video systems, ATVEF can take advantage of a wealth of previous work.

The specification defines two transports—one for broadcast data and one for data pulled through a return path. While the ATVEF specification has the capability to run on any video network, a complete specification requires a specific binding to each video network standard in order to ensure true interoperability.

Reference and example bindings also are specified in the document, although it is assumed that appropriate standards bodies will define the bindings for each video standard—PAL, SECAM, DVB, ATSC, and others.

There are many roles in the production and delivery of television enhancements. The ATFEF document identifies three key roles:

- **Content creator**. The content creator originates the content components of the enhancement including graphics, layout, interaction, and triggers.

- **Transport operator**. The transport operator runs a video delivery infrastructure (terrestrial, cable, or satellite) that includes a transport mechanism for ATVEF data.

- **Receiver**. The receiver is a hardware and software implementation (television, set-top box, or personal computer) that decodes and plays ATVEF content.

A particular group or company may participate as one, two or all three of these roles.

10.4.4   Digital Application Software Environment

The ATSC T3/S17 specialist group was charged with defining a *digital television application software environment* (DASE) for broadcast interactive applications. This environment contains two principal components: a *presentation engine* for declarative applications, and a Java-based set of interfaces for procedural applications [52].

The purpose of the presentation engine is to integrate so-called *declarative content* with streaming audio and video, and to deliver the resulting content to the television display. In partnership with the presentation engine is a set of Java interfaces that provide a means for content authors to develop procedural applications for drawing to the screen.

In defining the presentation engine, there are two distinct interfaces: the *content authoring* specification and the receiver specification. This split approach provides two benefits. First, it allows for content authoring tools to mature at a different rate than the installed base of client receivers. Second, it provides a means for authors to develop content that is delivered to different platforms with different capabilities (such as a television set-top box and a mobile phone).

The requirement for supporting different receiver profiles is based on the premise that receivers may be classified based on features that are discernible to the customer, support multiple price-point strategies, and are simple to implement and simple to understand. The presentation engine supports these requirements through modularization of the features, a layering scheme for delivering these features, and support for backward compatibility of content.

Predictive Rendering

Traditionally, television-based content producers have a great deal of control over how their product appears to the customer [52]. The conventional television paradigm would be problematic if there were no assurances that a content producer could predict how their content would be rendered on every receiver-display combination in use by consumers. The requirement for *predictive rendering*, then, is essentially a contract between the content developer and the receiver manufacturer that guarantees the following parameters:

- Content will be displayed at a specific time

- Content will be displayed in a specific sequence

- Content will look a certain way

The presentation engine supports these requirements through a well-defined model of operation, media synchronization, pixel-level positioning, and the fact that it is a *conformance specification*. The model of operation formally defines the relationship between broadcast applications, native applications, television programs, and on-screen display resources.

Pixel level positioning allows a content author to specify where elements are rendered on a display. It also allows content authors to specify elements in relation to each other or relative to the dimensions of the screen.

The presentation engine architecture consists of five principal components:

- *Markup language*, which specifies the content of the document

- *Style language*, which specifies how the content is presented to the user

- *Event model*, which specifies the relationship of events with elements in the document

- *Application programming interfaces*, which provide a means for external programs to manipulate the document

- *Media types*, which are simply those media formats that require support in a compliant receiver

For additional details on the DASE presentation engine see [52].

10.4.5 DTV Product Classification

On August 31, 2000, the Consumer Electronics Association (CEA) announced new definitions and labels for DTV products. The definitions were expected to be incorporated into manufacturers' television marketing materials as DTV receivers continued into the retail channels.

The CEA Video Division Board resolved that analog-only televisions (televisions/monitors with a scanning frequency of 15.75 kHz) should not be marketed or designated to consumers as having any particular DTV capabilities or attributes. In a second related resolution, the Board agreed that the new definitions for monitors and tuners should be used by all manufacturers and retailers to replace general, non-industry terminology like "DTV-ready" or "HDTV-ready." They also defined minimums for HDTV displays as those with active top-to-bottom scan lines of 720 progressive or 1080 interlaced, or higher. Manufacturers were also to disclose the number of active scan lines for a high-definition image within a 16:9 aspect ratio "letter boxed" image area if the unit has a 4:3 HDTV display.

The CEA digital television definitions are as follows [53]:

- **High-definition television** (HDTV): a complete product/system with the following minimum performance attributes: 1) receives ATSC terrestrial digital transmissions and decodes all ATSC Table 3 video formats; 2) has active vertical scanning lines of 720P, 1080I, or higher; 3) capable of displaying a 16:9 image; and 4) receives and reproduces, and/or outputs Dolby Digital audio.

- **High-definition television monitor**: a monitor or display with the following minimum performance attributes: 1) has active vertical scanning lines of 720P, 1080I, or higher; and 2) capable of displaying a 16:9 image. In specifications found on product literature and in owner's manuals, manufacturers were required to disclose the number of vertical scanning lines in the 16:9 viewable area, which must be 540P, 1080I or higher to meet the definition of HDTV.

- **High-definition television tuner:** an RF receiver with the following minimum performance attributes: 1) receives ATSC terrestrial digital transmissions and decodes all ATSC Table 3 video formats; 2) outputs the ATSC Table 3 720P and 1080P/I formats in the form of HD with minimum active vertical scanning lines of 720P, 1080I, or higher, and 3): receives and reproduces, and/or outputs Dolby Digital audio. Additionally, this tuner may output HD formats converted to other formats. The lower resolution ATSC Table 3 formats can be output at lower resolution levels. Alternatively, the output can be a digital bitstream with the full resolution of the broadcast signal.

- **Enhanced definition television** (EDTV): a complete product/system with the following minimum performance attributes: 1) receives ATSC terrestrial digital transmissions and decodes all ATSC Table 3 video formats; 2) has active vertical scanning lines of 480P or higher; and 3) receives and reproduces, and/or outputs Dolby Digital audio. The aspect ratio is not specified.

- **Enhanced definition television monitor**: a monitor or display with the following minimum performance attributes: 1) has active vertical scanning lines of 480P or higher. No aspect ratio is specified.

- **Enhanced definition television tuner**: an RF receiver with the following minimum performance attributes: 1) receives ATSC terrestrial digital transmissions and decodes all ATSC Table 3 video formats; 2) outputs the ATSC Table 3 720P and 1080I/P and 480P formats with minimum active vertical scanning lines of 480P; and 3) receives and reproduces, and/or outputs Dolby Digital audio. Alternatively, the output can be a digital bitstream output capable of transporting 480P, except the ATSC Table 3 480I format, which can be output at 480I.

- **Standard definition television** (SDTV): a complete product/system with the following performance attributes: 1) receives ATSC terrestrial digital transmissions and decodes all

ATSC Table 3 video formats, and produces a useable picture; 2) has active vertical scanning lines less than that of EDTV; and 3) receives and reproduces usable audio. No aspect ratio is specified.

- **Standard definition television tuner**: an RF receiver with the following minimum performance attributes: 1) receives ATSC terrestrial digital transmissions and decodes all ATSC Table 3 video formats; 2) outputs all ATSC table 3 formats in the form of NTSC output; and 3) receives and reproduces, and/or outputs Dolby Digital audio.

These industry standard definitions were intended to eliminate the confusion over product features and capabilities of television sets and monitors intended for DTV applications. The agreement promised to spur the sale of DTV-compliant sets by injecting a certain amount of logic into the marketing efforts of TV set manufacturers. The consumer electronics industry had come under fire early in their DTV product rollouts because of the use of confusing and—in many cases meaningless—marketing terms. For example, terms such as "DTV-ready" means different things to different people.

10.4.6   Cable/DTV Receiver Labeling

On the heels of the CEA DTV product classification agreement, the FCC adopted a Report and Order requiring standardized labeling of DTV receivers that are marketed for connection to cable television systems. The R&O, issued September 14, 2000, specified that such receivers offered for sale after July 1, 2001, must be permanently marked with a label on the outside of the product that reads: "Meets FCC Labeling Standard Digital Cable Ready (DCR) x," where x = 1, 2, or 3 [54].

The Commission prohibited marketing of receiving devices claimed to be fully compatible with digital cable services unless they have the functionality of one of the three categories defined. The new rules cover any consumer television receiving device with digital signal processing capability that is intended to be used with cable systems. The rules permit marketing devices with less capability, provided full compatibility is not claimed.

DCR 1 refers to a consumer electronics television receiving device capable of receiving analog basic, digital basic, and digital premium cable television programming by direct connection to a cable system providing digital programming. This device does not have an IEEE 1394 connector or other digital interlace. A security card or point of deployment module provided by the cable operator is required to view encrypted programming.

DCR 2, a superset of DCR 1, adds a 1394 digital interface connector. It is clear from the R&O that other digital interfaces also could be present. The FCC noted that connection of a DCR 2 receiver to a digital set-top box (also presumably a level 2 device) may support advanced and interactive digital services and programming delivered by the cable system to the set-top box.

The distinction asserted for DCR 3 is the addition of the capability to receive advanced and interactive digital services. A device with this label is not required to have a 1394 connector. As such services are not defined, the meaning of this distinction is unclear. The FCC did state that additional industry work wa required to design specifications for the DCR 3 category of receivers, and that it would keep the record open in this proceeding, allowing the option to incorporate these specifications into the rules at a later date.

The FCC required the consumer electronics and cable industries to report their progress on developing technical standards in two other areas: direct connection of digital TV receivers to digital cable television systems and the provision of tuning and program schedule information to support on-screen program guides for consumers. The FCC said these two issues had been substantially, but not completely, resolved in an earlier agreement. Reporting requirements were consolidated into a single reporting timetable that began on October 31, 2000, and every six months thereafter until October 2002.

The Commission determined that the copy protection issue raised in this proceeding, in fact, relates to its navigation device rules. For that reason, the copy protection licensing issue was incorporated in the Further Notice of Proposed Rulemaking and Memorandum Opinion and Order/Declaratory Ruling in the navigation devices docket (CS 97-80).

When these rules were announced by the Commission, it was unclear how much the R&O would aid in the delivery of DTV over a cable system to a DTV receiver, other than preventing some grossly misleading marketing practices. The actual connection of the cable systems coaxial cable to the DTV set is not covered. Presumably the products will use the relevant CEA or Society of Cable Television Engineers standards for the coax interface, which were being harmonized as this book went to press. The FCC also stopped short of selecting the standard for transport of signals over the IEEE-1394 physical interface (technically only requiring a connector, not any signaling through this connector). It is unlikely that a manufacturer will put a 1394 connector on a product without some implementation of a protocol using one of the 1394-based protocol standards. Unfortunately, TV receivers with the DCR 2 label are not assured to work with set-top boxes with the same label because of multiple 1394-based protocols.

10.5   References

1. "Receiver Planning Factors Applicable to All ATV Systems," Final Report of PS/WP3, Advanced Television Systems Committee, Washington, D.C., Dec. 1, 1994.

2. ATSC, "Guide to the Use of the ATSC Digital Television Standard," Advanced Television Systems Committee, Washington, D.C., Doc. A/54, Oct. 4, 1995.

3. Qureshi, Shahid U. H.: "Adaptive Equalization," *Proceedings of the IEEE*, IEEE, New York, vol. 73, no. 9, pp. 1349–1387, September 1985.

4. Ciciora, Walter, et. al.: "A Tutorial on Ghost Canceling in Television Systems," *IEEE Transactions on Consumer Electronics*, IEEE, New York, vol. CE-25, no. 1, pp. 9–44, February 1979.

5. Ungerboeck, Gottfried: "Fractional Tap-Spacing Equalizer and Consequences for Clock Recovery in Data Modems," *IEEE Transactions on Communications*, IEEE, New York, vol. COM-24, no. 8, pp. 856–864, August 1976.

6. Sgrignoli, Gary: "Preliminary DTV Field Test Results and Their Effects on VSB Receiver Design," *ICEE '99*.

7. Einolf, Charles: "DTV Receiver Performance in the Real World," *2000 Broadcast Engineering Conference Proceedings*, National Association of Broadcasters, Washington, D.C., pp. 478–482, 2000.

8. Ledoux, B, P. Bouchard, S. Laflèche, Y. Wu, and B. Caron: "Performance of 8-VSB Digital Television Receivers," *Proceedings of the International Broadcasting Convention*, IBC, Amsterdam, September 2000.

9. FCC/ACATS SSWP2-1306: *Grand Alliance System Test Procedures*, May 18, 1994.

10. Digital HDTV Grand Alliance System: *Record of Test Results*, October 1995.

11. *NAB TV TechCheck*: "Canadians Perform Indoor DTV Reception Tests," National Association of Broadcasters, Washington, D.C., October 9, 2000.

12. *NAB TV TechCheck*: "New Digital Receiver Technology Announced," National Association of Broadcasters, Washington, D.C., August 30, 1999.

13. Rhodes, Charles: "Terrestrial High-Definition Television," *The Electronics Handbook*, Jerry C. Whitaker (ed.), CRC Press, Boca Raton, Fla., pp.1599–1610, 1996.

14. Whitaker, Jerry C.: *Electronic Displays: Technology, Design and Application*, McGraw-Hill, New York, 1994.

15. Mazur, Jeff: "Video Special Effects Systems," *NAB Engineering Handbook*, 9th ed., Jerry C. Whitaker (ed.), National Association of Broadcasters, Washington, D.C., to be published 1998.

16. DeMarsh, LeRoy E.: "Displays and Colorimetry for Future Television," *SMPTE Journal*, SMPTE, White Plains, N.Y., pp. 666–672, October 1994.

17. Thorpe, Laurence J.: "High-Definition Production Systems," *Television Engineering Handbook*, rev. ed., K. B. Benson and Jerry C. Whitaker (eds.), McGraw-Hill, New York, pp. 24.1–24.28, 1992.

18. "SMPTE C Color Monitor Colorimetry," SMPTE Recommended Practice RP 145-1994, SMPTE, White Plains, N.Y., June 1, 1994.

19. Ajluni, Cheryl: "SID Conference Highlights: The Road to Invention," *Electronic Design*, Penton Publishing, Hasbrouck Heights, N.J., pp. 39–46, May 12, 1997.

20. Bohannon, W. K.: "Which Technology Will Make Digital TV a Household Name," *Electronic Design*, Penton Publishing, Hasbrouck Heights, N.J., pp. 48–51, Oct. 23, 1997.

21. Markandey, Vishal, and Robert J. Gove: "Digital Display Systems Based on the Digital Micromirror Device," *SMPTE Journal*, SMPTE, White Plains, N.Y., pp. 680–685, October 1995.

22. Tong, Hua-Sou: "HDTV Display—A CRT Approach," *Display Technologies*, Shu-Hsia Chen and Shin-Tson Wu (eds.), Proc. SPIE 1815, SPIE, Bellingham, Wash., pp. 2-8, 1992.

23. Hockenbrock, Richard: "New Technology Advances for Brighter Color CRT Displays," *Display System Optics II*, Harry M. Assenheim (ed.), Proc. SPIE 1117, SPIE, Bellingham, Wash., pp. 219-226, 1989.

24. Robinder, R., D. Bates, P. Green: "A High Brightness Shadow Mask Color CRT for Cockpit Displays," *SID Digest*, Society for Information Display, vol. 14, pp. 72-73, 1983.

25. Mitsuhashi, Tetsuo: "HDTV and Large Screen Display," *Large-Screen Projection Displays II*, William P. Bleha, Jr. (ed.), Proc. SPIE 1255, SPIE, Bellingham, Wash., pp. 2-12, 1990.

26. Benson, K. B., and D. G. Fink: *HDTV: Advanced Television for the 1990s*, McGraw-Hill, New York, 1990.

27. Morrell, A., et al.: *Color Picture Tubes*, Academic Press, New York, pp. 91-98, 1974.

28. Benson, K. B., and Jerry C. Whitaker (eds.): *Television Engineering Handbook*, rev. ed., McGraw-Hill, New York, p. 12.20, 1991.

29. Hutter, Rudolph G. E.: "The Deflection of Electron Beams," *Advances in Image Pickup and Display*, B. Kazen (ed.), vol. 1, Academic Press, New York, pp. 212-215, 1974.

30. Barkow, W. H., and J. Gross: "The RCA Large Screen 110° Precision In-Line System," ST-5015, RCA Entertainment, Lancaster, Pa.

31. Eccles, D. A., and Y. Zhang: "Digital-Television Signal Processing and Display Technology," *SID 99 Digest*, Society for Information Display, San Jose, Calif., pp. 108–111, 1999.

32. Benson, K. B., and Jerry C. Whitaker (eds.): *Television Engineering Handbook*, rev. ed., McGraw-Hill, New York, 1991.

33. Luxenberg, H., and R. Kuehn: *Display Systems Engineering*, McGraw-Hill, New York, 1968.

34. McKechnie, S.: Philips Laboratories (NA) report, 1981, unpublished.

35. Kikuchi, M., et al.: "A New Coolant-Sealed CRT for Projection Color TV," *IEEE Trans.*, vol. CE-27, IEEE, New York, no. 3, pp. 478-485, August 1981.

36. Yoshizawa, T., S. Hatakeyama, A. Ueno, M. Tsukahara, K. Matsumi, K. Hirota, "A 61-in High-Definition Projection TV for the ATSC Standard," *SID 99 Digest*, Society for Information Display, San Jose, Calif., pp. 112–115, 1999.

37. Pease, Richard W.: "An Overview of Technology for Large Wall Screen Projection Using Lasers as a Light Source," *Large Screen Projection Displays II*, William P. Bleha, Jr. (ed.), Proc. SPIE 1255, SPIE, Bellingham, Wash., pp. 93-103, 1990.

38. Takeuchi, Kazuhiko, et. al.: "A 750-TV-Line Resolution Projector Using 1.5 Megapixel a-Si TFT LC Modules," *SID 91 Digest*, Society for Information Display, pp. 415–418, 1991.

39. Fritz, Victor J.: "Full-Color Liquid Crystal Light Valve Projector for Shipboard Use," *Large Screen Projection Displays II*, William P. Bleha, Jr. (ed.), Proc. SPIE 1255, SPIE, Bellingham, Wash., pp. 59–68, 1990.

40. Phillips, Thomas E., et. al.: "1280 × 1024 Video Rate Laser-Addressed Liquid Crystal Light Valve Color Projection Display," *Optical Engineering*, Society of Photo-Optical Instrumentation Engineers, vol. 31, no. 11, pp. 2300–2312, November 1992.

41. Bleha, W. P.: "Image Light Amplifier (ILA) Technology for Large-Screen Projection," *SMPTE Journal*, SMPTE, White Plains, N.Y., pp. 710–717, October 1997.

42. Hornbeck, Larry J.: "Digital Light Processing for High-Brightness, High-Resolution Applications," *Projection Displays III*, Electronic Imaging '97 Conference, SPIE, Bellingham, Wash., February 1997.

43. Florence, J., and L. Yoder: "Display System Architectures for Digital Micromirror Device (DMD) Based Projectors," *Proc. SPIE*, SPIE, Bellingham, Wash., vol. 2650, *Projection Displays II*, pp. 193–208, 1996.

44. Gove, R. J., V. Markandey, S. Marshall, D. Doherty, G. Sextro, and M. DuVal: "High-Definition Display System Based on Digital Micromirror Device," International Workshop on HDTV (HDTV '94), International Institute for Communications, Turin, Italy (October 1994).

45. Sextro, G., I. Ballew, and J. Lwai: "High-Definition Projection System Using DMD Display Technology," *SID 95 Digest*, Society for Information Display, pp. 70–73, 1995.

46. Younse, J. M.: "Projection Display Systems Based on the Digital Micromirror Device (DMD)," SPIE Conference on Microelectronic Structures and Microelectromechanical Devices for Optical Processing and Multimedia Applications, Austin, Tex., *SPIE Proceedings*, SPIE, Bellingham, Wash., vol. 2641, pp. 64–75, Oct. 24, 1995.

47. Glenn, W. E., C. E. Holton, G. J. Dixon, and P. J. Bos: "High-Efficiency Light Valve Projectors and High-Efficiency Laser Light Sources," *SMPTE Journal*, SMPTE, White Plains, N.Y., pp. 210–216, April 1997.

48. Mitsuhashi, Tetsuo: "HDTV and Large Screen Display," *Large-Screen Projection Displays II*, William P. Bleha, Jr. (ed.), Proc. SPIE 1255, SPIE, Bellingham, Wash., pp. 2–12, 1990.

49. Hoffman, Gary A., "IEEE 1394: The A/V Digital Interface of Choice," 1394 Technology Association Technical Brief, 1394 Technology Association, Santa Clara, Calif., 1999.

50. *NAB TV TechCheck*, National Association of Broadcasters, Washington, D.C., February 1, 1999.

51. "Advanced Television Enhancement Forum Specification," Draft, Version 1.1r26 updated 2/2/99, ATVEF, Portland, Ore., 1999.

52. Wugofski, T. W.: "A Presentation Engine for Broadcast Digital Television," *International Broadcasting Convention Proceedings*, IBC, London, England, pp. 451–456, 1999.

53. *NAB TV TechCheck*: "CEA Establishes Definitions for Digital Television Products," National Association of Broadcasters, Washington, D.C., September 1, 2000.

54. *NAB TV TechCheck:* "FCC Adopts Rules for Labeling of DTV Receivers," National Association of Broadcasters, Washington, D.C., September 25, 2000.

10.6　Bibliography

Chalamala, B., and B. Gnade: "FED Up with Fat Tubes," *IEEE Spectrum*, IEEE, New York, N.Y., pp. 42–51, April 1998.

"Digital Cable–DTV Receiver Compatibility Standard Announced," *NAB TV TechCheck*, National Association of Broadcasters, Washington, D.C., November 15, 1999.

Goede, Walter F: "Electronic Information Display Perspective," *SID Seminar Lecture Notes*, Society for Information Display, San Jose, Calif., vol. 1, pp. M-1/3–M1/49, May 17, 1999.

The DVB Standard

11.1 Introduction

The DVB system is to Europe what the ATSC system is the United States: a technical achievement of considerable note that will propel television into its next level of service to the public. The DVB suite of standards was, essentially, an outgrowth of earlier work on the Eureka 95 project, just as the ATSC DTV standard was an outgrowth of now-discarded analog implementations of advanced television. Regardless of the road taken to DVB, the work is significant—even in North America—because of the widespread implementation of DVB that is planned, and—indeed—is now underway.

11.2 European System

The roots of European enhanced- and high-definition television can be traced back to the early 1980s, when development work began on *multiplexed analog component* (MAC) hardware. Designed from the outset as a direct broadcast satellite (DBS) service, MAC went through a number of infant stages, including *A*, *B*, *C*, and *E*; MAC finally reached maturity in its *D* and *D2* form. High-definition versions of the transmission formats also were developed with an eye toward European-based HDTV programming.

The PAL/SECAM transmission systems use *frequency-division multiplexing* (FDM) to transmit luminance, chrominance, and sound components. Cross-effects commonly are experienced between luminance and chrominance, and picture resolution is limited to about 120,000 pixels. MAC, on the other hand, used *time-division multiplexing* (TDM) of digital elements to eliminate cross-effects. Resolution of 180,000 pixels, 50 percent greater than PAL/SECAM, was the result. The sound component was digital, and the MAC system data capacity allowed up to eight sound channels to be transmitted with each program.

Although the *D*-MAC and *D2*-MAC systems failed to gain marketplace acceptance, it is worthwhile to briefly examine the core technologies involved, because they had a significant impact on subsequent European work on HDTV.

11.2.1   *D*-MAC/*D*2-MAC Systems

D-MAC and *D*2-MAC provided for enhanced television service in a 625/50 format and for multiple high-quality sound channels. Although the systems were technically superior to the existing PAL, a combination of marketplace forces, economic restraints, and technical problems limited the launch of *D*-MAC and *D*2-MAC to such a point that an interim step between PAL/SECAM and high-definition television was of questionable value. However, as the anticipated introduction timetable for HDTV service in Europe stretched further into the future, the concept of enhanced television, with an aspect ratio of 16:9, gained some measure of acceptance among broadcasters, if not consumers. Design philosophy for wide-screen enhanced television for Europe included double-line scanning and high-slew-rate output stages in domestic receivers to permit the display of externally fed HDTV signals at a later date.

The differing aspect ratios of PAL/SECAM and wide-screen enhanced TV presented formidable operational problems for European broadcasters. The *D*2-MAC/packet format, therefore, was designed to cope with both aspect ratios. Although *D*2-MAC initially was intended to be used exclusively for DBS, its field of applications was extended to serve cable networks and various forms of terrestrial transmission.

Under the MAC/packet concept, both the sound and the picture signals for a given program were transmitted in sequential packets of luminance, chrominance, digital sound, and data. Depending on the configuration of the 4:3 *D*2-MAC/packet receiver—with or without vertical amplitude switching—either letterbox or pan-scan options could be selected for display. Wide-screen 16:9 programming was to be displayed directly on a 16:9 *D*2-MAC/packet receiver without the need for image reconstruction. Because the transmitted signal format was identical in both aspect ratios, with the exception of a flag identifying 16:9 operation, no residual artifacts were anticipated on 4:3 *D*2-MAC/packet displays.

Parallel development efforts were conducted to implement an extended 16:9 aspect ratio PAL-compatible signal, referred to as *PALplus*. PALplus was to contain a compatible 4:3 element (transmitted on the existing terrestrial broadcast channel) and an auxiliary channel to carry supplementary information necessary to enable the PALplus receiver to reconstruct the original 16:9 image. The development efforts included improvements to PAL encoding and decoding to reduce artifacts while extending the resolution of the transmitted signal to the maximum extent possible. Designers hoped to include digital sound (*NICAM* or another system) within the enhanced signal spectrum.

11.2.2   Enhanced Television Objectives and Constraints

In Europe, the vast majority of television programs encoded in PAL are conveyed through 7 MHz channels (*PAL-G*). The final quality of any enhanced TV service, then, would be determined by the following elements:

* Source bandwidth

* Channel bandwidth

* Display bandwidth

The bandwidths of the source and the display are variables; the channel bandwidth is generally a constant. To provide a cost-efficient service, the quality level of each link in the chain must be matched. In the case of PALplus, the move to a 16:9 display caused the horizontal axis to be extended (for the same picture height) by 33 percent. To provide the same spatial resolution as an equivalent 4:3 display, the bandwidth of both the luminance and the chrominance channels would have to be expanded by a similar factor. For PAL-G, an expansion of luminance from 5 to 6.7 MHz was required. Given the same picture height and viewing distance for both aspect ratios, vertical resolution defined by 575 lines/2:1 interlace then could be maintained. With essentially the same horizontal and vertical resolution for PAL and PALplus, the appeal of the new service was based primarily on the wide-screen aspect ratio, not improved resolution.

Because signal defects in the horizontal direction (including chrominance/luminance delay, overshoots, and ringing) were more conspicuous in a 16:9 image, the entire transmission chain had to be carefully optimized to avoid picture-quality degradation.

PALplus letterbox encoding options included:

- 625/50 interlaced

- 625/50 progressive (alternatively 1250/50/2:1)

The algorithms required for splitting the 16:9 picture into compatible letterbox components were different in each case, as was the auxiliary (control) information that enabled the 16:9 receiver to reconstruct the original image. In both cases, the supplementary information was carried within 2×72 active lines above and below the compatible letterbox picture, properly encoded to make the conveyed information invisible to the 4:3 viewer. Two-stage PAL encoding and decoding was planned, switchable to optimize operation for stationary and moving picture content. Encoding options permitted optimization for film or electronic sources. For such functions, flags in the vertical blanking interval triggered complementary action in the receiver on a field-by-field basis.

11.2.3 Eureka Program

The Eureka program, EU-95, was established to formulate standards for a European HDTV system that would be compatible with the conventional *D*-MAC and *D*2-MAC/packet DBS systems. Because the project was one of the MAC family of DBS systems, it was commonly known as HD-MAC.

The source scanning employed twice the line rate of the 625-line MAC system at an aspect ratio of 16:9, interlaced. The luminance and chrominance basebands were 21 and 10.5 MHz, respectively. Through 3:2 and 3:1 compressions, respectively, a common sampling frequency of 20.5 MHz was achieved, imposing a maximum modulation bandwidth of 10.125 MHz. This constituted the uplink FM satellite signal, using the 27 MHz bandwidth available on European satellites. Additional data was carried through multiplexing in the vertical and horizontal blanking intervals.

HD-MAC was a hybrid digital/analog system. The analog portion of the format was the main visual signal, which was based on MAC concepts; the digital portion contained sound channels, teletext, conditional-access data, and *digitally assisted television* (DATV) data. To maintain compatibility with conventional MAC signals, the main data burst was set at

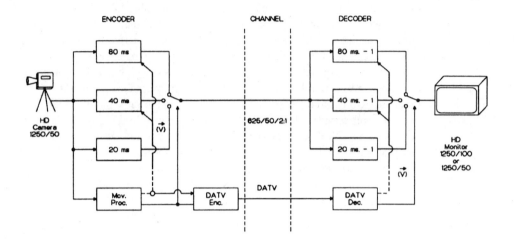

Figure 11.1 Elements and transmission paths of the Eureka EU-95 HD-MAC system. (*After* [1].)

10.125 Mbits/s. The DATV data was 20.25 Mbits/s, carried in the vertical blanking interval. The analog visual signal had a bandwidth of 10.125 MHz, compared with 8.4 MHz for conventional MAC. The wider bandwidth meant that HD-MAC was operating at the limits of what was then achievable in either a satellite or cable channel, in terms of noise and interference. Special measures, such as nonlinear preemphasis, were used to enhance the S/N performance.

To transmit the 21 MHz luminance baseband compatibly with *D*-MAC and *D*2-MAC receivers, bandwidth reduction by a factor of approximately 4 was required. This function was accomplished by the encoding and decoding process shown in Figure 11.1. The encoder included *branches* for three degrees of motion:

- 80 ms (4-field) branch for stationary and slow moving areas of the image

- 40 ms (2-field) branch for moving areas

- 20 ms (1-field) branch for rapid motion and sudden scene changes

These branches were switched to the transmission channel by a *motion processor* circuit. The switching signals were transmitted to the receiver via the DATV channel, where the branch in use at a particular time (after decoding) was connected to the receiver for processing and display at the 1250/50/2:1/16:9 rates of the source equipment at the transmitter. The chrominance signals, each 10.5 MHz baseband, were transmitted after similar 3-branch encoding, but without motion compensation.

Encoding in the 80 ms branch extended over four fields. Hence, the luminance bandwidth for stationary areas was reduced from 21 to 5.25 MHz. Because the 40 and 20 ms

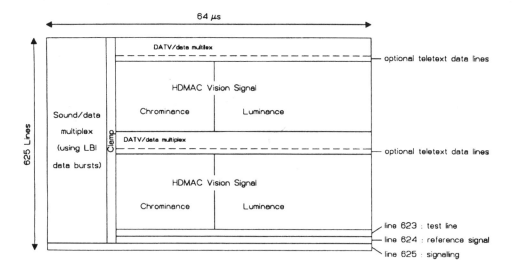

Figure 11.2 Luminance, chrominance, DATV, and sound/data components of the HD-MAC system. (After [2].)

branches extended over just two fields and one field, respectively, additional bandwidth reduction was required. This reduction was achieved through several processes, including:

- *Quincunx* scanning on alternate fields. This process, which scanned successive picture elements alternately from two adjacent lines, produced a "synthetic" interlace.

- *Line shuffling*, which interleaved high-definition samples so that two lines within a field were transmitted as one MAC/packet line. *D*-MAC and *D*2-MAC receivers responded compatibly to this process. The data required to perform the inverse operations at the receiver was transmitted over the DATV channel.

Figure 11.2 illustrates the HD-MAC signal structure in terms of the times occupied during the 64 μs line-scanning time and the image line structure. The left portion of the figure shows the sound and data signals multiplexed during the horizontal blanking interval. The DATV data occupies vertical (field) blanking times. Two sets of the HD-MAC vision signals are shown, transmitted on successive fields. Times for teletext data also are provided during the vertical blanking intervals. The luminance and chrominance signals were transmitted in time sequence, compatibly with the *D*-MAC and *D*2-MAC/packet systems.

While the video compression scheme used in the HD-MAC system was primitive by today's standards, it was reasonably effective. The fact that it was quickly replaced by digital bit-rate reduction techniques does not dilute its ingenuity.

The foregoing documentation of HD-MAC compression efforts illustrates the enormous challenges faced by design engineers attempting to transmit a high-definition signal within the limited constraints of conventional terrestrial and/or satellite channels without the benefit of modern compression systems, such as MPEG-2.

Standardization Efforts

A major thrust of the Eureka program was to forge an international—or at least a European—production standard based on the 1250/50 system. To accommodate differing applications, a hierarchy of standards was offered, including:

- *High-definition progressive scanning* (HDP), based on 1250 lines with a 50 Hz field rate and progressive scanning. Each active line had 1920 samples for luminance, and the sampling pattern was orthogonal. Color-difference samples were co-sited with each other and with the luminance samples. A total of 960 samples of each color difference were on every active line. The sampling pattern was quincunxial, so that there were 1920 color-differencing sampling positions horizontally. Luminance and color-difference sampling frequencies were 144 and 72 MHz, respectively.

- *High-definition progressive scanning and quincunxial sampling* (HDQ), similar to HDP, but designed for reduced-bandwidth applications. The color-difference sampling pattern was orthogonal with samples on every other line.

- *High-definition interlace scanning* (HDI), a reduced-bandwidth interlaced system that used orthogonal sampling patterns for both luminance and color differences.

HDQ and HDI luminance and color-difference sampling frequencies were the same, 72 and 36 MHz, respectively. Both systems could use the same digital interface.

The primary delivery system for HD-MAC was based on planned high-power satellites operating in the 11.7 to 12.5 GHz band (BSS band). An ERP of 62 dBW was provided over the (national) service area. This was to provide a received carrier-to-noise ratio (C/N) during 99 percent of the worst month of about 20 dB when using a 60 cm receive antenna. The S/N criterion for HDTV was more critical compared with conventional television. In addition, the observed S/N was reduced by the use of a closer viewing distance, an increased bandwidth, and a wider aspect ratio. Also, the effects of signal processing in the *bandwidth-reduction encoder* (BRE) caused a further small degradation in noise performance of the system. These effects were countered by the use of nonlinear preemphasis, which produced about 5 dB of noise suppression and reduced the adjacent-channel interference.

The analog visual signal consisted of data at a sampling rate of 20.25 MHz. To avoid intersample interference, a Nyquist channel was provided with a 10 percent rolloff factor. This value was chosen for compatibility with cable distribution systems. The signal was transmitted over a BSS satellite using frequency modulation with a nominal deviation of 13.5 MHz/V, using the same parameters used by conventional MAC. However, the optimum receiver bandwidth was less than 27 MHz (actually 25.5 MHz), in order to avoid adjacent-channel interference and to minimize inband distortion.

11.2.4 The End of Eureka

Although a great deal of developmental effort was poured into the Eureka EU-95 and *D*-MAC/*D2*-MAC systems, marketplace forces and unexpected advancements in digital processing combined to render the EU-95 system essentially obsolete. Acknowledging the realities of the situation, EU-95 quietly disappeared from the scene. In its place emerged the Digital Video Broadcasting (DVB) effort, which attempted to use as much of the research and knowledge gained from the EU-95 project as possible, but discarded essentially all of the analog-based hardware.

11.3 Digital Video Broadcasting (DVB)

In the early 1990s, it was becoming clear that the once state-of-the-art MAC systems would have to give way to all-digital technology. DVB provided a forum for gathering all the major European television interests into one group to address the issue [3]. The DVB project promised to develop a complete digital television system based on a unified approach.

As the DVB effort was taking shape, it was clear that digital satellite and cable television would provide the first broadcast digital services. Fewer technical problems and a simpler regulatory climate meant that these new technologies could develop more rapidly than terrestrial systems. Market priorities dictated that digital satellite and cable broadcasting systems would have to be developed rapidly. Terrestrial broadcasting would follow later.

From the beginning, the DVB effort was aimed primarily at the delivery of digital video signals to consumers. Unlike the 1125-line HDTV system and the European HDTV efforts that preceded DVB, the system was not envisioned primarily as a production tool. Still, the role that the DVB effort played—and continues to play—in the production arena is undeniable. The various DVB implementations will be examined in this chapter.

11.3.1 Technical Background of the DVB System

From the outset, it was clear that the sound- and picture-coding systems of ISO/IEC MPEG-2 should form the audio- and image-coding foundations of the DVB system [3]. DVB would need to add to the MPEG transport stream the necessary elements to bring digital television to the home through cable, satellite, and terrestrial broadcast systems. Interactive television also was examined to see how DVB might fit into such a framework for new video services of the future.

MPEG-2

The video-coding system for DVB is the international MPEG-2 standard. As discussed in Chapter 5, MPEG-2 specifies a data-stream *syntax*, and the system designer is given a "toolbox" from which to make up systems incorporating greater or lesser degrees of sophistication [3]. In this way, services avoid being overengineered, yet are able to respond fully to market requirements and are capable of evolution.

The sound-coding system specified for all DVB applications is the MPEG audio standard MPEG Layer II (MUSICAM), which is an audio coding system used for many audio products and services throughout the world. MPEG Layer II takes advantage of the fact that a given sound element will have a masking effect on lower-level sounds (or on noise) at nearby frequencies. (See Chapter 5.) This is used to facilitate the coding of audio at low data rates. Sound elements that are present, but would not be heard even if reproduced faithfully, are not coded. The MPEG Layer II system can achieve a sound quality that is, subjectively, very close to the compact disc. The system can be used for mono, stereo, or multilingual sound, and (in later versions) surround sound.

The first users of DVB digital satellite and cable services planned to broadcast signals up to and including MPEG-2 Main Profile at Main Level, thus forming the basis for first-generation European DVB receivers. Service providers, thus, were able to offer programs giving up to "625-line studio quality" (ITU-R Rec. 601), with either a 4:3, 16:9, or 20:9 aspect ratio.

Having chosen a given MPEG-2 *compliance point*, the service provider also must decide on the operating bit rates (variable or constant) that will be used. In general, the higher the bit rate, the greater the proportion of transmitted pictures that are free of coding artifacts. Nevertheless, the law of diminishing returns applies, so the relationship of bit rate to picture quality merits careful consideration.

To complicate the choice, MPEG-2 encoder design has a major impact on receiver picture quality. In effect, the MPEG-2 specification describes only syntax laws, thus leaving room for technical-quality improvements in the encoder. Early tests by DVB partners established the approximate relationship between bit rate and picture quality for the Main Profile/Main Level, on the basis of readily available encoding technology. These tests suggested the following:

- To comply with ITU-R Rec. 601 "studio quality" on all material, a bit rate of up to approximately 9 Mbits/s is required.

- To match current "NTSC/PAL/SECAM quality" on most television material, a bit rate of 2.5 to 6 Mbits/s is required, depending upon the program material.

- Film material (24/25 pictures/s) is easier to code than scenes shot with a video camera, and it also will look good at lower bit rates.

MPEG-2 Data Packets

MPEG-2 *data packets* are the basic building blocks of the DVB system [3]. The data packets are fixed-length containers with 188 bytes of data each. MPEG includes *program-specific information* (PSI) so that the MPEG-2 decoder can capture and decode the packet structure. This data, transmitted with the pictures and sound, automatically configures the decoder and provides the synchronization information necessary for the decoder to produce a complete video signal. MPEG-2 also allows a separate *service information* (SI) system to complement the PSI.

11.3.2 DVB Services

DVB has incorporated an *open service information system* to accompany the DVB signals, which can be used by the decoder and the user to navigate through an array of services offered. The following sections detail the major offerings.

DVB-SI

As envisioned by the system planners of DVB, the viewer of the future will be capable of receiving a multitude (perhaps hundreds) of channels via the DVB *integrated receiver decoder* (IRD) [3]. These services could range from interactive television to near video-on-demand to specialized programming. To sort out the available offerings, the DVB-SI provides the elements necessary for the development of an *electronic program guide* (EPG), a guide which, it is believed, is likely to become a feature of new digital television services.

Key data necessary for the DVB IRD to automatically configure itself is provided for in the MPEG-2 PSI. DVB-SI adds information that enables DVB IRDs to automatically tune to particular services and allows services to be grouped into categories with relevant schedule information. Other information provided includes:

- Program start time

- Name of the service provider

- Classification of the event (sports, news, entertainment, and so on)

DVB-SI is based on four tables, plus a series of optional tables. Each table contains descriptors outlining the characteristics of the services/event being described. The tables are:

- *Network information table* (NIT). The NIT groups together services belonging to a particular network provider. It contains all of the tuning information that might be used during the setup of an IRD. It also is used to signal a change in the tuning information.

- *Service description table* (SDT). The SDT lists the names and other parameters associated with each service in a particular MPEG multiplex.

- *Event information table* (EIT). The EIT is used to transmit information relating to all the events that occur or will occur in the MPEG multiplex. The table contains information about the current transport and optionally covers other transport streams that the IRD can receive.

- *Time and date table* (TDT). The TDT is used to update the IRD internal clock.

In addition, there are three optional SI tables:

- *Bouquet association table* (BAT). The BAT provides a means of grouping services that might be used as one way an IRD presents the available services to the viewer. A particular service can belong to one or more "bouquets."

- *Running status table* (RST). The sections of the RST are used to rapidly update the running status of one or more events. The running status sections are sent out only once—at the time the status of an event changes. The other SI tables normally are repetitively transmitted.

- *Stuffing table* (ST). The ST may be used to replace or invalidate either a subtable or a complete SI table.

With these tools, DVB-SI covers the range of practical scenarios, facilitating a seamless transition between satellite and cable networks, near video-on-demand, and other operational configurations.

DVB-S

DVB-S is a satellite-based delivery system designed to operate within a range of transponder bandwidths (26 to 72 MHz) accommodated by European satellites such as the Astra series, Eutelsat series, Hispasat, Telecom series, Tele-X, Thor, TDF-1 and 2, and DFS [3].

DVB-S is a single carrier system, with the *payload* (the most important data) at its core. Surrounding this core are a series of layers intended not only to make the signal less sensitive to errors, but also to arrange the payload in a form suitable for broadcasting. The video, audio, and other data are inserted into fixed-length MPEG transport-stream packets. This packetized data constitutes the payload. A number of processing stages follow:

- The data is formed into a regular structure by inverting synchronization bytes every eighth packet header.

- The contents are randomized.

- Reed-Solomon forward error correction (FEC) overhead is added to the packet data. This efficient system, which adds less than 12 percent overhead to the signal, is known as the *outer code*. All delivery systems have a common outer code.

- Convolutional interleaving is applied to the packet contents.

- Another error-correction system, which uses a *punctured convolutional code*, is added. This second error-correction system, the *inner code*, can be adjusted (in the amount of overhead) to suit the needs of the service provider.

- The signal modulates the satellite broadcast carrier using quadrature phase-shift keying (QPSK).

In essence, between the multiplexing and the physical transmission, the system is tailored to the specific channel properties. The system is arranged to adapt to the error characteristics of the channel. Burst errors are randomized, and two layers of forward error correction are added. The second level (inner code) can be adjusted to suit the operational circumstances (power, dish size, bit rate available, and other parameters).

DVB-C

The cable network system, known as DVB-C, has the same core properties as the satellite system, but the modulation is based on quadrature amplitude modulation (QAM) rather than QPSK, and no inner-code forward error correction is used [3]. The system is centered on 64-QAM, but lower-level systems, such as 16-QAM and 32-QAM, also can be used. In each case, the data capacity of the system is traded against robustness of the data.

Higher level systems, such as 128-QAM and 256-QAM, also may become possible, but their use will depend on the capacity of the cable network to cope with the reduced decoding margin. In terms of capacity, an 8 MHz channel can accommodate a payload capacity of 38.5 Mbits/s if 64-QAM is used, without spillover into adjacent channels.

DVB-MC

The DVB-MC digital multipoint distribution system uses microwave frequencies below approximately 10 GHz for direct distribution to viewers' homes [3]. Because DVB-MC is based on the DVB-C cable delivery system, it will enable a common receiver to be used for both cable transmissions and this type of microwave transmission.

DVB-MS

The DVB-MS digital multipoint distribution system uses microwave frequencies above approximately 10 GHz for direct distribution to viewers' homes [3]. Because this system is based on the DVB-S satellite delivery system, DVB-MS signals can be received by DVB-S satellite receivers. The receiver must be equipped with a small microwave multipoint distribution system (MMDS) frequency converter, rather than a satellite dish.

DVB-T

DVB-T is the system specification for the terrestrial broadcasting of digital television signals [3]. DVB-T was approved by the DVB Steering Board in December 1995. This work was based on a set of user requirements produced by the Terrestrial Commercial Module of the DVB project. DVB members contributed to the technical development of DVB-T through the DTTV-SA (Digital Terrestrial Television—Systems Aspects) of the Technical Module. The European Projects SPECTRE, STERNE, HD-DIVINE, HDTVT, dTTb, and several other organizations developed system hardware and produced test results that were fed back to DTTV-SA.

As with the other DVB standards, MPEG-2 audio and video coding forms the payload of DVB-T. Other elements of the specification include:

- A transmission scheme based on *orthogonal frequency-division multiplexing* (OFDM), which allows for the use of either 1705 carriers (usually known as *2k*), or 6817 carriers (*8k*). Concatenated error correction is used. The 2k mode is suitable for single-transmitter operation and for relatively small single-frequency networks with limited transmitter power. The 8k mode can be used both for single-transmitter operation and for large-area single-frequency networks. The guard interval is selectable.

- Reed-Solomon outer coding and outer convolutional interleaving are used, as with the other DVB standards.

- The inner coding (punctured convolutional code) is the same as that used for DVB-S.

- The data carriers in the *coded orthogonal frequency-division multiplexing* (COFDM) frame can use QPSK and different levels of QAM modulation and code rates to trade bits for ruggedness.

- Two-level hierarchical channel coding and modulation can be used, but hierarchical source coding is not used. The latter was deemed unnecessary by the DVB group because its benefits did not justify the extra receiver complexity that was involved.

- The modulation system combines OFDM with QPSK/QAM. OFDM uses a large number of carriers that spread the information content of the signal. Used successfully in DAB (digital audio broadcasting), OFDM's major advantage is its resistance to multipath.

Improved multipath immunity is obtained through the use of a *guard interval*, which is a portion of the digital signal given away for echo resistance. This guard interval reduces the transmission capacity of OFDM systems. However, the greater the number of OFDM carriers provided, for a given maximum echo time delay, the less transmission capacity is lost. But, certainly, a tradeoff is involved. Simply increasing the number of carriers has a significant, detrimental impact on receiver complexity and on phase-noise sensitivity.

Because of the multipath immunity of OFDM, it may be possible to operate an overlapping network of transmitting stations with a single frequency. In the areas of overlap, the weaker of the two received signals is similar to an echo signal. However, if the two transmitters are far apart, causing a large time delay between the two signals, the system will require a large guard interval.

The potential exists for three different operating environments for digital terrestrial television in Europe:

- Broadcasting on a currently unused channel, such as an adjacent channel, or broadcasting on a clear channel

- Broadcasting in a small-area *single-frequency network* (SFN)

- Broadcasting in a large-area SFN

One of the main challenges for the DVB-T developers is that the different operating environments lead to somewhat different optimum OFDM systems. The common 2k/8k specification has been developed to offer solutions for all (or nearly all) operating environments.

11.3.3 The DVB Conditional-Access Package

The area of *conditional access* has received particular attention within DVB [3]. Discussions were difficult and lengthy, but a consensus yielded a package of practical solutions. The seven points of the DVB conditional-access package are:

- Two routes to develop the market for digital television reception should be permitted to coexist: receivers incorporating a single conditional-access system (the *Simulcrypt* route), and receivers with a common interface, allowing for the use of multiple conditional-access systems (the *Multicrypt* route). The choice of route would be optional.

- The definition of a common scrambling algorithm and its inclusion, in Europe, in all receivers able to descramble digital signals. This provision enables the concept of the single receiver in the home of the consumer.

- The drafting of a Code of Conduct for access to digital decoders, applying to all conditional-access providers.

- The development of a common interface specification.

- The drafting of antipiracy recommendations.

- Agreement that the licensing of conditional-access systems to manufacturers should be on fair and reasonable terms and should not prevent the inclusion of the common interface.

- The conditional-access systems used in Europe should allow for simple *transcontrol*; for example, at cable headends, the cable operators should have the ability to replace the conditional-access data with their own data.

11.3.4 Multimedia Home Platform

In 1997, the DVB Project expanded its scope to encompass the *multimedia home platform*, MHP [4]. From a service and application point of view, enhanced broadcasting, interactive services, and Internet access were deemed to be important to the future of the DVB system. The intention was to develop standards and/or guidelines that would establish the basis for an unfragmented horizontal market in Europe with full competition in the various layers of the business chain. A crucial role was expected to be played by the *application programming interface* (API). A comprehensive set of user- and market-based commercial requirements were subsequently approved.

Since the DVB project was established in 1993, it has produced a large family of specifications for almost every aspect of digital broadcasting. These specifications were subsequently adopted through the European Telecommunications Standards Institute (ETSI) as formal European standards. In its first phase, the DVB project focused its standardization work on the broadcasting infrastructure. As documented previously in this chapter, a comprehensive set of standards was delivered, including broadcast transmission standards for different transport media, service information standards related to services and associated transport networks, and transport-related standards for interactive services using different types of return channels (cable, PSTN, ISDN, and so on). In early implementations, however, problems developed wherein different applications and set-top boxes used different APIs that were incompatible. An end-user wanting to have access to all the DVB services available would, thus, need to buy several set top boxes. This formed a considerable road block in building full confidence of consumers in the future of digital TV services.

The expansion of the DVB project focus to include the standardization of a multimedia home platform was a logical next step. Aimed squarely at achieving full convergence of consumer information and entertainment devices, the MHP comprises the home terminal (set top box, integrated television set, multimedia personal computer), its peripherals, and an in-home digital network. From an application point of view, such standardization should lead to advanced broadcasting with multimedia data applications arriving alongside conventional linear broadcasting, plus interactive services and Internet access capabilities.

For standard 6, 7 or 8 MHz TV channels, the DVB standard offers a data throughput potential of between 6 Mbits/s and 38 Mbits/s, depending on whether only a part of the channel or the full channel or transponder is used. DVB systems provide a means of delivering MPEG-2 transport streams via a variety of transmission media. These transport streams traditionally contain MPEG-2-compressed video and audio. The use by DVB of MPEG-based "data containers" opens the way for anything that can be digitized to occupy these containers.

Implementation Considerations

The DVB data broadcasting standard allows a wide variety of different, fully interoperable data services to be implemented [5]. Data-casting or Internet services would typically use a broadcasters' extra satellite transponder space to broadcast content into the home via the consumer's receiving dish. The desired content would then be directed to the consumer's PC via a coaxial cable interfaced with a DVB-compliant plug-in PC card. After decoding, it could be viewed on a browser, or saved on the PC's hard disk for later use.

Where there is a need to have two-way communications, the user could connect via the public network to a specific host computer, or to a specific Web site. At the subscriber end, conditional access components built into the PC card would integrate with the subscriber management system, allowing the broadcaster to track and charge for the data that each subscriber receives.

The wide area coverage offered by a single satellite footprint ensures that millions of subscribers could receive data in seconds from just one transmission. Because much of the infrastructure is already in place, little additional investment would be needed from both broadcasters and subscribers to take advantage of data broadcasts over satellite. With possible data rates of more than 30 Mbits/s per transponder, a typical CD-ROM could be transmitted to an entire continent in just under three minutes.

11.3.5 DVB Data Broadcast Standards

DVB defines a set of methods for encapsulating data inside MPEG-2 transport stream packets [6]. There are various methods defined, each designed to provide a flexible and efficient means for supporting a specific set of applications. For example, the *multiprotocol encapsulation* method is intended for interconnecting two networks operating under various protocols by providing the facility for addressing multiple receivers, as well as efficient segmentation and de-segmentation of packets with arbitrary sizes.

DVB defines the following basic protocols for data broadcasting:

- *Data piping*—provides a mechanism for inserting data directly into the payload of transport stream packets. The mechanisms for the fragmentation of data into packets, the reassembly, and data interpretation are privately defined by users.

- *Asynchronous datagrams*—a data streaming method to encapsulate data inside PES (packetized elementary stream) packets in which the data has neither intra-stream nor inter-stream timing requirements.

- *Synchronous streaming data*—a data streaming method to encapsulate data inside PES packets in which the data streams are characterized by a periodic interval between consecutive packets so that both maximum and minimum arrival times between packets are bounded. Synchronous streams have no strong timing association with other data streams.

- *Synchronized streaming data*—a data streaming method to encapsulate data inside PES packets in which the data stream has the same intra-stream timing requirements as the synchronous streaming protocol. In addition, synchronized streaming implies a strong timing association with other PES streams, such as video and audio streams.

- *Multi-protocol encapsulation*—a format that provides a mechanism for transporting packets defined by arbitrary protocols, such as IP, inside MPEG-2 transport stream packets. The addressing scheme covers *unicast*, *multicast*, and broadcast applications.

- *Data carousel*—a format for encapsulating data into MPEG-2 streams that allows the server to present a set of distinct data modules to a receiver by cyclically repeating the contents of the carousel. If the receiver wants to access a particular module from the data carousel, it simply waits for the next time that the data for the requested module is broadcast.

- *Object carousel*—a format for encapsulating data into MPEG-2 streams that provides the facility to transmit a structured group of objects from a broadcast server to a broadcast client using *directory objects*, *file objects*, and *stream objects*.

DAVIC Standards

The Digital Audio Visual Council (DAVIC) is a nonprofit organization created to develop standards for the delivery of interactive data services to cable modems and set-top boxes [6]. DAVIC specifications define the minimum tools and dynamic behavior required by digital audio-visual systems for end-to-end interoperability across countries, applications, and services. To achieve this interoperability, DAVIC specifications define the technologies and information flows to be used within and between major components of generic digital audio-visual systems.

DAVIC specification encompasses the entire architectural components needed for the delivery of interactive audio-visual services. These components are:

- The server

- Delivery system

- Service consumer systems (i.e., set top boxes)

The specification covers all information layers, from the physical layer, middle- and high-layer protocols, to the managed object classes. DAVIC specifies a *reference decoder model* that defines specific memory and behavior requirements of a set-top device without specifying the internal design of the unit. Other parameters specified by the standard include the following:

- A *virtual machine* for application execution.

- Standard for presentation.

- Set of API's that are accessible by applications to be executed at the set-top device.

- Interfaces, protocols, and tools for implementing security, billing, system control, system validation, and conformance/interoperability testing.

The DVB data broadcast standard complements the DAVIC documents by specifying MPEG-2 based transport and physical layers of the DAVIC standard. Taken together, the DVB data broadcast and DAVIC specifications provide a complete, end-to-end specifications for implementing data broadcasting and interactive services.

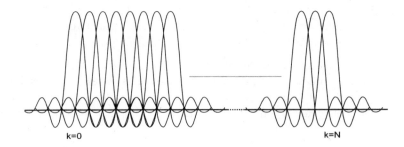

Figure 11.3 Frequency domain representation of orthogonal carriers. (*From* [7]. *Used with permission.*)

11.3.6   DVB and the ATSC DTV System

As part of the ATSC DTV specification package, a recommended practice was developed for use of the ATSC standard to ensure interoperability internationally at the transport level with the European DVB project (as standardized by the European Telecommunications Standards Institute, ETSI). Guidelines for use of the DTV standard are outlined to prevent conflicts with DVB transport in the areas of *packet identifier* (PID) usage and assignment of user private values for descriptor tags and table identifiers.

Adherence to these recommendations makes possible the simultaneous carriage of system/service information (SI) conforming to both the ATSC standard (ATSC A/65) and the ETSI ETS-300-486 standard. Such dual carriage of SI may be necessary when transport streams conforming to the ATSC standard are made available to receivers supporting only the DVB service information standard, or when transport streams conforming to the DVB standard are made available to receivers supporting only the ATSC SI standard.

11.4   COFDM Technical Principles

Orthogonal frequency demodulation multiplex (OFDM) is a wideband modulation technique that uses multiple orthogonal carriers, each modulated in phase and amplitude [7]. OFDM transmission has a slow signalling rate; the long symbol duration is the key property that makes its application attractive for terrestrial broadcasting.

For DVB-T transmissions, the OFDM modulator collects 2^{11} or 2^{13} code words into a single OFDM symbol. Depending on the modulation depth, one code word consists of 2, 4, or 6 bits. The code words are mapped onto orthogonal carriers. (See Figure 11.3.)

An inverse fast Fourier transform (IFFT) is used to convert the set of resulting modulated frequencies into a time-domain signal. Collecting multiple codewords for concurrent transmission on multiple subcarriers can be advantageous over transmitting these codewords serially on a single carrier. When comparing OFDM with the equivalent single-carrier system, the concurrent use of N subcarriers causes the signalling rate to slow-down N-times.

The long OFDM symbol duration protects the transmission against short-lived distortions. Because only a small part of a symbol is interfered, the distortion of the information

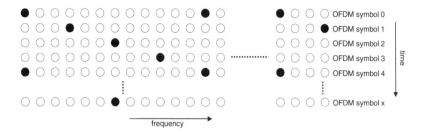

Figure 11.4 Positions of pilots in the OFDM symbol. (*From* [7]. *Used with permission.*)

bearing carriers is minimal. To further protect OFDM transmission from distortions, certain provisions are included in its application for DVB-T, including:

- The first and last carriers in the frequency domain are not modulated. The free parts in the spectrum enable filtering to protect the OFDM signal from adjacent channel interference.

- In the time-domain, the OFDM symbol is enlarged by inserting a copy of the last part of the symbol in front of it. This guard-interval protects the OFDM symbol from inter-symbol interference (ISI), caused by echoes with a distance smaller than the guard-interval length. Within the "stretched" time-domain signal there is always a complete OFDM symbol available without distortion from previous or consecutive symbols. The "zero-ISI" component of the enlarged symbol is only influenced by an echo of itself. This results in fading and amplification of its sub-carriers, an effect that can be remedied by the equalizer in the receiver system. Zero ISI reception by using a guard interval is only possible because of the long OFDM symbol duration. Only then, the unused time of the guard interval can be a modest fraction of the useful time.

- Reference signals, called *pilots*, are inserted in the frequency domain. The position, phase, and energy level of the pilots are pre-defined, as shown in Figure 11.4, enabling the receiver to reconstruct the shape of the channel. Knowing the fading, amplification, and phase-shift of all the individual sub-carriers, the receiver is able to equalize each subcarrier. This reverses effects of the transmission channel.

An *echo* is a copy of the original signal delayed in time. Problems can occur when one OFDM symbol overlaps with the next one. There is no correlation between two consecutive OFDM symbols and therefore interference from one symbol with the other will result in a disturbed signal. Because of the efficient spectrum usage, the interfering signal very much resembles white noise. The DVB-T standard adds a cyclic copy of the last part of the OFDM symbol in front of the symbol to overcome this problem. This guard-interval protects the OFDM symbol from being disturbed by its predecessor, as illustrated in Figure 11.5.

When ISI occurs, the resulting signal to noise ratio can be described as [7]

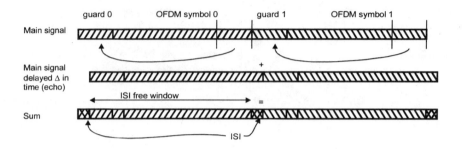

Figure 11.5 Protection of the OFDM symbol against echoes through the use of a guard-interval. (*From* [7]. *Used with permission.*)

$$\text{SNR} = 10 \cdot {}^{10}\log\left(\frac{A \cdot t}{A' \cdot Tu}\right)\text{dB}$$

(11.1)

Where:
A' = the power of the echo
A = the power of the original signal
t = the duration of interference in complex samples
Tu = the duration of one OFDM symbol without guard interval

The receiver must find a window of duration Tu within the Tu plus Tg time frame that suffers from minimal ISI.

The OFDM method uses N carriers. At least N complex discrete time samples are required to represent the OFDM symbol. The N complex time domain samples $(0\ldots N-1)$ resulting from a single sub-carrier k modulated with C_k within an OFDM symbol are

$$s_{ofdm_k}[n] = \frac{C_k}{N} \cdot e^{j\frac{2\pi}{N}k \cdot n}$$

(11.2)

Where:
N = number of sub-carriers and time-domain samples used
n = time-domain sample index $(0\ldots N-1)$
k = sub-carrier index $(0\ldots N-1)$
C_k = complex phase and amplitude information to be transmitted

Both k and C_k are constant for a single sub-carrier during the period of an OFDM-symbol. When examining Equation (11.2), it appears that the N complex samples for sub-carrier k rotate exactly k circles in the complex plane during the useful period of an OFDM-symbol.

The complete time-domain symbol is constructed from the N subcarriers by superimposing these waves onto each other

$$S_{ofdm}[n] = \sum_{k=0}^{N-1} S_{ofdm_k}[n]$$

(11.3)

Inside the DVB-T receiver, the OFDM signal is analyzed by applying an FFT to the time-domain signal. The originally transmitted information is reconstructed by comparing each subcarrier with a reference subcarrier of known phase and amplitude, and equal frequency

$$S_{ref_k}[n] = 1 \cdot e^{j\frac{2\pi}{N}k \cdot n}$$

(11.4)

As a result of the orthogonality of the N subcarriers, a zero result is obtained in the FFT for any other subcarrier than the reference

$$(l \neq k) \Rightarrow \sum_{n=0}^{N-1} \frac{S_{ofdm_l}[n]}{S_{ref_k}[n]} = 0$$

(11.5)

Therefore, to isolate the transmitted information for subcarrier k, the complete OFDM-symbol can be analyzed without error using an FFT

$$\sum_{n=0}^{N-1} \frac{S_{ofdm}[n]}{S_{ref_k}[n]} = C'_k$$

(11.6)

in which C'_k is the reconstructed phase and amplitude information.

If the receiver receives a delayed input signal, Equation (11.2) can be substituted by

$$S_{ofdm_k}[n] = \frac{C_k}{N} \cdot e^{j\frac{2\pi}{N}k \cdot (n-\Delta)}$$

(11.7)

where Δ = delay in units of complex samples. Then, this yields at the output of the FFT

$$\sum_{n=0}^{N-1} \frac{S_{ofdm}[n]}{S_{ref_k}[n]} = C'_k \cdot e^{j\frac{2\pi}{N}k \cdot \Delta}$$

(11.8)

Equation (11.8) shows that a delay in the input signal causes a rotation over the carriers in the frequency domain. Adding this delayed signal to the original will result in fading and amplification of different parts of the frequency domain. This effect is graphically shown in Figure 11.6.

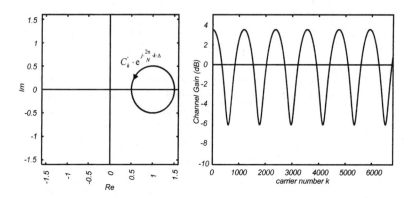

Figure 11.6 Fading effect of an echo. (*From* [7]. *Used with permission.*)

In order to reconstruct this distortion of the channel, the OFDM symbol contains reference signals (pilots). They are all modulated with known phase and amplitude information (C_k). After collecting pilots from four symbols, channel information is available at every third sub-carrier. The missing information for the two sub-carriers in between the reference signals is obtained through interpolation. Of course, this interpolation sets a limit to the maximum frequency of the distortion and therefore to the maximum delay(Δ) of the echo. Figure 11.7 shows this limitation when using a standard interpolation technique.

Delays in the range of $-N/6$ and $+N/6$ can be resolved. The OFDM symbol is protected by the guard interval from echoes up to a delay of $N/4$ [7]. Therefore, it makes sense to shift the interpolation filter range to cope with delay in the range of 0 to $N/3$ (Figure 11.8).

When a receiver locks to the strongest signal instead of the nearest in case of a pre-echo, it can be seen that this pre-echo cannot be resolved by the shifted interpolation filter. This pre-echo at $-\Delta$ is aliased into the interpolation filters range to an echo at $N/3 - \Delta$, which in the case of a strong pre-echo is disastrous. (See Figure 11.9.) Additionally, synchronized in this manner, the pre-echo causes ISI.

For the interpolation filter to reconstruct the distorted channel correctly, the input signal of the FFT must be delayed in time. The receiver then synchronizes to the pre-echo that will yield the channel distortion shown Figure 11.10. The interpolation filter can resolve this distortion and, therefore, the originally transmitted information can be retrieved without error.

In order synchronize to a weak pre-echo, the receiver must be able to detect the presence of it. Echoes can be detected by performing an inverse Fourier transform on the available reference carriers. This yields the channel impulse response. The impulse response can be scanned for pre-echoes.

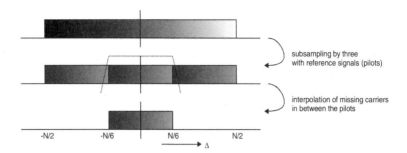

Figure 11.7 Limitation caused by a small number of reference signals. (*From* [7]. *Used with permission.*)

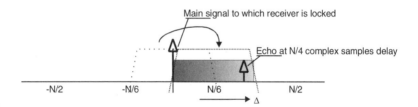

Figure 11.8 Shift of the interpolation filter towards 0 to $N/3$ distortion range. (*From* [7]. *Used with permission.*)

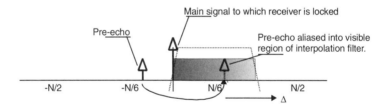

Figure 11.9 Aliasing of pre-echo. (From [7]. *Used with permission.*)

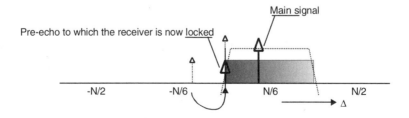

Figure 11.10 Delay input signal of FFT to shift pre-echo into interpolation filters range. (*From* [7]. *Used with permission.*)

11.5 DVB-T and ATSC Modulation Systems Comparison

As the battle over modulation systems in the U.S. began to emerge (8-VSB-vs.-COFDM), an ITU Radiocommunication Study Group released a detailed comparison of the 8-VSB and COFDM modulation systems that objectively compared the two systems under a variety of operating modes and conditions. That document, "Guide for the Use of Digital Television Terrestrial Broadcasting Systems Based on Performance Comparison of ATSC 8-VSB and DVB-T COFDM Transmission Systems," served as the basis for further consideration of each approach by regions and countries evaluating digital television transmission methods. A summary of the findings, adapted from the ITU document, follows.

11.5.1 Operating Parameters

Generally speaking, each system has its unique advantages and disadvantages [8]. Table 11.1 presents an overall transmission system performances comparison of the ATSC 8-VSB and DVB-T COFDM systems. For fair comparison, the DVB-T system convolutional coding rate of R = 2/3 with 64 QAM modulation, which is also called ITU *mode M3*, is selected.

The COFDM signal can be statistically modeled as a two-dimensional Gaussian process [9]. Its *peak to average power ratio* (PAR) is somewhat independent of filtering. On the other hand, the 8-VSB PAR is largely set by the roll-off factor of the spectrum shaping filter, i.e., 11.5 percent for the ATSC 8-VSB signal. Studies show that the DVB-T signal PAR (for 99.99 percent of the time) is about 2.5 dB higher than the ATSC [9–11].

For the same level of adjacent channel spill-over, which is the major source of adjacent channel interference, the DVB-T system requires a larger transmitter (2.5 dB or 1.8 times power) to accommodate the 2.5 dB additional output power back-off, or a better channel filter with additional side-lobe attenuation. However, the high PAR has no impact on system performance. It just adds to the start-up cost for the broadcaster, and to the on-going power supply costs.

Theoretically, from a modulation point of view, OFDM and single carrier modulation schemes, such as VSB and QAM, should have the same C/N threshold over *additive white Gaussian channel* (AWGN). It is the channel coding, channel estimation, and equalization schemes—as well as other implementation margins (phase noise, quantization noise, and intermodulation products)—that result in different C/N thresholds.

Both the DVB-T and ATSC systems use concatenated forward error correction and interleaving. The DVB-T outer code is a R-S(204, 188, t = 8) with 12 R-S block interleaving. The R-S(204, 188) code, which is shortened from R-S(255, 239) code, can correct 8-byte transmission errors and is consistent with the DVB-S (satellite) and DVB-C (cable) standards for commonality and easy inter-connectivity.

The ATSC system implemented a more powerful R-S (207, 187, t = 10) code, which can correct 10-byte errors, and uses a much larger 52 R-S block interleaver to mitigate impulse and co-channel NTSC interference. The differences of R-S code implementations will result in about 0.5 dB C/N performance benefit for the ATSC system. Meanwhile, the ATSC system implements R = 2/3 trellis coded modulation (TCM) as the inner code, while the DVB-T system uses a sub-optimal punctured convolutional code (the same as the one used in the DVB-S standard for commonality). There is up to 1 dB coding advantage in favor of the

Table 11.1 General System Comparison of ATSC 8-VSB and DVB-T COFDM (*After* [8].)

System Parameter	ATSC 8-VSB	DVB-T COFDM (ITU mode M3)	Comments
Signal peak-to-average power ratio	7 dB	9.5 dB	99.99% of time
E_b/N_o AWGN channel			
Theoretical	10.6 dB	11.9 dB	A 0.8 dB correction factor is used to compensate the measurement threshold difference
RF back-to-back test	11.0 dB	14.6 dB	
Static multipath distortion			
> 4 dB	Better	Worse	
< 4 dB	Worse	Better	
Dynamic multipath	Worse	Much better	
Mobile reception	No	2k-mode	
Spectrum efficiency	Better	Worse	
HDTV capability	Yes	Yes*	*6 MHz DVB-T might have difficulty, because of low data rate
Interference into analog TV system	Low	Medium	ATSC E_b/N_o is low, which require less transmission power
Single frequency networks			
Large scale SFN	No	Yes	DVB-T 8k mode.
On-channel repeater	Yes	Yes	ATSC and DVB-T 2k mode
Impulse noise	Better	Worse	
Tone interference	Worse	Better	
Co-channel analog TV interference into DTV	Same	Same	Assuming ATSC system has comb-filter
Co-channel DTV	Better	Worse	
Phase noise sensitivity	Better	Worse	
Noise figure	Same	Same	
Indoor reception	N/A	N/A	Needs more investigation
System for different channel bandwidth	Same	Same	ATSC might need different comb-filter, and DVB-T 6 MHz (8k mode) might be sensitive to phase noise

ATSC system. Therefore, the implementation difference in forward error correction gives the ATSC system a C/N advantage of about 1.5 dB. This 1.5 dB difference is unlikely to be reduced with the technical advances or system improvements.

The Grand Alliance prototype receiver implemented a *decision feedback equalizer* (DFE). The DFE causes very small noise enhancement, but it also results in a very sharp bit error rate (BER) threshold, because of the error feedback. On the other hand, the DVB-T will suffer a C/N degradation of about 2 dB as the system is utilizing in-band pilots for fast channel estimation and, until now (May 1999), implementing one-tap linear equalizers [12, 13]. The aggregate C/N performance difference, based on today's technology, is about 3.5 dB in favor of the ATSC system over AWGN channel [10, 14, 15].

From the transmitter implementation point of view, a DVB-T transmitter must be 6 dB (3.5 dB C/N difference plus 2.5 dB PAR), or 4 times, more powerful than an ATSC transmitter to achieve the same coverage and the same unwanted adjacent interference limit. However, it should be pointed out that the AWGN channel C/N performance is only one benchmark for a transmission system. It is an important performance indicator, but it might not represent a real-world channel model. Meanwhile, the equalization and automatic gain control (AGC) systems designed to perform well on a AWGN channel might be slow to response to moving echo or signal variations. The additional 2 dB implementation margin now found in the DVB-T system can be reduced in the future.

In Europe, the *Ricean channel model* is used in the DTTB spectrum planning process [13, 15, 17]. Computer simulation results show that the C/N threshold differences for Gaussian channel and Ricean channel (K = 10 dB) is about 0.5 to 1 dB, depending on the modulation and channel coding used [18]. The actual C/N threshold values recommended for the planning process factored in 2 dB noise degradation caused by equalization and the receiver noise floor [13]. However, the C/N differences between Gaussian channel and Ricean channel (i.e., 0.5 to 1 dB) are preserved.

The frequency planning with the ATSC system has been done with different approaches. In the US, the FCC uses Gaussian channel performance [11]. In Canada, a generous 1.3 dB C/N margin is allocated for multipath distortion (direct path to multipath power ratio K = 7.6 dB), which is much like the European approach [18].

11.5.2   Multipath Distortion

The COFDM system has a strong immunity against multipath distortion [8]. It can withstand echoes of up to 0 dB. The implementation of a guard interval can eliminate the intersymbol interference, but in-band fading will still exist. A strong inner error correction code and a good channel estimation system are mandatory for a DVB-T system to withstand 0 dB echoes. It also needs at least 7 dB more signal power to deal with the 0 dB echoes [10, 14]. *Soft decision decoding* using a so-called eraser technique can significantly improve performance [19]. For static echoes with levels less than 4 to 6 dB, the 8-VSB system, using a *decision feedback equalizer* (DFE), yields less noise enhancements [15]. The DVB-T system guard interval can be used to deal with both advanced or delayed multipath distortions.

This capability is important for single-frequency network (SFN) operation. The ATSC system can not handle long advanced echoes because it was designed for a multiple-frequency network (MFN) environment where they almost never happen.

11.5.3   Mobile Reception

COFDM can be used for mobile reception, but lower-order modulation on OFDM sub-carriers and a lower rate of convolutional coding have to be used for reliable reception [8]. Therefore, there is a significant penalty in data throughput for mobile reception in comparison to fixed reception. It is nearly impossible to achieve the 19 Mbits/s data capacity required for one HDTV program and associated multi-channel audio and data services.

Meanwhile, in the high UHF band, assuming a receiver traveling at 120 km/hr., the OFDM sub-carrier spacing should be larger than 2 kHz to accommodate Doppler effects. This indicates that only the DVB-T 2k mode is viable for mobile reception. The 2k mode was not intended to support a large scale SFN. If QPSK is used on OFDM sub-carriers, the data rate is up to 8 Mbits/s (BW = 8 MHz, R = 2/3, GI = 1/32) [18]. Using 16QAM modulation, the data rate is 16 Mbits/s (BW = 8 MHz, R = 2/3, GI = 1/32). With a higher order of modulation, the system will be sensitive to the fading and Doppler effects, which will demand more transmission power.

One potential problem to offering mobile service is spectrum availability. Because mobile reception requires different modulation and channel coding than fixed services, it might have to be offered in separate channels. Many countries have difficulties trying to allocate one fixed service DTV channel to every existing analog TV broadcasters. Finding additional spectrum for mobile service might prove difficult.

11.5.4 Spectrum Efficiency

OFDM, as a modulation scheme, is slightly more spectrum efficient than single carrier modulation systems because its spectrum can have a very fast initial roll-off, even without an output spectrum-shaping filter [8]. For a 6 MHz channel, the useful (3 dB) bandwidth is as high as 5.65 MHz (or 5.65/6 = 94 percent) [17] in comparison with the 5.38 MHz (or 5.38/6 = 90 percent) useful bandwidth of the ATSC system [20]. OFDM modulation has, therefore, a 4 percent theoretical advantage in spectrum efficiency.

However, the guard interval that is needed to mitigate the strong multipath distortions and the in-band pilots inserted for fast channel estimation significantly reduces the data capacity for the DVB-T system. For example, DVB-T offers a selection of system guard intervals, i.e., 1/4, 1/8, 1/16 and 1/32 of the active symbol duration. These are equivalent to data capacity reductions of 20 percent, 11 percent, 6 percent and 3 percent, respectively. The 1/12 in-band pilot insertion will result in an 8 percent loss of data rate. Overall, the data throughput losses are up to 28 percent, 19 percent, 14 percent, and 11 percent for the different guard intervals. Subtracting the previous 4 percent bandwidth efficiency advantage for the OFDM system, the total data capacity reductions for the DVB-T system, in comparison with the ATSC system, are 24 percent, 19 percent, 10 percent, and 7 percent, respectively. This means that, assuming equivalent channel coding scheme for both systems, the DVB-T system will suffer a 4.7, 3.7, 1.9, or 1.4 Mbits/s data capacity reduction for a 6 MHz system. The corresponding data rates are then approximately 14.7, 15.7, 17.5, and 17.9 Mbits/s respectively [17].

11.5.5 HDTV Capability

Research on digital video compression has demonstrated that, based on current technology, a data rate of at least 18 Mbits/s is required to provide a satisfactory HDTV picture for sports and fast-action programming [17]. Additional data capacity is required to accommodate multi-channel audio and ancillary data services [8]. Based on the DVB-T standard, with an equivalent channel coding scheme as the ATSC 8-VSB system (R = 2/3 punctured convolutional code, or ITU-mode M3 [13, 16]), the 6 MHz DVB-T system data throughput is between 14.7 Mbits/s and 17.90 Mbits/s, depending on the guard interval selection. Therefore, it is difficult for the DVB-T system to provide HDTV service within a 6 MHz channel, unless a weaker error correction coding is selected. For example, by increasing the convolutional coding rate to R = 3/4 and selecting GI = 1/16, the data rate becomes 19.6 Mbits/s, which is comparable with the ATSC system data rate of 19.4 Mbits/s. However, this approach will require at least 1.5 dB additional signal power [17]. Increasing the coding rate will also compromise the performance against the multipath distortions, especially for indoor reception and SFN environments.

Other techniques are available for decoding the COFDM signal without using the inband pilots [21, 22], which could significantly improve the spectrum efficiency. Unfortunately, those techniques were not fully developed when the DVB-T standard was finalized.

11.5.6   Single Frequency Network

The 8k mode DVB-T system was designed for large scale (nation-wide or region-wide) SFN, where a cluster of transmitters are used to cover a designated service area. It uses a small carrier spacing, which can support very long (up to 224 ms) guard intervals. It can also sustain 0 dB multipath distortion, if a strong convolutional code is selected (R < 3/4). However, at least 7 dB more signal power is required to deal with the 0 dB multipath distortion [10, 17]. This extra power requirement is in addition to the recommended 6 dB transmitter headroom. One alternative to reduce the excess transmission power is to use a directional receiving antenna, which would likely eliminate 0 dB multipath distortion. Such an antenna will also improve the reception of ATSC 8-VSB system.

Another problem that might impact a large-scale SFN implementation is co-channel and adjacent channel interference. In many countries, it might be difficult to allocate a DTV channel for large-scale SFN operation that will not generate substantial interference into existing analog TV services during the analog TV to DTV transition period. Finding additional tower sites at desired locations and the associated expenses (such as property, equipment, legal, construction, operation, and environmental studies) might not be practical or economically viable.

On the other hand, the SFN approach can provide stronger field strength throughout the core coverage area and can significantly improve service availability. The receivers have more than one transmitter to access (*diversity gain*); there is a better chance of having a line-of-sight path to a transmitter for reliable service. By optimizing the transmitter density, tower height, and location, as well as the transmission power, an SFN might yield better coverage and frequency economy, while maintaining a satisfactory level of interference to and from neighboring networks [23].

The ATSC system was not specifically designed for SFN implementation. Limited on-channel repeater and gap filler operation is possible, if enough isolation between the pick-up of the off-air signal and its retransmission can be achieved [24]. An another option is a full digital on-channel system, where the signal is demodulated, decoded, and re-modulated. The transmission error in the first hop can be corrected and the system does not need high level of isolation between pick-up and retransmission antennas.

The key difference between a DTV and an analog TV systems is that the DTV can withstand at least 20 dB co-channel interference, while the analog TV co-channel threshold of visibility is around 50 dB. In other words, DTV is up to 30 dB more robust than conventional analog TV, which provides more flexibility in repeater design and planning. For an ATSC system repeater implementation [24], using a directional receiving antenna will increase the location availability as well as reduce the impact of fast moving or long delay multipath distortions. The operational parameters will depend on the population distribution, terrain environment, and intended coverage area.

It should be pointed out that under any circumstances, ATSC or DVB-T, SFN or MFN, 100 percent location availability can not be achieved.

11.5.7 Impulse Noise

Theoretically, OFDM modulation should be more robust to time-domain impulse interference because the FFT process in the receiver can average out the short duration impulses [8]. However, as mentioned previously, the channel coding and interleaver implementation also play an important role. The stronger R-S(207, 187) code with 52-segment interleaver makes the ATSC system more immune to impulse interference than DVB-T using R-S(204, 188) code with 12-segment interleaver [15]. For the inner code, the shorter constraint length of 2 for ATSC (7 for DVB-T) also results in shorter error bursts, which are easier to correct by the outer code.

The impulse noise interference usually occurs in the VHF and low UHF bands, and is caused by industrial equipment and home appliances, such as microwave ovens, fluorescent lights, hair-dryers, and vacuum cleaners. High-voltage power transmission lines, which often generates arcing and corona, are also a common impulse noise source. The robustness of the carrier recovery and synchronization circuits against impulse noise can also limit the system performance.

11.5.8 Tone Interference

Because a COFDM system is a frequency domain technique, which implements a large number of sub-carriers for data transmission, a single tone or narrow band interference will destroy a few sub-carriers, but the lost data can be easily corrected by the error correction code [8]. On the other hand, tone interference will cause eye closing for the 8-VSB modulation. The adaptive equalizer could reduce the impact of the tone interference, but, in general, the DVB-T system should outperform the ATSC system on tone interference [10, 15]. However, tone interference is just another performance benchmark. In the real world, a DTTB system should never experience a tone interference-dominated environment, as a well engineered spectrum allocation plan is made to avoid that problem.

11.5.9 Co-Channel Analog TV Interference

As mentioned in the previous section, co-channel analog TV interference will destroy a limited number of COFDM sub-carriers on specific portions of the DTTB band. A good channel estimation system combined with soft decision decoding using eraser techniques should result in good performance against the analog TV interference [8]. The ATSC system used a much different approach. A carefully designed comb-filter is implemented to notch out the analog TV's video, audio and color sub-carriers to improve the system performance.

Both systems have similar performance benchmarks. It should be pointed out that the comb-filter was turned off in [15], where a 7 MHz analog TV interference signal was used to test a 6 MHz ATSC system. In the DTV spectrum planning process [18], the co-channel analog TV interference was not identified as the most critical factor. The DTV interference into the existing analog TV services is a more serious concern.

11.5.10 Co-Channel DTV Interference

Both DTV signals behave like an additive white Gaussian noise. Therefore, the co-channel DTV interference performance should be highly correlated with the C/N performance, which is largely dependent upon the channel coding and modulation used [8]. There is about 3 to 4 dB advantage for the ATSC system (shown in Table 11.2) because it benefits from a better forward error correction system. Good co-channel DTV C/I performance will result in less interference into the existing analog TV services. It will also mean more spectrum efficiency after the analog services are phased out.

11.5.11 Phase Noise Performance

Theoretically, the OFDM modulation is more sensitive to the tuner phase noise [8]. The phase noise impact can be modeled into two components [25, 26]:

- A common rotation component that causes a phase rotation of all OFDM sub-carriers

- A dispersive component, or *inter-carrier interference component*, that results in noise-like defocusing of sub-carrier constellation points.

The first component can easily be tracked by using in-band pilots as references. However, the second component is difficult to compensate. It will slightly degrade the DVB-T system noise threshold.

For a single carrier modulation system, such as 8-VSB, the phase noise generally causes constellation rotation that can mostly be tracked via a phase-locked-loop. A tuner with improved phase noise performance might be needed for the DVB-T system [27]. Using a single conversion tuner or double conversion tuner will also cause performance differences. Single conversion tuners have less phase noise, but are less tolerant to adjacent channel interference. A tuner that covers both VHF and UHF bands will be slightly worse than a single-band tuner.

11.5.12 Noise Figure

Generally speaking, noise figure is a receiver implementation issue. It is system independent [8]. A low noise figure receiver front end can be used for the ATSC or DVB-T system to reduce the minimum signal level required. A single-conversion tuner has low noise figure and low phase noise, but its noise figure is inconsistent over different TV channels. Some channels have better noise figure than others. Single-conversion tuners provide less sup-

Table 11.2 DTV Protection Ratios for Frequency Planning (*After* [8].)

System Parameters (protection ratios)	Canada [16]	USA [9]	EBU [12, 13] ITU-mode M3
C/N for AWGN Channel	+19.5 dB (16.5dB[1])	+15.19 dB	+19.3 dB
Co-Channel DTV into Analog TV	+33.8 dB	+34.44 dB	+34 ~ 37 dB
Co-Channel Analog TV into DTV	+7.2 dB	+1.81 dB	+4 dB
Co-Channel DTV into DTV	+19.5 dB (16.5 dB[1])	+15.27 dB	+19 dB
Lower Adjacent Channel DTV into Analog TV	−16 dB	−17.43 dB	−5 ~ −11 dB
Upper Adjacent Channel DTV into Analog TV	−12 dB	−11.95 dB	−1 ~ −10 dB
Lower Adjacent Channel Analog TV into DTV	−48 dB	−47.33 dB	−34 ~ −37 dB
Upper Adjacent Channel Analog TV into DTV	−49 dB	−48.71 dB	−38 ~ −36 dB
Lower Adjacent Channel DTV into DTV	−27 dB	−28 dB	N/A
Upper Adjacent Channel DTV into DTV	−27 dB	−26 dB	N/A
1 The Canadian parameter, $C/(N+I)$ of noise plus co-channel DTV interference should be 16.5 dB.			

pression of adjacent channel interference. They are also inconsistent over different channels. On the other hand, a double-conversion tuner has a high noise figure and high phase noise. However, it can achieve better adjacent channel suppression, and its noise figure and adjacent channel suppression are consistent over different frequencies.

Tuner performance is very much linked to cost (materials, components, frequency range, and related factors). With today's technology, for a low-cost consumer-grade tuner, the single-conversion tuner noise figure is about 7 dB; for double-conversion, it is around 9 dB. Tuner noise figure only impacts the system performance at the fringe of the coverage area, where signal strength is very low and there is no co-channel interference present. This situation might only represent a very small percentage of the intended coverage areas because most of the coverage is interference-limited. However, some countries do regulate receiver noise figure.

11.5.13 Indoor Reception

The DTTB system indoor reception issue needs more investigation [8]. At this writing, there was no published large scale field trial data to support a meaningful system comparison[1]. In general, an indoor signal has strong multipath distortion, resulting from reflection between indoor walls, as well as from outdoor structures. The movement of human bodies and even pets can significantly alter the distribution of indoor signal, which causes moving echoes and field strength variation.

The indoor signal strength and its distribution are related to many factors, such as building structure (concrete, brick, wood), siding material (aluminum, plastic, wood), insulation material (with or without metal coating), and window material (tinted glass, multi-layer glass). Measurements on indoor set-top antennas have shown that gain and directivity depend very much on frequency and location [16]. For a "rabbit ear" antenna, the measured gain varied from about –10 to –4 dB. For five-element logarithmic antenna, the gains were –15 to +3 dB [16]. Meanwhile, indoor environments often experience a high level of impulse noise interference from power line and home appliance sources.

11.5.14 Scaling for Different Channel Bandwidth

The DVB-T system was originally designed for 7 and 8 MHz channels [8]. By changing the system clock rate, the signal bandwidth can be adjusted to fit 6, 7 and 8 MHz channels. The corresponding hardware differences are the channel filter, IF unit, and system clock. On the other hand, the ATSC system was designed for 6 MHz channel. The 7/8 MHz systems can also be achieved by changing the system clock, as for the DVB-T case. However, the ATSC system implemented a comb-filter to combat the co-channel NTSC interference. The comb-filter might need to be changed to deal with different analog TV systems that it will encounter. The use of comb-filter is not mandatory and might not be needed, if co-channel analog TV interference is not a major concern. For instance, some countries might implement DTV on dedicated DTV channels where there is no analog co-channel interference.

Generally speaking, a narrower channel results in a lower data rate for both modulation systems because of the slower symbol rate. However, it also means longer guard interval for DVB-T system and longer echo correction capability for the ATSC system. One minor weak point for the 6 MHz DVB-T system is that its narrow sub-carrier spacing might cause the system to be more sensitive to phase noise.

11.6 References

1. "HD-MAC Bandwidth Reduction Coding Principles," Draft Report AZ-11, International Radio Consultative Committee (CCIR), Geneva, Switzerland, January 1989.

2. "Conclusions of the Extraordinary Meeting of Study Group 11 on High-Definition Television," Doc. 11/410-E, International Radio Consultative Committee (CCIR), Geneva, Switzerland, June 1989.

3. Based on technical reports and background information provided by the DVB Consortium.

1. The Zenith project outlined in Chapter 10 was underway at the time the ITU report was prepared (Sgrignoli, Gary: "Preliminary DTV Field Test Results and Their Effects on VSB Receiver Design," *ICEE '99*.)

4. Luetteke, Georg: "The DVB Multimedia Home Platform," DVB Project technical publication, 1998.

5. Jacklin, Martin: "The Multimedia Home Platform: On the Critical Path to Convergence," DVB Project technical publication, 1998.

6. Sariowan, H.: "Comparative Studies Of Data Broadcasting, *International Broadcasting Convention Proceedings*, IBC, London, England, pp. 115–119, 1999.

7. van Klinken, N., and W. Renirie: "Receiving DVB: Technical Challenges," *Proceedings of the International Broadcasting Convention*, IBC, Amsterdam, September 2000.

8. ITU Radiocommunication Study Groups, Special Rapporteur's Group: "Guide for the Use of Digital Television Terrestrial Broadcasting Systems Based on Performance Comparison of ATSC 8-VSB and DVB-T COFDM Transmission Systems," International Telecommunications Union, Geneva, Document 11A/65-E, May 11, 1999.

9. Chini, A., Y. Wu, M. El-Tanany, and S. Mahmoud: "Hardware Nonlinearities in Digital TV Broadcasting Using OFDM Modulation," *IEEE Trans. Broadcasting*, IEEE, New York, N.Y., vol. 44, no. 1, March 1998.

10. Wu, Y., M. Guillet, B. Ledoux, and B. Caron: "Results of Laboratory and Field Tests of a COFDM Modem for ATV Transmission over 6 MHz Channels," *SMPTE Journal*, SMPTE, White Plains, N.Y., vol. 107, February 1998.

11. ATTC: "Digital HDTV Grand Alliance System Record of Test Results," Advanced Television Test Center, Alexandria, Virginia, October 1995.

12. Salter, J. E.: "Noise in a DVB-T System," *BBC R&D Technical Note*, R&D 0873(98), February 1998.

13. ITU-R SG 11, Special Rapporteur—Region 1, "Protection Ratios and Reference Receivers for DTTB Frequency Planning," ITU-R Doc. 11C/46-E, March 18, 1999.

14. Morello, Alberto, et. al.: "Performance Assessment of a DVB-T Television System," *Proceedings of the International Television Symposium 1997*, Montreux, Switzerland, June 1997.

15. Pickford, N.: "Laboratory Testing of DTTB Modulation Systems," Laboratory Report 98/01, Australia Department of Communications and Arts, June 1998.

16. Joint ERC/EBU: "Planning and Introduction of Terrestrial Digital Television (DVB-T) in Europe," Izmir, Dec. 1997.

17. ETS 300 744: "Digital Broadcasting Systems for Television, Sound and Data Services: Framing Structure, Channel Coding and Modulation for Digital Terrestrial Television," ETS 300 744, 1997.

18. Wu, Y., et. al.: "Canadian Digital Terrestrial Television System Technical Parameters," *IEEE Transactions on Broadcasting*, IEEE, New York, N.Y., to be published in 1999.

19. Stott, J. H.: "Explaining Some of the Magic of COFDM", *Proceedings of the International TV Symposium 1997*, Montreux, Switzerland, June 1997.

20. ATSC: "ATSC Digital Television Standard", ATSC Doc. A/53, ATSC, Washington, D. C., September 16, 1995.

21. Chini, A., Y. Wu, M. El-Tanany, and S. Mahmoud: "An OFDM-based Digital ATV Terrestrial Broadcasting System with a Filtered Decision Feedback Channel Estimator," *IEEE Trans. Broadcasting*, IEEE, New York, N.Y., vol. 44, no. 1, pp. 2–11, March 1998.

22. Mignone, V., and A. Morello: "CD3-OFDM: A Novel Demodulation Scheme for Fixed and Mobile Receivers," *IEEE Trans. Commu.*, IEEE, New York, N.Y., vol. 44, pp. 1144–1151, September 1996.

23. Ligeti, A., and J. Zander: "Minimal Cost Coverage Planning for Single Frequency Networks", *IEEE Trans. Broadcasting*, IEEE, New York, N.Y., vol. 45, no. 1, March 1999.

24. Husak, Walt, et. al.: "On-channel Repeater for Digital Television Implementation and Field Testing," *Proceedings 1999 Broadcast Engineering Conference*, NAB'99, Las Vegas, National Association of Broadcasters, Washington, D.C., pp. 397–403, April 1999.

25. Stott, J. H: "The Effect of Phase Noise in COFDM", *EBU Technical Review*, Summer 1998.

26. Wu, Y., and M. El-Tanany: "OFDM System Performance Under Phase Noise Distortion and Frequency Selective Channels," *Proceedings of Int'l Workshop of HDTV 1997*, Montreux Switzerland, June 10–11, 1997.

27. Muschallik, C.: "Influence of RF Oscillators on an OFDM Signal", *IEEE Trans. Consumer Electronics*, IEEE, New York, N.Y., vol. 41, no. 3, pp. 592–603, August 1995.

11.7 Bibliography

Arragon, J. P., J. Chatel, J. Raven, and R. Story: "Instrumentation for a Compatible HD-MAC Coding System Using DATV," *Conference Record*, International Broadcasting Conference, Brighton, Institution of Electrical Engineers, London, 1989.

Basile, C.: "An HDTV MAC Format for FM Environments," International Conference on Consumer Electronics, IEEE, New York, June 1989.

ETS-300-421, "Digital Broadcasting Systems for Television, Sound, and Data Services; Framing Structure, Channel Coding and Modulation for 11–12 GHz Satellite Services," DVB Project technical publication.

ETS-300-429, "Digital Broadcasting Systems for Television, Sound, and Data Services; Framing Structure, Channel Coding and Modulation for Cable Systems," DVB Project technical publication.

ETS-300-468, "Digital Broadcasting Systems for Television, Sound, and Data Services; Specification for Service Information (SI) in Digital Video Broadcasting (DVB) Systems," DVB Project technical publication.

ETS-300-472, "Digital Broadcasting Systems for Television, Sound, and Data Services; Specification for Carrying ITU-R System B Teletext in Digital Video Broadcasting (DVB) Bitstreams," DVB Project technical publication.

ETS-300-473, "Digital Broadcasting Systems for Television, Sound, and Data Services; Satellite Master Antenna Television (SMATV) Distribution Systems," DVB Project technical publication.

Eureka 95 HDTV Directorate, *Progressing Towards the Real Dimension*, Eureka 95 Communications Committee, Eindhoven, Netherlands, June 1991.

European Telecommunications Standards Institute: "Digital Video Broadcasting; Framing Structure, Channel Coding and Modulation for Digital Terrestrial Television (DVB-T)", March 1997.

Lee, E. A., and D.G. Messerschmitt: *Digital Communication*, 2nd ed.,. Kluwer, Boston, Mass., 1994.

Lucas, K.: "B-MAC: A Transmission Standard for Pay DBS," *SMPTE Journal*, SMPTE, White Plains, N.Y., November 1984.

Muschallik, C.: "Improving an OFDM Reception Using an Adaptive Nyquist Windowing," *IEEE Trans. on Consumer Electronics,* no. 03, 1996

Pollet, T., M. van Bladel, and M. Moeneclaey: "BER Sensitivity of OFDM Systems to Carrier Frequency Offset and Wiener Phase Noise," *IEEE Trans. on Communications*, vol. 43, 1995.

Raven, J. G.: "High-Definition MAC: The Compatible Route to HDTV," *IEEE Transactions on Consumer Electronics*, vol. 34, pp. 61–63, IEEE, New York, February 1988.

Robertson, P., and S. Kaiser: "Analysis of the Effects of Phase-Noise in Orthogonal Frequency Division Multiplex (OFDM) Systems," ICC 1995, pp. 1652–1657, 1995.

Sabatier, J., D. Pommier, and M. Mathiue: "The *D*2-MAC-Packet System for All Transmission Channels," *SMPTE Journal*, SMPTE, White Plains, N.Y., November 1984.

Sari, H., G. Karam, and I. Jeanclaude: "Channel Equalization and Carrier Synchronization in OFDM Systems," *IEEE Proc.* 6th. Tirrenia Workshop on Digital Communications, Tirrenia, Italy, pp. 191–202, September 1993.

Schachlbauer, Horst: "European Perspective on Advanced Television for Terrestrial Broadcasting," *Proceedings of the ITS*, International Television Symposium, Montreux, Switzerland, 1991.

Story, R.: "HDTV Motion-Adaptive Bandwidth Reduction Using DATV," BBC Research Department Report, RD 1986/5.

Story, R.: "Motion Compensated DATV Bandwidth Compression for HDTV," International Radio Consultative Committee (CCIR), Geneva, Switzerland, January 1989.

Teichmann, Wolfgang: "HD-MAC Transmission on Cable," *Proceedings of the ITS*, International Television Symposium, Montreux, Switzerland, 1991.

Vreeswijk, F., F. Fonsalas, T. Trew, C. Carey-Smith, and M. Haghiri: "HD-MAC Coding for High-Definition Television Signals," International Radio Consultative Committee (CCIR), Geneva, Switzerland, January 1989.

Video Measurement Techniques

12.1 Introduction

As video technology steams full speed ahead into the digital domain, the shortcomings and degradations associated with analog technology—which video engineers have come to accept and deal with—are rapidly disappearing. In their place, however, are new problems. Digital devices and systems bring their own unique mix of issues that must be addressed, including:

- **Quantization.** The quantization process, by design, discards information. It takes an analog waveform with infinite variability and blocks it into a collection of bits, the number of which is determined by the bit length of the system.

- **Concatenation.** Defined as the connection of elements end-to-end, concatenation for video and audio describes the effects of chaining compression and decompression systems.

- **Video Processing.** It is commonly assumed that as long as a video clip is manipulated in the digital domain, it will not be degraded. In a general sense this is true, however, certain operations will discard information that can not be recreated downstream. Changes in sizing, adjustment of color hue and saturation, and adjustment of luminance values are just some of the operations that can result in degradation of the signal unless proper precautions are taken. Something as simple as improper gamma setup on monitors can result in a host of problems as the signal meanders through the production process. Once picture information is discarded, it cannot be completely recreated.

- **Transmission.** In order for a digital video signal to be useful, it usually must be moved from one location to another. This almost always involves codecs and a transmission medium. This medium may be coax, fiber, or a radio frequency link. With any of these systems, degradations are possible; some are more vulnerable than others. An RF link usually has the greatest level of exposure to interfering signals. Coax, on the other hand, is basically closed to outside influences but has a finite cable length over which reliable communications can take place.

The important message here is that digital is not always perfect and that the need for test equipment and quality control does not disappear simply because a room full of analog boxes is replaced with a computer workstation. Furthermore, just because the picture "looks good" on a local monitor does not mean that it will look good (or at least look the same) at the end of a terrestrial or satellite link.

New test instruments are rising to the challenge posed by the new technologies being introduced to the video production process. As the equipment used by broadcasters and video professionals becomes more complex, the requirements for advanced, specialized maintenance tools also increases. These instruments range from simply go/no-go status indicators to automated test routines with preprogrammed pass/fail limits. Video quality control efforts must focus on the overall system, not just a particular island.

The attribute that makes a good test instrument is really quite straightforward: accurate measurement of the signal under test. The attributes important to the user, however, usually involve the following:

- Affordability

- Ease of use

- Performance

Depending upon the application, the order of these criteria may be inverted (performance, ease of use, then affordability). Suffice it to say, however, that all elements of these specifications combine to translate into the user's definition of the ideal instrument.

The memory functions of the new breed of instruments provide important new capabilities, including archiving test setups and reference waveforms for ongoing projects and comparative tests. Hundreds of files typically can be saved for later use. With automatic measurement capabilities, even a novice technician can perform detail-oriented measurements quickly and accurately.

In the rush to embrace advanced, specific-purpose test instruments, it is easy to overlook the grandparents of all video test devices—the waveform monitor and vectorscope. Just because they are not new to the scene does not mean that they have outlived their usefulness.

The waveform monitor and vectorscope still fill valuable roles in the test and measurement world. Both, of course, have their roots in the general-purpose oscilloscope. This heritage imparts some important benefits. The scope is the most universal of all instruments, combining the best abilities of the human user and the machine. Electronic instruments are well equipped to quickly and accurately measure a given amplitude, frequency, or phase difference; they perform calculation-based tasks with great speed. The human user, however, is far superior to any machine in interpreting and analyzing an image. The waveform monitor and vectorscope—in an instant—presents to the user a wealth of information that allows rapid characterization and understanding of the signal under consideration.

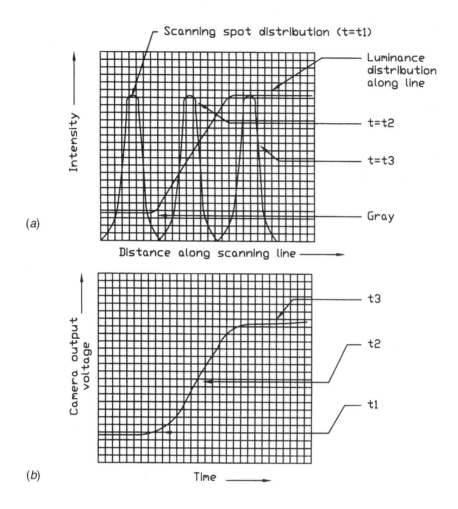

Figure 12.1 Video signal spectra: (*a*) camera scanning spot, shown with a Gaussian distribution, passing over a luminance boundary on a scanning line; (*b*) corresponding camera output signal resulting from the convolution of the spot and luminance distributions.

12.2 The Video Spectrum

The spectrum of the video signal arising from the scanning process in a television camera extends from a lower limit determined by the timed rate of change of the average luminance of the scene to an upper limit determined by the time during which the scanning spots cross the sharpest vertical boundary in the scene as focused within the camera. This concept is illustrated in Figure 12.1. The distribution of spectral components within these limits is determined by the following:

• The distribution of energy in the camera scanning system

- Number of lines scanned per second

- Percentage of line-scan time consumed by horizontal blanking

- Number of fields or frames scanned per second

- Rates at which the luminance and chrominance values of the scene change in size, position, and boundary sharpness

To the extent that the contents and dynamic properties of the scene cannot be predicted, the spectrum limits and energy distribution are not defined. However, the spectra associated with certain static and dynamic test charts and waveform generators may be used as the basis for video system design and testing. Among the configurations of interest are:

- Flat fields of uniform luminance and/or chrominance

- Fields divided into two or more segments of different luminance by sharp vertical, horizontal, or oblique boundaries

The case of the divided fields includes the horizontal and vertical wedges of test charts and the concentric circles of zone plate charts, illustrated in Figure 12.2. The reproductions of such patterns typically display diffuse boundaries and other degradations that may be introduced by the camera scanning process, the amplitude and phase responses of the transmission system, the receiver scanning system, and other artifacts associated with scanning, encoding, and transmission.

12.2.1 Minimum Video Frequency

To reproduce a uniform value of luminance from top to bottom of an image scanned in the conventional interlaced fashion, the video signal spectrum must extend downward to include the field-scanning frequency. This frequency represents the lower limit of the spectrum resulting from scanning an image whose luminance does not change. Changes in the average luminance are reproduced by extending the video spectrum to a lower frequency equal to the reciprocal of the duration of the luminance change. Because a given average luminance may persist for many minutes, the spectrum extends essentially to zero frequency (dc). Various techniques of preserving or restoring the dc component are employed in conventional television to extend the spectrum from the field frequency down to dc.

12.2.2 Maximum Video Frequency

In the analysis of the maximum operating frequency for a conventional video system, three values must be distinguished:

- The maximum output signal frequency generated by the camera or other pickup/generating device

- Maximum modulating frequency corresponding to: 1) the fully transmitted (radiated) sideband, or 2) the system used to convey the video signal from the source to the display

- Maximum video frequency present at the picture-tube (display) control electrodes

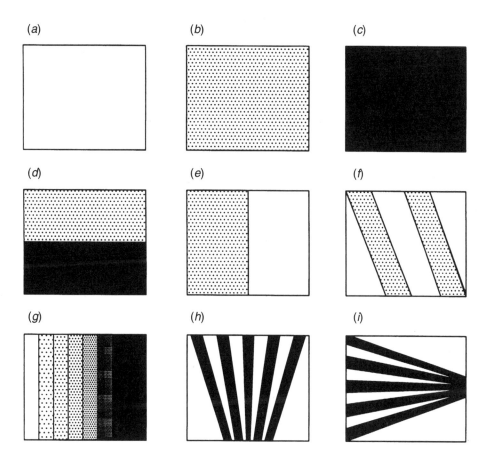

Figure 12.2 Scanning patterns of interest in analyzing conventional video signals: (*a*), (*b*), (*c*) flat fields useful for determining color purity and transfer gradient (gamma); (*d*) horizontal half-field pattern for measuring low-frequency performance; (*e*) vertical half field for examining high-frequency transient performance; (*f*) display of oblique bars; (*g*) in monochrome, a tonal wedge for determining contrast and luminance transfer characteristics; in color, a display used for hue measurements and adjustments; (*h*) wedge for measuring horizontal resolution; (*i*) wedge for measuring vertical resolution.

The maximum camera frequency is determined by the design and implementation of the imaging element. The maximum modulating frequency is determined by the extent of the video channel reserved for the fully transmitted sideband. The channel width, in turn, is chosen to provide a value of horizontal resolution approximately equal to the vertical resolution implicit in the scanning pattern. The maximum video frequency at the display is determined by the device and support circuitry of the display system.

12.2.3 Horizontal Resolution

The *horizontal resolution factor* is the proportionality factor between horizontal resolution and video frequency. It may be expressed as:

$$H_r = \frac{R_h}{\alpha} \times \iota \qquad (11.1)$$

Where:
H_r = horizontal resolution factor in lines per megahertz
R_h = lines of horizontal resolution per hertz of the video waveform
α = aspect ratio of the display
ι = active line period in microseconds

For NTSC, the horizontal resolution factor is:

$$78.8 = \frac{2}{4/3} \times 52.5 \qquad (11.2)$$

12.2.4 Video Frequencies Arising From Scanning

The signal spectrum arising from the scanning process comprises a number of discrete components at multiples of the scanning frequencies. Each spectrum component is identified by two numbers, m and n, which describe the pattern that would be produced if that component alone were present in the signal. The value of m represents the number of sinusoidal cycles of brightness measured horizontally (in the width of the picture) and n the number of cycles measured vertically (in the picture height). The 0, 0 pattern is the dc component of the signal, the 0, 1 pattern is produced by the field-scanning frequency, and the 1, 0 pattern is produced by the line-scanning frequency. Typical patterns for various values of m and n are shown in Figure 12.3. By combining a number of such patterns (including m and n values up to several hundred), in the appropriate amplitudes and phases, any image capable of being represented by the scanning pattern may be built up. This is a 2-dimensional form of the Fourier series.

The amplitudes of the spectrum components decrease as the values of m and n increase. Because m represents the order of the harmonic of the line-scanning frequency, the corresponding amplitudes are those of the left-to-right variations in brightness. A typical spectrum resulting from scanning a static scene is shown in Figure 12.4. The components of major magnitude include:

- The dc component

- Field-frequency component

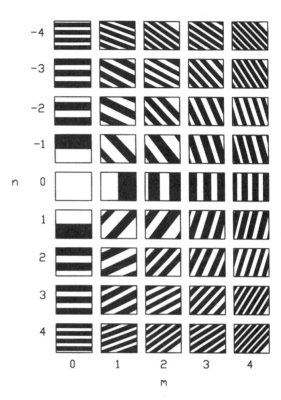

Figure 12.3 An array of image patterns corresponding to indicated values of *m* and *n*. (*After*[1].)

- Components of the line frequency and its harmonics

Surrounding each line-frequency harmonic is a cluster of components, each separated from the next by an interval equal to the field-scanning frequency.

It is possible for the clusters surrounding adjacent line-frequency harmonics to overlap one another. As shown in Figure 12.4, two patterns situated on adjacent vertical columns produce the same value of video frequency when scanned. Such "intercomponent confusion" of spectral energy is fundamental to the scanning process. Its effects are visible when a heavily striated pattern (such as that of a fabric with an accented weave) is scanned with the striations approximately parallel to the scanning lines. In the NTSC and PAL color systems, in which the luminance and chrominance signals occupy the same spectral region (one being interlaced in frequency with the other), such intercomponent confusion may produce prominent color fringes. Precise filters, which sharply separate the luminance and chrominance signals (comb filters), can remove this effect, except in the diagonal direction.

In static and slowly moving scenes, the clusters surrounding each line-frequency harmonic are compact, seldom extending further than 1 or 2 kHz on either side of the line-harmonic frequency. The space remaining in the signal spectrum is unoccupied and may be used to accommodate the spectral components of another signal having the same structure and frequency spacing. For scenes in which the motion is sufficiently slow for the eye to

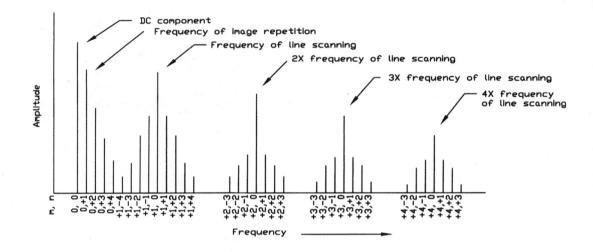

Figure 12.4 The typical spectrum of a video signal, showing the harmonics of the line-scanning frequency surrounded by clusters of components separated at intervals equal to the field-scanning frequency. (*After* [1].)

perceive the detail of moving objects, it may be safely assumed that less than half the spectral space between line-frequency harmonics is occupied by energy of significant magnitude. It is on this principle that the NTSC- and PAL-compatible color television systems are based. The SECAM system uses frequency-modulated chrominance signals, which are not frequency interlaced with the luminance signal.

The DTV system, thankfully, eliminates the built-in shortcomings of conventional composite video. As with most new technologies, however, the complex elements of the DTV system—most notably MPEG compression—offer up a whole new palette of issues that must be addressed.

12.3 Measurement of Color Displays

The chromaticity and luminance of a portion of a color display device may be measured in several ways. The most fundamental approach involves a complete spectroradiometric measurement followed by computation using tables of color-matching functions. Portable spectroradiometers with built-in computers are available for this purpose. Another method, somewhat faster but less accurate, involves the use of a photoelectric colorimeter. Because these devices have spectral sensitivities approximately equal to the CIE color-matching functions, they provide direct readings of tristimulus values.

For setting up the reference white, it is often simplest to use a split-field visual comparator and to adjust the display device until it matches the reference field (usually D_{65}) of the comparator. However, because a large spectral difference (large *metamerism*) usually exists between the display and the reference, different observers may make different settings by this method. Consequently, settings by one observer—or a group of observers—with normal color vision often are used simply to provide a reference point for subsequent photoelectric measurements.

An alternative method of determining the luminance and chromaticity coordinates of any area of a display involves measuring the output of each phosphor separately, then combining the measurements using the *center of gravity law*, by which the total tristimulus output of each phosphor is considered as an equivalent weight located at the chromaticity coordinates of the phosphor.

Consider the CIE chromaticity diagram shown in Figure 12.5 to be a uniform flat surface positioned in a horizontal plane. For the case illustrated, the center of gravity of the three weights (T_r, T_g, T_b), or the *balance point*, will be at the point C_o. This point determines the chromaticity of the mixture color. The luminance of the color C_o will be the linear sum of the luminance outputs of the red, green, and blue phosphors. The chromaticity coordinates of the display primaries may be obtained from the manufacturer. The total tristimulus output of one phosphor may be determined by turning off the other two CRT guns, measuring the luminance of the specified area, and dividing this value by the y chromaticity coordinate of the energized phosphor. This procedure then is repeated for the other two phosphors. From this data, the color resulting from given excitations of the three phosphors may be calculated as follows:

- Chromaticity coordinates of red phosphor = x_r, y_r

- Chromaticity coordinates of green phosphor = x_g, y_g

- Chromaticity coordinates of blue phosphor = x_b, y_b

- Luminance of red phosphor = Y_r

- Luminance of green phosphor = Y_g

- Luminance of blue phosphor = Y_b

$$\text{Total tristimulus value of red phosphor} = X_r + Y_r + Z_r = \frac{Y_r}{y_r} = T_r \tag{11.3}$$

$$\text{Total tristimulus value of green phosphor} = X_g + Y_g + Z_g = \frac{Y_g}{y_g} = T_g \tag{11.4}$$

$$\text{Total tristimulus value of blue phosphor} = X_b + Y_b + Z_b = \frac{Y_b}{y_b} = T_b \tag{11.5}$$

Consider T_r as a weight located at the chromaticity coordinates of the red phosphor and T_g as a weight located at the chromaticity coordinates of the green phosphor. The location of the chromaticity coordinates of color C_1 (blue gun of color CRT turned off) can be deter-

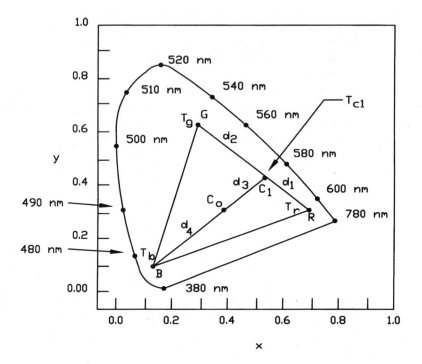

Figure 12.5 The CIE 1931 chromaticity diagram illustrating use of the *center of gravity law* ($T_r d_1 = T_g d_2$, $T_{c1} = T_r + T_g$, $T_{c1} d_3 = T_b d_4$).

mined by taking moments along line RG to determine the center of gravity of weights T_r and T_g:

$$T_r \times d_1 = T_g \times d_2 \tag{11.6}$$

The total tristimulus value of C_1 is equal to $T_r + T_g = T_{c1}$ \qquad (11.7)

Taking moments along line C_{1B} will locate the chromaticity coordinates of the mixture color C_o:

$$T_{c1} \times d_3 = T_b \times d_4 \tag{11.8}$$

The luminance of the color Co is equal to $Y_r + Y_g + Y_b$ \qquad (11.9)

12.3.1   Assessment of Color Reproduction

A number of factors may contribute to poor color rendition in a display system. To assess the effect of these factors, it is necessary to define system objectives, then establish a method of measuring departures from the objectives. Visual image display may be categorized as follows:

- *Spectral color reproduction*: The exact reproduction of the spectral power distributions of the original stimuli. Clearly, this is not possible in a video system with three primaries.

- *Exact color reproduction*: The exact reproduction of tristimulus values. The reproduction is then a metameric match to the original. Exact color reproduction will result in equality of appearance only if the viewing conditions for the picture and the original scene are identical. These conditions include the angular subtense of the picture, the luminance and chromaticity of the surround, and glare. In practice, exact color reproduction often cannot be achieved because of limitations on the maximum luminance that can be produced on a color monitor.

- *Colorimetric color reproduction*: A variant of exact color reproduction in which the tristimulus values are proportional to those in the original scene. In other words, the chromaticity coordinates are reproduced exactly, but the luminances all are reduced by a constant factor. Traditionally, color video systems have been designed and evaluated for colorimetric color reproduction. If the original and the reproduced reference whites have the same chromaticity, if the viewing conditions are the same, and if the system has an overall gamma of unity, colorimetric color reproduction is indeed a useful criterion. These conditions, however, often do not hold; then, colorimetric color reproduction is inadequate.

- *Equivalent color reproduction*: The reproduction of the original color appearance. This might be considered as the ultimate objective, but it cannot be achieved because of the limited luminance generated in a display system.

- *Corresponding color reproduction*: A compromise by which colors in the reproduction have the same appearance that colors in the original would have had if they had been illuminated to produce the same average luminance level and the same reference white chromaticity as that of the reproduction. For most purposes, corresponding color reproduction is the most suitable objective of a color video system.

- *Preferred color reproduction*: A departure from the preceding categories that recognizes the preferences of the viewer. It is sometimes argued that corresponding color reproduction is not the ultimate aim for some display systems, such as color television, and that it should be taken into account that people prefer some colors to be different from their actual appearance. For example, suntanned skin color is preferred to average real skin color, and sky is preferred bluer and foliage greener than they really are.

Even if corresponding color reproduction is accepted as the target, some colors are more important than others. For example, flesh tones must be acceptable—not obviously reddish, greenish, purplish, or otherwise incorrectly rendered. Likewise, the sky must be blue and the clouds white, within the viewer's range of acceptance. Similar conditions apply to other well-known colors of common experience.

12.3.2 Chromatic Adaptation and White Balance

With properly adjusted cameras and displays, whites and neutral grays are reproduced with the chromaticity of D65. Tests have shown that such whites (and grays) appear satisfactory in home viewing situations even if the ambient light is of quite different color temperature. Problems occur, however, when the white balance is slightly different from one camera to the next or when the scene shifts from studio to daylight or vice versa. In the first case, unwanted shifts of the displayed white occur, whereas in the other, no shift occurs even though the viewer subconsciously expects a shift.

By always reproducing a white surface with the same chromaticity, the system is mimicking the human visual system, which adapts so that white surfaces always appear the same, whatever the chromaticity of the illuminant (at least within the range of common light sources), as discussed briefly in Chapter 10. The effect on other colors, however, is more complicated. In video cameras, the white balance adjustment usually is made by gain controls on the R, G, and B channels. This is similar to the *von Kries* model of human chromatic adaptation, although the R, G, and B primaries of the model are not the same as the video primaries. It is known that the von Kries model does not accurately account for the appearance of colors after chromatic adaptation, and so it follows that making simple gain changes in a video camera is not the ideal approach. Nevertheless, this approach seems to work well in practice, and the viewer does not object to the fact, for example, that the relative increase in the luminances of reddish objects in tungsten light is lost.

12.3.3 Overall Gamma Requirements

Colorimetric color reproduction requires that the overall gamma of the system—including the camera, the display, and any gamma-adjusting electronics—be unity. This simple criterion is the one most often used in the design of a video color rendition system. However, the more sophisticated criterion of corresponding color reproduction takes into account the effect of the viewing conditions. In particular, several studies have shown that the luminance of the surround is important. For example, a dim surround requires a gamma of about 1.2, and a dark surround requires a gamma of about 1.5 for optimum color reproduction.

12.3.4 Perception of Color Differences

The CIE 1931 chromaticity diagram does not map chromaticity on a uniform-perceptibility basis. A just-perceptible change of chromaticity is not represented by the same distance in different parts of the diagram. Many investigators have explored the manner in which perceptibility varies over the diagram. The study most often quoted is that of MacAdam, who identified a set of ellipses that are contours of equal perceptibility about a given color, as shown in Figure 12.6 [2].

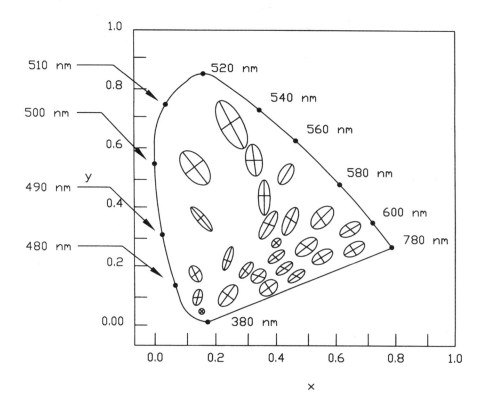

Figure 12.6 Ellipses of equally perceptible color differences (as identified by MacAdam).

From this and similar studies it is apparent, for example, that large distances represent relatively small perceptible changes in the green sector of the diagram. In the blue region, much smaller changes in the chromaticity coordinates are readily perceived.

Furthermore, viewing identical images on dissimilar displays can result in observed differences in the appearance of the image [3]. Several factors affect the appearance of the image:

- Physical factors, including display gamut, illumination level, and black point

- Psychophysical factors, including chromatic induction and color constancy

Each of these factors interact in such a way that prediction of the appearance of an image on a given display becomes difficult. System designers have experimented with colorimetry to facilitate the sharing of image data among display devices that vary in manufacture, calibration, and location. As pointed out previously in this book, colorimetric standardization was one of the goals of SMPTE 240M. Of particular interest today is the application of colorimetry to imaging in a networked window system environment, where it is often necessary to ensure that an image displayed remotely looks like the image displayed locally.

Studies have indicated that image context and image content are also factors that affect color appearance. The use of highly chromatic backgrounds in a windowed display system is popular, but it usually will affect the appearance of the colors in the foreground.

12.3.5 Display Resolution and Pixel Format

The pixel represents the smallest resolvable element of a display. The size of the pixel varies from one type of display to another. In a monochrome CRT, pixel size is determined primarily by the following factors:

- Spot size of the electron beam (the current density distribution)

- Phosphor particle size

- Thickness of the phosphor layer

The term *pixel* was developed in the era of monochrome television, and the definition was—at that time—straightforward. With the advent of color-triad-based CRTs and solid-state display systems, the definition is not nearly so clear.

For a color CRT, a single triad of red, green, and blue phosphor dots constitutes a single pixel. This definition assumes that the mechanical and electrical parameters of the CRT will permit each triad to be addressed without illuminating other elements on the face of the tube. Most display systems, however, will not meet this criterion. Depending on the design, a number of triads may constitute a single pixel in a CRT display. A more all-inclusive definition for the pixel is: the smallest spatial-information element as seen by the viewer [4].

Dot pitch is one of the principal mechanical criteria of a CRT that determines, to a large extent, the resolution of the display. Dot pitch is defined as the center-to-center distance between adjacent green phosphor dots of the red, green, blue triad.

The *pixel format* is the arrangement of pixels into horizontal rows and vertical columns. For example, an arrangement of 640 horizontal pixels by 480 vertical pixels results in a 640 × 480 pixel format. This description is not a resolution parameter in itself, simply the arrangement of pixel elements on the screen. *Resolution* is the measure of the ability to delineate picture detail; it is the smallest discernible and measurable detail in a visual presentation [5].

Pixel density is a parameter closely related to resolution, stated in terms of pixels per linear distance. Pixel density specifies how closely the pixel elements are spaced on a given display. It follows that a display with a given pixel format will not provide the same pixel density, or resolution, on a large-size screen (such as 27-in diagonal) as on a small-size screen (such as 12-in diagonal).

Television lines is another term used to describe resolution. The term refers to the number of discernible lines on a standard test chart. As before, the specification of television lines is not, in itself, a description of display resolution. A 525-line display on a 17-in monitor will appear to a viewer to have greater resolution than a 525-line display on a 30-inch monitor. Pixel density is the preferred resolution parameter.

12.3.6 Contrast Ratio

The purpose of a video display is to convey information by controlling the illumination of phosphor dots on a screen, or by controlling the reflectance or transmittance of a light source. The *contrast ratio* specifies the observable difference between a pixel that is switched on and one that is in its corresponding off state:

$$C_r = \frac{L_{on}}{L_{off}}$$

(11.9)

Where:
C_r = contrast ratio of the display
L_{on} = luminance of a pixel in the on state
L_{off} = luminance of a pixel in the off state

The area encompassed by the contrast ratio is an important parameter in assessing the performance of a display. Two contrast ratio divisions typically are specified:

• *Small area*: comparison of the on and off states of a pixel-sized area

• *Large area*: comparison of the on and off states of a group of pixels

For most display applications, the small-area contrast ratio is the more critical parameter.

12.3.7 Color Bar Test Patterns

Color bars are the most common pattern for testing encoders, decoders, and other video devices. The test pattern typically contains several bars filled with primary and complementary colors. There are many variants, which differ in color sequence, orientation, saturation and intensity.

The standard color bar sequence is white, yellow, cyan, green, magenta, red, blue, and black. This sequence can be produced in an RGB format by a simple 3-bit counter. The typical specification of color bar levels (ITU-R Rec. BT.471-1, "Nomenclature and Description of Color Bar Signals") is a set of four numbers separated by slashes or dots and giving RGB levels as a percentage of reference white in the following sequence:

white bar / black bar / max colored bars / min colored bars

For example, 100/0/100/25 means 100 percent R, G, and B on the white bar; 0 percent R, G, and B on the black bar; 100 percent maximum of R, G, and B on colored bars; and 25 percent minimum of R, G, and B on colored bars. (See Figure 12.7a.)

Some color bar patterns merit special names. For example, 100/0/75/0 bars are often called "EBU bars" or "75 percent bars", and 100/0/100/25 bars are known as "BBC bars." Nevertheless, the ITU four number nomenclature remains the only reliable specification system to designate the exact levels for color bar test patterns.

The SMPTE variation is a matrix test pattern according to SMPTE engineering guideline EG 1-1990, "Alignment Color Bar Test Signal for Television Picture Monitors." The guide-

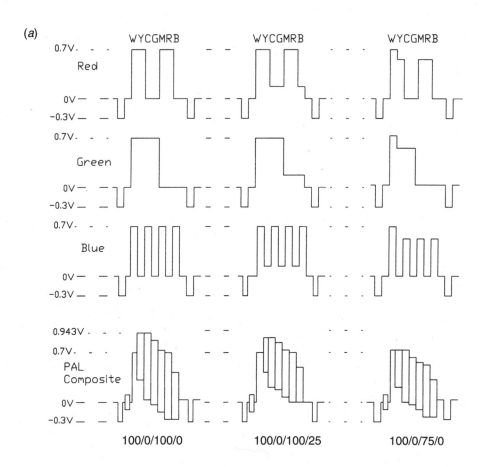

Figure 12.7 Examples of color test bars: (*a*) ITU nomenclature, (*b*, next page) SMPTE EG 1-1990 color bar test signal, (*c*, next page) EIA RS-189-A test signal. [*Drawing* (b) *after SMPTE EG 1-1990*, (c) *after EIA RS-189-A.*]

line specifies that 67 percent of the field shall contain seven (without black) 75 percent color bars, plus 8 percent of the field with the "new chroma set" bars (blue/black/magenta/black/cyan/black/gray), and the remaining 25 percent shall contain a combination of –*I*, white, *Q*, black, and the black set signal (a version of PLUGE). This arrangement is illustrated in Figure 12.7*b*.

Chroma gain and chroma phase for picture monitors are usually adjusted by observing the standard encoded color bar signal with the red and green CRT guns switched off. The four visible blue bars are set for equal brightness. The use of the chroma set feature greatly increases the accuracy of this adjustment because it provides a signal with the blue bars to be matched vertically adjacent to each other. Because the bars are adjacent, the eye can easily perceive differences in brightness. This also eliminates effects resulting from shading or purity from one part of the screen to another.

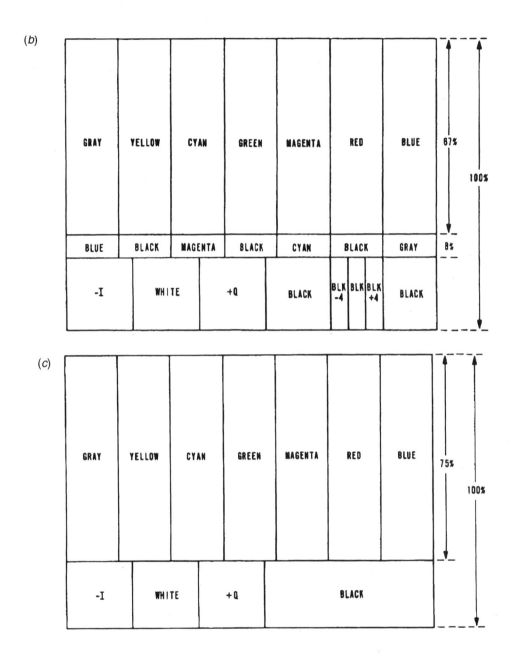

The EIA color bar signal is a matrix test pattern according to RS-189-A, which consists of 75 percent of the field containing seven (without black) 75 percent color bars (same as SMPTE) and the remaining 25 percent containing a combination of –I, white, Q, and black. (See Figure 12.7c.)

12.3.8   Conventional Video Measurements

Although there are a number of computer-based television signal monitors capable of measuring video, sync, chroma, and burst levels (as well as pulse widths and other timing factors), sometimes the best way to see what a signal is doing is to monitor it visually. The waveform monitor and vectorscope are oscilloscopes specially adapted for the video environment. The waveform monitor, like a traditional oscilloscope, operates in a voltage-versus-time mode. While an oscilloscope timebase can be set over a wide range of intervals, the waveform monitor timebase triggers automatically on sync pulses in the conventional TV signal, producing line- and field-rate sweeps, as well as multiple lines, multiple fields, and shorter time intervals. Filters, clamps, and other circuits process the video signal for specific monitoring needs. The vectorscope operates in an X-Y voltage-versus-voltage mode to display chrominance information. It decodes the signal in much the same way as a television receiver or a video monitor to extract color information and to display phase relationships. These two instruments serve separate, distinct purposes. Some models combine the functions of both types of monitors in one chassis with a single CRT display. Others include a communications link between two separate instruments.

Beyond basic signal monitoring, the waveform monitor and vectorscope provide a means to identify and analyze signal aberrations. If the signal is distorted, these instruments allow a technician to learn the extent of the problem and to locate the offending equipment.

Although designed for analog video measurement and quality control, the waveform monitor and vectorscope still serve valuable roles in the digital video facility, and will continue to do so for some time.

Basic Waveform Measurements

Waveform monitors are used to evaluate the amplitude and timing of video signals and to show timing relationships between two or more signals. The familiar color-bar pattern is the only signal required for these basic tests. Figure 12.8 shows a typical waveform display of color bars. It is important to realize that all color bars are not created equal. Some generators offer a choice of 75 percent or 100 percent amplitude bars. Sync, burst, and setup amplitudes remain the same in the two color-bar signals, but the peak-to-peak amplitudes of high-frequency chrominance information and low-frequency luminance levels change. The saturation of color, a function of chrominance and luminance amplitudes, remains constant at 100 percent in both modes. The 75 percent bar signal has 75 percent amplitude with 100 percent saturation. In 100 percent bars, amplitude and saturation are both 100 percent.

Chrominance amplitudes in 100 percent bars exceed the maximum amplitude that should be transmitted via NTSC. Therefore, 75 percent amplitude color bars, with no chrominance information exceeding 100 IRE, are the standard amplitude bars for conventional television. In the 75 percent mode, a choice of 100 IRE or 75 IRE white reference level may be offered. Figure 12.9 shows 75 percent amplitude bars with a 100 IRE white level. Either white level can be used to set levels, but operators must be aware of which signal has been selected. SMPTE bars have a white level of 75 IRE as well as a 100 IRE white flag.

The vertical response of a waveform monitor depends upon filters that process the signal in order to display certain components. The *flat response* mode displays all components of the signal. A *chroma response* filter removes luminance and displays only chrominance.

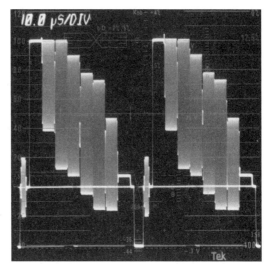

Figure 12.8 Waveform monitor display of a color-bar signal at the two-line rate. (*Courtesy of Tektronix.*)

The low-pass filter removes chrominance, leaving only low-frequency luminance levels in the display. Some monitors include an IRE filter, designed to average-out high-level, fine-detail peaks on a monochrome video signal. The IRE filter aids the operator in setting brightness levels. The IRE response removes most, but not all, of the chrominance.

If the waveform monitor has a dual filter mode, the operator can observe luminance levels and overall amplitudes at the same time. The instrument switches between the flat and low-pass filters. The line select mode is another useful feature for monitoring live signals.

Sync Pulses

The duration and frequency of the sync pulses must be monitored closely for conventional video. Most waveform monitors include 0.5 µs or 1 µs per division magnification (MAG) modes, which can be used to verify H-sync width between 4.4 µs and 5.1 µs. The width is measured at the –4 IRE point. On waveform monitors with good MAG registration, sync appearing in the middle of the screen in the 2-line mode remains centered when the sweep is magnified. Check the rise and fall times of sync, and the widths of the front porch and entire blanking interval. Examine burst and verify there are between 8 and 11 cycles of subcarrier.

Check the vertical intervals for correct format, and measure the timing of the equalizing pulses and vertical sync pulses. The acceptable limits for these parameters are shown in Figure 12.10.

Basic Vectorscope Measurements

The vectorscope displays chrominance amplitudes, aids hue adjustments, and simplifies the matching of burst phases of multiple signals. These functions require only the color-bar test signal as an input stimulus. To evaluate and adjust chrominance in the TV signal, observe color bars on the vectorscope. The instrument should be in its calibrated gain position. Adjust the vectorscope phase control to place the burst vector at the 9 o'clock position. Note

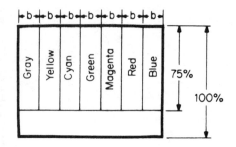

(a)

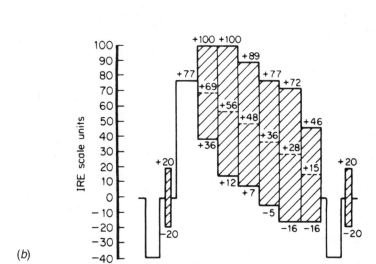

(b)

Figure 12.9 EIA RS-189-A color-bar displays: (a) color displays of gray and color bars, (b) waveform display of reference gray and primary/complementary colors, plus sync and burst.

the vector dot positions with respect to the six boxes marked on the graticule. If everything is correctly adjusted, each dot will fall on the crosshairs of its corresponding box, as shown in Figure 12.11.

The chrominance amplitude of a video signal determines the intensity or brightness of color. If the amplitudes are correct, the color dots fall on the crosshairs in the corresponding graticule boxes. If vectors overshoot the boxes, chrominance amplitude is too high; if they undershoot, it is too low. The boxes at each color location can be used to quantify the error. In the radial direction, the small boxes indicate a ±2.5 IRE error from the standard amplitude. The large boxes indicate a ±20 percent error.

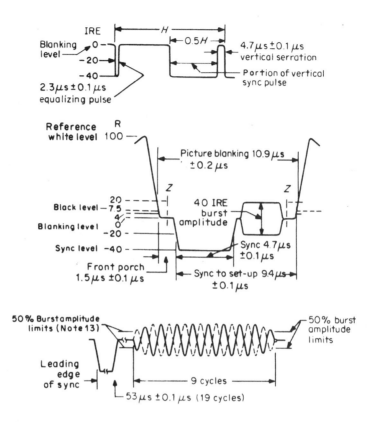

Figure 12.10 Sync pulse widths for NTSC.

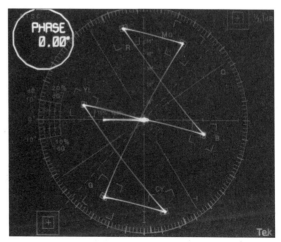

Figure 12.11 Vectorscope display of a color-bar signal. (*Courtesy of Tektronix.*)

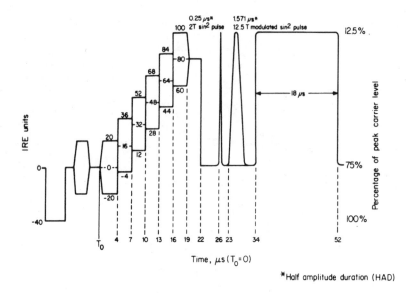

Figure 12.12 Composite vertical-interval test signal (VITS) inserted in field 1, line 18. The video level in IRE units is shown on the left; the radiated carrier signal is shown on the right.

Other test signals, including a modulated staircase or multiburst, are used for more advanced tests. It is important to closely observe how these signals appear at the output of the waveform generator. Knowing what the undistorted signal looks like simplifies the subsequent identification of distortions.

Line Select Features

Some waveform monitors and vectorscopes have line select capability. They can display one or two lines out of the entire video frame of 525 lines. (In the normal display, all of the lines are overlaid on top of one another.) The principal use of the single line feature is to monitor VITS (*vertical interval test signals*). VITS allows in-service testing of the transmission system. A typical VITS waveform is shown in Figure 12.12. A full-field line selector drives a picture monitor output with an intensifying pulse. The pulse causes a single horizontal line on a picture monitor to be highlighted. This indicates where the line selector is within the frame to correlate the waveform monitor display with the picture.

Distortion Mechanisms

The video system should respond uniformly to signal components of different frequencies. In an analog system, this parameter is generally evaluated with a waveform monitor. Different signals are required to check the various parts of the frequency spectrum. If the signals

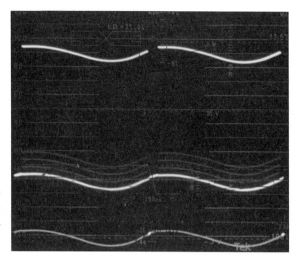

Figure 12.13 Waveform monitor display showing additive 60 Hz degradation. (*Courtesy of Tektronix.*)

are all faithfully reproduced on the waveform monitor screen after passing through the video system, it is safe to assume that there are no serious frequency response problems.

At very low frequencies, look for externally introduced distortions, such as power-line hum or power supply ripple, and distortions resulting from inadequacies in the video processing equipment. Low-frequency distortions usually appear on the video image as flickering or slowly varying brightness. Low-frequency interference can be seen on a waveform monitor when the dc restorer is set to the slow mode and a 2-field sweep is selected. Sine wave distortion from ac power-line hum may be observed in Figure 12.13. A *bouncing APL* signal can be used to detect distortion in the video chain. Vertical shifts in the blanking and sync levels indicate the possibility of low-frequency distortion of the signal.

Field-rate distortions appear as a difference in shading from the top to the bottom of the picture. A field-rate 60 Hz square wave is best suited for measuring field-rate distortion. Distortion of this type is observed as a tilt in the waveform in the 2-field mode with the dc restorer off.

Line-rate distortions appear as streaking, shading, or poor picture stability. To detect such errors, look for tilt in the bar portion of a pulse-and-bar signal. The waveform monitor should be in the 1H or 2H mode with the fast dc restorer selected for the measurement.

The *multiburst* signal is used to test the high-frequency response of a system. The multiburst includes packets of discrete frequencies within the television passband, with the higher frequencies toward the right of each line. The highest frequency packet is at about 4.2 MHz, the upper frequency limit of the NTSC system. The next packet to the left is near the color subcarrier frequency (3.58 MHz, approximately) for checking the chrominance transfer characteristics. Other packets are included at intervals down to 500 kHz. The most common distortion is high-frequency rolloff, seen on the waveform monitor as reduced amplitude packets at higher frequencies. This type of problem is shown in Figure 12.14. The television picture exhibits loss of fine detail and color intensity when such impairments are

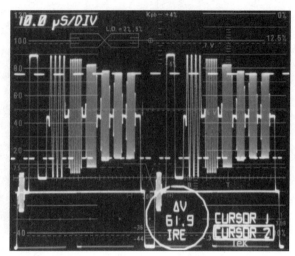

Figure 12.14 Waveform monitor display of a multiburst signal showing poor high-frequency response. (*Courtesy of Tektronix.*)

present. High frequency peaking, appearing on the waveform as higher amplitude packets at the higher frequencies, causes ghosting on the picture.

Differential Phase

Differential phase (*d*ϕ) distortion occurs if a change in luminance level produces a change in the chrominance phase. If the distortion is severe, the hue of an object will change as its brightness changes. A modulated staircase or ramp is used to quantify the problem. Either signal places chrominance of uniform amplitude and phase at different luminance levels. Figure 12.15 shows a 100 IRE modulated ramp. Because *d*ϕ can change with changes in APL, measurements at the center and at the two extremes of the APL range are necessary.

To measure *d*ϕ with a vectorscope, increase the gain control until the vector dot is on the edge of the graticule circle. Use the phase shifter to set the vector to the 9 o'clock position. Phase error appears as circumferential elongation of the dot. The vectorscope graticule has a scale marked with degrees of *d*ϕ error. Figure 12.16 shows a *d*ϕ error of 5°.

More information can be obtained from a swept R–Y display, which is a common feature of waveform monitor and vectorscope systems. If one or two lines of demodulated video from the vectorscope are displayed on a waveform monitor, differential phase appears as tilt across the line. In this mode, the phase control can be adjusted to place the demodulated video on the baseline, which is equivalent in phase to the 9 o'clock position of the vectorscope. Figure 12.17 shows a *d*ϕ error of 5° with the amount of tilt measured against a vertical scale. This mode is useful in troubleshooting applications. By noting where along the line the tilt begins, it is possible to determine at what dc level the problem starts to occur. In addition, field-rate sweeps enable the operator to look at *d*ϕ over the field.

A variation of the swept R–Y display may be available in some instruments for precise measurement of differential phase. Highly accurate measurements can be made with a vectorscope that includes a precision phase shifter and a double-trace mode. This method involves nulling the lowest part of the waveform with the phase shifter, and then using a sep-

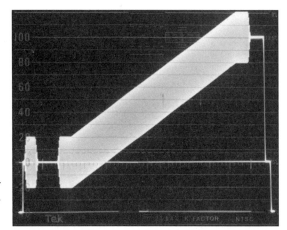

Figure 12.15 Waveform monitor display of a modulated ramp signal. (*Courtesy of Tektronix.*)

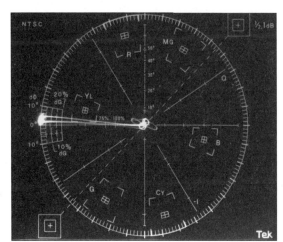

Figure 12.16 Vectorscope display showing 5° differential phase error. (*Courtesy of Tektronix.*)

arate calibrated phase control to null the highest end of the waveform. A readout in tenths of a degree is possible.

Differential Gain

Differential gain (dG) distortion refers to a change in chrominance amplitude with changes in luminance level. The vividness of a colored object changes with variations in scene brightness. The modulated ramp or staircase is used to evaluate this impairment with the measurement taken on signals at different APL points.

To measure differential gain with a vectorscope, set the vector to the 9 o'clock position and use the variable gain control to bring it to the edge of the graticule circle. Differential gain error appears as a lengthening of the vector dot in the radial direction. The dG scale at the left side of the graticule can be used to quantify the error. Figure 12.18 shows a dG error of 10 percent.

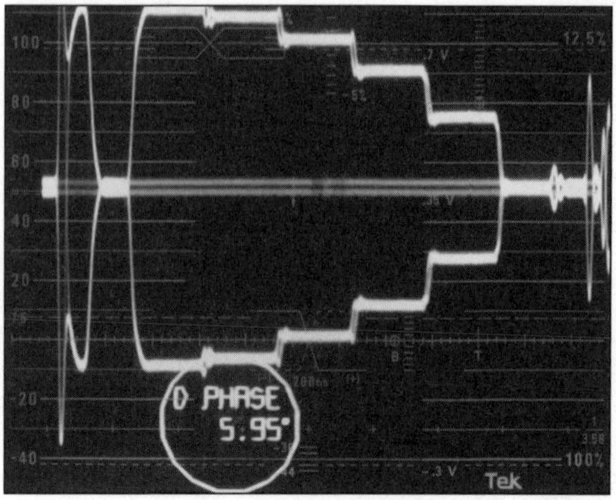

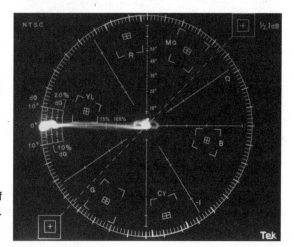

Figure 12.17 Vectorscope monitor display showing a differential phase error of 5.95° as a tilt on the vertical scale. (*Courtesy of Tektronix.*)

Figure 12.18 Vectorscope display of a 10 percent differential gain error. (*Courtesy of Tektronix.*)

Differential gain can be evaluated on a waveform monitor by using the chroma filter and examining the amplitude of the chrominance from a modulated staircase or ramp. With the waveform monitor in 1H sweep, use the variable gain to set the amplitude of the chrominance to 100 IRE. If the chrominance amplitude is not uniform across the line, there is dG error. With the gain normalized to 100 IRE, the error can be expressed as a percentage. Finally, dG can be precisely evaluated with a swept display of demodulated video. This is similar to the single trace R–Y methods for differential phase. The B–Y signal is examined for tilt when the phase is set so that the B–Y signal is at its maximum amplitude. The tilt can be quantified against a vertical scale.

12.3.9 Automated Video Signal Measurement

Video test instruments based on microcomputer systems provide the maintenance engineer with the ability to rapidly measure a number of parameters with exceptional accuracy. Automated instruments offer a number of benefits, including reduced setup time, test repeatability, waveform storage and transmission capability, and remote control of instrument/measurement functions. Typical features of this class of instrument include the following:

- Waveform monitor functions

- Vectorscope monitor functions

- Picture display capability

- Automatic analysis of an input signals

- RS-232 and/or GPIB I/O ports

Figure 12.19 shows a block diagram of a representative automated video test instrument. Sample output waveforms are shown in Figure 12.20.

Automated test instruments offer the end-user the ability to observe signal parameters and to make detailed measurements on them. The instrument depicted in Figure 12.19 offers a "Measure Mode," in which a captured waveform can be expanded and individual elements of the waveform examined. The instrument provides interactive control of measurement parameters, as well as graphical displays and digital readouts of the measurement results. Figure 12.21 illustrates measurement of sync parameters on an NTSC waveform.

Another feature of many automated video measurement instruments is the ability to set operational limits and parameters that—if exceeded—are brought to the attention of an operator. In this way, operator involvement is required only if the monitored signal varies from certain preset limits. Figure 12.22 shows an example error log.

12.3.10 Applications of the Zone Plate Signal

The increased information content of advanced high-definition display systems requires sophisticated processing to make recording and transmission practical [6]. This processing uses various forms of bandwidth compression, scan-rate changes, motion-detection and motion-compensation algorithms, and other techniques. Zone plate patterns are well suited to exercising a complex video system in the three dimensions of its signal spectrum: horizontal, vertical, and temporal. Zone plate signals, unlike most conventional test signals, can be complex and dynamic. Because of this, they are capable of simulating much of the detail and movement of actual video, exercising the system under test with signals representative of the intended application. These digitally generated and controlled signals also have other important characteristics needed in test waveforms for video systems.

A signal intended for meaningful testing of a video system must be carefully controlled, so that any departure from a known parameter of the signal is attributable to a distortion or other change in the system under test. The test signal also must be predictable, so that it can be accurately reproduced at other times or places. These constraints usually have led to test signals that are electronically generated. In a few special cases, a standardized picture has

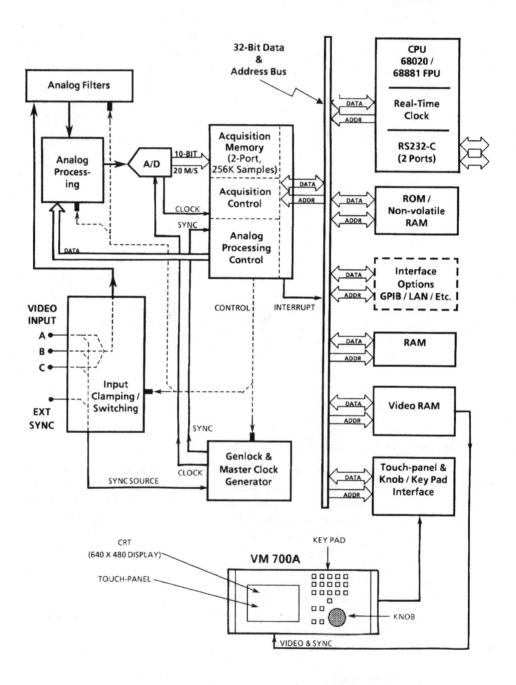

Figure 12.19 Block diagram of an automated video test instrument. (*Courtesy of Tektronix.*)

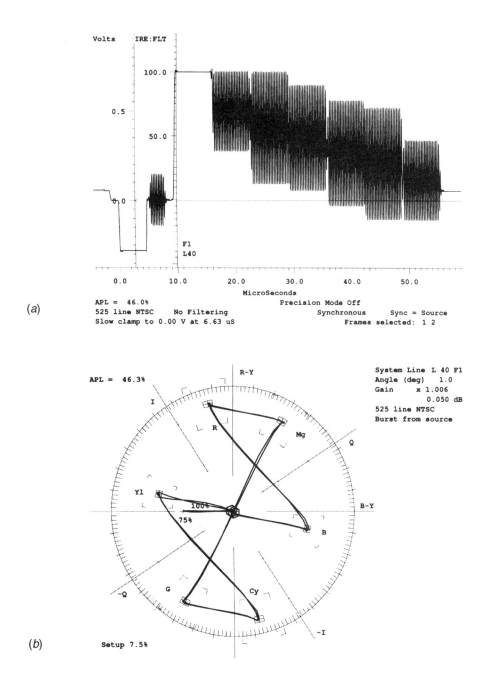

Figure 12.20 Automated video test instrument output charts: (*a*) waveform monitor mode display of color bars, (*b*) vectorscope mode display of color bars. (*Courtesy of Tektronix.*)

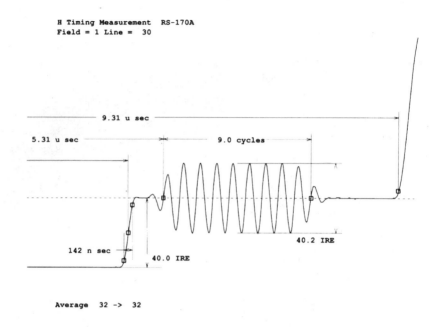

Figure 12.21 Expanded view of a sync waveform with measured parameters. (*Courtesy of Tektronix.*)

				Violated Limits		
				Lower	Upper	
Bar Top	—— % Carr	**		10.0	15.0	Bar Not Found
Blanking Level	—— % Carr	**		72.5	77.5	ZC Pulse Unselected
Bar Amplitude	—— IRE	**		96.0	104.0	Bar Not Found
Sync Variation	—— % Carr	**		0.0	5.0	ZC Pulse Unselected
VIRS Setup	—— % Bar	**		5.0	10.0	Not Found
VIRS Luminance Ref	—— % Bar	**		45.0	55.0	Not Found
VIRS Chroma Ampl	—— % Burst	**		90.0	110.0	Not Found
VIRS Chroma Ampl	—— % Bar	**		36.0	44.0	Not Found
VIRS Chroma Phase	—— Deg	**		-10.0	10.0	Not Found
Line Time Distortion	—— %	**		0.0	2.0	No Composite VITS
Pulse/Bar Ratio	—— %	**		94.0	106.0	No Composite VITS
2T Pulse K-Factor	—— % Kf	**		0.0	2.5	No Composite VITS
IEEE-511 ST Dist	—— % SD	**		0.0	3.0	No Composite VITS

Figure 12.22 Example error log for a monitored signal. (*Courtesy of Tektronix.*)

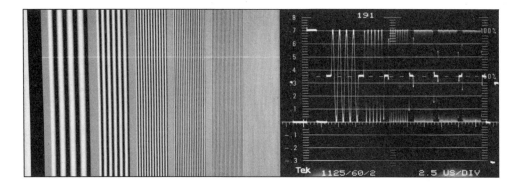

Figure 12.23 Multiburst video test waveform: (*a, left*) picture display, (*b, right*) multiburst signal as viewed on a waveform monitor (1H). (*Courtesy of Tektronix.*)

been televised by a camera or monoscope—usually for a subjective, but more detailed, evaluation of overall performance of the video system.

A zone plate is a physical optical pattern, which was first used by televising it in this way. Now that electronic generators are capable of producing similar patterns, the label "zone plate" is applied to the wide variety of patterns created by video test instruments.

Conventional test signals, for the most part limited by the practical considerations of electronic generation, have represented relatively simple images. Each signal is capable of testing a narrow range of possible distortions; several test signals are needed for a more complete evaluation. Even with several signals, this method may not reveal all possible distortions or allow study of all pertinent characteristics. This is true especially in video systems employing new forms of sophisticated signal processing.

Simple Zone Plate Patterns

The basic testing of a video communication channel historically has involved the application of several single frequencies—in effect, spot-checking the spectrum of interest [6]. A well-known and quite practical adaptation of this idea is the multiburst signal, shown in Figure 12.23[1]. This test waveform has been in use since the earliest days of video. The multiburst signal provides several discrete frequencies along a TV line.

The frequency-sweep signal is an improvement on multiburst. Although harder to implement in earlier generators, it was easier to use. The frequency-sweep signal, illustrated in Figure 12.24, varies the applied signal frequency continuously along the TV line. In some cases, the signal is swept as a function of the vertical position (field time). Even in these

1. Figure 12.23(*a*) and other photographs in this section show the "beat" effects introduced by the screening process used for photographic printing. This is largely unavoidable. The screening process is quite similar to the scanning or sampling of a television image—the patterns are designed to identify this type of problem.

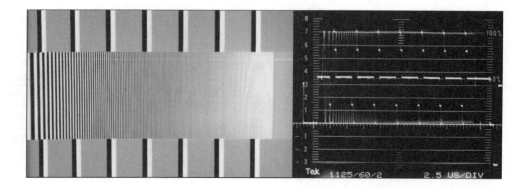

Figure 12.24 Conventional sweep-frequency test waveform: (*a, left*) picture display, (*b, right*) waveform monitor display, with markers (1H). (*Courtesy of Tektronix.*)

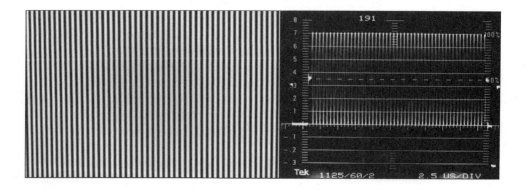

Figure 12.25 Single horizontal frequency test signal from a zone plate generator: (*a, left*) picture display, (*b, right*) waveform monitor display (1H). (*Courtesy of Tektronix.*)

cases, the signal being swept is appropriate for testing the spectrum of the horizontal dimension of the picture.

Figure 12.25 shows the output of a zone plate generator configured to produce a horizontal single-frequency output. Figure 12.26 shows a zone plate generator configured to produce a frequency-sweep signal. Electronic test patterns, such as these, may be used to evaluate the following system characteristics:

- Channel frequency response
- Horizontal resolution
- Moiré effects in recorders and displays
- Other impairments

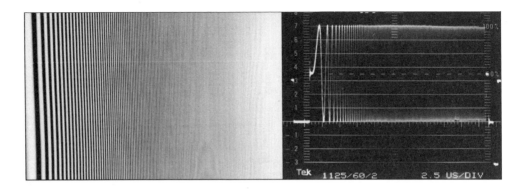

Figure 12.26 Horizontal frequency-sweep test signal from a zone plate generator: (*a, left*) picture display, (*b, right*) waveform monitor display (1H). (*Courtesy of Tektronix.*)

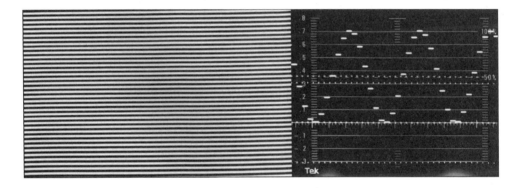

Figure 12.27 Single vertical frequency test signal: (*a, left*) picture display, (*b, right*) magnified vertical-rate waveform, showing the effects of scan sampling. (*Courtesy of Tektronix.*)

Traditionally, patterns that test vertical (field) response have been less frequently used. As new technologies implement conversion from interlaced to progressive scan, line-doubling display techniques, vertical antialiasing filters, scan conversion, motion detection, or other processes that combine information from line to line, vertical testing patterns will be more in demand.

In the vertical dimension, as well as the horizontal, tests can be done at a single frequency or with a frequency-sweep signal. Figure 12.27 illustrates a magnified vertical-rate waveform display. Each "dash" in the photo represents one horizontal scan line. Sampling of vertical frequencies is inherent in the scanning process, and the photo shows the effects on the signal waveform. Note also that the signal voltage remains constant during each line, changing only from line to line in accord with the vertical-dimension sine function of the signal. Figure 12.28 shows a vertical frequency-sweep picture display.

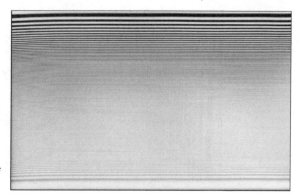

Figure 12.28 Vertical frequency-sweep picture display. (*Courtesy of Tektronix.*)

The horizontal and vertical sine waves and sweeps are quite useful, but they do not use the full potential of a zone plate signal source.

Producing the Zone Plate Signal

A zone plate signal is created in real time by a test signal generator [6]. The value of the signal at any instant is represented by a number in the digital hardware. This number is incremented as the scan progresses through the three dimensions that define a point in the video image: horizontal position, vertical position, and time.

The exact method by which these dimensions alter the number is controlled by a set of coefficients. These coefficients determine the initial value of the number and control the size of the increments as the scan progresses along each horizontal line, from line to line vertically, and from field to field temporally. A set of coefficients uniquely determines a pattern, or a sequence of patterns, when the time dimension is active.

This process produces a sawtooth waveform; overflow in the accumulator holding the signal number effectively resets the value to zero at the end of each cycle of the waveform. Usually, it is desirable to minimize the harmonic energy content of the output signal; in this case, the actual output is a sine function of the number generated by the incrementing process.

Complex Patterns

A pattern of sine waves or sweeps in multiple dimensions may be produced, using the unique architecture of the zone plate generator [6]. The pattern shown in Figure 12.29, for example, is a single signal sweeping both horizontally and vertically. Figure 12.30 shows the waveform of a single selected line (line 263 in the 1125/60/2 HDTV system). Note that the horizontal waveform is identical to the one shown in Figure 12.26(*b*), even though the vertical-dimension sweep is now also active. Actually, different lines will give slightly different waveforms. The horizontal frequency and sweep characteristics will be identical, but the starting phase must be different from line to line to construct the vertical signal.

Figure 12.29 Combined horizontal and vertical frequency-sweep picture display. (*Courtesy of Tektronix.*)

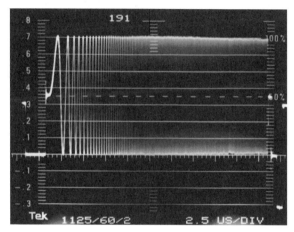

Figure 12.30 Combined horizontal and vertical frequency sweeps, selected line waveform display (1H). This figure shows the maintenance of horizontal structure in the presence of vertical sweep. (*Courtesy of Tektronix.*)

Figure 12.31 shows a 2-axis sweep pattern that is most often identified with zone plate generators, perhaps because it quite closely resembles the original optical pattern. In this circle pattern, both horizontal and vertical frequencies start high, sweep to zero (in the center of the screen), and sweep up again to the end of their respective scans. The concept of 2-axis sweeps actually is more powerful than it might first appear. In addition to purely horizontal or vertical effects, there are possible distortions or artifacts that are apparent only with simultaneous excitation in both axes. In other words, the response of a system to diagonal detail may not be predictable from information taken from the horizontal and vertical responses.

The Time (Motion) Dimension

Incrementing the number in the accumulator of the zone plate generator from frame to frame (or field to field in an interlaced system) creates a predictably different pattern for each vertical scan [6]. This, in turn, creates apparent motion and exercises the signal spectrum in the temporal dimension. Analogous to the single-frequency and frequency-sweep

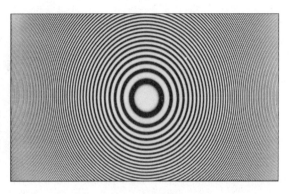

Figure 12.31 The best-known zone plate pattern, combined horizontal and vertical frequency sweeps with zero frequency in the center screen. (*Courtesy of Tektronix.*)

Figure 12.32 Vertical frequency-sweep picture display. (*Courtesy of Tektronix.*)

examples given previously, appropriate setting of the time-related coefficients will create constant motion or motion sweep (acceleration).

Specific motion-detection and interpolation algorithms in a system under test may be exercised by determining the coefficients of a critical sequence of patterns. These patterns then may be saved for subsequent testing during development or adjustment. In an operational environment, appropriate response to a critical sequence could ensure expected operation of the equipment or facilitate fault detection.

Although motion artifacts are difficult to portray in the still-image constraints of a printed book, the following example gives some idea of the potential of a versatile generator. In Figure 12.32, the vertical sweep maximum frequency has been increased to the point where it is zero-beating with the scan at the bottom of the screen. (The cycles/ph of the pattern matches the lines/ph per field of the scan.) Actually, in direct viewing, there is another noticeable artifact in the vertical center of the screen: a harmonic beat related to the gamma of the display CRT. Because of interlace, this beat flickers at the field rate. The photograph integrates the interfield flicker, thereby hiding the artifact, which is readily apparent when viewed in real time.

Figure 12.33 is identical to the previous photo, except for one important difference—upward motion of ½-cycle per field has been added to the pattern. Now the sweep pattern itself is integrated out, as is the first-order beat at the bottom. The harmonic effects in center screen no longer flicker, because the change of scan vertical position from field to field is

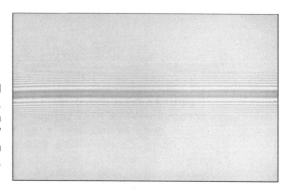

Figure 12.33 The same vertical sweep as shown in Figure 12.32, except that appropriate pattern motion has been added to "freeze" the beat pattern in the center screen for photography or other analysis. (*Courtesy of Tektronix.*)

Figure 12.34 A hyperbolic variation of the 2-axis zone plate frequency sweep. (*Courtesy of Tektronix.*)

compensated by a change in position of the image. The resulting beat pattern does not flicker and is easily photographed or, perhaps, scanned to determine depth of modulation.

A change in coefficients produces hyperbolic, rather than circular 2-axis patterns, as shown in Figure 12.34. Another interesting pattern, which has been used for checking complex codecs, is shown in Figure 12.35. This is also a moving pattern, which was altered slightly to freeze some aspects of the movement for the purpose of taking the photograph.

12.3.11 Display Measurement Techniques

A number of different techniques have evolved for measuring the static performance of picture display devices and systems [7]. Most express the measured device performance in a unique figure of merit or metric. Although each approach provides useful information, the lack of standardization in measurement techniques makes it difficult or even impossible to directly compare the performance of a given class of devices.

Regardless of the method used to measure performance, the operating parameters must be set for the anticipated operating environment. Key parameters include:

• Input signal level

• System/display line rate

• Luminance (brightness)

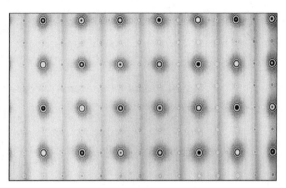

Figure 12.35 A 2-axis frequency sweep in which the range of frequencies is swept several times in each axis. Complex patterns such as this may be created for specific test requirements. (*Courtesy of Tektronix.*)

- Contrast

- Image size

- Aspect ratio

If the display is used in more than one environmental condition—such as under day and night conditions—a set of measurements is appropriate for each application.

12.3.12 Subjective CRT Measurements

Three common subjective measurement techniques are used to assess the performance of a CRT [7]:

- Shrinking raster

- Line width

- TV limiting resolution

Predictably, subjective measurements tend to exhibit more variability than objective measurements. Although they generally are not used for acceptance testing or quality control, subjective CRT measurements provide a fast and relatively simple means of performance assessment. Results usually are consistent when performed by the same observer. Results for different observers often vary, however, because different observers use different visual criteria to make their judgments.

The shrinking raster and line-width techniques are used to estimate the vertical dimension of the display (CRT) beam spot size (*footprint*). Several underlying assumptions accompany this approach:

- The spot is assumed to be symmetrical and Gaussian in the vertical and horizontal planes.

- The display modulation transfer function (MTF) calculated from the spot-size measurement results in the best performance envelope that can be expected from the device.

- The modulating electronics are designed with sufficient bandwidth so that spot size is the limiting performance parameter.

- The modulation contrast at low spatial frequencies approaches 100 percent.

Depending on the application, not all of these assumptions are valid:

- **Assumption 1.** Verona [7] has reported that the symmetry assumption is generally not true. The vertical spot profile is only an approximation to the horizontal spot profile; most spot profiles exhibit some degree of astigmatism. However, significant deviations from the symmetry and Gaussian assumptions result in only minor deviations from the projected performance when the assumptions are correct.

- **Assumption 2.** The optimum performance envelope assumption infers that other types of measurements will result in the same or lower modulation contrast at each spatial frequency. The MTF calculations based on a beam footprint in the vertical axis indicate the optimum performance that can be obtained from the display because finer detail (higher spatial frequency information) cannot be written onto the screen smaller than the spot size.

- **Assumption 3.** The modulation circuit bandwidth must be sufficient to pass the full incoming video signal. Typically, the video circuit bandwidth is not a problem with current technology circuits, which usually are designed to provide significantly more bandwidth than the display is capable of reproducing. However, in cases where this assumption is not true, the calculated MTF based purely on the vertical beam profile will be incorrect. The calculated performance will be better than the actual performance of the display.

- **Assumption 4.** The calculated MTF is normalized to 100 percent modulation contrast at zero spatial frequency and ignores the light scatter and other factors that degrade the actual measured MTF. Independent modulation-contrast measurements at a low spatial frequency can be used to adjust the MTF curve to correct for the normalization effects.

Shrinking Raster Method

The shrinking raster measurement procedure is relatively simple. Steps include the following [7]:

- The brightness and contrast controls are set for the desired peak luminance with an active raster background luminance (1 percent of peak luminance) using a stair-step video signal.

- While displaying a flat-field video signal input corresponding to the peak luminance, the vertical gain/size is reduced until the raster lines are barely distinguishable.

- The raster height is measured and divided by the number of active scan lines to estimate the average height of each scan line. The number of active scan lines is typically 92 percent of the line rate. (For example, a 525-line display has 480 active lines, an 875-line display has 817, and a 1025-line display has 957 active lines.)

The calculated average line height typically is used as a stand-alone metric of display performance.

The most significant shortcoming of the shrinking raster method is the variability introduced through the determination of when the scan lines are *barely distinct* to the observer. Blinking and other eye movements often enhance the distinctness of the scan lines; lines that were indistinct become distinct again.

Line-Width Method

The line-width measurement technique requires a microscope with a calibrated graticule [7]. The focused raster is set to a 4:3 aspect ratio, and the brightness and contrast controls are set for the desired peak luminance with an active raster background luminance (1 percent of peak luminance) using a stair-step video signal. A single horizontal line at the anticipated peak operating luminance is presented in the center of the display. The spot is measured by comparing its luminous profile with the graticule markings. As with the shrinking raster technique, determination of the line edge is subjective.

TV Limiting Resolution Method

This technique involves the display of 2-dimensional high-contrast bar patterns or lines of various size, spacing, and angular orientation [7]. The observer subjectively determines the limiting resolution of the image by the smallest set of bars that can be resolved. Following are descriptions of potential errors with this technique:

- A phenomenon called *spurious resolution* can occur that leads the observer to overestimate the limiting resolution. Spurious resolution occurs beyond the actual resolution limits of the display. It appears as fine structures that can be perceived as line spacings closer than the spacing at which the bar pattern first completely blurs. This situation arises when the frequency-response characteristics fall to zero, then go negative, and perhaps oscillate as they die out. At the bottom of the negative trough, contrast is restored, but in reverse phase (white becomes black, and black becomes white).

- The use of test charts imaged with a video camera can lead to incorrect results because of the addition of camera resolution considerations to the measurement. Electronically generated test patterns are more reliable image sources.

The proper setting of brightness and contrast is required for this measurement. Brightness and contrast controls are adjusted for the desired peak luminance with an active raster background luminance (1 percent of peak luminance) using a stair-step video signal. Too much contrast will result in an inflated limiting resolution measurement; too little contrast will result in a degraded limiting resolution measurement.

Electronic resolution pattern generators typically provide a variety of resolution signals from 100 to 1000 *TV lines/picture height* (TVL/ph) or more in a given multiple (such as 100). Figure 12.36 illustrates an electronically generated resolution test pattern for high-definition video applications.

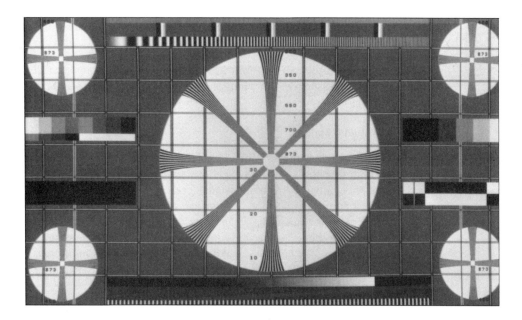

Figure 12.36 Wide aspect ratio resolution test chart produced by an electronic signal generator. (*Courtesy of Tektronix.*)

Application Considerations

The subjective techniques discussed in this section, with the exception of TV limiting resolution, measure the resolution of the *display* [7]. The TV pattern test measures *image* resolution, which is quite different.

Consider as an example a video display in which the scan lines can just be perceived—about 480 scan lines per picture height. This indicates a *display* resolution of at least 960 TV lines, counting light *and* dark lines, per the convention. If a pattern from an electronic generator is displayed, observation will show the image beginning to deteriorate at about 340 TV lines. This characteristic is the result of beats between the image pattern and the raster, with the beat frequency decreasing as the pattern spatial frequency approaches the raster spatial frequency. This ratio of 340/480 = 0.7 (approximately) is known as the *Kell factor*. Although debated at length, the factor does not change appreciably in subjective observations.

12.3.13 Objective CRT Measurements

Four common types of objective measurements may be performed to assess the capabilities of a CRT:

- Half-power width

- Impulse Fourier transform

- Knife-edge Fourier transform

- Discrete frequency

Although more difficult to perform than the subjective measurements discussed so far, objective CRT measurement techniques offer greater accuracy and better repeatability. Some of the procedures require specialized hardware and/or software.

Half-Power-Width Method

Under the half-power-width technique, a single horizontal line is activated with the brightness and contrast controls set to a typical operating level. The line luminance is equivalent to the highlight luminance (maximum signal level). The central portion of the line is imaged with a microscope in the plane of a variable-width slit. The open slit allows all the light from the line to pass through to a photodetector. The output of the photodetector is displayed on an oscilloscope. As the slit is gradually closed, the peak amplitude of the photodetector signal decreases. When the signal drops to 50 percent of its initial value, the slit width is recorded. The width measurement divided by the microscope magnification represents the *half-power width* of the horizontal scan line.

The half-power width is defined as the distance between symmetrical integration limits, centered about the maximum intensity point, which encompasses half of the total power under the intensity curve. The half-power width is not the same as the half-intensity width measured between the half-intensity points. The half-intensity width is theoretically 1.75 times greater than the half-power width for a Gaussian spot luminance distribution.

It should be noted that the half-power line-width technique relies on line width to predict the performance of the CRT. Many of the precautions outlined in the previous section apply here also. The primary difference, however, is that line width is measured under this technique objectively, rather than subjectively.

Fourier Transform Methods

The impulse Fourier transform technique involves measuring the luminance profile of the spot and then taking the Fourier transform of the distribution to obtain the MTF. The MTF, by definition, is the Fourier transform of the line spread function. Commercially available software may be used to perform these measurements using either an impulse or knife edge as the input waveform. Using the vertical spot profile as an approximation to the horizontal spot profile is not always appropriate, and the same reservations expressed in the previous section apply in this case as well.

The measurement is made by generating a single horizontal line on the display with the brightness and contrast set as discussed previously. A microphotometer with an effective slit-aperture width approximately one-tenth the width of the scan line is moved across the scan line (the long slit axis parallel to the scan line). The data taken is stored in an array, which represents the luminance profile of the CRT spot, distance vs. luminance. The microphotometer is calibrated for luminance measures and for distance measures in the object plane. Each micron step of the microphotometer represents a known increment in the object plane. The software then calculates the MTF of the display based on its line spread from the calibrated luminance and distance measurements. Finite slit-width corrections also may be made to the MTF curve by dividing it by a measurement-system MTF curve obtained from the luminance profile of an ideal knife-edge aperture or a standard source.

The knife-edge Fourier transform measurement may be conducted using a low-spatial-frequency vertical bar pattern (5 to 10 cycles) across the display with the brightness and contrast controls set as discussed previously. The frequency response of the square wave pattern generator and video pattern generator should be greater than the frequency response of the display system (100 MHz is typical). The microphotometer scans from the center of a bright bar to the center of a dark bar (left to right), measuring the width of the boundary and comparing it to a knife edge. The microphotometer slit is oriented vertically, with its long axis parallel to the bars. The scan usually is made from a light bar to a dark bar in the direction of spot movement. This procedure is preferred because waveforms from scans in the opposite direction may contain certain anomalies. When the beam is turned on in a square wave pattern, it tends to overshoot and oscillate. This behavior produces artifacts in the luminance profile of the bar edge as the beam moves from an off to an on state. In the on-to-off direction, however, the effects are minimal and the measured waveform does not exhibit the same anomalies that can corrupt the MTF calculations.

The bar-edge (knife-edge) measurement, unlike the other techniques discussed so far, uses the horizontal spot profile to predict display performance. All of the other techniques use the vertical profile as an approximation of the more critical horizontal spot profile. The bar-edge measurement will yield a more accurate assessment of display performance because the displayed image is being generated with a spot scanned in the horizontal direction.

Discrete Frequency Method

The discrete sine wave frequency-response measurement technique provides the most accurate representation of display performance. With this approach, there are no assumptions implied about the shape of the spot, the electronics bandwidth, or low-frequency light scatter. Discrete spatial frequency sine wave patterns are used to obtain a discrete spatial frequency MTF curve that represents the signal-in to luminance-out performance of the display.

The measurement is begun by setting the brightness and contrast as discussed previously, with black-level luminance at 1 percent of the highlight luminance. A sine wave signal is produced by a function generator and fed to a pedestal generator where it is converted to an RS-170-A or RS-343 signal, then applied to the CRT. The modulation and resulting spatial frequency pattern are measured with a scanning microphotometer. The highlight and black-level measurements are used to calculate the modulation constant for each spatial frequency from 5 cycles/display width to the point that the modulation constant falls to less than 1 percent. The modulation constant values then are plotted as a function of spatial frequency, generating a discrete spatial frequency MTF curve.

12.3.14 Viewing Environment Considerations

The environment in which a video display device is viewed is an important criterion for critical viewing situations. Applications in which color purity and adherence to set standards are important require a standardized (or at least consistent) viewing environment. For example, textile colors viewed on a display with a white surround will appear different than the same colors viewed with a black surround. By the same token, different types of ambient lighting will make identical colors appear different on a given display.

The SMPTE has addressed this issue with RP 166-1995, which specifies the environmental and surround conditions that are required in television or video program review areas for the "consistent and critical evaluation" of conventional television signals [8]. Additionally, the practice is designed to provide for repeatable color grading or correction. A number of important parameters are specified in RP 166-1995, including the following:

- The distance of the observer from the monitor screen should be 4 to 6 picture heights for SDTV displays.

- The observer should view the monitor screen at a preferred angle in both the horizontal and vertical planes of $0° \pm 5°$ and, in any event, no greater than $\pm 15°$ from the perpendicular to the midpoint of the screen.

- The viewing area decor should have a generally neutral matte impression, without dominant colors.

- Surface reflectances should be nonspecular and should not exceed 10 percent of the peak luminance value of the monitor white.

The Recommended Practice suggests placing the monitor in a freestanding environment 2.5 to 5 screen heights in front of the wall providing the visual surround. Another acceptable approach is to mount the monitor in a wall with its face approximately flush with the surface of the wall. It is further recommended that all light sources in use during picture assessment or adjustment have a color quality closely matching the monitor screen at reference white (i.e., D65).

It is often necessary to have black-and-white monitors surrounding one or more color monitors in a studio control room. According to RP 166-1995, the black-and-white monitors should be the same color temperature as the properly adjusted color monitor(s), 6500 K. Black-and-white monitors are normally equipped with P4 phosphors, at about 9300 K. This cooler color temperature prevents the background surrounding the color monitors from remaining neutral. Most black-and-white monitors can be ordered with 6500 K phosphors.

Picture Monitor Alignment

The proper adjustment and alignment of studio picture monitors is basic to video quality control. Uniform alignment throughout the production chain also ensures consistency in color adjustment, which facilitates the matching of different scenes within a program that may be processed at different times and in different facilities. The SMPTE has addressed this requirement for conventional video through RP 167-1995. The Recommended Practice offers a step-by-step process by which color monitors can be set. Key elements of RP 167-1995 include the following [9]:

- **Initial conditions.** Setup includes allowing the monitor to warm up and stabilize for 20 to 30 minutes. The room ambient lighting should be the same as it is when the monitor is in normal service, and several minutes must be allowed for visual adaptation to the operating environment.

- **Initial screen adjustments.** The monitor is switched to the setup position, in which the red, green, and blue screen controls are adjusted individually so that the signals are barely visible.

- **Purity.** Purity, the ability of the gun to excite only its designated phosphor, is checked by applying a low-level flat-field signal and activating only one of the three guns at a time. The display should have no noticeable discolorations across the face.

- **Scan size.** The color picture monitor application establishes whether the *overscan* or *underscan* presentation of the display will be selected. An underscanned display is one in which the active video (picture) area, including the corners of the raster, is visible within the screen mask. Normal scan brings the edges of the picture tangent to the mask position. Overscan should be no more than 5 percent.

- **Geometry and aspect ratio.** Display geometry and aspect ratio are adjusted with the crosshatch signal by scanning the display device with the green beam only. Correct geometry and linearity are obtained by adjusting the pincushion and scan-linearity controls so that the picture appears without evident distortions from the normal viewing distance.

- **Focus.** An ideal focus target is available from some test signal generators; if it is unavailable, multiburst, crosshatch, or white noise can be used as tools to optimize the focus of the displayed picture.

- **Convergence.** Convergence is adjusted with a crosshatch signal; it should be optimized for either normal scan or underscan, depending upon the application.

- **Aperture correction.** If aperture correction is used, the amount of correction can be estimated visually by ensuring that the $2T \sin^2$ pulse has the same brightness as the luminance bar or the multiburst signal when the 3 and 4.2 MHz bursts have the same sharpness and contrast.

- **Chrominance amplitude and phase.** The chrominance amplitude and phase are adjusted using the SMPTE color bar test signal and viewing only the blue channel. Switching off the comb filter, if it is present, provides a clear blue channel display. Periodically, the red and green channels should be checked individually in a similar manner to verify that the decoders are working properly. A detailed description of this procedure is given in [9].

- **Brightness, color temperature, and gray scale tracking.** The 100-IRE window signal is used to supply the reference white. Because of typical luminance shading limitations, a centrally placed PLUGE [10] signal is recommended for setting the monitor brightness control. The black set signal provided in the SMPTE color bars also can be used for this purpose.

- **Monitor matching.** When color matching two or more color monitors, the same alignment steps should be performed on each monitor in turn. Remember, however, that monitors cannot be matched without the same phosphor sets, similar display uniformity characteristics, and similar sharpness. The most noticeable deviations on color monitors are the lack of uniform color presentations and brightness shading. Color matching of monitors for these parameters can be most easily assessed by observing flat-field uniformity of the picture at low, medium, and high amplitudes.

For complete monitor-alignment procedures, see [9].

As more experience is gained with DTV-based systems, operating parameters such as those detailed in this section will no doubt be updated to take into consideration the unique attributes and requirements of HDTV.

12.4 Video Camera Performance Characterization and Verification

With tube-type cameras, it was almost mandatory to fine-tune the camera before a major shoot to achieve optimum performance [11]. The intrinsic stability of the CCD imager and the stability of the circuitry used in modern CCD cameras now make it possible to operate the camera for several months without internal readjustment. Physical damage to the camera or lens, in use or transport, is probably the most frequent cause of a loss in performance in a CCD-based device. With careful handling, the probability of malfunction is very small. It is nevertheless prudent to schedule a quick check-out of the camera before the start of a major shoot, when the high cost of talent and other aspects of the production are considered.

The following items are appropriate for inclusion in such a check-out procedure. If the test results show a significant deviation from the manufacturers specifications or from the data previously obtained for the same test, a more thorough examination of the camera, as prescribed in the camera service manual, is then indicated.

12.4.1 Visual Inspection and Mechanical Check

Visually inspect the camera and lens for evidence of physical damage as a clue to the possibility of more serious internal damage [11]. Carefully operate the lens adjustments—manual and servo zoom, focus, and manual iris. If there is evidence of binding or a rough spot in any of these adjustments, physical damage that may affect the optical performance of the lens must be suspected. Inspect the front and rear lens elements; clean if necessary using pure alcohol and soft, lint-free wipes. Fingerprints, in particular, should be removed as quickly as possible to avoid harm to the optical coating of the lens elements. Note that lens manufacturers generally discourage the use of silicon-impregnated wipes for cleaning high quality optics.

12.4.2 Confirmation of the Camera Encoder

A properly adjusted encoder is particularly useful because it provides a convenient window to look inside the camera and confirm proper operation of the remaining circuitry [11]. Encoder set-up is easy to confirm because the color bar generator, normally provided in a professional camera, offers a convenient self test of the encoder. To confirm proper operation of the camera encoder, perform the following steps:

- Apply the camera (encoder) output signal to a waveform monitor (WFM), vectorscope, and picture monitor (a high-resolution black and white monitor with 800 TVL or higher resolution is recommended).

- Terminate the WFM, vectorscope, and picture monitor using a discrete 75 Ω termination. The preferred tolerance for this termination is ±0.1 percent and no greater than ±1 percent. Internal terminations should not be used unless they have been tested and, if necessary replaced with terminations within the recommended tolerance.

- Select the color bar mode on the camera. Confirm on the vectorscope that the burst and I and Q vectors are of the correct phase and amplitude. Confirm that all of the color bar vectors fall within the tolerance boxes on the vectorscope.

If all of the foregoing vectors are within tolerance, correct operation and adjustment of the encoder is confirmed.

12.4.3 Confirmation of Auto Black

To confirm proper operation of the Auto Black circuit [11]:

- Activate the Auto Black circuitry of the camera.

- Confirm that the lens caps during this adjustment. Confirm the character display in the viewfinder indicates the Auto Black adjustment has been successfully executed.

- Select the 0 dB, +9 dB, and +18 dB gain settings in sequence and confirm the black level adjustment is correct for all three gain settings. This is most easily confirmed with the vectorscope; the output signal should be a dot at the center of the display with no shift in position as gain is switched. The only change should be an increase in noise at the higher gain settings.

12.4.4 Lens Back-Focus

The lens back-focus adjustment trims the lens to the specific optical dimensions of the camera [11]. Whenever a new lens is put on a camera, it is necessary to make this adjustment. Some lenses use a screwdriver lock, while others use a knurled knob to lock the lens back-focus adjustment in place. Accidental misadjustment in use has been known to occur, and it is therefore recommended to confirm this adjustment. To set the lens back-focus:

- Place a Siemens Star Chart (Figure 12.37) at least 10 ft from the camera. Place the chart in a location with low lighting such that the lens iris is wide open.

- Using a high-resolution picture monitor: 1) adjust the lens zoom for full close-up and adjust for best focus using the focus ring on the front of the lens, and 2) adjust the zoom for maximum wide angle position and adjust for optimum focus using the lens back-focus adjustment. Repeat both steps several times.

- Securely lock the lens back-focus adjustment in place.

- Confirm the lens stays in focus over the full zoom range.

Figure 12.37 Siemens star chart. (*After* [11].)

12.4.5 Black Shading

To confirm black shading, perform the following steps [11]:

- Cap the lens. Raise the master black level to about 10–12 IRE to avoid any clipping.

- Observe the waveform monitor in the vertical display mode, and then in the horizontal display mode. If the black level is a thin horizontal line, there is no black shading in any of the three color components.

- Restore the black level to its proper position.

12.4.6 Detail Circuit

Using the 11-step gray scale chart, confirm the amplitude of the detail signal as required for the application [11]. A stronger detail signal typically is required for a lower performance recorder and less for a higher performance recorder.

12.4.7 Optional Tests

If the camera system is capable of resolving fine detail close to the limiting resolution specified by the manufacturer, there is a strong assurance that the lens, camera optics, and overall camera signal processing circuits are working correctly [11]. Use a suitable chart with resolution wedges or a chart with a multiburst pattern and confirm that the overall resolution of the camera system is close to the manufacturers specification (Figures 12.36 and 12.38).

White Shading

To confirm white shading [11]:

- Set up a uniformly lit white test chart.

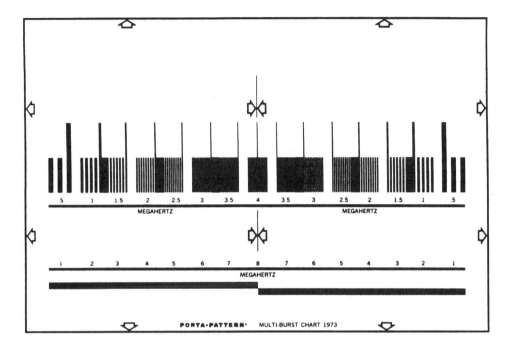

Figure 12.38 Multiburst chart (10 MHz). (*After* [11].)

- Using the waveform monitor, open the lens to obtain about 70 IRE units of video (confirm that the iris is in the range of $f/4.0$ to $f/5.6$; adjust the lighting if necessary), and confirm there is a minimum of vertical, then horizontal, shading.

- Adjust as necessary using the camera horizontal and vertical white shading controls.

Flare

The camera flare correction circuitry provides an approximate correction for flare or scattering of peripheral rays in the various parts of the optical system [11]. To confirm the adjustment of the flare correction circuit:

- Frame an 11-step gray scale chart that includes a very low reflectance strip of black velvet added to the chart.

- Adjust the iris from fully closed until the white chip is at 100 IRE units of video. The flare compensation circuitry is adjusted correctly if there is almost no rise in the black level of the velvet strip as the white level is increased to 100 IRE units and only a small rise in the black level, with no change in hue, when the iris is opened one more f-stop beyond the 100 IRE units point.

- Adjust the R, G, and B flare controls as defined in the camera service manual if the flare correction is not adjusted correctly.

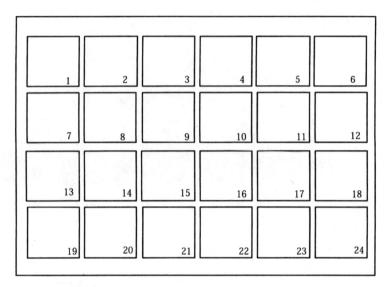

Figure 12.39 Color reference pattern layout specified in SMPTE 303M (proposed). (*From* [12]. *Used with permission.*)

Linear Matrix

When it is necessary to use two dissimilar cameras models in a multicamera shoot, and either of the two models provides an adjustable linear matrix, it is possible to use the variable matrix to obtain a better colorimetry match between cameras [11]. Specific matrix parameters and adjustments (if any) will be found in the camera service manuals.

12.4.8 Color Reference Pattern

SMPTE 303M, proposed at this writing, defines the electrical and physical representation of a television color reference pattern [12]. It also specifies colorimetry, geometry, and related parameters.

The color reference pattern is made up of 24 sample colors whose colorimetric designations are distributed throughout the color television gamut. As shown in Figure 12.39, these samples are square and are arranged in four rows of six samples per row. The first two rows consist of colors that are designed to simulate the color appearance of natural objects. The third row consists of colors that represent subtractive primaries (cyan, magenta, and yellow) as well as the binary combinations of these colors (red, green, and blue). The last row consists of a six-step neutral gray scale.

All video signal levels are calibrated using SMPTE RP 177, the standard television system D_{65} white point, and each of the following standard television primary colorimetry definitions:

- ITR-R BT.709

- SMPTE 170M/SMPTE 240M

- PAL (EBU)

- NTSC (1953 original values)

The standard further specifies illumination and camera positioning details.

12.5   Picture-Quality Measurements for Digital Television

Picture-quality measurement methods include subjective testing, which is always used—at least in an informal manner—and objective testing, which is most suitable for system performance specification and evaluation [13]. A number of types of objective measurement methods are possible for digital television pictures, but those using a human visual system model are the most powerful.

As illustrated in Figure 12.40, three key testing layers can be defined for the modern television system:

- **Video quality.** This consists of signal quality and picture quality (discussed in some detail previously).

- **Protocol analysis.** Protocol testing is required because the data formatting can be quite complex and is relatively independent of the nature of the uncompressed signals or the eventual conversion to interfacility transmission formats. Protocol test equipment can be both a source of signals and an analyzer that locates errors with respect to a defined standard and determines the value of various operational parameters for the stream of data.

- **Transmission system analysis.** To send the video data to a remote location, one of many possible digital data transmission methods may be used, each of which imposes its own analysis issues.

Table 12.1 lists several dimensions of video-quality measurement methods. Key definitions include the following:

- *Subjective* measurements: the result of human observers providing their opinions of the video quality.

- *Objective* measurements: performed with the aid of instrumentation, manually with humans reading a calibrated scale or automatically using a mathematical algorithm.

- *Direct* measurements: performed on the material of interest, in this case, pictures (also known as *picture-quality* measurements).

- *Indirect* measurements: made by processing specially designed test signals in the same manner as the pictures (also known as *signal-quality* measurements). Subjective measurements are performed only in a direct manner because the human opinion of test signal picture quality is not particularly meaningful.

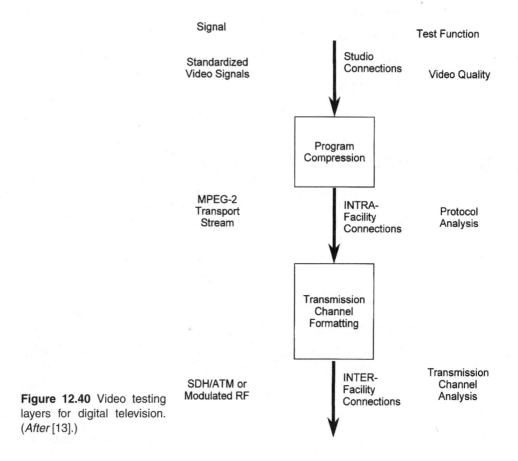

Figure 12.40 Video testing layers for digital television. (*After* [13].)

- *In-service* measurements: made while the program is being displayed, directly by evaluating the program material or indirectly by including test signals with the program material.

- *Out-of-service*: appropriate test scenes are used for direct measurements, and full-field test signals are used for indirect measurements.

In the mixed environment of compressed and uncompressed signals, video-quality measurements consist of two parts: signal quality and picture quality.

12.5.1 Signal/Picture Quality

Signal-quality measurements are made with a suite of test signals as short as one line in the vertical interval [13]. In a completely uncompressed system, such testing will give a good characterization of picture quality. This is not true, however, for a system with compression encoding/decoding because picture quality will change based on the data rate, complexity,

Table 12.1 Video Quality Definitions (*After* [13].)

Parameter	In-Service	Out-of-Service
Indirect measurement		
Objective signal quality	Vertical interval test signals	Full-field test signals
Direct measurement		
Subjective picture quality	Program material	Test scenes
Objective picture quality	Program material	Test scenes

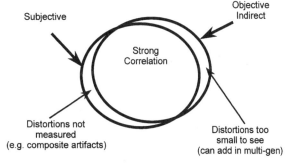

Figure 12.41 Functional environment for traditional video measurements. (*After* [13].)

and encoding algorithm. Picture-quality measurements, instead, require natural scenes (or some equivalent thereof) that are much more complex than traditional test signals. These complex scenes stress the capabilities of the encoder, resulting in nonlinear distortions that are a function of the picture content.

Out-of-service picture-quality measurements are similar to indirect signal-quality measurements in one aspect: the determination of the system response to a high-quality reference. However, they actually measure the degradation in reference picture quality rather than that of a synthetic test signal. In-service, such signals determine the response to program material and its degradation through the system. If the program material is "easy," the measurement may not have a great deal of practical value.

Objective indirect signal-quality measurements are a reasonably good way to determine the picture quality for uncompressed systems. That is, there is a good correlation between subjective measurements made on pictures from the system and objective measurements made on a suite of test signals using the same system. (See Figure 12.41.) The correlation is not perfect for all tests, however. There are distortions in composite systems, such as false color signals caused by poorly filtered high-frequency luminance information being detected as chroma. These distortions are not easily measured by objective means. Also, there are objective measurements that are so sensitive they do not directly relate to subjective results. However, such objective results often are useful because their effect will be seen by a human observer if the pictures are processed in the same way a number of times.

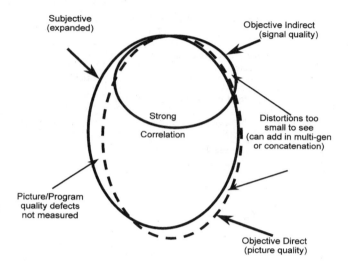

Figure 12.42 Functional environment for digital video measurements. (*After* [13].)

The use of digital compression has expanded the types of distortions that can occur in the modern television system. Because signal-quality measurements will not do the job, objective picture-quality measurements are needed, as illustrated in Figure 12.42. The total picture-quality measurement space has increased because of subjective measurements that now include multiminute test scenes with varying program material and variable picture quality. The new objective measurement methods must have strong correlation with subjective measurements and cover a broad range of applications.

Even with all the objective testing methods available for analog and full-bandwidth digital video, it is important to have human observation of the pictures. Some impairments are not easily measured, yet are obvious to a human observer. This situation certainly has not changed with the addition of modern digital compression. Therefore, casual or informal subjective testing by a reasonably expert viewer remains an important part of system evaluation and/or monitoring.

12.5.2 Automated Picture-Quality Measurement

Objective measurements can be made automatically with an instrument that determines picture degradation through the system [13]. Two somewhat mutually exclusive ways are available for classifying objective picture-quality measurement systems. Although there are several practical methods with a variety of algorithmic approaches, they may be divided into the following classes:

- *Feature extraction*, which does an essentially independent analysis of input and output pictures.

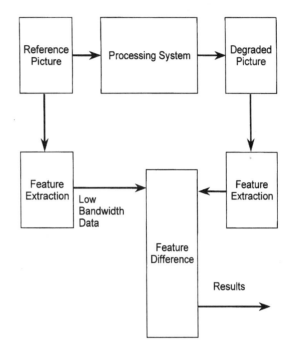

Figure 12.43 Block diagram of the feature-extraction method of picture-quality analysis. (*After* [13].)

- *Picture processing*, where the complete input and output pictures are directly compared in some manner and must be available at the measurement instrument (also known as *picture differencing*).

From a system standpoint, the *glass box* approach utilizes knowledge of the compression system to measure degradation. An example would be looking for blockiness in a DCT system. The *black box* approach makes no assumptions about operation of the system.

In feature extraction, analysis time is reduced by comparing only a limited set of picture characteristics. These characteristics are calculated, and the modest amount of data is embedded into the picture stream. Examples of compression impairments would be block distortion, blurring/smearing, or "mosquito noise." At the receiver, the same features of the degraded picture are calculated, providing a measure of the differences for each feature (Figure 12.43). The major weakness of this approach is that it does not provide correlation between subjective and objective measurements across a wide variety of compression systems or source material.

For the picture-processing scheme, the reference and degraded pictures are filtered in an appropriate manner, resulting in a data set that may be as large as the original pictures. The difference between the two data sets is a measure of picture degradation, as illustrated in Figure 12.44.

The need for a standard for objective measurements has been addressed in ANSI T1.801.03 [14]. Several variations and improvements to the basic ANSI toolbox are listed in

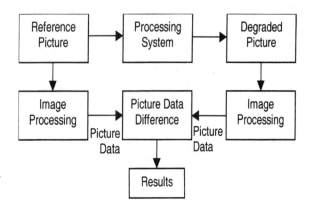

Figure 12.44 Block diagram of the picture-differencing method of picture-quality analysis. (*After* [13].)

[13]. Because of the limitations of the ANSI method, work continues with the goal of refining objective measurement methods.

A number of approaches have been proposed by researchers using the human visual system (HVS) model as a basis. Such a model provides an image-quality metric that is independent of video material, specific types of impairments, and the compression system used. The study of the HVS has been going on for decades, investigating such properties as contrast sensitivity, spatio-temporal response, and color perception. Perhaps one of the best-known derivatives of this work is the *JNDmetrix* (Sarnoff/Tektronix), described in [15]. Other work in this area involves the *Three-Layered Bottom-Up Noise Weighting Model* described in [16].

In-Service Measurements

In-service measurements are made simultaneously with regular program transmissions [17]. For traditional analog signal quality measurements, this is easily accomplished using vertical blanking interval test signals. In-service measurements also are important for systems incorporating bit-rate reduction and other forms of digital processing. Picture quality analysis must use the actual program material analyzed at the source and the processed output for the measurement.

Both picture differencing and feature extraction can be used for in-service measurements. The availability of the source sequence may not be a limitation for in-service monitoring in most applications, given that picture quality is limited primarily by the (transmission) compression process. Presuming this allows a defect monitoring solution to be designed to look for trace artifacts, rather than a loss of image fidelity.

Although this approach would seemingly relegate double-ended picture quality testing to the systems evaluations lab, it is conceivable that program producers may want to have absolute image quality confirmed, or even guaranteed.

Real-time measurements are desirable for in-service operation. In this context, *real-time* means that measurements are continuously available as the sequence passes through the system under test. There may be a nominal delay for processing, much as is done with the aver-

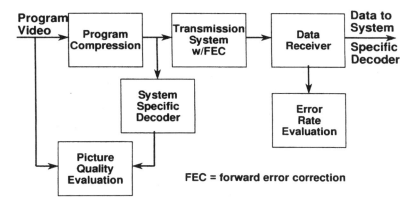

Figure 12.45 In-service picture quality measurement system. (*After* [17].)

aging function in sophisticated analog measurement systems. Generally speaking, every video field would be evaluated.

Non-real-time is the condition where the source and processed sequence information are captured, with results available after some nominal time for computation, perhaps a minute or two.

In-service picture quality measurement using the picture differencing method can be accomplished as shown in Figure 12.45. At the program source, a system-specific decoder is used to provide the processed sequence to a picture quality evaluation system. The results would be available in non-real-time, sampled real-time, or real-time, depending on the computational power of the measurement instrument. The system-specific decoder provides the same picture quality output as the decoder at the receive end of the transmission system.

Practical transmission systems for broadcast applications operate either virtually error-free (resulting in no picture degradation) or with an unrecoverable number of errors (degradation so severe that quality measurements are not particularly useful). Therefore, bit-error-rate evaluation based on calculations in the receiver may provide sufficient monitoring of the transmission system.

The foregoing is based on the assumption that the transmission path is essentially fixed. Depending upon the application, this may not always be the case. Consider a terrestrial video link, which may be switched to a variety of paths before reaching its ultimate destination. The path also can change at random intervals depending on the network arrangement. The situation gives rise to compression concatenation issues for which there are no easy solutions.

As identified by Uchida [18], video quality deterioration accumulates with each additional coding device added to a cascade connection. This implies that, although picture deterioration caused by a single codec may be imperceptible, cascaded codecs, as a whole, can cause serious deterioration that cannot be overlooked. Detailed measurements of concatenation issues are documented in [18].

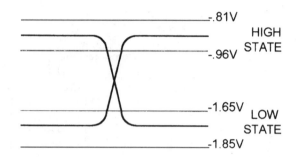

Figure 12.46 Valid data acceptance bands for a digital signal. (*After* [19].)

12.6 Serial Digital Bit Stream Analysis

The creation, processing, storage, and transmission of video in a digital form has numerous, well-documented advantages over analog signals. It is no surprise, therefore, that use of the serial digital interface (SDI) and serial digital transport interface (SDTI) extension to move signals within and among facilities is increasing every year, to the point now where it is commonplace. As with all good things, however, there are a few hidden problems lurking in the background.

Specifications for the interconnection of digital video signal paths are established with the purpose of setting limits for deviation that will permit proper operation under a variety of conditions. Like analog signal specifications, digital signal specs also establish the limits within which the sending equipment must operate and the receiving equipment must accept [19]. In the digital environment, small violations of the specified limits will usually cause no detrimental effect in the resulting image; this is the nature of digital information transfer, and one of its better-known attributes.

The serial digital interface, naturally, concerns itself with transfer of data from one point to another. It follows, then, that three elements must be considered in an analysis of system reliability:

- The transmitter

- The receiver

- The interconnecting medium

If we accept that professional digital video products meet the required SDI specifications (a fair assumption), then the interconnecting medium must be the focus of attention when SDI system design and maintenance is concerned. Given the foregoing, the *point* of attention is the receiver, or more correctly put, the signal delivered to the input terminals of the receiver.

A digital receiving device has a latitude of acceptable variation within which reliable recovery of data will occur. (See Figure 12.46.) Variations within the *high state* and *low state* (as shown) will cause no lost information. Variations greater than the stated specifications may also permit complete information recovery if the performance of the receiver is superior to the nominal (minimum and/or maximum) specification. However, as the state change excursions extend beyond the low state and high state tolerance bands, the performance of the receiver will become unpredictable. Eventually, catastrophic errors will be generated that cannot be masked or recovered by the system (the *digital cliff*).

Because the "cliff" is usually rather broad in a digital system, users often operate without much regard for the specified parameters. Still, reliable operation of a video facility dictates that safety margins be measured and documented, and—in some cases—monitored on a prescribed schedule. The latter case would apply to mission-critical links, such as inter-facility lines that would take the plant down if they failed. Also, systems subject to physical stress, such as remote truck equipment, would qualify for regular signal quality analysis.

12.6.1 SMPTE RP 259M

SMPTE recommended practice 259M, the 10 bit 4:2:2 component digital video protocol, produces a signal of 270 Mbits/s. RP 259M, thus, provides all of the bandwidth necessary to accommodate eight or ten-bit ITU-R Rec. 601 video, with audio if required, in real time and using conventional 75 Ω coaxial cable [20, 21]. The specifications for SMPTE 259M are given in [21]. Key among the specifications are the peak-to-peak value of 0.8 V and the rise time (*transition time*) of 0.75 ns to 1.5 ns. If the signal transmission path had infinite band-width and no group delay, the 259M signal would appear as a perfect square wave pulse train. No path is ideal, of course, and herein lies the potential for problems.

As stated previously, the weak link in an SDI system is the path from the receiver to the transmitter, typically several runs of coax interconnected through a router and/or patch bay. The cable itself represents the greatest problem potential when long path lengths are required. Coax can be modeled as an infinite network of inductive and resistive components in series, with distributed shunt capacitance. This, in effect, describes a low-pass filter whose poles increase in number and whose corner frequency moves closer to zero with extending length. Such attenuation with increasing frequency and distance can deteriorate the SDI signal to the point that it becomes unusable.

It is intuitive that different video signals result in different data patterns in a digital system. Consider the case of successive "1", as illustrated in Figure 12.47. This condition results in a true square wave of 50 percent duty cycle. Real-life video, of course, results in a unpredictable variety of ones and zeros, producing—in effect—*rectangular* square waves whose duty cycle is less than 50 percent [22]. This condition requires a more dense spectrum of harmonics to properly define. It follows that considerable low- and high-frequency harmonics will be present in the SDI signal (significant energy can extend to beyond 1 GHz).

12.6.2 Jitter

Jitter is a related distortion mechanism that may be observed in the SDI/SDTI signal. Jitter is defined as the difference in timing between where a data transition *should occur* and where it actually *does occur*. As illustrated in Figure 12.48, imperfections in the generation and transmission of the data stream can result in displacement of the transition points to either before or after their proper locations. This timing offset can remain relatively stable, or oscillate between two or more points in time. The latter case is what most engineers consider to be *jitter*. Minimizing jitter is critical to the performance of SDI-based systems.

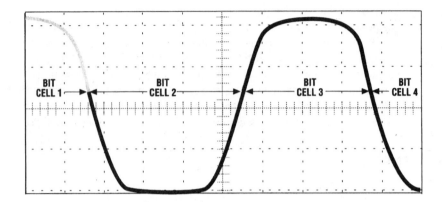

Figure 12.47 A portion of a SMPTE 259M datastream showing three successive ones. (*After* [22].)

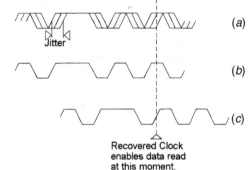

Figure 12.48 Representation of jitter: (*a*) jitter present on the measured waveform, (*b*) data correctly centered on the recovered clock, (*c*) signal instantaneously advanced as a result of jitter. (*After* [19].)

In order for the receiver to be able to decode the logic levels of the SDI signal, a clock is recovered from the data stream, and this recovered clock is used to facilitate decoding of the received data. Because the data transfer is asynchronous, sufficiently large variations in time can result in received data being incorrectly interpreted. Lost data is the result.

12.6.3 The Serial Digital Cliff

Generally speaking, the recovered signal from an SDI/SDTI link is either perfect or basically worthless. An SDI link that is experiencing zero errors, or just a few errors, is considered to be on the *operational plateau* of the SDI reliability curve. As the link is extended or the S/N otherwise degrades, the system moves forward to the *error cliff*. As the link progresses over the knee of this cliff, errors climb rapidly to a point sufficient to swamp error-control mechanisms built into the SDI system. The path, thus, becomes unusable.

Avoiding this well-documented cliff effect requires careful attention to system planning, installation, maintenance, and on-going quality control.

Measurement of the S/N of the principle spectral elements in the SDI bit stream is an effective way to determine how far a path is from the error cliff. For ITU-R Rec. 601 SDI, experience has shown that there are two spectral components whose S/N values are useful in determining the overall health of the link [22]. These are the fundamental frequency of 135 MHz and its third harmonic (405 MHz). It should be noted that because of the coding method used, the fundamental frequency for a component SDI link is not 270 Mbits/s, but one-half of that value (135 Mbits/s). A detailed discussion of this point is given in [22].

The third harmonic is easy to observe with a spectrum analyzer. At the output of most SDI drivers, the third harmonic starts approximately 35 dB above the noise floor. This compares with the fundamental 135 MHz frequency at 45 to 50 dB above noise. Tests demonstrate that after this signal has passed through approximately 300 meters (1,000 ft.) of high-quality coaxial cable, the third harmonic is typically down to 8 to 10 dB above the noise floor. As the third harmonic signal approaches 6 dB above the noise floor, clock recovery becomes unreliable and large numbers of errors begin to occur.

As the error cliff is approached, the displayed error rate at the receiver may increase from one per day to one per frame over a S/N range difference of just 3 dB or less. Such a link can become unusable even though the level of the fundamental experiences only modest attenuation through the path.

12.6.4 Pathological Testing

There are a number of tools that can be useful in determining how close a path is to the knee of the cliff. One readily-available method is the use of *pathological* test signals. As outlined previously, the receiver circuitry of an SDI/SDTI link must regenerate the clock signal. To facilitate decoding, most SDI receivers incorporate a signal equalizing circuit to boost the high frequencies of the incoming waveform. This permits easier clock regeneration and data-value determination. Pathological test signals produce bitstreams that stress these circuits.

There are a large number of signal forms that fall under the general category of pathological testing. One common signal stresses the clock regeneration and equalizing circuits by producing values for *C* and *Y* that force the SDI bit-scrambling circuits to produce a run of 19 zeros and a single *one* approximately every frame. Another common signal puts the values of *C* and *Y* such that a run of 20 *ones*, followed by 20 *zeros* periodically, is produced. There are—in fact—thousands of possible *C* and *Y* combinations that will stress the receiver.

Measurements under real-world conditions have found that a path will fail under testing with a pathological signal at received levels approximately 2 dB higher than where a typical program (non-pathological) signal will fail [22]. Such tests, therefore, can help to identify whether a given SDI path is at or near the error cliff.

The issue of data scrambling for in-service serial digital links is addressed in SMPTE Engineering Guideline EG 34-1999, "Pathological Conditions in Serial Digital Video Systems."

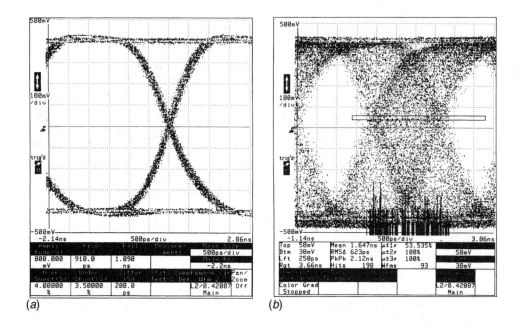

Figure 12.49 Measurement of jitter using a digitizing oscilloscope: (*a*) low jitter, absolute jitter = 200 ps p-p; (*b*) significant jitter present, absolute jitter = 2.12 ns p-p. (*Courtesy of Tektronix.*)

Measuring Jitter

Test equipment is available to measure and characterize the amount of jitter in a link. Several approaches can be taken. One, recommended in SMPTE RP-184, extracts a clock signal from the data stream, which is divided by a given value and used to trigger an *eye pattern* display. The divisor is typically the same value as the word size (i.e., 10 bits). This method will, thus, mask any word-related jitter, which is usually quite small. Figure 12.49 illustrates jitter measurement using a digitizing oscilloscope.

Specific jitter measurement procedures are specified in SMPTE RP 192-1996 [23]. This document describes methods for measuring jitter performance in bit-serial digital interfaces, and the techniques are specifically suited for jitter specifications that follow the form described in SMPTE RP 184.

In RP 192-1996, the principal elements of jitter are defined as follows:

- **Alignment jitter.** The variation in position of a signal's transitions relative to those of a clock extracted from that signal. The bandwidth of the clock extraction process determines the low-frequency limit for alignment jitter.

- **Input jitter tolerance**. The peak-to-peak amplitude of sinusoidal jitter that, when applied to an equipment input, causes a specified degradation of error performance.

- **Intrinsic jitter**. The jitter at an equipment output in the absence of input jitter.

Table 12.2 Error Frequency and Bit Error Rates (*After* [24].)

Time Between	NTSC, 143 Mbits/s	PAL, 177 Mbits/s	Component, 270 Mbits/s
1 television frame	2×10^{-7}	2×10^{-7}	1×10^{-7}
1 second	7×10^{-9}	6×10^{-9}	4×10^{-9}
1 minute	1×10^{-10}	9×10^{-11}	6×10^{-11}
1 hour	2×10^{-12}	2×10^{-12}	1×10^{-12}
1 day	8×10^{-14}	7×10^{-14}	4×10^{-14}
1 week	1×10^{-14}	9×10^{-15}	6×10^{-15}
1 month	3×10^{-15}	2×10^{-15}	1×10^{-15}
1 year	2×10^{-16}	2×10^{-16}	1×10^{-16}
1 decade	2×10^{-17}	2×10^{-17}	1×10^{-17}
1 century	2×10^{-18}	2×10^{-18}	1×10^{-18}

- **Jitter transfer**. Jitter in the output of equipment resulting from applied input jitter.

- **Jitter transfer function**. The ratio of the output jitter to the applied jitter as a function of frequency.

- **Output jitter**. Jitter at the output of equipment that is embedded in a system network. This quantity consists of *intrinsic jitter* and the *jitter transfer* of jitter at the equipment input.

- **Timing jitter**. The variations in position of a signal's transitions occurring at a rate greater than a specified frequency, typically 10 Hz or less. Variations occurring below this specified frequency are termed *wander*.

Four jitter measurement methods are described in RP 1992-1996. See [23] for details.

Quantifying Errors

Errors in digital systems usually are quantified by the bit error rate (BER), which is simply the ratio of bits in error to total bits. Table 12.2 gives the BER for one error over different lengths of time for various television systems [24]. BER is a useful measure of system performance where the S/N at the receiver is such that noise-produced random errors occur.

Bit scrambling is used in the SDI system to lower the dc content of the signal and to provide sufficient zero crossings for reliable clock recovery at the receiver [21]. It is the nature of the descrambler that a single bit error will cause an error in two words (samples). Furthermore, there is a 50 percent probability that the error will occur in one of the words being in the most significant, or next to the most significant, bit position. The resulting error rate of 1 error/frame will be noticeable by a reasonably patient observer. This situation, clearly, is unacceptable in professional video applications.

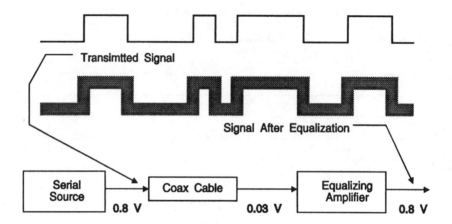

Figure 12.50 Basic serial digital transmitter/receiver system. Note typical voltage levels (for composite video) and the function of the equalizing amplifier. (*After* [24].)

Figure 12.50 shows the block diagram of a basic serial digital transmitter and receiver system. The intuitive method of testing the serial link is to add cable to the point where the link is unusable. Because coax is not itself a significant source of noise, it is the *noise figure* (NF) of the receiver that will determine the basic operating S/N of the system. Assuming an automatic equalizer in the receiver (which is usually the case), as more cable is added, eventually the signal level resulting from coax attenuation will cause the S/N in the receiver to degrade to the point that errors occur.

Based on the scrambled NRZI (non-return-to-zero-inverted) channel code used with SDI and assuming gaussian-distributed noise, a calculation using the *error-function* provides theoretical values for error rate as a function of S/N, as shown in Table 12.3. [24] An examination of the table will show that for composite (NTSC) serial digital transmission, a 4.7 dB increase in S/N changes the resulting condition from 1 error/frame to 1 error/century. For composite digital, the calibration point for the calculation is 400 meters of Belden 8281 coax. (Other types of cable can, of course, be used with adjustments made for the calculation point, if necessary.)

The same theoretical data can be expressed to show error rates as a function of cable length, as given in Table 12.4 [24, 25]. This data is reproduced in graph form in Figure 12.51. As you would expect, there is a sharp knee in the graph as cable length is extended beyond a certain *critical point* (380 meters, or approximately 1,250 ft., for composite digital video). Similar results are obtained for other standards; the critical point calculation is 360 meters for PAL and 290 meters for ITU-R Rec. 601 video. Cable lengths and headroom scale proportionally. Good engineering practice would suggest a minimum of 6–8 dB margin for reliable SDI transmission.

Practical systems include equipment that does not necessarily completely reconstitute the signal in terms of S/N. For example, sending the SDI signal through a distribution amplifier

Table 12.3 Error Rate as a Function of S/N for Composite Serial Digital (*After* [24].)

Time Between Errors	BER	SNR (dB)	S/N (volts ratio)
1 microsecond	7×10^{-3}	10.8	12
1 millisecond	7×10^{-6}	15.8	38
1 television frame	2×10^{-7}	17.1	51
1 second	7×10^{-9}	18.1	64
1 minute	1×10^{-10}	19.0	80
1 day	8×10^{-14}	20.4	109
1 month	3×10^{-15}	20.9	122
1 century	2×10^{-18}	21.8	150

Table 12.4 Error Rate as a Function of Cable Length Using 8281 Coax for Composite Serial Digital (*After* [24].)

Time Between Errors	BER	Cable Length (meters)	Attenuation (dB) at 1/2 Clock Frequency
1 microsecond	7×10^{-3}	484	36.3
1 millisecond	7×10^{-6}	418	31.3
1 television frame	2×10^{-7}	400	30.0
1 second	7×10^{-9}	387	29.0
1 minute	1×10^{-10}	374	28.1
1 day	8×10^{-14}	356	26.7
1 month	3×10^{-15}	350	26.2
1 century	2×10^{-18}	338	25.3

or routing switcher may result in a completely useful, but not completely "standard" signal that is sent to a receiving device. The characteristics of this "non-standardness" could include both jitter and noise. The sharp knee characteristic of the system, however, would remain, occurring at a different amount of signal attenuation.

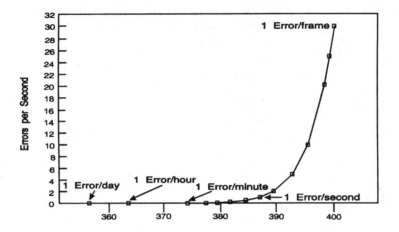

Figure 12.51 Error rate as a function of cable length for composite video signals on an SDI link. (*After* [24].)

12.6.5 Eye Diagram

Most engineers are familiar with an oscilloscope display of repetitive waveforms such as sine, square, or triangle waves. These are known as *single-value* displays because each point on the time axis has only a single voltage value associated with it [26].

When analyzing a digital telecommunications waveform, single-value displays are not very useful. Real communications signals are not repetitive, but instead consist of random or pseudorandom patterns. A single-value display can only show a few of the many different possible one/zero combinations. A number of pattern-dependent problems will be overlooked if they do not occur in the small segment of the waveform appearing on the display of the conventional oscilloscope.

The *eye diagram* overcomes the limitations of the single-value display by overlapping all of the possible one/zero combinations on the screen. Eye diagrams are *multivalued displays* in that each point on the time axis has multiple voltage values associated with it.

The eye diagram has become a valuable tool in analyzing digital communications systems because of the unique properties that it is capable of displaying.

12.7 Transmission Issues

For analog signals, some transmission impairments are tolerable because the effect at the receiver is often negligible, even for some fairly significant faults [27]. With DTV, however, an improperly adjusted transmitter could mean the loss of viewers in the Grade B coverage area (or worse). DTV reception, of course, does not degrade gracefully; it simply disappears. Attention to several parameters is required for satisfactory operation of the 8-VSB system. First, there is the basic FCC requirement against creating interference to other over-the-air services. To verify that there is no leakage into adjacent channels, out-of-band emission testing is required. Second, for NTSC, there is concern with S/N performance, and so for DTV the *desired-to-undesired signal ratio* (D/U) is measured.

As with analog transmitters, flat frequency response across the channel passband is required. A properly aligned DTV transmitter exhibits many of the same characteristics as a properly aligned analog unit: flat frequency response and group delay, with no leakage into adjacent channels. In the analog domain, group delay results in chroma/luma delay, which degrades the displayed picture but still leaves it viewable. Group delay problems in DTV transmitters, however, result in *intersymbol interference* (ISI) and a rise in the BER, causing the receiver to drop in and out of lock. Even low levels of ISI may cause receivers operating near the edge of the cliff to lose the picture completely. Amplitude and phase errors may cause similar problems, again, resulting in reduced viewer coverage.

Eye patterns and BER have become well-known parameters of digital signal measurement, although they may not always be the best parameters to monitor for 8-VSB transmission. The *constellation diagram* and *modulation error ratio* (MER), on the other hand, provide insight into the overall system health. RF constellations are displayed on the *I* and *Q* axes. Constellations of tight vertical dot patterns with no slanting or bending indicate proper operation, as illustrated in Figure 12.52*a*. The 8-VSB levels are the in-phase signal, so they are displayed left to right.

An 8-VSB signal is a single sideband waveform with a pilot carrier added. In a single sideband signal, phase does not remain constant. Therefore, the constellation points (dots) occur in a vertical pattern. As long as the dot pattern is vertical and the points form narrow lines of equal height, the signal is considered good and can be decoded.

Figure 12.52*b* shows an 8-VSB signal that has noise and phase shift. Noise is indicated by the spreading of the dot pattern. Phase problems are indicated by the slant along the *Q*-axis.

Although BER is a valid measurement for 8-VSB, a better approach involves monitoring the MER, which usually will reveal problems before the BER is affected. In many cases, MER provides enough warning time to correct problems that would result in an increase in the BER. MER provides an indication of how far the points in the constellation have migrated from the ideal. A considerable amount of migration might occur before boundary limits are exceeded. Degradation in the BER is apparent only when those limits have been exceeded.

12.7.1 Transmission System Measurements

The key measurements required for a DTV transmitter proof of performance include output power, pilot frequency, intermodulation products (IMD), *error vector magnitude* (EVM), and harmonic power levels [27, 28]. The average or RMS output power can be measured with a calorimeter, or a precision probe and average-reading power meter. The calorimeter measures flow rate and temperature rise of the transmitter test load liquid coolant, from which average power may be computed. Because this is the quantity of interest for DTV applications, no other conversion factors are necessary. Note that there is no need to measure peak power so long as the average power is known and EVM and *spectral regrowth* requirements are met.

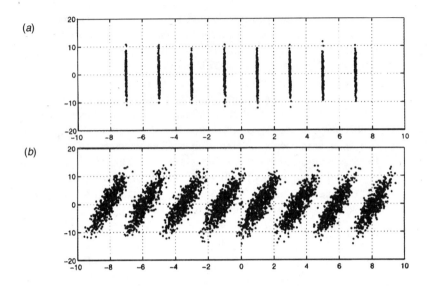

Figure 12.52 An 8-VSB constellation diagram: (*a*) a near-perfect condition, (*b*) constellation diagram with noise and phase shift (the spreading of the pattern is caused by noise; the slant is caused by phase shift). (*After* [17].)

The pilot frequency can be measured with a frequency counter or spectrum analyzer. The results should be the frequency of the lower channel edge plus 309,440.6 Hz, ±200 Hz, unless precise frequency control is required.

In practice, it is difficult to measure compliance with the RF emissions mask (discussed in the next section) directly. For near-in spectral components, within the adjacent channel, a common procedure is as follows [28]:

- Measure the transmitter IMD level with a resolution bandwidth of 30 kHz throughout the frequency range of interest. This results in an adjustment to the standard FCC mask by 10.3 dB. Under this test condition, the measured shoulder breakpoint levels should be at least –36.7 dB from the mid band level.

- Measure the transmitter output spectrum without the filter using a spectrum analyzer.

- Measure the filter rejection vs. frequency using a network analyzer.

- Add the filter rejection to the measured transmitter spectrum. The sum should equal the transmitter spectrum with the filter.

Output harmonics may be determined in the same manner as the rest of the output spectrum. They should be at least –99.7 dB below the mid band power level. EVM is checked with a vector signal analyzer or similar instrument.

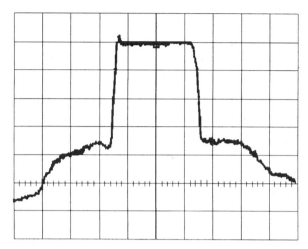

Figure 12.53 Measured DTV transmitter output and sideband splatter. (*After* [29]. *Courtesy of Harris.*)

The output power, pilot frequency, inband frequency response, and adjacent channel spectrum should be measured periodically to assure proper transmitter operation. These parameters can be measured while the transmitter is in service with normal programming.

Sideband Splatter

Interference from a DTV signal on either adjacent channel into NTSC or another DTV signal will be primarily due to *sideband splatter* from the DTV channel into the adjacent channel [28]. The limits to this out-of-channel emission are defined by the RF mask as described in the "Memorandum Opinion and Order on Reconsideration of the Sixth Report and Order," adopted February 17, 1998, and released February 23, 1998.

For all practical purposes, high-power television transmitters will invariably generate some amount of intermodulation products as a result of non-linear distortion mechanisms. Intermodulation products appear as spurious sidebands that fall outside the 6 MHz channel at the output of the transmitter. (See Figure 12.53.) Intermodulation products appear as noise in receivers tuned to either first adjacent channel, and this noise adds to whatever noise is already present. The overall specifications for the FCC mask are given in Figure 12.54.

Salient features of the RF mask include the following [28]:

- The *shoulder level*, at which the sideband splatter first appears outside the DTV channel, is specified to be 47 dB below the effective radiated power (ERP) of the radiated DTV signal. When this signal is displayed on a spectrum analyzer whose resolution bandwidth

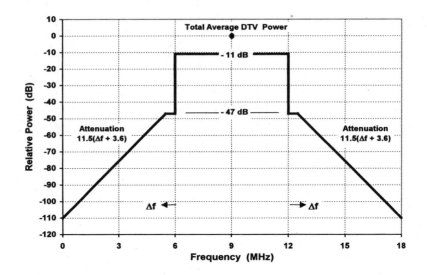

Figure 12.54 RF spectrum mask limits for DTV transmission. The mask is a contour that illustrates the maximum levels of out-of-band radiation from a transmitted signal permitted by the FCC. This chart is based on a measurement bandwidth of 500 kHz. (*From* [30]. *Used with Permission.*)

is small compared to the bandwidth of the signal to be measured, it is displayed at a lower level than would be the case in monitoring an unmodulated carrier (one having no sidebands). If the analyzer resolution bandwidth is 0.5 MHz, and the signal power density is uniform over 5.38 MHz (as is the case for DTV), then the analyzer would display the DTV spectrum within the DTV channel 10.3 dB below its true power. The correction factor is 10 log (0.5/5.38) = 10.3 dB. Thus, the reference line for the in-band signal shown across the DTV channel is at –10.3 dB, relative to the ERP of the radiated signal.

- The shoulder level is specified as –47 dB, relative to the ERP. The shoulder level is at 36.7 dB below the reference line at –10.3 dB.

- The RF mask is flat for the first 0.5 MHz from the DTV channel edges, at –47 dB relative to the ERP, and is shown to be 36.7 dB below the reference level, which is 10.3 dB below the ERP.

- The RF mask from 0.5 MHz outside the DTV channel descends in a straight line from a value of –47 dB to –110 dB at 6.0 MHz from the DTV channel edges.

- Outside of the first adjacent channels, the RF mask limits emissions to –110 dB below the ERP of the DTV signal. No frequency limits are given for this RF mask. This limit on out-of-channel emissions extends to 1.8 GHz in order to protect the 1.575 MHz GPS signals.

- The total power in either first adjacent channel permitted by the RF mask is 45.75 dB below the ERP of the DTV signal within its channel.

- The total NTSC weighted noise power in the lower adjacent channel is 59 dB below the ERP.

- The total NTSC weighted noise power in the upper adjacent channel is 58 dB below the ERP.

12.7.2 In-Band Signal Characterization

The quality of the in-band emitted signal of a DTV transmitter can be specified and measured by determining the departure from the 100 percent *eye opening* [29]. This departure, or error, has three components:

- Circuit noise (*white noise*)

- Intermodulation noise caused by various nonlinearities

- Intersymbol interference

The combination of all of these effects can be specified and measured by the error vector magnitude. This measurement is described in [29].

12.7.3 Power Specification and Measurement

Conventional NTSC broadcast service allows a power variation ranging from 80 to 110 percent of authorized power [29]. These values correspond to –0.97 and +0.41 dB, respectively. Because of the cliff effect at the fringes of the service coverage area for a DTV signal, the allowable lower power value will have a direct impact on the DTV reception threshold. A reduction of 0.97 dB in transmitted power will change the DTV threshold of 14.9 dB (which has been determined to yield a 3×10^{-6} error rate) to 15.87 dB, or approximately a 1-mile reduction in coverage distance from the transmitter. Therefore, the average operating power of the DTV transmitted signal is of significant importance.

The ATSC in [28] recommends a lower allowed power value of 95 percent of authorized power and an upper allowed power value of 105 percent of authorized power.

12.8 References

1. Mertz, P.: "Television and the Scanning Process," *Proc. IRE*, vol. 29, pp. 529–537, October 1941.

2. MacAdam, D. L.: "Visual Sensitivities to Color Differences in Daylight," *J. Opt. Soc. Am.*, vol. 32, pp. 247–274, 1942.

3. Bender, Walter, and Alan Blount: "The Role of Colorimetry and Context in Color Displays," *Human Vision, Visual Processing, and Digital Display III*, Bernice E. Rogowitz (ed.), *Proc. SPIE* 1666, SPIE, Bellingham, Wash., pp. 343–348, 1992.

4. Tannas, Lawrence E., Jr.: *Flat Panel Displays and CRTs*, Van Nostrand Reinhold, New York, pg. 18, 1985.

5. Standards and definitions committee, Society for Information Display.

6. "Broadening the Applications of Zone Plate Generators," Application Note 20W7056, Tektronix, Beaverton, Oreg., 1992.

7. Verona, Robert: "Comparison of CRT Display Measurement Techniques," *Helmet-Mounted Displays III*, Thomas M. Lippert (ed.), Proc. SPIE 1695, SPIE, Bellingham, Wash., pp. 117–127, 1992.

8. "Critical Viewing Conditions for Evaluation of Color Television Pictures," SMPTE Recommended Practice RP 166-1995, SMPTE, White Plains, N.Y., 1995.

9. "Alignment of NTSC Color Picture Monitors," SMPTE Recommended Practice RP 167-1995, SMPTE, White Plains, N.Y., 1995.

10. Quinn, S. F., and C. A. Siocos: "PLUGE Method of Adjusting Picture Monitors in Television Studios—A Technical Note," *SMPTE Journal*, SMPTE, White Plains, N.Y., vol. 76, pg. 925, September 1967.

11. Gloeggler, Peter: "Video Pickup Devices and Systems," in *NAB Engineering Handbook*," 9th Ed., Jerry C. Whitaker (ed.), National Association of Broadcasters, Washington, D.C., 1999.

12. "Proposed SMPTE Standard: For Television—Color Reference Pattern," SMPTE 303M, SMPTE, White Plains, N.Y., 1999.

13. Fibush, David K.: "Picture Quality Measurements for Digital Television*," Proceedings of the Digital Television '97 Summit*, Intertec Publishing, Overland Park, Kan., December 1997.

14. ANSI Standard T1.801.03-1996, "Digital Transport of One-Way Video Signals: Parameters for Objective Performance Assessment," ANSI, Washington, D.C., 1996.

15. Fibush, David K.: "Practical Application of Objective Picture Quality Measurements," *Proceedings IBC '97*, IEE, pp. 123–135, Sep. 16, 1997.

16. Hamada, T., S. Miyaji, and S. Matsumoto: "Picture Quality Assessment System by Three-Layered Bottom-Up Noise Weighting Considering Human Visual Perception," *SMPTE Journal*, SMPTE, White Plains, N.Y., pp. 20–26, January 1999.

17. Bishop, Donald M.: "Practical Applications of Picture Quality Measurements," *Proceedings of Digital Television '98*, Intertec Publishing, Overland Park, Kan., 1998.

18. Uchida, Tadayuki, Yasuaki Nishida, and Yukihiro Nishida: "Picture Quality in Cascaded Video-Compression Systems for Digital Broadcasting," *SMPTE Journal*, SMPTE, White Plains, N.Y., pp. 27–38, January 1999.

19. Finck, Konrad: "Digital Video Signal Analysis for Real-World Problems," in *NAB 1994 Broadcast Engineering Conference Proceedings*, National Association of Broadcasters, Washington, D.C., pg. 257, 1994.

20. Haines, Steve: "Serial Digital: The Networking Solution?," in *NAB 1994 Broadcast Engineering Conference Proceedings*, National Association of Broadcasters, Washington, D.C., pg. 270, 1994.

21. SMPTE Standard: SMPTE 259M-1997, "Serial Digital Interface for 10-bit 4:2:2 Components and $4F_{sc}$ NTSC Composite Digital Signals," SMPTE, White Plains, N.Y., 1997.

22. Boston, J., and J. Kraenzel: "SDI Headroom and the Digital Cliff," *Broadcast Engineering*, Intertec Publishing, Overland Park, Kan., pg. 80, February 1997.

23. SMPTE Recommended Practice: SMPTE RP 192-1996, "Jitter Measurement Procedures in Bit-Serial Digital Interfaces," SMPTE, White Plains, N.Y., 1996.

24. Fibush, David K.: "Error Detection in Serial Digital Systems," *NAB Broadcast Engineering Conference Proceedings*, National Association of Broadcasters, Washington, D.C., pp. 346-354, 1993.

25. Stremler, Ferrel G.: "Introduction to Communications Systems," *Addison-Wesley Series in Electrical Engineering*, Addison-Wesley, New York, December 1982.

26. "Eye Diagrams and Sampling Oscilloscopes," *Hewlett-Packard Journal*, Hewlett-Packard, Palo Alto, Calif., pp. 8–9, December 1996.

27. Reed-Nickerson, Linc: "Understanding and Testing the 8-VSB Signal," *Broadcast Engineering*, Intertec Publishing, Overland Park, Kan., pp. 62–69, November 1997.

28. *DTV Express Training Manual on Terrestrial DTV Broadcasting*, Harris Corporation, Quincy, Ill., September 1998.

29. ATSC, "Transmission Measurement and Compliance for Digital Television," Advanced Television Systems Committee, Washington, D.C., Doc. A/64, Nov. 17, 1997.

30. ATSC, "Transmission Measurement and Compliance for Digital Television," Advanced Television Systems Committee, Washington, D.C., Doc. A/64-Rev. A, May 30, 2000.

12.9 Bibliography

SMPTE Engineering Guideline EG 1-1990, "Alignment Color Bar Test Signal for Television Picture Monitors," SMPTE, White Plains, N.Y., 1990.

Pank, Bob (ed.): *The Digital Fact Book*, 9th ed., Quantel Ltd, Newbury, England, 1998.

DTV Implementation Issues

13.1 Introduction

As with any new technology, the DTV system has experienced growing pains. Some system requirements and tradeoffs—not anticipated during initial design of the DTV standard—inevitably must be resolved while the technology is being implemented. Such issues include logo insertion, switching, and scanning/conversion techniques.

The larger question of how to plan and construct a new DTV facility is a point of great concern as stations begin their DTV rollouts. Fortunately, a number of technical guide-lines—templates, if you will—are available as a starting point.

13.2 MPEG Bit Stream Splicing

In today's editing environment, audio, video, or some combination of the two is spliced into or onto existing material. In the uncompressed domain, this is a simple procedure [1]. It is relatively easy to synchronize two or more video streams. Vertical intervals occur regularly, allowing switches to be performed as required (Figure 13.1). Digital audio is similar to video in this regard, and analog audio is even easier because it requires no synchronization whatsoever. However, in the compressed domain of an MPEG bit stream, several factors must be considered. Among them are:

• The varying number of bits per frame

• The use of motion prediction

• The fact that frames may not be sent in the order in which they will be displayed

I- or *P*-frames that are displayed after *B*-frames need to be sent before the *B*-frames so that the *B*-frames can be properly assembled. (See Figure 13.2.)

Because the number of bits per frame varies, it is virtually impossible to synchronize two MPEG bit streams. However, bit streams can be loaded into RAM, and memory pointers then can be manipulated. If two bit streams were loaded into RAM, the pointer used to read the data could be shifted such that after one stream is output, it is followed immediately by a

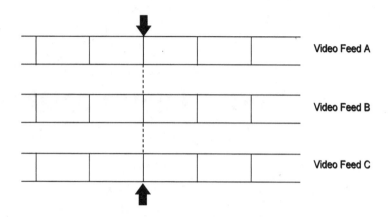

Figure 13.1 Splicing procedure for uncompressed video. (*After* [2].)

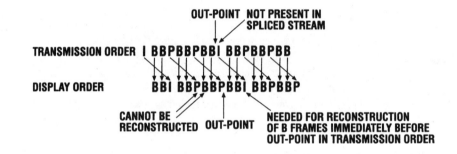

Figure 13.2 Splice point considerations for the MPEG bit stream. (*After* [1].)

section of the other bit stream. This process is much like edits performed by many nonlinear desktop editing systems, except that in the editing systems the data is not read from RAM, but from a hard drive using pointers that are essentially lists of frames and their locations.

Most nonlinear editors use JPEG compression, which is comparable to an MPEG bit stream composed entirely of *I*-frames. Jumping from the end of one *I*-frame to the beginning of another is relatively simple. Bit streams made up of all *I*-frames are fine for editing, but inefficient when it comes to storage or transmission. Bit streams that are far more efficient to store and transport make considerable use of *P*- and *B*-frames, but these elements complicate the editing process.

Significant effort has been applied to develop tools within the MPEG structure that would allow for bit stream splicing in the compressed domain. One tool thought to be needed was an encoder that would mark potential splice points within the stream. One requirement of a splice point would be that the first frame after the splice be an *I*-frame. Among other things, an *I*-frame would ensure that no previous frames were needed for

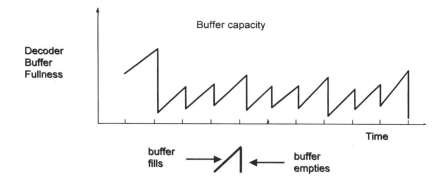

Figure 13.3 Typical operational loading of the buffer. (*After* [2].)

proper decoding. A second requirement would be that the last frame before a splice point be an *I*-frame or a *P*-frame, guaranteeing that all the needed *B*-frames could be decoded.

Another requirement of a splice point involves the state of the buffer in the decoder. This state could be anything from nearly full but emptying out to nearly empty but filling up. Decoder *buffer fullness* is a dynamic parameter of the encoding process, and as long as the splicing is within a particular bit stream, it is not likely to cause a problem (Figure 13.3). However, splicing one bit stream onto another could cause the receiver's buffer to underflow or overflow. This could occur if the stream to be switched from leaves the buffer fairly empty, and the stream to be switched into assumes a nearly full buffer and expects to empty it shortly, resulting in a buffer *underflow*, as illustrated in Figure 13.4. Flushing the decoder buffer is one way to deal with the problem, but this probably would result in display disruption on the viewer's screen. Another splicing method is to constrain splice points so that they occur only when the decoder buffer is 50 percent full. However, this could make potential splice points few and far between.

Meeting the previously mentioned requirements would go a long way toward solving the splicing problem, but it would not be a complete solution. Within the data stream are variables such as time stamps that must be updated to prevent problems at the decoder. Additional data-stream processing is required to update these variables properly.

Up to this point, only video splicing has been discussed. Within an MPEG program or transport stream, the audio and video are sent as separate packets. Because the packets are sent serially in a single stream, audio packets end up being sent before or after the video packets with which they are associated. To resynchronize the audio and video, each packet has a *presentation time stamp* (PTS) that allows the decoder to present the various audio and video packets in a synchronized manner. Both the audio and video signal paths include buffers. However, because of the amount of calculation required to reassemble the video, the video buffer is much larger than the audio buffer. To some extent, the larger the buffer, the larger the signal delay. Because of the additional delay in the video buffer, a bit-stream

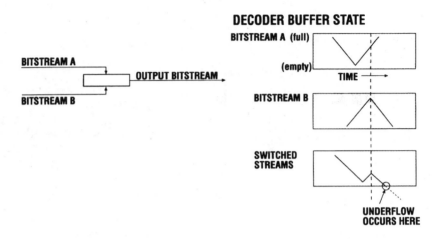

Figure 13.4 MPEG decoder buffer state issues for bit stream switching. (*After* [1].)

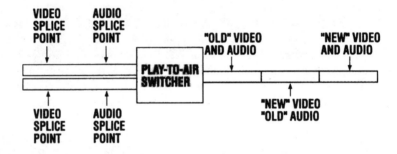

Figure 13.5 Relative timing of video and audio splice points and the end result. (*After* [1].

splice that contains "old" audio and video before the splice and "new" video and audio after the splice probably will be presented to the viewer as two separate splices. The first splice will affect the audio and, because of the longer buffer, the second splice will affect the video. This process is illustrated in Figure 13.5.

13.2.1 Splice Flags

One proposed solution to the MPEG bit stream splicing problem is the insertion of *splice flags* [2]. These flags are inserted during encoding at defined points of buffer occupancy. As illustrated in Figure 13.6, the flags identify allowable switch points in the MPEG bit stream. The benefits of this approach include:

- No artifacts from the switching process

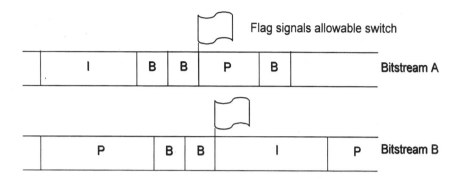

Figure 13.6 The use of splice flags to facilitate the switching of MPEG bit streams. (*After* [2].)

- Continuity of video and audio at the receiver display

- No unpredictable behavior of the video decoder

In addition to conventional cuts editing, splice flag-based operation allows for switching between progressive and interlaced images and from one picture resolution to another. The drawback of this approach is that the splice points must first be identified at the origination center. This requirement limits the usefulness of the process somewhat. However, for most network-to-affiliate feed operations, the required splice/insertion points are well known and clearly defined.

13.2.2 SMPTE 312M

In response to the problems posed by the bit stream splicing issue, the SMPTE examined possible solutions. The result of this work was SMPTE 312M.

SMPTE 312M defines constraints on the encoding of and syntax for MPEG-2 transport streams such that they may be spliced without modifying the PES (packetized elementary stream) payload [3]. Generic MPEG-2 transport streams that do not comply with the constraints in the standard may require more sophisticated techniques for splicing.

The constraints specified are applied individually to programs within transport streams, a *program* being defined as a collection of video, audio, and data streams that share a common timebase. The presence of a video component is not assumed. The standard enables splicing of programs within a multiprogram transport stream either simultaneously or independently. Splice points in different programs may be presentation-time-coincident, but do not have to be. The standard also may be used with single-program transport streams.

The 312M document specifies constraints for both *seamless* and *nonseamless* splice points. Seamless splice points must adhere to all the stated constraints; nonseamless splice points adhere to a simplified subset of constraints.

In addition to constraints for creating spliceable bit streams, the standard specifies the technique for carrying notification of upcoming splice points in the transport stream. A *splice information table* is defined for notifying downstream devices of splice events, such as a network break or return from a network break. The splice information table that pertains to a given program is carried in a separate PID (program identifier) stream referred to by the program map table. In this way, splice event notification can pass through transport stream remultiplexers without the need for special processing.

Buffer Issues

As addressed previously, the splicing of MPEG bit streams requires careful management of buffer fullness. When MPEG bit streams are encoded, there is an inherent buffer occupancy at every point in time [3]. The buffer fullness corresponds to a delay, the amount of time that a byte spends in the buffer. When splicing two separately encoded bit streams, the delay at the splice point usually will not match. This mismatch in delay can cause the buffer to over-flow or underflow. The seamless splicing method requires that the MPEG encoder match the delay at splicing points to a given value. The nonseamless method does not require the encoder to match the delay. Instead, the splicing device is responsible for matching the delay of the new material and the old material as well as it can. In some cases, this will result in a controlled decoder buffer underflow. This underflow can be masked in the decoder by holding the last frame of the outgoing video and muting the audio until the first access unit of the new stream has been decoded. In the worst case, this underflow may last for a few frames.

Both splicing methods may cause an underflow of the audio buffer, and consequently a gap in the presentation of audio at the receiver. The perceived quality of the splice in both cases will benefit from audio decoders that can handle such a gap in audio data gracefully.

Splice Points

To enable the splicing of compressed bit streams, SMPTE 312M defines *splice points* [3]. Splice points in an MPEG-2 transport stream provide opportunities to switch from one program to another. They indicate a safe place to switch: a place in the bit stream where a switch can be made and result in good visual and audio quality. In this way, they are analogous to the vertical interval used to switch uncompressed video. Unlike uncompressed video, frame boundaries in an MPEG-2 bit stream are not evenly spaced. Therefore, the syntax of the transport packet itself is used to convey where these splice points occur. Transport streams are created by multiplexing PID streams. Two types of splice points for PID streams are defined:

- *In Points,* places in the bit streams where it is safe to enter and start decoding the data

- *Out Points*, places where it is safe to exit the bit stream

Methods are defined that can be used to group In Points of individual PID streams into Program In Points to enable the switching of entire programs (video with audio). Program Out Points for exiting a program also are defined.

Out Points and In Points are imaginary benchmarks in the bit stream located between two transport stream packets. An Out Point and an In Point may be co-located; that is, a single packet boundary may serve as both a safe place to leave a bit stream and a safe place to enter it.

13.2.3 Transition Clip Generator

Another approach to bit stream splicing is the *transition clip generator* (TCG) scheme, developed by Sarnoff and Silicon Graphics (SGI). TCG is a suite of software tools intended to solve the MPEG transport stream splicing problem. Designed for use in a video server environment, TCG performs frame-accurate seamless splicing from one MPEG-2 transport stream to another.

The TCG creates a *transition clip*—a new sequence of MPEG transport stream packets that replace (in the resulting spliced video stream) a portion of each of the MPEG transport streams of the video being spliced together (clip 1 and clip 2) around the point where the splice is to be made. Creation of this transition clip involves a small amount of decoding and subsequent re-encoding of compressed video frames (from the streams being spliced), but only of frames from those regions being replaced.

The fact that only a few frames need to be decoded and then re-encoded to create a seamless splice is one of the significant benefits of the TCG approach. A detailed discussion of the TCG system can be found in [Ward, et. al. (see the Bibliography)].

13.2.4 SMPTE 328M

The SMPTE 328M standard defines the MPEG video *elementary stream* (ES) information to facilitate seamless edits under defined circumstances [4]. The video ES, as defined by the MPEG standards, is supplemented with additional information for professional studio applications. This supplementary information is carried within the sequence header and the user data area of the video ES. SMPTE 328M defines the data to be carried and the location of that data.

Seamless, frame-accurate editing of compressed video is most easily accomplished with the use of short GOP structures. Longer GOP structures can be edited by decoding and reencoding, by transcoding to shorter GOP structures, or (with more involved processing) edited directly. The best approach is determined by a range of application-specific considerations.

ISO/IEC 13818-2 does not define the repetition frequency of the sequence header. To be compliant with SMPTE 328M, the sequence header must exist at every *I* frame.

As specified in SMPTE 328M, the following syntax elements and functional descriptions are inserted in the MPEG ES in the user data area:

- **V/H coding phase**. The basic implementation of MPEG does not specify the horizontal and vertical coding phase. SMPTE 328M requires that the vertical and horizontal coding phase be known in order for decoding and peripheral equipment to correctly process the signal. V and H coding information are included only for SDTV signals where the coding phase is not compliant with SMPTE RP 202. For HDTV signals, H/V coding phase information is defined by SMPTE RP 202.

- **Time code**. Provision is made for the insertion of two time codes complying with SMPTE 12M. At least one time code, the reference date time stamp (as defined in SMPTE 326M) is carried as a means of maintaining synchronization with other content or metadata streams. Carriage of a second time code is optional. Compliant decoders must have the capability to decode both time codes.

- **Picture order**. Picture order information specifies the picture duration and is the equivalent to the PTS/DTS present in the MPEG transport stream. The picture order value is counted by field units. In some cases, the latency of the system will be minimized using the picture order information.

- **Video index**. Video index, as defined by SMPTE RP 186, is carried (if present) on the baseband signal. Information carried by the video index should be preserved during any coding, recoding, editing, or transcoding process. It was envisioned that the data described in the SMPTE metadata dictionary (SMPTE 335M, proposed) would be handled by the transport mechanism described in SMPTE 326M. These parametric data will include all of the parameters currently coded in the video index, although the data representation of some items may be different.

- **Ancillary data**. Data that is carried in the vertical interval of the baseband signal should be preserved. Ancillary data may consist of more than 23 consecutive zeros. To prevent this condition, a marker is inserted every 22 bits.

- **History data**. History data, consisting of original and subsequent encoding parameters that may be useful in transcoding or reencoding, can be carried by the bit stream. SMPTE 327M defines the content of the history data information. History data may consist of more than 23 consecutive zeros. To prevent this condition, a marker is inserted every 22 bits.

- **User data**. User data is defined by ISO/IEC 13818-2.

13.3 MPEG-2 Recoding

The MPEG-2 video *recoding data set* allows a full description of the MPEG-2 parameter set that characterizes any MPEG-2 encoding process. According to this recoding data set, it is accepted, from a theoretical point of view, that any MPEG-2 encoding equipment could generate an identical MPEG-2 bitstream from a given digital video signal. However, the MPEG-2 recoding data set may be used in practical recoding applications where the environment can introduce additional, unpredictable constraints. For example, the bit-rate of any recoding stage may or may not differ from that of the previous encoding stage. This means that the recoding data set will not necessarily be fully reused at any further stage.

Another critical aspect of the recoding process is the bandwidth available for transport of the MPEG-2 recoding data set. In many practical applications, a reduced set of the generic MPEG-2 recoding data set must be addressed.

The issue of MPEG-2 recoding is an important one as the need increases to modify previously encoded programs. To address this need, the SMPTE developed a suite of tools that includes the following standards:

- SMPTE 327M, MPEG-2 Video Recoding Data Set

- SMPTE 329M, MPEG-2 Video Recoding Data Set—Compressed Stream Format

- SMPTE 319M, Transporting MPEG-2 Recoding Information Through 4:2:2 Component Digital Interfaces

- SMPTE 351M, Transporting MPEG-2 Recoding Information through High-Definition Digital Interfaces

- SMPTE 353M, Transport of MPEG-2 Recoding Information as Ancillary Data Packets

These standards are discussed in the following sections.

13.3.1 SMPTE 327M

SMPTE 327M specifies the content of the picture-related recoding data set for the representation of ISO/IEC 13818-2 MPEG coding information for the purpose of optimally cascading decoders and recoders at any bit rate or GOP structure [5]. The coding information is derived from an ISO/IEC 13818-compliant MPEG bit stream during the picture decoding process, as described in ISO/IEC 13818-2. The scope and operation of this standard are the definition of the content of a *sufficient recoding data set* that may be derived in decoders complying with ISO/IEC 13818-2, including all nonscalable profiles defined in ISO/IEC 13818-2.

To allow the resynchronization of the video and its associated audio or data after processing, a mechanism using some additional information derived from ISO/IEC 13818-1 is also included in SMPTE 327M. This sufficient data set can be transported by various means (defined in other SMPTE standards).

The principal application of this standard is to preserve the quality of the video signal when cascading MPEG-2 decoders and coders (including transcoding) by feeding forward previous coding decisions. The MPEG-2 recoding data set is described as *sufficient* when it contains the data required that, in combination with an MPEG-2 decoded or partially decoded picture, allows bit-accurate recreation of the previously coded bit stream. The information required in the sufficient MPEG-2 recoding data set can be broken down into three parts.

- The picture rate information

- Macroblock rate information

- Additional housekeeping data

13.3.2 SMPTE 329M

SMPTE 329M specifies the stream format of the MPEG-2 recoding data set for the representation of compressed ISO/IEC 13818-2 MPEG coding information, as used in applications requiring transport systems of reduced data capacity [6]. The coding information is derived from an ISO/IEC 13818-2 compliant MPEG bit stream during the decoding process, as described in ISO/IEC 13818-2. The information based on this stream format can be transported by various means; for example, the elementary stream format defined in SMPTE 328M.

There are applications in which the transmission of all the recoding data set is not possible. Some legacy equipment may have restricted capacity for the transmission of the recoding data. This limitation has an impact on subsequent compression stages that can make use of the MPEG-2 recoding process. In order to decrease the bit rate for the recoding data set, the MPEG-2 recoding data set is converted into an MPEG-like stream, which is called the *compressed stream format* of the MPEG-2 recoding set. SMPTE 329M defines this stream format. The stream format is independent of application, and all the transport information in the reduced bandwidth recoding data transportation system should be based on this stream. The transport mechanism depends on the application, which is defined in other standards documents.

13.3.3 SMPTE 319M

SMPTE 319M specifies an embedded transport mechanism for the MPEG-2 recoding data set as defined in SMPTE 327M for the representation of MPEG-2 recoding information in ITU-R BT.656, 4:2:2 component digital interfaces [7]. The recoding data set is derived from an ISO/IEC 13818-1 and 2 compliant MPEG bit stream during the decoding process, as described in ISO/IEC 13818-1 and -2.

For the minimum operation of this standard, the MPEG-2 recoding data set is spatially and temporally aligned to each decoded macroblock mapped into an ITU-R BT.656 interface. The standard specifies the spatially and temporally aligned transport of the MPEG-2 recoding data set within the active picture area on ITU-R BT.656 interfaces for equipment that complies with ISO/IEC 13818-1 and -2, including 422P@ML and MP@ML for both the 625/50 and 525/60 video standards.

The information contained in the MPEG-2 recoding data set is defined in SMPTE 327M. This recoding information is temporally locked to the decoded (or partially decoded) video to the nearest MPEG-2 frame or field depending on the picture structure of the coded MPEG-2 bit stream. It is also spatially locked with the decoded video to the nearest MPEG-2 macroblock within the decoded frame/field. It is necessary for the recoding information to be aligned with the decoded MPEG macroblocks in the decoded pictures, both spatially and temporally.

13.3.4 SMPTE 351M

SMPTE 351M (proposed at this writing) specifies an embedded transport mechanism for the MPEG-2 recoding data set as defined in SMPTE 327M for the representation of MPEG-2 recoding information on a SMPTE 274M interface and subsequently upon a SMPTE 292M bit-serial digital interface [8]. The recoding data set is derived from an ISO/IEC 13818-1/2 compliant MPEG bitstream during the decoding process, as described in the ISO/IEC 13818-1/2 standards. For the minimum operation of this standard, the MPEG-2 recoding data set is spatially and temporally aligned to each decoded macroblock mapped into a SMPTE 274M/292M interface.

The standard specifies the spatially and temporally aligned transport of the MPEG-2 recoding data set within the active picture area on SMPTE 274M/292M interfaces for equipment that complies with the ISO/IEC 13818-1/2 standards, including 4:2:2P@HL and MP@HL for 60- and 50-Hz interlaced and 60-, 30-, 25-, and 24-Hz progressive video standards.

The recoding information is temporally locked to the decoded (or partially decoded) video to the nearest MPEG-2 frame or field depending on the picture structure of the coded MPEG-2 bitstream. It is also spatially locked with the decoded video to the nearest MPEG-2 macroblock within the decoded frame/field. To accrue the full benefits of the recoding information when cascading via a digital baseband interface, the following recommendations must be adhered to:

- The transport mechanism must preserve at least the 8 most significant bits of active video. The mechanism outlined uses the least significant bit of each 10-bit chrominance sample to transmit the data through the SMPTE 274M or 292M interface.

- The recoding information must be aligned with the decoded MPEG macroblocks in the decoded pictures, both spatially and temporally.

SMPTE 351M is based on producing a SMPTE 274M/292M compliant output to cover HD-MPEG bitstreams up to and including 4:2:2P@HL and MP@HL for 1920×1080 60 (59.94) or 50 2:1; 1920×1080 30 (29.97), 25 or 24 (23.98) 1:1; and 1280×720 60 (59.94) 1:1 systems.

13.3.5 SMPTE 353M

SMPTE 353M (proposed at this writing) specifies the mechanism for the transport of MPEG-2 video recoding information as ancillary data packets in an ancillary data space—for example, through ITU-R BT.656/SMPTE 259M interfaces [9]. The video recoding information transported through this mechanism is used to preserve picture quality at re-encoding stages when cascading MPEG-2 decoders and encoders. The transport mechanism specified in the standard has been designed so that it can work with digital video systems in which operation is limited to 8-bit resolution.

Principle parameters of SMPTE 353M include:

- The transported MPEG-2 video recoding information is compliant with the MPEG-2 video recoding data set as defined in SMPTE 327M

- The data set is formatted according to the stream format defined in SMPTE 329M

- The formatted data set is transported in the form of ancillary (ANC) data packets as specified in SMPTE 291M

- The transport mechanism specified in the standard is compliant with SMPTE 291M

Part of both the *vertical blanking ancillary data space*, (V-ANC) and the *horizontal blanking ancillary data space* (H-ANC) are used. The V-ANC space carries picture rate information only; this is the most basic, highest priority element of the recoding data set. For low bit rate, long GOP applications, this typically brings the greatest picture quality improvement; further refinements are achieved when more information is available. The H-ANC space is used to carry the other part of the recoding data set. Use of the reduced bandwidth indicator, as specified in SMPTE 329M, allows the transmission—more or less—of this part of the recoding data set, depending upon the transmission capacity available in the H-ANC space.

13.4 Planning the DTV Infrastructure

The implementation of digital television broadcasting launched an industry-wide upgrade program—both on the RF side and the studio side. Planning how the DTV facility will function is a difficult exercise, and one that will have far-reaching effects.

13.4.1 Considerations Regarding Interlaced and Progressive Scanning

The issues relating to interlaced and progressive scanning originate in the video camera itself [10]. The implications of the choice of scanning method echo throughout the entire television production, transmission, and display system. It is important to remember that, unlike the photoconductive pickup tube, the CCD itself does not determine the scanning mode at the point of optical-to-electronic conversion. All sensors of the CCD are simultaneously engaged in the conversion of light to electronic charges every 1/60-second. The decision to scan the video in a progressive or interlaced manner is subsequently made within the charge readout mechanism.

Figure 13.7 compares progressive and interlaced scanning as executed at the CCD level. Note that in the interlaced case, practical implementation of interlaced scanning overlaps the scanning rows of two fields, in turn creating an important linkage between the spatio-temporal domains. This overlap of the scanning rows affords the following advantages:

- It increases the sensitivity of the imager

- It increases the overall temporal resolution for a given bandwidth

A comparison of DTV transmission bandwidth requirements is useful in considering the relative merits of progressive and interlaced scanning. For an MPEG-2 bit stream at 4:2:0 (per ATSC A/53), 480I requires 4 to 6 Mbits/s (depending upon the complexity of the program material), resulting in a maximum of four SDTV programs per 6 MHz terrestrial RF channel. A similar bit stream at 480P, however, requires 8 to 13 Mbits/s (again, depending upon the complexity of the program material), resulting in a maximum of two SDTV programs within a 6 MHz channel. The visual appearance of the two images is, of course, substantially different.

Consequently, the scanning choice is as much a business decision as it is a technical one. This decision entails a business model, one that works for the available number of DTV services that can be transmitted simultaneously as well as for the long-term plant-conversion scenario.

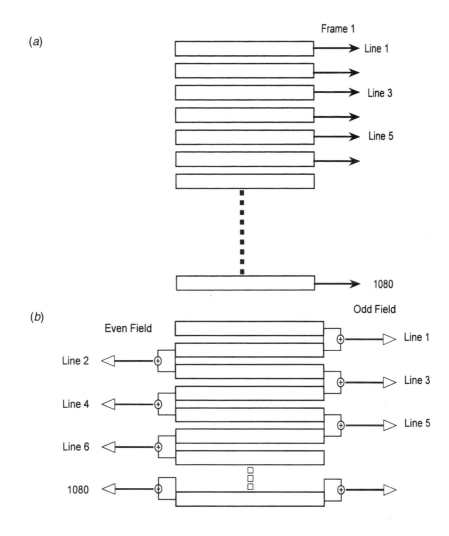

Figure 13.7 Raster scanning methods: (*a*) progressive, (*b*) interlaced. (*After* [10].)

13.4.2 Network Contribution Options

The form in which material is delivered to broadcast affiliates is a point of some interest and concern as a stations plan and build their DTV plants. The simplest approach has the station taking a broadcast-ready 19.4 Mbits/s ATSC data stream from the network. This feed, however, is basically acceptable for pass-through functions only because of the compression already taken on the signal. For applications requiring some form of manipulation of the incoming signal, a higher bit-stream rate is required. Options include the following:

- 45 Mbits/s over satellite or telco fiber links

- 68 Mbits/s from satellite systems using the latest in modem technology

A higher bit stream also carries several potentially negative implications:

- It requires one satellite transponder per signal
- It requires a contribution-grade encoder/decoder
- Affiliates who simply pass through still require a 19.4 Mbits/s bit stream
- Concatenation artifacts may arise from different compression systems

In-Plant Distribution

Mezzanine level distribution is one possible solution to plant infrastructure issues. In this scenario, video signals are compressed from 1.5 Gbits/s down to 200 to 300 Mbits/s using *higher level* MPEG-2 (higher level = less compression). SDTI is used as a carrier for the resulting compressed stream. This permits the continued use of the existing SDI plant infrastructure, which may consist of:

- Routers
- DAs
- Cabling
- DVTRs and servers (>200 Mbits/s raw-data-rate playback and record)

In an ideal arrangement, new DTV devices would support mezzanine I/O. The data rate could vary from one device to the next, as it is assumed that encoders would support a variety of rates and be user-configurable. The higher the data rate, the greater the flexibility of the compressed bit stream. The maximum data rate would be established based on the lowest of the following two parameters:

- The SDI router and available DAs (270 or 360 Mbits/s)
- The uncompressed (*bit bucket*) capability of the recorders being used

Mezzanine encoding/decoding is attractive because it is much less severe in terms of compression than 19.4 Mbits/s. A mezzanine encoder also is considerably less expensive than a 19.4 Mbits/s ATSC encoder. Figure 13.8 illustrates one possible mezzanine implementation.

SMPTE 310M

The SMPTE 310M-1998 standard describes the physical interface and modulation characteristics for a synchronous serial interface to carry MPEG-2 transport bit streams at rates of up to 40 Mbits/s [11]. The interface is a point-to-point scheme intended for use in a low-noise environment, "low noise" being defined as a noise level that would corrupt no more than one MPEG-2 data packet per day at the transport clock rate. When other transmission systems, such as an STL microwave link, are interposed between devices employing this interface, higher noise levels may be encountered. Appropriate error correction methods

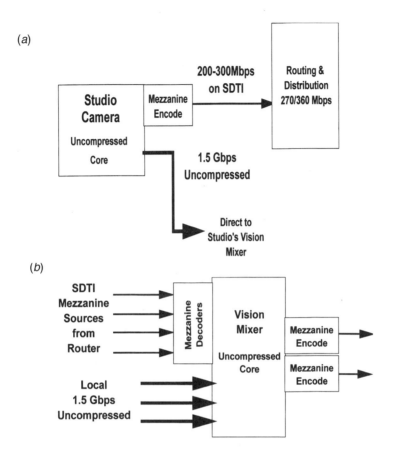

Figure 13.8 One possible implementation for mezzanine level compression in the studio environment: (*a*) camera system, (*b*) vision mixer. (*Courtesy of Leitch.*)

then must be used. Figure 13.9 shows how SMPTE 310M fits into the overall DTV station infrastructure.

13.4.3 DTV Implementation Scenarios

The transition from a conventional NTSC plant to DTV is dictated by any number of factors, depending upon the dynamics of a particular facility. The choices, however, usually can be divided into four overall groups [12]:

- Simple pass-through

- Pass-through with limited local insert

- Local HDTV production

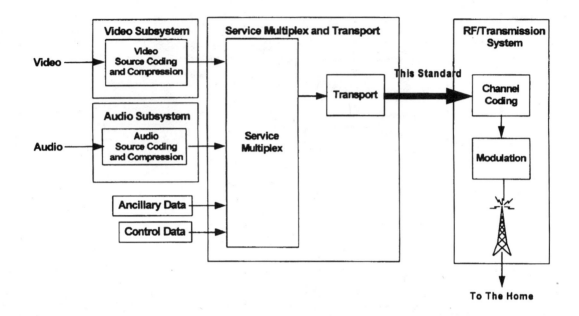

Figure 13.9 A typical application of the SMPTE 310M synchronous serial interface. (*From* [10]. *Used with permission.*)

- Multicasting

Simple Pass-Through

The intent of this scenario is to simply pass through network DTV programming. Important capabilities that are missing include:

- No local identity
- No local value added
- Little revenue potential from the DTV service

Pass-Through With Limited Local Insert

The intent here is to pass through network DTV programming and, additionally, facilitate the following capabilities:

- Local insertion of interstitials, promotional announcements, and commercials
- Local insertion of full-screen logos
- Network delay and program time shifting

- Insertion of local 525 programming

Significant drawbacks to this approach include the following:

- Insertion capabilities are limited

- The bit-stream splicer cannot switch on every frame

- Network contributions must be processed to include splice points

- To keep costs down, local insert material probably would be precompressed off-site, not live

Local HD Production

The basic capabilities of this approach include:

- Pass-through of network DTV programming

- Local insertion of interstitials and promos

- Unconstrained switch-point selection

- Local insertion of full-screen or keyed logos

- Local insertion of weather crawls or alerts

- Full-bandwidth audio processing

- Enhanced video capabilities

The addition of an upconverter allows the insertion of live local programming and news originating from the SDTV facility. The primary challenges with this scenario are:

- The necessary equipment may be cost-prohibitive

- It is practical only on a relatively small scale

- The short-term payback is small in light of the sizable expense

A block diagram of a full-featured HDTV/SDTV television facility is given in Figure 13.10. Note the tandem HD/SD routing switchers and the use of both SDI and SDTI to distribute data around the plant.

Multicasting

The most flexible of the four primary scenarios, multicasting permits the station to adjust its program schedule as a function of day parts and/or specific events. The multicasting environment is configured based on marketplace demands and corporate strategy. To preserve the desired flexibility, the facility should be designed to permit dynamic reconfiguration. As the audience for HDTV builds, and as the equipment becomes available, the station would move with the audience toward the "new viewing experience."

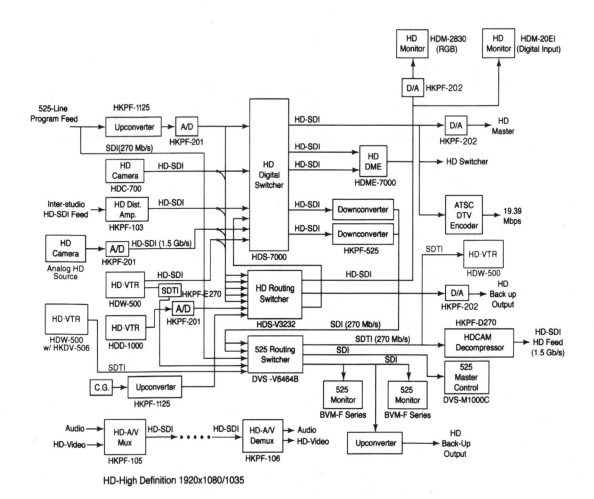

Figure 13.10 Example television plant incorporating support for high-definition and standard-definition video capture, production, and distribution. (*Courtesy of Sony.*)

13.4.4 Top Down System Analysis

In an effort to facilitate the transition to digital television broadcasting, the Implementation Subcommittee (IS) of the ATSC undertook an effort to inventory the various systems and their interfaces that could potentially exist in a typical station, regardless of the implementation scenario. This inventory was intended to serve as a guide to point to the standards that exist for equipment interfaces, identity potential conflicts between those standards, and identity areas where standards and/or technology need further development.

The resulting report and referenced system maps were generalized blueprints for not only construction of an early DTV facility, but also a joint informal agreement among a number of industry manufacturers and end-user consultants, all of whom have worked in digital television for some time. For the station engineer, the maps provide a basic blueprint for their facilities. No station would build a facility as shown in the main system map, but rather each would take portions of the map as a foundation upon which to build to meet their local requirements.

By establishing a single system map where all of the likely system elements and interfaces could exist, a commonality in design and functionality throughout the industry was established that not only pointed the way for the early DTV adopters, but also established a framework upon which to build and expand systems in the future.

One of the most interesting aspects of the committee's efforts was the single format philosophy of the *plant native format*.

Plant Native Format

Several DTV plant architectures were considered by the ATSC IS [13]. The most basic function, pass through of a pre-compressed stream, with or without local bit-stream splicing, is initially economical but lacks operational flexibility. To provide the operational flexibility required, digital video levels could be added to an existing analog NTSC facility, but managing multiple video formats would likely become complex and expensive. An alternative is to extend the existing facility utilizing a single-format plant core infrastructure surrounded by appropriate format conversion equipment. This single-format philosophy of facility design is based on application of one format, called the *plant native format*, to as much of the facility as practical.

The chosen native format would be based on numerous factors, including the preferred format of the station's network, economics, existing legacy equipment and migration strategy, equipment availability, and station group format.

A native format assumes that all material will be processed in the plant in a single common standard to allow production transitions such as keying, bug insertions, and so on. The format converters at the input of the plant would convert the contribution format to the native plant format. A format converter would not be necessary if the native format is the same as the contribution format. Legacy streams will also likely require format conversion. The native plant infrastructure could be traditional analog or digital routers, production switchers, master control switchers, or new concepts such as servers and digital networks.

Choosing a native plant format has a different set of criteria than choosing a broadcast emission format. Input signals arrive in a variety of formats, necessitating input signal format conversion. Similarly, changes in broadcast emission format will necessitate output format conversion. Therefore the native plant format should be chosen to help facilitate low-cost, low-latency, and high-quality format conversions. It is well understood that format conversions involving interlaced scan formats require careful attention to de-interlacing, while progressive scan formats offer fewer conversion challenges.

The native plant format could be chosen from a variety of candidates, including:

- 480I 4:3 or 16:9 carried by SMPTE 259M-1997

- 1080I or 720P using SMPTE 292M

- 480P using SMPTE 293M-1996

• Intermediate compressed formats

Although the full benefits of adopting a plant native format are most apparent with an implementation in which "plant native format" means a single plant native format, many stations will not be able to afford to adopt a single plant native format immediately. For them "plant native format" will, at least on an interim basis, mean a practical mix of perhaps two formats. One of the formats might be the legacy format, which is being phased out while the other is the preferred format for future operations.

13.4.5 Advanced System Control Architecture

The early experiences of DTV facility construction demonstrated the need for a comprehensive control and monitoring environment for studio equipment [14]. Because of differing product lines and market directions, various systems were derived from different sets of operational requirements. Furthermore, varied enabling technologies within the network and computing industries were incorporated into these systems. In recognition of this somewhat hap-hazard situation, and the need for broadcasters and audio/video professionals to chart the course, rather than just follow, a working group within the SMPTE was formed to study advanced system control architectures.

The focus of the S22.02 Advanced System Control Architectures Working Group was to unify existing systems and proposed new systems. This work was intended to be a living document, specifically:

• To provide a comprehensive system overview

• Identify specific areas for additional SMPTE efforts

• Solicit input from interested parties

At this writing, the SMPTE was preparing to embark on standardization efforts in this area.

In terms of overall architecture, the Working Group's efforts were intended to meet the following requirements:

• To be sufficiently flexible to accommodate varying enabling technologies where possible

• Independent of specific technologies where possible

• Scalable to a range of platforms and environments

• Extensible to adapt to emerging technologies

• Modular in nature, allowing functional pieces of the system to be used as building blocks

• Offer a viable migration path for existing facilities

Functional Planes

Under the S22.02 Working Group plan, *functional planes* exist as the front line interface between the studio applications and the functionality of the control system [14]. Four basic functional planes are defined:

- **Content management**: Manages content in the studio; understands physical storage allocation for the content; performs activities including content distribution, content creation, scheduled operations, and storage management; presents a view of content as required for *data streaming*

- **Service**: Combines multiple content streams into complete services; maps services onto resources available in the *path plane*; abstracts the mapping of content to individual pieces of equipment; uses content sources to create paths

- **Path**: Facilitates the physical connection between devices for the purposes of data streaming; manages the physical links between devices; manages the resources required for these connections

- **Device**: Contains the interfaces used to access studio equipment; provides specific information for device I/O capability (ports); is based on a SMPTE-defined hierarchy of functional classifications; provides for extensible interfaces

The planes are organized into a form of *usage abstraction* from content, to service, to path, to device. Adjacent planes cooperate to carry out individual operations as generated by the studio applications. Though this abstraction is beneficial in most cases, it is not explicitly required. For example, applications can access functionality at any plane, or planes can access any other plane in order to accomplished a given task.

These functional planes represent logical partitioning within the control system. From an implementation point of view, the devices themselves will likely be responsible for running local software that will contribute to the functionality of one or more of these planes. For example, within a network router, software exists at runtime that is responsible for some path plane functionality.

Content Management Plane

The *content plane* is the locus of all the "higher level" applications ("business" applications) in the studio and is the first line of support for these applications. [14] It implements *content management* classes of objects, including:

- **Library server**. A studio will have zero or more objects of this class, whose purpose is to provide a *library service*. If there are multiple such objects, they will link together to provide a unified, though distributed, library service.

- **Streaming element**. Objects of this class are created to present a view of content files as endpoints for streaming file transfers. These objects will be created as necessary to support scheduled operations, and be deleted after the operations end.

Service Plane

The *service plane* is responsible for aggregating multiple content streams into complete services and then mapping the implementation of these services onto aggregated resources exposed by the path plane [14]. The service plane manages multiple logical paths within the network that together represent a program, and therefore need to be treated together.

Path Plane

The *path plane* represents the end-to-end connectivity from one device to another, or through chained devices, over a network of routers [14]. The responsibility of the path plane is to transport the media content from one point to another to accomplish a particular task in a timely fashion and with a guaranteed quality of service.

A *physical connectivity* path is a connection between a source device and a destination device via a number of intermediate devices, i.e. a number of chained devices. The adjacent devices are connected through a network connection that is routed over the routers in the network. Control and management of the physical connectivity path involves allocation, setup, and monitoring of the components (devices and links between adjacent devices) that make up the path.

A logical and abstract representation of the connectivity path is desirable to simplify the interface for control of the content data flow. The concept of a *stream* is used to represent such a logical view and the controllable attributes of the content data transfer process. A stream is a logical construction (object) corresponding to the transient state of content data being passed from one device to another through the physical connectivity path.

The path plane contains a collection of logical entities (objects) that collectively build, operate, and manage the connectivity. These entities can be categorized into two functional groups:

- Streams that control the content data flow over the path

- *Studio resource management* (SRM) that facilitates and manages the allocation and reservation of shared studio resources to meet the resource requirements of various tasks and thereby achieve better resource utilization. Manageable studio resources include the devices that are registered with the device *registration repository* and the network connections and bandwidth. SRM can also provide a central point for resource *access control*. The SRM functions may be carried out collaboratively by a number of logical entities.

Device Plane

The *device plane* represents the collection of object classes that implement the interfaces used to access equipment throughout the studio [14]. Devices can also represent the interfaces to software-only objects that perform specific functions within the control network. These device objects need to support the following generic functionality:

- *Dynamic discovery* of the various *interface elements* used in the interface. This does not necessarily mean that the interface description information travels over any network; it may be derived from a local information database for the devices.

- Interfaces for later versions of a device are compatible with earlier versions of the interface.

- Interface elements are based on a consistent set of SMPTE-defined broadcast types.

- All information that can be used to define the unambiguous operational state of a device is available in the device interface. That is, a device state can be completely represented (and therefore restored) by a snapshot of its current interface state.

- Device interfaces are made up of a hierarchical collection of sub-functions (functional blocks).

- Interfaces are extensible and therefore inherently backward compatible.

Functional Layers

Functional layers exist as the foundation on which the planes are built [14]. They provide a homogeneous set of functions that facilitate the transport of control information throughout the network. The three basic functional layers are:

- **System**: Provides common reusable system services as follows: studio time, configuration, security, identification and naming, fault recovery, and status and alarming.

- **Communication**: Provides a common facility for the exchange of information among communicating endpoints within a distributed network, allows the use of multiple protocols simultaneously, and facilitates the transfer of time-synchronized operations.

- **Transport**: Provides for the reliable transmission of information throughout a network, provides an abstraction for physical network access, and allows for adaptation to legacy networks.

13.5   24-Frame Mastering

About 75 percent of prime time television programming is shot on film at 24 f/s. These programs are then downconverted to the 525- or 625-line broadcast formats of conventional television. The process of standards conversion (via telecine) is, thus, a fundamental philosophy inherent to the total scheme of programming for conventional television. A 24-frame celluloid medium is converted to a 25-frame or a 29.97-frame TV medium—depending on the region of the globe—for conventional television broadcast.

This paradigm changes, of course with the introduction of the ATSC digital television standard. Multiple frame rates are possible, including the 24 f/s rate native to film. The question then becomes, why convert at all? Instead, simply broadcast the material in its native format. This simplifies the production process, simplifies the encoding process, and results in higher displayed quality. With regard to the last point, it must be understood that no standards converter (telecine or otherwise) is going to make a film-based program look better than the film product itself, scratch removal and other corrective measures notwithstanding.

To solve—or at least lessen—conversion problems between the multiple transmission standards in use and the original material, most of which originate from the ubiquitous 24 f/s film standard, a new mastering format has been proposed to the video community [15]. The concept is to master at 1080×1920, 24P. This simplifies the process of converting to the various video standards in use worldwide by networks and broadcasters, relative to the inter-format video conversions that now take place. Such an approach makes the final video program spatially and temporally compatible with 480I, 480P 60, 720P 60, and 1080I 30.

A working group of the SMPTE was studying the issue as this book went to press. The proposal included two important characteristics:

- A 1080 × 1920, 16:9, 24P mastering format standard for film-originated TV programs.

- A new 24P standard referred to as 48*sF*, where each frame is progressively scanned and output as two *segmented frames* (*sF*). Under this scheme, segmented Frame 1 contains all the odd lines of the frame, and segmented Frame 2 contains all even lines of the frame, derived from the progressively scanned film image.

Perhaps the most important reason that 24 f/s post production is gaining interest is because of the many different scanning formats that different broadcasters are likely to use in the DTV era. In Hollywood, it is clear that post should be performed at the highest resolution that any client will want the product delivered [16]. The next step, then, is to downconvert to the required lower-resolution format(s). Posting film-shot productions at 24 f/s with 1920 × 1080 resolution, intuitively, makes a great deal of sense. It is a relatively simple matter to extract a 1920 × 1080 60I signal, a 1280 × 720 60P or 24P signal, a 480P signal, or a 480I signal. Perhaps equally important, a 50 Hz signal can be extracted by running the 24 Hz tape at 25 Hz. This makes a PAL copy that is identical to the end result of using a 50 Hz high-definition telecine and running the film 4 percent fast at 25 f/s.

Motion picture producers also are interested in 24 Hz electronic shooting. Producers want to be able to seamlessly mix material shot on film and material shot electronically, both for material that will end up as video, and for material that will end up as film. The easiest way to accomplish these objectives is to shoot video at 24 f/s.

The use of a 1080 × 1920, 24P mastering format means that all telecine-to-tape transfers would be done at the highest resolution possible under the ATSC transmission standard [15]. Just as important, all other lower-quality formats can be derived from this 24P master format. This means that NTSC, PAL, or even the HDTV 720 × 1280 formats with their slightly lower resolution can be downconverted from a high-definition 24 f/s master. With the 24 f/s rate, higher-frame or segmented frame rates can easily be generated.

In order to simplify the transfer process to make a 1080 × 1920 image, 24P becomes the more economical format of choice. It complies with ATSC Table 3 and is progressive. Being a film rate and progressive, it compresses more efficiently than interlaced formats; a progressively scanned 24 f/s image running through an MPEG encoder can reduce the needed bit rate from 25 to 35 percent. Or viewed from another perspective, image quality may be improved for a given bit rate. With a lower frame rate and adjacent picture elements from the progressive scan, motion vector calculations become more efficient. Furthermore, in a 24 f/s system, the post production process—being 24-frame-based—is not burdened with 3:2 tracking issues throughout the process.

Figure 13.11 shows the telecine concept applied to the all-electronic system of HDTV studio mastering at 24 f/s, followed by conversion to the existing 525/625 television media and to the DTV transmission formats. As illustrated in the figure, the conventional telecine has been replaced with electronic digital standards converters. The principle, however, remains identical. This concept is the very essence of the long search for a single worldwide HDTV format for studio origination and international program exchange.

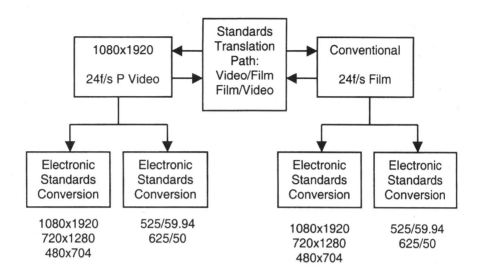

Figure 13.11 A practical implementation of the integration of HDTV video and film using a 24 f/s mastering system.

13.5.1 SMPTE 211

Recommended Practice RP211 (proposed at this writing) defines changes to SMPTE 274M to implement various 1920 × 1080 progressive systems in their *segmented frame* format: 24*sF*, 25*sF*, and 30*sF* [17]. Only the changes to the appropriate clauses of SMPTE 274M are contained in the document. (The same clause, table, and figure numbering system, as used in SMPTE 274M, is employed in RP211.) Table 13.1 lists the additions to Table 1 of SMPTE 274M.

In a segmented frame system, the frame is scanned as a first field then as a second field, in which the scan lines of each field have twice the vertical spatial sampling pitch of the frame. Scanning lines in the second field are displaced vertically by the vertical sampling pitch, but the scanning time is the same temporally as that of scanning lines in the first field. The first field conveys 540 active picture lines, starting with the top picture line of the frame. The second field conveys 540 active picture lines, ending with the bottom picture line of the frame.

All of the scanning systems defined in RP211 and in SMPTE 274M use a total of 1125 lines per picture. In an analog-only system, this would normally imply that the interlaced versions would divide this total into two equal-length fields of 562 1/2 lines each. However, because a digital interface must also be supported, only whole numbers of lines in each field are allowed, in order to permit unambiguous identification of lines by the digital timing reference sequences. Therefore, the interlaced and segmented frame versions define integer, and hence unequal, numbers of lines (563 and 562) in each of the two fields comprising one frame. Analog vertical sync sequences, however, must remain equally spaced in time and are therefore not fully aligned to the fields as defined for the digital interface. This results in

Table 13.1 Additions to SMPTE 274M Table 1 as Specified in RP211 (Proposed) (*After* [17]

System Nomenclature	Samples/ Active Line (S/AL)	Active Lines/ Frame	Frame Rate (Hz)	Scanning Format	Interface Sampling Freq. f_s (MHz)	Samples/ Total Line (S/TL)	Total Lines/ Frame
12	1920 × 1080/ 30 (*sF*)	1920	1080	30	P (*sF*)	74.25	2200	1125
13	1920 × 1080/ 29.97 (*sF*)	1920	1080	30/1.001	P (*sF*)	74.25/1.001	2200	1125
14	1920 × 1080/ 25 (*sF*)	1920	1080	25	P (*sF*)	74.25	2640	1125
15	1920 × 1080/ 24 (*sF*)	1920	1080	24	P (*sF*)	74.25	2750	1125
16	1920 × 1080/ 23.98 (*sF*)	1920	1080	24/1.001	P (*sF*)	74.25/1.001	2750	1125

the analog vertical sync for the second digital field beginning one half-line before the end of the first digital field.

As specified in RP211, in a segmented frame system, the assignment of lines within a frame are the same as that of an interlaced system. More specifically, the assignment of each even line of a progressive system corresponds to lines 1 through 562 of a segmented frame system, and each odd line of a progressive system corresponds to lines 563 through 1125 of a segmented frame system.

Ancillary signals may be conveyed in a progressive system during lines 7 through 41 inclusive, and in an interlaced and segmented frame system during lines 7 through 20 inclusive and lines 569 through 583 inclusive. The portion within each of these lines that may be used for ancillary data is defined in the Recommended Practice. Ancillary signals do not convey picture information although they may be employed to convey other related or unrelated signals, coded similarly to picture information.

13.5.2 Global HDTV Program Origination Standard

In June 1999, the ITU finally brought to closure a long 15-year quest for a global standard for HDTV production and international program exchange [18]. Formerly wrestling with the technical vagaries that continued to separate the 50 Hz and 60 Hz regions of the world, the organization became galvanized by the new surge of interest swirling around the emergence of the 24 frame progressive HD format. In a matter of weeks, all 172 countries represented in the ITU had removed all of the differences formerly blocking a convergence to a singular digital production standard. All regions finally agreed to a unique set of numbers that completely prescribed the active still picture in that standard:

• Aspect ratio

- Digital sampling structure

- Total number of lines

- Colorimetry

- Transfer characteristic

The concept of an international *common image format* had arrived. Viewed from the vantage point of history, the agreement was one of monumental proportions. In a brilliant stroke, the ITU recognized the implementation of a worldwide digital picture within a framework of different capture rates. Specifically, both progressive and interlace scanning were recognized and 50 and 60 Hz were equally acknowledged. The 24 f/s rate emerged as the all-important electronic emulation of the global defacto 24 f/s film standard; and for good measure, they also added 25P and 30P.

The implications of this new world production standard for HD are far-reaching. Digital HD and motion picture film can now be on a common platform of 24 f/s—making transfers between the two media that much simpler and with little attendant quality loss. This will affect movie-making as profoundly as it will prime-time television production. The use of a single 24P high-definition master in postproduction (whether it is originated on 24 f/s film or electronic 24P) will greatly reduce the technical burden (and costs) associated with preparing different distribution masters to service the diverse DTV needs of broadcasters.

13.6 References

1. Epstein, Steve: "Editing MPEG Bitstreams," *Broadcast Engineering*, Intertec Publishing, Overland Park, Kan., pp. 37–42, October 1997.

2. Cugnini, Aldo G.: "MPEG-2 Bitstream Splicing," *Proceedings of the Digital Television '97 Conference*, Intertec Publishing, Overland Park, Kan., December 1997.

3. SMPTE Standard: SMPTE 312M-1999, *Splice Points for MPEG-2 Transport Streams*, SMPTE, White Plains, N.Y., 1999.

4. SMPTE Standard: SMPTE 328M-2000: "MPEG-2 Video Elementary Stream Editing Information," SMPTE, White Plains, N.Y., 2000.

5. SMPTE 327M-2000, "MPEG-2 Video Recoding Data Set," SMPTE, White Plains, N.Y., 2000.

6. SMPTE 329M-2000, "MPEG-2 Video Recoding Data Set—Compressed Stream Format," SMPTE, White Plains, N.Y., 2000.

7. SMPTE 319M-2000, "Transporting MPEG-2 Recoding Information Through 4:2:2 Component Digital Interfaces," SMPTE, White Plains, N.Y., 2000.

8. SMPTE 351M (Proposed), "Transporting MPEG-2 Recoding Information through High-Definition Digital Interfaces," SMPTE, White Plains, N.Y., 2000.

9. SMPTE 353M (Proposed), "Transport of MPEG-2 Recoding Information as Ancillary Data Packets," SMPTE, White Plains, N.Y., 2000.

10. Thorpe, Larry: "The Great Debate: Interlaced Versus Progressive Scanning," *Proceedings of the Digital Television '97 Conference*, Intertec Publishing, Overland Park, Kan., December 1997.

11. SMPTE Standard: SMPTE 310M-1998, "Synchronous Serial Interface for MPEG-2 Digital Transport Stream," SMPTE, White Plains, N.Y., 1998.

12. Course notes, "DTV Express," PBS/Harris, Alexandria, Va., 1998.

13. ATSC: "Implementation Subcommittee Report on Findings," Draft Version 0.4, ATSC, Washington, D.C., September 21, 1998.

14. SMPTE: "System Overview—Advanced System Control Architecture, S22.02, Revision 2.0," S22.02 Advanced System Control Architectures Working Group, SMPTE, White Plains, N.Y., March 27, 2000.

15. Mendrala, Jim: "Mastering at 24P," *Broadcast Engineering*, Intertec Publishing, Overland Park, Kan., pp. 92–94, February 1999.

16. Hopkins, Robert: "What We've Learned from the DTV Experience," *Proceedings of Digital Television '98*, Intertec Publishing, Overland Park, Kan., 1998.

17. SMPTE RP 211 (Proposed): "Implementation of 24P, 25P, and 30P Segmented Frames for 1920 × 1080 Production Format," SMPTE, White Plains, N.Y., 2000.

18. Thorpe, Laurence,: "A New Global HDTV Program Origination Standard: Implications for Production and Technology," *Proceedings of DTV99*, Intertec Publishing, Overland Park, Kan., 1999.

13.7 Bibliography

Ward, Christopher, C. Pecota, X. Lee and G. Hughes: "Seamless Splicing for MPEG-2 Transport Stream Video Servers," *Proceedings, 33rd SMPTE Advanced Motion Imaging Conference*, SMPTE, White Plains, N.Y., 2000.

14.1 Terms Employed

For the purposes of the ATSC digital television standard, the following definition of terms apply [1–6]:

16 VSB Vestigial sideband modulation with 16 discrete amplitude levels.

8 VSB Vestigial sideband modulation with 8 discrete amplitude levels.

access unit A coded representation of a presentation unit. In the case of audio, an access unit is the coded representation of an audio frame. In the case of video, an access unit includes all the coded data for a picture, and any *stuffing* that follows it, up to but not including the start of the next access unit.

anchor frame A video frame that is used for prediction. *I*-frames and *P*-frames are generally used as anchor frames, but *B*-frames are never anchor frames.

asynchronous transfer mode (ATM) A digital signal protocol for efficient transport of both constant-rate and variable-rate information in broadband digital networks. The ATM digital stream consists of fixed-length packets called *cells*, each containing 53 8-bit bytes (a 5-byte header and a 48-byte information payload).

bidirectional pictures (B-pictures or **B-frames)** Pictures that use both future and past pictures as a reference. This technique is termed *bidirectional prediction*. *B*-pictures provide the most compression. *B*-pictures do not propagate coding errors as they are never used as a reference.

bit rate The rate at which the compressed bit stream is delivered from the channel to the input of a decoder.

block An 8-by-8 array of DCT coefficients representing luminance or chrominance information.

byte-aligned A bit stream operational condition. A bit in a coded bit stream is byte-aligned if its position is a multiple of 8-bits from the first bit in the stream.

channel A medium that stores or transports a digital television stream.

coded representation A data element as represented in its encoded form.

compression The reduction in the number of bits used to represent an item of data.

constant bit rate The operating condition where the bit rate is constant from start to finish of the compressed bit stream.

conventional definition television (CDTV) This term is used to signify the *analog* NTSC television system as defined in ITU-R Rec. 470. (*See also standard definition television* and ITU-R Rec. 1125.)

CRC Cyclic redundancy check, an algorithm used to verify the correctness of data.

decoded stream The decoded reconstruction of a compressed bit stream.

decoder An embodiment of a decoding process.

decoding (process) The process defined in the ATSC digital television standard that reads an input coded bit stream and outputs decoded pictures or audio samples.

decoding time-stamp (DTS) A field that may be present in a PES packet header which indicates the time that an access unit is decoded in the system target decoder.

***D*-frame** A frame coded according to an MPEG-1 mode that uses dc coefficients only.

DHTML (dynamic HTML) A term used by some vendors to describe the combination of HTML, style sheets, and scripts that enable the animation of web pages.

DOM (document object model) A platform- and language-neutral interface that allows programs and scripts to dynamically access and update the content, structure, and style of documents. The document can be further processed and the results of that processing can be incorporated back into the presented page.

digital storage media (DSM) A digital storage or transmission device or system.

discrete cosine transform (DCT) A mathematical transform that can be perfectly undone and which is useful in image compression.

editing A process by which one or more compressed bit streams are manipulated to produce a new compressed bit stream. Conforming edited bit streams are understood to meet the requirements defined in the ATSC digital television standard.

elementary stream (ES) A generic term for one of the coded video, coded audio, or other coded bit streams. One elementary stream is carried in a sequence of PES packets.

elementary stream clock reference (ESCR) A time stamp in the PES stream from which decoders of PES streams may derive timing.

encoder An embodiment of an encoding process.

encoding (process) A process that reads a stream of input pictures or audio samples and produces a valid coded bit stream as defined in the ATSC digital television standard.

entitlement control message (ECM) Private conditional access information that specifies control words and possibly other stream-specific, scrambling, and/or control parameters.

entitlement management message (EMM) Private conditional access information that specifies the authorization level or the services of specific decoders. They may be addressed to single decoders or groups of decoders.

entropy coding The process of variable-length lossless coding of the digital representation of a signal to reduce redundancy.

entry point A point in a coded bit stream after which a decoder can become properly initialized and commence syntactically correct decoding. The first transmitted picture after an entry point is either an *I*-picture or a *P*-picture. If the first transmitted picture is not an *I*-picture, the decoder may produce one or more pictures during acquisition.

event A collection of elementary streams with a common time base, an associated start time, and an associated end time.

field For an interlaced video signal, a *field* is the assembly of alternate lines of a frame. Therefore, an interlaced frame is composed of two fields, a top field and a bottom field.

frame Lines of spatial information of a video signal. For progressive video, these lines contain samples starting from one time instant and continuing through successive lines to the bottom of the frame. For interlaced video, a frame consists of two fields, a top field and a bottom field. One of these fields will commence one field later than the other.

group of pictures (GOP) One or more pictures in sequence.

high-definition television (HDTV) An imaging system with a resolution of approximately twice that of conventional television in both the horizontal (H) and vertical (V) dimensions, and a picture aspect ratio (H × V) of 16:9. ITU-R Rec. 1125 further defines "HDTV quality" as the delivery of a television picture that is subjectively identical with the interlaced HDTV studio standard.

high level A range of allowed picture parameters defined by the MPEG-2 video coding specification that corresponds to high-definition television.

HTML (hypertext markup language) A collection of tags typically used in the development of Web pages.

HTTP (hypertext transfer protocol) A set of instructions for communication between a server and a World Wide Web client.

Huffman coding A type of source coding that uses codes of different lengths to represent symbols which have unequal likelihood of occurrence.

intra-coded pictures (*I*-pictures or *I*-frames) Pictures that are coded using information present only in the picture itself and not depending on information from other pictures. *I*-pictures provide a mechanism for random access into the compressed video data. *I*-pictures employ transform coding of the pixel blocks and provide only moderate compression.

layer One of the levels in the data hierarchy of the DTV video and system specifications.

level A range of allowed picture parameters and combinations of picture parameters.

macroblock In the advanced television system, a macroblock consists of four blocks of luminance and one each C_r and C_b block.

main level A range of allowed picture parameters defined by the MPEG-2 video coding specification, with maximum resolution equivalent to ITU-R Rec. 601.

main profile A subset of the syntax of the MPEG-2 video coding specification that is supported over a large range of applications.

MIME (multipart/signed, multipart/encrypted content-types) A protocol for allowing e-mail messages to contain various types of media (text, audio, video, images, etc.).

motion vector A pair of numbers that represent the vertical and horizontal displacement of a region of a reference picture for prediction purposes.

MPEG Standards developed by the ISO/IEC JTC1/SC29 WG11, *Moving Picture Experts Group*. MPEG may also refer to the Group itself.

MPEG-1 ISO/IEC standards 11172-1 (Systems), 11172-2 (Video), 11172-3 (Audio), 11172-4 (Compliance Testing), and 11172-5 (Technical Report).

MPEG-2 ISO/IEC standards 13818-1 (Systems), 13818-2 (Video), 13818-3 (Audio), and 13818-4 (Compliance).

pack A header followed by zero or more packets; a layer in the ATSC DTV system coding syntax.

packet A header followed by a number of contiguous bytes from an elementary data stream; a layer in the ATSC DTV system coding syntax.

packet data Contiguous bytes of data from an elementary data stream present in the packet.

packet identifier (PID) A unique integer value used to associate elementary streams of a program in a single or multi-program transport stream.

padding A method to adjust the average length of an audio frame in time to the duration of the corresponding PCM samples by continuously adding a slot to the audio frame.

payload The bytes that follow the header byte in a packet. The transport stream packet header and adaptation fields are not payload.

PES packet The data structure used to carry elementary stream data. It consists of a packet header followed by PES packet payload.

PES stream A stream of PES packets, all of whose payloads consist of data from a single elementary stream, and all of which have the same stream identification.

picture Source, coded, or reconstructed image data. A source or reconstructed picture consists of three rectangular matrices representing the luminance and two chrominance signals.

pixel "Picture element" or "pel." A pixel is a digital sample of the color intensity values of a picture at a single point.

predicted pictures (*P*-pictures or *P*-frames) Pictures that are coded with respect to the nearest *previous I* or *P*-picture. This technique is termed *forward prediction*. *P*-pictures provide more compression than *I*-pictures and serve as a reference for future *P*-pictures or *B*-pictures. *P*-pictures can propagate coding errors when *P*-pictures (or *B*-pictures) are predicted from prior *P*-pictures where the prediction is flawed.

presentation time-stamp (PTS) A field that may be present in a PES packet header that indicates the time that a presentation unit is presented in the system target decoder.

presentation unit (PU) A decoded audio access unit or a decoded picture.

profile A defined subset of the syntax specified in the MPEG-2 video coding specification.

program A collection of program elements. Program elements may be elementary streams. Program elements need not have any defined time base; those that do have a common time base and are intended for synchronized presentation.

program clock reference (PCR) A time stamp in the transport stream from which decoder timing is derived.

program element A generic term for one of the elementary streams or other data streams that may be included in the program.

program specific information (PSI) Normative data that is necessary for the demultiplexing of transport streams and the successful regeneration of programs.

quantizer A processing step that intentionally reduces the precision of DCT coefficients

random access The process of beginning to read and decode the coded bit stream at an arbitrary point.

scrambling The alteration of the characteristics of a video, audio, or coded data stream in order to prevent unauthorized reception of the information in a clear form.

slice A series of consecutive macroblocks.

source stream A single, non-multiplexed stream of samples before compression coding.

splicing The concatenation performed on the system level or two different elementary streams. It is understood that the resulting stream must conform totally to the ATSC digital television standard.

standard definition television (SDTV) This term is used to signify a *digital* television system in which the quality is approximately equivalent to that of NTSC. This equivalent quality may be achieved from pictures sourced at the 4:2:2 level of ITU-R Rec. 601 and subjected to processing as part of bit rate compression. The results should be such that when judged across a representative sample of program material, subjective equivalence with NTSC is achieved. Also called standard digital television.

start codes 32-bit codes embedded in the coded bit stream that are unique. They are used for several purposes, including identifying some of the layers in the coding syntax.

STD input buffer A first-in, first-out buffer at the input of a system target decoder (STD) for storage of compressed data from elementary streams before decoding.

still picture A video sequence containing exactly one coded picture that is intra-coded. This picture has an associated PTS and the presentation time of succeeding pictures, if any, is later than that of the still picture by at least two picture periods.

system clock reference (SCR) A time stamp in the program stream from which decoder timing is derived.

system header A data structure that carries information summarizing the system characteristics of the ATSC digital television standard multiplexed bit stream.

system target decoder (STD) A hypothetical reference model of a decoding process used to describe the semantics of the ATSC digital television standard multiplexed bit stream.

time-stamp A term that indicates the time of a specific action such as the arrival of a byte or the presentation of a presentation unit.

transport stream packet header The leading fields in a transport stream packet.

UHTTP (unidirectional hypertext transfer protocol) A is a simple, robust, one-way resource transfer protocol that is designed to efficiently deliver resource data in a one-way broadcast-only environment. This resource transfer protocol is appropriate for IP multicast over the television vertical blanking interval (IP-VBI), in an IP multicast carried in MPEG-2, or in other unidirectional transport systems.

UUID (universally unique identifier) An identifier that is unique across both space and time, with respect to the space of all UUIDs. Also known as GUID (globally unique identifier).

variable bit rate An operating mode where the bit rate varies with time during the decoding of a compressed bit stream.

video buffering verifier (VBV) A hypothetical decoder that is conceptually connected to the output of an encoder. Its purpose is to provide a constraint on the variability of the data rate that an encoder can produce.

video sequence An element represented by a sequence header, one or more groups of pictures, and an end of sequence code in the data stream.

14.2 Acronyms and Abbreviations

A/D analog to digital converter

ACATS Advisory Committee on Advanced Television Service

AES Audio Engineering Society

ANSI American National Standards Institute

ATEL Advanced Television Evaluation Laboratory

ATM asynchronous transfer mode

ATSC Advanced Television Systems Committee

ATTC Advanced Television Test Center

ATV advanced television

bps bits per second

bslbf bit serial, leftmost bit first

CAT conditional access table

CDT carrier definition table

CDTV conventional definition television

CRC cyclic redundancy check

DCT discrete cosine transform

DSM digital storage media

DSM-CC digital storage media command and control

DTS decoding time-stamp

DVCR digital video cassette recorder

ECM entitlement control message

EMM entitlement management message

ES elementary stream

ESCR elementary stream clock reference

FPLL frequency- and phase-locked loop

GA Grand Alliance

GMT Greenwich mean time

GOP group of pictures

GPS global positioning system

HDTV high-definition television

IEC International Electrotechnical Commission

IRD integrated receiver-decoder

ISO International Organization for Standardization

ITU International Telecommunication Union

JEC Joint Engineering Committee of EIA and NCTA

MCPT multiple carriers per transponder

MMT modulation mode table

MP@HL main profile at high level

MP@ML main profile at main level

MPEG Moving Picture Experts Group

NAB National Association of Broadcasters

NTSC National Television System Committee

NVOD near video on demand

PAL phase alternation each line

PAT program association table

PCR program clock reference

pel pixel

PES packetized elementary stream

PID packet identifier

PMT program map table

PSI program specific information

PTS presentation time stamp

PU presentation unit

SCR system clock reference

SDTV standard definition television

SECAM sequential couleur avec mémoire (sequential color with memory)

SIT satellite information table

SMPTE Society of Motion Picture and Television Engineers

STD system target decoder

TAI international atomic time

TDT transponder data table

TNT transponder name table

TOV threshold of visibility

TS transport stream

UTC universal coordinated time

VBV video buffering verifier

VCN virtual channel number

VCT virtual channel table

14.3 References

1. *ATSC Digital Television Standard*, Doc. A/53, Advanced Television Systems Committee, Washington, D.C., 1996.

2. *Digital Audio Compression (AC-3) Standard*, Doc. A/52, Advanced Television Systems Committee, Washington, D.C., 1996.

3. *Guide to the Use of the ATSC Digital Television Standard*, Doc. A/54, Advanced Television Systems Committee, Washington, D.C., 1996.

4. *Program Guide for Digital Television*, Doc. A/55, Advanced Television Systems Committee, Washington, D.C., 1996.

5. *System Information for Digital Television*, Doc. A/56, Advanced Television Systems Committee, Washington, D.C., 1996.

6. "Advanced Television Enhancement Forum Specification," Draft, Version 1.1r26 updated 2/2/99, ATVEF, Portland, Ore., 1999.

Reference Documents

15.1 General

The following references provide additional information on digital television in general, and the ATSC standard in particular.

15.1.1 Video

ISO/IEC IS 13818-1, International Standard (1994), MPEG-2 Systems

ISO/IEC IS 13818-2, International Standard (1994), MPEG-2 Video

ITU-R BT.601-4 (1994), Encoding Parameters of Digital Television for Studios

SMPTE 274M-1995, Standard for Television, 1920×1080 Scanning and Interface

SMPTE 293M-1996, Standard for Television, 720×483 Active Line at 59.94 Hz Progressive Scan Production, Digital Representation

SMPTE 294M-1997, Standard for Television, 720×483 Active Line at 59.94 Hz Progressive Scan Production, Bit-Serial Interfaces

SMPTE 295M-1997, Standard for Television, 1920×1080 50 Hz, Scanning and Interface

SMPTE 296M-1997, Standard for Television, 1280×720 Scanning, Analog and Digital Representation, and Analog Interface

15.1.2 Audio

ATSC Standard A/52 (1995), Digital Audio Compression (AC-3)

AES 3-1992 (ANSI S4.40-1992), AES Recommended Practice for digital audio engineering—Serial transmission format for two-channel linearly represented digital audio data

ANSI S1.4-1983, Specification for Sound Level Meters

IEC 651 (1979), Sound Level Meters

IEC 804 (1985), Amendment 1 (1989), Integrating/Averaging Sound Level Meters

15.2 ATSC DTV Standard

The following documents form the basis for the ATSC digital television standard.

15.2.1 Service Multiplex and Transport Systems

ATSC Standard A/52 (1995), Digital Audio Compression (AC-3)

ISO/IEC IS 13818-1, International Standard (1994), MPEG-2 Systems

ISO/IEC IS 13818-2, International Standard (1994), MPEG-2 Video

ISO/IEC CD 13818-4, MPEG Committee Draft (1994), MPEG-2 Compliance

15.2.2 System Information Standard

ATSC Standard A/52 (1995), Digital Audio Compression (AC-3)

ATSC Standard A/53 (1995), ATSC Digital Television Standard

ATSC Standard A/80 (1999), Modulation And Coding Requirements For Digital TV (DTV) Applications Over Satellite

ISO 639, Code for the Representation of Names of Languages, 1988

ISO CD 639.2, Code for the Representation of Names of Languages: alpha-3 code, Committee Draft, dated December 1994

ISO/IEC 10646-1:1993, Information technology—Universal Multiple-Octet Coded Character Set (UCS) — Part 1: Architecture and Basic Multilingual Plane

ISO/IEC 11172-1, Information Technology—Coding of moving pictures and associated audio for digital storage media at up to about 1.5 Mbit/s—Part 1: Systems

ISO/IEC 11172-2, Information Technology—Coding of moving pictures and associated audio for digital storage media at up to about 1.5 Mbit/s—Part 2: Video

ISO/IEC 11172-3, Information Technology—Coding of moving pictures and associated audio for digital storage media at up to about 1.5 Mbit/s—Part 3: Audio

ISO/IEC 13818-3:1994, Information Technology—Coding of moving pictures and associated audio—Part 3: Audio

ISO/CD 13522-2:1993, Information Technology—Coded representation of multimedia and hypermedia information objects—Part 1: Base notation

ISO/IEC 8859, Information Processing—8-bit Single-Octet Coded Character Sets, Parts 1 through 10

ITU-T Rec. H. 222.0 / ISO/IEC 13818-1:1994, Information Technology—Coding of moving pictures and associated audio—Part 1: Systems

ITU-T Rec. H. 262 / ISO/IEC 13818-2:1994, Information Technology—Coding of moving pictures and associated audio—Part 2: Video

ITU-T Rec. J.83:1995, Digital multi-programme systems for television, sound, and data services for cable distribution

ITU-R Rec. BO.1211:1995, Digital multi-programme emission systems for television, sound, and data services for satellites operating in the 11/12 GHz frequency range

15.2.3 Receiver Systems

47 CFR Part 15, FCC Rules

EIA IS-132, EIA Interim Standard for Channelization of Cable Television

EIA IS-23, EIA Interim Standard for RF Interface Specification for Television Receiving Devices and Cable Television Systems

EIA IS-105, EIA Interim Standard for a Decoder Interface Specification for Television Receiving Devices and Cable Television Decoders

15.2.4 Program Guide

ATSC Standard A/53 (1995), ATSC Digital Television Standard

ANSI/EIA-608-94 (1994), Recommended Practice for Line 21 Data Service

ISO/IEC IS 13818-1, International Standard (1994), MPEG-2 Systems

15.2.5 Program/Episode/Version Identification

ATSC Standard A/53 (1995), Digital Television Standard

ATSC Standard A/65 (1998), Program and System Information Protocol for Terrestrial Broadcast and Cable

ATSC Standard A/70 (1999), Conditional Access System for Terrestrial Broadcast

ATSC Standard A/90 (2000), ATSC Data Broadcast Standard

ISO/IEC IS 13818-1, International Standard (1994), MPEG-2 systems

15.3   DVB

The following documents form the basis of the DVB digital television standard.

15.3.1 General

Digital Satellite Transmission Systems, ETS 300 421

Digital Cable Delivery Systems, ETS 300 429

Digital Terrestrial Broadcasting Systems, ETS 300 744

Digital Satellite Master Antenna Television (SMATV) Distribution Systems, ETS 300 473

Specification for the Transmission of Data in DVB Bitstreams, TS/EN 301 192

Digital Broadcasting Systems for Television, Sound and Data Services; Subtitling Systems, ETS 300 743

Digital Broadcasting Systems for Television, Sound and Data Services; Allocation of Service Information (SI) Codes for Digital Video Broadcasting (DVB) Systems, ETR 162

15.3.2 Multipoint Distribution Systems

Digital Multipoint Distribution Systems at and Above 10 GHz, ETS 300 748

Digital Multipoint Distribution Systems at or Below 10 GHz, ETS 300 749

15.3.3 Interactive Television

Return Channels in CATV Systems (DVB-RCC), ETS 300 800

Network-Independent Interactive Protocols (DVB-NIP), ETS 300 801

Interaction Channel for Satellite Master Antenna TV (SMATV), ETS 300 803

Return Channels in PSTN/ISDN Systems (DVB-RCT), ETS 300 802

Interfacing to PDH Networks, ETS 300 813

Interfacing to SDH Networks, ETS 300 814

15.3.4 Conditional Access

Common Interface for Conditional Access and Other Applications, EN50221

Technical Specification of SimulCrypt in DVB Systems, TS101 197

15.3.5 Interfaces

DVB Interfaces to PDH Networks, prETS 300 813

DVB Interfaces to SDH Networks, prETS 300 814

15.4 SMPTE Documents Relating to Digital Television

The following documents relating to digital television have been approved (or are pending at this writing) by the Society of Motion Picture and Television Engineers.

15.4.1 General Topics

AES/EBU Emphasis and Preferred Sampling Rate, EG 32

Alignment Color Bar Signal, EG 1

Audio: Linear PCM in MPEG-2 Transport Stream, SMPTE 302M

Camera Color Reference Signals, Derivation of, RP 176-1993

Color, Equations, Derivation of, RP 177

Color, Reference Pattern, SMPTE 303M

Wide-Screen Scanning Structure, SMPTE RP 199

15.4.2 Ancillary

AES/EBU Audio and Auxiliary Data, SMPTE 272M

Camera Positioning by Data Packets, SMPTE 315M

Data Packet and Space Formatting, SMPTE 291M

DTV Closed-Caption Server to Encoder Interface, SMPTE 333M

Error Detection and Status Flags, RP 165

Format for Non-PCM Audio and Data in an AES3 Serial Digital Audio Interface, SMPTE 337M

Format for Non-PCM Audio and Data in an AES3 Serial Digital Audio Interface—ATSC A/52 (AC-3) Data Type, SMPTE 340M

Format for Non-PCM Audio and Data in an AES3 Serial Digital Audio Interface—Captioning Data Type, SMPTE 341M

Format for Non-PCM Audio and Data in an AES3 Serial Digital Audio Interface—Data Types, SMPTE 338M

Format for Non-PCM Audio and Data in an AES3 Serial Digital Audio Interface—Generic Data Types, SMPTE 339M

HDTV 24-bit Digital Audio, SMPTE 299M

LTC and VITC Data as HANC Packets, RP 196

Time and Control Code, RP 188

Transmission of Signals Over Coaxial Cable, SMPTE 276M

15.4.3 Digital Control Interfaces

Common Messages, RP 172

Control Message Architecture, RP 138

Electrical and Mechanical Characteristics, SMPTE 207M

ESlan Implementation Standards, EG 30

ESlan Remote Control System, SMPTE 275M

ESlan Virtual Machine Numbers, RP 182

Glossary, Electronic Production, EG 28

Remote Control of TV Equipment, EG 29

Status Monitoring and Diagnostics, Fault Reporting, SMPTE 269M

Status Monitoring and Diagnostics, Processors, RP 183-1995

Status Monitoring and Diagnostics, Protocol, SMPTE 273M

Supervisory Protocol, RP 113

System Service Messages, RP 163

Tributary Interconnection, RP 139

Type-Specific Messages, ATR, RP 171

Type-Specific Messages, Routing Switcher, RP 191

Type-Specific Messages, VTR, RP 170

Universal Labels for Unique ID of Digital Data, SMPTE 298M

Video Images: Center, Aspect Ratio and Blanking, RP 187

Video Index: Information Coding, 525- and 625-Line, RP 186

15.4.4 Edit Decision Lists

Device Control Elements, SMPTE 258M

Storage, 3-1/2-in Disk, RP 162

Storage, 8-in Diskette, RP 132

Transfer, Film to Video, RP 197

15.4.5 Image Areas

8 mm Release Prints, TV Safe Areas, RP 56

16 mm and 35 mm Film and 2 × 2 slides, SMPTE 96

Review Rooms, SMPTE 148

Safe Areas, RP 27.3

15.4.6 Interfaces and Signals

12-Channel for Digital Audio and Auxiliary Data, SMPTE 324M

Checkfield, RP 178

Development of NTSC, EG 27

Key Signals, RP 157

NTSC Analog Component 4:2:2, SMPTE 253M

NTSC Analog Composite for Studios, SMPTE 170M

Pathological Conditions, EG 34

Reference Signals, 59.94 or 50 Hz, SMPTE 305M

Bit-Parallel Interfaces

1125/60 Analog Component, RP 160

1125/60 Analog Composite, SMPTE 240M

1125/60 High-Definition Digital Component, SMPTE 260M

NTSC Digital Component, SMPTE 125M

NTSC Digital Component, 16×9 Aspect Ratio, SMPTE 267M

NTSC Digital Component 4:4:4:4 Dual Link, RP 175

NTSC Digital Component 4:4:4:4 Single Link, RP 174

NTSC Digital Composite, SMPTE 244M

Bit-Serial Interfaces

4:2:2p and 4:2:0p Bit Serial, SMPTE 294M

540 Mbits/s Serial Digital Interface, SMPTE 344M

Digital Component 4:2:2 AMI, SMPTE 261M

Digital Component S-NRZ, SMPTE 259M

Digital Composite AMI, SMPTE 261M

Digital Composite, Error Detection Checkwords/Status Flag, RP 165

Digital Composite, Fiber Transmission System, SMPTE 297M

Digital Composite, S-NRZ, SMPTE 259M

Element and Metadata Definitions for the SDTI-CP, SMPTE 331M

Encapsulation of Data Packet Streams over SDTI (SDTI-PF), SMPTE 332M

HDTV, SMPTE 292M

High Data-Rate Serial Data Transport Interface (HD-SDTI), SMSPTE 348M

HDTV, Checkfield, RP 198

Jitter in Bit Serial Systems, RP 184

Jitter Specification, Characteristics and Measurements, EG 33

Jitter Specification, Measurement, RP 192

SDTI Content Package Format (SDTI-CP), SMPTE 326M

Serial Data Transport Interface, SMPTE 305M

Time Division Multiplexing Video Signals and Generic Data over High-Definition Interfaces, SMPTE 346M

Vertical Ancillary Data Mapping for Bit-Serial Interface, SMPTE 334M

Scanning Formats

1280×720 Scanning, SMPTE 296M

1920×1080 Scanning, 60 Hz, SMPTE 274M

1920×1080 Scanning, 50 Hz, SMPTE 295M

720×483 Digital Representation, SMPTE 293M

15.4.7 Monitors

Alignment, RP 167

Colorimetry, RP 145

Critical Viewing Conditions, RP 166

Receiver Monitor Setup Tapes, RP 96

15.4.8 MPEG-2

4:2:2 Profile at High Level, SMPTE 308M

4:2:2 P@HL Synchronous Serial Interface, SMPTE 310M

Alignment for Coding, RP 202

MPEG-2 Video Elementary Stream Editing Information, SMPTE 328M

MPEG-2 Video Recoding Data Set, SMPTE 327M

MPEG-2 Video Recoding Data Set—Compressed Stream Format, SMPTE 329M

Opportunistic Data Broadcast Flow Control, SMPTE 325M

Splice Points for the Transport Stream, SMPTE 312M

Transport of MPEG-2 Recoding Information as Ancillary Data Packets, SMPTE 353M

Transporting MPEG-2 Recoding Information Through 4:2:2 Component Digital Interfaces, SMPTE 319M

Transporting MPEG-2 Recoding Information Through High-Definition Digital Interfaces, SMPTE 351M

Unique Material Identifier (UMID), SMPTE 330M

15.4.9 Test Patterns

Alignment Color Bars, EG 1

Camera Registration, RP 27.2

Deflection Linearity, RP 38.1

Mid-Frequency Response, RP 27.5

Operational Alignment, RP 27.1

Safe Areas, RP 27.3

Telecine Jitter, Weave, Ghost, RP 27.4

15.4.10 Video Recording and Reproduction

Audio Monitor System Response, SMPTE 222M

Channel Assignments, AES/EBU Inputs, EG 26

Channel Assignments and Magnetic Masters to Stereo Video, RP 150

Cassette Bar Code Readers, EG 31-1995

Data Structure for DV-Based Audio, Data, and Compressed Video, SMPTE 314M

Loudspeaker Placement, HDEP, RP 173

Relative Polarity of Stereo Audio Signals, RP 148

Telecine Scanning Capabilities, EG 25

Tape Care, Handling, Storage, RP 103

Time and Control Code

Binary Groups, Date and Time Zone Transmissions, SMPTE 309M

Binary Groups, Storage and Transmission, SMPTE 262M

Directory Index, Auxiliary Time Address Data, RP 169

Directory Index, Dialect Specification of Page-Line, RP 179

Specifications, TV, Audio, Film, SMPTE 12M

Time Address Clock Precision, EG 35

Vertical Interval, 4:2:2 Digital Component, SMPTE 266M

Vertical Interval, Encoding Film Transfer Information, 4:2:2 Digital, RP 201

Vertical Interval, Location, RP 164

Vertical Interval, Longitudinal Relationship, RP 159

Vertical Interval, Switching Point, RP 168

Tape Recording Formats

SMPTE Documents Relating to Tape Recording Formats (*Courtesy of SMPTE.*)

	B	C	D-1	D-2	D-3	D-5	D-6	D-7 (1)	D-9 (2)	E (3)	G (4)	H (5)	L (6)	M-2
Basic system parameters														
525/60	15M	18M	EG10	EG20	264M	279M	277M	306M	316M	21M			RP144	RP158
625/50					265M	279M	277M	306M	316M					
Record dimensions	16M	19M	224M	245M	264/5M	279M	277M			21M		32M	229M	249M
Characteristics														
Video signals	RP84	RP86								RP87		32M	230M	251M
Audio and control signals	17M	20M	RP155	RP155	264/5M	279M	278M			RP87		32M	230M	251M
Data and control record			227M	247M	264/5M	279M	278M							
Tracking control record	RP83	RP85				279M	277M							
Pulse code modulation audio														252M
Time and control recording	RP93		228M	248M	264/5M	279M	278M						230M	251M
Audio sector time code, equipment type information			RP181											
Nomenclature		18M	EG21	EG21						21M		32M		
Index of documents				EG22										
Stereo channels	RP142	RP142								RP142	RP142	RP142	RP142	
Relative polarity	RP148	RP148	RP148	RP148						RP148	RP148	RP148	RP148	
Tape	25M	25M	225M	246M	264/5M	279M	277M				35M	32M	238M	250M
Reels	24M	24M												
Cassettes			226M	226M	263M	263M	226M	307M	317M	22M	35M	32M	238M	250M
Small										31M				
Bar code labeling			RP156	RP156										
Dropout specifications	RP121	RP121												
Reference tape and recorder														
System parameters	29M													
Tape	26M	26M												

Notes:
1 DVCPRO, 2 Digital S, 3 U-matic, 4 Beta, 5 VHS, 6 Betacam

15.5 SCTE Standards

The following documents relating to digital television have been adopted by the Society of Cable Telecommunications Engineers.

DVS 011 Cable and Satellite Extensions to ATSC System Information Standards

DVS 018 ATSC Digital Television Standard

DVS 019 Digital Audio Compression (AC-3) Standard

DVS 020 Guide to the Use of the ATSC Digital Television Standard

DVS 022 System Information for Digital Information

DVS 026 Subtitling Method for Broadcast Cable

DVS 031 Digital Video Transmission Standard for Cable Television

DVS 033 SCTE Video Compression Formats

DVS 043 QPSK Tools for Forward and Reverse Data Paths

DVS 046 Specifications for Digital Transmission Technologies

DVS 047 National Renewable Security Standard (NRSS)

DVS 051 Methods for Asynchronous Data Services Transport

DVS 053 VBI Extension for ATSC Digital Television Standards

DVS 055 EIA Interim Standard IS-679 B of National Renewable Security Standard (NRSS)

DVS 057 Usage of A/53 Picture (Video) User Data

DVS 061 SCTE Cable and Satellite Extensions to ATSC System Information (SI)

DVS 064 National Renewable Security Standards (NRSS) Part A and Part B

DVS 068 ITU-T Recommendation J.83–"Digital Multi-Programme Systems for Television, Sound and Data Services for Cable Distribution"

DVS 071 Digital Multi Programming Distribution by Satellite

DVS 076 Digital Cable Ready Receivers: Practical Considerations

DVS 077 Requirements for Splicing of MPEG-2 Transport Streams

DVS 080 Digital Broadband Delivery Phase 1.0 Functions

DVS 082 Broadband File System Product Description Release 1.2s

DVS 084 Common Interface for DVB Decoder Interface

DVS 085 DAVIC 1.2 Basic Security

DVS 092 Draft System Requirements for ATV Channel Navigation

DVS 093 Draft Digital Video Service Multiplex and Transport System

DVS 097 Program and System Information Protocol for Terrestrial Broadcast and Cable

DVS 098 IPSI Protocol for Terrestrial Broadcast with examples

DVS 110 Response to SCTE DVS CFI Cable Headend and Distribution Systems

DVS 111 Digital Headend and Distribution CFI Phase 1.0 System Description

DVS 114 SMPTE Splice point for MPEG-2 Transport

DVS 131 Draft Point-of-Development (POD) Proposal on Open Cable

DVS 132 Methods for Isochronous Data Services Transport

DVS 147 Revision to DVS 022 (Standard System Information)

DVS 151 Operational Impact on Currently Deployed Systems

DVS 153 ITU-T Draft Recommendation J.94

DVS 154 Digital Program Insertion Control API

DVS 157 SCTE Proposed Standard Methods for Carriage of Closed Captions and Non-Real Time Sampled Video

DVS 159 Optional Extensions for Carriage of NTSC VBI Data in Cable Digital Transport Streams

DVS 161 ATSC Data Broadcast Specification

DVS 165 DTV Interface Specification

DVS 166 Draft Corrigendum for Program ad System Information Protocol for Terrestrial Broadcast and Cable (A/65

DVS 167 Digital Broadband Delivery System: Out of Band Transport–Quadrature Phase Shifting Key (QPSK) Out of Band Channels Based on DAVIC, first draft

DVS 168 Emergency Alert System Interface to Digital Cable Network

DVS-178 Cable System Out-of-Band Specifications (GI)

DVS-179 MPAA Response to DVS CFI

DVS-181 Service Protocol

DVS-190 Standard for Conveying VBI Data in MPEG-2 Transport Streams

DVS-191 Draft Standard API to Splicing Equipment for MPEG-2 Transport StreamsDVS-192 Splicer Application Programmer's Interface Definition Overview

DVS-194 Home Digital Network Interface Specification Proposal with Copy Protection

DVS-195 Home Digital Network Interface Specification Proposal without Copy Protection

DVS-208 Proposed Standard: Emergency Alert Message for Cable

DVS-209 DPI System Physical Diagram

DVS-211 Service Information for Digital Television

DVS-213 Copy Protection for POD Module Interface

15.6   References Cited in this Book

ACATS: "ATV System Description: ATV-into-NTSC Co-channel Test #016," Grand Alliance Advisory Committee on Advanced Television, p. I-14-10, Dec. 7, 1994.

ATSC: "Comments of The Advanced Television Systems Committee, MM Docket No. 00-39," ATSC, Washington, D.C., May, 2000.

"Advanced Television Enhancement Forum Specification," Draft, Version 1.1r26 updated 2/2/99, ATVEF, Portland, Ore., 1999.

Aitken, S., D. Carr, G. Clayworth, R. Heppinstall, and A. Wheelhouse: "A New, Higher Power, IOT System for Analogue and Digital UHF Television Transmission," *Proceedings of the 1997 NAB Broadcast Engineering Conference*, National Association of Broadcasters, Washington, D.C., p. 531, 1997.

Ajluni, Cheryl: "SID Conference Highlights: The Road to Invention," *Electronic Design*, Penton Publishing, Hasbrouck Heights, N.J., pp. 39–46, May 12, 1997.

"Alignment of NTSC Color Picture Monitors," SMPTE Recommended Practice RP 167-1995, SMPTE, White Plains, N.Y., 1995.

Alkin, Oktay: "Digital Coding Schemes," *The Electronics Handbook*, J. C. Whitaker (ed.), CRC Press, Boca Raton, Fla., pp. 1252–1258, 1996.

ANSI Standard T1.801.03-1996, "Digital Transport of One-Way Video Signals: Parameters for Objective Performance Assessment," ANSI, Washington, D.C., 1996.

Appelquist, P.: "The HD-Divine Project: A Scandinavian Terrestrial HDTV System," *1993 NAB HDTV World Conference Proceedings*, National Association of Broadcasters, Washington, D.C., p. 118, 1993.

Arragon, J. P., J. Chatel, J. Raven, and R. Story: "Instrumentation for a Compatible HD-MAC Coding System Using DATV," *Conference Record*, International Broadcasting Conference, Brighton, Institution of Electrical Engineers, London, 1989.

Arvind, R., et al.: "Images and Video Coding Standards," *AT&T Technical J.*, p. 86, 1993.

ATSC: "Amendment No. 1 to ATSC Standard: Program and System Information Protocol for Terrestrial Broadcast and Cable," Doc. A/67, ATSC, Washington, D.C, December 17, 1999.

ATSC: "Conditional Access System for Terrestrial Broadcast," Advanced Television Systems Committee, Washington, D.C., Doc. A/70, July 1999.

ATSC: "Data Broadcast Standard," Advanced Television Systems Committee, Washington, D.C., Doc. A/90, July 26, 2000.

ATSC: "Digital Audio Compression Standard (AC-3)," Advanced Television Systems Committee, Washington, D.C., Doc. A/52, Dec. 20, 1995.

ATSC: "Digital Audio Compression Standard (AC-3), Annex A: AC-3 Elementary Streams in an MPEG-2 Multiplex," Advanced Television Systems Committee, Washington, D.C., Doc. A/52, Dec. 20, 1995.

ATSC: "Digital Television Standard," Advanced Television Systems Committee, Washington, D.C., Doc. A/53, 1995.

ATSC: "Guide to the Use of the ATSC Digital Television Standard," Advanced Television Systems Committee, Washington, D.C., Doc. A/54, Oct. 4, 1995.

ATSC: "Implementation of Data Broadcasting in a DTV Station," Advanced Television Systems Committee, Washington, D.C., Doc. IS/151, November 1999.

ATSC: "Implementation Subcommittee Report on Findings," Draft Version 0.4, ATSC, Washington, D.C., September 21, 1998.

ATSC: "Modulation And Coding Requirements For Digital TV (DTV) Applications Over Satellite," Doc. A/80, ATSC, Washington, D.C., July, 17, 1999.

ATSC: "Program and System Information Protocol for Terrestrial Broadcast and Cable," Advanced Television Systems Committee, Washington, D.C., Doc. A/65, February 1998.

ATSC: "Technical Corrigendum No.1 to ATSC Standard: Program and System Information Protocol for Terrestrial Broadcast and Cable," Doc. A/66, ATSC, Washington, D.C., December 17, 1999.

ATSC: "Transmission Measurement and Compliance for Digital Television," Advanced Television Systems Committee, Washington, D.C., Doc. A/64, Nov. 17, 1997.

ATSC, "Transmission Measurement and Compliance for Digital Television," Advanced Television Systems Committee, Washington, D.C., Doc. A/64-Rev. A, May 30, 2000.

ATTC: "Digital HDTV Grand Alliance System Record of Test Results," Advanced Television Test Center, Alexandria, Virginia, October 1995.

"ATV System Recommendation," 1993 NAB HDTV World Conference Proceedings, National Association of Broadcasters, Washington, D.C., pp. 253-258, 1993.

Baldwin, M. W., Jr.: "The Subjective Sharpness of Simulated Television Images," *Proc. IRE*, vol. 28, p. 458, July 1940.

Ballard, Randall C.: U.S. Patent 2,152,234, filed July 19,1932.

Barkow, W. H., and J. Gross: "The RCA Large Screen 110° Precision In-Line System," ST-5015, *RCA Entertainment*, Lancaster, Pa.

Baron, Stanley: "International Standards for Digital Terrestrial Television Broadcast: How the ITU Achieved a Single-Decoder World," *Proceedings of the 1997 BEC*, National Association of Broadcasters, Washington, D.C., pp. 150–161, 1997.

Basile, C.: "An HDTV MAC Format for FM Environments," *International Conference on Consumer Electronics*, IEEE, New York, June 1989.

Battison, John: "Making History," *Broadcast Engineering*, Intertec Publishing, Overland Park, Kan., June 1986.

Bauer, Richard W.: "Film for Television," *NAB Engineering Handbook*, 9th ed., J. C. Whitaker (ed.), National Association of Broadcasters, Washington, D.C., 1998.

Belton, J.: "The Development of the CinemaScope by Twentieth Century Fox," *SMPTE Journal*, vol. 97, pp. 711–720, September 1988.

Bender, Walter, and Alan Blount: "The Role of Colorimetry and Context in Color Displays," *Human Vision, Visual Processing, and Digital Display III*, Bernice E. Rogowitz (ed.), *Proc. SPIE* 1666, SPIE, Bellingham, Wash., pp. 343–348, 1992.

Bendov, Oded: "Coverage Contour Optimization of HDTV and NTSC Antennas," *Proceedings of the 1996 NAB Broadcast Engineering Conference*, National Association of Broadcasters, Washington, D.C., p. 69, 1996.

Bennett, Christopher: "Three MPEG Myths," *Proceedings of the 1996 NAB Broadcast Engineering Conference*, National Association of Broadcasters, Washington, D.C., pp. 129–136, 1996.

Benson, K. B., and D. G. Fink: "Gamma and Its Disguises," *SMPTE Journal*, SMPTE, White Plains, N.Y., vol. 102, no. 12, pp. 1009–1108, December 1993.

Benson, K. B., and D. G. Fink: *HDTV: Advanced Television for the 1990s*, McGraw-Hill, New York, 1990.

Benson, K. B., and Jerry C. Whitaker (eds.): *Television and Audio Handbook for Engineers and Technicians*, McGraw-Hill, New York, 1989.

Benson, K. B., and Jerry C. Whitaker (eds.): *Television Engineering Handbook*, rev. ed., McGraw-Hill, New York, p. 12.20, 1991.

Benson, K. Blair (ed.): *Television Engineering Handbook*, McGraw-Hill, New York, pp. 21.57–21.72, 1986.

Bishop, Donald M.: "Practical Applications of Picture Quality Measurements," *Proceedings of Digital Television '98*, Intertec Publishing, Overland Park, Kan., 1998.

Bleha, W. P.: "Image Light Amplifier (ILA) Technology for Large-Screen Projection," *SMPTE Journal*, SMPTE, White Plains, N.Y., pp. 710–717, October 1997.

Bohannon, W. K.: "Which Technology Will Make Digital TV a Household Name," *Electronic Design*, Penton Publishing, Hasbrouck Heights, N.J., pp. 48–51, Oct. 23, 1997.

Bonomi, Mauro: "The Art and Science of Digital Video Compression," *NAB Broadcast Engineering Conference Proceedings*, National Association of Broadcasters, Washington, D.C., pp. 7–14, 1995.

Boston, J., and J. Kraenzel: "SDI Headroom and the Digital Cliff," *Broadcast Engineering*, Intertec Publishing, Overland Park, Kan., pg. 80, February 1997.

Brandenburg, K., and Gerhard Stoll: "ISO-MPEG-1 Audio: A Generic Standard for Coding of High Quality Digital Audio," *92nd AES Convention Proceedings*, Audio Engineering Society, New York, N.Y., 1992, revised 1994.

"Broadening the Applications of Zone Plate Generators," Application Note 20W7056, Tektronix, Beaverton, Oreg., 1992.

Cadzow, James A.: *Discrete Time Systems*, Prentice-Hall, Inc., Englewood Cliffs, N. J., 1973.

CCIR Document PLEN/69-E (Rev. 1), "Minutes of the Third Plenary Meeting," pp. 2-4, May 29, 1990.

CCIR Report 122-4, 1990.

CCIR Report 801-3, "The Present State of High-Definition Television," p. 37, 46, June 1989.

Chini, A., Y. Wu, M. El-Tanany, and S. Mahmoud: "Hardware Nonlinearities in Digital TV Broadcasting Using OFDM Modulation," *IEEE Trans. Broadcasting*, IEEE, New York, N.Y., vol. 44, no. 1, March 1998.

Chairman, ITU-R Task Group 11/3, "Report of the Second Meeting of ITU-R Task Group 11/3, Geneva, Oct. 13-19, 1993," p. 40, Jan. 5, 1994.

Chalamala, B., and B. Gnade: "FED Up with Fat Tubes," *IEEE Spectrum*, IEEE, New York, N.Y., pp. 42–51, April 1998.

Chambers, J. A., S. Tantaratana, and B. W. Bomar: "Digital Filters," *The Electronics Handbook*, Jerry C. Whitaker (ed.), CRC Press, Boca Raton, Fla., pp. 749–772, 1996.

Ciciora, Walter, et. al.: "A Tutorial on Ghost Canceling in Television Systems," *IEEE Transactions on Consumer Electronics*, IEEE, New York, vol. CE-25, no. 1, pp. 9–44, February 1979.

Chini, A., Y. Wu, M. El-Tanany, and S. Mahmoud: "An OFDM-based Digital ATV Terrestrial Broadcasting System with a Filtered Decision Feedback Channel Estimator," *IEEE Trans. Broadcasting*, IEEE, New York, N.Y., vol. 44, no. 1, pp. 2–11, March 1998.

"Conclusions of the Extraordinary Meeting of Study Group 11 on High-Definition Television," Doc. 11/410-E, International Radio Consultative Committee (CCIR), Geneva, Switzerland, June 1989.

Course notes, "DTV Express," PBS/Harris, Alexandria, Va., 1998.

Craig, Donald: "Network Architectures: What does Isochronous Mean?," *IBC Daily News*, IBC, Amsterdam, September 1999.

Cugnini, Aldo G.: "MPEG-2 Bitstream Splicing," *Proceedings of the Digital Television '97 Conference*, Intertec Publishing, Overland Park, Kan., December 1997.

Dare, Peter: "The Future of Networking," *Broadcast Engineering*, Intertec Publishing, Overland Park, Kan., p. 36, April 1996.

DeMarsh, LeRoy E.: "Displays and Colorimetry for Future Television," *SMPTE Journal*, SMPTE, White Plains, N.Y., pp. 666–672, October 1994.

DeWith, P. H. N.: "Motion-Adaptive Intraframe Transform Coding of Video Signals," *Philips J. Res.*, vol. 44, pp. 345–364, 1989.

Digital HDTV Grand Alliance System: *Record of Test Results*, October 1995.

DTV Express Training Manual on Terrestrial DTV Broadcasting, Harris Corporation, Quincy, Ill., September 1998.

Dubois, E., and W. F. Schreiber: "Improvements to NTSC by Multidimensional Filtering," *SMPTE J.*, SMPTE, White Plains, N.Y., vol. 97, pp. 446–463, July 1988.

Ehmer, R. H.: "Masking of Tones Vs. Noise Bands," *J. Acoust. Soc. Am.*, vol. 31, pp. 1115–1256, September 1959.

Einolf, Charles: "DTV Receiver Performance in the Real World," *2000 Broadcast Engineering Conference Proceedings*, National Association of Broadcasters, Washington, D.C., pp. 478–482, 2000.

Epstein, Steve: "Editing MPEG Bitstreams," *Broadcast Engineering*, Intertec Publishing, Overland Park, Kan., pp. 37–42, October 1997.

Ericksen, Dane E.: "A Review of IOT Performance," *Broadcast Engineering*, Intertec Publishing, Overland Park, Kan., p. 36, July 1996.

ETS-300-421, "Digital Broadcasting Systems for Television, Sound, and Data Services; Framing Structure, Channel Coding and Modulation for 11–12 GHz Satellite Services," DVB Project technical publication.

ETS-300-429, "Digital Broadcasting Systems for Television, Sound, and Data Services; Framing Structure, Channel Coding and Modulation for Cable Systems," DVB Project technical publication.

ETS-300-468, "Digital Broadcasting Systems for Television, Sound, and Data Services; Specification for Service Information (SI) in Digital Video Broadcasting (DVB) Systems," DVB Project technical publication.

ETS-300-472, "Digital Broadcasting Systems for Television, Sound, and Data Services; Specification for Carrying ITU-R System B Teletext in Digital Video Broadcasting (DVB) Bitstreams," DVB Project technical publication.

ETS-300-473, "Digital Broadcasting Systems for Television, Sound, and Data Services; Satellite Master Antenna Television (SMATV) Distribution Systems," DVB Project technical publication.

ETS 300 744: "Digital Broadcasting Systems for Television, Sound and Data Services: Framing Structure, Channel Coding and Modulation for Digital Terrestrial Television," ETS 300 744, 1997.

Eureka 95 HDTV Directorate, Progressing Towards the Real Dimension, Eureka 95 Communications Committee, Eindhoven, Netherlands, June 1991.

European Telecommunications Standards Institute: "Digital Video Broadcasting; Framing Structure, Channel Coding and Modulation for Digital Terrestrial Television (DVB-T)", March 1997.

"Eye Diagrams and Sampling Oscilloscopes," *Hewlett-Packard Journal*, Hewlett-Packard, Palo Alto, Calif., pp. 8–9, December 1996.

Favreau, M., S. Soca, J. Bajon, and M. Cattoen: "Adaptive Contrast Corrector Using Real-Time Histogram Modification," *SMPTE J.*, SMPTE, White Plains, N.Y., vol. 93, pp. 488–491, May 1984.

FCC/ACATS SSWP2-1306: *Grand Alliance System Test Procedures*, May 18, 1994.

Federal Communications Commission: Notice of Proposed Rule Making 00-83, FCC, Washington, D.C., March 8, 2000.

Federal Communications Commission Report and Order: "Closed Captioning Requirements for Digital Television Receivers," FCC, Washington, D.C., ET Docket 99-254 and MM Docket 95-176, adopted July 21, 2000.

Fibush, David K., *A Guide to Digital Television Systems and Measurement*, Tektronix, Beaverton, Ore., 1994.

Fibush, David K., *A Guide to Digital Television Systems and Measurements*, Tektronix, Beaverton, Ore., 1997.

Fibush, David K.: "Error Detection in Serial Digital Systems," *NAB Broadcast Engineering Conference Proceedings*, National Association of Broadcasters, Washington, D.C., pp. 346-354, 1993.

Fibush, David K.: "Picture Quality Measurements for Digital Television*," Proceedings of the Digital Television '97 Summit*, Intertec Publishing, Overland Park, Kan., December 1997.

Fibush, David K.: "Practical Application of Objective Picture Quality Measurements," *Proceedings IBC '97*, IEE, pp. 123–135, Sept. 16, 1997.

Fibush, David K.: "Testing MPEG-Compressed Signals," *Broadcast Engineering*, Overland Park, Kan., pp. 76–86, February 1996.

Finck, Konrad: "Digital Video Signal Analysis for Real-World Problems," in *NAB 1994 Broadcast Engineering Conference Proceedings*, National Association of Broadcasters, Washington, D.C., pg. 257, 1994.

Fink, D. G, et. al.: "The Future of High Definition Television," *SMPTE Journal*, vol. 9, SMPTE, White Plains, N.Y., February/March 1980.

Fink, D. G.: "Perspectives on Television: The Role Played by the Two NTSCs in Preparing Television Service for the American Public," *Proc. IEEE*, vol. 64, pp. 1322–1331, September 1976.

Fink, D. G.: *Color Television Standards*, McGraw-Hill, New York, pp. 108-111, 1955.

Florence, J., and L. Yoder: "Display System Architectures for Digital Micromirror Device (DMD) Based Projectors," *Proc. SPIE*, SPIE, Bellingham, Wash., vol. 2650, Projection Displays II, pp. 193–208, 1996.

Freed, Ken: "Video Compression," *Broadcast Engineering*, Overland Park, Kan., pp. 46–77, January 1997.

Fritz, Victor J.: "Full-Color Liquid Crystal Light Valve Projector for Shipboard Use," Large Screen Projection Displays II, William P. Bleha, Jr. (ed.), *Proc. SPIE 1255*, SPIE, Bellingham, Wash., pp. 59–68, 1990.

Fujio, T., J. Ishida, T. Komoto, and T. Nishizawa: "High-Definition Television Systems—Signal Standards and Transmission," *SMPTE Journal*, vol. 89, pp. 579–584, August 1980.

Gaggioni, H., M. Ueda, F. Saga, K. Tomita, and N. Kobayashi, "The Development of a High-Definition Serial Digital Interface," Sony Technical Paper, Sony Broadcast Group, San Jose, Calif., 1998.

Gallo and Hancock: *Networking Explained*, Digital Press, pp. 191–235, 1999.

Garrod, S., and R. Borns: *Digital Logic: Analysis, Application, and Design*, Saunders College Publishing, Philadelphia, p. 919, 1991.

Garrod, Susan A. R.: "D/A and A/D Converters," *The Electronics Handbook*, Jerry C. Whitaker (ed.), CRC Press, Boca Raton, Fla., pp. 723–730, 1996.

Gilder, George: "IBM-TV?," *Forbes*, Feb. 20, 1989.

Gilge, M.: "Region-Oriented Transform Coding in Picture Communication," VDI-Verlag, Advancement Report, Series 10, 1990.

Gilmore, A. S.: *Microwave Tubes*, Artech House, Dedham, Mass., pp. 196–200, 1986.

Glenn, W. E., C. E. Holton, G. J. Dixon, and P. J. Bos: "High-Efficiency Light Valve Projectors and High-Efficiency Laser Light Sources," *SMPTE Journal*, SMPTE, White Plains, N.Y., pp. 210–216, April 1997.

Gloeggler, Peter: "Video Pickup Devices and Systems," in *NAB Engineering Handbook*, 9th Ed., Jerry C. Whitaker (ed.), National Association of Broadcasters, Washington, D.C., 1999.

Gove, R. J., V. Markandey, S. Marshall, D. Doherty, G. Sextro, and M. DuVal: "High-Definition Display System Based on Digital Micromirror Device," *International Workshop on HDTV* (HDTV '94), International Institute for Communications, Turin, Italy (October 1994).

Haines, Steve: "Serial Digital: The Networking Solution?," in *NAB 1994 Broadcast Engineering Conference Proceedings*, National Association of Broadcasters, Washington, D.C., pg. 270, 1994.

Hamada, T., S. Miyaji, and S. Matsumoto: "Picture Quality Assessment System by Three-Layered Bottom-Up Noise Weighting Considering Human Visual Perception," *SMPTE Journal*, SMPTE, White Plains, N.Y., pp. 20–26, January 1999.

Hamasaki, Kimio: "How to Handle Sound with Large Screen," *Proceedings of the ITS*, International Television Symposium, Montreux, Switzerland, 1991.

"HD-MAC Bandwidth Reduction Coding Principles," Draft Report AZ-11, International Radio Consultative Committee (CCIR), Geneva, Switzerland, January 1989.

Heymans, Dennis: "Channel Combining in an NTSC/ATV Environment," *Proceedings of the 1996 NAB Broadcast Engineering Conference*, National Association of Broadcasters, Washington, D.C., p. 165, 1996.

Hockenbrock, Richard: "New Technology Advances for Brighter Color CRT Displays," *Display System Optics II*, Harry M. Assenheim (ed.), Proc. SPIE 1117, SPIE, Bellingham, Wash., pp. 219-226, 1989.

Hoffman, Gary A., "IEEE 1394: The A/V Digital Interface of Choice," 1394 Technology Association Technical Brief, 1394 Technology Association, Santa Clara, Calif., 1999.

Holman, Tomlinson: "Psychoacoustics of Multi-Channel Sound Systems for Television," *Proceedings of HDTV World*, National Association of Broadcasters, Washington, D.C., 1992.

Holman, Tomlinson: "The Impact of Multi-Channel Sound on Conversion to ATV," *Perspectives on Wide Screen and HDTV Production*, National Association of Broadcasters, Washington, D.C., 1995.

Hopkins, R.: "Advanced Television Systems," *IEEE Transactions on Consumer Electronics*, vol. 34, pp. 1-15, February 1988.

Hornbeck, Larry J.: "Digital Light Processing for High-Brightness, High-Resolution Applications," *Projection Displays III*, Electronic Imaging '97 Conference, SPIE, Bellingham, Wash., February 1997.

Hubel, David H.: *Eye, Brain and Vision*, Scientific American Library, New York, 1988.

Hulick, Timothy P.: "60 kW Diacrode UHF TV Transmitter Design, Performance and Field Report," *Proceedings of the 1996 NAB Broadcast Engineering Conference*, National Association of Broadcasters, Washington, D.C., p. 442, 1996.

Hunold, Kenneth: "4:2:2 or 4:1:1—What are the Differences?," *Broadcast Engineering*, Intertec Publishing, Overland Park, Kan., pp. 62–74, October 1997.

Husak, Walt, et. al.: "On-channel Repeater for Digital Television Implementation and Field Testing," *Proceedings 1999 Broadcast Engineering Conference*, NAB'99, Las Vegas, National Association of Broadcasters, Washington, D.C., pp. 397–403, April 1999.

Hutter, Rudolph G. E.: "The Deflection of Electron Beams," *Advances in Image Pickup and Display*, B. Kazen (ed.), vol. 1, Academic Press, New York, pp. 212-215, 1974.

IEEE Standard Dictionary of Electrical and Electronics Terms, ANSI/IEEE Standard 100-1984, Institute of Electrical and Electronics Engineers, New York, 1984.

"IEEE Standard Specifications for the Implementation of 8 × 8 Inverse Discrete Cosine Transform," std. 1180-1990, Dec. 6, 1990.

Internal Report of the Ad Hoc Group on Colorimetry and Transfer Characteristic (under the SMPTE Working Group on High-Definition Electronic Production, N15.04/ 05), SMPTE, White Plains, N.Y., October 24, 1987.

Isnardi, M. A.: "Exploring and Exploiting Subchannels in the NTSC Spectrum," *SMPTE J.*, SMPTE, White Plains, N.Y., vol. 97, pp. 526–532, July 1988.

Isnardi, M. A.: "Multidimensional Interpretation of NTSC Encoding and Decoding," *IEEE Transactions on Consumer Electronics*, vol. 34, pp. 179–193, February 1988.

Isnardi, M., and T. Smith: "MPEG Tutorial," *Proceedings of the Advanced Television Summit*, Intertec Publishing, Overland Park, Kan., 1996.

ITU-R Document TG11/3-2, "Outline of Work for Task Group 11/3, Digital Terrestrial Television Broadcasting," June 30, 1992.

ITU-R Recommendation BS-775, "Multi-channel Stereophonic Sound System with and without Accompanying Picture."

ITU Radiocommunication Study Groups, Special Rapporteur's Group: "Guide for the Use of Digital Television Terrestrial Broadcasting Systems Based on Performance Comparison of ATSC 8-VSB and DVB-T COFDM Transmission Systems," International Telecommunications Union, Geneva, Document 11A/65-E, May 11, 1999.

ITU-R SG 11, Special Rapporteur—Region 1, "Protection Ratios and Reference Receivers for DTTB Frequency Planning," ITU-R Doc. 11C/46-E, March 18, 1999.

Jacklin, Martin: "The Multimedia Home Platform: On the Critical Path to Convergence," DVB Project technical publication, 1998.

Joint ERC/EBU: "Planning and Introduction of Terrestrial Digital Television (DVB-T) in Europe," Izmir, Dec. 1997.

Jones, Ken: "The Television LAN," *Proceedings of the 1995 NAB Engineering Conference*, National Association of Broadcasters, Washington, D.C., p. 168, April 1995.

Judd, D. B.: "The 1931 C.I.E. Standard Observer and Coordinate System for Colorimetry," *J. Opt. Soc. Am.*, vol. 23, pp. 359–374, 1933.

Kell, R. D., A. V. Bedford, and M. Trainer: "Scanning Sequence and Repetition of Television Images," *Proc. IRE*, vol. 24, p. 559, April 1936.

Keller, Thomas B.: "Proposal for Advanced HDTV Audio," *1991 HDTV World Conference Proceedings*, National Association of Broadcasters, Washington, D.C., April 1991.

Kelly, K. L.: "Color Designation of Lights," *J. Opt. Soc. Am.*, vol. 33, pp. 627–632, 1943.

Kelly, R. D., A. V. Bedbord, and M. Trainer: "Scanning Sequence and Repetition of Television Images," *Proceedings of the IRE*, vol. 24, April 1936.

Kikuchi, M., et al.: "A New Coolant-Sealed CRT for Projection Color TV," *IEEE Trans.*, vol. CE-27, IEEE, New York, no. 3, pp. 478-485, August 1981.

Krivocheev, Mark I., and S. N. Baron: "The First Twenty Years of HDTV: 1972–1992," *SMPTE Journal*, SMPTE, White Plains, N.Y., p. 913, October 1993.

Lagadec, Roger, Ph.D.: "Audio for Television: Digital Sound in Production and Transmission," *Proceedings of the ITS*, International Television Symposium, Montreux, Switzerland, 1991.

Lakhani, Gopal: "Video Compression Techniques and Standards," *The Electronics Handbook*, J. C. Whitaker (ed.), CRC Press, Boca Raton, Fla., pp. 1273–1282, 1996.

Ledoux, B, P. Bouchard, S. Laflèche, Y. Wu, and B. Caron: "Performance of 8-VSB Digital Television Receivers," *Proceedings of the International Broadcasting Convention*, IBC, Amsterdam, September 2000.

Lee, E. A., and D. G. Messerschmitt: *Digital Communications*, 2nd ed., Kluwer, Norell, Mass., 1994.

Legault, Alain, and Janet Matey: "Interconnectivity in the DTV Era—The Emergence of SDTI," *Proceedings of Digital Television '98*, Intertec Publishing, Overland Park, Kan., 1998.

Libin, Louis: "The 8-VSB Modulation System," *Broadcast Engineering*, Intertec Publishing, Overland Park, Kan., p. 22, December 1995.

Ligeti, A., and J. Zander: "Minimal Cost Coverage Planning for Single Frequency Networks", *IEEE Trans. Broadcasting*, IEEE, New York, N.Y., vol. 45, no. 1, March 1999.

Lincoln, Donald: "TV in the Bay Area as Viewed from KPIX," *Broadcast Engineering*, Intertec Publishing, Overland Park, Kan., May 1979.

Lucas, K.: "B-MAC: A Transmission Standard for Pay DBS," *SMPTE Journal*, SMPTE, White Plains, N.Y., November 1984.

Luetteke, Georg: "The DVB Multimedia Home Platform," DVB Project technical publication, 1998.

Luxenberg, H., and R. Kuehn: *Display Systems Engineering*, McGraw-Hill, New York, 1968.

Lyman, Stephen, "A Multichannel Audio Infrastructure Based on Dolby E Coding," *Proceedings of the NAB Broadcast Engineering Conference*, National Association of Broadcasters, Washington, D.C., 1999.

MacAdam, D.L.: "Visual Sensitivities to Color Differences in Daylight," *J. Opt. Soc. Am.*, vol. 32, pp. 247–274, 1942.

Markandey, Vishal, and Robert J. Gove: "Digital Display Systems Based on the Digital Micromirror Device," *SMPTE Journal*, SMPTE, White Plains, N.Y., pp. 680–685, October 1995.

Mathias, H.: "Gamma and Dynamic Range Needs for an HDTV Electronic Cinematography System," *SMPTE J.*, SMPTE, White Plains, N.Y., vol. 96, pp. 840–845, September 1987.

Mazur, Jeff: "Video Special Effects Systems," *NAB Engineering Handbook*, 9th ed., J. C. Whitaker (ed.), National Association of Broadcasters, Washington, D.C., to be published 1998.

McCroskey, Donald: "Setting Standards for the Future," *Broadcast Engineering*, Intertec Publishing, Overland Park, Kan., May 1989.

McKechnie, S.: Philips Laboratories (NA) report, 1981, unpublished.

Mertz, P.: "Television and the Scanning Process," *Proc. IRE*, vol. 29, pp. 529–537, October 1941.

Mignone, V., and A. Morello: "CD3-OFDM: A Novel Demodulation Scheme for Fixed and Mobile Receivers," *IEEE Trans. Commu.*, IEEE, New York, N.Y., vol. 44, pp. 1144–1151, September 1996.

Miller, Howard: "Options in Advanced Television Broadcasting in North America," *Proceedings of the ITS*, International Television Symposium, Montreux, Switzerland, 1991.

Mitsuhashi, Tetsuo: "HDTV and Large Screen Display," *Large-Screen Projection Displays II*, William P. Bleha, Jr. (ed.), Proc. SPIE 1255, SPIE, Bellingham, Wash., pp. 2-12, 1990.

Morello, Alberto, et. al.: "Performance Assessment of a DVB-T Television System," *Proceedings of the International Television Symposium 1997*, Montreux, Switzerland, June 1997.

Moore, B. C. J., and B. R. Glasberg: "Formulae Describing Frequency Selectivity as a Function of Frequency and Level, and Their Use in Calculating Excitation Patterns," *Hearing Research*, vol. 28, pp. 209–225, 1987.

Morrell, A., et al.: *Color Picture Tubes*, Academic Press, New York, pp. 91-98, 1974.

Muschallik, C.: "Improving an OFDM Reception Using an Adaptive Nyquist Windowing," *IEEE Trans. on Consumer Electronics,* no. 03, 1996

Muschallik, C.: "Influence of RF Oscillators on an OFDM Signal*", IEEE Trans. Consumer Electronic*s, IEEE, New York, N.Y., vol. 41, no. 3, pp. 592–603, August 1995.

NAB: "An Introduction to DTV Data Broadcasting," *NAB TV TechCheck*, National Association of Broadcasters, Washington, D.C., August 2, 1999.

NAB: "Canadians Perform Indoor DTV Reception Tests," *NAB TV TechCheck*, National Association of Broadcasters, Washington, D.C., October 9, 2000.

NAB: "CEA Establishes Definitions for Digital Television Products," *NAB TV TechCheck*, National Association of Broadcasters, Washington, D.C., September 1, 2000.

NAB: "Digital TV Closed Captions," *NAB TV TechCheck*, National Association of Broadcasters, Washington, D.C., August 7, 2000.

NAB: "FCC Adopts Rules for Labeling of DTV Receivers," *NAB TV TechCheck*, National Association of Broadcasters, Washington, D.C., September 25, 2000.

NAB: "Navigation of DTV Data Broadcasting Services," *NAB TV TechCheck*, National Association of Broadcasters, Washington, D.C., November 1, 1999.

NAB: "New Digital Receiver Technology Announced," *NAB TV TechCheck*, National Association of Broadcasters, Washington, D.C., August 30, 1999.

NAB: "Pay TV Services for DTV," *NAB TV TechCheck*, National Association of Broadcasters, Washington, D.C., October 4, 1999.

NAB TV TechCheck, National Association of Broadcasters, Washington, D.C., February 1, 1999.

NAB TV TechCheck, National Association of Broadcasters, Washington, D.C., January 4, 1999.

NAB TV TechCheck, National Association of Broadcasters, Washington, D.C., June 7, 1999.

Nelson, Lee J.: "Video Compression," *Broadcast Engineering*, Intertec Publishing, Overland Park, Kan., pp. 42–46, October 1995.

Netravali, A. N., and B. G. Haskell: *Digital Pictures, Representation, and Compression*, Plenum Press, 1988.

"Networking and Internet Broadcasting," Omneon Video Networks, Campbell, Calif, 1999.

"Networking and Production," Omneon Video Networks, Campbell, Calif., 1999.

Nyquist, H.: "Certain Factors Affecting Telegraph Speed," *Bell System Tech. J.*, vol. 3, pp. 324–346, March 1924.

Ogomo, M., T. Yamada, K. Ando, and E. Yamazaki: "Considerations on Required Property for HDTV Displays," *Proc. of HDTV 90 Colloquium*, vols. 1, 2B, 1990.

Ostroff, Nat S.: "A Unique Solution to the Design of an ATV Transmitter," *Proceedings of the 1996 NAB Broadcast Engineering Conference*, National Association of Broadcasters, Washington, D.C., p. 144, 1996.

Owen, Peter: "Gigabit Ethernet for Broadcast and Beyond," *Proceedings of DTV99*, Intertec Publishing, Overland Park, Kan., November 1999.

Pank, Bob (ed.): *The Digital Fact Book*, 9th ed., Quantel Ltd, Newbury, England, 1998.

Parks, T. W. and J. H. McClellan: "A Program for the Design of Linear Phase Infinite Impulse Response Filters," *IEEE Trans. Audio Electroacoustics*, AU-20(3), pp. 195–199, 1972.

Pease, Richard W.: "An Overview of Technology for Large Wall Screen Projection Using Lasers as a Light Source," *Large Screen Projection Displays II*, William P. Bleha, Jr. (ed.), Proc. SPIE 1255, SPIE, Bellingham, Wash., pp. 93-103, 1990.

Peterson, R., R. Ziemer, and D. Borth: *Introduction to Spread Spectrum Communications*, Prentice-Hall, Englewood Cliffs, N. J., 1995.

Phillips, Thomas E., et. al.: "1280 × 1024 Video Rate Laser-Addressed Liquid Crystal Light Valve Color Projection Display," *Optical Engineering*, Society of Photo-Optical Instrumentation Engineers, vol. 31, no. 11, pp. 2300–2312, November 1992.

Pickford, N.: "Laboratory Testing of DTTB Modulation Systems," Laboratory Report 98/01, Australia Department of Communications and Arts, June 1998.

Piercy, John: "ATM Networked Video: Moving From Leased-Lines to Packetized Transmission," *Proceedings of the Transition to Digital Conference*, Intertec Publishing, Overland Park, Kan., 1996.

Pitts, K. and N. Hurst: "How Much Do People Prefer Widescreen (16 x 9) to Standard NTSC (4 x 3)?," *IEEE Transactions on Consumer Electronics*, IEEE, New York, August 1989.

Plonka, Robert J., "Bandpass Filter and Linearization Requirements for the New FCC Mask," *Proceedings of the NAB Broadcast Engineering Conference*, National Association of Broadcasters, Washington, D.C., pp. 387–396, 1999.

Plonka, Robert J.: "Can ATV Coverage Be Improved With Circular, Elliptical, or Vertical Polarized Antennas?" *Proceedings of the 1996 NAB Broadcast Engineering Conference*, National Association of Broadcasters, Washington, D.C., p. 155, 1996.

Plonka, Robert J.: "Planning Your Digital Television Transmission System," *Proceedings of the 1997 NAB Broadcast Engineering Conference*, National Association of Broadcasters, Washington, D.C., p. 89, 1997.

Pointer, R. M.: "The Gamut of Real Surface Colors, *Color Res. App.*, vol. 5, 1945.

Pollet, T., M. van Bladel, and M. Moeneclaey: "BER Sensitivity of OFDM Systems to Carrier Frequency Offset and Wiener Phase Noise," *IEEE Trans. on Communications*, vol. 43, 1995.

Porter, J. C.: *Proc. Roy. Soc.*, London, vol. 86, p. 945, 1912.

Powers, Kerns H.: "High Definition Production Standards—Interlace or Progressive?," *Implementing HDTV: Television and Film Applications*, S. Baron (ed.), SMPTE, White Plains, N.Y., pp. 27–35, 1996.

Priest, D. H., and M. B. Shrader: "The Klystrode—An Unusual Transmitting Tube with Potential for UHF-TV," *Proc. IEEE*, vol. 70, no. 11, pp. 1318–1325, November 1982.

Quinn, S. F., and C. A. Siocos: "PLUGE Method of Adjusting Picture Monitors in Television Studios—A Technical Note," *SMPTE Journal*, SMPTE, White Plains, N.Y., vol. 76, p. 925, September 1967.

Qureshi, Shahid U. H.: "Adaptive Equalization," *Proceedings of the IEEE*, IEEE, New York, vol. 73, no. 9, pp. 1349–1387, September 1985.

Raven, J. G.: "High-Definition MAC: The Compatible Route to HDTV," *IEEE Transactions on Consumer Electronics*, vol. 34, pp. 61–63, IEEE, New York, February 1988.

"Receiver Planning Factors Applicable to All ATV Systems," Final Report of PS/WP3, Advanced Television Systems Committee, Washington, D.C., Dec. 1, 1994.

Reed-Nickerson, Linc: "Understanding and Testing the 8-VSB Signal," *Broadcast Engineering*, Intertec Publishing, Overland Park, Kan., pp. 62–69, November 1997.

Reimers, U. H.: "The European Perspective for Digital Terrestrial Television, Part 1: Conclusions of the Working Group on Digital Terrestrial Television Broadcasting," *1993 NAB HDTV World Conference Proceedings*, National Association of Broadcasters, Washington, D.C., p. 117, 1993.

Rhodes, Charles W.: "Terrestrial High-Definition Television," *The Electronics Handbook*, Jerry C. Whitaker (ed.), CRC Press, Boca Raton, Fla., pp. 1599–1610, 1996.

Robertson, P., and S. Kaiser: "Analysis of the Effects of Phase-Noise in Orthogonal Frequency Division Multiplex (OFDM) Systems," ICC 1995, pp. 1652–1657, 1995.

Robin, Michael: "Digital Resolution," *Broadcast Engineering*, Intertec Publishing, Overland Park, Kan., pp. 44–48, April 1998.

Robinder, R., D. Bates, P. Green: "A High Brightness Shadow Mask Color CRT for Cockpit Displays," *SID Digest*, Society for Information Display, vol. 14, pp. 72-73, 1983.

Sabatier, J., D. Pommier, and M. Mathiue: "The D2-MAC-Packet System for All Transmission Channels," *SMPTE Journal*, SMPTE, White Plains, N.Y., November 1984.

Salter, J. E.: "Noise in a DVB-T System," *BBC R&D Technical Note*, R&D 0873(98), February 1998.

Sari, H., G. Karam, and I. Jeanclaude: "Channel Equalization and Carrier Synchronization in OFDM Systems," *IEEE Proc.* 6th. Tirrenia Workshop on Digital Communications, Tirrenia, Italy, pp. 191–202, September 1993.

Sariowan, H.: "Comparative Studies Of Data Broadcasting, *International Broadcasting Convention Proceedings*, IBC, London, England, pp. 115–119, 1999.

Schachlbauer, Horst: "European Perspective on Advanced Television for Terrestrial Broadcasting," *Proceedings of the ITS*, International Television Symposium, Montreux, Switzerland, 1991.

Schow, Edison: "A Review of Television Systems and the Systems for Recording Television," *Sound and Video Contractor*, Intertec Publishing, Overland Park, Kan., May 1989.

Schreiber, W. F., A. B. Lippman, A. N. Netravali, E. H. Adelson, and D. H. Steelin: "Channel-Compatible 6-MHz HDTV Distribution Systems," *SMPTE Journal*, SMPTE, White Plains, N.Y., vol. 98, pp. 5-13, January 1989.

Schreiber, W. F., and A. B. Lippman: "Single-Channel HDTV Systems—Compatible and Noncompatible," Report ATRP-T-82, Advanced Television Research Program, MIT Media Laboratory, Cambridge, Mass., March 1988.

Sextro, G., I. Ballew, and J. lwai: "High-Definition Projection System Using DMD Display Technology," *SID 95 Digest*, Society for Information Display, pp. 70–73, 1995.

Sklar, B.: *Digital Communications: Fundamentals and Applications*, Prentice-Hall, Englewood Cliffs, N. J., 1988.

Slamin, Brendan: "Sound for High Definition Television," *Proceedings of the ITS*, International Television Symposium, Montreux, Switzerland, 1991.

Smith, Paul D.: "New Channel Combining Devices for DTV," *Proceedings of the 1997 NAB Broadcast Engineering Conference*, National Association of Broadcasters, Washington, D.C., p. 218, 1996.

Smith, Terry: "MPEG-2 Systems: A Tutorial Overview," Transition to Digital Conference, *Broadcast Engineering,* Overland Park, Kan., Nov. 21, 1996.

SMPTE and EBU, "Task Force for Harmonized Standards for the Exchange of Program Material as Bitstreams," *SMPTE Journal*, SMPTE, White Plains, N.Y., pp. 605–815, July 1998.

SMPTE Engineering Guideline: EG 1-1990, *Alignment Color Bar Test Signal for Television Picture Monitors*, SMPTE, White Plains, N.Y., 1990.

SMPTE Engineering Guideline: EG 36 (Proposed), *Transformations Between Television Component Color Signals*, SMPTE, White Plains, N.Y., 1999.

SMPTE Recommended Practice: RP 145-1994, *SMPTE C Color Monitor Colorimetry*, SMPTE, White Plains, N.Y., 1994.

SMPTE Recommended Practice: RP 165-1994, *Error Detection Checkwords and Status Flags for Use in Bit-Serial Digital Interfaces for Television*, SMPTE, White Plains, N.Y., 1994.

SMPTE Recommended Practice: RP 166-1995, *Critical Viewing Conditions for Evaluation of Color Television Pictures*, SMPTE, White Plains, N.Y., 1995.

SMPTE Recommended Practice: RP 192-1996, *Jitter Measurement Procedures in Bit-Serial Digital Interfaces*, SMPTE, White Plains, N.Y., 1996.

SMPTE Recommended Practice: RP 199-1999, *Mapping of Pictures in Wide-Screen (16:9) Scanning Structure to Retain Original Aspect Ratio of the Work*, SMPTE, White Plains, N.Y., 1999.

SMPTE Recommended Practice: RP 202 (Proposed), *Video Alignment for MPEG-2 Coding*, SMPTE, White Plains, N.Y., 1999.

SMPTE Recommended Practice RP 203 (Proposed): *Real Time Opportunistic Data Flow Control in an MPEG-2 Transport Emission Multiplex*, SMPTE, White Plains, N.Y., 1999.

SMPTE Recommended Practice RP 204 (Proposed): *SDTI-CP MPEG Decoder Templates*, SMPTE, White Plains, N.Y., 1999.

SMPTE Recommended Practice RP 211 (Proposed): *Implementation of 24P, 25P, and 30P Segmented Frames for 1920 × 1080 Production Format*, SMPTE, White Plains, N.Y., 2000.

SMPTE Standard: SMPTE 170M-1999, *Composite Analog Video Signal NTSC for Studio Applications*, SMPTE, White Plains, N.Y., 1999.

SMPTE Standard: SMPTE 240M-1988, *Signal Parameters—1125/60 High-Definition Production Systems* SMPTE, White Plains, N.Y., 1988.

SMPTE Standard: SMPTE 240M-1995, *Signal Parameters—1125-Line High-Definition Production Systems,"* SMPTE, White Plains, N.Y., 1995.

SMPTE Standard: SMPTE 259M-1997, *Serial Digital Interface for 10-bit 4:2:2 Components and $4F_{sc}$ NTSC Composite Digital Signals,* SMPTE, White Plains, N.Y., 1997.

SMPTE Standard: SMPTE 260M-1992, *Digital Representation and Bit-Parallel Interface— 1125/60 High-Definition Production System*, SMPTE, White Plains, N.Y., 1992.

SMPTE Standard: SMPTE 260M-1999, *Digital Representation and Bit-Parallel Interface— 1125/60 High-Definition Production System*, SMPTE, White Plains, N.Y., 1999.

SMPTE Standard: SMPTE 274M-1998, *1920 × 1080 Scanning and Analog and Parallel Digital Interfaces for Multiple-Picture Rates,"* SMPTE, White Plains, N.Y., 1998.

SMPTE Standard: SMPTE 291M-1998, *Ancillary Data Packet and Space Formatting*, SMPTE, White Plains, N.Y., 1998.

SMPTE Standard: SMPTE 292M-1998, *Bit-Serial Digital Interface for High-Definition Television Systems*, SMPTE, White Plains, N.Y., 1998.

SMPTE Standard: SMPTE 293M-1996, *720 × 483 Active Line at 59.94 Hz Progressive Scan Production—Digital Representation*, SMPTE, White Plains, N.Y., 1996.

SMPTE Standard: SMPTE 294M-1997, *720 × 483 Active Line at 59.94-Hz Progressive Scan Production Bit-Serial Interfaces*, SMPTE, White Plains, N.Y., 1997.

SMPTE Standard: SMPTE 295M-1997, *1920 × 1080 50 Hz Scanning and Interfaces*, SMPTE, White Plains, N.Y., 1997.

SMPTE Standard: SMPTE 296M-1997, *1280 × 720 Scanning, Analog and Digital Representation and Analog Interface*, SMPTE, White Plains, N.Y., 1997.

SMPTE Standard: SMPTE 297M-1997, *Serial Digital Fiber Transmission System for ANSI/ SMPTE 259M Signals*, SMPTE White Plains, N.Y., 1997.

SMPTE Standard: SMPTE 302M-1998, *Mapping of AES3 Data into MPEG-2 Transport Stream*, SMPTE, White Plains, N.Y., 1998.

SMPTE Standard: SMPTE 303M (Proposed), *Color Reference Pattern*, SMPTE, White Plains, N.Y., 1999.

SMPTE Standard: SMPTE 304M-1998, *Broadcast Cameras: Hybrid Electrical and Fiber-Optic Connector*, SMPTE, White Plains, N.Y., 1998.

SMPTE Standard: SMPTE 305M-1998, *Serial Data Transport Interface*, SMPTE, White Plains, N.Y., 1998.

SMPTE Standard: SMPTE 308M-1998, *MPEG-2 4:2:2 Profile at High Level*, SMPTE, White Plains, N.Y., 1998.

SMPTE Standard: SMPTE 311M-1998, *Hybrid Electrical and Fiber-Optic Camera Cable*, SMPTE, White Plains, N.Y., 1998.

SMPTE Standard: SMPTE 312M-1999, *Splice Points for MPEG-2 Transport Streams*, SMPTE, White Plains, N.Y., 1999.

SMPTE Standard: SMPTE 315M-1999, *Camera Positioning Information Conveyed by Ancillary Data Packets*, SMPTE, White Plains, N.Y., 1999.

SMPTE Standard: SMPTE 319M-2000, *Transporting MPEG-2 Recoding Information Through 4:2:2 Component Digital Interfaces*, SMPTE, White Plains, N.Y., 2000.

SMPTE Standard: SMPTE 320M-1999, *Channel Assignments and Levels on Multichannel Audio Media*, SMPTE, White Plains, N.Y., 1999.

SMPTE Standard: SMPTE 324M (Proposed), *12-Channel Serial Interface for Digital Audio and Auxiliary Data*, SMPTE, White Plains, N.Y., 1999.

SMPTE Standard: SMPTE 325M-1999, *Opportunistic Data Broadcast Flow Control*, SMPTE, White Plains, N.Y., 1999.

SMPTE Standard: SMPTE 326M, *SDTI Content Package Format* (*SDTI-CP*), SMPTE, White Plains, N.Y., 2000.

SMPTE Standard: SMPTE 327M-2000, *MPEG-2 Video Recoding Data Set*, SMPTE, White Plains, N.Y., 2000.

SMPTE Standard: SMPTE 328M-2000, *MPEG-2 Video Elementary Stream Editing Information*, SMPTE, White Plains, N.Y., 2000.

SMPTE Standard: SMPTE 329M-2000, *MPEG-2 Video Recoding Data Set—Compressed Stream Format*, SMPTE, White Plains, N.Y., 2000.

SMPTE Standard: SMPTE 331M, *Element and Metadata Definitions for the SDTI-CP*, SMPTE, White Plains, N.Y., 2000.

SMPTE Standard: SMPTE 332M, *Encapsulation of Data Packet Streams over SDTI (SDTI-PF)*, SMPTE, White Plains, N.Y., 2000.

SMPTE Standard: SMPTE 333M, *DTV Closed-Caption Server to Encoder Interface*, SMPTE, White Plains, N.Y., 1999.

SMPTE Standard: SMPTE 334M (Proposed), *Vertical Ancillary Data Mapping for Bit-Serial Interface*, SMPTE, White Plains, N.Y., 2000.

SMPTE Standard: SMPTE 344M (Proposed), *540 Mb/s Serial Digital Interface,* SMPTE, White Plains, N.Y., 2000.

SMPTE Standard: SMPTE 346M (Proposed), *Signals and Generic Data over High-Definition Interfaces,*" SMPTE, White Plains, N.Y., 2000.

SMPTE Standard: SMPTE 348M (Proposed), *High Data-Rate Serial Data Transport Interface (HD-SDTI)*, SMPTE, White Plains, N.Y., 2000.

SMPTE Standard: SMPTE 351M (Proposed), *Transporting MPEG-2 Recoding Information through High-Definition Digital Interfaces*, SMPTE, White Plains, N.Y., 2000.

SMPTE Standard: SMPTE 353M (Proposed), *Transport of MPEG-2 Recoding Information as Ancillary Data Packets*, SMPTE, White Plains, N.Y., 2000.

SMPTE: "System Overview—Advanced System Control Architecture, S22.02, Revision 2.0," S22.02 Advanced System Control Architectures Working Group, SMPTE, White Plains, N.Y., March 27, 2000.

Smyth, Stephen: "Digital Audio Data Compression," *Broadcast Engineering*, Intertec Publishing, Overland Park, Kan., February 1992.

Solari, Steve. J.: *Digital Video and Audio Compression*, McGraw-Hill, New York, 1997.

Sgrignoli, Gary: "Preliminary DTV Field Test Results and Their Effects on VSB Receiver Design

Stallings, William: *ISDN and Broadband ISDN*, 2nd Ed., MacMillan, New York.

Standards and definitions committee, Society for Information Display.

Story, R.: "HDTV Motion-Adaptive Bandwidth Reduction Using DATV," BBC Research Department Report, RD 1986/5.

Story, R.: "Motion Compensated DATV Bandwidth Compression for HDTV," International Radio Consultative Committee (CCIR), Geneva, Switzerland, January 1989.

Stremler, Ferrel G.: "Introduction to Communications Systems," *Addison-Wesley Series in Electrical Engineering*, Addison-Wesley, New York, December 1982.

Stott, J. H.: "Explaining Some of the Magic of COFDM", *Proceedings of the International TV Symposium 1997*, Montreux, Switzerland, June 1997.

Stott, J. H: "The Effect of Phase Noise in COFDM", *EBU Technical Review*, Summer 1998.

Suitable Sound Systems to Accompany High-Definition and Enhanced Television Systems: Report 1072. Recommendations and Reports to the CCIR, 1986. Broadcast Service—Sound. International Telecommunications Union, Geneva, 1986.

Symons, R., M. Boyle, J. Cipolla, H. Schult, and R. True: "The Constant Efficiency Amplifier—A Progress Report," *Proceedings of the NAB Broadcast Engineering Conference*, National Association of Broadcasters, Washington, D.C., pp. 77–84, 1998.

Symons, Robert S.: "The Constant Efficiency Amplifier," *Proceedings of the NAB Broadcast Engineering Conference*, National Association of Broadcasters, Washington, D.C., pp. 523–530, 1997.

Takeuchi, Kazuhiko, et. al.: "A 750-TV-Line Resolution Projector Using 1.5 Megapixel a-Si TFT LC Modules," *SID 91 Digest*, Society for Information Display, pp. 415–418, 1991.

Tanaka, H., and L. J. Thorpe: "The Sony PCL HDVS Production Facility," *SMPTE J.*, SMPTE, White Plains, N.Y., vol. 100, pp. 404–415, June 1991.

Tannas, Lawrence E., Jr.: *Flat Panel Displays and CRTs*, Van Nostrand Reinhold, New York, pg. 18, 1985.

Tardy, Michel-Pierre: "The Experience of High-Power UHF Tetrodes," *Proceedings of the 1993 NAB Broadcast Engineering Conference*, National Association of Broadcasters, Washington, D.C., p. 261, 1993.

Taylor, P.: "Broadcast Quality and Compression," *Broadcast Engineering*, Intertec Publishing, Overland Park, Kan., p. 46, October 1995.

"Technology Brief—Networking and Storage Strategies," Omneon Video Networks, Campbell, Calif., 1999.

Teichmann, Wolfgang: "HD-MAC Transmission on Cable," *Proceedings of the ITS*, International Television Symposium, Montreux, Switzerland, 1991.

Terry, K. B., and S. B. Lyman: "Dolby E—A New Audio Distribution Format for Digital Broadcast Applications," *International Broadcasting Convention Proceedings*, IBC, London, England, pp. 204–209, September 1999.

Thorpe, Laurence J., E. Tamura, and T. Iwasaki: "New Advances in CCD Imaging," *SMPTE J.*, SMPTE, White Plains, N.Y., vol. 97, pp. 378–387, May 1988.

Thorpe, Laurence J., et. al.: "New High Resolution CCD Imager," *NAB Engineering Conference Proceedings*, National Association of Broadcasters, Washington, D.C., pp. 334–345, 1988.

Thorpe, Laurence J.: "A New Global HDTV Program Origination Standard: Implications for Production and Technology," *Proceedings of DTV99*, Intertec Publishing, Overland Park, Kan., 1999.

Thorpe, Laurence J.: "Applying High-Definition Television," *Television Engineering Handbook*, rev. ed., K. B. Benson and Jerry C. Whitaker (eds.), McGraw-Hill, New York, p. 23.4, 1991.

Thorpe, Laurence J.: "HDTV and Film—Digitization and Extended Dynamic Range," 133rd SMPTE Technical Conference, Paper no. 133-100, SMPTE, White Plains, N.Y., October 1991.

Thorpe, Laurence J.: "High-Definition Production Systems," *Television Engineering Handbook*, rev. ed., K. B. Benson and Jerry C. Whitaker (eds.), McGraw-Hill, New York, p. 23.4, 1991.

Thorpe, Laurence J.: "The Great Debate: Interlaced Versus Progressive Scanning," *Proceedings of the Digital Television '97 Conference*, Intertec Publishing, Overland Park, Kan., December 1997.

Thorpe, Laurence J.: "The HDTV Camcorder and the March to Marketplace Reality," *SMPTE Journal*, SMPTE, White Plains, N.Y., pp. 164–177, March 1998.

Todd, C., et. al.: "AC-3: Flexible Perceptual Coding for Audio Transmission and Storage," AES 96th Convention, Preprint 3796, Audio Engineering Society, New York, February 1994.

Tong, Hua-Sou: "HDTV Display—A CRT Approach," *Display Technologies*, Shu-Hsia Chen and Shin-Tson Wu (eds.), Proc. SPIE 1815, SPIE, Bellingham, Wash., pp.2-8, 1992.

Torick, Emil L.: "HDTV: High Definition Video—Low Definition Audio?," *1991 HDTV World Conference Proceedings*, National Association of Broadcasters, Washington, D.C., April 1991.

Turow, Dan: "SDTI and the Evolution of Studio Interconnect," *International Broadcasting Convention Proceedings*, IBC, Amsterdam, September 1998.

Uchida, Tadayuki, Yasuaki Nishida, and Yukihiro Nishida: "Picture Quality in Cascaded Video-Compression Systems for Digital Broadcasting," *SMPTE Journal*, SMPTE, White Plains, N.Y., pp. 27–38, January 1999.

Ungerboeck, G.: "Trellis-Coded Modulation with Redundant Signal Sets," parts I and II, *IEEE Comm. Mag.*, vol. 25 (Feb.), pp. 5-11 and 12-21, 1987.

Ungerboeck, Gottfried: "Fractional Tap-Spacing Equalizer and Consequences for Clock Recovery in Data Modems," *IEEE Transactions on Communications*, IEEE, New York, vol. COM-24, no. 8, pp. 856–864, August 1976.

van Klinken, N., and W. Renirie: "Receiving DVB: Technical Challenges," *Proceedings of the International Broadcasting Convention*, IBC, Amsterdam, September 2000.

Verona, Robert: "Comparison of CRT Display Measurement Techniques," *Helmet-Mounted Displays III*, Thomas M. Lippert, (ed.), Proc. SPIE 1695, SPIE, Bellingham, Wash., pp. 117–127, 1992.

Venkat, Giri, "Understanding ATSC Datacasting—A Driver for Digital Television," *Proceedings of the NAB Broadcast Engineering Conference*, National Association of Broadcasters, Washington, D.C., pp. 113–116, 1999.

Vreeswijk, F., F. Fonsalas, T. Trew, C. Carey-Smith, and M. Haghiri: "HD-MAC Coding for High-Definition Television Signals," *International Radio Consultative Committee (CCIR)*, Geneva, Switzerland, January 1989.

Ward, Christopher, C. Pecota, X. Lee and G. Hughes: "Seamless Splicing for MPEG-2 Transport Stream Video Servers," *Proceedings, 33rd SMPTE Advanced Motion Imaging Conference*, SMPTE, White Plains, N.Y., 2000.

Whitaker, Jerry C., and H. Winard (eds.): *The Information Age Dictionary*, Intertec Publishing/ Bellcore, Overland Park, Kan., 1992.

Whitaker, Jerry C.: "Microwave Power Tubes," *Power Vacuum Tubes Handbook*, Van Nostrand Reinhold, New York, p. 259, 1994.

Whitaker, Jerry C.: "RF Combiner and Diplexer Systems," *Power Vacuum Tubes Handbook*, Van Nostrand Reinhold, New York, p. 368, 1994.

Whitaker, Jerry C.: "Solid State RF Devices," *Radio Frequency Transmission Systems: Design and Operation*, McGraw-Hill, New York, p. 101, 1990.

Whitaker, Jerry C.: *Electronic Displays: Technology, Design and Application*, McGraw-Hill, New York, 1994.

Wilkinson, J. H., H. Sakamoto, and P. Horne: "SDDI as a Video Data Network Solution," *International Broadcasting Convention Proceedings,* IBC, Amsterdam, September 1997.

Wu, Tsong-Ho: "Network Switching Concepts," *The Electronics Handbook*, Jerry C. Whitaker (ed.), CRC Press, Boca Raton, Fla., p. 1513, 1996.

Wu, Y., et. al.: "Canadian Digital Terrestrial Television System Technical Parameters," *IEEE Transactions on Broadcasting*, IEEE, New York, N.Y., to be published in 1999.

Wu, Y., and M. El-Tanany: "OFDM System Performance Under Phase Noise Distortion and Frequency Selective Channels," *Proceedings of Int'l Workshop of HDTV 199*7, Montreux Switzerland, June 10–11, 1997.

Wu, Y., M. Guillet, B. Ledoux, and B. Caron: "Results of Laboratory and Field Tests of a COFDM Modem for ATV Transmission over 6 MHz Channels," *SMPTE Journal*, SMPTE, White Plains, N.Y., vol. 107, February 1998.

Wugofski, T. W.: "A Presentation Engine for Broadcast Digital Television*,*" *International Broadcasting Convention Proceedings*, IBC, London, England, pp. 451–456, 1999.

Wylie, Fred: "Audio Compression Techniques," *The Electronics Handbook*, Jerry C. Whitaker (ed.), CRC Press, Boca Raton, Fla., pp. 1260–1272, 1996.

Wylie, Fred: "Audio Compression Technologies," in *NAB Engineering Handbook*, 9th ed., Jerry C. Whitaker (ed.), National Association of Broadcasters, Washington, D.C., 1998.

Yoshizawa, T., S. Hatakeyama, A. Ueno, M. Tsukahara, K. Matsumi, K. Hirota, "A 61-in High-Definition Projection TV for the ATSC Standard," *SID 99 Digest*, Society for Information Display, San Jose, Calif., pp 112–115, 1999.

Younse, J. M.: "Projection Display Systems Based on the Digital Micromirror Device (DMD)," SPIE Conference on Microelectronic Structures and Microelectromechanical Devices for Optical Processing and Multimedia Applications, Austin, Tex., *SPIE Proceedings*, SPIE, Bellingham, Wash., vol. 2641, pp. 64–75, Oct. 24, 1995.

Ziemer, R., and W. Tranter: *Principles of Communications: Systems, Modulation, and Noise*, 4th ed., Wiley, New York, 1995.

Ziemer, Rodger E.: "Digital Modulation," *The Electronics Handbook*, J. C. Whitaker (ed.), CRC Press, Boca Raton, Fla., pp. 1213–1236, 1996.

Zwicker, E.: "Subdivision of the Audible Frequency Range Into Critical Bands (Frequen-zgruppen)," *J. Acoust. Soc. of Am.*, vol. 33, p. 248, February 1961.

Index of Tables and Figures

Index of Tables

Table 1.1 Channel Designations for VHF- and UHF-TV Stations in the United States 9

Table 1.2 Basic Characteristics of Video Signals Based on an 1125/60 System 19

Table 1.3 Basic Characteristics of Video Signals Based on a 1250/50 System 19

Table 2.1 Common Computer System Display Formats 55

Table 2.2 Basic Characteristics Of IBM-Type Computer Graphics Displays 56

Table 3.1 Spatial Characteristics of HDTV (NHK) and Conventional Television Systems, Based on Luminance or Equivalent *Y* Signal) 84

Table 3.2 Typical Day and Night Illumination Levels 88

Table 3.3 Typical Artificial Illumination Levels 89

Table 3.4 Bandwidth Characteristics of HDTV (NHK) and Conventional Television Systems 96

Table 3.5 Picture Element Parameters in Various Television Systems 97

Table 3.6 Luma Equations for Scanning Standards 101

Table 3.7 System Colorimetry for SMPTE 274M 103

Table 3.8 Typical ATSC Compression Ratios 106

Table 3.9 Typical ATSC Channel Capacities as a Function of Compression Rate 106

Table 4.1 Binary Values of Amplitude Levels for 8-Bit Words 111

Table 4.2 Example of the Differential Encoding Process 127

Table 4.3 SMPTE 272M Mode Definitions 142

Table 4.4 Fibre Channel General Performance Specifications 149

Table 4.5 Basic Specifications of Ethernet Performance 149

Table 5.1 Participants in Early MPEG Proceedings 164

Table 5.2 Layers of the MPEG-2 Video Bit-Stream Syntax 167

Table 5.3 Common MPEG Profiles and Levels in Simplified Form 169

Table 5.4 Recommended MPEG-2 Coding Ranges for Various Video Formats 192

Table 5.5 Operational Parameters of Subband APCM Algorithm 199

Table 5.6 Target Applications for ITU-R Rec. BS.1116 PEAQ 211

Table 6.1 Quantizing S/N Associated with Various Quantization Levels 231

Table 6.2 Encoding Parameter Values for SMPTE-260M 234

Table 6.3 Scanned Raster Characteristics of SMPTE-240M-1995 239

Table 6.4 Analog Video Signal Levels 240

Table 6.5 Scanning System Parameters for SMPTE-274M 241

Table 6.6 Scanning Systems Parameters for SMPTE-295M-1997 243

Table 6.7 Scanning System Parameters for SMPTE 296M-1997 244

Table 6.8 Scanning System Parameters for SMPTE 293M-1996 244

Table 6.9 SMPTE 294M-1997 Interface Parameters 245

Table 6.10 SDI Reference Source Format Parameters 248

Table 6.11 Source Format Parameters for SMPTE 292M 249

Table 6.12 Summary of SD Video Formats Referenced in SMPTE 346M 258

Table 6.13 Different Terminology Used in the Video and Film Industries for Comparable Imaging parameters 274

Table 7.1 Theoretical S/N as a Function of the Number of Sampling Bits 280

Table 7.2 Language Code Table for AC-3 310

Table 7.3 Typical Bit Rates for Various Services 311

Table 8.1 Standardized Video Input Formats 317

Table 8.2 ATSC DTV Compression Format Constraints 317

Table 8.3 Picture Formats Under the DTV Standard 319

Table 8.4 Partial Trellis Coding Interleaving Sequence 332

Table 8.5 ATM Cell Header Fields 352

Table 9.1 Parameters for VSB Transmission Modes 374

Table 9.2 System Interfaces Specified in A/80 391

Table 9.3 Input Data Stream Structures 391

Table 10.1 Receiver Planning Factors Used by the FCC 414

Table 10.2 DTV Interference Criteria 414

Table 10.3 Target DTV Receiver Parameters 435

Table 10.4 Susceptibility to Random Noise Test Results 436

Table 10.5 Susceptibility to Random Noise in the Presence of Static Multipath 436

Table 10.6 Color CRT Diagonal Dimension vs. Weight 463

Table 10.7 Comparative Resolution of Shadow-Mask Designs 464

Table 10.8 Performance Levels of Video and Theater Displays 476

Table 10.9 Screen Gain Characteristics for Various Materials 478

Table 10.10 Minimum Projector Specifications for Consumer and Theater Displays 509

Table 10.11 Appearance of Scan Line Irregularities in a Projection Display 510

Table 11.1 General System Comparison of ATSC 8-VSB and DVB-T COFDM 545

Table 11.2 DTV Protection Ratios for Frequency Planning 551

Table 12.1 Video Quality Definitions 609

Table 12.2 Error Frequency and Bit Error Rates 619

Table 12.3 Error Rate as a Function of S/N for Composite Serial Digital 621

Table 12.4 Error Rate as a Function of Cable Length Using 8281 Coax for Composite Serial Digital 621

Table 13.1 Additions to SMPTE 274M Table 1 as Specified in RP211 656

SMPTE Documents Relating to Tape Recording Formats 678

Index of Figures

Figure 1.1 The interlaced-scanning pattern (raster) of the television image. 6

Figure 1.2 The principal components of the NTSC color system. 11

Figure 1.3 Sync pulse widths for the NTSC color system. 11

Figure 1.4 Major elements of a mechanically scanned laser-beam telecine. 15

Figure 1.5 Block diagram of a mechanically scanned laser-beam video-to-film recorder. 16

Figure 2.1 Applications for high-definition imaging in business and industrial facilities. 52

Figure 2.2 An illustration of the differences in the scene capture capabilities of conventional video and HDTV. 59

Figure 2.3 Viewing angle as a function of screen distance for HDTV. 60

Figure 2.4 Viewing angle as a function of screen distance for conventional video systems. 60

Figure 2.5 Comparison of the aspect ratios of television and motion pictures. 62

Figure 2.6 The effects of listener positioning on center image shift. 64

Figure 2.7 Optimum system speaker placement for HDTV viewing. 66

Figure 2.8 Conventional telecine conversion process. 69

Figure 2.9 Standards conversion between 35 mm 24-frame film, HDTV, and conventional broadcast systems. 70

Figure 2.10 The primary components of a telecine. 72

Figure 2.11 Block diagram of cathode-ray tube flying-spot scanner. 73

Figure 2.12 Block diagram of CCD line array telecine. 74

Figure 2.13 Modulation transfer function (MTF) of typical films and scanned laser beams. 76

Figure 2.14 Block diagram of the basic tape-to-film electronic-beam recording system. 76

Figure 2.15 Relative timing of components in the EBR step printing process. 77

Figure 3.1 The geometry of the field of view occupied by a television image. 83

Figure 3.2 Visual sharpness as a function of the relative values of horizontal and vertical resolution. The liminal unit given in the figure is the least perceptible difference. 85

Figure 3.3 Simplified block diagram of an end-to-end video imaging system. 87

Figure 3.4 The C.I.E. Chromaticity Diagram, showing the color sensations and the triangular gamuts of the C.I.E. primaries. 90

Figure 3.5 The gamuts of NTSC and common CRT phosphors (P22). 91

Figure 3.6 CRT light intensity output as a function of video signal input. 97

Figure 3.7 Linearity characteristics of five common phosphors. P22 is commonly used in tricolor television displays. The red, green, and blue phosphors are denoted by P22R, P22G, and P22B, respectively. 98

Figure 3.8 DCT 2-dimensional spectrum sorting. 105

Figure 4.1 Basic elements of an analog-to-digital converter. 111

Figure 4.2 Video waveform quantized into 8-bit words. 112

Figure 4.3 Successive approximation A/D converter block diagram. 113

Figure 4.4 Block diagram of a flash A/D converter. 114

Figure 4.5 The Nyquist volume of the luminance signal: (*a*) progressive scanning, (*b*) interlaced scanning. 117

Figure 4.6 Sampled intervals in the vertical-picture and field-time axes for progressive scanning. 118

Figure 4.7 Sampled intervals in the vertical-picture and field-time axes for interlaced scanning. 119

Figure 4.8 Sampling rates corresponding to the intervals in Figure 4.6 and Figure 4.7: (*a*) progressive scanning, (*b*) interlaced scanning. 120

Figure 4.9 The Nyquist volume of the NTSC signal space: (*a*) location of the chrominance subcarrier, (*b*) cross section showing field- and line-frequency components. 121

Figure 4.10 Nyquist volumes of the chrominance space positioned within the luminance volume. 122

Figure 4.11 Subcarrier locations on the faces of the Nyquist volume: (*a*) relative positions, (*b*) cross-sectional view. 122

Figure 4.12 One-dimensional filtering between luminance and chrominance, as provided by a typical NTSC receiver. 125

Figure 4.13 Receiver systems for noncoherent detection of binary signals: (*a*) ASK, (*b*) FSK. 126

Figure 4.14 The Huffman coding algorithm. 129

Figure 4.15 The magnitude and phase response of the simple moving average filter with $M = 7$. 134

Figure 4.16 The impulse and magnitude response of an optimal 40th-order half-band FIR filter. 135

Figure 4.17 Direct form realizations of IIR filters: (*a*) direct form I, (*b*) direct form II, (*c*) transposed direct form I, (*d*) transposed direct form II.137

Figure 4.18 The basic SDI bitstream. 142

Figure 4.19 SMPTE 305M system block diagram. 144

Figure 4.20 The application of IEEE 1394 in a large scale internet broadcast environment. 146

Figure 4.21 The application of IEEE 1394 in a nonlinear postproduction environment. 147

Figure 5.1 Block diagram of a sequential DCT codec: (*a*) encoder, (*b*) decoder. 156

Figure 5.2 A simplified search of a best-matched block. 157

Figure 5.3 The mechanics of motion-compensated prediction. Shown are the pictures for a planar 4×4 DCT. Element C_{11} is located at row 1, column 1; element C_{23} is located at row 2, column 3. Note that picture C_{11} values are constant, referred to as dc coefficients. The changing values shown in picture C_{23} are known as ac coefficients. 158

Figure 5.4 Overall block diagram of a DPCM system: (*a*) encoder, (*b*) decoder. 159

Figure 5.5 Block diagram of a DCT-based image-compression system. Note how the 8×8 source image is processed through a forward-DCT (FDCT) encoder and related systems to the inverse-DCT (IDCT) decoder and reconstructed into an 8×8 image. 162

Figure 5.6 A typical MPEG-1 codec: (*a*) encoder, (*b*) decoder. 165

Figure 5.7 A simplified search of a best-matched block. 166

Figure 5.8 Illustration of I-frames, P-frames, and B-frames. 167

Figure 5.9 Sequence of video frames for the MPEG-2/ATSC DTV system. 173

Figure 5.10 Simplified encoder prediction loop. 176

Figure 5.11 Scanning of coefficient blocks: (*a*) alternate scanning of coefficients, (*b*) zigzag scanning of coefficients. 185

Figure 5.12 ATSC DTV video system decoder functional block diagram. 187

Figure 5.13 Generalized frequency response of the human ear. Note how the PCM process captures signals that the ear cannot distinguish. 194

Figure 5.14 Example of the masking effect of a high-level sound. 195

Figure 5.15 Variation of subband gain as a function of the number of subbands. 197

Figure 5.16 apt-X100 audio coding system: (*a*) encoder block diagram, (*b*) decoder block diagram. 202

Figure 5.17 ISO/MPEG-1 Layer 2 system: (*a*) encoder block diagram, (*b*) decoder block diagram. 204

Figure 5.18 ISO/MPEG-1 Layer 2 data frame structure. 206

Figure 5.19 Functional block diagram of the MPEG-2 AAC coding system. 207

Figure 5.20 Basic frame structure of the Dolby E coding system. 209

Figure 5.21 Overall coding scheme of Dolby E. 210

Figure 6.1 Color gamut curve for SMPTE 240M. 221

Figure 6.2 Colorimetry process under the initial (step one) guidelines of SMPTE 240M. 222

Figure 6.3 Colorimetry process under the secondary (step two) guidelines of SMPTE 240M. 223

Figure 6.4 Overall pixel count for the digital representation of the 1125/60 HDTV production standard. 232

Figure 6.5 Basic sampling structure for the digital representation of the 1125/60 HDTV production standard. 233

Figure 6.6 Line numbering scheme, first digital field ($F = 0$). 236

Figure 6.7 Line numbering scheme, second digital field ($F = 1$). 237

Figure 6.8 Production and clean aperture recommendations in SMPTE-260M. 238

Figure 6.9 SMPTE 274M analog interface vertical timing. 242

Figure 6.10 SMPTE 274M digital interface vertical timing. 243

Figure 6.11 Block diagram of an HD-SDI transmission system. 249

Figure 6.12 The HD-SDI signal format from the coprocessor IC to the P/S device. 250

Figure 6.13 Block diagram of an HD-SDI receiver system. 251

Figure 6.14 Attenuation characteristics of 5C-2V coaxial cable. 252

Figure 6.15 Eye diagram displays of transmission over 5C-2V coaxial cable: (*a*) 3 m length, (*b*) 100 m length. 253

Figure 6.16 Recovered signal clocks: (*a*) 1.485 GHz 60 Hz system, (*b*) 1.4835 GHz 59.94 Hz system. 254

Figure 6.17 SMPTE 259M principal video timing parameters. 260

Figure 6.18 Header data packet structure. 261

Figure 6.19 SDTI header data structure. 262

Figure 6.20 SDTI data structure: (*a*) fixed block size, (*b*) variable block size. 262

Figure 6.21 SDI/SDTI implementation for a computer-based editing/compositing system. 263

Figure 6.22 The basic content package structure of SMPTE 326M. 265

Figure 6.23 The elements that combine to define the quality of a video camera. 267

Figure 6.24 Optical/sampling path for a high-performance HDTV camera. 269

Figure 6.25 Pre-knee approach to reducing the dynamic range of a camera front-end. 269

Figure 6.26 Video camera signal flow: (*a*) analog system, (*b*) digital system. 271

Figure 6.27 Block diagram of the digital processing elements of an HDTV camera. 272

Figure 7.1 AES audio data format structure. 281

Figure 7.2 Example application of the AC-3 audio subsystem for satellite audio transmission. 283

Figure 7.3 Overview of the AC-3 audio-compression system encoder. 284

Figure 7.4 Overview of the AC-3 audio-compression system decoder. 285

Figure 7.5 The audio subsystem in the DTV standard. 286

Figure 7.6 Overview of the AC-3 audio-compression system. 289

Figure 7.7 The AC-3 synchronization frame. 293

Figure 7.8 Generalized flow diagram of the AC-3 encoding process. 296

Figure 7.9 Generalized flow diagram of the AC-3 decoding process. 298

Figure 8.1 ITU-R digital terrestrial television broadcasting model. 314

Figure 8.2 High-level view of the DTV encoding system. 315

Figure 8.3 Sample organization of functionality in a transmitter-receiver pair for a single DTV program. 318

Figure 8.4. Placement of luma/chroma samples for 4:2:0 sampling. 322

Figure 8.5 DTV system block structure: (*a*) blocks, (*b*) macroblocks. 323

Figure 8.6 Simplified block diagram of an example VSB transmitter. 326

Figure 8.7 The basic VSB data frame. 327

Figure 8.8 Nominal VSB channel occupancy. 328

Figure 8.9 Randomizer polynomial for the DTV transmission subsystem. 329

Figure 8.10 Reed-Solomon (207,187) $t = 10$ parity-generator polynomial.330

Figure 8.11 Convolutional interleaving scheme (byte shift register illustration). 330

Figure 8.12 An 8-VSB trellis encoder, precoder, and symbol mapper. 331

Figure 8.13 Trellis code interleaver. 332

Figure 8.14 The 8-VSB data segment. 333

Figure 8.15 Nominal VSB system channel response (linear-phase raised-cosine Nyquist filter). 333

Figure 8.16. Sample organization of a transmitter-receiver system for a single digital television program. 336

Figure 8.17 Basic transport packet format for the DTV standard. 338

Figure 8.18 Example program insertion architecture. 341

Figure 8.19 Illustration of the multiplex function in the formation of a program transport stream. 342

Figure 8.20 The system-level bit stream multiplex function. 343

Figure 8.21 Overview of the transport demultiplexing process for a program. 344

Figure 8.22 Structural overview of a PES packet. 345

Figure 8.23 PES header flags in their relative positions (all sizes in bits). 346

Figure 8.24 Organization of the PES header. 347

Figure 8.25 Typical data segment for the 16-VSB mode. 347

Figure 8.26 Functional block diagram of the 16-VSB transmitter. 348

Figure 8.27 A 16-VSB mapping table. 348

Figure 8.28 The typical packing of an internodal ATM trunk. 350

Figure 8.29 The ATM cell format. 351

Figure 8.30 Comparison of the ATM cell structure and the MPEG-2 transport packet structure: (*a*) structure of the ATM cell, (*b*) structure of the transport packet. 353

Figure 8.31 Overall structure for the PSIP tables. 354

Figure 8.32 Extended text tables in the PSIP hierarchy. 355

Figure 8.33 The principle elements of a conditional access system as defined in ATSC A/70. 359

Figure 8.34 The opportunistic flow control environment of SMPTE 325M. 367

Figure 8.35 ATSC data broadcast system diagram. 369

Figure 9.1 Segment-error probability for 8-VSB with 4-state trellis coding; RS (207,187). 376

Figure 9.2 The cumulative distribution function of 8-VSB peak-to-average power ratio. 376

Figure 9.3 The error probability of the 16-VSB signal. 378

Figure 9.4 Cumulative distribution function of the 16-VSB peak-to-average power ratio. 379

Figure 9.5 Typical spectrum of a digital television signal. 380

Figure 9.6 Measured co-channel interference, DTV-into-NTSC. 381

Figure 9.7 Spectrum of the 8-VSB DTV signal on channel 12. Note the pilot carrier at 204.3 MHz. 382

Figure 9.8 The spectrum of the 8-VSB DTV signal at the output of a VHF transmitter after bandpass filtering. Note the shoulders 38 dB down, which are intermodulation products. 383

Figure 9.9 The spectrum of the 8-VSB DTV signal at the output of a UHF transmitter after bandpass filtering. Note the shoulders 35 dB down, which are intermodulation products, as in Figure 9.8. 384

Figure 9.10 The cumulative distribution function transient peaks, relative to the average power of the 8-VSB DTV signal at the output of the exciter (IF). 385

Figure 9.11 Overall system block diagram of a digital satellite system. The ATSC standard described in document A/80 covers the elements noted by the given reference points. 389

Figure 9.12 Block diagram of the baseband and modulator subsystem. 390

Figure 9.13 Comparison of the DTV peak transmitter power rating and NTSC peak-of-sync. 393

Figure 9.14 DTV RF envelope at a spectral spread of –43 dB. 393

Figure 9.15 DTV RF envelope at a spectral spread of –35 dB. 394

Figure 9.16 Schematic diagram of a 600 W VHF amplifier using eight FETs in a parallel device/parallel module configuration. 395

Figure 9.17 Cutaway view of the tetrode (left) and the Diacrode (right). Note that the RF current peaks above and below the Diacrode center, but the tetrode has only one peak at the bottom. 398

Figure 9.18 The elements of the Diacrode, including the upper cavity. Double current, and consequently, double power is achieved with the device because of the current peaks at the top and bottom of the tube, as shown. 398

Figure 9.19 Mechanical design of the multistage depressed collector assembly. Note the "V" shape of the 4-element system. 400

Figure 9.20 Functional schematic diagram of an IOT. 401

Figure 9.21 Mechanical configuration of a high-power IOT (EEV). 402

Figure 9.22 Schematic overview of the MSDC IOT or constant efficiency amplifier. 403

Figure 9.23 An example 8-VSB sideband intermodulation distortion corrector functional block diagram for the broadcast television IF band from 41 to 47 MHz for use with DTV television transmitters. 405

Figure 9.24 After amplification, the out-of-band signal is corrected or suppressed by an amount determined by the ability of the correction circuit to precisely match the delay, phase, and amplitude of the cancellation signal. 406

Figure 9.25 FCC linearity emissions requirements: (*a*) typical transmitter output with linearity such that the IMD shoulder levels are at –37 dB, (*b*) DTV mask filter on transmitter output to be fully mask compliant. 407

Figure 10.1 Simplified block diagram of a VSB receiver. 415

Figure 10.2 Block diagram of the tuner subsystem. 416

Figure 10.3 Block diagram of the tuner IF FPLL system. 417

Figure 10.4 Data segment sync waveforms. 419

Figure 10.5 Segment sync and symbol clock recovery with AGC. 420

Figure 10.6 The process of data field sync recovery. 421

Figure 10.7 Receiver filter characteristics: (*a*) primary NTSC components, (*b*) comb filter response, (*c*) filter band edge detail, (*d*) expanded view of band edge. 422

Figure 10.8 Block diagram of the NTSC interference-rejection filter. 423

Figure 10.9 Simplified block diagram of the VSB receiver equalizer. 425

Figure 10.10 The phase-tracking loop system. 426

Figure 10.11 Functional diagram of the trellis code de-interleaver. 427

Figure 10.12 Block diagram of the 8-VSB receiver segment sync suspension system. 428

Figure 10.13 Trellis decoding with and without the NTSC rejection filter. 428

Figure 10.14 Functional diagram of the convolutional de-interleaver. 429

Figure 10.15 Data field sync waveform. 431

Figure 10.16 The CIE 1931 chromaticity diagram showing three sets of phosphors used in color television displays. 444

Figure 10.17 Comparison of the Y, C_b, C_r and R, G, B color spaces. Note that about one-half of the Y, C_b, C_r values are outside of the R, G, B gamut. 446

Figure 10.18 Color gamut for SMPTE 240M. 447

Figure 10.19 Basic concept of a shadow-mask color CRT: (*a*) overall mechanical configuration, (*b*) delta-gun arrangement on tube base, (*c*) shadow-mask geometry. 452

Figure 10.20 Basic concept of the Trinitron color CRT: (*a*) overall mechanical configuration, (*b*) in-line gun arrangement on tube base, (*c*) mask geometry. 454

Figure 10.21 Shadow-mask CRT using in-line guns and round mask holes: (*a*) overall tube geometry, (*b*) detail of phosphor dot layout. 455

Figure 10.22 Shadow-mask CRT using in-line guns and vertical stripe mask holes. 456

Figure 10.23 Delta-gun, round-hole mask negative guard band tri-dot screen. The taper on the mask holes is shown in the detail drawing only. 457

Figure 10.24 The relationship between beam shift at local doming and the effective radius of the faceplate with various mask materials. 460

Figure 10.25 Comparison of small-area-mask doming for a conventional CRT and a flat-tension-mask CRT. 461

Figure 10.26 Mechanical structure of a taut shadow-mask CRT. 462

Figure 10.27 Relationship between screen size and CRT weight. 463

Figure 10.28 Simplified mechanical structure of a bipotential color electron gun. 465

Figure 10.29 Electrode arrangement of a UniBi gun. 467

Figure 10.30 Electrode arrangement of a BiUni gun. 467

Figure 10.31 Electrode arrangement of the Trinitron gun. 468

Figure 10.32 Spot-size comparison of high-resolution guns vs. commercial television guns: (*a*) 13-in vertical display, (*b*) 19-in vertical display. 469

Figure 10.33 Astigmatism errors in a color CRT. 470

Figure 10.34 Astigmatism and coma errors. 470

Figure 10.35 Misconvergence in the four corners of the raster in a color CRT. 471

Figure 10.36 Field configurations suitable for correcting misconvergence in a color CRT. 471

Figure 10.37 Self-converging deflection field in an in-line gun device: (*a*) horizontal deflection field, (*b*) vertical deflection field. 472

Figure 10.38 Principal elements of a video projection system. 475

Figure 10.39 Projection-system screen characteristics: (*a*) front projection, (*b*) rear projection. 477

Figure 10.40 High-contrast rear-projection system. 478

Figure 10.41 Optical projection configurations: (*a*) single-tube, single-lens rear-projection system; (*b*) crossed-mirror trinescope rear projection; (*c*) 3-tube, 3-lens rear-projection system; (*d*) folded optics, front projection with three tubes in-line and a dichroic mirror; (*e*) folded optics, rear projection. 480

Figure 10.42 A 3-tube in-line array. The optical axis of the outboard (red and blue) tubes intersect the axis of the green tube at screen center. Red and blue rasters show trapezoidal (keystone) distortion and result in misconvergence when superimposed on the green raster, thus requiring electrical correction of scanning geometry and focus. 482

Figure 10.43 Mechanical construction of a liquid-cooled projection CRT. 484

Figure 10.44 Projection CRT with integral Schmidt optics. 484

Figure 10.45 Mechanical configuration of the Eidophor projector optical system. 486

Figure 10.46 Functional operation of the General Electric single-gun light-valve system. 487

Figure 10.47 Functional operation of the 2-channel HDTV light-valve system. 489

Figure 10.48 Block diagram of a laser-scanning projection display. 490

Figure 10.49 Configuration of a color laser projector. 491

Figure 10.50 Rotating polygon line scanner for a laser projector. 492

Figure 10.51 Laser-screen projection CRT. 493

Figure 10.52 Structure of a liquid-cooled LC panel. 494

Figure 10.53 LC panel temperature dependence on lamp power in a liquid crystal projector. 495

Figure 10.54 Layout of a homeotropic liquid crystal light valve. 496

Figure 10.55 Block diagram of an LCLV system. 497

Figure 10.56 Functional optical system diagram of an LCLV projector. 498

Figure 10.57 Layout of the fluid-filled prism assembly for an LCLV projector. 499

Figure 10.58 Block diagram of a color laser-addressed light-valve display system. 500

Figure 10.59 Cross-sectional diagram of the Image Light Amplifier light valve. 502

Figure 10.60 Optical channel of the ILA projector. 502

Figure 10.61 Full-color optical system for the ILA projector (Series 300/400). 503

Figure 10.62 Color gamut of the ILA projector (Series 300/400) compared with SMPTE 240M. 504

Figure 10.63 A pair of DMD pixels with one mirror shown at $-10°$ and the other at $+10°$. 505

Figure 10.64 The basics of DMD optical switching. 506

Figure 10.65 A scanning electron microscope view of the DMD yoke and spring tips (the mirror has been removed). 506

Figure 10.66 Optical system of the DLP 3-chip implementation. 507

Figure 10.67 DMD display resolution as a function of chip diagonal. 508

Figure 10.68 The relationship between misconvergence and display resolution. 510

Figure 11.1 Elements and transmission paths of the Eureka EU-95 HD-MAC system. 526

Figure 11.2 Luminance, chrominance, DATV, and sound/data components of the HD-MAC system. 527

Figure 11.3 Frequency domain representation of orthogonal carriers. 538

Figure 11.4 Positions of pilots in the OFDM symbol. 539

Figure 11.5 Protection of the OFDM symbol against echoes through the use of a guard-interval. 540

Figure 11.6 Fading effect of an echo. 542

Figure 11.7 Limitation caused by a small number of reference signals. 543

Figure 11.8 Shift of the interpolation filter towards 0 to N/3 distortion range. 543

Figure 11.9 Aliasing of pre-echo. 543

Figure 11.10 Delay input signal of FFT to shift pre-echo into interpolation filters range. 543

Figure 12.1 Video signal spectra: (a) camera scanning spot, shown with a Gaussian distribution, passing over a luminance boundary on a scanning line; (b) corresponding camera output signal resulting from the convolution of the spot and luminance distributions. 559

Figure 12.2 Scanning patterns of interest in analyzing conventional video signals: (a), (b), (c) flat fields useful for determining color purity and transfer gradient (gamma); (d) horizontal half-field pattern for measuring low-frequency performance; (e) vertical half field for examining high-frequency transient performance; (f) display of oblique bars; (g) in monochrome, a tonal wedge for determining contrast and luminance transfer characteristics; in color, a display used for hue measurements and adjustments; (h) wedge for measuring horizontal resolution; (i) wedge for measuring vertical resolution. 561

Figure 12.3 An array of image patterns corresponding to indicated values of *m* and *n*. 563

Figure 12.4 The typical spectrum of a video signal, showing the harmonics of the line-scanning frequency surrounded by clusters of components separated at intervals equal to the field-scanning frequency. 564

Figure 12.5 The CIE 1931 chromaticity diagram illustrating use of the center of gravity law. 566

Figure 12.6 Ellipses of equally perceptible color differences (as identified by MacAdam). 569

Figure 12.7 Examples of color test bars: (*a*) ITU nomenclature, (*b*) SMPTE EG 1-1990 color bar test signal, (*c*) EIA RS-189-A test signal. 572

Figure 12.8 Waveform monitor display of a color-bar signal at the two-line rate. 575

Figure 12.9 EIA RS-189-A color-bar displays: (*a*) color displays of gray and color bars, (*b*) waveform display of reference gray and primary/complementary colors, plus sync and burst. 576

Figure 12.10 Sync pulse widths for NTSC. 577

Figure 12.11 Vectorscope display of a color-bar signal. 577

Figure 12.12 Composite vertical-interval test signal (VITS) inserted in field 1, line 18. The video level in IRE units is shown on the left; the radiated carrier signal is shown on the right. 578

Figure 12.13 Waveform monitor display showing additive 60 Hz degradation. 579

Figure 12.14 Waveform monitor display of a multiburst signal showing poor high-frequency response. 580

Figure 12.15 Waveform monitor display of a modulated ramp signal. 581

Figure 12.16 Vectorscope display showing 5° differential phase error. 581

Figure 12.17 Vectorscope monitor display showing a differential phase error of 5.95° as a tilt on the vertical scale. 582

Figure 12.18 Vectorscope display of a 10 percent differential gain error. 582

Figure 12.19 Block diagram of an automated video test instrument. 584

Figure 12.20 Automated video test instrument output charts: (*a*) waveform monitor mode display of color bars, (*b*) vectorscope mode display of color bars. 585

Figure 12.21 Expanded view of a sync waveform with measured parameters. 586

Figure 12.22 Example error log for a monitored signal. 586

Figure 12.23 Multiburst video test waveform: (*a*) picture display, (*b*) multiburst signal as viewed on a waveform monitor (1H). 587

Figure 12.24 Conventional sweep-frequency test waveform: (*a*) picture display, (*b*) waveform monitor display, with markers (1H). 588

Figure 12.25 Single horizontal frequency test signal from a zone plate generator: (*a*) picture display, (*b*) waveform monitor display (1H). 588

Figure 12.26 Horizontal frequency-sweep test signal from a zone plate generator: (*a*) picture display, (*b*) waveform monitor display (1H). 589

Figure 12.27 Single vertical frequency test signal: (*a*) picture display, (*b*) magnified vertical-rate waveform, showing the effects of scan sampling. 589

Figure 12.28 Vertical frequency-sweep picture display. 590

Figure 12.29 Combined horizontal and vertical frequency-sweep picture display. 591

Figure 12.30 Combined horizontal and vertical frequency sweeps, selected line waveform display (1*H*). This figure shows the maintenance of horizontal structure in the presence of vertical sweep. 591

Figure 12.31 The best-known zone plate pattern, combined horizontal and vertical frequency sweeps with zero frequency in the center screen. 592

Figure 12.32 Vertical frequency-sweep picture display. 592

Figure 12.33 The same vertical sweep as shown in Figure 12.32, except that appropriate pattern motion has been added to "freeze" the beat pattern in the center screen for photography or other analysis. 593

Figure 12.34 A hyperbolic variation of the 2-axis zone plate frequency sweep. 593

Figure 12.35 A 2-axis frequency sweep in which the range of frequencies is swept several times in each axis. Complex patterns such as this may be created for specific test requirements. 594

Figure 12.36 Wide aspect ratio resolution test chart produced by an electronic signal generator. 597

Figure 12.37 Siemens star chart. 604

Figure 12.38 Multiburst chart (10 MHz). 605

Figure 12.39 Color reference pattern layout specified in SMPTE 303M (proposed). 606

Figure 12.40 Video testing layers for digital television. 608

Figure 12.41 Functional environment for traditional video measurements. 609

Figure 12.42 Functional environment for digital video measurements. 610

Figure 12.43 Block diagram of the feature-extraction method of picture-quality analysis. 611

Figure 12.44 Block diagram of the picture-differencing method of picture-quality analysis. 612

Figure 12.45 In-service picture quality measurement system. 613

Figure 12.46 Valid data acceptance bands for a digital signal. 614

Figure 12.47 A portion of a SMPTE 259M datastream showing three successive ones. 616

Figure 12.48 Representation of jitter: (*a*) jitter present on the measured waveform, (*b*) data correctly centered on the recovered clock, (*c*) signal instantaneously advanced as a result of jitter. 616

Figure 12.49 Measurement of jitter using a digitizing oscilloscope: (*a*) low jitter, absolute jitter = 200 ps p-p; (*b*) significant jitter present, absolute jitter = 2.12 ns p-p. 618

Figure 12.50 Basic serial digital transmitter/receiver system. Note typical voltage levels (for composite video) and the function of the equalizing amplifier. 620

Figure 12.51 Error rate as a function of cable length for composite video signals on an SDI link. 622

Figure 12.52 An 8-VSB constellation diagram: (*a*) a near-perfect condition, (*b*) constellation diagram with noise and phase shift (the spreading of the pattern is caused by noise; the slant is caused by phase shift). 624

Figure 12.53 Measured DTV transmitter output and sideband splatter. 625

Figure 12.54 RF spectrum mask limits for DTV transmission. The mask is a contour that illustrates the maximum levels of out-of-band radiation from a transmitted signal permitted by the FCC. This chart is based on a measurement bandwidth of 500 kHz. 626

Figure 13.1 Splicing procedure for uncompressed video. 632

Figure 13.2 Splice point considerations for the MPEG bit stream. 632

Figure 13.3 Typical operational loading of the buffer. 633

Figure 13.4 MPEG decoder buffer state issues for bit stream switching. 634

Figure 13.5 Relative timing of video and audio splice points and the end result. 634

Figure 13.6 The use of splice flags to facilitate the switching of MPEG bit streams. 635

Figure 13.7 Raster scanning methods: (*a*) progressive, (*b*) interlaced. 643

Figure 13.8 One possible implementation for mezzanine level compression in the studio environment: (*a*) camera system, (*b*) vision mixer. 645

Figure 13.9 A typical application of the SMPTE 310M synchronous serial interface. 646

Figure 13.10 Example television plant incorporating support for high-definition and standard-definition video capture, production, and distribution. 648

Figure 13.11 A practical implementation of the integration of HDTV video and film using a 24 f/s mastering system. 655

Subject Index

Numerics

10 BASE 150
1080-line format 319
10-bit quantization 230
1125/60 HDTV signal parameters 228
1125/60 NHK standard 14
1125/60 scanning format 217
1125-line HDTV 215
1250/50 Eureka HDTV 18
16:9 format 62
16-VSB 659
16-level slicer 431
22:11:11 sampling 233
22:22 sampling 233
24 f/s mastering 653
2D transform 157
35 mm film 67, 99
4:2:2:4 video sampling 139
480-line format 320
720-line format 320
8:8:8 video sampling 140
8-VSB 30, 659
8-level slicer 431

A

AB operation 395
ac plasma display pane 448
ac thin film display 449
AC-3 bit stream 284
AC-3 decoding 297
AC-3 surround sound system 279
AC-3 sync frame 288, 295

access control 652
access unit 659
acoustic-optical modulator (AOM) 75
active loop antenna 437
active picture area 172
active TV lines 224
ac-to-RF efficiency 407
ACTV-I 26
ACTV-II 26
adaptive differential PCM 196
adaptive equalizer 433
adaptive PCM 196
adder overflow limit cycle 136
additive white Gaussian channel 544
additive white Gaussian noise 36
addressable section encapsulation 364
adjacent channel interference 381, 544
advanced audio coding (AAC) 206
advanced systems 24
Advanced Television Enhancement Forum 513
Advanced Television Technology Center 386
Advanced Television Test Center (ATTC) 25
Advisory Committee on Advanced Television Service (ACATS) 25
AES audio 255, 279
AES data format 281
afterglow correction 73
AGC delay 418
aggregate capacity 150
aliasing 109
aliasing effects 110
alignment jitter 618
alpha channel 139

alphanumeric and graphics 448
AM/PM conversion 385
Amendment No.1 to ATSC Standard A/65 355
American Electronics Association 23
American National Standards Institute (ANSI) 16
amplitude-shift keying (ASK) 125
analog-to-digital conversion 55, 109
analog-to-digital converter 110
analysis filterbank 284
anamorphic CinemaScope 68
anchor frame 659
ancillary data 314, 638
ancillary data space 260
anisochronous 148
antenna pattern 409
aperture correction 601
aperture grille 473
application programming interface 515, 535
apt-X100 200
artifacts 190
ASPEC 198
aspect ratio 18, 43, 50, 319
aspect ratio management 266
aspheric corrector lens 483
associated service 306
asynchronous 148
asynchronous data transfer 511
asynchronous datagram 536
asynchronous transfer mode (ATM) 315, 512, 659
ATSC 34
ATSC A/80 387
ATSC satellite transmission standard 388
ATSC T3/S17 specialist group 514
audio bit rates 199
audio compression 193
audio elementary stream 286
Audio Engineering Society 279
audio item 264
audio sample bit 282
aural monitoring 67
auto black circuit 603
automatic frequency control (AFC) 417
automatic gain control (AGC) 415
automatic phase control (APC) 418
auxiliary item 264

average luminance 560
average picture level (APL) 384
average power 392

B

back focal distance 268
backward prediction 163
backward-compatible 40
balance point 565
bandwidth 128
bandwidth efficiency 127
bandwidth-reduction encoder (BRE) 528
Bell Telephone Laboratories 84
B-frame 173, 319
bidirectional pictures 659
binary exponent 284
binding 366
bipotential lens 466
bit bucket 644
bit error rate (BER) 413, 619
bit rate 659
bit scrambling 619
bit stream 110, 111
bit-allocation 301
bit-allocation algorithm 291
bit-allocation computation 299
bit-allocation pointers (BAPs) 292, 299
bit-rate-reduction systems 193
bit-stream information (BSI) 293
black matrix shadow-mask 458
blind equalization 424, 431, 440
block 659
block coding 130
block effects 188
block layer 172
blocking effect 155
block-matching 164
bouquet association table (BAT) 531
B-picture 166
brightness 86, 601
broadcast applications for HDTV 52
broadcast architecture 150
Broadcast Television Association (BTA) 218
BTSC (Broadcast Television Systems Committee) 382

buffer fullness 633
buffer overflow 367
buffer underflow 633
business applications for HDTV 51
byte-aligned 659

C

C data space 258
C service 308
C.I.E. diagram 89
C.I.E. illuminant C 89
C.I.E. Standard Observer 89
C/N threshold 544
cable equalizer 251
cable virtual channel table 356
camera black shading 604
camera detail circuit 604
camera encoder measurement 602
camera flare correction 605
camera performance mesurement 602
camera positioning information 272
camera technology 266
camera white shading 604
carrier-to-noise ratio (C/N) 528
carryover smear 94
cathode-ray tube (CRT) 98, 440, 448
CBS color system 6
CCD (charge-coupled device) 71, 268
CCD line array telecine 74
CCIR 43
CCIR-601 168
CEMA 512
center of deflection 453
center of gravity law 565
channel 660
channel buffer 184
channel coding 128, 129
channel equalizer 415
channel filtering 414
channel impairments 432, 434
channel status block 282
channel status data block 281
channel tuning 416
channel-combining systems 408
checkwords 143

chromatic adaptation 568
chromaticity coordinates 565
chrominance 88, 102
chrominance amplitude and phase 601
CIE 1931 chromaticity diagram 568
CIE chromaticity coordinates 238
CIE color-matching 238
CIE illuminant D65 238, 442
CinemaScope 62, 84
circular polarization (CP) 409
clean aperture 236
clock recovery circuit 419
clock regeneration 617
closed captions 362
CM service 306
co-channel interference 423
code rate 390
coded audio field 209
coded orthogonal frequency-division multiplexing 22, 533
coded representation 660
coding gain 130, 284
coding matrix 100
coefficient design 134
coefficient quantization error 136
COFDM 544
coherent AGC 418
coherent receiver 126
color acuity 92
color bars 571
color channel coding matrix 100
color CRT 457
color depth 50
color gamut 89, 446
color information representation 43
color primaries 100
color reference pattern 606
color reproduction 267
color science 89
color separation system 268
color space 441
color temperature 601
color-bar pattern 574
color-difference components 140, 225
color-difference signals 89

colorimetric color reproduction 567
colorimetric match 442
colorimetry 274
common image format 100, 657
common line combiner 408
Communications Research Center of Canada 434
compact disc (CD) 192
compatibility 42
compatibility in HDTV systems 40
compressed stream format 640
compression 160, 660
compression artifacts 187
compressionist 190
computer applications for HDTV 54
concatenated forward error correction 544
concatenation 186, 557, 613
conditional access 357, 534
conformance specification 515
constant bit rate 660
constant efficiency amplifier 401
constant impedance diplexer 408
constant modulus algorithm (CMA) 433
constellation diagram 623
content authoring 514
content management 651
content package 264
content plane 651
continuity of motion 93
continuous phase modulation (CPM) 127
continuous-wave laser 491
contrast 50, 86
contrast handling range 87
contrast ratio 571
contribution service 388
control data 365
conventional definition television 660
conventional systems 24
convergence 601
convergence yoke 485
conversion time 115
convolutional coding 130
convolutional coding rate 544
convolutional de-interleaver 429
cooperative processing 124
copy protection issues 518

corresponding color reproduction 567
coupling coordinates 293
CRC 660
CRC word checks 299
cross-polarization 381
CRT flying-spot scanner (FSS) 71
CRT projection display 474
cubic transforms 158
curvature of field 470

D

D service 307
D/A converter 114
D2-MAC 523
darkfield performance 489
data broadcast 57
data broadcasting 364
data carousel 537
data de-interleaver 415
data derandomizer 415
data field synchronization 415
data information table 365
data payload capability 131
data piping 364, 536
data projector 474
data segment sync 418
data service table 365
data sink 389
data source 389
data streaming 364
datacasting 57
data-stream syntax 529
David Sarnoff 3
dc coefficient 157
dc plasma display panel 448
decision feedback equalizer 545, 546
declarative content 514
decode time stamp (DTS) 315
decoded stream 660
decoder 660
decoding 660
decoding time-stamp 660
deflection plane 453
deflection yoke 485
delta bit-allocation code 292

delta-gun CRT 451
depth perception 86
de-rotator (complex multiplier) 426
desired to undesired (D/U) signal power ratio 380, 421, 623
device plane 652
D-frame 660
DHTML 660
Diacrode 394
dialogue normalization 302
dichroic interference filter 507
difference color set 239
difference pulse-code-modulation (DPCM) 158
differential coding 301
differential gain distortion 581
differential linearity 115
differential phase distortion 580
differential vector 179
differentially coherent PSK (DPSK) modulation 126
diffraction grating 488
digital active line 258
digital audio data compression 192
Digital Audio Visual Council (DAVIC) 537
digital cable 356
digital cable system 439
digital cliff 615
digital coding 128
digital compact cassette (DCC) 200
digital convergence system 485
digital filters 131
digital headroom 445
Digital Light Processing system 504
digital line blanking 258
digital micromirror device (DMD) 449
digital modulation 125
digital signal processing (DSP) 138
digital storage media 660
Digital Television Application Software Environ-ment 57, 514
digital television audio encoder 287
digital terrestrial television broadcasting 44
Digital Video Broadcasting (DVB) 20, 528
digitally assisted television 525
digital-to-analog converter (D/A) 300

diode-gun impregnated-cathode Saticon (DIS) 14
diplexer 408
direct form filter realization 136
directed channel change 357
directory object 537
direct-view CRT 479
discovery 366
discrete cosine transform (DCT) 105, 156, 181, 660
display colorimetry 218
display device 447
display format 55
display measurement 593
display system convergence 509
displayed white 442
distribution service 388
distribution systems 24
diversity gain 548
D-MAC 523
Dolby AC-3 65
Dolby E coding 208
DOM 660
dot pitch 570
downmixing 300
DSP devices 138
DTV Closed Captions 362
DTV core spectrum 386
DTV Implementation Subcommittee (IS) 35
DTV peak envelope power (PEP) 392
DTV receiver 413, 439
DTV receiver labeling 517
DTV receiving subsystem 414
DTV RF envelope 392
DTV scanning format 101
DTV system implementation 631
DTVCC 362
dual-link interface 245
dual-mode channel combiner 408
DVB data broadcasting 536
DVB project 21
DVB-C 532
DVB-MC 533
DVB-MS 533
DVB-S 532
DVB-T 533

dynamic bit allocation 198
dynamic range compression 300, 303
dynamic range control value (DynRng) 294

E

E service 308
EAV 258
EDH 143
editing 660
EDTV-II 244
effective radiated power 379
EG 36 100
EIA color bar signal 573
EIA-708-B 362
Eidophor system 486
electroluminescent display 448
electron gun 453, 465
electron-beam deflection 462
electron-beam recording (EBR) 77
electronic pin registration (EPR) 73
electronic program guide 361
electronic program guide (EPG) 531
electronic-intermediate (EI) digital video system 22
Electronics Industries Association 4
elementary stream 637, 660
elementary stream clock reference 660
elementary streams 255
embedded audio 141
embedded DSP 138
Emergency Alert System 439
encapsulated key 360
encoder 660
encoding 660
encoding decision list 191
encoding tools 189
encrypted keys 358
encryption 358
end of active video 260
end-of-block (EOB) marker 184
Energy Saving Collector IOT 404
entitlement control message 359, 661
entitlement management message 359, 661
entropy 129
entropy coding 160, 661

entry point 661
equalizer/ghost canceller 424
equalizing circuit 617
equivalent color reproduction 567
error cliff 616
error correction 130
error detection and handling 143
error detection data packet 143
error flag 143
error vector magnitude 623, 627
ET Docket No. 99-25 362
Ethernet protocol 149
Eureka 1187 Advanced Digital Television Technologies (ADTT) 21
Eureka EU-95 17, 525
European Broadcasting Union 279
European HDTV systems 17
European Telecommunications Standards Institute 535
event 661
event information table 365, 531
event model 515
exact color reproduction 567
exponent coding 291
exponent strategies 301
exposure latitude 267
extrafoveal region 82
eye diagram 622
eye pattern 618

F

Farnsworth, Philo 2
Fast Ethernet 150
fast Fourier transform (FFT) 105, 203
FCC emissions mask 406
feature extraction 610
Fibre Channel 148
Fibre Channel Arbitrated Loop 148
field 94, 661
file object 537
film formats 68
film look 68
film-to-video transfer 71
filterbank tool 206

filtering 190
finite impulse response (FIR) filter 131
FireWire 145
five-halves power law 96
fixed-reference D/A converter 114
fixed-reference oscillator 417
flash A/D conversion 113
flat CRT 448
flat face display 472
flat- panel displays (FPDs) 450
flat-panel PDP (plasma display panel) 475
flicker 83, 94
flying-spot scanner 71
focus 601
format 43
format developmemt 61
forward error correction 532
forward prediction 163
four-quadrant multiplying D/A converter 115
fovea 82
foveal vision 82
frame 94, 661
frequency and phase-locked loop (FPLL) 417
frequency sampling 134
frequency-division multiplexing (FDM) 523
frequency-shift keying (FSK) 125
front projection system 476
functional layers 653
functional planes 650

G

G.722 196
gamma 97
gamma correction 86, 97, 221, 443
gamut 447
geometric nesting 458
geometry and aspect ratio 601
Gibb's phenomena 384
Gigabit Ethernet 149
glitter 83
Goldmark, Peter 6
GOP layer 166
Grand Alliance 20, 29, 31, 101
graphics projectors 474
gray scale 274

gray scale tracking 601
group of pictures (GOP) 166, 661
group-delay 133
guard band (CRT) 456
guard interval 534, 539, 546

H

Hadamard transform 156
half-intensity width 598
half-power-width 598
HANC 141
HD-DIVINE 21
HD-MAC (high-definition multiplexed analog
    component) 17, 525
HDTV 24, 60, 661
HDTV and computer graphics 54
HDTV applications 49
HDTV audio 62
HDTV film-to-video transfer 68
HDTV image content 60
HDTV image size 60
HDTV launch 34
HDTV projector 450, 474
HDTV standardization 18
HDTV studio origination standard 219
HDTV-6-7-8 system 20
hearing perception 63
HI service 307
high level 661
high-band VHF 8
high-data-rate cable mode 131, 430
high-definition serial digital interface (HD-SDI)
    247
high-definition television (see *HDTV*)
high-gain receive antenna 413
high-side injection 416
Hilbert transform 426
history data 638
homeotropic light valve 495
homing 360
horizontal ancillary data 141
horizontal blanking ancillary data space 641
horizontal resolution 50, 55, 84, 225
horizontal resolution factor 562
HTML 57, 661

HTTP 661
Huffman coding 128, 183, 661
human auditory system 193, 203
human visual system (HVS) 161
HyperText Mark-up Language 57

I

Iconoscope 3
IDCT mismatch 180
IEEE 1212 511
IEEE 1394 145, 510, 512
I-frames 319
iLINK 145
image depth 273
Image Light Amplifier 501
image refresh 180
image response channels 379
Implementation Subcommittee 648
improved-definition television (IDTV) 24
impulse Fourier transform 597
impulse interference 549
impulse PPV 358
incidental carrier phase modulation (ICPM) 385
infinite focus 492
infinite impulse response (IIR) filter 131, 136
input jitter tolerance 618
in-service measurement 612
integral linearity 115
integrated receiver decoder (IRD) 530
interaural amplitude differences 63
interaural time delay 63
inter-carrier interference component 550
intercomponent confusion 563
interface elements 652
interference-rejection 420
interference-rejection filter 415, 419
interframe coding 158, 161
interlace factor 83, 225
interlaced scanning 83, 94, 642
interline flicker 94
intermediate frequency (IF) 131
intermediate power amplifier (IPA) 392
intermodulation distortion 404
intermodulation noise 627
intermodulation products 544, 625

International Radio Consultative Committee (CCIR) 13, 43
International Standards Organization 104, 163
International Telecommunication Union 43, 313
Internet Protocol 151
inter-picture encoding 166
interpolation filter 542
inter-symbol interference 539, 546, 623, 627
intra-coded pictures 661
intraframe coding 161, 171, 173
intrinsic jitter 618
inverse discrete cosine transform (IDCT) 177
inverse fast Fourier transform 538
inverse transform 300
inverse-quantized coefficients 178
inverse-quantizer 178
IOT (inductive output tube) 394
I-picture 166
ISO/IEC 13818-2 246, 637
ISO/MPEG-1 Layer 2 200, 203
isochronous 147
isochronous data transfer 511
isochronous delivery 145
ITU 43
ITU mode M3 544
ITU-R 601 140
ITU-R BT.601 100
ITU-R BT.709 100
ITU-R Rec. 601 139
ITU-R Rec. BS.1116 210
ITU-R Study Group 11 44

J

JAVA 57
jitter 615
jitter transfer 619
jitter transfer function 619
JNDmetrix 612
joint stereo coding 198
JPEG (Joint Photographic Experts Group) 159

K

Karhunen-Loeve transform (KLT) 156
Kell factor 83, 223

key signal 139
keystone scanning 479
kinescope 3
K-L-V 265
K-L-V format 255
Klystrode 394
klystron 394, 399
knife-edge Fourier transform 598

L

lambertian surface 477
large signal overload 416
laser beam transfer systems 75
laser projection display 490
laser-addressed liquid crystal light valve 499
lateralization 63
layer 662
LDMOS 394, 396
least significant bit (LSB) 115
least-mean-square (LMS) algorithm 424
lens 268
lens back-focus 603
letter-box format 61
level 662
level of impairment 210
library service 651
light-valve projection display 474, 486
line select 578
line shuffling 527
linear gamma 99
linear PCM 192
linear phase response 133
linear-beam device 399
linear-phase FIR filter 133
liquid crystal display 448
liquid crystal light valve (LCLV) 493
LN/CRC 258
local oscillator (LO) 416
local program insertion 646
local programming 647
localization 63
long transform block 295
lossless compression 162
loudspeaker placement 63
low-band VHF 8

low-frequency effects (LFE) 279
low-noise amplifier (LNA) 413
low-power television (LPTV) 10
luminance 88, 102
luminance equation 100
luminous efficiency 447

M

macroblock 179, 662
Main Level 168, 662
Main Profile 168, 662
Main Profile/Low Level 169
Main Profile/Main Level 169
main service 306
major channel number 356, 361
markup language 515
M-ary modulation 126
masking curve 292
masking threshold 198
Maximum Service Telecasters 23
ME service 306
mean time between failure (MTBF) 407
mean time to repair (MTTR) 407
media type 515
metadata 265, 365
metadata field 209
metal-oxide-semiconductor field-effect transistor
    (MOSFET) 394
metameric reproduction 238
meter field 209
mezzanine encoding/decoding 644
mezzanine level MPEG 644
microelectromechanical systems (MEMS) 505
microwave multipoint distribution system
    (MMDS) 533
MIME 662
MiniDisc 200
minimum shift keying (MSK) 127
minor channel number 362
MM Docket 87-268 386
modified discrete cosine transform 200
modified reduced constellation algorithm
    (MRCA) 433
modulation error ratio 623
modulation transfer function 594

modulation transfer function (MTF) 479
moiré 459
moiré patterns 110
monophonic sound 279
mosquito noise 188
motion compensation 160, 163
motion estimation 174
motion JPEG 162
motion picture theater projection-system 508
motion vector 174, 662
motion-compensated hybrid discrete cosine trans-
    form coding 22
motion-detection and interpolation 592
moving average operation 132
Moving Picture Experts Group (MPEG) 163, 662
MPEG bit stream splicing 631
MPEG Layer II 529
MPEG-1 160, 167, 662
MPEG-2 30, 160, 168, 662
MPEG-2 4:2:2 profile at high level 246
MPEG-2 advanced audio coding 206
MPEG-2 compliance point 530
MPEG-2 data packets 530
MPEG-2 TS packet 266
MPEG-3 168
MPEG-4 168, 207
MSDC (multistage depressed collector) klystron
    394
multiburst 578
multicast 537
multicasting 647
multichannel audio media 311
multi-frequency network 37, 546
multilingual service 309
multimedia home platform 535
multipath distortion 546
multipath immunity 534
multiplexed analog component (MAC) 523
multiplexer 408
multiplying D/A converter 114
multi-protocol encapsulation 537
multiprotocol encapsulation 536
multivalued displays 622
MUSE (multiple sub-Nyquist encoding) 16
MUSICAM 200

N

national information infrastructure (NII) 31
National Television System Committee (NTSC) 5
near video on demand 358
negative guard band 456
network contribution 643
network delay and program time shifting 646
Network File System 151
network information table (NIT) 531
network interface 150
network interface card 151
network resource table 365
new viewing experience 51
Nippon Hoso Kyokai (NHK) 12
noise 189
noise figure (NF) 413, 550, 620
noise weighting function 105
non return to zero inverted 141
noncoherent detection 126
noncoherent receiver 126
nonintracoded picture 246
non-linear distortion 625
normalized mantissa 290
NPRM 00-83 38
NRSS-A 358
NRSS-B 358
NRZI 141
NRZI (non return to zero inverted) 250
NTSC 7, 10
NTSC service 8
NTSC-rejection filter 428
Nyquist rule 109
Nyquist slope 423
Nyquist volume 116

O

object carousel 537
offset QPSK 127
on-channel repeater 548
Open architecture 57
open service information system 530
Open Systems Interconnect model 150
operational plateau 616
opportunistic data broadcast 366

optical filter 268
optical system 268
optical system inspection 602
optimal viewing distance 82
optimal viewing ratio 82
order-ahead PPV 357
orthogonal diffraction 488
orthogonal frequency-division multiplexing
  (OFDM) 533, 538
orthogonal sampling structure 233
oscilloscope 574
outer code 532
out-of-band 356
out-of-channel emission 625
out-of-service picture-quality measurements 609
output concentration 157
output jitter 619
overflow oscillation 136
oversampling A/D converter 115
overscan 601

P

pack 662
packet 662
packet data 662
packet identifier 662
packet jitter 265
packetized elementary stream 255
packetized elementary stream (PES) 317
padding 662
PAL-G 524
PALplus 524
pan and scan 61
parallel amplification 394
parallel-stripe CRT 453
Parks-McClellan algorithm 134
partial response process 421
passband corner frequency 133
pass-through of DTV programming 647
path plane 651, 652
pathological test signals 617
payload 532, 662
pay-per-view 358
PC99 56
PCM output buffer 300

peak to average power ratio 544
perceived audio quality 210
perceptual coding 193
Perceptual Evaluation of Audio Quality 211
perceptual weighting 182
periodic filter 124
periodic subscription service 357
peripheral vision 82
persistence of vision 92
PES packet 662
PES stream 662
Peter Goldmark 6
P-frame 173, 319
phantom image 63
phase alternation line (PAL) 8
phase noise 544
phase-alternate-frequency (PAF) carriers 121
phase-shift keying (PSK) 125
phase-tracking loop 425
Philo Farnsworth 2
phosphor chromaticities 444
phosphor screen 453
phosphor thermal quenching 483
photomultiplier tube 72
physical connectivity path 652
physical layer 315
picture 662
picture differencing 611
picture element (pixel) 83
picture item 264
picture layer 166
picture order 638
picture rate 319
picture sharpness 267
pilot carrier 382
pilot frequency 624
pixel 663
pixel density 570
pixel format 570
pixel format for DTV 30
pixel rate 104
pixels 319
planar transform 157
plant native format 649
plasma/gas discharge display 448

point of unusability 438
point-of-deployment 439
point-to-point architecture 150
polarization diversity 409
polarizing beam-splitter (PBS) 501
positive guard band 456
power efficiency 127
power spectral density (PSD) 302
P-picture 166
predicted pictures 663
prediction errors 175
predictive rendering 515
preferred color reproduction 567
pre-knee 268
presentation engine 514
presentation time stamp (PTS) 315, 633, 663
presentation unit 663
primary color correction 75
primary color set 239
prime lens 268
production HDTV systems 24, 50
production telecine 71
profile 663
program 663
Program Association Table 361
program clock reference 663
program clock reference base 315
program clock reference extension 315
program element 663
program ID (PID) 306
program specific information 663
programmable color-correction 75
program-specific information (PSI) 317, 530
progressive DCT 161
progressive scanning 174, 642
projection CRT 483
projector luminance output 477
proof of performance 623
propagation delay 198
protocol analysis 607
pseudorandom (PN) sequence 420, 430
Public Broadcasting System (PBS) 31
pulse code modulation 110, 192, 230
punctured convolutional code 532
purity 601

Q

Q-spacing 457
quadrature amplitude modulation (QAM) 532
quadrature mirror filter 201
quadrature modulation 123
quadrature phase-shift keying 532
quadrature-amplitude-shift keying (QASK) 127
quadriphase-shift keying (QPSK) 127
quality of service 151
quantization 177, 194, 557
quantization compression 161
quantization error 194
quantization levels 230
quantization noise 109, 194, 544
quantizer 663
quincunx scanning 527

R

Radio Manufacturers Association 4
random access 663
random noise 435
raw exponent 290
rear projection system 476
receiver loop acquisition sequencing 415
recoding 638
recoding data set 638
Recommended Practice RP211 655
reduced constellation algorithm (RCA) 433
redundancy reduction 160
Reed Solomon encoding/decoding 329
Reed-Solomon decoder 415
Reed-Solomon outer coding 533
reference primaries 238
reference white 100
refresh rates 55
registration repository 652
relative accuracy 115
rematrixing 292, 300
resolution 49, 115, 459
resonant loop combiner 408
retina 82
return-beam Saticon 14
RF mask 625
RF space current 399

RF/transmission 315
Ricean channel model 546
roundoff noise 136
RP 165-1994 143
RP 202 191
RP 203 368
RP 204 264, 266
RP199-1999 62
run-length encoding 184
running status table (RST) 531

S

S22.02 Advanced System Control Architectures
    Working Group 650
sample word 279
sampling frequencies 231
sampling rate 115
Sarnoff, David 3
satellite link 388
satellite transmission 387
SAV 258
scalable decoder 206
scale factor 203
scale factor select information 205
scaleable sampling rate 206
scanning 95
scanning standard 43
Schmidt reflective optical system 483
scrambling 358, 663
scrambling keys 360
screen gain 477
screen size 463
SCTE 512
SDI bitstream 141
SDI/SDTI adapter 263
SDTI 260
SDTI content package format 263
SDTI transport 264
SDTI-CP 264
SECAM 8
second audio program 382
secondary color correction 75
security module 358, 439
segment sync 414
self-converging yoke 471

separable transforms 157
separate service information (SI) 530
sequence layer 171
sequential DCT 161
serial data transport interface 259
serial digital interface 141, 247, 614
service description framework 365
service description table (SDT) 531
service multiplex and transport 314
service plane 651
settling time 115
shading correction 73
shadow-mask CRT 451
shadow-mask geometry 453
shimmer 83
short transform block 295
sideband splatter 383, 625
Siemens Star Chart 603
signal characterization 210
signal conditioning 190
signal dynamic range 87
signal levels 43
signal-to-mask ratio 203
signal-to-mask threshold 203
signal-to-thermal noise power ratio 104
signal-to-white-noise 378
simulcast systems 24
Simulcrypt 360
Sinclair Broadcast Group 35
single format philosophy 649
single frequency network 37, 534, 546
single-link interface 245
Sixth Report and Order 386
skin effect 485
Slant transform 156
slice 663
slice layers 171
smear 93
SMPTE 170M 102, 245, 320
SMPTE 240M 17, 59, 99, 219, 226
SMPTE 259M 141, 247, 615
SMPTE 260M 228
SMPTE 272M 142
SMPTE 274M 102, 240, 320
SMPTE 291M 261

SMPTE 292M 247
SMPTE 293M-1996 244
SMPTE 294M-1997 244
SMPTE 295M-1997 242
SMPTE 296M-1997 242
SMPTE 299M-1997 254
SMPTE 302M 255
SMPTE 303M 606
SMPTE 304M-1998 270
SMPTE 305M 144, 260
SMPTE 308M 246
SMPTE 310M 644
SMPTE 311M-1998 270
SMPTE 312M 635
SMPTE 315M 272
SMPTE 319M 640
SMPTE 320M 311
SMPTE 324M 282
SMPTE 325M 366
SMPTE 326M 264
SMPTE 327M 639
SMPTE 328M 637
SMPTE 329M 639
SMPTE 331M 264
SMPTE 332M 264, 265
SMPTE 333M 364
SMPTE 334M 255
SMPTE 336M 255
SMPTE 344M 259
SMPTE 346M 257
SMPTE 348M 257
SMPTE 351M 640
SMPTE 353M 641
SMPTE C phosphors 222
SMPTE C reference primary 222
SMPTE color bars 571
SMPTE dynamic metadata dictionary 265
SMPTE EG 34-1999 617
SMPTE RP 173 311
SMPTE RP 192 618
SMPTE RP-184 618
SMPTE RP211 655
SMPTE WG-HDEP 217
SMPTE working group on high-definition elec-
    tronic production 217, 446

Society of Motion Picture and Television Engi-
    neers (SMPTE) 16
soft decision decoding 546
solid-state PA systems 395
sound field 279
sound system design 66
sound-pressure level (SPL) 302
source coding 128, 313
source coding and compression 313
source stream 663
spatial correlation 160
spatial resolution 274
spatial transform 175
spatio-temporal analysis 116
spectral color reproduction 567
spectral correlation 161
spectral envelope 284
spectral regrowth 623
SPECTRE 21
spectroradiometric measurement 564
spectrum efficiency 547
splice flag 634
splice information table 636
splice point 635, 636
splicing 663
spurious resolution 596
spurious sidebands 625
standard definition television 663
standards conversion 43, 69
starpoint combiner 408
start codes 664
start of active video 260
statistical compression 193
STD input buffer 664
steepest-descent technique 432
stigmatism distortion 470
still picture 664
stop-band corner frequency 133
stream clock reference 257
stream object 537
streaming file transfer 651
structured audio decoder 207
structured inputs 207
Studio Profile/Main Level 169
stuffing table (ST) 531

style language 515
subband coding 196
subband gain 196
subcarrier frequencies 121
subframe 280
subjective CRT measurements 594
sub-optimal punctured convolutional code 544
successive approximation A/D converter 112
successive approximation register (SAR) 112
sufficient recoding data set 639
surface acoustic wave (SAW) oscillator 417
surround sound 63
suspension of disbelief 66
symbol clock recovery 414
symbol clock synchronization 418
sync pulse measurement 575
synchronization field 209
synchronization frames 293
synchronization information (SI) header 293
synchronized streaming data 536
synchronous 147
synchronous detector 417
synchronous streaming data 536
system clock reference 664
system header 664
system item 264
system target decoder 664

T

T3/S14 387
Talaria system 487
Task Force for Harmonized Standards for the Exchange of Program Material as Bitstreams 44
taut shadow-mask CRT 461
TCP/IP 151
Technical Corrigendum No.1 to ATSC Standard A/65 355
Technology Group on Distribution (T3) 35
Technology Group on Production (T4) 35
telecine 69
Telecommunications Decoder Circuitry Act 362
television lines 570
television service 42
television transmission standards 4

temporal factors in vision 92
temporal noise shaper (TNS) 206
tension mask CRT 461
terrestrial broadcast mode 131
terrestrial virtual channel table 356
test instruments 558
test pattern 571
text-to-speech 207
the "film look" 275
Three-Layered Bottom-Up Noise Weighting Model 612
threshold of visibility 413, 435, 549
time and date table (TDT) 531
time code 638
time division multiplexing 257, 523
time smearing 290
time-domain aliasing cancellation (TDAC) 290
time-invariant FIR filter 132
time-stamp 664
timing jitter 619
tonal reproduction 267
total internal reflection (TIR) prism 507
transcode 42
transcoding 639
transcoding systems 42
transfer function 443
transform coding 155
transform filterbank 289
transition bandwidth 133
Transition Clip Generator 637
translators 10
Transmission Control Protocol 151
Transmission Signal ID 361
transport packets 287
Transport Stream Identifier 360
transport stream packet header 664
trellis coded modulation 544
Trellis decoder 415
trellis-coded modulation (TCM) 130
trellis-decoded byte data 429
Trinitron 453
Trinitron gun 467
tripotential electron lens 466
tristimulus values 442
TSID 360

tuner 414
TV limiting resolution 594

U

UHF band 8
UHF taboos 379
UHF tetrode 394, 396
UHTTP 664
underscan 601
unicast 537
unidirectional Internet protocol 266
Uni-IP 266
unipotential gun 466
universal reset 430
unpacking (demultiplexing) 299
usage abstraction 651
user data 638
UUID 664

V

V/H coding phase 637
validity bit 282
VANC 141
variable bit rate 664
variable-length codes (VLC) 175
variable-length decoder (VLD) 185
vectorscope 574, 575
vertical ancillary data 141
vertical ancillary data space 255
vertical blanking ancillary data space 641
vertical lines 319
vertical resolution 50, 55, 84, 224
very large scale integration (VLSI) 139
VI service 307
video bandwidth 95
video buffering verifier 664
video compression 104
video encoding 189
video index 638

video processing 557
video projection systems 473
video recoding information 641
video sampling 139
video sequence 166, 172, 664
video standards 43
video system distortion mechanisms 578
vignetting effects 480
virtual image 63
visual acuity 89
visual field 81, 82
Vladimir Zworykin 3
VO service 308
vocoder 207
VSB carrier recovery 414

W

wave filter 124
waveform monitor 574
white balance 568
white noise 418, 627
window function 290
windowing 134
wire speed 150

X

xenon lamp 507

Y

Y data space 258

Z

zone plate 587
zone plate signal 590
zoom lens 268
Zworykin, Vladimir 3

Afterword

What We've Learned from the DTV Experience

Dr. Robert Hopkins, Sony Pictures High-Definition Center

Before closing this book packed with over 700 pages of standards, figures, tables, and related details, it is perhaps appropriate to step back and consider, what have we learned from the DTV experience? I will concentrate my comments on our key experiences related to digital terrestrial broadcasting during the last fifteen years.

As this book goes to press, about 150 digital stations are on the air, more than what was expected back in 1997 when the build-out began This is great news. Perhaps one thing we have learned from our DTV experience, contrary to what we seem to read in the newspapers, is that broadcasters really do want to do something with digital broadcasting.

So, where are we now. What have we learned.?

1. New standards are difficult to complete, especially when government regulations are involved.

Perhaps the most significant thing we have learned is that a totally new technical standard is exceedingly difficult to complete if the standard is regulated by the government. Terrestrial broadcasting is subject to strict technical regulation. Then, you have to add the fact that television broadcasting affects 100 percent of our citizens. Perhaps the wonder is that we were able to finish the standard during my lifetime.

This difficulty is not present in technical standards for pre-recorded media. Look at DVD as an example. That was not nearly so tough to complete. This difficulty is not present in technical standards for satellite broadcasting, or cable. Did you hear such long debates about DBS standards? Or digital cable standards?

Because of the technical regulation, more people have to be involved. That means more voices. Different opinions. Many different opinions. You have to get widespread agreement. General agreement. Consensus. Because of this, the DTV standard came about very slowly. When a direction seemed to appear, new people became involved and they, of course, had new views. And so the direction changed many times.

And, as if the job was not already tough enough, national politics intervened. I mean politics with a capital "P". We all know that politics with a little "p" is involved in technical standards because different companies try to obtain an advantage over their competitors. But, in this case, national politics became a factor. The Advisory Committee was formed by a Chairman who expressed interest in HDTV. After a few months, a new Chairman was appointed, and he showed even greater interest in HDTV. Then the Administration changed from one political party to the other. A temporary Chairman was appointed. After some delay, another Chairman was appointed.

Most participants in the standardization process, especially at the beginning, had a goal of all media following one standard. That view seemed to persist for a long time, almost to the end. As we approached the end, though, I think too much time had passed. Digital satellite broadcasting had begun using slightly different technical parameters. The DVD group released their standard with slightly different technical parameters. Yet, the FCC still had not ruled on the digital terrestrial broadcasting standard.

Today we continue to see that kind of fragmentation. You still hear debates about "computers" and "broadcasting". Convergence. Convergence, to me, means convergence of the technology, not convergence of the products, or applications. I have a four wheel drive sports utility vehicle. It uses tires, four of them. They are big and fat. I have a motorcycle. It uses tires, two of them. They are not big and fat. All six tires are made of rubber, and they are all round, with a big hole in the center. But, they are not interchangeable. They all use the same technology, but they are not the same product. They are used in different applications. That's my view of convergence.

Back to the fragmentation. We still hear debates on progressive scan and interlace scan. And colorimetry. Different broadcasters will use different scanning formats. There continue to be debates on the number of pixels per line, for both standard-definition and high-definition. High-definition DBS may use still different scanning formats. It continues to be unclear what will happen with cable.

2. Bandwidth of 6 MHz is important, but NTSC compatibility is not.

Fifteen years ago, when the ATSC was formed, it was conventional wisdom that more than 6 MHz would be required for HDTV, that terrestrial broadcasters would not be able to broadcast HDTV, and that a new standard must be backward compatible with NTSC.

My how things changed in fifteen years. The earliest proposals were for satellite broadcasting only. And they required two channels. One channel carried the normal 525-line signal, the second channel carried an augmentation signal which, when added to the first signal, produced a high-definition picture. Some time later, there were proposals for satellite broadcasting that used wider bandwidth channels, and were not NTSC-compatible.

The earliest proposals for terrestrial broadcasting assumed NTSC compatibility. If the system required only 6 MHz, it really was an enhanced system, not a high-definition system. Early terrestrial proposals for high-definition assumed two channels, like the satellite example mentioned previously. The first channel was simply an NTSC signal, perhaps an enhanced NTSC signal. The second channel provided an augmentation signal which, when added to the first signal, produced a high-definition signal.

In my opinion, the big breakthrough came late in 1988 when Zenith proposed their "Digital Spectrum Compatible" high-definition system. This was a non-NTSC compatible 6 MHz system. For the first time, people began to seriously consider a non-compatible system. There seemed to be a recognition that NTSC was terribly bandwidth inefficient, and that carrying an NTSC signal forevermore would be a terrible waste of bandwidth.

3. Digital broadcasting is a winner.

Fifteen years ago we thought that digital broadcasting would not be possible for many, many years into the future. Even ten years ago it was conventional wisdom. That changed in 1990, when General Instrument proposed a 100 percent digital system to the FCC. Within six months, all proposals, save one, were modified to be all-digital. The one remaining analog system was dropped two years later.

Let me make a clarification for people who have not been involved in this area. Digital television is not new, not by any means. I began my digital television career in 1970. The IBA in Great Britain designed a digital standards converter around 1970. The first commercial digital time base corrector was available around 1972. The first commercial digital television frame store was available around 1974 or 1975. Broadcast quality digital tape recorders were being demonstrated around 1980. This form of digital television is not new.

Digital bit rate reduction, on the other hand, is new. Or, let me say, bit rate reduction to the degree that permits digital broadcasting to use less bandwidth than analog broadcasting is certainly much newer. The work of MPEG in the late 1980s and early 1990s made this happen.

Broadcasting bits, rather than broadcasting an analog signal, also is newer technology. This development, more or less, coincides with the bit rate reduction development. Finding techniques to broadcast more bits per second, coupled with reduction of the number of bits required to represent pictures, proved to be a powerful combination.

This change in technology also affects how companies use other technologies. Consider, for example, my own company, the Sony Pictures High-Definition Center. Our biggest business is converting movies to high-definition video. As I am sure you know, there is not an overwhelming demand for HD software today. So, we downconvert the movies to standard-definition for broadcasting and for VHS and Laserdisc releases, and for DVD releases. Because of the great amount of bit rate reduction that is needed for DVD, other elements of the process begin to take on greater importance than ever before. Some of these important items are elimination of weave, low noise, and proper maintenance of the 3:2 pulldown. And this will be just as important for digital broadcasting, whether standard-definition or high-definition. Weave and noise gobble up bits.

Think for a moment just how important bits are. If you start with 1920×1080 pixels, 24 times each second, with a 4:2:2 representation, meaning 16 bits per pixel, you have almost 800 million bits per second. For HD broadcasting, this must be reduced to about 18 million bits per second. Or, about 0.36 bits per pixel. We go from 16 bits per pixel to only 1/3 of a bit per pixel. You certainly do not want to waste bits under such constraints. So, we use pin registration to remove the weave and low noise full-frame cameras to minimize the noise. We have designed our own telecine equipment and have transferred around 350 movies with it, gaining an Emmy in the meantime.

4. Material shot on film should be posted and broadcast at 24 frames per second.

Fifteen years ago we did not think of broadcasting film-originated programming at 24 frames per second. Today, it is becoming the rage.

About 75 percent of prime time programming is shot on film. That means 24 frames per second. Generally, each film shot is transferred to video, and then treated as if it were 60 Hz interlaced video rather than 24 frames per second film. Edit timing is based on 60 Hz fields, not 24 Hz frames. This means that the 24 frames per second character of the film probably is maintained within the shot, but there is a 3 out of 4 chance that it is broken between shots. And, when the length of the show is changed by dropping or repeating a field, it is lost within the shot. This wastes those precious bits. If the 24 Hz character is maintained, fewer bits are required for a given level of quality.

A second reason that 24 frames per second post production is becoming so popular is because of the different scanning formats that different broadcasters plan to use. In Hollywood, it is clear that post should be performed at the highest resolution that any client will want the product delivered. Then you downconvert to lower resolution formats. Posting film-shot productions at 24 frames per second with 1920 × 1080 resolution makes sense. You can easily extract a 1920 × 1080 60I signal, or a 1280 × 720 60P or 24P signal, or a 480P signal, or a 480I signal. Perhaps equally important, you can easily extract a 50 Hz signal by running the 24 Hz tape at 25 Hz. This makes a PAL tape that is identical to what you would have if you used a 50 Hz high-definition telecine and ran the film 4 percent fast at 25 frames per second.

The HD Center has been aware of this issue for some time. We have been maintaining proper 3:2 pulldown in our transfers from the beginning. This has been important to us because of our down-conversions to a DVD master, and because of our downconversions to PAL. To make a PAL tape, we simply drop the redundant one-out-of-five 60 Hz field, downconvert from 1920 × 1080 pixels to 576 × 720 pixels, and record at 48 Hz. Then we play back that tape at 50 Hz. As described before, this produces a flawless PAL tape.

We also are interested in 24 Hz electronic shooting. We want to be able to seamlessly mix material shot on film and material shot electronically. This is for material that will end up as video, and for material that will end up as film. We believe the easiest way to accomplish this is to shoot video at 24 frames per second.

5. The distinction between high-definition and standard-definition has blurred.

Fifteen years ago, the ATSC formed three different technology groups. The first group was dedicated to improved-NTSC systems. The second group was dedicated to enhanced 525-line systems. The third group was dedicated to HDTV systems. Around 1990, the ATSC changed that organizational arrangement. It had become clear that the more important breakdown was production and distribution.

Today, we continue to see this as a continuum, not three different standards, or sets of standards. Yes, the ATSC Digital Television Standard incorporates three quality levels, 480 lines, 720 lines, and 1080 lines. But we also believe receivers should decode and display a picture regardless of which quality level was broadcast. This is consistent with my earlier comments that productions will likely be done at the highest quality level, then distributed at whatever quality level seems appropriate for that show at that time on that service.

This approach that we took in America is different from the approach that has been taken in Europe. In America, we have embraced HD and SD in the original standard, and we have encouraged the manufacture of all-format receivers. In this approach, viewers select the resolution they desire when purchasing a television set. The highest quality will cost the most. Even the lowest resolution television set, though, will be able to display pictures from a high-definition broadcast. The European approach has been to incorporate only SD in the original standard. Then, whenever HD is desired, HD broadcasts can be made to HD receivers. I am an advocate of the American approach because I am concerned that it will be too difficult to take that second step up to HD broadcasting. This means two transitions, not one. I am afraid that one transition is difficult enough!

Today we do not know if the American approach is a winner or a loser. As I have said, I believe it is the right approach. But, only time will tell.

The future

What do I see for the future. Or, what might I have written if these comments were published five or ten years from now?

I am sold on quality. That's why I am a big advocate of HD. I believe our standard of 1920×1080 will be able to stand the test of time. Let me give you an example of what this quality level means. The HD Center is involved in digital image restoration. Often this means scanning a 35 mm film that has been damaged, repairing the scanned images, then replacing the damaged images. A recent project was the restoration of the 1969 Columbia Pictures classic "Easy Rider." We replaced two missing negative reels, accomplished by scanning (at 1920×1080) black and white separation prints that were made from the original negatives back in 1969. Then we combined the three black and white images to make a color image, repaired any damage with computers, then made a 35 mm negative from the HD master. The two replacement reels were then included in the movie, along with the other original reels.

So, what will I say five or ten years from now? All those pesky little issues I mentioned at the beginning of this essay will go away. At some point, we will stop worrying about the number of pixels, the colorimetry, the scanning formats; yes, even progressive and interlaced scanning. All forms of "open" media will use the ATSC Digital Television Standard. And the all-format receiver approach will be seen as the appropriate manner to have begun the new service. And consumers will embrace HD. They will love it just as I do.

Robert Hopkins

October 2000

About the Author

Jerry C. Whitaker is President of Technical Press, a consulting company based in the San Jose (CA) area. Mr. Whitaker has been involved in various aspects of the electronics industry for over 25 years, with specialization in communications. Current book titles include the following:

- Editor-in-chief: *Standard Handbook of Video and Television Engineering*, 3rd ed., McGraw-Hill, 2000

- Editor: *Video and Television Engineers' Field Manual*, McGraw-Hill, 2000

- *Radio Frequency Transmission Systems: Design and Operation*, McGraw-Hill, 1990

- *Electronic Displays: Technology, Design, and Applications*, McGraw-Hill, 1993

- *Maintaining Electronic Systems*, CRC Press, 1991

- Editor-in-chief, *The Electronics Handbook*, CRC Press, 1996

- *AC Power Systems Handbook*, 2nd Ed., CRC Press, 1999

- *Power Vacuum Tubes Handbook*, 2nd Ed., CRC Press, 1999

- *The Communications Facility Design Handbook*, CRC Press, 2000

- *The Resource Handbook of Electronics*, CRC Press, 2000

- Co-author, *Television and Audio Handbook for Engineers and Technicians*, McGraw-Hill, 1989

- Co-author, *Communications Receivers*, 2nd edition, McGraw-Hill, 1996

- Co-editor, *Information Age Dictionary*, Intertec Publishing/Bellcore, 1992

Mr. Whitaker has lectured extensively on the topic of electronic systems design, installation, and maintenance. He is the former editorial director and associate publisher of *Broadcast Engineering* and *Video Systems* magazines, and a former radio station chief engineer and television news producer.

Mr. Whitaker is a Fellow of the Society of Broadcast Engineers and an SBE-certified professional broadcast engineer. He is also a fellow of the Society of Motion Picture and Television Engineers, and a member of the Institute of Electrical and Electronics Engineers.

Mr. Whitaker has twice received a Jesse H. Neal Award *Certificate of Merit* from the Association of Business Publishers for editorial excellence. He has also been recognized as *Educator of the Year* by the Society of Broadcast Engineers.

Mr. Whitaker resides in Morgan Hill, California.

On the CD-ROM

The enclosed CD includes valuable background information on the ATSC DTV standards. These files are provided courtesy of the Advanced Television Systems Committee, Washington, D.C. For more information on the activities of the ATSC, visit the organization's web site at **www.atsc.org**.

- ATSC Document **A/52**: Digital Audio Compression (AC-3) Standard, December 20, 1995.

- ATSC Document **A/53**: ATSC Digital Television Standard, September 16, 1995.

- ATSC Document **A/54**: Guide to the Use of the ATSC Digital Television Standard, October 4, 1995.

- ATSC Document **A/57**: Program/Episode/Version Identification, August 30, 1996.

- ATSC Document **A/58**: Harmonization with DVB SI in the use of the ATSC Digital Television Standard, September 16, 1996.

- ATSC Document **A/63**: Standard for Coding 25/50 Hz Video, May 2, 1997.

- ATSC Document **A/64**: Transmission Measurement and Compliance for Digital Television, November 17, 1997. Also included is Revision A of the document, dated May 30, 2000. The A/64-Rev. A document includes the revised FCC DTV emission mask, and specifies that the standard is for use in the U.S. and not world-wide.

- ATSC Document **A/65**: Program and System Information Protocol for Terrestrial Broadcast and Cable, December 23, 1997.

- ATSC Document **A/66**: Technical Corrigendum No.1 to ATSC Standard: Program and System Information Protocol for Terrestrial Broadcast and Cable, December 17, 1999.

- ATSC Document **A/67**: Amendment No. 1 to ATSC Standard: Program and System Information Protocol for Terrestrial Broadcast and Cable, December 17, 1999.

- ATSC Document **A/70**: Conditional Access System for Terrestrial Broadcast, July 17, 1999.

- ATSC Document **A/80**: Modulation and Coding Requirements for Digital TV (DTV) Applications Over Satellite, July 17, 1999.

- ATSC Document **A/90**: Data Broadcast Standard, July 26, 2000.

The ATSC documents are available in printed form from the Society of Motion Picture and Television Engineers (**www.smpte.org**) and the National Association of Broadcasts (**www.nab.org**).

The following documents are provided courtesy of the Federal Communications Commission. For additional information and resources, consult the FCC web site at **www.fcc.gov**.

- FCC: **Fifth Report and Order on Digital Television**, adopted April 3, 1997
- FCC: **Memorandum Opinion and Order on Reconsideration of the Fifth Report and Order**, adopted February 17, 1998
- FCC: **Second Memorandum Opinion and Order on Reconsideration of the Fifth and Sixth Report and Orders on Digital Television**, adopted November 24, 1998
- FCC: **Sixth Report and Order on Digital Television**, adopted April 3, 1997
- FCC: **Sixth Report and Order Appendix A: Technical Data**
- FCC **Memorandum Opinion and Order on Reconsideration of the Sixth Report and Order**, adopted February 17, 1998
- FCC **Report and Order on Closed Captioning Requirements for Digital Television Receivers**, adopted July 21, 2000.

Other resources provided on the CD-ROM include the following:

- MSTV: **Proposed TSID Assignments**, released December 12, 1998 (**www.mstv.org**)
- ATVEF: **Advanced Television Enhancement Forum Specifications**, draft version 1.1r26, released February 2, 1999. Copyright© ATVEF. (**www.atvef.com**)
- SCTE Standards relating to digital television (**www.scte.org**)

Important Note

Two file formats are used for CD-ROM documents: Adobe Acrobat 4.0 Portable Document Format and Microsoft Word 2000 for the PC. The Acrobat reader is available at no cost from Adobe Systems. See the company's web site at **www.adobe.com**. Microsoft offers a Word reader for various platforms. Check the Microsoft web site at **www.microsoft.com**.

The enclosed CD-ROM is supplied "as is." No warranty or technical support is available for the CD.

On-Line Updates

Additional updates relating to DTV in general, and this book in particular, can be found at the *Standard Handbook of Video and Television Engineering* web site:

www.tvhandbook.com

The tvhandbook.com web site supports the professional video community with news, updates, and product information relating to the broadcast, post production, and business/ industrial applications of digital video.

Check the site regularly for news, updated chapters, and special events related to video engineering. The technologies encompassed by *DTV: The Revolution in Digital Video* are changing rapidly, with new standards proposed and adopted each month. Changing market conditions and regulatory issues are adding to the rapid flow of news and information in this area.

Specific services found at **www.tvhandbook.com** include:

- **Video Technology News**. News reports and technical articles on the latest developments in digital television, both in the U.S. and around the world. Check in at least once a month to see what's happening in the fast-moving area of digital television.

- **Television Handbook Resource Center**. Check for the latest information on professional and broadcast video systems. The Resource Center provides updates on implementation and standardization efforts, plus links to related web sites.

- **tvhandbook.com Update Port**. Updated material for *DTV: The Revolution in Digital Video* is posted on the site each month. Material available includes updated sections and chapters in areas of rapidly advancing technologies.

- **tvhandbook.com Book Store**. Check to find the latest books on digital video and audio technologies. Direct links to authors and publishers are provided. You can also place secure orders from our on-line bookstore.

In addition to the resources outlined above, detailed information is available on other books in the McGraw-Hill Video/Audio Series.

CD-ROM WARRANTY

This software is protected by both United States copyright law and international copyright treaty provision. You must treat this software just like a book. By saying "just like a book," McGraw-Hill means, for example, that this software may be used by any number of people and may be freely moved from one computer location to another, so long as there is no possibility of its being used at one location or on one computer while it also is being used at another. Just as a book cannot be read by two different people in two different places at the same time, neither can the software be used by two different people in two different places at the same time (unless, of course, McGraw-Hill's copyright is being violated).

LIMITED WARRANTY

McGraw-Hill takes great care to provide you with top-quality software, thoroughly checked to prevent virus infections. McGraw-Hill warrants the physical CD-ROM contained herein to be free of defects in materials and workmanship for a period of sixty days from the purchase date. If McGraw-Hill receives written notification within the warranty period of defects in materials or workmanship, and such notification is determined by McGraw-Hill to be correct, McGraw-Hill will replace the defective CD-ROM. Send requests to:

McGraw-Hill
Customer Services
P.O. Box 545
Blacklick, OH 43004-0545

The entire and exclusive liability and remedy for breach of this Limited Warranty shall be limited to replacement of a defective CD-ROM and shall not include or extend to any claim for or right to cover any other damages, including but not limited to, loss of profit, data, or use of the software, or special, incidental, or consequential damages or other similar claims, even if McGraw-Hill has been specifically advised of the possibility of such damages. In no event will McGraw-Hill's liability for any damages to you or any other person ever exceed the lower of suggested list price or actual price paid for the license to use the software, regardless of any form of the claim.

McGRAW-HILL SPECIFICALLY DISCLAIMS ALL OTHER WARRANTIES, EXPRESS OR IMPLIED, INCLUDING, BUT NOT LIMITED TO, ANY IMPLIED WARRANTY OF MERCHANTABILITY OR FITNESS FOR A PARTICULAR PURPOSE.

Specifically, McGraw-Hill makes no representation or warranty that the software is fit for any particular purpose and any implied warranty of merchantability is limited to the sixty-day duration of the Limited Warranty covering the physical CD-ROM only (and not the software) and is otherwise expressly and specifically disclaimed.

This limited warranty gives you specific legal rights; you may have others which may vary from state to state. Some states do not allow the exclusion of incidental or consequential damages, or the limitation on how long an implied warranty lasts, so some of the above may not apply to you.